Understanding Food Science and Technology

Understanding Food Science and Technology

Peter S. Murano
Texas A & M University

THOMSON
———✳———
WADSWORTH

Australia • Canada • Mexico • Singapore • Spain
United Kingdom • United States

THOMSON

WADSWORTH

Publisher: Peter Marshall
Development Editor: Elizabeth Howe
Assistant Editor: John Boyd
Editorial Assistant: Madinah Chang
Marketing Manager: Jennifer Somerville
Marketing Assistant: Mona Weltmer
Print/Media Buyer: Tandra Jorgensen
Permissions Editor: Bob Kauser
Production: Mary Douglas, Rogue Valley Publications
Production Project Managers: Sandra Craig, Dianne Toop

Text Designer: Burl Sloan
Art Editor: Pat Brewer
Photo Researchers: Lili Weiner, Myrna Engler
Copy Editor: Pat Brewer
Illustrators: Lotus Art, New England Typographic Service
Cover Designer: Stephen Rapley
Cover Images: Getty Images
Cover Printer: Phoenix Color Corporation
Compositor: New England Typographic Service
Printer: Phoenix Color Corporation

COPYRIGHT © 2003 Wadsworth, a division of Thomson Learning, Inc. Thomson Learning™ is a trademark used herein under license.

ALL RIGHTS RESERVED. No part of this work covered by the copyright hereon may be reproduced or used in any form or by any means—graphic, electronic, or mechanical, including but not limited to photocopying, recording, taping, Web distribution, information networks, or information storage and retrieval systems—without the written permission of the publisher.

Printed in the United States of America
1 2 3 4 5 6 7 06 05 04 03 02

For more information about our products, contact us at:
**Thomson Learning Academic Resource Center
1-800-423-0563**
For permission to use material from this text, contact us by:
Phone: 1-800-730-2214 **Fax:** 1-800-730-2215
Web: http://www.thomsonrights.com

Library of Congress Control Number: 2002110525

Student Edition with InfoTrac College Edition: ISBN 0-534-54486-X
Student Edition without InfoTrac College Edition: ISBN 0-534-54487-8

Instructor's Edition: ISBN 0-534-54492-4

Wadsworth/Thomson Learning
10 Davis Drive
Belmont, CA 94002-3098
USA

Asia
Thomson Learning
5 Shenton Way #01-01
UIC Building
Singapore 068808

Australia
Nelson Thomson Learning
102 Dodds Street
South Melbourne, Victoria 3205
Australia

Canada
Nelson Thomson Learning
1120 Birchmount Road
Toronto, Ontario M1K 5G4
Canada

Europe/Middle East/Africa
Thomson Learning
High Holborn House
50/51 Bedford Row
London WC1R 4LR
United Kingdom

Latin America
Thomson Learning
Seneca, 53
Colonia Polanco
11560 Mexico D.F.
Mexico

Spain
Paraninfo Thomson Learning
Calle/Magallanes, 25
28015 Madrid, Spain

BRIEF CONTENTS

CHAPTER 1	Introduction to Food Science and Technology	1
CHAPTER 2	Food Categories and Composition	21
CHAPTER 3	Human Nutrition and Food	57
CHAPTER 4	Food Chemistry I: Functional Groups and Properties, Water, and Acids	91
CHAPTER 5	Food Chemistry II: Carbohydrates, Lipids, Proteins	120
CHAPTER 6	Food Chemistry III: Color, Flavor, and Texture	154
CHAPTER 7	Food Additives, Food Laws, and Dietary Supplements	178
CHAPTER 8	Understanding Dimensions of Food Processing and Preservation: Animal Products	210
CHAPTER 9	Understanding Fat, Sugar, Beverage, and Plant Product Processing	249
CHAPTER 10	Food Microbiology and Fermentation	282
CHAPTER 11	Food Safety	303
CHAPTER 12	Food Toxicology	329
CHAPTER 13	Food Engineering	356
CHAPTER 14	Food Biotechnology	392
CHAPTER 15	Sensory Evaluation and Food Product Development	419
GLOSSARY	G-1	
INDEX	I-1	

CONTENTS

CHAPTER 1
Introduction to Food Science and Technology 1

1.1 The Dimensions of Food Science 2
Food Science and Nutrition 3
Biology and Food Science 4
Chemistry and Food Science 4
Physics and Food Science 5
Engineering and Food Science 5

1.2 The Food Processing Industry 5
History of the Food Industry 6
The Work of the Food Scientist Today 8

1.3 Major Classes of Food Components 10
The Food Guide Pyramid 10

1.4 Food Science Education 11
Earning a Food Science Degree 11
Food Science FAQ 13

Key Points, Study Questions, References, Additional Resources, Web Sites 15

Challenge! Reading the Research 17

CHAPTER 2
Food Categories and Composition 21

2.1 Food Composition Tables 22
Food Categories and the Food Guide Pyramid 23
Serving Sizes 23

2.2 Beverages 24
The Degrees Brix of Beverages 26
The Importance of the °Brix/Acid Ratio 26

2.3 Cereals, Grains, and Baked Products 27
Leavening of Baked Products 28

2.4 Fruits and Vegetables 28
Health Benefits 30
Maturity and Ripeness 31
Quantitative and Qualitative Quality Considerations 31
Dried Fruits 32

2.5 Legumes and Nuts 32
Common Soy Products 33
Tree Nuts 33

2.6 Meat and Meat Products 35
The Structure of Muscle Tissue 35
Meat Emulsions 37

2.7 **Seafood 37**
Why Are Fish So Perishable? 38

2.8 **Eggs 39**
Egg Quality 40

2.9 **Milk and Dairy Products 41**
The Varieties of Milk 42
The Milk Proteins: Casein and Whey 42
Ice Cream—A Complex Colloidal System 44
Butter and Margarine 44
Cheese 45

2.10 **Chocolate and Confectionery 46**
Noncrystalline and Crystalline Confectionery 47

Key Points, Study Questions, References, Additional Resources, Web Sites 48

Challenge! Phytochemicals as Food Components 51

CHAPTER 3
Human Nutrition and Food 57

3.1 **Proper Nutrition—Making the Right Food Choices 58**
The Dietary Guidelines and the Food Guide Pyramid 58
Digestion, Absorption, and Transport of Food 60
Regulating Digestion and Absorption 61
Reading Food Labels 62

3.2 **The Nutritional Role of the Macronutrients 64**
Water, Water, Everywhere 64
Carbohydrates—Our Basic Fuel 66
Lipids—Those Essential Fatty Acids 68
Proteins—Essential Body Builders 73

3.3 **The Micronutrients—Vitamins and Minerals 75**
Vitamins 75
Essential Minerals 75

3.4 **Substitutions for Sugar and Fat 77**
Alternative Sweeteners 77
Fat Replacers 79

3.5 **Energy Metabolism 82**

3.6 **Ergogenic Substances 84**

Key Points, Study Questions, References, Additional Resources, Web Sites 85

Challenge! Can a High-Protein, High-Fat Diet Work? 88

CHAPTER 4
Food Chemistry I: Functional Groups and Properties, Water, and Acids 91

4.1 **The Nature of Matter 92**
Chemical Symbols, Formulas, and Equations 92

Electron Orbits and Chemical Bonds 93
Chemical Bonds in Foods 94

4.2 Chemical Reactions in Foods 96
Enzymatic Reactions 96
Nonenzymatic Reactions 98

4.3 Functional Groups 99
Alcohol Group 99
Aldehyde Group 100
Amino Group 100
Carboxylic Acid Group 100
Ester Group 101
Ketone Group 101
Methyl Group 101
Phosphate Group 101
Sulfhydryl Group 102
Ionic Groups 102

4.4 The Chemical and Functional Properties of Water 102
Water Molecule Structure 102
Solvation and Dispersing Action 103
Water Activity and Moisture 103
Water as a Component of Emulsions 104
Water and Heat Transfer 105
Water as an Ingredient 105
Water as a Plasticizer 105

4.5 The Chemical and Functional Properties of Food Acids 106
Food Acid Structure 106
Acid Strength 106
Fumaric Acid and Dough Softening 107
Salts of Organic Acids 107
Buffers 108
Leavening 108

4.6 Food Acidity 110
pH and the pH Scale 110
Titratable Acidity 110
pH and Acid Foods 111

Key Points, Study Questions, References, Additional Resources, Web Sites 112

Challenge! Food Systems 115

CHAPTER 5
Food Chemistry II: Carbohydrates, Lipids, Proteins 120

5.1 Food Carbohydrates 121
The Structures of Sugars 121
The Functional Properties of Sugars 122
The Functional Properties of Polysaccharides 126

5.2 Food Lipids 133
Structures and Types of Lipids 133
Chemical Reactions of Lipids 137
The Functional Properties of Lipids 140

5.3 Food Proteins 143
The Structure of Proteins 143
The Chemical Reactions and Functional Properties of Proteins 145

Key Points, Study Questions, References, Additional Resources, Web Sites 148

Challenge! Milk Protein Chemistry 151

CHAPTER 6
Food Chemistry III: Color, Flavor, and Texture 154

6.1 Food Color Chemistry 155
What Is Color? 155
Pigment Molecules 156
The Color Chemistry of Red Meat 156
The Color Chemistry of Fruits and Vegetables 158
The Color Chemistry of Food Colorants 160

6.2 Food Flavor Chemistry 162
Chemical Structure and Taste 162
Process and Reaction Flavors 164
Flavor Enhancers 166
The Chemistry of Flavor Deterioration 166

6.3 Food Texture 167
Texture Classification 168
The Influence of Chemical Forces in Water and Fat Systems on Texture 168
The Chemistry of Food Texturizing Agents 169

Key Points, Study Questions, References, Additional Resources, Web Sites 172

Challenge! The Chemistry of Sweeteners and Sweetness 175

CHAPTER 7
Food Additives, Food Laws, and Dietary Supplements 178

7.1 What Is a Food Additive? 179
The Uses of Food Additives 180
The Major Types of Food Additives 180

7.2 Food Laws and Regulations in the United States 183
Differentiating Laws and Regulations 183
Early Events and Legislation 184
The 1938 FFDCA and Amendments 185
Other Legislation and Significant Regulatory Actions 187

7.3 The Enforcers of Food Laws 190
International Food Agencies 190
National Food Agencies 190

7.4 The Approval Process for Food Additives 193
Testing Additive Safety 194
The Ames Test 194

7.5 The Nutrition Labeling and Education Act of 1990 195
  The Nutrition Facts Label 196
  General Product Labeling 197
  Nutrition Content Descriptors 197
  Health Claims 198

7.6 The Dietary Supplement Health and Education Act of 1994 200
  Safety Testing of Dietary Supplements 201
  Dietary Supplement or Drug? 201

Key Points, Study Questions, References, Additional Resources, Web Sites 202

Challenge! Regulation of Functional Foods, Bioengineered Foods, and Organic Foods 205

CHAPTER 8
Understanding Dimensions of Food Processing and Preservation: Animal Products 210

8.1 Food Processing—From Field and Farm to Consumers 211
  The Unit Operations of Food Processing 212
  The Basic Principles of Food Processing 214

8.2 What Is Heat Transfer? 219
  Heat Transfer in a Retort Canner 219
  Heat Transfer Within a Can 219
  Vacuum Canning 220

8.3 Food Preservation—Preventing Food Spoilage 220
  Thermal Processing for Food Preservation 220
  Traditional Nonthermal Processing for Food Preservation 223
  Innovative Nonthermal Methods of Food Preservation 223

8.4 Dairy Products Processing 225
  Milk 225
  Cheese 226
  Ice Cream 226
  Yogurt 227

8.5 Egg Processing 227
  Egg Product Processing 228
  Egg Substitute 230

8.6 Meat Processing 230
  Meat Quality 230
  Meat Preservation and Processing 231
  Red Meat Processing 232
  Fish Processing 235
  Poultry Processing 236

Key Points, Study Questions, References, Additional Resources, Web Sites 237

Challenge! Food Irradiation 241

CHAPTER 9
Understanding Fat, Sugar, Beverage, and Plant Product Processing 249

- **9.1 Processing of Fats and Oils 250**
  Processing of Specific Fats 252
  Chemical and Physical Testing of Fats 253

- **9.2 Sugar Processing 253**
  Extraction 254
  Neutralization and Clarification 254
  Concentration and Crystallization 254
  Separation and Drying 255
  Sugar Refining 255

- **9.3 Beverage Processing 255**
  Water Beverages 255
  Soft Drink Beverages 256
  Special Beverage Categories 258

- **9.4 Processing of Cereal Grains 258**
  Wheat Milling 259
  Breadmaking 260
  Breakfast Cereal 261
  Pasta Processing 262
  Snack Foods 262

- **9.5 Fruit and Vegetable Processing 262**
  Ripening 263
  Processing 264
  Packaging 265
  Storage 266
  Freezing 266
  Manufacture of Fruit Juice 267
  Canning 267
  Pickling 268
  Dehydration 269
  Hot-Break/Cold-Break Processing 269

- **9.6 Soybean Processing 270**

- **9.7 Chocolate Processing 272**

Key Points, Study Questions, References, Additional Resources, Web Sites 275

Challenge! Enzymes in Food Processing—What Are Protein Hydrolysates? 279

CHAPTER 10
Food Microbiology and Fermentation 282

- **10.1 What Are Microorganisms? 283**
- **10.2 Factors Affecting Microbial Growth 285**
  Nutrient Availability 285
  Water Activity 285
  Acidity/Alkalinity 286

Oxygen 286
Temperature 287
Food Effects 287
Using the Hurdle Concept 288

10.3 Foodborne Microorganisms 288
Sources of Microorganisms 288
Types of Microorganisms Found in Food 289

10.4 Food Spoilage by Microorganisms 291
Metabolizing or Producing Carbohydrates 291
Metabolizing Proteins 292
Mold Growth 292

10.5 Microbial Fermentation 292
The Fermentation of Milk 293
The Fermentation of Meat 295
Fermenting Fruit and Vegetable Products 296
Fermenting Cereal Grains 297

Key Points, Study Questions, References, Additional Resources, Web Sites 298

Challenge! Microbial Sampling to Verify Food Quality 301

CHAPTER 11
Food Safety 303

11.1 What Is Foodborne Illness? 304

11.2 Types of Biological Hazards in Food 305
Bacterial Causes 305
Mycotoxins from Molds 306
Virus Transmission 307
Ingestion of Parasites 307

11.3 The Most Common Biological Hazards in Food 308
Bacteria—The Main Culprits 308
Molds—Ergotism and Aleukia 311
Viruses in Foods 312
Parasites—Protozoa and Worms 313

11.4 What Is Mad Cow Disease? 315
Transmissible Spongiform Encephalopathies (TSEs) 315
Causes of TSEs 316
Is PrPsc Transmitted to Humans Through Consumed Beef? 316

11.5 Preventing Foodborne Illness 317
Preventing Food Contamination 317
Preventing Proliferation of Foodborne Microorganisms 318
Eliminating or Reducing Biological Hazards 320

11.6 HACCP—A Preventive Approach 320
Principles of a HACCP System 321
GMPs and SOPs 322

Key Points, Study Questions, References, Additional Resources, Web Sites 323

Challenge! Risk Assessment for Biological Hazards 327

CHAPTER 12
Food Toxicology 329

12.1 What Is a Food Toxicant? 330

12.2 Risk Assessment for Chemical Hazards 331
Modes of Action of Toxicant 332
Assessing Dose-Response 332
Carcinogen Testing 333

12.3 Endogenous Toxicants 334
Flavonoids in Fruits and Vegetables 334
Goitrogens in Cruciferous Vegetables 335
Coumarins in Citrus 335
Cyanide Compounds 335
Herbal Extracts 336
Toxic Mushrooms 338

12.4 Naturally Occurring Toxicants 340
Marine Toxins 340
Microbial Toxins 341

12.5 Synthetic Toxicants 343
Antimicrobial Agents 343
Growth Promotants 345
Pesticides 346

Key Points, Study Questions, References, Additional Resources, Web Sites 350

Challenge! Food Allergies and Food Intolerances 353

CHAPTER 13
Food Engineering 356

13.1 Food Engineering—Basic Terms and Principles 357
Characteristics of Temperature and Heat 357
Conservation of Mass 362
Thermodynamics 363
Heat Exchangers 364

13.2 Deep-fat Frying—An Illustration of Heat Transfer, Mass Transfer, and Boundary Layers 365
Changes to Frying Food and Frying Oil 365
Mass Transfer 365

13.3 Food Materials Science—A Physicochemical Approach 365
Physicochemical States, Glass Transition, and Water Mobility 366
Mobility of Food Molecules and Ingredients 367

13.4 Food Microstructure—Influencing Physical and Sensory Qualities 368

13.5 Psychrometrics—Looking at Air and Food Processing 369
Latent and Sensible Heating and Cooling 370
Psychrometric Charts 370
Food Dehydration and Psychrometry 373

13.6 Rheology—Studying Flow and Deformation 374
Mechanical and Rheological Measurements 375
Stresses, Elastic Solids, and Plastic Solids 375

Newtonian and Non-Newtonian Foods 376
Viscosity in Food and in Processing 377

13.7 Extrusion Technology in the Food Industry 379
Extrusion and Extruded Products 379
How an Extruder Works 379

Key Points, Study Questions, References, Additional Resources, Web Sites 381

Challenge! Food Packaging 385

CHAPTER 14
Food Biotechnology 392

14.1 What Is Food Biotechnology? 393
Surmounting the Species Barrier 393
The Benefits of Food Biotechnology 394

14.2 Genetic Engineering 395
Protein Synthesis 396
Manipulating DNA in Food Production 396

14.3 Regulations Controlling the Application of Food Biotechnology 400
FDA Policy 401
Determining the Safety of Biotechnology-Derived Foods 403

14.4 Improving Plant Products Through Biotechnology 405
Insect-, Virus-, and Drought-Resistant Plants 405
Crop Improvement 400

14.5 Improving Animal Products Through Biotechnology 407

14.6 Improving Food Processing Aids Through Biotechnology 408
Amino Acids Used in Food Production 409
Gums Used in Food Production 409
Enzymes Used in Food Production 409
Microorganisms Used in Food Production 409

14.7 Applying Biotechnology-Derived Foods in Food Safety 410
Nisin: Antimicrobial Agent 410
Diagnostic Biotechnology 410

14.8 Major Concerns About Biotechnology-Derived Foods 410
Concern 1: Are Biotech Foods Harmful? 410
Concern 2: Do Biotech Foods Harm the Ecosystem Balance? 411
Concern 3: Will Big Companies Dictate Food Biotech Practices? 411
Concern 4: Does Genetic Engineering Create Unnatural Consequences? 411
Some Answers 411
The International Food Biotechnology Council 412
The StarLink Controversy 412

Key Points, Study Questions, References, Additional Resources, Web Sites 413

Challenge!   Bioengineering of β-lactoglobin 416

CHAPTER 15
Sensory Evaluation and Food Product Development 419

15.1 What Is Sensory Evaluation? 420
A Scientific Method 420
A Quantitative Science 421
Sensory Science in the Food Industry 421

15.2 Sensory Odor, Flavor, and Mouthfeel Perception 422
Taste 422
Transduction and Sensitivity 423
Odor 423
Flavor 423
Mouthfeel 423

15.3 Sensory Texture and Color Perception 425
Sensory Texture 425
Sensory Interactions 426
Color 427

15.4 Responses Contributing to Sensory Perception 428
Objectivity and Subjectivity 429
Intensity 429
Threshold 430

15.5 Sensory Tests 430
Classification of Test Methods 430
Affective Test Methods 430
Discrimination Test Methods 431
Descriptive Test Methods 431
Selection of Test Method 432

15.6 The Role of the Sensory Evaluation Specialist in Product Development 432
The Sensory Specialist 433
The Sensory Environment 434

15.7 Product Development 434
The Scientific Method in Product Development 435
The Stages of Product Development 435

15.8 The Role of Marketing in Food Product Development 438
Marketing Steps 438
Meeting Market Need and Product Marketability 440

15.9 Product Probability, Life Cycle, and ANN 440
Product Probability 440
Product Life Cycle 441
Artificial Neural Network? 441

Key Points, Study Questions, References, Additional Resources, Web Sites 442

Challenge! Experimental Design in Product Development 445

GLOSSARY G-1

INDEX I-1

PREFACE

Food processing is one of the largest industries in the United States. The demand for food scientists is ever increasing because of the need to maintain a wholesome food supply and to improve the quantity, quality, safety, and variety of foods. To accomplish this task, the food scientist integrates knowledge of biology, chemistry, engineering, physics, and nutrition and applies them to the study of foods.

This comprehensive introductory food science and technology text introduces students to important food science concepts. This text meets the needs of undergraduate students in departments of food science, food technology, and nutrition at two- and four-year colleges and universities. *Understanding Food Science and Technology* is intended for use as an introductory course for food science and technology majors, or for a "science of foods" course for nutrition majors.

A unique fifteen-chapter organization makes this text stand out from other food science and technology texts. Traditional texts are compilations of information written in an encyclopedic style with little care to pedagogy. This fully illustrated textbook has been written specifically to facilitate student learning through the use of special features. Its design enables students to master the material in a traditional semester course. More than 250 easy-to-understand charts, photos, and illustrations enhance the text and provide visual reinforcement of chapter concepts.

This book provides thorough and up-to-date information (for example, bioterrorism and the food supply is mentioned in the food safety chapter), covering a broad range of topics in food science and technology and creates an awareness of the wide range of employment opportunities that exist for trained professionals. The text begins with an overview of food science, describes the interdisciplinary nature of the field, and presents avenues of advanced study and career opportunities in the field.

It then explores key food groups and composition and the functional properties of the major food components. Three chapters on food chemistry cover the chemical and physical properties of foods and include many figures, tables, and illustrations to enhance learning. The next chapters include an overview of food law that provides historical perspective as well as the latest information on food additives, nutrition labeling, and food regulation. Thorough coverage of processing methods is included for all major food commodities as well as a background in microbiology and fermentation, food handling and safety, food contamination, HACCP principles, and toxicology. The final chapters cover food engineering concepts and applications, biotechnology, and the field of sensory evaluation and food product development, including marketing principles.

Pedagogy Overview

- Each chapter opens with an **Outline** of main topics and series of **Chapter Objectives,** helping students preview chapter material and begin thinking critically about the core topics.
- Chapters are summarized with a **Key Points** summary to help students identify major areas for review.
- Each chapter includes introductory and advanced level **study questions.** An answer key is located on the book's Web site, www.nutrition.wadsworth.com.
- Key food science **Web sites** are identified at the end of each chapter and link students to the various online resources.
- A **Challenge!** feature following each chapter provides in-depth coverage of topics related to content of every chapter. For example, a feature on functional, GMO, and organic foods accompanies Chapter 7, "Food Additives, Food Laws, and Dietary Supplements."

Challenge Topics
Reading the research
Phytochemicals as food products
Can a high-protein, high-fat diet work?
Food systems
Milk protein chemistry
Chemistry of sweeteners and sweetness
Regulation of functional foods, bioengineered foods, and organic foods
Food irradiation
Enzymes in food processing—what are protein hydrolysates?
Microbial sampling to verify food quality
Risk assessment for biological hazards
Food allergies and food intolerance
Food packaging
Bioengineering of β-lactoglobin
Experimental design in product development

- **Challenge questions** follow each Challenge section to help students think critically about the topics covered.
- **Three food chemistry chapters** (Chapters 4, 5, 6) start with a preview of functional properties and conclude with discussions of food textures, flavors, and colors.
- Chapter 15, "Sensory Evaluation and Product Development," **synthesizes knowledge** gained from the study of food additives, food chemistry, food safety, and food processing spanning functional foods, marketing, and sensory evaluation.

The world of food science is changing rapidly as new science and new research develops. *Understanding Food Science and Technology* is designed to meet the needs of students and instructors in an evolving and expanding discipline.

Acknowledgments

This text owes much of its form and substance to the following manuscript reviewers for their valuable comments and expert suggestions throughout the development of this text.

James Acton, *Clemson University*
Alfred Bushway, *University of Maine*
Daniel E. Carroll, *North Carolina State University*
Katherine L. Cason, *The Pennsylvania State University*
Suzanne R. Curtis, *University of Maryland*
Douglas L. Holt, *University of Missouri—Columbia*
Trish Lobenfeld, *New York University*
Mark H. Love, *Iowa State University*
Michael Mangino, *The Ohio State University*
Joe Montecalvo, *California Polytechnic State University*
Karen Pesaresi Penner, *Kansas State University Food Science Institute*
Shridhar K. Sathe, *The Florida State University*
Shelley J. Schmidt, *University of Illinois at Urbana-Champaign*
Martha B. Stone, *Colorado State University*
Martha Verghese, *Alabama A&M University*
Charles H. White, *Mississippi State University*

I dedicate this book to my wife Elsa, and I am especially thankful for the expertise of Pat Brewer and Mary Douglas, and the support of Janet Johnson, Duane Suter, Ralph Waniska, Merle Pierson, and of course the Good Lord.

Peter S. Murano

CHAPTER 1

Introduction to Food Science and Technology

1.1 The Dimensions of Food Science

1.2 The Food Processing Industry

1.3 Major Classes of Food Components

1.4 Food Science Education

Challenge!
Reading the Research

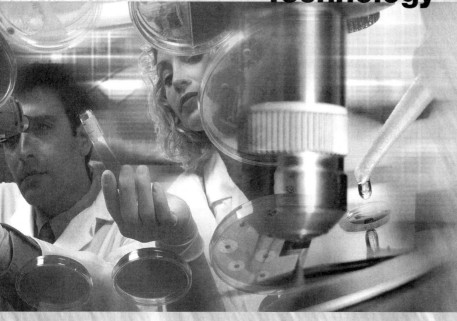

CHAPTER OBJECTIVES

After completing this chapter, you will be able to:
- Define the scope of food science and food technology.
- Distinguish between food science and nutrition.
- List the dimensions of food science.
- Describe the historical development of the food industry.
- List the seven major classes of food components.
- Discuss the ways in which experiments in food science are carried out.
- Describe the importance of teamwork in the food processing industry.
- Describe the major research focus areas in food science and technology today.

Welcome to *Understanding Food Science and Technology*. In this opening chapter, we will consider the educational discipline of food science and explore the connections that food science and technology has with the food manufacturing industry. We will discuss the background necessary to become a food scientist and look at the work that food scientists actually do. The chapter opens with pertinent definitions and clarifies the distinction between "food science" and "nutrition." Toward the end of this chapter is a preview of the major components of foods—the "stars" of the remainder of the text. A special learning feature called Challenge! is at the end of each chapter, and for Chapter 1 the feature is devoted to five specific examples of food research performed by food scientists.

1.1 THE DIMENSIONS OF FOOD SCIENCE

Food is a critical human need. Human physiology responds with a growl, a grumble, and a gnawing emptiness when deprived of it. Our optimum physical and mental functioning, collectively termed *health*, depends upon the nutritional quality, safety, and balance of the foods we eat. To say that the foods we have access to have been influenced by technology is an understatement. Technology makes possible the staggering variety and availability of frozen, baked, dried, pickled, canned, fresh, fermented, refrigerated, and packaged foods. Part of the technological equation is related to the numerous nonfood substances that are formulated into manufactured foods as processing aids. These food **additives** include everything from color stabilizers to anti-staling agents, emulsifiers and fat substitutes to preservatives.

Practical definitions for food science have been developed, not surprisingly, by organizations whose members are food scientists. By considering the two definitions in Figure 1.1, we can propose our own definition: **Food science** is the scientific study of raw food materials and their behavior during formulation, processing, packaging, storage, and evaluation as consumer food products. This definition clearly differentiates food science

Institute of Food Technologists (IFT) of the United States:
Food science is the discipline in which biology, physical sciences, and engineering are used to study the nature of foods, the causes of their deterioration, and the principles underlying food processing.

Institute of Food Science and Technology (IFST) of the United Kingdom:
Food science is a coherent and systematic body of knowledge and understanding of the nature and composition of food materials, and their behavior under the various conditions to which they may be subject.

FIGURE 1.1 Definitions of food science from the Institute of Food Technologists (IFT) in the United States and the Institute of Food Science & Technology (IFST) in the United Kingdom.

FIGURE 1.2 How are food science and technology involved in your breakfast? *The egg white has formed an opaque gel—due to a change in the albumen proteins that denature and coagulate. The toasted bread contains flour components that undergo a browning reaction when heated. Orange juice is a processed fruit beverage that provides a certain sweetness and tartness due to the ratio of naturally present sugars and acids in the oranges. Milk is a processed beverage also: homogenized, pasteurized, and processed to contain a specific level of fat; it is slightly sweet due to lactose sugar.*

from the science of nutrition. In a nutshell, food science is concerned with all quality and safety aspects of foods before a person consumes them (Figure 1.2), while nutrition is related to how the body uses foods after we eat them to promote and maintain our health.

As a field, food science and technology is concerned with the following aspects of agricultural, avian, mammalian, and marine food materials:

- Food processing and manufacture
- Food preservation and packaging
- Food wholesomeness and safety
- Food quality evaluation
- Food distribution
- Consumer food preparation and use

IFT identifies **food technology** as "the application of food science to the selection, preservation, processing, packaging, distribution, and use of safe, nutritious, and wholesome food," and *food manufacturing* as "the mass production of food products from raw animal and plant materials utilizing the principles of food technology."

In actual practice the differences between what are called food science and food technology are so subtle that they are often ignored. The difference between science and technology in general is that technology is what people can make happen through today's technological expertise whereas an understanding of the science and consequences of those technologies (e.g., long-term effects on people and the environment) may belong to the future. The development of nutraceutical food products is an example.

Food Science and Nutrition

The American Medical Association (AMA) has defined **nutrition** as the nutrients and other substances in food, "their action, interaction, and balance in relation to health and disease, and the processes by which the organism (body) ingests, digests, absorbs, transports, utilizes and excretes food substances."

Consider that food science deals with food manipulations and their consequences on the food components themselves, whereas nutrition deals with the consequences the food components have upon the humans who consume them.

Food science overlaps not only nutrition but other knowledge areas such as biology, chemistry, engineering, and physics. For this reason, food science is not a traditional scientific discipline but is rather an application or combination of these other sciences, meaning that food science is an interdisciplinary field of study (Figure 1.3).

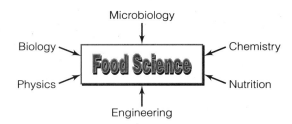

FIGURE 1.3 Food science, an interdisciplinary field of study.

Biology and Food Science

Biology is the study of living things and their life-sustaining systems. Foods are derived from living organisms, either animal or plant in origin. The food scientist must know how the edible tissue of these food sources is organized on the cellular level. Figure 1.4 shows important organelles for animal and plant cells. This foundation is needed to understand changes that occur during processing, storage, and so forth. Living organisms such as bacteria, yeast, and molds can contaminate living tissue, creating a variety of consequences. The microbiological action of these organisms can be harmful (causing foodborne illness through infections and intoxications), can be detrimental to food (microbial enzymatic reactions causing it to rot and smell), or can be useful (purposely added yeast to bread flour, causing the desired leavening effect in baking). One branch of biology, genetics, is the scientific foundation of food biotechnology.

Chemistry and Food Science

Chemistry is the study of the atoms and molecules that make up the substances present in the universe, their arrangement into structures, and the reactions they participate in. Sometimes students are put off by the abstract nature of chemistry and fail to see its relevance. But we can appreciate the fact that foods such as meats, fruits and vegetables, and breads and cereals contain carbon, hydrogen, and oxygen atoms. These and other atoms listed in the periodic table of the elements exist in foods as combined structures forming the characteristic proteins, carbohydrates, vitamins, minerals, pigments, and fat molecules found in every food. Food scientists want to know the molecular structures of these food components and to study the changes to them that might affect food quality. These changes can be represented by chemical equations.

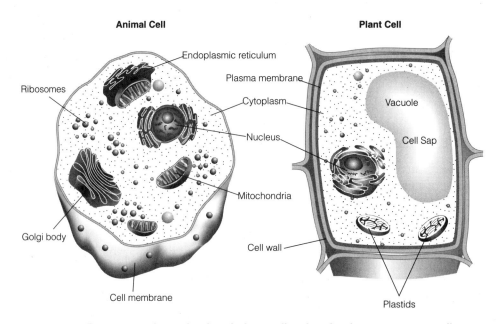

FIGURE 1.4 Representative animal and plant cells, showing important organelles.

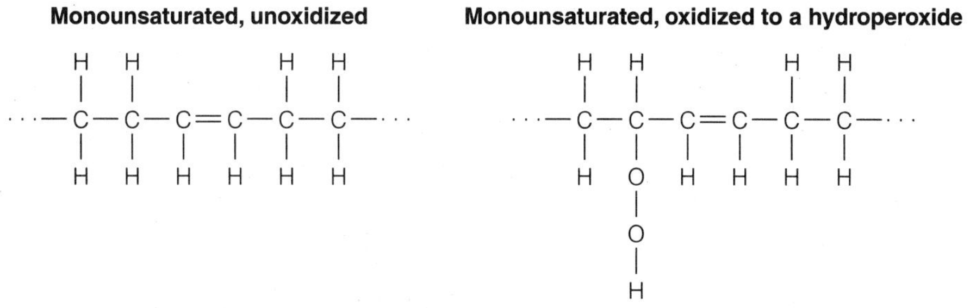

Monounsaturated, unoxidized

···—C—C—C=C—C—C—···

Monounsaturated, oxidized to a hydroperoxide

···—C—C—C=C—C—C—···
          |
          O
          |
          O
          |
          H

FIGURE 1.5 The chemistry of an opened jar of peanut butter. *Monounsaturated fatty acids like oleic acid in peanuts can become rancid. The chemical change that leads to the unpleasant smell is shown in the second chemical structure.*

For example, if an opened jar of peanut butter sits on the shelf long enough, a chemical reaction will occur to the peanut oil in the product that results in an undesirable odor. The food scientist would recognize the reaction as an oxidation of unsaturated fatty acids in the peanut butter and could use chemical structures to describe it (Figure 1.5).

Physics and Food Science

Physics is the study of matter and energy and is concerned with the physical changes that matter experiences under certain conditions. Many branches of physics are important to food scientists. Thermodynamics seeks to explain the properties of matter such as the solid or liquid state of food. For example, in the manufacture of solid chocolate, it is critical to understand the concept of melting point behavior and how it relates to the final product texture. Food scientists use the physics of the electromagnetic energy spectrum as a basis for understanding how food color is detected, how microwave cooking works, and how the application of irradiation makes foods safe.

Engineering and Food Science

Engineering is a discipline traditionally devoted to the study of momentum, heat, and mass transfer. Food scientists apply the science of engineering to understand the interaction of physical and energy transfer operations with food materials. The unit operations of the food industry (Chapter 8) invoke engineering principles that are applied to food processing (heat transfer during thermal processing), packaging, refrigeration, freezing, evaporation, and drying.

The manner in which a raw potato slice dropped into hot oil is transformed into a hot french fry or a crisp potato chip is explained by discussing the concept of heat and mass transfer from the cooking oil to the potato. Despite our everyday familiarity with such food products, explaining "how" a food is transformed presents a challenge to the food scientist, who must be simultaneously knowledgeable in the basics of biology, chemistry, physics, and engineering.

1.2 THE FOOD PROCESSING INDUSTRY

Food processing technology, in which raw (animal or plant) food ingredients are converted into specific foods for consumption, has had a long history (Table 1.1). For thousands of years people have been salting, drying, smoking, pickling, and chilling their foods. Canned and frozen foods are a fairly recent development of the food processing industry. Researchers over the years have found new and better ways to process, package, and preserve foods from the time of the harvest to the time they get to the consumer.

Historically the human race has met challenges via breakthrough efforts—new developments and discoveries that in turn generate further discoveries and developments with the gradual passage of time. However, with the industrial era, the speed of change was startlingly and dramatically swift. A minor technological improvement in one field often has resulted in an avalanche of unforeseen progress in others. For example, advances with optics and lenses enabled the development of microscopes, which permitted the

TABLE 1.1 Examples of food technology throughout history

Technology Developed	Approximate Date
Milling of cereal grains into flour	10,000 BC
Baking unleavened bread	10,000 BC
Meat and fish smoking, salting, drying	4,000 BC
Grape and barley fermentation	3,000 BC
Yogurt fermentation	200 AD
Canning	1800s
Iron roller milling of flour	1800s
Milk pasteurization	1800s
Freeze-drying	1900s
Modified atmosphere packaging	1900s
Food irradiation	1900s

birth of the science of bacteriology, which profoundly influenced the areas of health, medicine, food safety, and food preservation.

Today, the food processing industry is vital to the global economy. More nutritious food is available to the American public, to consumers in developing countries, and to countries suffering from famine, thanks to modern food processing. Food processing is the largest manufacturing industry in the United States. We will talk about these food processing areas in more detail in Chapters 2, 8, and 9:

- Food ingredients of all kinds
- Milk and dairy products
- Meats, poultry, seafood, eggs, and legumes
- Fruits and vegetables
- Cereals, grains, and baked goods
- Beverages
- Sugar and alternative sweeteners
- Fats, oils, and fat substitutes
- Confection and chocolate

History of the Food Industry

Food, clothing, and shelter are commonly regarded as the necessities of life. Of the three, food (including water and other water-based beverages) is the most essential. People simply cannot live without food, and the need for a continuous food supply has led to the development of the modern, organized food industry that feeds the world. But how did this industry arise? In Europe during the mid 1700s, machines and power tools replaced the crude and simple hand tools used in agriculture and in manufacture. This advance allowed for the development of large-scale industrial production of consumer goods and had profound economic and societal effects. This Industrial Revolution had a bearing on the development of the food industry as well.

Historical problems such as food shortages and starvation became less frequent with the advent of more efficient mechanized food production and distribution. A specialized dairy industry developed in Holland, which became famous for exporting Edam and Gouda cheeses during the eighteenth century. In Europe, **scientific agriculture**, involving crop rotation and improved soil fertilization, meshed with improved livestock feeding and management.

Later, technological advances, the science of genetics, and the use of chemical fertilizers further expanded agricultural output. The growth of towns into cities and the transition of paths into roads and railways paved the way for greater goods transport and marketing opportunities. In the decades of the 1800s and early 1900s, the urbanization and industrialization of Europe and America allowed for a rise in the standard of living for more people, who were then better able to afford to feed themselves and their families.

Canning of Foods. The idea of **canning** grew from the knowledge (circa 1700) that cooked meats could be preserved for a limited time if they were covered with a layer of fat, which kept the meat from contact with the air. One hundred years later, Europeans were storing substances like fruit syrups and preserves in glass bottles. In response to Napoleon's need to feed his troops, Frenchman Nicolas Appert (b. 1752) combined these ideas. He placed foods such as cooked meat, vegetables, and even milk into glass bottles that were then closed off (corked) and heated, the idea being to drive the air off by heating. A series of food spoilage mishaps taught preparers that the heating step had to be standardized for different can sizes. It was the heat that killed the spoilage bacteria and thus prevented spoilage, not driving off the air as originally thought.

The Englishman Brian Donkin took the essential next step, substituting an unbreakable "tin" can for the breakable glass bottle. The canning industry was born in 1812 with Donkin's canning factory. In six years it boasted a canned product lineup that included beef, carrots, mutton, vegetable stew, veal, and a soup. Gail Borden in the United States developed a canned version of milk that used added sugar to help extend the shelf life. This product was so popular with the armies during the Civil War that the men carried their taste for it back to the civilian market. By the end of the 1800s, the Massachusetts Institute of Technology created a chart showing the effective pasteurization times

FIGURE 1.6 Cans extend the shelf life of foods.

and temperatures for a wide variety of foods. The positive results of the canning revolution thus included consumer convenience, low food cost, variety, and safety (Figure 1.6).

Refrigeration Technology. In the days before **refrigeration** was developed, "icehouses" were used to maintain a supply of ice. In the United States, the Shakers in particular were icehouse adherents. An icehouse was a heavily insulated room with a carpeted floor and insulated roof designed to store blocks of natural ice obtained from cold northern lakes.

By the mid 1800s ice-making technology had arrived, and ice machines were patented. An improved compressor designed by Scotsman James Harrison allowed for the development of an ice factory and a refrigerating machine. The latter was exploited by the Australian brewing industry. As you can imagine, manufactured ice was an enormous help to the fishing industry too. An unheard of dinner event took place in Australia in 1873—at a public banquet, meat, poultry, and fish that had been frozen for six months using the new technology were cooked and served to the highly impressed guests.

The Discovery of Vitamins. Our brief survey of the development of the food industry would be incomplete without mention of the discovery of vitamins and other nutrients and their use in food production along with other food additives. In 1846 Justus von Liebig, writing in his report *Chemistry and Its Applications to Agriculture and Physiology*, established the foundation of modern human nutrition by demonstrating that living tissue (and foods) were composed of carbohydrates, fats, and proteins, which were of differing biological value.

Dutch physician Christiaan Eijkman's investigations into beriberi treatment using rice polishings and the work of certain European scientists including A. C. Pekelharing and Wilhelm Stepp paved the way for the isolation and discovery of the vitamins. In the 1930s British doctor Harriet Chick and colleagues proved beyond all doubt the value of vitamin D in relieving the symptoms of rickets in a large number of artificially fed babies in pre-World War II Austria.

Adulteration of Foods. Some wholesalers and retailers in the 1800s extended food profits and supply by adding cheap bulk ingredients to certain foodstuffs. The practice was termed **adulteration**. Examples were adding "pepper dust" to pepper, which may have consisted of pea flour, mustard husks, or even floor sweepings! Tea was adulterated with a substance called "smouch," which was made from dried ash leaves and mixed in with the tea leaves. Used tea leaves were even recycled by unscrupulous merchants, who revitalized them with gum solutions and colored them with black lead.

FIGURE 1.7 Adulteration. *During the 1800s tea was adulterated with "smouch," and used tea leaves were resold.*

FIGURE 1.8 Assorted products developed by food scientists.

Commercial bread flour was extended with alum, a mineral-salt whitening agent; coffee was diluted with chicory or acorns; and cocoa powder often contained an addition of brick dust. These ingenious but deceitful and potentially harmful practices were exposed in the 1850s in England, which drafted its first British Food and Drugs Acts in 1860 and 1872 to regulate food purity. In the United States as early as 1785, Massachusetts enacted the first general food adulteration law. A 1905 book penned by Upton Sinclair, entitled *The Jungle*, detailed unsanitary practices in the meat industry; its publication led to enactment of the 1906 Food and Drugs Act and the Meat Inspections Act.

The Work of the Food Scientist Today

A **food scientist** or food technologist is a person who applies scientific knowledge and technological principles to the study of foods and their components, either in a research setting (university, industry, or government) or a manufacturing setting. Research food scientists are skilled at making observations, and observations lead to an understanding of how biological, chemical, and physical factors affect food.

In the manufacturing area, food scientists are concerned with analyzing product quality and safety, creating new products, and investigating new manufacturing methods. Where did the ideas for fruit roll-ups, fat-free yogurt, coffee singles, microwave brownies, flavored water, or high-fiber cereal come from? The answer: from the minds of food scientists.

The need for trained food scientists has grown in pace with consumer demands for convenient, safe, and nutritious foods and beverages. More emphasis is presently being placed on the safety of the food supply. Manufacturers are undergoing a transition to adopting food to new processing/quality assurance strategies.

Energy and cost-efficient technologies are being investigated in the processing of foods, including **microwave pasteurization, ultrafiltration,** and **reverse osmosis.** Health-conscious consumers and food companies with marketing savvy are also paying much attention to the development of nutraceutical foods—foods that provide health benefits beyond those due to the essential nutrients (discussed below). The skilled food scientist will therefore need to exhibit a combination of technological expertise as well as consumer awareness.

Food scientists use the laws of science and engineering to produce, process, evaluate, package, and distribute foods. The kind of work they do depends on the company they work for and the products they develop. Food scientists may concentrate on basic research, product development, quality assurance, processing, packaging, labeling, technical sales, or market research. They may work in production or technical management. They check on food standards, laws, and safety (Figure 1.9).

Basic Research. Food scientists in basic research study the chemical, microbiological, physical, and sensory properties of foods and their ingredients. An example of a chemical property is the color change in canned vegetables during storage. A microbiological study might check on the stability of baked muffins to spoilage by mold. The hardening of gummy bear candies is an example of an important physical property. The identification of altered flavor in an ultra-high temperature heat-treated dairy beverage is an example of a critical sensory property.

Basic food scientists obtain fundamental information from food research, while *applied* food scientists apply the information to food product development and testing. Research is an activity driven by questions that are answered through experimental testing and observation. It is considered an iterative process, in that a single question that receives an answer can lead to a multitude of successive questions requiring answers. Through this systematic process of inquiry, new information is constantly obtainable, and previous knowledge can be refined.

Product Development. Most large food processing companies have test kitchens, and food scientists work closely with test kitchen staff. Food scientists also work with engineers, microbiologists, flavor experts, sensory evaluation experts, packaging specialists,

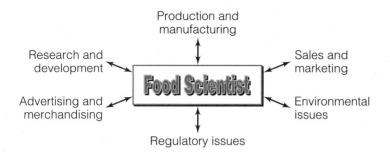

FIGURE 1.9 Food scientists do not work in isolation.

statisticians, and marketing professionals. They also interact with field buyers, production line workers, and warehouse staff.

Quality Assurance. Food scientists in **quality assurance** (QA) sample and check raw products to see if they are fresh and conform to purchasing specifications. Quality assurance personnel ensure that operators monitor each processing step to make sure the products meet government, company, and industry standards. They check warehouses and storage space for sanitation, temperature, and humidity. Of major importance is food safety testing, including microbial analysis. Foods are sampled and tested for the presence of specific biological, chemical, and physical hazards, including microorganisms of significance in terms of food safety.

Research Tools and Scope. The tools of the research food scientist are similar to that of other research scientists, in that well-established analytical methodologies and instrumental techniques are useful in the study of changes to foods and food components especially during processing and storage. Research food scientists are usually specialists in one of the focus areas mentioned previously, such as food chemistry, food engineering, and so on. Experiments may be related to flavor research, product formulation revisions, or innovations.

Scientists are skilled in making observations, which lead to the development of **scientific principles.** Food research and development (R&D) scientists are trained in the **scientific method.** Although the complete, stepwise scientific method approach may not be feasible in certain instances due to time, resource, or other limitations, it is important to learn and apply it whenever possible and practical.

This scientific approach begins with a statement of the experimental problem and is often in the form of a question ("how can we formulate a fat-free vanilla ice cream?"). The scientist next proposes a possible solution (hypothesis) to solve the problem and proceeds to design the necessary experiments to test the hypothesis. This process requires careful planning; designing appropriate experimental strategies; conducting experiments under specific controlled conditions; recording observations; analyzing, interpreting, and reporting results; and drawing conclusions related to the original hypothesis.

Research food scientists may specialize in studying one food group and its components, for example, cereal grains or dairy products. Some food scientists conduct research into ingredients only—**alternative sweeteners, fat substitutes, stabilizers, antioxidants, acidulants,** flavors, and food additives. The *Journal of Food Science*, a publication of the IFT, prints food research articles in five specific focus areas: food chemistry/biochemistry; food engineering/processing; food microbiology; nutrition; and sensory evaluation. Another IFT publication, *Food Technology*, features a specific topic each month, as well as food industry news and analysis, and information on new food products and technologies.

Nutraceuticals. In discussing the difference between science and technology, we indicated that knowing the impact that new technologies have on people and the environment requires long-term study. Today we are experiencing such an event with the development of *nutraceutical* food products. The term **nutraceutical** has been given to a proposed new regulatory category of food components—a food that, in whole or in part, is able to provide health improvement or disease prevention. The technology exists to design and produce nutraceutical food products now. However little scientific evidence is in regarding which food products and which specific chemical nutraceutical substances would be the most useful in the diets of people. Information is also lacking regarding the consequence of chronic ingestion of a certain level of a particular nutraceutical.

1.3 MAJOR CLASSES OF FOOD COMPONENTS

The molecules that make up foods can be classified into those that are needed in the diet in relatively greater amounts: water, carbohydrates, lipids, and proteins, and those that, although critically important, we require less of: vitamins and minerals. These six represent the traditional food components and are termed *nutrients*, in that the human body requires them to sustain life and maintain health. These nutrients—the "stars" of food science—will be discussed in much detail in upcoming chapters.

A seventh class of food component comprises certain substances present in plants that are thought to be potentially beneficial to health. These plant-derived chemicals, called **phytochemicals**, are not proteins or parts of proteins, not carbohydrates, not lipids, not vitamins, and not minerals. Some scientists consider them nonessential and nonnutrient. However, their benefit may be in providing protection against cardiovascular disease, certain cancers, and other age-related diseases, and our understanding of their potential benefits is growing. The present challenge of nutritional science is to discover their specific health effects. It is also clear that not all of a plant's natural chemicals are beneficial; some in fact are proven toxicants. To the toxicologist, the challenge is to establish dose tolerances. The challenge to the food scientist is to create new foods incorporating these substances that will offer an enjoyable as well as safe and healthy eating experience.

Phytochemicals are ingested in varying degrees when we consume foods like fruits, vegetables, cereals, and grains. Certain plant foods are concentrated sources of particular phytochemicals (e.g., *alliin* in garlic, and *genistein* in soybeans). Phytochemicals occur naturally in foods, or as food extracts, or as additions to designed or engineered foods.

The Food Guide Pyramid

We make food choices daily. These choices may be wise when we consume a variety of foods, with the proper balance of nutrients, or unwise, when we eat out of balance, such as eating too much fat, too many calories, too much protein, not enough vitamin D, or not enough calcium.

🍴 **The seven types of food components:**

Major components

water
carbohydrates
lipids
proteins

Minor components

vitamins
minerals
phytochemicals

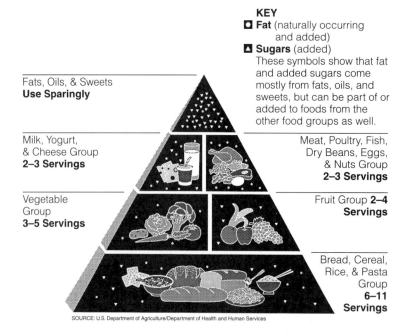

FIGURE 1.10 The food guide pyramid. *The pyramid shows the relative amounts of foods that should be consumed to promote health. The foods that need to be consumed the most (grains, fruits, and vegetables) are found in the bottom half of the pyramid, while those that should be consumed in moderate amounts (meats and dairy) or less frequently (fats and sweets) are found in the top half.*

Unwise eating may occur when people respond to food in an emotional manner. We occasionally indulge in an extra helping of dessert to reward ourselves after a hard day's work, or for comfort after an upsetting day at school. Although people occasionally eat without conscious regard to variety, balance, nutrients, and calories, most realize that one of the keys to health is proper selection of foods from among several groupings. The Food Guide Pyramid provides a useful representation of these food groupings.

The **Food Guide Pyramid** has been developed as an educational tool to remind people which sources of food components (as nutrients) we need to consume on a daily basis, and in what relative amounts. The food groups represented by the Pyramid starting at its base (most to least recommended consumption), are: breads, cereal, rice and pasta; vegetables; fruits; meat, poultry, fish and dry beans, nuts, eggs and egg products; milk, yogurt and cheese; fats and oil products and sugars, sweeteners, and confections (Figure 1.10).

1.4 FOOD SCIENCE EDUCATION

Behind the thousands of kinds of foods and food products on supermarket shelves are university-trained researchers. The initial development of food science as a field of study paralleled the development of the food processing industry. The food science programs in North American universities evolved largely from the dairy science programs that were common, particularly at agricultural colleges, during the early to mid-portion of the twentieth century.

Spoilage problems in the early days of the canning industry (circa 1900) led to the involvement of university scientists, H. L. Russell of the University of Wisconsin and Samuel C. Prescott of Massachusetts Institute of Technology. These researchers solved the problem of microbial spoilage of canned corn by identifying the organisms involved and developing more effective **sterilization procedures.** Prescott and others went on to organize lectures, publish research findings, and develop courses combining areas such as chemistry and microbiology into the newly developing field of food science.

The establishment of pioneering courses in food chemistry, food preservation, food microbiology, and food processing can be traced to individuals such as W. V. Cruess, a chemist at the University of California, and Ernest Weigand, a horticulturist at Oregon State University. Cruess initially worked in the area of **enology** (the science of winemaking), studying ways to improve the quality of California wines to make them competitive with European products. When Prohibition brought his research to an end, he expanded his interests and developed teaching and research programs in other areas of food science. In many states, dairy science programs evolved into the food science and technology programs we recognize today.

Earning a Food Science Degree

Food science has had an identity problem. Although many universities have food science programs, college students interested in general science are frequently unaware of the food science major or the opportunities awaiting them if they earn a degree at any level (B.S., M.S., or Ph.D.) in food science.

Students wishing to study food science at the undergraduate level generally take courses in food chemistry, food engineering and processing, food microbiology, nutrition, and food marketing, as well as courses in the supporting disciplines, general sciences, and commodity areas according to interest. Topics covered include engineering aspects of food processing, packaging, and storage; measurement of physical properties including food texture; food industry unit operations; evaluation of changes in nutrient content of foods during processing and storage; protein and enzyme technology; food product development; functional properties of food components; quality assurance; and current research.

Graduates with a B.S. degree can work in a wide variety of jobs in the food industry and in government. The primary job areas include food analysis and testing, food pro-

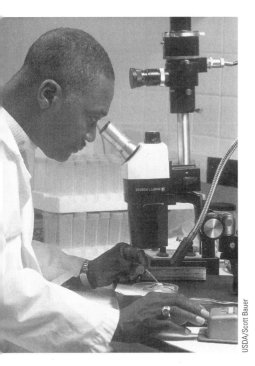

FIGURE 1.11 Food microbiologists investigate the incidence and growth of microorganisms associated with foods.

☞ HACCP is a systematic approach to ensure food safety in a food-processing environment (find more details about HACCP in the food microbiology and food safety chapters, 10 and 11).

☞ Hurdle strategy is using a combination of preservation methods to inhibit microorganisms (find more details about the hurdle concept in the food processing and food safety chapters, 8 and 11).

cessing, production supervision, product development, quality assurance, and sales and marketing. Many of the job titles related to these areas include the word *technician*. For example, a product development technician's responsibilities might include preparing product concepts and testing product samples to determine optimal production processes, participating in the product design process, performing product line trials, documenting the methods of pilot plant equipment, and coordinating pilot plant operations. Job areas range from quality assurance on testing of raw materials and finished products, to nutrition on labeling requirements and regulations, to the biotechnology area in new food development.

Graduate degrees (M.S. or Ph.D.) are usually needed to perform research in academic laboratories. For students interested in studying the advanced topics in food science—food chemistry, food engineering, food microbiology, and sensory evaluation—a graduate degree (M.S. or Ph.D.) is essential.

Food Microbiology. Microbiology impacts food production, processing, preservation, storage, and safety. Food microbiology students may use a wide variety of state-of-the-art technologies from fields including immunology, bacteriology, and molecular biology in their research (Figure 1.11). Microbes such as yeast, molds, and bacteria are being used for the production of foods and food ingredients such as proteins, enzymes, vitamins, and flavors. Beneficial microbes are exploited in the fermentative production, processing, and preservation of many foods and beverages. Spoilage microorganisms cost food producers, processors, and consumers millions of dollars annually in lost products. Lost productivity resulting from illness caused by foodborne microorganisms is an enormous economic burden throughout the world.

The study of **food microbiology** includes understanding not only the factors influencing the growth of microorganisms in food systems but also the means of controlling them. Knowledge of Hazard Analysis and Critical Control Points programs **(HACCP)** (pronounced ha-sip) and microbial hurdle strategies is essential. Students who specialize in food microbiology usually have sound undergraduate training in biochemistry, chemistry, microbiology, organic chemistry, and physics.

Food Chemistry. Foods are essentially chemicals in various arrangements. Food flavor, color, texture, and nutrition are the result of a food's chemical composition. **Food chemistry** is concerned with analytical, biochemical, chemical, nutritional, physical, and toxicological aspects of foods and food ingredients. Food chemistry research focuses on understanding relationships between the structure and functional properties of food molecules and improving the nutrition, safety, and sensory aspects of food. Product development and flavor research require students to become skilled, knowledgeable food chemists. Students studying food chemistry should have a good background in the basic sciences and should specialize in one or more of the following minor areas: biochemistry, chemical engineering, nutritional biochemistry, organic chemistry, physical chemistry, toxicology, or analytical chemistry.

Food Engineering. **Food engineering** applies engineering principles to food processes and food processing equipment. Because engineering is a quantitative discipline, the food engineer's fundamental tool is mathematics. Chemistry and microbiology are also important disciplines because processes of concern to food engineers may involve chemical reactions and microbial interactions. Food engineering students gain a thorough understanding of thermodynamics, reaction kinetics, and transport phenomena applied to food processes. Knowledge of computer programming, microprocessor applications, statistics, and engineering economics is encouraged. Courses in thermal processing and other unit operations, physical and engineering properties of foods, rheology, and food packaging are vital for the student of food engineering.

Sensory Evaluation. **Sensory evaluation** uses test methods that provide information on how products are perceived through the human senses of sight, smell, taste, touch,

FIGURE 1.12 Sensory evaluation panelists often input their data directly into computers.

and hearing. The importance of sensory perception to food quality is widely appreciated in the food industry. There is an ongoing need for sensory specialists (Figure 1.12).

Like other quantitative disciplines, sensory evaluation attempts to provide precise and accurate measurements. However, data collected from human beings present a special challenge. Proper experimental design is critical and sensory specialists should also be statistical specialists. Learning the principles of human judgment and perception means that courses in the behavioral sciences are also important.

Food Science

- *Who should study food science?* If you are generally interested in foods, nutrition, and health, or if you enjoy studying science, then food science may be for you. It provides a fundamental background in science so it opens a variety of career doors. Students of food science have time to explore related fields and may choose food science as a minor as career objectives become focused. Applied courses, such as cereal, meat, or dairy sciences, allow students to explore many areas of the food industry. We are all involved with this industry, either directly as employees, or as consumers. A continuous need for trained personnel to guide and run the food industry means many opportunities, and students can prepare for a wide range of careers or for graduate school.
- *Can food science students use their major in preparation for professional school or medical school?* Yes. Food science provides enough rigor in the basic sciences for further educational advancement in graduate schools or professional schools. Demand has increased for students educated in the area of food safety, due in part to outbreaks of diseases associated with foodborne infections. But the other food science areas are needed as well.
- *Will food science students be immediately employable after completion of a Bachelor of Science (B.S.) degree?* Yes. Most students have no difficulty obtaining employment throughout the food industry. Well-known companies such as General Mills, Coca-Cola, Pillsbury, Dean Foods, Nestle and Campbell's, as well as state and federal government agencies, hire at the B.S. level.
- *Can food science students finish the B.S. degree in four years?* Yes. Some students, however, do opt to extend their programs to take advantage of various internship opportunities or minor studies that require an extra semester or two for completion.
- *Do food science students need a minor?* No, but in general a minor is selected based on interest and gives your education added depth. In certain cases, a minor may

> **Some membership divisions of IFT available to students:**
> *Biotechnology*
> *Carbohydrate*
> *Citrus Products*
> *Dairy Foods*
> *Food Chemistry*
> *Food Engineering*
> *Food Laws & Regulations*
> *Food Microbiology*
> *Food Packaging*
> *Foodservice*
> *Fruits & Vegetable Products*
> *Marketing & Management*
> *Muscle Foods*
> *Nonthermal Processing*
> *Nutraceuticals & Functional Foods*
> *Nutrition*
> *Product Development*
> *Quality Assurance*
> *Refrigerated & Frozen Foods*
> *Religious & Ethnic Foods*
> *Seafood Technology*
> *Sensory Evaluation*
> *Toxicology & Safety Evaluation*

increase your marketability to a prospective employer. Also, a minor (such as business) allows a student to explore and discover a field not necessarily limited to the area of the major field. For example, it is not uncommon for students of food science to minor in business, or chemistry, or microbiology.

- *What activities outside of coursework in food science are important?* Participation in a student food science club and IFT activities provides an opportunity to learn leadership skills by serving as officers or by organizing events. The IFT provides various divisions of food science specialization that food science students can join.
- *Are opportunities in food science limited geographically?* No. Graduates obtain positions across the U.S. and also internationally, since the food industry is worldwide.
- *Do food science students receive a traditional specialist education?* Yes, the typical specialist-oriented educational food science degree programs are offered through traditional American colleges and universities. These build a foundation of basic knowledge in the sciences and further expertise in essential food science and technology areas. However, this approach has been criticized for training students to be competitive individualists rather than the team players needed in a corporate environment. **Teamsmanship** describes an ability to work well with others (as part of a group or team) toward a common goal and requires interpersonal skills such as listening, communicating, and providing feedback and accountability. Team players utilize the resources available to the team in an efficient and effective manner to solve problems and meet deadlines.

In response to this criticism, many undergraduate programs and courses offer team exercises that allow students to work together cooperatively toward common goals. In addition, food companies often hire undergraduate student interns, providing opportunities for learning not only specific job tasks but also the process of top-down, bottom-up planning, executing, and measuring of functions that are key to the success of a company.

Key Points

- The scope of food science and technology includes the application of biology, chemistry, engineering, microbiology, and physics to the study of food composition, preservation, processing, distribution, storage, and use.
- The difference between food science and nutrition is that nutrition deals with the effects of foods in the person who consumes them, while food science is concerned with changes in foods during processing, manufacture, and, storage.
- The development of the food industry was related to changes occurring in Europe during the 1700s, while the study of food science at universities originated much later during the early part of the twentieth century.
- The Food Guide Pyramid offers an easy way for consumers to remember which foods to eat the most of and which to eat the least of, by arranging food groups within a pyramid-shaped three-dimensional display chart.
- The seven classes of food components are water, carbohydrate, lipid, protein, vitamins, minerals, and phytochemicals. The first six are nutrients. The seventh is a category of nonnutrient plant-derived substances that may promote health.
- The *Journal of Food Science*, an IFT research publication, has identified five food research focus areas: food chemistry/biochemistry, food engineering/processing, food microbiology, nutrition, and sensory evaluation.
- The experimental method of food scientists involves a sequence of steps that scientists apply when solving problems and during experimentation, including making a hypothesis, designing experiments, recording observations, data analysis and interpretation, and revisiting the hypothesis in light of the results.

Study Questions

1. Which could *not* be an example of a processed food developed with the aid of technology?
   a. dried pineapple slices
   b. fresh peas
   c. bottled pickled herring
   d. fermented sausage sticks
   e. frozen microwave scrambled eggs

2. Which statement is false?
   a. Food science is the mass production of food products from raw animal and plant materials utilizing the principles of engineering.
   b. The IFT definition for food science implies that it is a discipline based upon biology, physical sciences, and engineering.
   c. Nutrition involves the study of what happens to foods after consumption to promote and maintain health.
   d. Food science and technology is concerned with aspects of agricultural, avian, mammalian, and marine food materials.
   e. Food science involves the study of quality aspects of foods prior to consumption.

3. This early example of the use of technology would have been practiced long before canning was developed.
   a. smoking fish
   b. food irradiation
   c. pasteurization of milk
   d. freeze drying
   e. modified atmosphere packaging

4. Which is an example of the relationship of chemistry to food science?
   a. the fact that fruits and vegetables contain carbon, hydrogen, and oxygen atoms
   b. plant crop genetics
   c. monounsaturated fatty acid oxidation
   d. both a and b
   e. choices a, b, and c

5. In understanding the processing of food via deep-fat frying, knowledge of _____ is critical.
   a. bioengineering
   b. heat and mass transfer
   c. microwave energy
   d. irradiation
   e. fermentation

6. The manufacture of an HFCS (high-fructose corn syrup) sweetened soft drink involves the _____ processing industry.
   a. beverage
   b. fats and oil
   c. corn
   d. both a and b
   e. both a and c

7. Research into the change in green color intensity of canned and stored peas represents a _____ study.
   a. microbiological
   b. engineering
   c. chemical
   d. sensory flavor
   e. sensory texture

8. This food scientist/technologist work area involves sampling raw products to ensure conformity to purchasing specifications.
   a. product development
   b. experimentation
   c. quality control
   d. marketing
   e. sensory research

9. Which sequence most accurately represents the scientific method?
   a. problem → hypothesis → experiment → publish
   b. problem → hypothesis → experiment → conclusion
   c. experiment → plan → interpret → hypothesis
   d. plan → hypothesis → experiment → report
   e. plan → hypothesis → experiment → publish

10. Which statement is false regarding nutraceuticals?
    a. All potential nutraceutical food substances are known.
    b. Nutraceutical products may enhance health.
    c. Food companies can design nutraceutical foods now.
    d. Nutraceutical are foods.
    e. Consequences of chronic nutraceutical ingestion are not entirely known.

11. Which organelle, present in a plant cell but not an animal cell, affects tissue structure?
    a. vacuole
    b. mitochondria
    c. cell membrane
    d. plastid
    e. cell wall

12. Which statement is true?
    a. The physical state of water depends on the motion of oxygen molecules.
    b. Water molecules in steam are more closely aligned than in ice.
    c. The physical state of water depends on the motion of H_2O molecules.
    d. Water molecules in liquid water are more closely aligned than in ice.
    e. The physical state of water depends on the motion of hydrogen molecules.

13. Which choice is accurate regarding the milk typically served for breakfast?
    a. It contains albumen proteins.
    b. It has received homogenization and pasteurization processing.
    c. It is slightly sweet due to the presence of milk fat.
    d. both a and b
    e. choices a, b, and c

14. Which item does *not* belong with the others in the group?
    a. food biotechnology
    b. organization of cellular tissue
    c. mold growth on cheese
    d. chemical oxidation of fat
    e. yeast action in bread dough

15. Which practice, characterized by crop rotation, applied genetics, and the use of fertilizers, aided the development of the food industry?

a. livestock management
b. industrialization
c. scientific agriculture
d. mechanized food distribution
e. waste-water management

16. Food adulteration is related to all of the following except
    a. reduced profitability.
    b. addition of cheap bulk ingredients.
    c. lessening product quality.
    d. deceiving the consumer.
    e. altered nutritional quality.

17. Who is responsible for early canning techniques?
    a. Gail Borden
    b. Nicholas Appert
    c. Upton Sinclair
    d. James Harrison
    e. Christian Eijkman

18. In Australia during the late 1800s, which new technology was behind the success of the serving of meat and fish entrees that had been stored for several months?
    a. ice houses
    b. thermal packaging
    c. nonthermal packaging
    d. fermentation
    e. refrigeration

19. Which branch of food science focuses on the functional properties of proteins?
    a. food microbiology
    b. food engineering
    c. food chemistry
    d. food processing
    e. sensory evaluation

20. Which statement regarding phytochemicals is true?
    a. Phytochemicals are substances derived from animals and plants that are thought to be beneficial to health.
    b. The vitamin "tocopherol" (vitamin E) is considered a phytochemical.
    c. Dose tolerances of all phytochemicals are studied by nutritionists.
    d. Food scientists are at present unable to incorporate phytochemicals into processed foods.
    e. Certain phytochemicals can produce toxic symptoms.

Internet Search Topic

To answer questions 21 to 26, go to the web site of the Institute of Food Technologists (http://www.ift.org), and under Continuing Education & Professional Development, find "Introduction to the Food Industry." Read Lesson 4: "Integrated Resource Management." Focusing on managing resources within a food processing manufacturing environment, this lesson explains how forecasting and master production scheduling affect material requirements planning, stresses the importance of employee teamwork, and introduces the role computers play in the manufacturing process.

21. In your own words, what is the meaning of integrated resource management (IRM)?

22. Why is integrated resource management important to implement in the food industry?

23. Create a chart or table to show how planning, execution, and measurement are used in process manufacturing.

24. What important resources would need to be managed in a facility that manufactures snack crackers?

25. What is the importance of teamwork in food product development?

26. How might a student avoid becoming a traditional "specialist loner" graduate?

References

Chen, Y. P., Andrews, L. S., and Grodner, R. M. 1996. Sensory and microbial quality of irradiated crab meat products. *Journal of Food Science* 61(6):1239.

He, Y., and Sebranek, J. G. 1996. Frankfurters with lean finely textured tissue as affected by ingredients. *Journal of Food Science* 61(6):1275.

Mitchell, G. V., Grundel, E., and Jenkins, M. Y. 1996. Bioavailability for rats of vitamin E from fortified breakfast cereal. *Journal of Food Science* 61(6):1257.

Mrak, E. H. 1978. Food science and technology: past, present, and future—the W. O. Atwater memorial lecture. In: *Agricultural and Food Chemistry: Past, Present, and Future*, pp. 47–57, R. Teranishi, ed. AVI Publishing Company, Westport, CT.

Potter, N. H., and Hotchkiss, J. H. 1995. *Food Science*, 5th ed. Chapman & Hall, New York.

Shieh, C.-J., Akoh, C. C., and Koehler, P. E. 1996. Optimizing low fat peanut spread containing sucrose polyester. *Journal of Food Science* 61(6):1227.

Toran, A. A., Barbera, R., Farre, R., Lagarda, M. J., and Lopez, J. C. 1996. HPLC method for cysteine and methionine in infant formulas. *Journal of Food Science* 61(6):1132.

Additional Resources

Boyle, M. A. 2001. *Personal Nutrition*, 4th ed. Wadsworth/Thomson Learning, Belmont, CA.

Brown, A. 2000. *Understanding Food: Principles and Preparation*. Wadsworth/Thomson Learning, Belmont, CA.

Hegarty, V. H. 1995. *Nutrition, Food, and the Environment*. Eagan Press, St. Paul, MN.

Kittler, P. G., and Sucher, K. P. 2001. *Food and Culture*, 3rd ed. Wadsworth/Thomson Learning, Belmont, CA.

Penfield, M. P., and Campbell, A. M. 1990. *Experimental Food Science*, 3rd ed. Academic Press, San Diego, CA.

Web Sites

http://www.easynet.co.uk/ifst/ifstfaq.htm
Institute of Food Science & Technology (UK)

http://www.ift.org/
Institute of Food Technologists

READING THE RESEARCH

The scope of food science research is broad—prepare to take a brief glimpse into five food research articles to see just what studies the scientists are performing. Since much of the terminology may be new, supplemental information is included to help you understand each summary. After examining the five research summaries, you should have a clearer insight into what food science is about and what research food scientists do. Each article summary and supplemental information is based upon specific *Journal of Food Science* published abstracts. The complete article citations are provided in the References for those interested in reading the full articles. (The *Journal* is one of two publications of IFT, the other is *Food Technology* magazine.)

The challenge is for you to see if you can answer these questions:

- What was the purpose of the research?
- Why was the research important?
- Which food or beverage component was studied?
- What was discovered?
- Did the study fit its category only (the designated research area) or was there overlap with another category?

By approaching these summaries within the context of the questions, you can begin to develop an understanding of basic concepts related to each topic and gain valuable food science vocabulary.

Article 1: "HPLC method for cysteine and methionine in infant formulas"
Research Area: Food Chemistry/Biochemistry

Supplemental Information. Commercial infant formula may be cow's milk-based (whey protein or casein protein) or plant protein (soy-based). Infant formula contains many nutrients including proteins, which are composed of amino acids. Two sulfur-containing amino acids, methionine (MET) and cysteine(CYS), are important nutritionally for infants. Although infant formulas are designed to resemble human breast milk nutritionally, these products are not required to carry amino acid content information on their labels. HPLC stands for high performance liquid chromatography and is a technique used to separate discrete compounds from a complex mixture (in this example, MET and CYS from milk formula samples). The researchers used this technique to answer the question: "How much methionine and cysteine are actually in these different infant formulas?"

Summary. The stated objective of this research was to use a particular instrumental analytical technique called HPLC in measuring the MET and CYS content of various infant formulas. Before performing HPLC, three important sample preparation steps (oxidation, hydrolysis, and derivitization) had to be completed per sample. These were accomplished using chemical reagents and a vacuum pump and nitrogen trap apparatus described in the article. Once prepared, the samples were easily analyzed using HPLC. The technique was found to be a reliable and relatively inexpensive procedure for measuring cysteine and methionine in infant formulas, though slightly time consuming compared to an automated system. The researchers discovered that the content of MET and CYS in three different products (whey, casein, or soy formulas) was comparable to that of human milk, which has approximately 4.2 grams of cysteine plus methionine combined per 100 grams of milk protein.

Article 2: "Optimizing low fat peanut butter spread containing sucrose polyester"
Research Area: Food Engineering/Processing

Supplemental Information. Standard commercial peanut butter is a high-calorie, high-fat (50%) food. Reducing fat content by removing peanut oil and replacing it with a low-calorie alternative must be accomplished without negatively affecting spreadability. SPE (sucrose polyester) is a sugar molecule with 4–8 fatty acids attached. Because of its structure, it is poorly absorbed and acts as a calorie-free fat replacer. In 1996 the FDA approved its use in snack foods. TPA stands for texture profile analysis, an instrumental method used to assess food texture. Modeling refers to the use of mathematical quadratic equations. In this experiment, modeling was used to help optimize the low-fat formulation in terms of product performance. The effects of peanut oil and emulsifier levels (which are called "independent variables") on each TPA parameter (which are called "dependent variables") were studied.

Summary. The stated objectives of this research were to model the physical properties of low-fat peanut butter spread formulated with varying proportions of peanut oil, SPE, and emulsifier and to determine the effects of these components on physical properties. TPA analyzed twelve formulations of varying SPE, peanut oil, and emulsifier content. TPA parameters were peanut butter hardness, adhesiveness, cohesiveness, and gumminess. The researchers discovered that SPE did not affect physical properties other than oiliness. Analysis of predicted models revealed that mainly the level of peanut oil in the formulation affected the TPA parameters. The optimum low-fat formulation was 26.6% fat content. It was made with 1% salt, 6% corn syrup solids, 2.6% emulsifier, 23.9% SPE, and 66.5% ground defatted peanuts.

Article 3: "Sensory and microbial quality of irradiated crab meat products"
Research Area: Microbiology

Supplemental Information. Irradiation is a food processing step that applies ionizing energy to foods at regulated doses. Commercial irradiation could be applied to foods such as crab meat that need to remain frozen to preserve freshness, although the process is not yet approved for seafood use. Irradiation neither raises the food temperature nor makes it radioactive. In this experiment a medium dose (2 kiloGrays) was used. Bacteria that contaminate food are called aerobes if they require oxygen in order to survive. They are called pathogens if they cause illness in people. Since crabs are hand processed, workers and unsanitary processing conditions can be sources of pathogenic organisms on the crabmeat. The "good" odors associated with freshly picked crabs are replaced by unpleasant "off" odors due to bacterial action during storage. Other chemical reactions caused by enzymes in the food and by oxygen in the air can interact with food molecules to produce undesirable changes. Volatile (gaseous) compounds such as ammonia are generated from protein breakdown, and rancid odor can result from a chemical reaction called lipid oxidation in crab meat.

Summary. The stated objectives of this research were to irradiate fresh crab meat products to reduce the risk from potential bacterial pathogens and to increase shelf life. Crab meat (white lump, claw, and finger) received up to 2 kGy ionizing energy. The products were analyzed over a 14-day ice-storage period for microbial and sensory quality and compared to nonirradiated controls. Microbial evaluation included total numbers of aerobic organisms, plus presumptive identification of pathogenic bacteria (*Staphylococcus*, *Listeria*, *Vibrio*, and *E. coli*). Sensory parameters included fresh ("good") crab odor and flavor, off-odor, and overall acceptability. The researchers discovered that irradiation was effective in reducing bacterial counts to below detectable levels and extended shelf life by more than 3 days. Off-odors and off-flavors developed more rapidly in nonirradiated controls. Overall acceptability was higher throughout the 14 days of storage for the irradiated samples.

Article 4: "Bioavailability for rats of vitamin E from fortified breakfast cereal"
Research Area: Nutrition

Supplemental Information. Vitamin E is referred to as tocopherol, and its four forms —alpha, beta, delta, and gamma—have different levels of antioxidant activity. In this experiment, the most potent form, alpha, was fed to rats that had previously received a deficient diet. The rat depletion-repletion model was used because it would be unethical to experimentally create a vitamin E deficiency in humans. Data from animal studies can often help scientists hypothesize how a nutrient might function in humans. So, findings from rat studies are cautiously extrapolated to humans. Bioavailability refers to the degree to which a particular form of a nutrient is able to be digested and utilized to maintain health. Cell, tissue, and blood plasma storage levels of vitamin E are indicative of bioavailability. Pyruvate kinase is an enzyme, meaning it is a substance that speeds up reactions that take place in living systems. Breakfast cereals are frequently fortified with vitamin E so they can serve as good sources of this nutrient. A food matrix effect implied that a substance (fiber) in the bran flakes reached a level that may have interfered with or confounded the effect of vitamin E during this feeding study.

Summary. The stated objective of this research was to extend the application of a vitamin E depletion-repletion animal (rat) model system by examining the response of selected biological parameters to graded levels of alpha tocopherol (a vitamin E form) and estimating the bioavailability of vitamin E in three vitamin E-fortified breakfast cereals. The parameters measured included alpha tocopherol levels in blood plasma, red blood cells, liver, heart, adipose tissue, and pyruvate kinase activity. The researchers discovered that the bioavailability of vitamin E when fed to rats in whole grain cereal and corn flakes was comparable to that when a pure standard form of the vitamin (tocopherol acetate) was used. Bioavailability in bran flakes was lower than that of the standard based on four of the five criteria, due to a food matrix effect.

Article 5: "Frankfurters with lean finely textured tissue as affected by ingredients"
Research Area: Sensory Evaluation

Supplemental Information. A frankfurter is considered a meat emulsion food system. As such, franks contain a fat phase and a watery phase that must mix and not separate. Various nonmeat food additives can be included in franks and other meat products to stabilize them as emulsions and to improve their quality. Salt (NaCl) and STPP, sodium tripolyphosphate, were used in this experiment as emulsion stabilizers. Kappa-carrageenan is a type of food gum. Food gums are classified as carbohydrates, so are related to starches. They improve meat texture by forming complexes with protein and water within meat products. ISP, which stands for isolated soy protein, acts as a fat and water binder to improve meat product juiciness and stability. LFTT—"lean finely textured tissue"—is tissue recovered from trimmed beef or pork. It was used in this experiment because it offered a low-cost, low-fat, high-quality protein alternative to standard beef and pork tissue.

Summary. The stated objective of this research was to evaluate the use of NaCl, STPP, kappa carrageenan and ISP for improving protein functionality and product sensory characteristics of frankfurters made with LFTT from beef and/or pork. Frankfurters were evaluated by 11 panelists for color, firmness, and cohesiveness. The researchers discovered that frankfurters made with LFTT were softer in texture, which was considered desirable since excessive toughness is a frequent problem with high-protein, low-fat meat products. ISP produced somewhat negative sensory results. Improved product stability and yields occurred when LFTT was used as a 50% replacement for lean meat in franks formulated with 2.5% NaCl, 0.25% STPP, and 0.5% kappa-carrageenan.

Finished?

This activity was intended to challenge you with new concepts and new terminology, illustrate the scope and breadth of real world food science research activities, and demonstrate the interdisciplinary nature of food science. Did it accomplish these three goals?

Challenge Questions

As a follow-up challenge, answer the following questions. Because they test learning and comprehension, completing them now is a good idea. Later, after covering the rest of the material in the course, revisit this material and see if it is easier for you to understand.

1. Define or explain the following terms without re-reading the material: TPA, ISP, HPLC, MET, CYS, SPE, STPP, LFTT, tocopherol.
2. List three types of infant formulas.
3. Describe one difference and one similarity between kappa-carrageenan and ISP.
4. Explain what is meant by bioavailability.
5. Define what is meant by extrapolation.
6. State the significance of a food matrix effect.
7. Describe two causes of off-odors in crab meat.
8. List two benefits of food irradiation.
9. Suggest how a low-fat peanut butter might be formulated and tested for success.
10. Provide the Latin (scientific) names of two potential crab meat pathogens. You only need the genus name (for instance, the "E" in *E. coli* stands for the genus *Escherichia*).

CHAPTER 2

Food Categories and Composition

2.1 Food Composition Tables

2.2 Beverages

2.3 Cereals, Grains, and Baked Products

2.4 Fruits and Vegetables

2.5 Legumes and Nuts

2.6 Meat and Meat Products

2.7 Seafood

2.8 Eggs

2.9 Milk and Dairy Products

2.10 Chocolate and Confectionery

Challenge!
Phytochemicals as Food Components

CHAPTER OBJECTIVES

After completing this chapter, you will be able to:
- Name the food categories used in the food industry and those in the Food Guide Pyramid.
- Explain the information in food composition tables.
- Define the concept of bioavailability.
- Define technical terms related to food composition and processing, including degrees Brix, leavening, sucrose inversion, comminuted meat emulsion, trimethylamine, isoelectric pH, and sugar crystallization.
- Explain the concept of nutrient density.
- Describe the structure of muscle tissue.
- Relate collagen content of meat to meat tenderness.
- Explain how Standards of Identity for milk products relate to compositional differences.
- Explain the difference in composition of crystalline and noncrystalline confectionery.
- Distinguish between the terms botanical, functional food, nutraceutical, and phytochemical.

In this chapter we focus on food composition and the major food categories. Data from food composition tables will be presented for each category, so you can compare the relative proportions of nutrients. The categories include beverages; cereals, grains, and baked products; fruits and vegetables; legumes and nuts; meat, poultry, eggs, and seafood; milk and dairy products; and confectionery and chocolate. The chapter closes with a challenge that looks at phytochemicals and distinguishes phytochemicals from botanicals, functional foods, and nutraceuticals.

2.1 FOOD COMPOSITION TABLES

Food composition refers to the substances or components found in a beverage or food. By identifying and quantifying the nutrients in foods, scientists have obtained the data needed to develop tables of food composition. These tables show uniformity as well as diversity among foods of the various groups. The key nutrients that compose foods include the larger molecular substances, such as protein, fat (lipid), and carbohydrate (starch, sugars, and fiber), as well as the smaller molecules of water, vitamins, minerals, and phytochemicals.

Raw foods are composed of naturally occurring substances, while processed foods often contain specific, functional additives. The relative amount of each nutrient or substance in a food depends on the category, and there is variation within categories as well. For instance, dairy products contain relatively high amounts of protein, carbohydrate, fat, and calcium. However, one cup of cottage cheese has much more protein and much less carbohydrate compared to ice cream, although both are derived from milk. Comparing dairy products to a different food group, typical fruits contain much less protein, fat, and calcium, but have a higher water content, more vitamins A and C, and more fiber.

Tables of food composition are printed by the U.S. Department of Agriculture (USDA) in handbooks and are accessible via the Internet, http://www.nal.usda.gov/fnic/foodcomp/. They provide nutritionists and consumers with information regarding the specific nutrient and calorie (kcal) content of beverages and foods. The Consumer Nutrition Center of the USDA has one of the most accessed and complete food composition (or nutrient database) systems in the world, the National Nutrient Data Bank (NDB). Consumers also have access to food composition information from food labels, a topic discussed in Chapter 3.

With food composition information, comparing the nutrients or calories in foods is easy. For example, Table 2.1 shows a wide range in iron and potassium content in sweeteners. Nutrient database systems enable international organizations to access food composition data for calculating food supplies. Epidemiologists (researchers who look at the relationship between diet and disease in human populations) use them as well.

Food groups/categories
Beverages
Cereals, grains, baked products
Confectionery and chocolate
Fruits and vegetables
Legumes and nuts
Meat, poultry, eggs, seafood
Milk and dairy

TABLE 2.1 Food composition—common sweeteners that make up foods and beverages *Tables of food composition may list calories, % water, protein, fiber, fat, carbohydrate, and specific vitamins and minerals. This example of the food composition of sugary foods demonstrates that, with the possible exception of molasses, each provides next to nothing in terms of required nutrients listed in the table. "Daily Values" refers to the nutrient needs of the average person per day for the nutrients listed.*

Food	Energy (kcal)	Protein (g)	Fiber (g)	Calcium (mg)	Iron (mg)	Magnesium (mg)	Potassium (mg)	Zinc (mg)	Vitamin A (RE)	Thiamin (mg)	Riboflavin (mg)	Niacin (mg)	Vitamin B_6 (mg)	Folate (ug)	Vitamin C (mg)
Sugar (1 tbs)	46	0	0	0	0.0	0	0	0.0	0	0	0	0.0	0	0	0
Honey (1 tbs)	64	0	0	1	0.1	0	11	0.0	0	0	0	0.0	0	<1	0
Molasses (1 tbs)	48	0	0	34	3.6	50	188	0.1	0	0	0	0.4	1.3	0	0
Jelly (1 tbs)	49	0	0	1	0.0	1	12	0.0	0	0	0	0.0	0	0	<1
Brown sugar (1 tbs)	34	0	0	8	0.2	3	31	0.0	0	0	0	0.0	0	0	0
Daily Values	2,000	56	25	1,000	18.0	400	3,500	15.0	1,000	1.5	1.7	20.0	2.0	400	60

Food Categories and the Food Pyramid

The classification of food groups depends on context—let's look at a few. A *commodity* is defined as "a useful comsumer good, a product of agriculture, produced and delivered for shipment." Although *commodity* usually refers to raw product, the USDA lists 14 commodities, some of which are processed foods: red meats, poultry, fish and shellfish, eggs, dairy products, beverage milks, fats and oils, fruits (fresh, canned, dried, frozen, and juices), vegetables (fresh, canned, dried, frozen, and pulses), shelled peanuts and tree nuts, flour and cereal products, caloric sweeteners, coffee, and cocoa. We can say that *processed commodities* are value-added commodities derived from agricultural commodities that offer convenience, longer shelf life, and sometimes added nutrients.

The food categories we will discuss in this chapter and throughout the book are those used within the food industry: beverages; cereals, grains, and baked products; fruits and vegetables; legumes and nuts; meat and poultry; seafood; eggs; milk and dairy products; and chocolate and confectionery.

In turn, the foods in the nutritionist's Food Guide Pyramid overlap the food categories important to the food scientist and the food industry but are not quite the same. For instance, processed meats like sausage and bacon are part of the Pyramid. However, special processing aids, such as stabilizers and emulsifiers, as well as flavors, perservatives, and other additives, are important components of their manufacture. Although consumed in the diet, these additives are not part of the Food Guide Pyramid, but they are important in the food industry. The student of food science must understand how these additives function and their regulated limits of addition to foods.

Serving Sizes

In planning our meals, it is not just *what* food we eat that matters, but how much of it. This is the idea behind the concepts of servings and **serving size.** The quantity of food recommended by the Food Guide Pyramid is termed servings; product food labels specify serving sizes; tables of food composition employ measures.

In the Food Guide Pyramid, the amount of food that counts as one serving depends on the food group. For example, one serving is equivalent to about 3 ounces of cooked lean meat, 1 cup of milk, and one medium apple, banana, or orange—see Figure 2.1 for more examples.

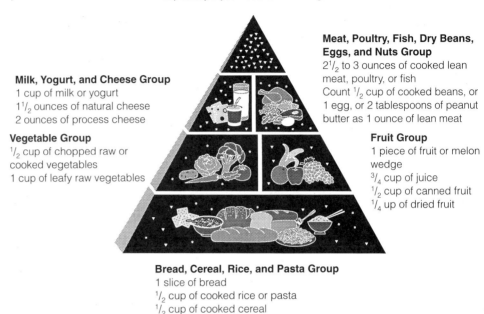

FIGURE 2.1 Serving size. *Just what amount of food constitutes a serving depends on the specific food group item. For example, from the bread group, one 42-gram slice of white bread is equivalent to one serving. On the other hand, ¾ cup of fresh orange juice, weighing 186 grams, is one serving from the fruit group.*

On a food label the serving size is the basis for reporting each food's nutrient content. In the past the serving size was up to the discretion of the food manufacturer. Serving sizes now are more uniform and reflect the amounts people actually eat. The Nutrition Labeling and Education Act (NLEA) defines serving size as the amount of food customarily eaten at one time. The serving sizes that appear on food labels are based on FDA-established lists of "Reference Amounts Customarily Consumed Per Eating Occasion." They must be expressed in both common household and metric measures—grams (g) and milliliters (mL).

Tables of food composition do not suggest a serving size, but instead provide a *measure* of food, identified by weight, followed by the nutrient composition found in that amount. For example, when a table says for a pickle: measure (1 each) and weight in grams (65), it is saying that the measure is for one pickle, at a weight of 65 grams.

2.2 BEVERAGES

A beverage is a drinkable liquid, consumed for a variety of reasons, including (1) thirst quenching, (2) stimulant effect, (3) alcoholic content, (4) health value, and (5) enjoyment. Examples of each type are (1) water, (2) caffeinated drinks (cocoa, coffe, tea), (3) beer and wine, (4) milk and fruit and vegetable juices, and (5) carbonated soft drinks.

Nutrients are substances in foods and beverages that, when absorbed into the body, are used for specific functions like growth, maintenance, and repair of tissue. *Nutrient density* is a concept that applies to beverages and to foods. An item is said to be nutrient dense if it supplies a variety of protein, complex carbohydrates, vitamins, and minerals without excess fat and calories. Figure 2.2 shows that orange juice and milk contain greater nutrient density compared to flavored soft drinks, which are primarily water, sweetener, and flavor. Tables 2.2–2.4 compare the nutrient content, caffeine content, and alcohol content of different types of beverages.

Although water represents the major constituent of beverages, this percentage varies according to product. While instant coffee is close to 100 percent water, orange juice is below 90 percent water and alcoholic beverages can contain even less water—see Table 2.5.

TABLE 2.2 Comparative compositions of five thirst-quenching beverages. *Although none of the beverages listed provide fiber, more vitamin C, potassium, and even protein are provided by orange juice than a typical sports drink. Club soda would be the dieter's choice, since it is calorie-free.*

Beverage	Energy (kcal)	Protein (g)	Carbohydrate (g)	Fat (g)	Dietary Fiber (g)	Calcium (mg)	Potassium (mg)	Sodium (mg)	Iron (mg)	Vitamin A* (RE)	Vitamin C (mg)
Carbonated club soda	0	0	0	0	0	18	7	7.5	0.04	0	0
Sports drink (1 cup)	60	1	5	0	1	18	64	11	0.14	0	0
Cola soft drink (12 fl oz)	152	0	38	0	0	11	4	15	0.11	0	0
Apple juice (1 cup)	116	<1	29	<1	<1	17	295	7	0.92	<1	2
Orange juice (1 cup fresh)	112	2	26	<1	<1	27	496	2	0.5	50	124

SOURCE: Based on USDA Nutrient Database for Standard Reference 13, www.nalusda.gov/fnic/foodcomp/.
*Vitamin A units are RE (retinol equivalents), μg (micrograms), and IU (International Units). The conversion: 1 μg retinol=1RE=3.33 IU.

TABLE 2.3 Comparative compositions of two stimulant beverages. *Tea and coffee are not devoid of nutrients; however, the caffeine content is responsible for the stimulant effect on the central nervous system. An identical serving size of instant coffee provides more than twice the caffeine as iced tea.*

Beverage	Energy (kcal)	Protein (g)	Carbohydrate (g)	Fat (g)	Dietary Fiber (g)	Calcium (mg)	Potassium (mg)	Sodium (mg)	Iron (mg)	Vitamin A (RE)	Vitamin C (mg)
Iced tea (8 fl oz unsweetened; caffeine content approx. 53 mg)	2	0	<1	0	0	5	47	7	0.05	0	0
Instant coffee (8 fl oz; caffeine content approx. 120 mg)	5	<1	1	<1	0	7	86	7	0.12	0	0

SOURCE: Based on USDA Nutrient Database for Standard Reference 13, www.nalusda.gov/fnic/foodcomp/.

TABLE 2.4 Comparative compositions of four alcoholic beverages. *Light beer is lower in calories compared to beer, mainly because of a lower carbohydrate content. However, alcohol contents are similar. Wine and vodka contain much higher levels of alcohol per fluid ounce than beer.*

Beverage	Energy (kcal)	Protein (g)	Carbohydrate (g)	Fat (g)	Dietary Fiber (g)	Calcium (mg)	Potassium (mg)	Sodium (mg)	Iron (mg)	Vitamin A* (RE)	Vitamin C (mg)
Beer (12 fl oz, approx. 13 g of alcohol)	146	1	13	0	3	18	89	18	0.11	0	0
Light beer (12 fl oz, approx. 11 g of alcohol)	78–131	1	5	0	1	18	64	11	0.14	0	0
Red wine (3.5 fl oz, approx. 9 g of alcohol)	74	<1	2	0	0	8	116	5	0.44	0	0
Vodka 90 proof (1.5 fl oz, approx. 16 g of alcohol)	110	0	0	0	0	8	1	<1	0.02	0	0

SOURCE: Based on USDA Nutrient Database for Standard Reference 13, www.nalusda.gov/fnic/foodcomp/.

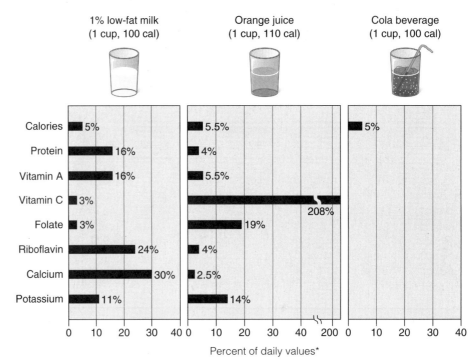

FIGURE 2.2 Comparison of the nutrient density of three beverages. *The bars indicate that milk and orange juice contain key nutrients in addition to calories, while cola beverage lacks nutrients. The milk and orange juice can be considered nutrient dense, while the cola provides only "empty" calories.*

TABLE 2.5 Water content of selected beverages. *All beverages are not created equal in the amount of water they provide. As the presence of dispersed substances in a beverage increases, the water content decreases. The presence of specific solute molecules such as citric acid, vitamins, and simple sugars in orange juice decreases the water content. In alcoholic beverages, the presence of alcohol and dissolved solids decreases the water content.*

Beverage	Water content (%)
Club soda	100
Iced tea (unsweetened)	100
Coffee (instant)	99
Light beer	95
Beer	92
Cola soda pop	89
Apple juice	88
Orange juice (fresh)	88
Red wine	88
Vodka (90 proof)	62

SOURCE: Based on USDA Nutrient Database for Standard Reference 13, www.nalusda.gov/fnic/foodcomp/.

The Degrees Brix of Beverages

Beverages can be naturally sweetened or have sweeteners added during processing. Commonly used beverage sweeteners are sugar (sucrose) and high-fructose corn syrup. Sweetened beverages include carbonated ones (soft drinks, pop, or sodas), fruit and vegetable drinks, dairy-based beverages, and powdered drink mixes.

Determining and maintaining the proper sweetener content is an important aspect of beverage quality. As a solid particle in a solution, sucrose is part of the *total soluble solids* content of a beverage. The sucrose concentration of beverages is measured as **degrees Brix** (°Brix), which is equal to the weight percent of sucrose in solution (grams of sucrose per 100 grams of sample). What this means is that in any sweetened beverage formulation using sucrose as the only sweetener, the °Brix reading is directly proportional to the amount of sucrose in solution. Analytically, the measurement of °Brix is made by refractometry, using either a refractometer or a hydrometer device, calibrated in degrees Brix, with corrections applied according to fruit type and acid content.

Sucrose is a special carbohydrate molecule called a disaccharide, meaning that it is composed of two small sugar molecules bonded together. See Figure 2.3. In a process called *sucrose inversion*, sucrose molecules in solution come apart and yield the two component sugars, glucose and fructose. Because fructose is a sweeter sugar than either sucrose or glucose, sucrose inversion can affect the final sweetness and flavor of a beverage. Any time fructose is present in a beverage due to inversion, sweetness increases.

Some of the factors that promote sucrose inversion are low storage pH (which means high acidity) and high storage temperature. In some products, inversion is desirable and is created by the action of an enzyme, *invertase*, or by action of an acidulant. In addition to beverages and juices, food products that are analyzed for °Brix include syrups such as honey, maple syrup and molasses, jams and jellies, and tomato products.

The Importance of the °Brix/Acid Ratio

We know that the percent sucrose in a solution is called the Brix concentration, or degrees Brix. A 50 percent sucrose-in-water solution is equivalent to 50° Brix on a weight/weight basis (weight of sucrose as a soluble solid in a weight of water). Since fruit juice beverage

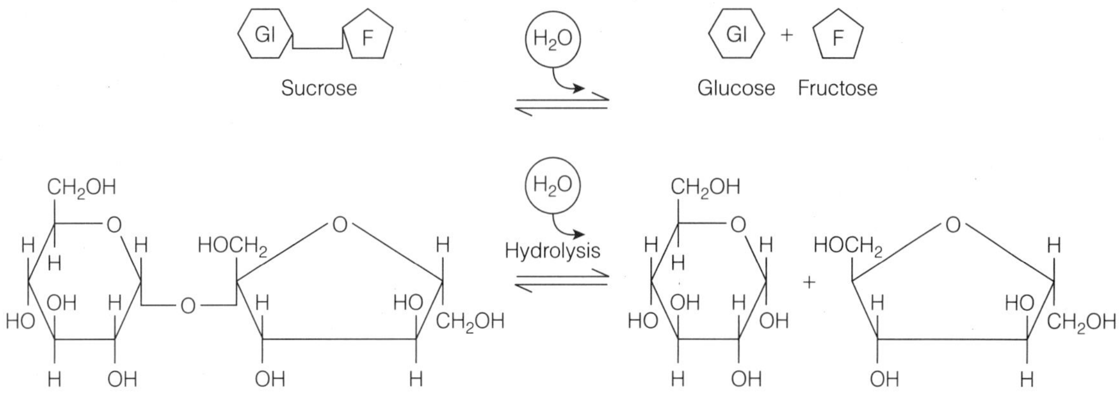

FIGURE 2.3 The inversion of sucrose—its hydrolysis (breakdown) into glucose and fructose. *Hydrolysis reactions are promoted by the presence of acid, enzyme, or high temperature. If this occurs in a beverage during storage, sweetness increases because fructose is generated by the reaction, and fructose is 1.7 times sweeter than sucrose.*

flavor is a function not only of sugar content but also of natural acid content, flavor depends upon the °Brix/acid ratio. This ratio varies as the relative proportion of sugar to acid content, and it is critical in selecting certain fruit and vegetable types to make juices.

Oranges are typically harvested at 12°Brix. However the °Brix/acid ratio can vary according to region. For California oranges, the °Brix/acid ratio may have a value of 8, while for Florida oranges it may be 13. Different °Brix/acid ratios is one reason why consumers prefer one orange variety over another.

2.3 CEREALS, GRAINS, AND BAKED PRODUCTS

Cereals are among the world's major crops. They include the **cereal grains** corn, rice, wheat, barley, millet, rye, sorghum, and oats. Cereal grains are high in carbohydrate content, such as starch, and the sugars glucose, maltose, and fructose, as well as fiber. See Table 2.6 for a comparison of selected grain products. Cereal grains can be consumed directly or milled into flours, and processed into a wide variety of food products, such as breakfast cereals, breads, pasta, snack chips, and syrups (Figure 2.4). Some grains, such as corn, are utilized in livestock feeds as well as being used by humans; some become components of pet foods. Two noncereal crops, sugar cane and beets, are also processed into sugars and syrups.

TABLE 2.6 **Fiber, protein, B vitamin, and iron content of selected grain products (processed, ready to eat) per one-cup serving size.** *Enriched cereal grain products are processed to contain added vitamins and minerals. However, the layer of the kernel used also has an impact on nutrient composition. For example, bran flakes contain significantly more protein and iron compared to corn flakes.*

FIGURE 2.4 **Breakfast cereals: processed grains such as corn, oats, rice, and wheat.**

Cereal	Dietary Fiber (g)	Protein (g)	Iron (mg)	Thiamin (Vitamin B_1) (mg)	Riboflavin (Vitamin B_2) (mg)	Niacin (mg)
Corn flakes	0.8	2	0.4	10	4	2
Macaroni (cooked, enriched)	1.8	6	1.4	0.23	0.14	1.8
White rice (cooked, enriched)	0.7	4	2	0.44	0.03	2.5
Bran flakes	6.9	5.6	7.8	0.64	0.74	8.6

SOURCE: Based on USDA Nutrient Database for Standard Reference 13, www.nalusda.gov/fnic/foodcomp/.

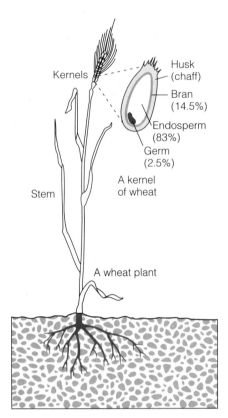

FIGURE 2.5 General structure of a cereal grain kernel.

Wheat is an example of a typical cereal grain seed or *kernel*. Cereal grains are composed of three nutritious parts or structures: the **endosperm,** which comprises about 83 percent of the kernel, the **bran** layer, which is about 15 percent of the kernel, and the remaining kernel **germ,** which is the embryonic or sprouting part. See Figure 2.5.

Energy and nutrient values vary among processed grain products. On average, unprocessed cereals are about 75 percent carbohydrate, 10 percent protein, and up to 2 percent fat. However, the protein is of a lower biological value than protein found in animal products and has generally lower bioavailability. **Biological value (BV)** refers to the amount of nitrogen derived from food protein that is used in the body to promote growth. BV is related to the amino acid content of a protein. For comparative purposes the hen's egg is used as a reference standard as a food complete in high-quality protein.

A food high in BV is said to possess high-quality protein, meaning all of the essential amino acids are present. For the most part, plant protein sources, such as cereals and legumes, are deficient in lysine and methionine, respectively, and they are termed "incomplete" protein sources (see Table 2.7). However, a diet containing a mix of cereal and legume protein can be adequate in terms of these key amino acids.

Bioavailability has a slightly different meaning. **Bioavailability** is the degree to which nutrients are digested and absorbed in the body. It is influenced by such factors as the food source (animal vs. plant) and food processing (certain nutrients, such as the B vitamins, are destroyed by heating).

Leavening of Baked Products

Leavening refers to the production of gases in dough that contributes to the volume achieved during baking ("leavening effect") and the final aerated texture (Figure 2.6). Typical leavening agents added to foods are *yeast,* an example of a biological leavening agent, and *baking soda* ($NaHCO_3$, sodium bicarbonate), *baking powder,* and *ammonium bicarbonate* (NH_4HCO_3). The chemistry of leavening agents will be presented in Chapter 4 (Food Chemistry I). Leavening agents function to produce carbon dioxide as the specific leavening gas. Steam (heated water vapor) is also produced during baking and contributes to "oven spring."

Yeast-raised baked products include crusty breads, specialty breads (multigrain, bran, and oatmeal), national breads (ethnic breads), rolls, bagels, pastries, croissants, and sweet buns. Baked products that are chemically leavened include cakes, cookies, and doughnuts, while some baked products, like crackers, tortillas, and certain cookies, are not leavened. Table 2.8 provides an overview of several baked products.

2.4 FRUITS AND VEGETABLES

Practical definitions of fruits and vegetables emphasize the use of the particular plant foods in meals. These definitions are useful to consumers, food processors, and certain regulatory agencies. For instance, a *fruit* is a fleshy or pulpy plant part commonly eaten as

TABLE 2.7 Approximate amino acid content of selected cereal grains. *Lysine and methionine are two amino acids that are essential to health. Cereal foods can be deficient in either lysine, methionine, or both. For example, although white flour contains some lysine, not enough is present in white bread made from it to support protein synthesis. In contrast, an egg contains a sufficient amount of both lysine and methionine to support protein synthesis.*

Food item	Lysine (mg of amino acid/gram of nitrogen)	Methionine (mg of amino acid/gram of nitrogen)
Cornmeal	160	210
Oats	230	270
Rice	220	220
White wheat flour	130	250
Hen's egg (standard)	430	360

SOURCE: Based on USDA Nutrient Database for Standard Reference 13 www.nalusda.gov/fnic/foodcomp/.

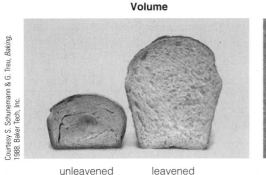

FIGURE 2.6 Leavening effect in bread. *Leavening refers to the expansion of a bread dough. Proper leavening results in a bread loaf of high volume and an open, aerated crumb texture.*

a dessert due to its sweetness. A *vegetable* is a plant or plant part that is served either raw or cooked as part of the main course of a meal. A *vegetable-fruit* is the fruit part of a plant that is not sweet, usually served with the main course of a meal, like cucumbers, squashes, and tomatoes.

A **fruit,** in botanical terms, is defined as the ripened ovary of a plant, which means that it contains the seeds. A **vegetable** is an herbaceous plant containing an edible portion such as a leaf, shoot, root, tuber, flower, or stem. Table 2.9 groups some common vegetables based upon plant parts. Because the botanical definition of fruit is defined as the "ripened ovary of a plant," some plants traditionally referred to as vegetables, such as tomatoes, squash, and avocados do fall within the fruit category.

Fruits and vegetables have many similarities in composition, methods of cultivation and harvesting, storage properties, and processing. Fruits and vegetables can be processed into juices, or they can be dried, frozen, or canned. Fruits are also processed into syrups, jams, and jellies.

Most fresh fruits and vegetables, though there are exceptions, are high in water content (up to 90%), low in protein (up to 3.5%), and low in fat (up to 0.5%)—see Table 2.10.

TABLE 2.8 Percentage composition of several baked products. *There are two divisions of baked products, those that are relatively low in fat calories, such as bagels, biscuits, bread, and tortillas, and those providing significant fat calories, including cakes, cookies, and doughnuts.*

Baked product	Energy (kcal)	Moisture (%)	Carbohydrate (g)	Protein (g)	Fat (g)	Dietary fiber (g)
Bagel (68 g)	187	33	36	7	1	2
Biscuit (28 g)	94	29	14	2	2	1
Bread (25-g slice, cracked wheat)	65	36	12	2	1	2
Cake (28-g piece, pound cake)	109	25	14	2	6	<1
Cookies (4 pieces, 42 g, chocolate chip variety)	192	12	25	1	10	1
Croissant (57 g)	231	23	26	5	12	1
Doughnut (60 g, glazed)	242	25	27	4	14	1
Toasted English muffin (37 g)	128	37	25	2	2	1
Roll or bun (40 g, hot dog bun)	114	34	20	3	2	1
Tortilla (57 g)	115	27	20	3	2	1

SOURCE: Based on USDA Nutrient Database for Standard Reference 13, www.nalusda.gov/fnic/foodcomp/.

TABLE 2.9 Vegetables based upon plant part.

Leaves	Seeds	Roots
lettuce	corn	carrot
spinach	peas	radish

Tubers	Bulbs	Flowers
potato	onion	broccoli
yam	garlic	cauliflower

"Veg fruits"	Stems/Shoots
tomato	celery
squash	asparagus

TABLE 2.10 Comparison of composition of several fruits and vegetables and wheat flour. *Although each food listed is a plant food, cereal grains are much lower in moisture and are highly concentrated sources of carbohydrate and protein compared to fresh fruits and vegetables.*

Food type	% Carbohydrate	% Protein	% Fat	% Water
Wheat flour	70	10	2	12
Potato	20	2	0.1	78
Carrot	9	1.1	0.2	89
Lettuce	2.8	1.3	0.2	95
Banana	24	1.3	0.4	74
Apple	15	0.3	0.4	87

SOURCE: Based on USDA Nutrient Database for Standard Reference 13, www.nalusda.gov/fnic/foodcomp/.

They contain various minerals, and are good sources of both digestible carbohydrates (sugars and starches) and indigestible carbohydrates (fiber, including cellulose and pectic substances). Fruits and vegetables can be good sources of specific vitamins: the vitamin A precursor, *beta-carotene* (in green leafy vegetables and yellow-orange fruits and vegetables), and vitamin C (in citrus fruits, green leafy vegetables, and tomatoes). Table 2.11 provides a comparison of the nutrient composition of several fruits and vegetables.

Health Benefits

The USDA's Food Guide Pyramid depicts two to four daily servings of fruits and three to five servings of vegetables recommended in the Dietary Guidelines. At present, the typical American consumes only about half the recommendation. However, new research is discovering a link between fruit and vegetable consumption and health. These foods often contain high levels of vitamin A and its precursor, beta-carotene, vitamin C, and fibers, such as pectin, and these substances may provide a protective effect against heart disease and certain cancers.

For instance, fermented grape juice (as red wine) contains *resveratrol*, a substance that may lower blood cholesterol, prevent heart disease, and act as an anticancer substance (Figure 2.7). For the food scientist, opportunities exist not only to identify key substances

TABLE 2.11 Nutrient compositions of several fruits and vegetables (100 gram portions). *Perhaps Popeye was right—spinach shines among the fruits and vegetables in this table, as an excellent source of vitamin A, fiber, and minerals. Sweet potatoes and carrots are highly concentrated sources of vitamin A, while broccoli, cantaloupe, oranges, and strawberries provide ample amounts of vitamin C.*

Food	Water (%)	Energy (kcal)	Calcium (mg)	Phosphorus (mg)	Iron (mg)	Vitamin A (IU)	Vitamin C (mg)	Thiamin (mg)	Dietary Fiber (g)
Apple	84	59	7	7	0.2	53	5.7	0.02	2.2
Apricot	86	48	14	19	0.5	2,612	10	0.11	<1
Avocado	74	161	11	41	1	612	8	0.07	<1
Banana	74	92	6	20	0.3	81	9	0.05	1.6
Broccoli	91	28	48	66	0.9	1,542	93	0.07	2.8
Cantaloupe	90	35	11	17	0.2	3,224	42	0.04	0.8
Carrot	88	43	27	44	0.5	28,129	9	0.1	3.2
Corn	76	86	2	89	0.5	281	7	0.2	3.2
Grapefruit	87	32	15	7	0.1	10	37	0.04	0.6
Lettuce	96	13	19	20	0.5	330	4	0.01	1
Onion	90	38	20	33	0.2	0	6	0.04	1.6
Orange	87	47	40	14	0.1	205	53	0.09	2.4
Peach	88	43	5	12	0.1	535	7	0.02	1.6
Spinach	92	22	99	49	2.7	6,715	28	0.08	2.6
Strawberry	92	30	14	19	0.4	27	58	0.02	2.6
Sweet potato	73	105	22	28	0.6	20,063	23	0.07	3
Tomato	94	21	5	24	0.5	623	19	0.06	1.3

SOURCE: Based on USDA Nutrient Database for Standard Reference 13, www.nalusda.gov/fnic/foodcomp/.

FIGURE 2.7 The skins of red grapes and resveratrol.
Resveratrol is found mainly in the skins of red grapes and came to scientific attention as a possible explanation for the "French Paradox"—the low incidence of heart disease among the French, despite a relatively high-fat diet. The resveratrol content of wine is related to the length of time the grape skins are present during the fermentation process (so red wine has higher resveratrol content than white wine). Grape juice, which is not a fermented beverage, is not a significant source of resveratrol.

naturally occurring in fruits and vegetables, but to incorporate them into new food products. More consumer awareness regarding the health benefits of fruit and vegetables should result in increased consumption of fruits, vegetables, and products made from them. Also see Table 2.21 in the Challenge section for phytochemicals found in fruits and vegetables.

Maturity and Ripeness

The state of ripeness, which is very important in fruits, is not the same as maturity. **Ripeness** is the optimum or peak condition of flavor, color, and texture for a particular fruit. **Maturity** is the condition of a fruit when it is picked. The fruit may be at or just before the ripened stage. Some fruits are picked when they are mature but not yet ripe. For instance, cherries if picked when fully ripe, would be damaged because of their softness. Fruits that continue to ripen after they are picked may become overripe if picked at peak ripeness.

Harvesting is the collection of fruits and vegetables at the specific time of peak quality in terms of color, texture, and flavor in order to market them. A fruit or vegetable may quickly pass beyond the period of peak quality (independent of microbial spoilage, which is another topic) after harvest. As an example, the sweetness of corn decreases dramatically during storage because up to 50 percent of sugars may convert back to starch, which is not sweet. This decline in the quality of stored, respiring fruits and vegetables that occurs after harvesting is called **senescence**.

Quantitative and Qualitative Quality Considerations

Fruit and vegetable quality depends upon such qualitative characteristics as size, color, blemishes and bruises, flavor, firmness, and presence of extraneous matter such as unwanted stems, pits, or leaves. Quantifiable quality data include °Brix, microbiological information, moisture content, pH, and viscosity. In fruit and vegetable juices, pectin levels, starch levels, pulp, and solids content are often measured. As an industry example, apple juice concentrate specifications include color, pH, soluble solids, and percent malic acid.

Some fruit and vegetable quality indicators:

- *Viscosity* can be measured if the fruit in any given form can flow (e.g., applesauce, fruit spread, puree). Viscosity can be an indication of solids content, particle size, and pectin or starch content.
- *Color* should be typical of the fruit or vegetable variety in the ripened state. Consistent color is usually desirable over the surface. The color of dried fruit depends on whether it was treated to prevent browning reactions.

- *pH* and *titratable acidity* (discussed in Chapter 4) are both indicators of the amount of acid present. Acidity is significant because it influences flavor as well as the potential for microbial spoilage. Most fruits are acid, with a pH of 4.6 or less (exceptions are bananas, figs, and papayas). Many are classified as high acid, with pH under 3.7 (most apples and berries). In the fruit juice industry, acid levels are frequently expressed as the number of grams of titratable acid per sample volume or weight.
- *Flavor and odor* should be representative of the fruit or vegetable in the fresh, unspoiled state, with no off-odors or off-flavors due to fermentation or other microbial activity.
- *Degrees Brix*, refers to the percent soluble solids, primarily sugar. Other substances that can influence °Brix readings include dissolved acids or salts.

Dried Fruits

Dried fruits are a processed category of fruits created by dehydration. **Dehydration** removes moisture from fruits to prevent microbial and enzymatic deterioration. Dehydration also concentrates the nutrients in a dried fruit product—see Table 2.12. With the exception of fragile-skinned fruits, most can be dehydrated, including grapes, dates, plums, figs, apricots, and apples. Fruits may be sun-dried or dried in special dehydrator chambers. Fruits to be dehydrated must be in a very ripened state so they contain the highest possible natural sugar levels. A process known as **infusion** can also be used to create dehydrated fruit. Using heat and pressure, fructose is forced into a fruit piece, and this sugar replaces the water in the fruit. Infusion also provides a means to add flavor, add color, and achieve a certain texture. Infused fruits can be chewy or soft, depending on the moisture level and variety of fruit used.

2.5 LEGUMES AND NUTS

The **legumes**, which are edible seeds and pods (as beans and peas) of certain flowering plants, include beans, lentils, soybeans, and peas. Beans are available to consumers in the dried, canned, or fresh state. Fresh beans include green, lima, and wax. Commonly available canned and dried beans are red, kidney, navy, pinto, black, white, and garbanzo (also called chickpeas).

Most legumes offer good-quality protein compared to other plants and are low in fat (see Table 2.13). Meat and dairy products are still the major sources of dietary protein in the United States, but legumes are becoming more popular due to their low-fat, low-sodium, high-fiber, vitamin, mineral, and protein content. The Food Guide Pyramid of the USDA includes legumes ("dry beans") with meat, poultry, fish, eggs, and nuts, at 2–3 recommended servings per day. However, Asian and Mediterranean food guide pyramids have been developed that differ slightly in dietary recommendations—meat and dairy products are de-emphasized while legumes are emphasized as protein sources.

TABLE 2.12 Nutrient compositions of several dried fruits, indicating how they can serve as good sources of vitamins, minerals, and fiber.

Food	Water (%)	Energy (kcal)	Protein (g)	Carbohydrate (g)	Iron (mg)	Vitamin A (RE)	Vitamin C (mg)	Dietary Fiber (g)
Date (178 g)	22	490	4	130	2	1,160	0	13
Fig (187 g)	28	477	6	122	4	1,331	2	18
Prune (168 g)	64	400	4	106	4	1,250	6	12
Raisin (145 g)	15	435	5	115	3	1,088	5	6

SOURCE: Based on USDA Nutrient Database for Standard Reference 13, www.nalusda.gov/fnic/foodcomp/.

Legume	Energy (kcal)	Protein (g)	Carbohydrate (g)	Fat (g)	Dietary Fiber (g)	Calcium (mg)	Potassium (mg)	Magnesium (mg)	Iron (mg)	Phosphorus (mg)	Niacin (mg)	Vitamin C (mg)
Chickpea (1 cup, 164 g)	269	15	45	4	8	80	477	79	4.7	276	0.9	2
Kidney bean (1 cup, canned, 256 g)	256	13	40	1	16	61	658	72	3.2	240	1.2	3
Lentil (1 cup, cooked, 198 g)	230	18	40	<1	10	32	730	36	6.6	356	2.1	2
Pea (1 cup, cooked, 160 g)	124	8	22	<1	8	38	268	46	2.5	144	2.4	16
Soybean (1 cup tofu, 248 g)	188	20	4	12	2	260	300	252	13.2	240	0.5	<1

SOURCE: Based on USDA Nutrient Database for Standard Reference 13, www.nalusda.gov/fnic/foodcomp/.

TABLE 2.13 Nutrient composition of several legumes—good sources of protein, minerals, and fiber. *Although soy foods like tofu contain a relatively high level of fat, the fat is low in saturated fatty acids and devoid of cholesterol.*

Higher fat legumes include soybeans, cottonseed, sesame seed, sunflower seed, and peanut seed—these are termed **oilseeds.** Peanuts (which are not true nuts!) are ground to produce peanut butter, are extremely popular as an edible "nut," and are pressed to extract the peanut oil. Peanuts come in three major varieties. Spanish peanuts are rather small and are used in nut clusters and peanut brittle. The Runner peanut is medium sized and is incorporated into confectionery products. Virginia peanuts are larger and longer and are the variety used for roasted nuts (in or out of shell).

Common Soy Products

Soybeans are a legume with a dual role as an "oil bean" and a "food bean" because of the fat and protein content. Soybeans are processed into a broad variety of soy products (Figure 2.8). Whole soybeans are fermented to produce products such as miso, tempeh, and soy sauce. When soybeans are pressed, the soybean oil can be removed and is used as a food-grade oil and in margarine formulations. The beans that remain after removal of the oil can be processed into soy flour, which is relatively high in protein (about 50%).

The protein in defatted soybean flakes can be concentrated to produce even more soy products. Soy protein concentrate (about 70% protein) or soy protein isolate (about 90% protein) are two examples. The isolate can be processed into several textured (spun or extruded) forms and used as meat analogs and extenders. Soy milk is made by soaking soybeans in water, followed by grinding to produce a slurry, which is cooked and filtered. Some of the early processed soy milks had a flavor defect termed "beany" or "grassy." An enzyme called **lipoxygenase** present in the bean was responsible because it converted soy fatty acids to odorous compounds. However, this enzyme is deactivated by a heat processing step, which eliminates the off-flavors.

Some soyfoods provide not only protein but a class of phytochemical called isoflavones, compounds that may be effective in the prevention and treatment of cancer and certain chronic diseases. Isoflavones are present in tofu, soy milk, and soy flour but not in soy oil or soy sauce (see list in margin).

Isoflavone content.

	Serving Size	Isoflavones (mg)
Green soy beans, edamame	1/2 cup	50
Soy beans, roasted	1/2 cup	110
Miso	2 tbsp	15
Soy milk	1 cup	24
Soy flour, roasted	1/4 cup	42
Tempeh	1/2 cup	36
Tofu	1/2 cup	24
Texturized soy protein (TVP), dry	1/4 cup	29
Soy burger (check label)	3 oz	38–55
Peanuts	100 g	0.26
Navy beans	100 g	0.21
Granola bar	100 g	0.15

Tree Nuts

Nuts are a well-balanced, nutritious food, providing protein, fiber, vitamins, minerals, unsaturated fatty acids, and phytochemicals—see Table 2.14. They are consumed out of the shell, often dry or oil roasted, or are consumed as pastes or incorporated into confections. Nuts are derived from various ground plants and trees. Popular tree nuts include almonds, hazelnuts (also called filberts), pecans, pistachios, macadamia, and walnuts (Figure

FIGURE 2.8 Soyfoods. *HVP (hydrolyzed vegetable protein) can be made from soybeans. It is used to add flavor as well as provide essential amino acids in soups, meat and poultry products, and canned vegetables.*

FIGURE 2.9 Nuts—many essential nutrients in attractive, crunchy packages. *The savvy consumer is also aware of the high fat content in nuts (approximately 5 grams of fat in a handful of peanuts).*

2.9). The peanut, though considered a nut snack food, is technically a member of the pea and bean family.

Almonds come in two varieties: sweet and bitter. The bitter almond is poisonous if consumed—it contains a substance called *amygdalin* that can be broken down to produce hydrogen cyanide. The sweet variety includes the long Jordan almond from Spain and the soft-shell "nonpareil" almond, grown extensively in California. Almonds are consumed raw or roasted, can be ground into almond paste, or can be incorporated into toffee brittles, which when ground into pastes are called nougat or praline.

Hazelnuts are used in chocolate coatings and other chocolate products because the flavor of the roasted hazelnut combines well with chocolate. They are grown primarily in Southern Europe, Turkey, and Oregon. Pecans are native to America, and the trees are grown in Texas, Oklahoma, and the Mississippi valley.

The pale green color of pistachios distinguishes them from other tree nuts. The shells are light tan in color, although some processors dye them bright red. Pistachio ice cream is manufactured with a bright green color. Pistachio nuts are a hardy nut, grown throughout the Middle East and India, even under conditions of poor soil and low rainfall.

Walnuts include the *regia* variety and the black walnut (*nigra*). Regia are native to Iran and are grown in the United States and various countries of Europe, Australia, South Africa, and China, while black walnuts are native to the United States. The walnut is primarily used as an addition to chocolates, cakes and confection.

Macadamia nuts, which are said to possess the most difficult shell to crack, come from an Australian evergreen tree and are named after Dr. John Macadam of Victoria, Australia. After macadamia trees were exported to Hawaii, the macadamia nut developed into a large industry—although at first it was difficult to produce nuts of uniform quality. Scientists at the University of Hawaii worked for twenty years to produce the three varieties available today.

Nut	Energy (kcal)	Protein (g)	Carbohydrate (g)	Fat (g)	Dietary Fiber (g)	Calcium (mg)	Iron (mg)	Magnesium (mg)	Potassium (mg)	Phosphorus (mg)	Zinc (mg)	Niacin (mg)
Almond (1 cup, dry roasted, 138 g)	810	22	33	71	14	389	5.2	420	1062	756	6.8	3.9
Cashew (1 cup, dry roasted, 137 g)	786	21	45	64	4	62	8.2	356	774	671	7.7	1.9
Hazelnut (1 cup, chopped, 115 g)	726	15	18	72	9	216	3.8	327	511	358	2.8	1.3
Macadamia (1 cup, oil roasted, 134 g)	962	17	12	102	12	60	2.4	157	440	268	1.5	2.7
Peanut (1 cup, oil roasted, 144 g)	837	38	27	71	10	126	2.6	266	982	744	9.6	20.4
Pecan (1 cup, dry roasted, 112 g)	780	8	24	72	4	40	2.5	152	420	344	6.4	1
Walnut (1 cup, chopped, 112 g)	770	17	22	74	5	113	2.9	203	602	380	3.3	1.3

TABLE 2.14 Nutrient composition of several nut varieties. *Nuts are a good choice for a quick snack, providing minerals, fiber, and protein. However, the high fat content of nuts means that they are high in calories as well.*

*The peanut is not a true nut, but a legume.
SOURCE: Based on USDA Nutrient Database for Standard Reference 13, www.nalusda.gov/fnic/foodcomp/.

2.6 MEAT AND MEAT PRODUCTS

Meat can be defined as the edible flesh and organs of animals and fowls. Red meat describes the flesh of cattle, pigs, and sheep. The flesh of poultry (chicken, turkey, duck) is considered white meat. Meat purchased by consumers can be either whole muscle with bone (a broiler chicken or turkey), whole muscle (a ribeye steak), or some form of processed meat. Meat is often processed in some way to create products that appeal to consumers. Processing can include physical methods such as grinding, chemical treatment such as curing, and smoking. Processed meats may be canned and thermally treated, as with corned beef. Drying strips of meat produces jerky. Frozen ground beef and turkey are available as chubs or formed into patties. Cured meats include bacon (cured pork) and ham. Meat by-products include organs and glands such as liver, heart, brain, thymus, and intestines ("carcass meat").

Meat is a source of high-quality protein and is valued for its cooked flavor and tender texture. It also provides various B vitamins, iron, magnesium, and other minerals. The fat content of red meat ranges widely, up to 40 percent, and is a source of saturated as well as unsaturated fatty acids and cholesterol (Table 2.15). Poultry meat contains more protein and less fat and cholesterol than red meat on a weight basis. Since the 1990s, the poultry meat industry has expanded the variety of products for sale to consumers, offering counterparts to beef and pork products—chicken products include chicken and turkey franks, chicken nuggets, chicken patties, and turkey bacon. For these reasons, poultry meat consumption has increased over recent years while that of red meat has declined.

The Structure of Muscle Tissue

The basic unit of muscle tissue is the muscle cell or muscle fiber, which is surrounded by a cytoplasmic substance called sarcoplasm. The muscle fiber consists of many units called muscle fibrils or **myofibrils.** These myofibrils contain contractile proteins organized in a specialized array of thick and thin rod-like filaments. The thick filament protein is called *myosin*, and the thin filament protein is called *actin* (Figure 2.10). When muscles contract, these filaments form crosslinkages with each other—bonds that produce a new protein called **actinomyosin**. Visible muscle is therefore a collection of muscle fibers, which are held together by connective tissue as bundles.

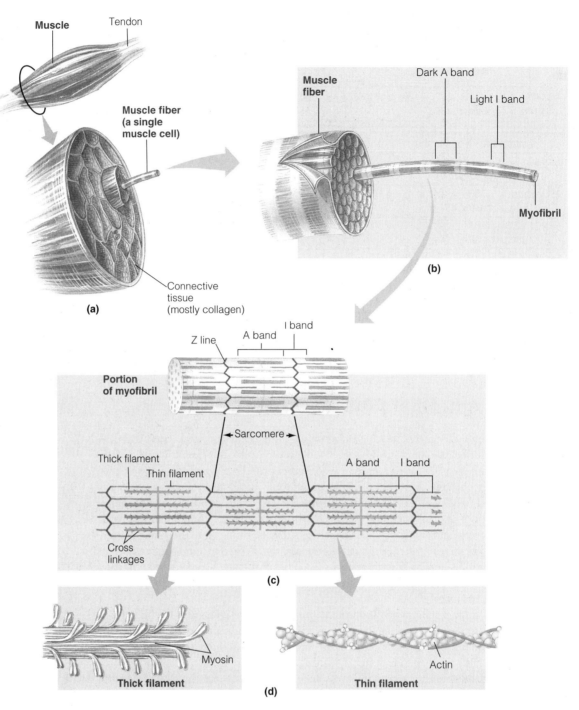

FIGURE 2.10 The organization of skeletal muscle. *(a) Cross-section of skeletal tissue. (b) Enlarged view of one myofibril within a single muscle fiber. (c) Regions between and within sarcomeres and the arrangement of actin and myosin filaments. (d) Thick and thin filaments. Muscle tissue is made up of muscle fibers, and muscle fibers are made up of myofibrils. Myofibrils in turn are composed of sarcomeres. Sarcomere units contain actin and myosin, which chemically bond together to form actinomyosin during muscle contraction. Each sarcomere is defined as the space between two Z lines.*

Connective tissue is composed of a watery dispersion of material called stromal protein matrix plus adipose (fat) tissue. Two kinds of connective stromal proteins are common in meats: elastin and collagen. Connective tissue sheets that surround muscle bundles are collagen sheets. **Collagen** is known as white connective tissue. The toughness of meat is directly related to the amount of collagen it contains. Older animals tend to have high levels of collagen and yield tougher meat than young animals.

TABLE 2.15 Nutrient composition of meats and poultry. *Poultry meats and pork are lower in fat and calories compared to red meat, although they are not lower in cholesterol.*

Food	Energy (kcal)	Protein (g)	Carbohydrate (g)	Fat (g)	Cholesterol (mg)	Iron (mg)	Magnesium (mg)	Potassium (mg)	Phosphorus (mg)	Folacin (µg)	Niacin (mg)
Beef (3 oz lean ground beef patty, 85 g)	238	24	0	15	86	2.1	20	297	155	9	5
Chicken (3 oz without skin, 86 g)	142	27	0	3	73	0.9	25	220	196	3	11.8
Frankfurter (Beef and pork, 57 g)	183	6	1	17	29	0.7	7	95	49	0.1	1.5
Pork (3 oz lean cutlet, 85 g)	175	24	0	8	78	1.7	22	293	179	19	5.4
Turkey (3 oz light meat roasted, 85 g)	133	25	0	7	107	1.1	24	186	186	5	5.8
Veal (3 oz braised cutlet, 85 g)	166	27	0	6	100	1	24	288	213	344	7.2

SOURCE: Based on USDA Nutrient Database for Standard Reference 13, www.nalusda.gov/fnic/foodcomp/.

Meat Emulsions

Certain meat products are examples of emulsions. An emulsion is a two-phase system in which one phase is dispersed in the other, in the form of finely divided droplets. Emulsions in which the main components are lipid (oil or liquid fat) and water can be classified as either oil-in-water (O/W) or water-in-oil (W/O). The first term refers to the dispersed phase and the second term to the external or continuous phase. Thus, in an O/W emulsion, a lesser amount of oil is dispersed as tiny droplets in a greater amount of water (Figure 2.11). Over time, without an emulsifying agent, emulsions separate into their component phases because they are thermodynamically unstable.

Comminuted meat emulsions contain finely chopped meat mixed with water, fat, and sometimes additives such as preservatives and water-binding agents (Figure 2.12). Some meat proteins are termed soluble because they remain dispersed within the watery phase, while others are insoluble and can form gels. Frankfurters and sausages are comminuted meat emulsions where the O/W emulsion is entrapped in a gel formed by the insoluble proteins and muscle fibers of the meat tissue plus water. The fat content is typically 20–45 percent by weight, and it is the dispersed phase in the emulsion.

2.7 SEAFOOD

"Seafood" refers to more than just fish. Seafood includes clams, oysters, lobsters, scallops, and other food animals derived from oceans, lakes, and streams. Fish can be divided into two groups: finfish and shellfish. **Finfish** are fish with a backbone and fins (e.g., freshwa-

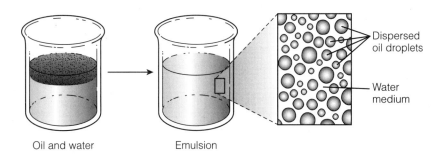

FIGURE 2.11 An oil-in-water emulsion. *In an O/W emulsion, oil is the dispersed phase distributed throughout the continuous water phase.*

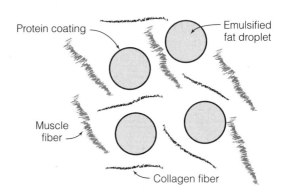

FIGURE 2.12 A comminuted meat emulsion, showing protein matrix of muscle fibers, collagen fibers, and emulsified fat droplets surrounded by a film of myofibrillar protein. *Myofibrillar proteins are "extractable" from meat, usually through the use of added salt. These proteins function in the formation of the comminuted meat emulsion by stabilizing fat droplets and by contributing to the formation of a viscous gel matrix, which stabilizes the emulsion still further by minimizing fat globule movement. The two important myofibrillar proteins are actin and myosin.*

ter trout and saltwater cod). **Shellfish** include *crustaceans*, which are sea invertebrates with a hard upper shell and a soft under shell (e.g., crab, shrimp, lobster), and *mollusks* with two enclosing shells (e.g., clams, oysters, scallops).

The composition of seafood is quite variable, as it depends on species, age, and season of capture or "harvest." Typically, finfish composition falls into the range of 14–20 percent protein and 0.2–20 percent fat. (See Table 2.16.) The protein is highly digestible and a good source of all the essential amino acids. Fish fat (e.g., cod liver oil) is rich in unsaturated fatty acids, as well as fat-soluble vitamins A and D. Fish oil is highly perishable; that is, it is susceptible to oxidation, rancidity, and off-odor and off-flavor development. The B vitamins are found in the fish muscle. Fish can be rich sources of iodine, and some magnesium, calcium, and iron are often present.

Why Are Fish So Perishable?

In general, fish flesh is more perishable than other kinds of animal flesh. This perishability has three basic reasons: fish microbiology, fish physiology, and fish fat chemistry. On the microbiology side, fish flesh is sterile, but surface slime is present as a mass of bacteria covering the fish body, and bacteria reside internally in the digestive tract. When fish are caught in nets and die, these bacteria rapidly invade all the tissue constituents. Fish microorganisms are **psychrotrophic,** which means they live and multiply in colder temperatures, even refrigeration temperatures (4–15°C). So, this process occurs even in chilled, refrigerated fish.

In fish physiology, fish store energy in muscle tissue as a compound called glycogen. When glycogen is used to produce energy, lactic acid is also produced from the glycogen breakdown. Since fish struggle when caught, they use up muscle glycogen stores, so none remains to be converted to lactic acid after they die. Because an acid environment inhibits bacterial growth and acid acts as a preservative for the fish muscle, the absence of lactic acid allows microorganisms to multiply, and spoilage takes place.

Looking at fish fat chemistry, fish lipids are easily *oxidized* since they are highly unsaturated molecules. Oxidation is a chemical reaction in which oxygen reacts with fatty acids to produce odorous products. The phospholipids associated with a fish's fat deposits are rich in a substance called *trimethylamine oxide (TMAO)*. When TMAO is split from the phospholipid molecule by fish or bacterial enzymes after capture, it is changed to

TABLE 2.16 **Nutrient composition of various seafoods.** *Fish flesh is low in fat and calories and high in protein and minerals. Crustaceans contain significant amounts of cholesterol.*

Food	Energy (kcal)	Protein (g)	Carbohydrate (g)	Fat (g)	Cholesterol (mg)	Iron (mg)	Magnesium (mg)	Potassium (mg)	Phosphorus (mg)	Folacin (μg)	Niacin (mg)
Cod (3.5 oz baked w/ butter, 100 g)	132	23	0	3	60	0.5	42	245	140	10	2.5
Crab (1 cup cooked, 135 g)	138	27	0	2	135	1.2	45	437	278	10	3
Shrimp (3.5 oz boiled, 100 g)	99	21	0	1	195	3.1	35	179	198	4	2.6
Tuna (3 oz canned in water, 85 g)	111	25	0	1	28	2.7	26	267	158	5	13.2

SOURCE: Based on USDA Nutrient Database for Standard Reference 13, www.nalusda.gov/fnic/foodcomp/.

trimethylamine (TMA). TMAO is not odorous, but TMA has a strong "fishy" odor—see Figure 2.13. Such odor is evidence of some deterioration.

Fish oil contains components referred to as omega fatty acids, which are essential in the diet. All fatty acids contain a CH_3 (or methyl group) at one end of the chemical structure and a COOH (or acid group) at the other. In between the ends are carbon atoms linked by single bonds, (C—C) and in the case of unsaturated fatty acids, with double bonds (C=C). (We will discuss these fatty acids further in Chapter 3.)

Linolenic, docosahexaenoic acid (DHA), and eicosapentaenoic acid (EPA) are examples of **omega-3 fatty acids.** Fish oil contains about 30 percent omega-3 fatty acids. These are polyunsaturated fatty acids having the first double bond (or site of unsaturation) three carbon atoms away from the methyl end of the molecule. Consumption of omega-3 fatty acids is associated with a decreased tendency to develop cholesterol-clogged arteries. The down side to unsaturated fish fat is that more double bonds mean greater instability to oxygen during storage. Thus, rancidity is more likely to develop as a result of oxidation of the double bonds in the fatty acids.

2.8 EGGS

Eggs are valued as one of the most nutritious foods available and have been a staple food item since the earliest civilizations. Eggs are composed of an outer protective shell and an inner yolk and white, consisting of proteins, lipids, a small amount of carbohydrate, vitamins, and minerals—see Figure 2.14. Nutrients are distributed unequally between the egg yolk and the egg white. Some consumers are aware that the yolks contain the fat and cholesterol, while the white is a rich source of high-quality protein.

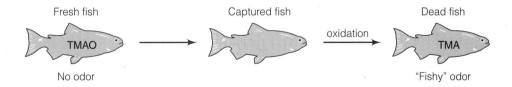

FIGURE 2.13 **Trimethylamine and fish odor.** *The fatty part of fish, especially sea fish, contains trimethylamine oxide (TMAO), which regulates osmotic pressure in living fish cells. After death, TMAO is converted by bacterial enzymes to a "fishy" smelling liquid substance, trimethylamine (TMA). Trimethylamine in the liquid state is odorous. However, if TMA liquid can be changed into a gel, such as by the addition of acid, the odor is greatly reduced.*

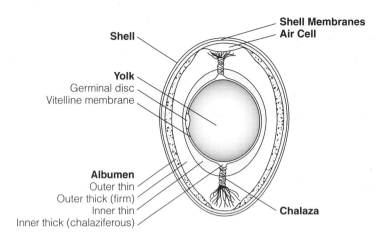

FIGURE 2.14 The organization of an egg.

Egg white contains water and proteins including *ovalbumin*, *conalbumin*, *ovomucoid*, and *lysozyme*. Another egg white protein called **avidin** is found in smaller amounts. Avidin is of interest because it binds to a particular B vitamin called biotin and renders it unavailable as a nutrient. However, cooking an egg destroys avidin and its ability to bind biotin.

Egg yolk constitutes about one-third of the edible portion and provides roughly 75 percent of the calories of an egg. The fatty portion of the egg yolk is divided into the *neutral lipids*, *phospholipids*, and *cholesterol*. The primary phospholipid is called phosphatidyl choline, or lecithin. However, the yolk is not composed only of lipids but also has a significant protein content. There are four classes of egg yolk proteins: *lipovitellin* (a high-density lipoprotein), *lipovitellenin* (a low-density lipoprotein), *phosvitin*, and *livetin*. The yolk also contains iron and other minerals and vitamins. The mineral content of eggs depends on the fortification of the animal feed provided the hens.

The published food composition of hen eggs represents an average of hen egg data from various flocks, taken at various times throughout the year, and from diverse geographic localities—see Table 2.17. The composition of a raw egg and a boiled egg are essentially identical, although an egg prepared with added milk and butter would differ. The nutrient composition of frozen and dried egg products is about the same as for fresh eggs.

Egg Quality

A freshly laid, quality egg has a uniform yolk of good color and size, and a thick egg white portion that resists spreading when the shell is cracked and egg is released onto a flat surface. The height of the egg white can be measured by a method proposed by Raymond Haugh in 1937. **Haugh units** relate egg weight and the height of the thick white, and the

TABLE 2.17 **Nutrient composition of hen eggs.** *Hen eggs provide the standard for high-quality protein among foods. The egg yolk contributes more protein on a weight basis than does the egg white but is also high in fat and cholesterol.*

Food	Energy (kcal)	Protein (g)	Carbohydrate (g)	Fat (g)	Cholesterol (mg)	Calcium (mg)	Iron (mg)	Phosphorus (mg)	Potassium (mg)	Riboflavin (mg)
One egg (50 g)	74	6	<1	5	213	25	0.7	89	61	0.25
Egg white (33 g)	17	4	<1	0	0	2	<0.1	4	47	0.15
Egg yolk (17 g)	59	3	<1	5	218	23	0.6	83	16	0.11

SOURCE: Based on USDA Nutrient Database for Standard Reference 13, www.nalusda.gov/fnic/foodcomp/.

higher the Haugh unit value, the better the albumen egg quality. The method known as candling is used to assess egg quality without breaking the shell and inspecting the contents. **Candling** involves examining the appearance of the egg as it is rotated in front of a light source (candles were often used). Important factors are size and position of the air cell, clearness of the white, yolk position and mobility, and shell condition.

2.9 MILK AND DAIRY PRODUCTS

Whole cow's milk is typically 88 percent water, 3.3 percent protein, 3.3 percent fat, 4.7 percent carbohydrate, and 0.7 percent ash. Milk is an emulsion in that microscopic droplets of milk fat are dispersed in the aqueous milk serum. Each droplet is coated with a complex of protein, carbohydrate, and lipid. Milkfat is mainly triglyceride, with short-chain fatty acids (fatty acids having 12 or fewer carbons in the structure) saturated with hydrogen atoms. Butyric acid, with four carbon atoms in its structure, is an example of a short-chain fatty acid in milkfat.

Milkfat contains not only cholesterol and phospholipids, but also fat-soluble vitamins A, D, E, and K and yellow carotenoid pigments, all carried in the fat fraction of milk. The primary milk carbohydrate is lactose, commonly referred to as "milk sugar." Lactose is a disaccharide that yields glucose and galactose upon hydrolysis. It is the least sweet and least soluble common sugar, and due to its low solubility, it may crystallize out of and give a sandy texture to some food products. This occurs when too much nonfat dry milk is added to an ice cream formulation. Table 2.18 provides an overview of the nutrient composition of selected milk and dairy products.

Many adults have difficulty digesting milk and milk products because they do not produce enough of the enzyme *lactase* that digests lactose. For these lactase-intolerant individuals, lactase-treated milks have been developed. The enzyme lactase is added to pasteurized milk and stored for 24 hours. Typically the reaction is allowed to proceed until the lactose content has been reduced by 70 percent (lactose-reduced milk). Milk that has 99.9 percent of its lactose hydrolyzed is labeled "lactose free."

Food	Energy (kcal)	Protein (g)	Carbohydrate (g)	Fat (g)	Cholesterol (g)	Calcium (mg)	Iron (mg)	Phosphorus (mg)	Potassium (mg)	Riboflavin (mg)
American cheese (1 oz, 28 g slice)	106	6	<1	9	27	174	0.11	211	405	0.1
Cream cheese (1 oz, low-fat, 28 g)	65	3	2	5	16	32	0.48	47	84	0.8
Cottage cheese (1 cup low-fat 2%, 226 g)	203	31	8	4	19	155	0.36	341	217	0.42
Ice cream (1 cup soft serve, 175 g, 3% fat)	221	9	38	5	21	275	0.1	212	123	0.35
Milk, whole (1 cup, 245 g)	88	8	11	8	33	290	0.12	371	120	0.39
Milk, skim (1 cup, 245 g)	86	8	12	<1	4	301	0.1	247	126	0.34
Sour cream (1 cup, 230 g)	492	7	10	48	104	267	0.14	195	122	0.34
Yogurt plain, low-fat (1 cup, 227 g)	144	12	16	4	14	415	0.18	327	159	0.49

TABLE 2.18 Nutrient composition of various milk and dairy products. *In general, milk products all offer high-quality protein, calcium, and varying levels of fat and cholesterol.*

SOURCE: Based on USDA Nutrient Database for Standard Reference 13, www.nalusda.gov/fnic/foodcomp/.

The Varieties of Milk

Fluid milk is classified as either *whole milk, reduced fat, low-fat* or *fat-free milk*. Whole milk contains not less than 3.25% fat and not less than 8.25% milk solids-not-fat (MSNF). Reduced fat milk contains 2% milkfat, and low-fat milk contains 1% milkfat. Fat-free (nonfat) milk, the current designation for what was formerly skim milk, must have a maximum fat content of 0.5% and at least 8.25% MSNF. All of these milks are usually homogenized, pasteurized, and fortified by the addition of 2,000 units vitamin A and 400 units vitamin D.

Other varieties of milk include:

- *Cultured milk*, also termed "fermented," includes yogurt, buttermilk, and sour cream. Buttermilk is made from skim milk, with added lactic acid bacteria, and may contain about 1% fat with the addition of small flecks of butter for improved palatability.
- *Evaporated milk* is a canned homogenized and vitamin D fortified milk product in which about half the water content has been evaporated prior to canning. Since canning destroys spoilage bacteria, milk can remain at room temperature in an unopened can.
- *Sweetened condensed milk* is a canned, homogenized, and fortified (vitamin D) milk product in which about half the water has been evaporated prior to canning. A high (44%) percentage of sucrose and/or glucose has been added to help retard bacterial growth, to increase flavor, and to increase viscosity.
- UHT (ultra-high temperature processed) milk is fluid milk heated at temperatures of about 300°F for 2–6 seconds. Packaged aseptically into presterilized containers to eliminate bacterial contamination, UHT milk can be stored for long periods of time, although after several months, "off-flavors" can develop due to residual enzyme activity and chemical changes.
- *Nonfat dry milk (NFDM)* "instant milk" powder is made from fresh, pasteurized skim milk by removing about two-thirds of the water content under a vacuum and then spray drying the concentrated milk into a chamber of hot filtered air. It is the least expensive form of milk. The ability of NFDM to whip into a foam is regarded as an important **functional property** of NFDM.

Different types of milk products are identified according to specific **Standards of Identity.** These federal standards are required for labeling purposes and identify the product name and the type and amount of ingredients that a specific food type must contain, as well as processing requirements. For example, canned condensed milk must be pasteurized, has the option to be homogenized, and must be formulated to contain 7.5% fat and 25.5% milk solids-not-fat (MSNF) which includes protein solids and lactose. See Table 2.19 for a listing of Standards of Identity.

The Milk Proteins: Casein and Whey

Freshly drawn cow's milk has a pH of 6.6 units. The major milk protein (roughly 80% of total milk protein) is called **casein.** It occurs in several forms—alpha-, beta-, and kappa-casein—and is dispersed as *calcium caseinate* in the watery part of milk, which is called the milk *serum*. An important characteristic of casein is its precipitation as a curd at pH of 4.6, which can be accomplished by action of rennin or "rennet". Rennin is also known as *chymosin*, an enzyme widely used in cheesemaking; "rennet" describes any enzymatic preparation used in the cheese manufacturing industry to coagulate milk. In milk, the kappa-casein form keeps the alpha- and beta-casein from precipitating by stabilizing the caseins in a particular configuration. Chymosin interacts with kappa-casein to disrupt this configuration and causes the caseins to precipitate in the presence of milk calcium, forming a curd. A by-product of the chymosin reaction is called *casein glycomacropeptide*. Casein glycomacropeptide is a protein known for its effect on intestinal microflora—the microorganisms that influence intestinal function and health.

TABLE 2.19 Standards of identity for milk products.

		Pasteurization[a] / Ultrapasteurization[b] / UHT Processing[c]	Homogenization	Fat % (min./range)	MSNF % min.	Vitamin D	Vitamin A
Fluid	Milk	M [a,b,c]	Opt.	3.25	8.25	Opt.	Opt.
	Reduced-fat milk	M [a,b,c]	Opt.	0.5-2.0	8.25	Opt.	M
	Fat-free milk	M [a,b,c]	Opt.	<0.5	8.25	Opt.	M
Cream	Half-and-half	M [a,b,c]	Opt.	10.5–18.0			
	Light cream (coffee cream or table cream)	M [a,b,c]	Opt.	18–30			
	Light whipping cream	M [a,b,c]	Opt.	30-36			
	Heavy whipping cream	M [a,b,c]	Opt.	36			
Can	Evaporated milk		M	7.5	25.0	M	Opt.
	Condensed milk	M [a]	Opt.	7.5	25.5	Opt.	
	Sweetened condensed milk	M	Opt.	8.0	28		
Dry	Fat-free dry milk	M	Opt.	Max 1.5			
	Fat-free dry milk fortified with vitamins A and D	M	Opt.	Max 1.5		M	M
Cultured	Sour Cream	M	Opt.	18.			
	Acidified sour cream	M	Opt.	18			
	Sour half-and-half	M	Opt.	10.5-18.0			
	Acidified sour half-and-half	M	Opt.	10.5-18.0			
	Yogurt	M [a,c]	Opt.	3.25	8.25	Opt.	Opt.
	Reduced-fat yogurt	M [a,c]	Opt.	0.5-2.0	8.25	Opt.	Opt.
	Fat-free yogurt	M [a,c]	Opt.	<0.5	8.25	Opt.	Opt.

M = Mandatory
Opt. = Optional

[a] Pasteurization is mandatory but declaration of term *pasteurized* is optional.
[b] "Ultra-pasteurization" is an optional process—declaration of term *ultra-pasteurized* is mandatory on principal display panel, if applicable.
[c] UHT declaration is on label although not included in standards at this time.
SOURCE: Courtesy of the National Dairy Council.

The other milk proteins are called **whey proteins** (specifically *lactalbumin, lactoferrin, lysozyme, lactoperoxidase,* and *lactoglobulin*). These proteins are not precipitated by acid or rennin, but can be heat coagulated. In whole milk, coagulated casein plus entrapped fat globules make up the **curd.** Whey is composed of the whey proteins plus the disaccharide lactose, all dispersed in the fluid serum. Two milk products that are high protein sources are whey protein concentrate (WPC) and whey protein isolate (WPI). WPC, roughly 75 percent protein content, is produced by heating milk and filtering through a very fine membrane (a process called ultrafiltration). WPC is used as an additive in various food products. WPI, roughly 90 percent protein content, is a more highly purified or concentrated form of protein. WPI is used in baked products, mixes, soups, and in confections as a gelling agent.

Ice Cream—A Complex Colloidal System

Premium quality ice cream is made from cream, not milk. **Cream** is the high-fat, liquid product that is separated from whole milk. According to federal Standards of Identity, cream must be at least 18% milkfat. **Ice cream** contains an emulsion of dispersed and clustered fat droplets. Clustering is due to the cold temperature and the presence of saturated fatty acids. Fat clusters exist within a concentrated mixture of sugars and proteins, with the additional presence of dispersed ice crystals, a polysaccharide gel, and air bubbles.

Ice cream is actually a *complex colloidal system*, meaning it contains several different molecules of a large enough size that prevents them from being soluble. Instead they are suspended throughout the ice cream system, which contains liquid, solid, and gas phases (Figure 2.15). In addition to air, milkfat, and water, ice cream may contain flavorings, sweeteners, stabilizers, emulsifiers, MSNF (milk solids-not-fat), and coloring.

The goal in producing quality ice cream is a product of smooth consistency. Because of the cold temperature of the product, ice crystals tend to form and can create a coarse consistency. Therefore, minimizing the development of large ice crystals through the proper addition and processing of ingredients is required to produce quality ice cream.

Butter and Margarine

Butter is a dairy spread made from either sweet or sour cream. The fat of cream is separated from other milk constituents by agitation or churning. The mechanical rupture of the protein film that surrounds each of the fat globules in cream allows separate globules to join (coalesce). Butter formation is an example of breaking an oil-in-water emulsion (in the original milk) by agitation. The resulting emulsion of butter itself is a water-in-oil emulsion, with about an 18:80 W/O ratio.

Margarine is made from various fat ingredients (e.g., animal fat, or soy or corn oils) that are churned with cultured, pasteurized nonfat milk or whey. It is a water-in-oil emulsion and must not contain less than 80 percent fat, according to its Standard of Identity as set by the USDA. By blending naturally polyunsaturated liquid oils from plants with partially hydrogenated oils, the total PUFA (polyunsaturated fatty acid) content of margarine can be relatively high. A reduced-calorie margarine can be produced by using less fat (50–75%) and adding water and emulsifier. Such a product does not meet the specific USDA Standards of Identity, so it cannot be labeled as margarine but instead is called a "vegetable/corn oil spread."

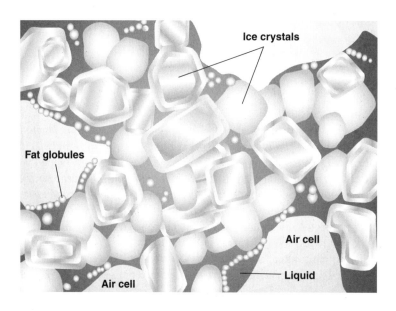

FIGURE 2.15 Ice cream. *The ice cream system can be thought of as a foam that contains air bubbles trapped within a frozen liquid phase of lactose, sugar, polysaccharides, and milk solids.*

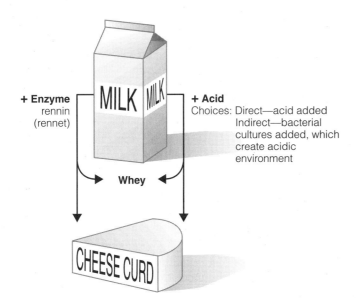

FIGURE 2.16 Cheesemaking. *An enzyme (rennin or rennet) or an acid (lactic acid) coagulates milk to produce a solid cheese curd.*

Cheese

Cheese is a concentrated dairy food defined as the fresh or matured product obtained by draining the whey (the moist serum from the original milk) after coagulation of casein, the major milk protein (Figure 2.16). Casein is clotted by coagulating enzymes **(rennin)** or by low pH, such as that produced by added bacterial cultures. Cheese has many nutritional similarities with milk; thus, cheese is a source of high-quality (complete) protein, calcium, phosphorus, and vitamin A. Because a great deal of water has been removed, cheese is a fairly concentrated source of nutrients, and also of saturated fat.

Many cheeses are high in fat and calories, although cottage cheese and those made from skim milk are low in fat. Adding salt for flavor makes cheese a high-sodium food, although reduced sodium varieties are available. Most cheeses need to undergo a ripening process. **Ripening** refers to the changes in physical and chemical composition of cheese that take place between the time of curd precipitation and the time when cheese develops the desired characteristics of its own particular type. Characteristics of flavor, aroma, texture, and appearance are therefore developed in each particular cheese.

Two principal cheese types are natural and process. Natural cheeses are categorized as hard or soft.

Natural Soft Cheeses. Examples of soft natural cheeses include cottage cheese, cream cheese, and Roquefort. Cottage cheese is made from skim milk allowed to clot by the action of the lactic acid-producing bacteria *Streptococcus lactis*. The casein precipitates at its **isoelectric point**—see Figure 2.17. As the pH drops due to lactic acid production by the bacteria in the milk, a curd forms.

Since a portion of the milk's calcium will be left behind in the watery whey portion as calcium lactate, the calcium content of cottage cheese made in this way is slightly lower than the milk it was made from. When rennin is used as the means to clot the casein, the calcium will not be lost to the whey, because the calcium forms an insoluble salt with the casein as calcium caseinate.

Cream cheese, another soft natural cheese, is made from whole milk plus added cream. Curd formation is by lactic acid. The Neufchatel variety is lower in cream content. The mold *Penicillium roqueforti* is injected into the rennin-clotted cheese called Roquefort. The mold will grow impressively throughout the cheese during the 2- to 12-month ripening period. Authentic Roquefort cheese is made from sheep's milk, and ripening must occur within the moist caves near Roquefort, France, which provide ideal atmospheric conditions for the ripening of this distinctly firmer cheese.

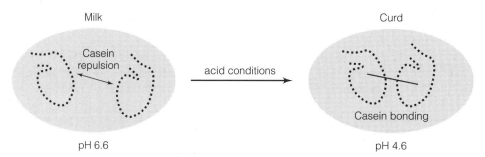

FIGURE 2.17 The importance of isoelectric point in cheesemaking. *Milk proteins (caseins, but not whey) coagulate to form a solid clot or curd under acidic conditions. They behave this way because of the effect of pH on the protein molecules. Fresh milk, at a pH of about 6.6, is nearly neutral. At this pH, the proteins carry overall repulsive electrical charges. Charge repulsion keeps the casein proteins dispersed in fluid milk. By acidifying milk to a pH below 4.6, which is the isoelectric point for casein, the charges are neutralized, so that there is no net charge on the proteins. Therefore, instead of repelling, they drift toward each other. This means that their solubility and stability in milk have decreased. As they touch, the proteins form chemical bonds with each other, and this creates a curd.*

Natural Hard Cheeses. Cheddar cheese, named for its English town of origin, is produced by the dual action of lactic acid bacteria and rennin. The characteristic yellowish orange color is due to a seed pod extract called *anatto*. Repeated cutting of the curd and draining of the whey, a process termed **cheddaring,** is carried out to achieve the desired moisture content before salting and ripening. Various stages of ripening (related to storage times) are used to create different flavors, ranging from mild (least expensive) to sharp (most expensive).

Regarding Swiss cheese, two microorganisms, the same two used for yogurt making, *Streptococcus thermophilus* and *Lactobacillus bulgaricus*, are added to start the processing by precipitating the curd. A third microorganism, *Propionibacterium shermanii*, causes the holes in ripened Swiss cheese because this bacterium produces gas during aging.

Process Cheese. **Process cheese** is produced from a mixture of natural cheeses and an emulsifying salt, which are blended together during controlled heating. The emulsifier can be either sodium citrate or disodium phosphate, which binds the high fat content with the water added to the mix. By heating this mixture to between 145°F and 165°F while stirring, a cheese with excellent consistency, keeping qualities, and mild flavor is produced. Heating prolongs shelf life because it destroys bacterial and enzymatic action, thus preventing ripening. The process yields a pasteurized product, so the accurate name for this type of cheese is *pasteurized process cheese*. American cheese is perhaps the best known example of a process cheese.

2.10 CHOCOLATE AND CONFECTIONERY

Is chocolate good for you? New research indicates that there is more to chocolate than saturated fat, calories, and flavor. Chocolate contains antioxidant flavonoid compounds that may benefit health. Find out more by visiting http://www.chocolateinfo.com/.

Although chocolates and confectionery "sweets" or candies are not consumed for reasons of good nutrition, they do have certain caloric and nutritive value—see Table 2.20. In most chocolates and confectioneries, sugars are the main ingredient (sucrose, glucose, glucose syrup, honey, invert sugar, lactose) and source of carbohydrate.

Cocoa butter supplies most of the fat calories in chocolates, although milkfat is present in milk chocolate and in caramel candy. Milk is the important source of protein in fudge, caramels, and milk chocolate. The vegetable proteins found in cocoa, nuts, and soy ingredients are used in certain chocolate and confectionery products. Refined white sugar is pure carbohydrate and does not supply other nutrients. Vitamins and minerals are present in confectionery and chocolate products made with added dried fruit pieces (e.g., raisins), milk solids, cocoa beans, and other natural products.

TABLE 2.20 Nutrient composition of chocolate and confectionery.

Product (100g)	Energy (kcal)	Protein (g)	Carbohydrate (g)	Fat (g)	Calcium (mg)	Iron (mg)	Magnesium (mg)	Potassium (mg)	Phosphorus (mg)
Dark chocolate	544	5.6	52.5	35.2	63	2.9	131	257	138
Milk chocolate	588	8.7	246	37.6	246	1.7	59	349	218
Milk chocolate bar	447	5.3	66.5	18.9	163	1.1	35	249	154
Hard candies	327	—	87.3	—	4.8	0.4	2.4	8	11.6
Cocoa powder	452	20.4	35.0	25.6	51	14.3	192	534	685
Toffee	399	0.2	90.8	6.2	11	0.6	4.0	91	9.7

SOURCE: Based on USDA Nutrient Database for Standard Reference 13, www.nalusda.gov/fnic/foodcomp/.

Noncrystalline and Crystalline Confectionery

Confectionery (candies) are classified as either crystalline or noncrystalline, depending on the manner in which they are manufactured. In general, both types require that a mixture of sugar (sucrose) and water be boiled to form a syrup solution. The syrup is a concentrated system of dispersed sugar molecules. In crystalline candies, the sugar molecules are allowed to form crystals through rapid cooling with agitation. This favors the production of small, fine microscopic sugar crystals and contributes to the smooth texture of products like fudge.

In the manufacture of jelly beans or gummy bears, two types of noncrystalline candies, the goal is to prevent sugar crystals from forming. This is usually accomplished by slowly cooling the hot syrup, without agitation. Adding substances such as corn syrup and cream of tartar also help to inhibit crystal formation—see Figure 2.18. Corn syrup lowers the water activity and concentrates the sugar solution. Cream of tartar lowers the pH of the syrup and causes sucrose inversion. The presence of the products of sucrose inversion, glucose and fructose molecules, physically interferes with the development of the nucleated sucrose crystal aggregates needed for crystallization.

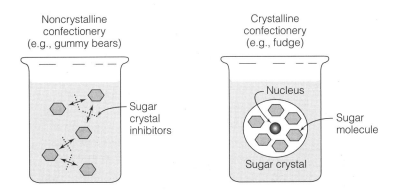

FIGURE 2.18 Comparing noncrystalline and crystalline confectionery. *A crystal forms when a group of sugar molecules collect around a nucleus. A nucleus is a limited aggregation of sugar molecules in a syrup solution. In noncrystalline manufacture, sugar crystal formation is prevented through the addition of interfering substances that block movement of sucrose molecules toward each other and toward nucleation sites.*

Key Points

- The food categories include beverages; cereals, grains, and baked products; fruits and vegetables; legumes and nuts; meat and meat products; seafood; eggs; milk products; and chocolate and confectionery.
- The important nutrients listed in tables of food composition include protein, fat, carbohydrate, dietary fiber, cholesterol, and specific vitamin and minerals.
- A cereal grain kernel is composed of three layers: a fibrous outer bran layer, a starchy endosperm layer, and a nutritious germ layer.
- Leavening is responsible for the expansion in volume of oven baked products, due to the production of specific gases.
- Fruit and vegetable quality is based not only on qualitative external characteristics but also on quantitative characteristics such as product moisture, viscosity, and pH.
- Soybeans are important foods due to their oil content for margarine production and protein content for producing meat-like and milk-like products.
- Meat tissue contains collagen, a tough, fibrous protein that can be softened through the application of moist heat.
- Trimethylamine is a substance found in fish that has not been properly stored with respect to chill temperature and is responsible for an unpleasant "fishy" odor.
- An egg yolk provides protein, fat, cholesterol, and calcium; an egg white provides protein and potassium.
- Ice cream is a complex colloidal system with characteristics of both an emulsion and a foam.
- In cheesemaking, the protein content of milk, specifically the protein casein, is responsible for the formation of a curd, due to decreased solubility of the casein at isoelectric pH.
- Candies are of two types, crystalline and noncrystalline, each requiring different processing to achieve crystal formation or not.

Study Questions

1. Food composition
    a. includes vitamins and minerals but not phytonutrients.
    b. of processed applesauce would be identical to raw apples.
    c. varies more in terms of quantity than quality from food group to food group.
    d. is reported in the Food Guide Pyramid.
    e. varies between and within food groups.

2. Which is *not* a group from the Food Guide Pyramid?
    a. milk, yogurt, cheese
    b. food ingredients and additives
    c. vegetables
    d. fruits
    e. bread, cereal, rice, and pasta

3. The average percentage of iron absorbed is higher for beef than for spinach, which is an illustration of
    a. biological value.
    b. Brix.
    c. bionutrient value.
    d. bioavailability.
    e. biochemical index.

4. Evaluate the following data from a food composition table of restaurant food, and select the correct answer choice.
    - Fish and chicken dinner, 1 each: 431 g, 55% water, 950 kcal, 36 g protein, 102 g carbohydrate, 49 g fat, 75 mg cholesterol, 200 mg calcium, 4.5 mg iron
    - Milk shake, 1 each: 461 g, 71% water, 600 kcal, 13 g protein, 101 g carbohydrate, 16 g fat, 50 mg cholesterol, 450 mg calcium, 1.44 mg iron

    a. The protein biological value of the dinner is much greater than for the milk shake.
    b. Milk is a better source of iron than fish and chicken.
    c. The fish and chicken dinner is probably fried rather than boiled.
    d. Two fish and chicken dinners provide more calcium than one milk shake.
    e. Both milk shake and fish and chicken dinner carbohydrates are primarily glucose.

5. Which would *not* cause a loss in quality?
    a. avidin
    b. trimethylamine
    c. senescence
    d. trimethylamine oxide
    e. crystallization of sugar in jelly beans

6. In a comminuted meat esmulsion,
   a. actin and myosin stabilize fat droplets.
   b. actin and myosin contribute to the formation of a gel matrix.
   c. actin and myosin minimize fat globule movement.
   d. choices (a) and (b).
   e. choices (a), (b), and (c).

7. Which statement regarding nutrient density is most accurate?
   a. Apple juice is less nutrient dense than carbonated water.
   b. Nutrient density is determined by the vitamin content of a food.
   c. Carrots and cantaloupes are roughly equivalent.
   d. Peanuts are more nutrient dense than peanut oil.
   e. Neither cooked peas nor raisins are particularly nutrient dense.

8. Which statement regarding collagen is most accurate?
   a. Low meat collagen content is associated with toughness.
   b. A collagen molecule is composed of three amino acids in total.
   c. Collagen is an important myofibrillar protein.
   d. Collagen's physical structure and ability to form crosslinks contribute to toughness.
   e. Collagen is composed of myofibrils.

9. Cheese
   a. is part of the meat alternate group of the Food Guide Pyramid.
   b. can be manufactured when milk proteins reach their isoelectric point.
   c. is a poor source of calcium and phosphorus compared to sour cream.
   d. is formed from milk proteins that repel each other at pH below 4.6.
   e. is formed by the coagulation of whey protein.

10. Regarding the Standards of Identity for milk products,
    a. fat-free milk and heavy whipping cream, but not yogurt, must be pasteurized.
    b. homogenization is required for all fluid milks.
    c. MSNF is greater in canned evaporated milk than for fluid milks.
    d. pasteurization, homogenization, and percent fat are the only regulated categories.
    e. only cheeses have published standards of identity.

11. How is margarine different from butter?
    a. Margarine potentially has a higher unsaturated fat content.
    b. Butter is a more concentrated source of protein.
    c. Margarine but not butter is an oil-in-water type of emulsion.
    d. Butter but not margarine is part of the dairy group of the Food Guide Pyramid.
    e. Butter is higher in vegetable oil content.

12. Foods manufactured with added HVP
    a. have soybeans added.
    b. must contain 100 mg of isoflavone per serving size.
    c. are used primarily as animal feed.
    d. will be deficient in lysine and methionine.
    e. contain essential amino acids.

13. Suppose you measure the °Brix/acid ratio of the same type of fruit from two suppliers (different parts of the country) and find supplier A = 7 and supplier B = 11. What might you conclude regarding the flavor of the juice resulting from A vs. B?

14. List four microorganisms and their use in creating specific foods.

15. How are qualitative and quantitative factors in fruits and vegetables related to their quality?

16. In a sweetened beverage that undergoes inversion, what would happen to the °Brix reading in comparison to the true solids content?

17. Lactose is the main sugar in milk. Since some people have lactose intolerances, the dairy industry makes reduced-lactose and lactose-free milk available. However, reduced-lactose milk tastes sweeter than milk with lactose. Why?

18. Complete the table for the number of calories and grams of alcohol in 6 fluid ounces of beer and 6 ounces of red wine.

Drink (6 fl oz)	Calories (kcal)	Alcohol (g)
Beer		
Red wine		

19. Explain how an O/W emulsion is different from a W/O emulsion.

20. Match the following proteins with their food source(s).
    - **Proteins:** actin, avidin, casein, collagen, conalbumin, elastin, lactalbumin, lactoferrin, lactoglobulin, lactoperoxidase, lipovitellenin, lipovitellin, lipoxygenase, livetin, lysozyme, ovalbumin, ovomucoid, and phosvitin
    - **Food source:** egg white, egg yolk, red meat, milk, and soybean

References

Brown, A. 2000. *Understanding Food: Principles and Preparation.* Wadsworth/Thomson Learning, Belmont, CA.

Chandan, R. 1997. *Dairy-Based Ingredients.* Eagan Press, St. Paul, MN.

Charley, H., and Weaver, C. 1998. *Foods: A Scientific Approach.* Prentice Hall, Upper Saddle River, NJ.

Demetrakakes, P. 1996. Sweet success: the latest options for in-line measurement of sugar solids. *Food Processing* 57(9):83.

Freeman, T. P., and Shelton, D. R. 1991. Microstructure of wheat starch: from kernel to bread. *Food Technology* 45(3):162.

Giese, J. H. 1992. Hitting the spot: beverages and beverage technology. *Food Technology* 46(7):70.

Homsey, C. 1999. Functional foods and phytochemicals. *Food Product Design* 8(7).

Liu, K. 1997. *Soybeans: Chemistry, Technology, and Utilization*. Chapman & Hall, New York.

Platzman, A. D. 1999. Functional foods: figuring out the facts. *Food Product Design* 8(11):32.

Vieira, E. R. 1996. *Elementary Food Science*, 4th ed. Chapman & Hall, New York.

Whitney, E. N., and Rolfes, S. R. 2002. *Understanding Nutrition*, 9th ed. Wadsworth/Thomson Learning, Belmont, CA.

Yamamoto, T., Juneja, L. R., Hatta, H., and Kim, M. 1997. *Hen Eggs: Their Basic and Applied Science*. CRC Press, Boca Raton.

Additional Resources

Bennion, M. 1995. *Introductory Foods*. Prentice Hall, Upper Saddle River, NJ.

Best, D. 2000. Foods of tomorrow: sports beverages follow road to recovery. *Food Processing* 61(5):85.

Boyle, M. A. 2001. *Personal Nutrition*, 4th ed. Wadsworth/Thomson Learning, Belmont, CA.

Brandt, L. 1996. Nuts: satisfying the nuttier side. *Food Product Design* 5(5):92.

Hedrick, H. B., Aberle, E. D., Forrest, J., Judge, M. D., Merkel, R. A., Gerrard, D., and Mills, E. 2001. *Principles of Meat Science*. 4th ed. Kendall/Hunt Publishing, Dubuque, IA.

Hegarty, V. H. 1995. *Nutrition, Food, and the Environment*. Eagan Press, St. Paul, MN.

Johnson, Q. W. 1997. What's in-store for bakeries. *Food Product Design* 6(3):72.

Karmas, E., and Harris, R. S. 1988. *Nutritional Evaluation of Food Processing*, 3rd ed. Van Nostrand Reinhold, New York.

McWilliams, M. 1997. *Foods: Experimental Perspectives*. Prentice Hall, Upper Saddle River, NJ.

Minifie, B. W. 1989. *Chocolate, Cocoa, and Confectionery: Science and Technology*, 3rd ed. Van Nostrand Reinhold, New York.

Morrow, B. 1991. The rebirth of legumes. *Food Technology* 45(9):96.

Penfield, M. P., and Campbell, A. M. 1990. *Experimental Food Science*, 3rd ed. Academic Press, San Diego.

Potter, N. H., and Hotchkiss, J. H. 1995. *Food Science*, 5th ed. Chapman & Hall, New York.

Sizer, F., and Whitney, E. N. 2000. *Nutrition: Concepts and Controversies*, 8th ed. Wadsworth/Thomson Learning, Belmont, CA.

Uhl, S. 1996. Spilling the beans on legumes. *Food Product Design* 5(6): 57.

Vaclavik, V. A. 1998. *Essentials of Food Science*. Chapman & Hall, New York.

Web Sites

www.ag.uiuc.edu/~ffh
The University of Illinois Functional Foods for Health Program

www.tamu.edu/food-protein/divisions/separ_sc/nutffpro.html
Nutraceutical/Functional Foods Program at Texas A&M University. This site provides accurate information to public, media, and governmental organizations on nutraceuticals and functional foods.

www.ift.org/divisions/nutraceuticals
IFT's Nutraceutical and Functional Foods Division's links page

http://phytochemicals.tamu.edu/citrus/introduction.html
Citrus Physiology & Nutraceutical Program. This site provides information on citrus nutraceuticals lutein, lycopene, pectin, zeaxanthine plus research and distance educational program.

http://www.agr.ca/food/nff/enutrace.html
Agriculture and Agri-Food Canada, Functional Foods & Nutraceuticals Division. This site contains industry and marketing, regulatory, research, and consumer awareness information on functional foods.

http://foodsci.rutgers.edu/nci/index.htm
Nutraceuticals Institute USA. This site is a joint venture between Rutgers, the State University of New Jersey, and St. Joseph's, Philadelphia's Jesuit University.

http://www.chocolateinfo.com/
Mars, Inc. chocolate and health information, including new research findings.

http://www.nal.usda.gov/fnic/foodcomp/
Release 14 of the USDA Nutrient Database for Standard Reference; provides downloadable data sets prepared by the USDA Nutrient Data Laboratory with nutrient values of foods.

PHYTOCHEMICALS AS FOOD COMPONENTS

Phytochemicals, medicinal herbs, functional foods—what's the difference? And should you add them to your diet? Read on.

Phytochemicals and Phytonutrients

A phytochemical ("phyto" means plant) is any plant-derived substance that is thought to function in the body to prevent certain diseases. Phytochemicals provide health benefits *beyond* what the standard nutrients provide, and they are not members of the standard nutrient groupings, because they are not required for health. An example of a phytochemical is lycopene, which is present in tomatoes. The term *phytonutrient* is sometimes used interchangeably with phytochemical, but the former term should be reserved specifically for plant-derived nutrients (starch, beta-carotene, etc.).

Phytochemical language can be confusing because a particular phytochemical can be classified according to its chemical structure and according to its health effects. For example, *genistein* is the name of a highly studied phytochemical present in soybeans. The structure of genistein places it in the broad chemical family of the *flavonoids*; specifically it is an *isoflavone* flavonoid. However, genistein exhibits hormone (specifically, estrogen like) effects in the body and can also be functionally classified as a *phytosterol* or *phytoestrogen*.

How do phytochemicals function? These substances are biologically active, meaning they enter into metabolic reactions in the body and affect specific physical or mental functions—see Table 2.21. Certain phytochemicals in carrots can help maintain the functioning of the immune system, while others in soy protein can reduce blood pressure.

Herbs and Botanicals

Phytochemical is a relatively new term, but the word *herb* has been in use for a long time. Medicinal **herbs** are plants or plant extracts that contain pharmacological or medicinal substances. This means that herbs can function as drugs. The practice of applying herbs to treat disease is referred to as herbal medicine or *phytotherapy*. Herbal medicines or herbal pharmaceuticals are known as **botanicals** and have been used as traditional medicines in older civilizations and as alternative medicines in Western culture. Examples of botanicals are aloe, echinacea, garlic, ginseng, licorice, and peppermint—see Table 2.22.

Nutraceuticals and Functional Foods

Nutraceuticals. The earliest recognized use of the term *nutraceutical* was for foods containing added calcium, fiber, and fish oil; however, there is no universally adopted definition at present. The Health Protection Branch of Health Canada defines nutraceutical as any healthful food ingredient produced from foods but sold in a pill or other concentrated form, demonstrated to have a medical or physiological benefit, not purely a nutritional one. The National Institutes of Health says that *Nutraceutical foods* include "any modified food or food ingredient that may provide a health benefit beyond the traditional nutrients it contains." This sets nutraceutical apart from phytochemical, in that a nutraceutical substance can be of animal origin.

Functional Foods. Health Canada has suggested that the term **functional food** refers to any food similar to a conventional food, consumed as part of a regular diet, that

TABLE 2.21 Specific phytochemicals, how they function, and food sources.

Phytochemical	Comments	Food Sources	Proposed Action
Allicin	Health benefits are associated with as little as a half clove of garlic per day on average.	Garlic, onions, leeks, shallots, chives, scallions	Interferes with the replication of cancer cells; decreases the production of cholesterol by the liver.
Carotenoids (beta-carotene, lutein, zeaxanthin, cryptoxanthin, lycopene)	Plants contain more than 600 types of carotenoids. An orange contains at least 20 types. Dark green vegetables are often good sources; the green chlorophyll obscures the colors of carotenoids in these plants.	Dark green vegetables, orange, yellow, and red vegetables and fruits	Neutralize oxidation reactions that can damage eye and other tissues.
Flavonoids	Foods low in vitamins and minerals may be rich sources of flavonoids (e.g., grape skins).	Apples, celery, strawberries, grapes, onions, green and black tea, red wine	Protect cells from oxidation; decrease plaque formation; increase HDL cholesterol; decrease DNA damage related to cancer development.
Indoles, isothiocynates (sulforaphanes)	These sulfur-containing compounds may be particularly protective against breast cancer.	Cruciferous vegetables (broccoli, brussels sprouts, cabbage, cauliflower)	Interfere with the cancer growth-promoting effects of genes on cells; increase the body's ability to neutralize cancer-causing substances.
Isoflavones (phytoestrogens, or plant estrogens)	Genistein and diadzein from soy may be particularly beneficial for the prevention of cancer.	Soybeans, soy food, chickpeas, other dried beans, peas, peanuts	Interfere with cancer growth-promoting effects of estrogen; block the action of cancer-causing substances; lower blood cholesterol; decrease menopausal symptoms and bone loss.
Isoprenoids	Refining wheat eliminates the germ and bran and causes a 200- to 300-fold loss in phytochemical content.	Fruits, vegetables, whole grains, flaxseed	Decrease cholesterol production by the liver; lower blood cholesterol.
Lignans (phytoestrogens)	Flaxseed is an extrememly rich source of lignans. In the digestive tract, lignans are converted to substances that may help prevent breast cancer.	Flaxseed, flaxseed oil, seaweed, soybeans and other dried beans, bran	Interfere with the action of estrogen.
Saponins	Nearly every type of dried beans is an excellent source of saponins.	Dried beans, whole grains, apples, celery, strawberries, grapes, onions, green and black tea, red wine	Neutralize certain potentially cancer-causing enzymes in the digestive tract.
Terpenes (monoterpenes, limonene)	These substances give citrus fruits a slightly bitter taste. Limonene is from the same family of compounds as tamoxifen, a drug used to treat breast cancer.	Oranges, lemons, grapefruit, and their juices	Facilitate the excretion of cancer-causing substances; decrease tumor growth.

has demonstrated physiological benefits and/or reduces the risk of chronic disease beyond basic nutritional functions. As with phytochemicals, functional foods and beverages contain specific ingredients that can enhance a specific physical or mental function. Fruits, vegetables, cereals, and legumes are functional foods because of the phytochemicals they contain.

The functional food and beverage industry is growing in the United States. Functional foods and beverages are sold like their conventional counterparts as over-the-counter (OTC) processed and packaged products like cereals, margarine, and yogurt. Another popular category of OTC functional food is snack food such as chewing gum, chips, and confections. Functional beverages include bottled water, herbal drinks, juices, milk-based beverages, sports and energy drinks, and teas.

TABLE 2.22 Commonly available herbal preparations, functions, and warnings regarding their use.*

Herb	Why It Is Used	How It Works	Cautions
Black cohosh	Reduces symptoms of premenstrual syndrome, painful menstruation, and the hot flashes associated with menopause.	It appears to function as an estrogen substitute and a suppressor of luteinizing hormone.	Occasional stomach pain or intestinal discomfort. Since no long-term studies have been done, use should be limited to six months.
Capsicum	Best known as the hot red peppers cayenne and chili. Applied topically for chronic pain from conditions such as shingles and trigeminal neuralgia.	The active ingredient, capsaicin, works as a counterirritant and decreases sensitivity to pain by depleting substance P, a neurotransmitter that facilitates the transmission of pain impulses to the spinal cord.	Overuse can result in prolonged insensitivity to pain. More concentrated products can cause a burning sensation. Users must avoid contact with eyes, genitals, and other mucous membranes.
Echinacea	These species of purple coneflower appear to shorten the intensity and duration of colds and flus, may help to control urinary tract infections, and, when applied topically, speed the healing of wounds.	Though echinacea lacks direct antibiotic activity, it helps the body muster up its own defenses against invading microorganisms.	Experts warn against using echinacea for more than eight weeks at a time, and against its use by people with autoimmune diseases like multiple sclerosis and rheumatoid arthritis or by those infected with HIV.
Feverfew	The dried leaves of this plant have been shown to reduce the frequency and severity of migraine as well as its frequently associated symptoms of nausea and vomiting.	The active ingredient, parthenolide, appears to act on the blood vessels of the brain making them less reactive to certain compounds.	Most commercial preparations recommend doses that are much too high. 250 μg a day of parthenolide—or 125 mg of the herb—is an adequate dose.
Garlic	Active against viruses, fungi, and parasites. It may also lower cholesterol and inhibit the formation of blood clots, actions that might help to prevent heart attacks.	When fresh garlic is crushed, enzymes convert alliin to allicin, a potent antibiotic. Garlic tablets and capsules containing alliin and the enzyme can be absorbed when dissolved in the intestines and not in the stomach.	More than five cloves of garlic a day can result in heartburn, flatulence, and other gastrointestinal problems. People taking anticoagulants should be cautious about taking garlic.
Ginger	A time-honored remedy for settling an upset stomach, ginger has been shown in clinical studies to prevent motion sickness and nausea following surgery.	Components in the aromatic oil and resin of ginger have been found to strengthen the heart and to promote secretion of saliva and gastric juices.	May prolong postoperative bleeding, aggravate gallstones, and cause heartburn. There is also debate about its safety when used to treat morning sickness.
Ginko	Used medicinally in China for hundreds of years, Ginko biloba was recently reported to improve short-term memory and concentration in people with early Alzheimer's disease.	It appears to work by increasing the brain's tolerance for low levels of oxygen and by enhancing blood flow to the brain and extremities.	Possible side effects include indigestion, headache, and allergic skin reactions.
Kava	This South Pacific herb, also known as kava-kava, is used to reduce anxiety, stress, and restlessness and to facilitate sleep without developing problems of tolerance, dependence, and withdrawal.	The active ingredients in the root and stem act as muscle relaxants and anticonvulsants.	Should not be used with alcohol or other depressants by people who are clinically depressed or by women who are pregnant or nursing. Should not be taken for longer than three months without medical supervision.
Milk thistle	One of the few herbs that has been demonstrated to protect the liver against toxins. Also encourages regeneration of liver cells.	The seeds contain a compound called silymarin, which helps liver cells keep out toxins and may promote formation of new liver cells.	When used as capsules containing 200 mg of concentrated extract (140 mg of silymarin), no harmful effects have been reported.
Psyllium	A laxative that doesn't undermine the natural action of the intestinal tract. It may reduce the risk of colorectal cancer and reduce blood levels of cholesterol.	The seeds of this herb have husks that are filled with mucilage, a fiber that swells with water in the intestines to add bulk and lubrication to the stool.	Increase in flatulence in some people, especially if a large amount is consumed.

(continued)

TABLE 2.22 *Continued*

Herb	Why It Is Used	How It Works	Cautions
Saw palmetto	Studies on patients with an enlarged prostate have shown that extracts of this palm tree can reduce urinary symptoms even though the gland may not shrink.	Nonhormonal chemicals in saw palmetto appear to work through their antiandrogen and anti-inflammatory activity.	Some experts are concerned that those taking the herb may have inaccurate PSA readings, used as an early warning sign of prostate cancer.
St John's wort	Reports attest to this herb's ability to relieve mild depression. It may also have sedating and antianxiety activity.	Though products are standardized for hypericin, the chemicals responsible for the herb's activity have not yet been identified.	Based on the sun-induced toxicity of hypericin in animals, people are advised to avoid exposure to bright sunlight.
Valerian	Perhaps best characterized as a mild tranquilizer.	Parts of this plant have antianxiety effects making it useful in treating nervousness and insomnia.	Long-term use can cause headache, restlessness, sleeplessness, and heart function disorders.

SOURCE: From *The New York Times*, Feb. 9, 1999. Reprinted by permission of the publisher.

Advice to Consumers: Safety First!

What constitutes sound, scientific advice for the consumer regarding botanicals, phytochemicals, nutraceuticals, and functional foods? First, differentiate between products sold and consumed as a food (e.g., whole garlic) versus an isolated food component sold separately as a pill, powder, or concentrate (e.g., garlic oil capsule). The consumed dose is likely to be higher with isolated components, which may correspond to a higher risk of harm to the user.

Second, since scientific evidence is still unfolding, add these components to your diet by consuming reasonable amounts of appropriate food sources—a variety of fruits, vegetables, grains, and plants, which are already emphasized in the Food Guide Pyramid. Another option is to select from manufactured foods and beverages formulated to contain useful amounts of beneficial ingredients. Although the food and pharmaceutical industries are developing new food products with specific herbal or phytochemical components, the health benefits remain to be seen.

The Food and Drug Administration (FDA) does allow companies to add ingredients to foods that have been proven safe in order to create functional foods. Examples of such foods include calcium-enriched orange juice, salmon burgers with omega-3 fatty acids, cookies with antioxidants, margarine that lowers LDL cholesterol, and cakes with added fiber. However, herbal supplements are unregulated dietary supplements; as such, they cannot make health claims on their labels. The FDA has warned several companies that their herb-containing products violate the law if they make illegal health claims.

The Role of the Food Scientist in Functional Food Product Development

Food scientists involved in the development of new functional foods must consider several key issues related to such foods and their active ingredients. The action of phytochemicals in the body must be understood. The efficacy of the phytochemical must be determined through research. Most phytochemicals do not have a recommended daily intake. Therefore epidemiological, animal, or clinical human studies need to be evaluated to suggest a guideline for safety and efficacy. Biologically active food compounds must be identified, isolated, and quantified. The task of product development is to incorporate appropriate levels of phytochemicals into manufactured foods and beverages. These must be present in a uniform quantity from batch to batch without variation in amount or activity. The effects of phytochemical ingredients on flavor, functional properties, and color of food products must be determined. For example, a beet extract might provide beneficial phenolic compounds in a functional food or beverage but also impart a deep red color.

Phytochemicals may affect other food properties. For instance, a given phytochemical may affect product viscosity in an undesirable manner by thickening a beverage excessively. Or a phytochemical may be sensitive to pH, heat, light, or oxygen and break down and lose its biological activity if exposed to those conditions. A phytochemical ingredient may exhibit different solubility in water versus oil, which may limit its use in certain food applications.

To become fully distributed throughout a food product, the substance may require an emulsifier or stabilizer. Also, the stability of a given phytochemical may vary depending on the botanical source and the processing conditions used. In sum, the food scientist must consider all of these factors in addition to safety and legality issues concerning botanical, nutraceutical, and phytochemical use in functional foods.

Challenge Questions

1. Functional foods, while promising, are somewhat controversial. Consider the following statements a, b, and c. Do you support any of them, and if so, why?

    a. "Functional foods are conventional foods, and there are ample regulations to make sure they're safe. Consumers are not being misled through deceptive claims and labeling."

    b. "Functional foods are a 'cash cow' for the food industry. Food companies are spiking fruit drinks, breakfast cereals, and snack foods with illegal ingredients and then misleading consumers about their health benefits."

    c. "If you get too much of a functional ingredient, it could be damaging to your health; on the other hand if you don't get what you think you're getting, you're wasting your money."

2. Construct a concept map showing the relationships between the following terms: cancer, cancer prevention, flavonoid, genistein, hormone, isoflavone, phytochemical, phytoestrogen, phytosterol, soybean.

Internet Search Exercises

3. Visit the University of Illinois Functional Foods for Health Program web site at www.ag.uiuc.edu/~ffh. Unscramble the following information and create a table matching six different examples of specific functional foods to their key phytochemical component (one per food), and the associated health benefits: soy foods, omega-3-fatty acids, garlic, reduce risk for cancer, sulforaphane, broccoli, catechin, reduce cholesterol, fish, alliin, black and green tea, soluble beta glucan fiber, genistein, reduce risk for heart disease, oats.

4. Using an Internet search engine, (a) find out what galactomannan is, as well as its reported health benefits, (b) describe its relationship to fenugreek seed, and (c) find out if any functional products or extracts containing the biologically active ingredient are presently available to food manufacturers.

5. Web team exercise: Nutraceuticals and Funtional Foods

    - Access http://www.agr.ca/food/nff/enutrace.html
    - As a team, begin this assignment by defining nutraceutical and functional food, using this site as a source for the definitions (cite the specific reference used).
    - The main part of the assignment is to have each team member prepare a ½–1 page single-spaced summary/review of the most important information from any reference listed under the five sections that follow. (Remember—it must be in your own words, to avoid plagiarism.) Begin by clicking on one of the following sections of the web site:

    Documents   *Industry/Market Environment*
    Regulatory Environment   *Research*
    Consumer Information

a. Each team member must investigate a different section. Have each team member read the other's work for clarity and content prior to submitting for a grade.

b. The team is then responsible for organizing the separate elements prepared by the team members into a single report. Include a title page, an overview of team report content, the team definitions for nutraceutical and functional food, the separate summaries, and a reference page.

IMPORTANT: *Include documentation of your group interaction.* This documentation may be in the form of a "transcript" of the main points discussed during any face-to-face meetings, phone calls, or e-mail conversations. The individual team member contributions to the final team report will also function to document participation.

Goals for the team regarding completion of this assignment:

- Complete assignment on time.
- Encourage each other.
- Cooperate rather than compete within your group.
- Everyone in the group should contribute her/his share.
- Learn and remember the material through this experience.
- Enjoy the opportunity to complete group work!

CHAPTER 3

Human Nutrition and Food

3.1 Proper Nutrition—Making The Right Food Choices

3.2 The Nutritional Role of the Macronutrients

3.3 The Micronutrients—Vitamins and Minerals

3.4 Substitutions for Sugar and Fat

3.5 Energy Metabolism

3.6 Ergogenic Substances

Challenge!
Can a High-Protein, High-Fat Diet Work?

CHAPTER OBJECTIVES

After completing this chapter, you will be able to:
- Define proper nutrition and describe ways to achieve it.
- Describe the Dietary Guidelines for Americans and the Food Guide Pyramid.
- Identify the nutrients considered essential for the human body.
- Explain how the digestion, absorption, and transport of the various nutrients are accomplished.
- Explain how to read a food label.
- Discuss the functions of the important nutrients in human nutrition.
- Calculate the energy value of any food.
- State the nutritional value of alternative sweeteners and fat replacers.
- Identify ergogenic substances and their functions.
- Discuss how to critically evaluate a weight-loss diet.

The study of human nutrition is both fascinating and important for students of food science because the two disciplines are closely related. Food scientists apply technology to keep whole and processed foods fresh, nutritious, and safe, and accomplishing this requires a fundamental knowledge of nutrition. This chapter focuses on identifying the essential nutrients and adequate nutrition, understanding the energy needs of the body, and describing nutrient digestion, absorption, and transport.

The food industry is developing new products to appeal to health and weight-conscious consumers, so food scientists must understand the role of the energy nutrients in foods and energy metabolism. The section on alternative sweeteners, fat replacers, and ergogenic substances illustrates the food science within these new products. Food scientists working in the functional food industry must consider the role of phytochemicals in promoting health. The Challenge section examines food intake, dieting, and weight loss in light of a popular weight-loss diet.

3.1 PROPER NUTRITION—MAKING THE RIGHT FOOD CHOICES

Nutrition is the study of foods and their contribution to health and disease. Because consumers have such a wide variety of food types to choose from, they need to know how their food choices contribute to proper nutrition. *Proper nutrition* implies an adequate and balanced consumption of foods and utilization of all the essential nutrients an individual needs.

The human body requires close to 50 specific nutrients to maintain health. It can synthesize certain nutrients, but essential nutrients must be obtained from food since the body cannot synthesize them. The nutrients that are needed in the largest amounts are **macronutrients** and include organic biomolecules like carbohydrates, lipids, and proteins. Those the body requires in lesser amounts are **micronutrients** and include vitamins and inorganic minerals. Nutrients have three general functions:

- To form body structures, such as organs, teeth, bones, and skin
- To serve as regulators of body processes, on a molecular and biological level (e.g., enzymes, hormones, coenzymes)
- To provide energy by being broken apart during metabolism.

Malnutrition implies an imbalance (too much or too little intake of essential nutrients), resulting in poor nutrition and health. In **undernutrition,** a dietary deficit for one or more nutrients occurs, and deficiency diseases result. In **overnutrition,** excessive intake of one or more nutrient may lead to toxic response and overdose disease.

The Dietary Guidelines and the Food Guide Pyramid

Americans are enthusiastic about pursuing a healthy lifestyle, and foods and nutrition play a critical role in personal health. Scientists and others working in governmental agencies

Too much or too little of a good thing: malnutrition can result from consuming too many or too few nutrients and calories.

have developed criteria for achieving proper nutrition based upon available epidemiological and scientific data. *The Dietary Guidelines for Americans*, first published in 1980 by the USDA and Department of Health and Human Services (HHS), was revised in 1990, and again in 2000—see Table 3.1. It provides recommendations that focus on the promotion of general health and the prevention of certain chronic diseases.

The Food Guide Pyramid was introduced by the USDA as a tool to illustrate the Dietary Guidelines. Recall that Chapter 2 examined food categories and compared them to the Food Guide Pyramid, with the emphasis on food composition. Here we recognize the use of the pyramid to achieve a balanced and nutritious diet. Figure 3.1 describes the nutrients within each food group within the pyramid, gives number of servings and typical serving sizes, and recommends the best choices for a healthy diet. The Dietary Guidelines and Food Guide Pyramid stress variety and moderation in the diet.

Once food is eaten, real nutrition begins. The human body works to obtain nourishment from the food through the processes of digestion, absorption, transport, and metabolism.

TABLE 3.1 The dietary guidelines for Americans, 2000.

Aim for Fitness

Aim for a healthy weight.
Be physically active each day.

Build a Healthy Base

Let the Food Guide Pyramid guide your food choices.
Choose a variety of grains daily, especially whole grains.
Choose a variety of fruits and vegetables daily.
Keep food safe to eat.

Choose Sensibly

Choose a diet that is low in saturated fat and cholesterol and moderate in total fat.
Choose beverages and foods that limit your intake of sugars.
Choose and prepare foods with less salt.
If you drink alcoholic beverages, do so in moderation.

SOURCE: *Dietary Guidelines for Americans* (5th ed.), (2000). Washington, DC, USDA.

FIGURE 3.1 The Food Guide Pyramid—a visual representation of a healthy diet.

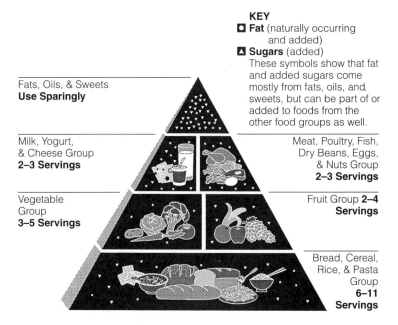

Bread, Cereal, Rice, and Pasta
These foods provide complex carbohydrates and fiber, vitamins (riboflavin, thiamin, niacin, and folic acid), minerals (iron and magnesium), and protein.
6 to 11 servings per day.
Serving = 1 slice bread; $1/2$ c cooked cereal, rice, or pasta; 1 oz ready-to-eat cereal; $1/2$ bun, bagel, or English muffin; 1 small roll, biscuit, or muffin; 5 to 6 small or 2 to 3 large crackers.
Choose most often foods that are made with little fat or sugars (bread, English muffins, rice, and pasta). Aim for at least 6 servings of whole grains per day (brown rice, bulgur, cracked wheat, oatmeal, whole barley, cornmeal, oats, rye, and wheat). Choose less often baked goods high in fat and sugars (cookies, croissants, pastries).

Vegetables
These foods provide fiber, vitamin A, vitamin C, folate, potassium, iron, and magnesium.
3 to 5 servings per day (use dark green, leafy vegetables and legumes [dried beans] several times a week).
Serving = $1/2$ c cooked or raw vegetables; 1 c. leafy raw vegetables; $1/2$ c cooked legumes, $3/4$ c vegetable juice.
Go easy on the fat you add to vegetables at the table or during cooking. French fries and olives are in this group but are high in fat.

Milk, Yogurt, and Cheese
These foods provide calcium, riboflavin, protein, vitamin B_{12}, and, when fortified, vitamin D and vitamin A.
2 servings per day.
3 servings per day for teenagers and young adults, pregnant/lactating women, women past menopause.
4 servings per day for pregnant/lactating teenagers.
Serving = 1 c milk or yogurt; 2 oz process cheese food; $1^{1}/_{2}$ oz cheese; 2 c cottage cheese[b]
Choose nonfat milk and nonfat yogurt often—they are lowest in fat. Go easy on high fat cheese and ice cream. They can add a lot of saturated fat to your diet.

Fats, Oils, and Sweets
Fat and added sugars supply calories but little or no vitamins and minerals (with the exception of vitamin E and some vitamin K). Alcoholic beverages are also included in this category.
Choose low-fat dairy and lean meats, fruits, vegetables, and grains to reduce fat intake. Go easy on fats and sugars added to foods in cooking or at the table (butter, margarine, gravy, salad dressing, sugar, and jelly). Choose fewer foods that are high in sugars such as candy, sweet desserts, and soft drinks.

Fruits
These foods provide fiber, vitamin A, vitamin C, and potassium.
2 to 4 servings per day.
Serving = typical portion (such as 1 medium apple, banana, or orange; $^{1}/_{2}$ grapefruit; 1 melon wedge; $^{3}/_{4}$ c juice; $^{1}/_{2}$ c berries; $^{1}/_{2}$ c diced, cooked, or canned fruit; $^{1}/_{4}$ c dried fruit.
Eat whole fruits often—they are higher in fiber than fruit juices. Have citrus fruits, melons, and berries regularly, because they are rich in vitamin C. Punches, ades, and most fruit "drinks" contain little juice and lots of added sugars.

Meat, Poultry, Fish, Dry Beans, Eggs, and Nuts
These foods contribute protein, phosphorus, vitamin B_6, vitamin B_{12}, zinc, magnesium, iron, niacin, and thiamin.
2 to 3 servings per day.
Serving = 2 to 3 oz lean, cooked meat, poultry, or fish (total 5 to 7 oz per day); count 1 egg, $^{1}/_{2}$ c cooked legumes,[a] 2 tbs peanut butter, $^{1}/_{2}$ c tofu, or $^{1}/_{3}$ c nuts as 1 oz meat (or about $^{1}/_{3}$ c serving).
Choose lean meat, poultry without skin, fish, and dry beans and peas often, because they are lowest in fat. Prepare meats in low-fat ways (trim visible fat; broil, roast, or boil rather than fry). Nuts, seeds, and egg yolks are high in fat, so eat them in moderation.

[a]The Food Guide Pyramid presents legumes (dried beans, lentils, and peas) both under vegetables, for their starch, fiber, and vitamins, and under meats, for their protein and minerals.
[b]Cottage cheese is lower in calcium than most cheeses; 1 cup cottage cheese counts as $^{1}/_{2}$ serving from the Milk, Yogurt, and Cheese group.
Note: Pregnant women may require additional servings of fruits, vegetables, meats, and breads to meet their higher needs for energy, vitamins, and minerals.
SOURCE: Photos by PhotoDisk

Digestion, Absorption, and Transport of Food

Digestion. Substances called *enzymes* are largely responsible for the digestion or breakdown of food molecules into absorbable units or pieces. Enzymes are protein molecules that cause chemical reactions to occur, without being altered in the process—see Figure 3.2. Specific enzyme names can be spotted as words ending in *-ase*.

Digestion and enzymatic reactions break food molecules apart through a process called *hydrolysis*. Food hydrolytic enzymes include pancreatic amylase for starch, pancreatic lipase for fat, and peptidases for protein. Table 3.2 provides an overview of the chemical digestion of starches, sugars, fiber, fats, proteins, and water.

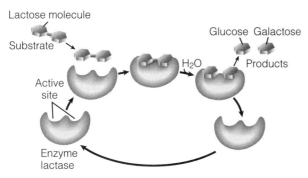

FIGURE 3.2 Enzyme action converting molecules into smaller units. *In enzyme reactions, it is possible to identify substrate, enzyme, and product names, like lactose, lactase, glucose, and galactose.*

Nutrient **bioavailability** refers to the degree to which nutrients are digested by human enzymes in the digestive tract and absorbed by the body. Digestibility is one of many factors that influences a nutrient's bioavailability. Others include adequate nutrient consumption, the nutrient's chemical form, and drug-nutrient, or nutrient-nutrient interactions that can negatively affect absorption.

Absorption. Absorption occurs when digested nutrients enter the bloodstream through the capillaries of the small intestine. Nutrients must pass through absorptive (mucosal) cells of the digestive system. Nutrients enter the mucosal cells in one of three ways:

1. By *simple diffusion*, small lipids and water freely cross into the mucosal cells.
2. Through *facilitated diffusion*, water-soluble vitamins utilize a carrier to transport them into the mucosal cells.
3. The process of *active transport* causes nutrients such as glucose and the amino acids to be absorbed into the mucosal cells—with the cost of energy input.

Transport. Nutrients must be released for distribution to the rest of the body from the mucosal cells of the intestine. The vascular (blood or circulatory) system delivers nutrients and oxygen to all of the body's cells; this is **transport**. One section of the system loops between the heart and lungs, while the other circles between the heart and all other body areas. The exchange of nutrients and wastes takes place across the walls of the smallest blood vessels, the capillaries.

Molecular size is important to nutrient transport. Small molecules of fat, carbohydrate, and protein digestion enter the blood directly via capillaries and are routed to the liver via the portal vein. These nutrients travel to all cells of the liver before moving on to the hepatic vein, and on to the heart. Large molecules of fat digestion cannot travel directly into the capillaries and to the heart. Instead, these form substances called *chylomicrons* that travel via the lymphatic system to the heart. Nutrients to be transported through the lymphatic system are picked up from the mucosal cells by small vessels called *lacteals*.

Regulating Digestion and Absorption

Before a meal, the body is said to be in a state of chemical and metabolic equilibrium called **homeostasis** ("staying the same"). Blood glucose is at a resting level, as is the amount of lipid in the blood. Digestive enzymes are not being secreted by the specific glands. Eating a meal temporarily changes glucose, lipid, and enzyme levels, but the body eventually returns to homeostasis through internal regulation.

Hormones and nerve pathways act as feedback regulators to coordinate digestion and absorption actions. They are part of the body's homeostatic mechanism. **Hormones** are chemical messengers secreted by a variety of glands, which travel to one or more specific target organs to produce an effect such as restoring normal conditions.

TABLE 3.2 Summary of chemical digestion.

	Mouth	Stomach	Small Intestine, Pancreas, Liver, and Gallbladder	Large Intestine (Colon)
Sugar and Starch	The salivary glands secrete saliva to moisten and lubricate food; chewing crushes and mixes it with a salivary enzyme that initiates starch digestion.	Digestion of starch continues while food remains in the upper storage area of the stomach. In the lower digesting area of the stomach, hydrochloric acid and an enzyme of the stomach's juices halt starch digestion.	The pancreas produces a starch-digesting enzyme and releases it into the small intestine. Cells in the intestinal lining possess enzymes on their surfaces that break sugars and starch fragments into simple sugars, which then are absorbed.	Undigested carbohydrates reach the colon and are partly broken down by intestinal bacteria.
Fiber	The teeth crush fiber and mix it with saliva to moisten it for swallowing.	No action.	Fiber binds cholesterol and some minerals.	Most fiber excreted with the feces; some fiber digested by bacteria in colon.
Fat	Fat-rich foods are mixed with saliva. The tongue produces traces of a fat-digesting enzyme that accomplishes some breakdown, especially of milk fats. The enzyme is stable at low pH and is important to digestion in nursing infants.	Fat tends to rise from the watery stomach fluid and foods and float on top of the mixture. Only a small amount of fat is digested. Fat is last to leave the stomach.	The liver secretes bile; the gallbladder stores it and releases it into the small intestine. Bile emulsifies the fat and readies it for enzyme action. The pancreas produces fat-digesting enzymes and releases them into the small intestine to split fats into their component parts (primarily fatty acids), which then are absorbed.	Some fatty materials escape absorption and are carried out of the body with other wastes.
Protein	Chewing crushes and softens protein-rich foods and mixes them with saliva.	Stomach acid works to uncoil protein strands and to activate the stomach's protein-digesting enzyme. Then the enzyme breaks the protein strands into smaller fragments.	Enzymes of the small intestine and pancreas split protein fragments into smaller fragments or free amino acids. Enzymes on the cells of the intestinal lining break some protein fragments into free amino acids, which then are absorbed. Some protein fragments are also absorbed.	The large intestine carries undigested protein residue out of the body. Normally, almost all food protein is digested and absorbed.
Water	The mouth donates watery, enzyme-containing saliva.	The stomach donates acidic, watery, enzyme-containing gastric juice.	The liver donates a watery juice containing bile. The pancreas and small intestine add watery, enzyme-containing juices; pancreatic juice is also alkaline.	The large intestine reabsorbs water and some minerals.

Examples of gastrointestinal hormones (enterogastrones) are **gastrin** and **secretin**. **Cholecystokinin (CCK)** is the name of the intestinal hormone that causes the gallbladder to release bile, and also slows gastrointestinal tract motility. Nerve cells called neurons are present as receptors in the digestive organs such as the stomach, and they can "sense" when food is eaten. They act to stimulate the flow of the gastric juices and also stimulate the intestinal contractions that move nutrient material through the digestive tract.

Reading Food Labels

Nutrition labeling was established by the FDA in 1973 to make consumers aware of the nutrient content of foods. At that time, the focus of the label was the micronutrient content of foods to help prevent deficiency diseases. However, in 1990 the FDA revised the

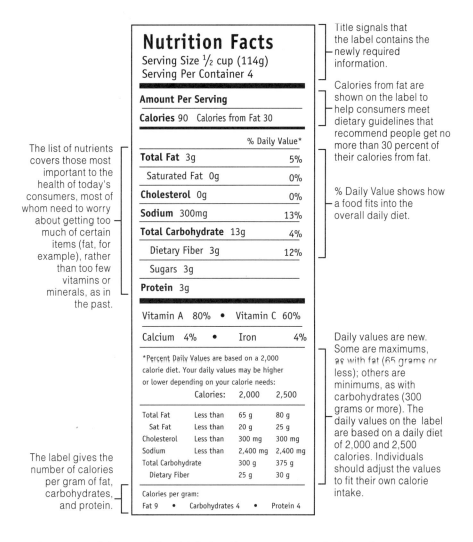

FIGURE 3.3 Example of a nutrition label.

layout and content of the nutritional label to focus more on calorie and macronutrient content. This change reflected dietary concerns about the role these factors play in chronic disease prevention (such as consuming a low-fat diet balanced in saturated and unsaturated fat content) and in contributing to chronic disease (e.g., high protein can contribute to osteoporosis and poor kidney function).

The standard label "Nutrition Facts" format contains information about serving size and quantitative amount per serving of each nutrient except vitamins and minerals. See Figure 3.3. The amount of fat, saturated fat, cholesterol, sodium, total carbohydrate, and fiber as a percent of the Daily Value for a 2000 calorie (kcal) diet is shown. Also provided is a footnote with Daily Values for selected nutrients based on 2000 and 2500 calorie diets (representing average female and male calorie intakes).

To understand Daily Value, you should know the two new terms, Reference Daily Intake (RDI) and Daily Reference Values (DRV), related to it. The term RDI replaces the older RDA (Recommended Dietary Allowance) and gives reference intakes of vitamins A and C and of minerals calcium (Ca) and iron (Fe). The DRV gives reference intake suggestions for fat, saturated fat, cholesterol, sodium, potassium, total carbohydrate, and fiber.

In addition to the nutrition information contained in the "Nutrition Facts" panel, the following information must be provided:

- Product name and place of business
- Product net weight
- Product ingredient content
- Company name and address
- Product code (UPC bar code)
- Product dating if applicable

Nutrient Daily Values are standard values for use on food labels. They are the basis upon which label claims like "low in sodium," "a good source of iron," or "high in calcium" are made possible. Any food providing ten percent or more of the Daily Value for a nutrient is a "good" source, while 20 percent is termed "high" for that nutrient. When consumers make Daily Value comparisons from product to product or label to label, they can make better informed food selections.

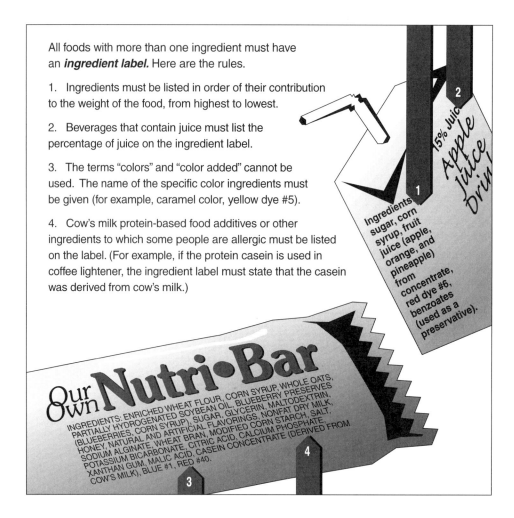

FIGURE 3.4 Ingredient label. *The ingredient list enables the consumer to identify food additives present in a product.*

- Religious symbols if applicable
- Safe handling instructions if applicable (e.g., raw meats)
- Special warning instructions if applicable (e.g., aspartame)

Full ingredient labeling is required for all foods containing more than one ingredient. Ingredients are always listed in descending order of amount present (most abundant to least)—see Figure 3.4.

Terms such as "free," "reduced," and "high" may appear on labels, but are only permitted if they meet specific definition standard—see Table 3.3. For example, if a label uses the descriptor "high" to indicate that a food product contains a high amount of a specific nutrient, this must equal 20 percent or more of the Daily Value for that nutrient.

3.2 THE NUTRITIONAL ROLE OF THE MACRONUTRIENTS

Proper nutrition requires the intake, digestion, absorption, and transport of the essential nutrients. Which nutrient do you think is the major component of the human body—protein, fat, carbohydrate, or water?

Water, Water, Everywhere

Water is the one nutrient we can least afford to go without and depending on age and fat level, comprises over 60 percent of body weight. It is the major constituent of the human body—every cell contains water (*intracellular water*) and is surrounded by water (*extracellular water*). Lean body (muscle) tissue contains more than twice the water content of

TABLE 3.3 Descriptor terms allowed on food labels and their definitions.

Description	Amount/Serving
Free	No nutrient or in very small amounts: calories <5, fat <$1/2$ g, saturated fat <$1/2$ g, cholesterol <2 mg, sugar <$1/2$ g, and sodium <5 mg
Low	<40 calories
	<3 g fat
	<1 g saturated fat
	<20 mg cholesterol
	<140 mg sodium (very low sodium <35 mg)
Extra Lean	<5 g fat
	<2 g saturated fat
	<95 mg cholesterol
Lean	<10 g fat
	<4 g saturated fat
	<95 mg cholesterol
Light/Lite	$1/3$ fewer calories
	Half the fat
	Can also describe color or texture
Reduced or Less	25% less than regular product
More	10% or more of Daily Value compared to reference food
High	20% or more of Daily Value for the nutrient
Good Source	10–19% or more of Daily Value for the nutrient
More Than	10% or more of Daily Value than comparison food
Healthy	<3 g fat
	<1 g saturated fat
	<480 mg sodium
	<60 mg cholesterol

adipose (fat) tissue. In lean tissue, about 60 percent of the water is intracellular water, and the remaining 40 percent is extracellular water, in interstitial and intervascular spaces.

Although water does not provide energy, it is the most critical nutrient because without it, the body becomes dehydrated, and death results. Water has five major functions in the body.

- Water helps regulate body temperature.
- Water lubricates the eyes, spinal cord, gastrointestinal tract, and joints and provides for the saliva in the mouth.
- Water, the medium for metabolic chemical reactions, allows reactants and products to come in contact with each other and with enzymes when necessary.
- Water participates in many reactions (e.g., hydrolysis).
- Water serves as a carrier and aids in the movement of nutrients during digestion, absorption, and circulation, as well as the excretion of wastes and toxins.

Metabolic Water. Water is not only present in foods and beverages, it is produced during the metabolism of nutrient molecules in body cells. When food is metabolized, *metabolic water* is produced. One hundred grams of glucose generates about 60 mL of metabolic water, while 100 grams of protein generates less (about 40 mL) and 100 grams of fat, more (about 107 mL).

Electrolytes and Dehydration. **Electrolytes** are specific mineral elements inside and outside of body cells that conduct electrical charges. Sodium is the major extracellular electrolyte in the body, and potassium is the major intracellular electrolyte. Electrolytes dissolve in water and dissociate (come apart) into *ions*—charged atoms that can attract water molecules:

$$NaCl \rightarrow Na^+ + Cl^- \qquad KCl \rightarrow K^+ + Cl^-$$

Water is attracted to these ions and is pulled along with their movement. In a healthy, well-nourished person, the correct concentration of electrolytes is maintained inside and outside of cells. This establishes a proper water balance on either side of the cell

TABLE 3.4 Nutritional overview of carbohydrates.

Classification

Monosaccharides: glucose, fructose, galactose
Disaccharides: sucrose, lactose, maltose
Polysaccharides: glycogen, starch, fiber

Energy Value 4 kcal/g

Functions

Provide energy
Maintain normal blood glucose level
Aid in normal functioning of gastrointestinal tract (insoluble fiber)
Reduce blood lipids (soluble fiber)

Food Sources

Breads, cereals, and grains
Vegetables and fruits
Milk and dairy
Legumes and nuts
Sweets and sweeteners

Food Sources of Fiber

Insoluble; wheat bran, fruits and vegetables, legumes, seeds
Soluble fruits and vegetables, legumes, oats, oat bran, barley, psyllium seeds

Recommended Intake

Lower limit: 55% of caloric intake*
Upper limit: 75% of caloric intake*
Sugars: ≤10% of caloric intake*
Fiber: 20–35 g/day (adults)[†]

*SOURCE: WHO Study Group on Diet, Nutrition, and Prevention of Noncommunicable Diseases (1991). Diet, nutrition and the prevention of chronic diseases; A report of the WHO Study Group on Diet, Nutrition, and Prevention of Noncommunicable Diseases, *Nutrition Reviews*, 49 (10), 291–301.
[†]SOURCE: American Dietetic Association (1997). Health implications of dietary fiber. *Journal of the American Dietetic Association*, 97, 1157.

membrane. To maintain this intra- and extracellular water balance, cells constantly move electrolytes in and out, via a mechanism called the sodium-potassium (Na–K) pump. Potassium attracts water into cells to maintain water balance, while sodium and chloride (Cl^-) serve as extracellular regulators of fluid balance.

When water intake is less than water output, water balance is upset and causes dehydration and an imbalance in the electrolytes. Insufficient fluid intake or illnesses, such as food poisoning that causes vomiting and diarrhea (resulting in water and electrolyte loss), can cause dehydration. To compensate, body cells release intracellular water, along with intracellular electrolytes, into the extracellular space. If too much water and electrolytes leave their cells, kidney and heart failure can result. Rehydration—restoring water balance to a dehydrated person through intravenous therapy—can be a lifesaver.

Carbohydrates—Our Basic Fuel

To the food scientist, carbohydrates are important for their functional properties. To the nutritionist carbohydrates are important as an energy nutrient. See Table 3.4 for an overview of the functions and sources of the carbohydrates.

Understanding nutrients at the molecular level is important for food scientists—including knowing a molecule's chemical formula, how nutrient molecules are shaped, and how size and shape impact their functioning in the body. All carbohydrates are composed of the elements carbon (C), hydrogen (H), and oxygen (O). The arrangement of atoms and the bonds between them distinguish the various carbohydrates, which are classified as simple carbohydrates (the sugars) and the complex carbohydrates (starch, glycogen, cellulose, and fiber).

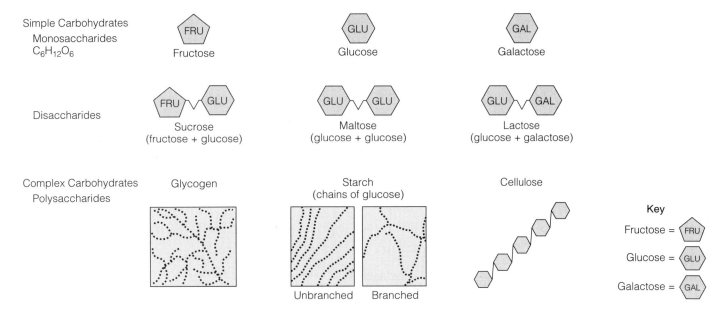

FIGURE 3.5 Classification of carbohydrates.

Simple Carbohydrates. In the simple sugar glucose, atoms of carbon, hydrogen, and oxygen bond to make a ring-shaped molecule. **Glucose** occurs in plants and is an example of a **monosaccharide,** or single sugar ring molecule (Figure 3.5). **Fructose** (levulose or fruit sugar—commonly found in fruits and other plant foods) and **galactose** are also monosaccharides.

Glucose is our primary energy nutrient or fuel. It can provide energy to the body immediately, be stored as glycogen to be used as energy later, or be converted to fat. Glucose is carried through the blood to different parts of the body (brain, nerves, red blood cells) to be broken down for energy. Apart from energy, glucose also serves as the backbone for synthesis of new compounds, such as amino acids.

Disaccharides ("di" meaning two) are carbohydrates formed by the joining of two monosaccharide units. A glucose molecule attached to fructose forms **sucrose,** or common table sugar. Galactose occurs in nature bound to glucose, and the resulting disaccharide is **lactose** (milk sugar). **Maltose** is composed of two glucose units bonded together.

Complex Carbohydrates. Polysaccharides or complex carbohydrates are long chain-like linkages of sugar units (Figure 3.5). Examples are starch and cellulose in plants, vegetable fibers (bran from grains), pectins from apple and other fruits, hydrocolloids or food gums (carrageenan and guar gum), plant starches, and glycogen or animal starch (although glycogen is not a dietary source of carbohydrate). Plant starches are enzymatically converted in the body into simple sugars like glucose to be used for energy.

Fiber is the nonstarch polysaccharide carbohydrate portion of plants (i.e., cellulose) that helps maintain structural rigidity. Fiber has no single chemical structure because fiber encompasses a diverse group of compounds. **Dietary fiber** is the residue of plants left undigested after consumption of edible fiber. Dietary fibers are of two main types: insoluble fiber, which does not dissolve in water, but will absorb water and swell up; and soluble fiber, which forms gel-like solutions as they dissolve in water.

Dietary fiber has been heavily studied owing to specific health effects when ingested above a critical level. The positive health effects of fiber are generally linked to their solubility and viscosity in water. Soluble fibers tend to delay food transit through the intestinal tract. Viscous fiber systems delay glucose uptake in the blood and help lower blood cholesterol. Insoluble fibers prevent constipation and are believed to decrease the incidence of colon cancer. (See Table 3.5.) Recommended fiber consumption for adults is 25–35 grams per day. The long-term health effects of consuming excessive amounts (50–100 grams or more per day) are unknown.

TABLE 3.5 Health benefits and dietary sources of soluble and insoluble fiber.

Foods Rich in Insoluble Fiber		Foods Rich in Soluble Fiber	
apples	pears	apples	green beans
bananas	peaches	apricots	green peas
beets	plums	bananas	legumes
brown rice	rice bran	barley	oat bran
cabbage family	seeds	berries	oatmeal
cauliflower	skins/peels of fruits and vegetables	black-eyed peas	pears
corn bran	strawberries	broccoli	potatoes
green beans	tomatoes	cabbage	prunes
green peas	wheat bran	carrots	rye
legumes	whole-grain breads and cereals	cherries	seeds
mature vegetables		citrus fruits	sweet potatoes
nuts		corn	zucchini
		grapes	

Health Problems	Fiber Type	Possible Health Benefits
Obesity	Soluble Insoluble	Replaces calories from fat, provides satiety, and prolongs eating time because of chewiness of food
Digestive tract disorders: Constipation Diverticulosis Hemorrhoids	Insoluble	Provides bulk and aids intestinal motility; binds bile acids
Colon cancer	Insoluble	Speeds transit time through intestines and may protect against prolonged exposure to carcinogens*
Diabetes	Soluble	May improve blood sugar tolerance by delaying glucose absorption
Heart disease	Soluble	May lower blood cholesterol by slowing absorption of cholesterol and binding bile

*This effect is based on epidemiologic studies and is usually observed along with a reduced-fat intake; it is unclear whether the protection comes from fiber or from other components that accompany fiber in foods—such as vitamins (e.g. folate), minerals (e.g., calcium), or phytochemicals.

Wheat bran, an insoluble fiber, can bind to minerals and remove them from the body before they can be digested, which constitutes a negative health effect. However, we must weigh the negatives and positives of such dietary substances.

Glycemic Effect. Carbohydrate food sources affect blood glucose levels. The ability of a food to cause a sharp increase in blood glucose is termed **glycemic effect**. Generally, the easier a food carbohydrate is digested and absorbed, the greater the glycemic effect. Simple sugars generate a high glycemic effect when consumed alone—see Table 3.6. Mixed meals containing a variety of foods can lessen the effect. Persons who have disorders of glucose metabolism, such as diabetes and hypoglycemia, should consume foods associated with low glycemic effect. The metabolism of these individuals cannot easily cope with the rapid increase in blood glucose caused by high glycemic foods.

Lipids—Those Essential Fatty Acids

Fat holds a negative connotation to most Americans. Cholesterol, saturated fat, and heart disease are closely associated. However, fat consumption is beneficial in the appropriate quantities. Fats are actually a subcategory of fat-soluble substances known as **lipids**. Lipids are the most concentrated source of energy fuel for the body.

Dietary lipids supply the essential fatty acids needed by the body to maintain proper health and functioning. Lipids include all types of fats, oils, phospholipids, sterols, and waxes. Fats are solid at room temperature, whereas oils are liquid. Butter, lard, and margarine are solid fats, while corn oil, canola oil, and soybean oil are liquid oils. Table 3.7 provides the classification, energy value, functions, food sources, and recommended intake of lipids.

TABLE 3.6 Trends among foods regarding glycemic effect. *The glycemic effect or index is a useful tool that measures how fast a particular food can elevate the blood sugar. Numbers are based on glucose, which is given the top value of 100. Foods in the "high" column raise low blood sugar levels quickly. Foods in the "low" column raise low blood sugar levels slowly.*

Low	Medium	High
apple	banana	carrots
beans	high-fructose corn syrup	dry dates
chickpeas	honey	glucose
grapefruit	ice cream	instant white rice
green leafy vegetables	oatmeal cookies	jelly beans
soy milk	sucrose	potatoes
strawberries	sweet corn	white bread

TABLE 3.7 Nutritional overview of lipids.

Classification

Triglycerides—95% of dietary fat
   1 glycerol + 3 fatty acids
      Fatty acids
         Short-, medium-, or long-chain   Polyunsaturated (>one double bond)
         Saturated (no double bonds)   Omega-3 fatty acids
         Monounsaturated (one double bond)   Omega-6 fatty acids
Cholesterol
Phospholipids

Energy Value 9 kcal/g

Functions

Provide energy	Serve as source of essential fatty acids
Serve as energy reserve	Insulate and protect organs
Assist in transport and absorption of fat-soluble vitamins	Maintain cell walls and play role in making Vitamin D, bile, and hormones (cholesterol)

Food Sources

Oils, sauces, spreads, butter
Meat, poultry, fish, eggs, nuts (fish oil for omega-3)
Milk, cheese, dairy
Breads, cereals, and grains with added fats

Recommended Intake*

Less than or equal to 30% of calories from total fat
Less than or equal to 10% of calories from saturated fat
No more than 300 mg cholesterol

*SOURCE: Expert Panel on Detection, Evaluation, and Treatment of High Blood Cholesterol in Adults (1993). Summary of the second report of the National Cholesterol Education Program (NCEP). Expert Panel on Detection, Evaluation, and Treatment of High Blood Cholesterol in Adults. (Adult Treatment Panel II). *Journal of the American Medical Association*, 269 (23), 3015–3023.

Lipid Composition. Like carbohydrates, lipids are chemical compounds containing primarily C, H and O atoms, but they contain more hydrogen and less oxygen than carbohydrates. The bonding arrangements also differ. Carbon atoms in lipids are usually bonded to hydrogen, while in carbohydrates carbons are often bound to oxygen. More chemical energy is stored in C—H bonds (lipids) than in C—O bonds (carbohydrates), which explains why lipids supply more energy calories than carbohydrates or proteins.

Lipid molecules contain two basic components: glycerol and fatty acids. Glycerol forms the backbone of lipid molecules and contains binding sites to hold up to 3 fatty acids. This allows for the formation of either monoglycerides, diglycerides, or triglycerides. Most food lipids are triglycerides—that is, 3 fatty acids are attached to each glycerol molecule—see Figure 3.6.

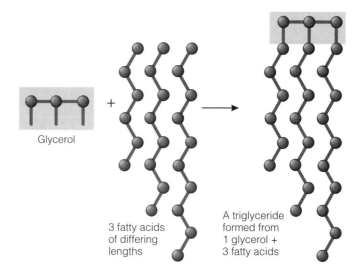

FIGURE 3.6 The triglyceride (chemical synonym: triacylglycerol). *Most food lipids are triglycerides, composed of 3 fatty acids attached to one glycerol molecule.*

Lipid Properties. Lipid molecules do not dissolve in water—they are insoluble molecules. Lipids are also termed *hydrophobic*, which literally means "water-fearing." In practical terms, lipids are unable to relate chemically with water to become dissolved because of their electrical nature. Lipids are electrically neutral (uncharged) compounds, whereas water is a *dipole*, meaning it has a negatively charged oxygen end and a positively charged hydrogen end—see Figure 3.7. Since the ends (the poles) are of opposite charge, they cancel each other electrically, and the overall molecule is neutral, or uncharged. However, the poles can attract other molecules that are polar, and this is the chemical basis for solubility—whether or not a substance will dissolve in water. Since lipid molecules lack this polar condition, they do not dissolve in water.

The insolubility of lipids affects the way they are digested and absorbed into the body. As hydrophobic compounds, lipids tend to separate themselves from the watery digestive juices and enzymes. Digestive enzymes are considered *hydrophilic* compounds—having an affinity for water. This means that hydrophilic substances can become *hydrated*, and dissolve or disperse in water. When a meal containing lipids is eaten, the fat sits on top of the watery mix of food and enzymes. The enzymes cannot digest lipids until they are part of the watery phase.

To accomplish this, the gallbladder secretes bile, a molecule with both a hydrophilic portion and a hydrophobic portion. Bile surrounds fats and "squeezes" them into small droplets from the original lipid matter. The hydrophilic end of bile is directed out into the aqueous environment, and the hydrophobic end is in contact with lipid material. This action of bile is called **emulsification.** Emulsification allows fat to mix into the watery layer of food where the digestive enzymes are waiting. Digestion separates triglycerides into glycerol and three free fatty acids.

Fatty Acids and Health. The fatty acids that are absorbed from food are often discussed in light of the chemical structure of the parent molecule (triglyceride), meaning whether or not they are composed of saturated or unsaturated fatty acids. Fatty acids that contain only single bonds between carbon atoms (C—C) are **saturated fatty acids.** *Saturated* means that each carbon pair, in addition to bonding to each other, is bonded to 4 hydrogen atoms (2 each). The fatty acid is said to be saturated with hydrogen.

By contrast, unsaturated fats are triglycerides that contain glycerol and between one and three unsaturated fatty acids. An **unsaturated fatty acid** contains one or more double bonds (C=C) between adjacent carbon atoms. Each double-bonded carbon pair, in addition to bonding to each other, is bonded to only 2 hydrogen atoms (1 each). Since this type of fatty acid is not saturated with hydrogen, it is said to be unsaturated. See Figure 3.8.

Essential fatty acids (EFAs), are essential to human health but cannot be made by the body. The EFAs are the building blocks of important lipids and lipid-containing materials. They are required as cell membrane components, strengthen immune cells, and lu-

FIGURE 3.7 Lipids are insoluble in water. *Lipids are neutral molecules, but water, although neutral, is polar and only dissolves other polar molecules. In a water (H_2O) molecule, two positively charged hydrogen atoms are chemically bonded to one negatively charged oxygen atom, creating a positive hydrogen pole and a negative oxygen pole. Water is thus called a dipole.*

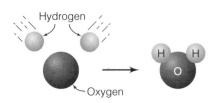

Saturated

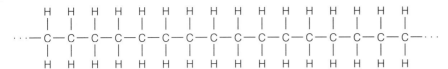

Primarily animal sources: meat, poultry, milk, butter, cheese, egg yolk, lard
Plant sources: chocolate, coconut, coconut oil, palm oil, vegetable shortening

Monounsaturated

Sources: avocado, peanuts, peanut butter, olives/olive oil

Polyunsaturated

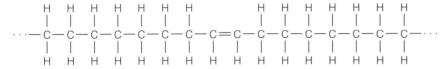

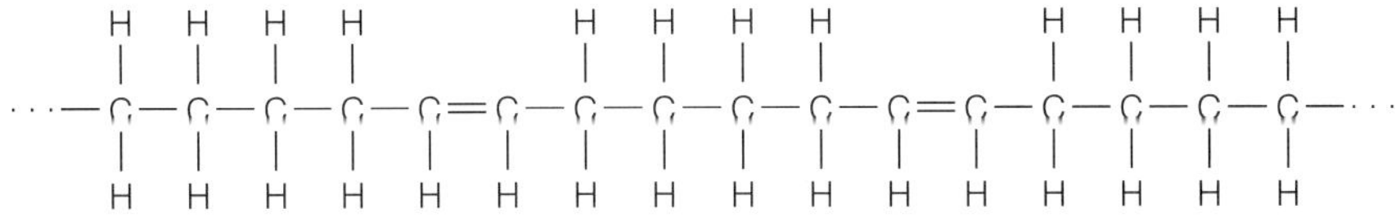

Primarily plant sources: vegetable oils (corn, safflower, soybean, sunflower, canola, etc.)
margarine (most), mayonnaise, certain nuts (almonds, filberts, pecans, walnuts)
Animal sources: fish

FIGURE 3.8 Structure of fatty acids. *Fatty acids are either saturated, monounsaturated (contain one $C=C$ double bond), or polyunsaturated (contain two or more $C=C$ double bonds).*

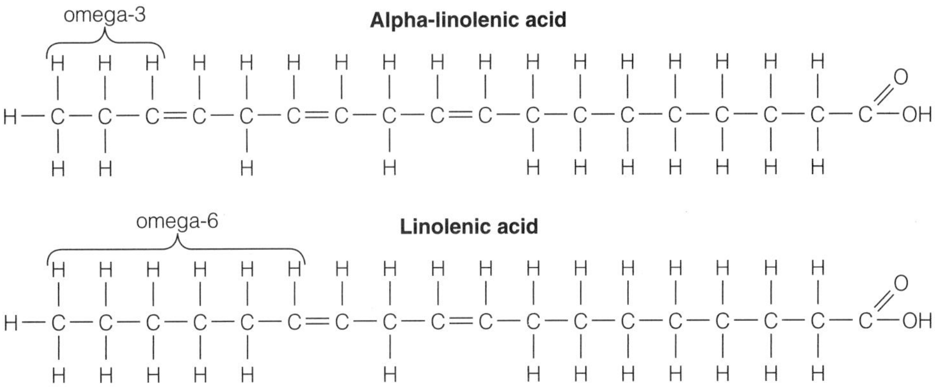

Alpha-linolenic acid (omega-3)

Linolenic acid (omega-6)

FIGURE 3.9 Structure of omega-3 fatty acids and omega-6 fatty acids. *These fatty acids are defined by the location of the first double bond.*

bricate body joints. Essential fatty acids also insulate the body against heat loss, prevent the skin from drying, provide energy, and are involved in prostaglandin synthesis. *Prostaglandins* are lipid-derived substances needed for energy metabolism, as well as cardiovascular and immune system functioning.

All of the EFAs are unsaturated and must be provided from food. Essential fatty acids occur in fish oils and a variety of seed oils. The essential fatty acids are either omega-3 fatty acids or omega-6 fatty acids (Figure 3.9). The two omega-3 essential fatty acids are eicosapentaenoic acid (EPA) and docosahexaenoic acid (DHA). Coldwater fish are the best sources of omega-3 EFAs, and some plant sources contain precursors to EPA and DHA. Flax seed, hemp, walnut, and soybean oil contain the omega-3 essential fatty acid called alpha-linolenic acid (ALA). ALA can be converted through several enzymatic steps into EPA, and EPA into DHA.

Nutritionists use chemical abbreviations when referring to fatty acids; in fatty acid nomenclature, C18:2 ω-6, identifies linoleic acid:
C18 = the number of carbon atoms in the molecule
2 = the number of double carbon bonds
ω-6 = the position of the first double bond nearest the methyl (CH_3) end of a fatty acid

> *Research into LCPUFAs (long chain polyunsaturated fatty acids) in pre- and postnatal human brain and eye development has led to their incorporation into infant formulas. In 2002, arachidonic acid (ARA) and docosahexaenoic acid (DHA) were approved by the FDA as additives to infant formulas.*

The omega-6 essential fatty acids include linoleic acid (LA) and gamma-linolenic acid (GLA). These are found in the seeds of borage, evening primrose, sunflower, and black currant. A nonessential omega-6 fatty acid that is in the news is conjugated linoleic acid (CLA). CLA is found mostly in dairy products (not to a great extent in the fat-free ones, however). Studies have hinted that CLA may play a role as a weight-gain deterrent by affecting the hormone insulin to help incorporate fatty acids and glucose preferentially into muscle rather than adipose tissue.

The omega-3 fatty acids of particular interest because of their effects on lipoprotein metabolism are alpha-linoleic acid (ALA), eicosapentaenoic acid (EPA), and docosahexaenoic acid (DHA). Studies have shown that lower total cholesterol can result from ingestion of these oils. Although elevated levels of total blood cholesterol is a major risk factor for coronary disease, dietary cholesterol may not play the major role once thought because cholesterol intake from food alone is not responsible for cholesterol levels. The liver also manufactures two lipoproteins, VLDL and LDL, as sources of cholesterol.

Lipoproteins are water-soluble substances that transport fat and cholesterol in the blood. There are three major classes of lipoproteins: low-density lipoprotein (LDL), very low-density lipoprotein (VLDL), and high-density lipoprotein (HDL). LDL carries cholesterol from the liver to body tissues. VLDL carries cholesterol and triglycerides from the liver to body tissues. HDL carries cholesterol back to the liver for disposal. Factors that raise LDL and /or lower HDL cholesterol levels in the blood are to be avoided, including a high-fat diet, a high–saturated-fat diet, and a diet high in trans-fatty acids.

Trans-fatty Acids. Trans-fatty acids are found in meat, poultry, and processed dairy products. (The term *trans* describes the geometric shape of the fatty acid and will be discussed in Chapter 5.) Foods that contain partially hydrogenated vegetable oils also contain trans-fatty acids. In the hydrogenation process, hydrogen atoms are added to liquid oils to harden them and provide a longer shelf life, creating a percentage of trans-fatty acids in the conversion. Table 3.8 lists the grams of trans-fatty acids and total fat grams found in some common foods. Trans-fatty acids are a concern because they may raise blood levels of LDL cholesterol in the same way that saturated fats do. In addition trans-fatty acids may decrease HDL levels.

TABLE 3.8   Trans-fatty acid content of selected foods.

Food	Grams of Trans-Fatty Acids*	Grams of Total Fat
Animal Foods		
5 oz beef	0.9	27.7
1 tsp butter	0.1	4.1
5 oz chicken	0.1	10.4
5 oz pork	0.1	21.7
Vegetable Fats		
1 tsp vegetable shortening	0.63	4.3
1 tsp stick margarine	0.62	4
1 tsp soft margarine	0.27	4
1 tsp vegetable oil	0.02	4.5
Packaged Foods and Fast Foods		
4 oz deep-fat fried french fries	5.5	18.5
1 cake doughnut	3.19	10.8
1 oz corn chips	1.42	9.5
1 piece yellow cake with chocolate frosting	1.04	11.2
1 slice apple pie	1	13.8
1 oatmeal raisin cookie	0.86	3.3
1 slice cheese pizza	0.13	6.1
1 oz potato chips	0.11	9.8
1 blueberry muffin	0.09	3.7

*Figures represent the average values derived from several brands or varieties of the foods.

SOURCE: Adapted from L. Litin and F. Sacks, Trans-fatty-acid content of common foods, *New England Journal of Medicine* 329 (1994): 1969–1970.

Proteins—Essential Body Builders

Anthropologists believe that early societies obtained dietary protein primarily from food gathering—that is, collecting and consuming plants and grains. Hunting animals for meat came somewhat later. Today animal products hold a central position in the North American diet, although a healthy vegetarian diet is possible too.

A lean (non-obese) healthy human body is about 16 percent protein by weight. Proteins are highly functional, involved in nearly every biological process in the body. Proteins contain atoms of carbon, oxygen, and hydrogen as do the carbohydrates and lipids, but in addition contain nitrogen (N) atoms and sometimes sulfur (S).

Proteins are composed of subunits called **amino acids.** Typical amino acids consists of a nitrogen-containing amino (NH_2) group, an acid or carboxyl (COOH) group, and a side chain (R group) opposite a hydrogen atom—see Figure 3.10. All of these groups are attached to a single carbon atom. Amino acids are identical in structure except for their side chains. This fundamental difference accounts for the 20 distinct naturally occurring amino acids known to be required by the human body. Of these twenty, only eleven are synthesized by the body. Therefore, nine must be supplied by the diet—referred to as the *essential amino acids*. Table 3.9 lists the amino acids and presents the functions, food sources, and recommended intake for protein.

The body not only synthesizes nonessential amino acids, it can join two or more amino acids together with **peptide bonds** to create more complex peptides (peptide = protein). Large proteins, which are made up of chains of amino acids connected by these peptide bonds, are referred to as polypeptides:

FIGURE 3.10 Structure of an amino acid. *The R group makes each amino acid different.*

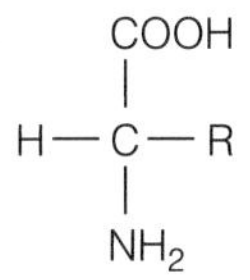

$$\begin{array}{c} COOH \\ | \\ H-C-R \\ | \\ NH_2 \end{array}$$

TABLE 3.9 Nutritional overview of proteins.

Classification
Proteins are made up of amino acids, which are classified as —

Essential Amino Acids (Indispensable)	*Nonessential Amino Acids (Dispensable)**
Histidine	Alanine
Isoleucine	Arginine
Leucine	Asparagine
Lysine	Aspartic acid
Methionine	Cysteine
Phenylalanine	Glutamic acid
Threonine	Glutamine
Tryptophan	Glycine
Valine	Proline
	Serine
	Tyrosine

Energy Value 4 kcal/g
Functions

Growth and maintenance of body tissues	Maintain fluid and electrolyte balance
Formation of enzymes, hormones, antibodies, and other essential components	Maintain acid-base balance
	Energy

Food Sources
Meat, poultry, fish, eggs, legumes, nuts
Milk, cheese, dairy
Breads, cereals, grains
Vegetables

Recommended Intake
0.8 g/kilogram of body weight per day (female adults); 0.9 g/kilogram of body weight per day (male adults)
10%–15% of caloric intake

*During some disease states, certain nonessential amino acids may become essential.
SOURCE: WHO Study Group on Diet, Nutrition, and Prevention of Noncommunicable Diseases. (1991). Diet, nutrition and the prevention of chronic diseases. A report of the WHO Study Group on Diet, Nutrition, and Prevention of Noncommunicable Diseases, *Nutrition Reviews* 49 (10), 291–301.

- A protein consisting of 2 amino acids is a *dipeptide*.
- A protein consisting of 3 amino acids is a *tripeptide*.
- A protein consisting of 4 or more amino acids is a *polypeptide*.

Protein Quality. After the body digests food protein into its component amino acids, it reassembles (synthesizes) the amino acids into specific proteins that the body needs for specific functions—this process is called **protein synthesis.** Protein quality is measured by the presence of all the essential amino acids in a protein-containing food, in the amounts required to support protein synthesis. Ideally, a protein is *complete* and contains all of the essential amino acids needed to support protein synthesis and is highly *digestible*. Digestible proteins are easily converted by enzymes into single amino acids.

Since the body does not store amino acids to a great extent, protein synthesis needs to occur at regular intervals. It will not occur unless all of the essential amino acids are present in sufficient quantity upon absorption. Absorbing all of the essential amino acids can be accomplished in one of two ways from a dietary standpoint:

1. Consistently, meaning several times per week, a person's diet should contain foods that are high in quality—that provide a source of complete, digestible protein (such as from the meat, poultry, fish, dry beans, eggs, and nuts group, and/or the milk, yogurt, and cheese group).
2. A person can consume a variety of incomplete protein foods together in a meal, several times a week, so that the body "thinks" a complete protein has been consumed. Incomplete proteins are those typically found in foods from the bread, rice, cereal, and pasta group. Beans and nuts represent incomplete protein as well. The practice of combining grains and legumes in this manner to make a high-quality protein meal is referred to as **protein complementation** or *mutual supplementation*.

For instance, the consumption of beans and rice is a tradition in many cultures. Beans are generally deficient in the amino acid lysine, whereas rice is deficient in the amino acid methionine. Both of these amino acids are termed *limiting amino acids*, in that their absence limits protein synthesis. Both are essential amino acids, and consumption of beans alone or rice alone would result in essential amino acid deficiency. (Look back at Table 2.7.)

However, by consuming these foods together at a meal, the combination provides high-quality protein, since beans contain adequate methionine and rice contains adequate lysine. Each "complements" the other, making up for their respective amino acid deficiencies. Succotash, which is a mixed meal of corn and lima beans, and a peanut butter sandwich (Figure 3.11) are examples of food combinations that provide complete protein.

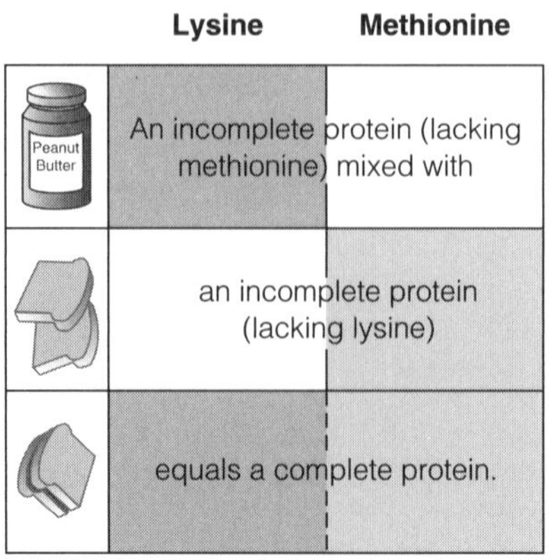

This discussion raises the issue of vegetarian diets. Clearly, vegetarians can meet their protein and other nutrient needs. Foods must be selected to ensure adequate intake of complete protein, vitamins (particularly vitamin B_{12}), and minerals, especially calcium and iron.

Biological value (BV) is also a measure of protein quality. Egg protein, one of the most complete and digestible of all food proteins, is the reference protein for BV with a rating of 100 (the maximum possible). In general, foods of animal origin tend to be of higher biological value than foods of plant origin. Beef protein has a BV of 80, milk protein is 71, and rice protein is 59. The amount of protein in foods can differ markedly. Animal sources provide roughly 15–20 grams of protein per 100 grams of food, while cereals, grains, nuts, and legumes provide only 1–2 grams of protein per 100 grams of food.

To measure a food's biological value, it is fed to test animals in controlled experiments to determine the amount of nitrogen retained by the animal in its tissues versus the amount of nitrogen excreted. If a food is composed of high BV protein, protein synthesis will occur after the animal consumes the meal, and most of the nitrogen will be retained as amino acids in the structure of new protein in the animal's body. If a food protein BV is low, excretion of nitrogen will be high, and protein synthesis will not be supported.

Nitrogen Balance. It is critical to health that the body be kept in what is called *nitrogen balance*. Because only dietary protein supplies nitrogen, nitrogen balance can be

FIGURE 3.11 Protein complementation. *The plant proteins in peanut butter and bread are incomplete due to* limiting amino acids, *but in combination, a peanut butter sandwich is a complete protein meal.*

equated with protein balance. In performing its primary functions, protein tends to be retained in the body, which is said to be in a state of positive nitrogen balance.

Consumption of adequate protein achieves nitrogen balance, wherein nitrogen intake equals nitrogen output. The recommended intake for an adult is 0.8 g/kilogram of body weight per day. However, in a growing child, during pregnancy and lactation, or during a prolonged illness or recovery from trauma, a person's dietary intake of protein should exceed this recommendation to ensure a positive protein balance.

In conditions of starvation or dietary imbalance, protein is used for energy. Low-calorie diets, low-protein diets, and disease states called kwashiorkor or P.E.M. (protein energy malnutrition) are other examples of situations in which protein breakdown for energy can occur. In these scenarios, the nitrogen from protein is excreted in the urine, and the body can go into negative nitrogen balance because urinary losses of nitrogen exceed dietary intake.

3.3 THE MICRONUTRIENTS—VITAMINS AND MINERALS

Vitamins

Vitamins are small organic compounds that must be supplied in relatively low amounts in the diet to maintain health. Nutritional functions of vitamins include acting as antioxidants and in disease prevention. Certain vitamins are important for *erythropoesis*, or red blood cell formation. Others function with enzymes to facilitate the metabolism of the energy nutrients, acting as coenzymes. A **coenzyme** is a substance that activates enzymes to promote enzymatic reactions.

Vitamins are classified according to their solubility. The *water-soluble vitamins* include the eight vitamins in the B-complex and vitamin C. The *fat-soluble vitamins* that cannot dissolve in water are A, D, E, and K. Table 3.10 lists the vitamins and food sources, functions, and deficiency and toxicity symptoms for each.

Substances you may hear or read about that are not true vitamins include citrus bioflavonoids, rutin, calcium pangamate ("vitamin B_{15}"), lipoic acid, DHEA, creatine, carnitine, PABA, and choline.

Essential Minerals

Minerals are inorganic substances, as listed in the periodic table of the elements. Of these, fewer than twenty are important for human health—see Table 3.11. The minerals are classified according to prevalence in the diet and in body tissues into the major minerals and the trace minerals. Figure 3.12 demonstrates the amounts of major and trace minerals in a

Water-soluble vitamins
vitamin C
B complex
  thiamin
  riboflavin
  niacin
  vitamin B_6
  folate
  vitamin B_{12}
  pantothenic acid
  biotin

Fat-soluble vitamins
vitamin A
vitamin D
vitamin E
vitamin K

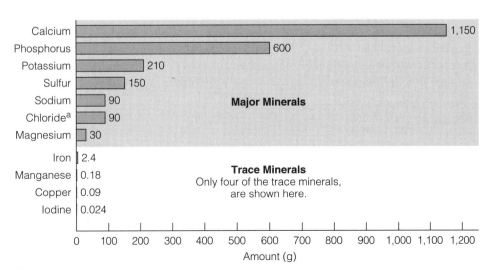

FIGURE 3.12 Body content of the seven major minerals and four trace minerals in a 60 kg (132 lb) person. *The major minerals are those present in amounts larger than 5g (a teaspoon). A pound is about 454g; thus, only calcium and phosphorus appear in amounts larger than one pound.*

[a]Chlorine appears in the body as a chloride ion.

TABLE 3.10 A vitamin guide.

Vitamin (Chemical Name)	Best Sources	Chief Roles	Deficiency Symptoms	Toxicity Symptoms
Water-soluble Vitamins Thiamin	Meat, pork, liver, fish, poultry, whole-grain and enriched breads, cereals, pasta, nuts, legumes, wheat germ, oats	Helps enzymes release energy from carbohydrate; supports normal appetite and nervous system function	Beriberi; edema, heart irregularity, mental confusion, muscle weakness, low morale, impaired growth	None reported
Riboflavin	Milk, leafy green vegetables, yogurt, cottage cheese, liver, meat, whole-grain or enriched breads and cereals	Helps enzymes release energy from carbohydrate, fat, and protein; promotes healthy skin and normal vision	Eye problems, skin disorders around nose and mouth	None reported
Niacin	Meat, eggs, poultry, fish, milk, whole-grain and enriched breads and cereals, nuts, legumes, peanuts	Helps enzymes release energy from energy nutrients; promotes health of skin, nerves, and digestive system	Pellagra; skin rash on parts exposed to sun, loss of appetite, dizziness, weakness, irritability, fatigue, mental confusion, indigestion	Flushing, nausea, headaches, cramps, ulcer irritation, heartburn, abnormal liver function, rapid heartbeat with doses above 500 mg per day; UL equals 35 mg/day from fortified foods or supplements*
Vitamin B_6 (pyridoxine)	Meat, poultry, fish, shellfish, legumes, whole-grain products, green leafy vegetables, fruits	Protein and fat metabolism; formation of antibodies and red blood cells; helps convert tryptophan to niacin	Nervous disorders, skin rash, muscle weakness, anemia, convulsions, kidney stones	Depression, fatigue, irritability, headaches, numbness, damage to nerves, difficulty walking, UL equals 100 mg/day
Folate (folacin, folic acid)	Green leafy vegetables, liver, legumes, seeds, citrus fruits, melons, enriched breads and grain products	Red blood cell formation; protein metabolism; new cell division	Anemia, heartburn, diarrhea, smooth tongue, depression, poor growth, neural tube defects, increased risk of heart disease, stroke, and certain cancers	Diarrhea, insomnia, irritability, may mask a vitamin B_{12} deficiency; UL equals 1 mg/day
Vitamin B_{12} (cobalamin)	Animal products: meat, fish, poultry, shellfish, milk, cheese, eggs; fortified cereals	Helps maintain nerve cells; red blood cell formation; synthesis of genetic material	Anemia, smooth tongue, fatigue, nerve degeneration progressing to paralysis	None reported
Pantothenic acid	Widespread in foods	Coenzyme in energy metabolism	Rare; sleep disturbances, nausea, fatigue	None reported
Biotin	Widespread in foods	Coenzyme in energy metabolism; fat synthesis; glycogen formation	Loss of appetite, nausea, depression, muscle pain, weakness, fatigue, rash	None reported
Vitamin C (ascorbic acid)	Citrus fruits, cabbage-type vegetables, tomatoes, potatoes, dark green vegetables, peppers, lettuce, cantaloupe, strawberries, mangos, papayas	Synthesis of collagen (helps heal wounds, maintains bone and teeth, strengthens blood vessels); antioxidant; strengthens resistance to infection; helps body absorb iron	Scurvy; anemia, depression, frequent infections, bleeding gums, loosened teeth, pinpoint hemorrhages, muscle degeneration, rough skin, bone fragility, poor wound healing, hysteria	Intakes of more than 1 g per day may cause nausea, abdominal cramps, diarrhea, and increased risk for kidney stones; UL equals 2,000 µg/day
Fat-soluble Vitamins Vitamin A	*Retinol:* fortified milk and margarine, cream, cheese, butter, eggs, liver *Beta-carotene:* Spinach and other dark leafy greens, broccoli, deep orange fruits (apricots, peaches, cantaloupe), and vegetables (squash, carrots, sweet potatoes, pumpkin)	Vision; growth and repair of body tissues; maintenance of mucous membranes; reproduction; bone and tooth formation; immunity; hormone synthesis; antioxidant (in the form of beta-carotene only)	Night blindness, rough skin, susceptibility to infection, impaired bone growth, abnormal tooth and jaw alignment, eye problems leading to blindness, impaired growth	Red blood cell breakage, nosebleeds, abdominal cramps, nausea, diarrhea, weight loss, blurred vision, irritability, loss of appetite, bone pain, dry skin, rashes, hair loss, cessation of menstruation, liver disease, birth defects

(continued)

TABLE 3.10 (continued)

Vitamin (Chemical Name)	Best Sources	Chief Roles	Deficiency Symptoms	Toxicity Symptoms
Vitamin D (cholecalciferol)	Self-synthesis with sunlight; fortified milk, fortified margarine, eggs, liver, fish	Calcium and phosphorus metabolism (bone and tooth formation); aids body's absorption of calcium	Rickets in children; osteomalacia in adults; abnormal growth, joint pain, soft bones	Deposits of calcium in organs such as the kidneys, liver, or heart, mental retardation, abnormal bone growth. UL equals 50 µg/day or 2,000 IU
Vitamin E	Vegetable oils, green leafy vegetables, wheat germ, whole-grain products, liver, egg yolk, salad dressings, mayonnaise, margarine, nuts, seeds	Protects red blood cells; antioxidant (protects fat-soluble vitamins); stabilization of cell membranes	Muscle wasting, weakness, red blood cell breakage, anemia, hemorrhaging	Doses over 800 IU/day may increase bleeding (blood clotting time); UL equals 1,000 mg/day
Vitamin K	Bacterial synthesis in digestive tract, liver, green leafy and cabbage-type vegetables, milk, grain products	Synthesis of blood-clotting proteins and a blood protein that regulates blood calcium	Hemorrhaging, decreased calcium in bones	Interference with anticlotting medication; synthetic forms may cause jaundice

*UL = Tolerable Upper Intake Level

TABLE 3.11 A mineral guide.

Mineral	Best Sources	Chief Roles	Deficiency Symptoms	Toxicity Symptoms
Major Minerals				
Calcium	Milk and milk products, small fish with (with bones), tofu, certain green vegetables, legumes, fortified juices	Principal mineral of bones and teeth; involved in muscle contraction and relaxation, nerve function, blood clotting, blood pressure	Stunted growth in children; bone loss (osteoporosis) in adults	Excess calcium is usually excreted except in hormonal imbalance states; Tolerable Upper Intake Level (UL) is 2,500 mg
Phosphorus	Meat, poultry, fish, dairy products, soft drinks, processed foods	Part of every cell; involved in acid-base balance and energy transfer	Muscle weakness and bone pain (rarely seen)	May cause calcium excretion; UL is 4,000 mg
Magnesium	Nuts, legumes, whole grains, dark green vegetables, seafoods, chocolate, cocoa	Involved in bone mineralization, protein synthesis, enzyme action, normal muscular contraction, nerve transmission	Weakness, confusion, depressed pancreatic hormone secretion, growth failure, behavioral disturbances, muscle spasms	Excess intakes (from overuse of laxatives) has caused low blood pressure, lack of coordination, coma, and death; UL is 350 mg
Sodium	Salt, soy sauce; processed foods: cured, canned, pickled, and many boxed foods	Helps maintain normal fluid and acid-base balance; nerve impulse transmission	Muscle cramps, mental apathy, loss of appetite	High blood pressure (in salt-sensitive persons)
Chloride	Salt, soy sauce; processed foods	Part of hydrochloric acid found in the stomach, necessary for proper digestion, fluid balance	Growth failure in children, muscle cramps, mental apathy, loss of appetite	Normally harmless (the gas chlorine is a poison but evaporates from water); disturbed acid-base balance; vomiting
Potassium	All whole foods: meats, milk, fruits, vegetables, grains, legumes	Facilitates many reactions, including protein synthesis, fluid balance, nerve transmission, and contraction of muscles	Muscle weakness, paralysis, confusion; can cause death; accompanies dehydration	Causes muscular weakness; triggers vomiting; if given into a vein, can stop the heart
Sulfur	All protein-containing foods	Component of certain amino acids; part of biotin, thiamin, and insulin	None known; protein deficiency would occur first	Would occur only if sulfur amino acids were eaten in excess; this (in animals) depresses growth

(continued)

TABLE 3.11 (continued)

Mineral	Best Sources	Chief Roles	Deficiency Symptoms	Toxicity Symptoms
Trace Minerals				
Iodine	Iodized salt, seafood, bread	Part of thyroxine, which regulates metabolism	Goiter, cretinism	Depressed thyroid activity
Iron	Red meats, fish, poultry, shellfish, eggs, legumes, dried fruits, fortified cereals	Hemoglobin formation; part of myoglobin; energy utilization	Anemia: weakness, pallor, headaches, reduced immunity, inability to concentrate, cold intolerance	Iron overload: infections, liver injury, acidosis, shock
Zinc	Protein-containing foods: meats, fish, shellfish, poultry, grains, vegetables	Part of many enzymes; present in insulin; involved in making genetic material and proteins, immunity, vitamin A transport, taste, wound healing, making sperm, fetal development	Growth failure in children, delayed development of sexual organs, loss of taste, poor wound healing	Fever, nausea, vomiting, diarrhea, kidney failure
Copper	Meats, drinking water	Helps make hemoglobin; part of several enzymes	Anemia, bone changes (rare in human beings)	Nausea, vomiting, diarrhea
Fluoride	Drinking water (if naturally fluoride containing or fluoridated), tea, seafood	Formation of bones and teeth; helps make teeth resistant to decay and bones resistant to mineral loss	Susceptibility to tooth decay	Fluorosis (discoloration of teeth); nausea, vomiting, diarrhea, UL is 10 mg
Selenium	Seafood, meats, grains, vegetables (depending on soil conditions)	Helps protect body compounds from oxidation; works with vitamin E	Fragile red blood cells, cataracts, growth failure, heart damage	Nausea, abdominal pain; nail and hair changes; liver and nerve damage; UL equals 400 μg
Chromium	Meats, unrefined foods, vegetable oils	Associated with insulin and required for the release of energy from glucose	Abnormal glucose metabolism	Occupational exposures damage skin and kidneys
Molybdenum	Legumes, cereals, organ meats	Facilitates, with enzymes, many cell processes	Unknown	Enzyme inhibition
Manganese	Widely distributed in foods	Facilitates, with enzymes, many cell processes	In animals: poor growth, nervous system disorders, abnormal reproduction	Poisoning, nervous system disorders
Cobalt	Meats, milk, and milk products	As part of vitamin B_{12}, involved in nerve function and blood formation	Unknown except in vitamin B_{12} deficiency	Unknown as a nutrition disorder

60 kg (132 lb) human body. Nutritional functions of minerals include assisting in bone growth, vision, blood clotting, maintaining electrolyte balance, and nerve functioning.

3.4 SUBSTITUTIONS FOR SUGAR AND FAT

Futurists are people who make predictions about the future. It would be interesting to know if they had predicted the current interest in fat reduction, sugar reduction, and calorie and fat elimination in foods, especially in the United States. This consumer interest, for health reasons or for weight loss, led to the development of alternative sweeteners and fat replacers.

Alternative Sweeteners

People with diabetes, consumers concerned with body weight, and those wanting to prevent tooth decay often wish to avoid sugar and sugary foods, which has led to the development of sweetener alternatives to sucrose. Sweeteners are considered nutritive (provide some calories) or nonnutritive (provide no calories). Sugar alcohols are nutritive alternative sweeteners. *Artificial sweeteners* are nonnutritive substances that provide the sweetening ability of sucrose but are metabolized differently and contribute few or no calories to foods. Asparatame, acesulfame-K, saccharin, and sucralose are nonnutritive artificial sweeteners (sometimes referred to as "intense" sweeteners).

Sugar Alcohols. Sugar alcohols, examples of which are isomalt, mannitol, sorbitol, and xylitol, are found naturally in fruits. They can be manufactured by hydrogenating sugars and provide between 1.5 and 3 kcal per gram (compared to 4 kcal/gram for sucrose). Sugar alcohols provide a sweet taste and a cooling sensation in the mouth without contributing to tooth decay. Some research has even indicated that xylitol may prevent the production of acids in the mouth that promote decay.

Acesulfame-K. The substance known as acesulfame potassium (Sunette) is heat stable and is used in a variety of foods, including baked products. Since acesulfame-K is not metabolized by the body, it is excreted without contributing any calories. Although highly sweet, acesulfame K does impart a bitter aftertaste.

Aspartame. Known to consumers by the brand names Nutrasweet and Equal, aspartame is composed of two amino acids, aspartic acid and phenylalanine. As such, it contributes 4 kcal of energy per gram like other amino acids and proteins. However, since it is intensely sweet (almost 200 times sweeter than sucrose), the amount added to foods and beverages is so low that it does not contribute calories. Although safe when consumed at reasonable levels (50 milligrams per kilogram, or 2.2 pounds, of body weight), a segment of the population is apparently sensitive to this substance even when consumed in small amounts.

Saccharin. The first artificial sweetener was saccharin, developed in 1878 and available as sodium or calcium saccharin. It is an intensely sweet substance, with a pronounced bitter aftertaste. Saccharin was taken off the market in 1977 because high doses were found to cause cancer in laboratory rats. At the time it was the only artificial sweetener available, and the ban caused public outcry that prompted Congress to respond and saccharin was returned to market by a congressional mandate. Products containing saccharin, and packets of it (Sweet'N Low) had to contain a consumer warning on the label.

Sucralose. A relatively new (1998) FDA-approved artificial sweetener is the sucrose derivative sucralose (Splenda). This substance has a sweetness level 600 times that of sucrose, with no aftertaste. From a molecular standpoint, sucralose is a sucrose molecule in which two alcohol (OH) groups have been replaced by chlorine (Cl) atoms. This change makes it so sweet that only a small amount is needed in food applications. Like other artificial sweeteners, sucralose does not provide calories or contribute to tooth decay.

Table 3.12 provides an overview of nonnutritive sweetener alternatives to sucrose.

Fat Replacers

The concept of fat reduction through the careful blending of ingredients and the development of fat replacement substances is a relatively new area of food product development. It is an important area both in food technology and in human health and nutrition.

The fat content of foods can be reduced by simply decreasing the amount of fat present, or through addition of water to the product (dilution effect). Fluid dairy products, reduced-fat cheese, and baked chips are examples of this approach. Replacement of fat

TABLE 3.12 The comparative sweetness, food applications, and characteristics of nonnutritive sweetener alternatives to sucrose.

Name (Trade Name)	Sweeteners Compared with Sucrose	Typical Uses	Characteristics*
Sweeteners approved for use in the United States			
Acesulfame-K (Sunette/Sweet One)	200 times sweeter	Tabletop sweetener, puddings, gelatins, chewing gum, candies, baked goods, desserts, diet drink mixes	Stable in high temperatures; soluble in water. A new blend of acesulfame-K and aspartame synergizes the sweetening power of these sweeteners.
Aspartame (NutraSweet/Equal)	180 times sweeter	General purpose sweetener in foods, beverages, chewing gum; tabletop sweetener	Loses sweetness at high temperatures and may lose sweetness over time. New forms can increase its sweetening power in cooking and baking
Saccharin (Sweet'N Low)	300 times sweeter	Diet soft drinks, tabletop sweetener	Stable at high temperatures
Sucralose (Splenda)	600 times sweeter	Soft drinks, jams, frozen desserts, dairy products, baked goods, chewing gum, salad dressings, syrups, tabletop sweetener	Stable at various temperatures
Sweeteners for which U.S. approval is pending			
Alitame (Novasweet)	2,000 times sweeter	Baked goods, beverages, frozen desserts, tabletop sweetener	Stable at high temperatures and in both acidic and nonacidic foods; highly soluble in water
Cyclamate[†]	30 times sweeter	Tabletop sweetener, baked goods	Stable at high temperatures; long shelf life

*In addition to the characteristics listed, all of the sweeteners have a synergistic effect when combined with other sweeteners. In other words, together they enhance each other's sweetness, yielding a combined sweetness greater than the sum of all the substances' sweetness.

†Cyclamate was banned in the United States in 1970 because studies suggested that it may cause cancer in rats. The validity of the studies has been questioned, however. Currently, a petition is before the FDA to reapprove cyclamate for use in the United States. In Canada, cyclamate has been approved for use in tabletop sweeteners and as a sweetening agent in drugs.

SOURCE: Position of the American Dietetic Association: Use of nutritive and nonnutritive sweeteners, *Journal of the American Dietetic Association* 98 (1998): 580–587.

with other foods including fruit purees and nonfat yogurt can result in products possessing sensory characteristics that resemble those contributed by fat.

Often, food manufacturers achieve fat reduction through the use of fat replacer ingredients. These fall into one of several categories: carbohydrate-based fat replacers, protein-based fat replacers, and fat-based fat replacers. The foods typically targeted for fat replacement include baked goods, butter and margarine, cake frosting, dairy products (cheese, coffee creamer, ice cream, sour cream, yogurt), frozen desserts, mayonnaise, meat products, puddings, salad dressing, and snack foods and chips. Table 3.13 provides a sampling of common fat replacers.

Carbohydrate-based Fat Replacers. Carbohydrate-based fat replacer ingredients are plant polysaccharides. They thicken foods and add bulk, providing a mouthfeel similar to that of fat. Some polysaccharides (e.g., dextrins, modified starches) are digested and provide 4 kcal/gram. Others are not digested to an appreciable extent (e.g., cellulose) and are essentially noncaloric. Although they cannot be used for frying, many carbohydrate-based fat replacers can withstand heat and can be used in meat products. Examples of fat replacer substances based upon carbohydrates are carrageenan, cellulose gels, corn maltodextrins, dextrins, fibers, gums (hydrocolloids), Oatrim (hydrolyzed oat flour), pectin, polydextrose, and starch gels. Since these are derived from natural plant products, they are considered safe for human consumption.

Protein-based Fat Replacers. Proteins can be blended with gums to form gels. These provide structure and functionality similar to that of fat. In addition, proteins of low molecular weight (microparticulated proteins) may act like fats to alter the texture of products like cheese. They contribute 1.3 to 4 kcal/gram, and their biological value depends on the component amino acids. **Microparticulation** is a food processing method that reduces

TABLE 3.13 Types of fat replacer ingredients. *The energy contributions for the various fat replacers range from 0 to 5 calories (k/cal) per gram.*

Fat Replacers	Energy (cal/g)	Properties	Uses in Foods
Carbohydrate-Based Fat Replacers			
Fruits purees and pastes of apples, bananas, cherries, plums, or prunes; add bulk and tenderness to baked goods	0–4	Replace bulk of fat; add moisture and tenderness	Baked goods, candy, dairy products
Gels derived from cellulose or starch to mimic the texture of fats in regular margarine and other products	0–4*	Replaces bulk; lends thickness	Fat-free margarines, salad dressing, frozen desserts
Gums extracted from beans, sea vegetables or other sources	0–4	Adds bulk; thickens salad dressings	Salad dressing, processed meats, desserts
Maltodextrins made from corn; is powdered and flavored to resemble the taste of butter	1–4	Adds "buttery" flavor	Butter-flavored "sprinkles" for melting on hot foods
Oatrim derived form oat fiber; has the added advantage of providing satiety	4	Creamy, replaces bulk of fat; can be used in baking but not frying	Dips, dressings, baked goods
Z-trim a modified form of insoluble fiber; is powdered and feels like fat in the mouth	0	Creamy, replaces bulk of fat; can be used in baking but not frying	Cheese, ground beef, chocolates, baked goods
Fat-Based Fat Replacers			
Olestra (OLEAN) a noncaloric artificial fat made from sucrose and fatty acids; formerly called *sucrose polyester*	0	Same properties as fats; heat stable in frying, cooking, and baking	Potato chips, tortilla chips, crackers
Salatrim (Benefat) derived from fat and contains short- and long-chain fatty acids	5	Same properties as fats; can be used in baking but not frying	Chocolate coatings, dairy products, spreads
Protein-Based Fat Replacers			
Microparticulated protein (Simplesse and K-Blazer) processed from the proteins of milk or egg white into mistlike particles that roll over the tongue, making it feel and taste like fat	4	Creamy, heat stable in some cooking and baking but not frying	Ice cream, dairy products

*Energy made available by action of colonic bacteria.

the particle size of a substance (in the case of protein, to about 1 micrometer in diameter). Microparticulated egg and milk proteins contribute the creamy mouthfeel of fats with less than half the calories. The microparticulated protein Simplesse, made from whey protein concentrate, is an example. Isolated soy protein (ISP) has been used for years as a fat replacer in ground meat products. Most protein-based replacers cannot be used at high temperatures because the protein coagulates and loses its functionality. Since protein fat replacers are derived from food proteins, they are safe for human consumption.

Fat-Based Fat Replacers. Fat-based fat replacers contain the same fatty acids found in regular food fats and can mimic their characteristics of flavor, baking and frying stability, and shelf life. They are produced by creating fat molecules that have a shorter fatty acid chain length, meaning fewer carbon atoms in the structure. Modifying the fatty acid composition of triglycerides in this way results in calorie reduction. For example, a long-chain fatty acid like oleic acid contains 18 carbon atoms and provides about 9 kcal per gram. By contrast, butyric acid, a short-chain fatty acid, is only four carbon atoms in length and provides only about half the calories of oleic.

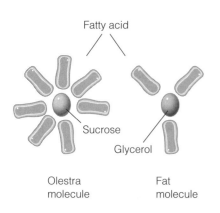

FIGURE 3.13 An olestra molecule. *Olestra is a sucrose molecule to which between 6 and 8 fatty acids have been attached. These fatty acids on the molecule block digestive enzymes, so the molecule travels through the digestive tract undigested and unabsorbed.*

A substance called caprenin, which contains short- and medium-chain fatty acids, is used as a substitute for cocoa butter in certain chocolate bars. Salatrim (an acronym for Short And Long chain Acid TRIglyceride Molecule) is another fat replacer used in chocolate product manufacture that contains short-chain fatty acids in its structure. Both caprenin and Salatrim contribute approximately 5 kcal per gram.

Olestra is a synthetic fat molecule that contains 6, 7, or 8 fatty acids. Unlike a triglyceride fat molecule in which the fatty acids are attached to glucose, in olestra the fatty acids are attached to sucrose. This inhibits the working of the digestive enzymes, with the result that olestra passes from the body without being absorbed or contributing calories—see Figure 3.13. Olestra is stable at high temperatures and can be used in cooking and frying. It is approved for use as a fat replacer in savory snack food items such as corn and potato chips.

Though nontoxic and functional, olestra is not without problems. A warning is required on all olestra-containing foods: "This product contains olestra. Olestra may cause abdominal cramping and loose stools. Olestra inhibits the absorption of some vitamins and other nutrients. Vitamins A, D, E, and K have been added." The consumer must weigh the benefits of consuming snacks containing olestra with the potential to experience gastrointestinal symptoms and decreased fat-soluble vitamin and phytonutrient absorption.

The Place of Fat Replacers in the Diet.   The final word on fat replacement is not out. Although new low-fat foods are increasingly available to the consumer, nutritionists caution that the population is not getting thinner. It can be argued that some individuals may consume more of a low-fat or fat-free food, which in turn can contribute to weight gain rather than weight loss, since no fat-free food is calorie-free. On the other hand, proponents point out that consumers want food choices that are lower in fat and calories, which creates a role for foods made with fat replacers.

From a health and nutrition standpoint, foods formulated with fat replacers contribute fewer fat grams and calories to the diet. However, fat replacers are often incorporated into foods that are not good sources of other nutrients, such as snack foods. Skeptics view such foods as contributing to a nutritionally inferior diet. This viewpoint would be correct if snack foods are repeatedly consumed at the expense of fruits, vegetables, and whole grains.

3.5 ENERGY METABOLISM

What happens to the nutrients after food is digested, absorbed, and transported to the cells of the body? The answer is they are metabolized. **Metabolism** refers to the chemical reactions that take place within the body. These reactions are of two types. *Anabolic reactions* serve to form (synthesize) chemical substances, while *catabolic reactions* are involved in the breakdown of chemical substances. The release of energy from the breakdown of the energy nutrients is an example of a catabolic process. **Energy metabolism** refers to the way the body uses energy nutrients (carbohydrates, fats, and proteins) as fuel sources—see Figure 3.14.

Vitamins, minerals, and water do not provide energy directly and are not energy nutrients. Although the human body can burn (oxidize) all of the energy nutrients (including alcohol), it prefers to burn carbohydrate and fat, while sparing protein. The catabolism of nutrients inside cells provides energy. The unit of measure for the amount of energy foods supply is the calorie.

One **calorie** equals the amount of energy required to raise the temperature of 1 kilogram of water 1 degree on the Celsius scale. This is equivalent to 4.2 kilojoules (United States uses calories, some countries use kilojoules). The calorie contribution of each energy nutrient is given in kilocalories (kcal). The energy values for each nutrient are 4 kcal/g for protein, 4 kcal/g for carbohydrate, 9 kcal/g for fat, and 7 kcal/g for alcohol.

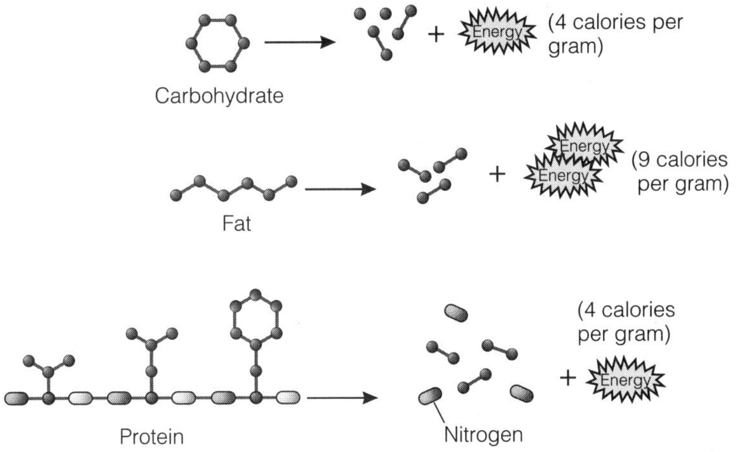

FIGURE 3.14 **Nutrients used by the body to provide energy.** *The compounds at the left yield the 2-carbon fragments shown at the right. These fragments oxidize quickly in the presence of oxygen to yield carbon dioxide, water, and energy.*

Knowing the energy values for the nutrients allows you to calculate the caloric value, and percentages of the nutrients in foods. Consider Figure 3.15. A fruit-flavored yogurt has a label that says it supplies 10 grams of protein, 3 grams of fat, and 43 grams of carbohydrate per serving. Although the serving sample size is 1 cup and may weigh over 200 grams, much of the weight is water, which supplies no energy. To determine the amount of energy from each nutrient, simply multiply the amount of each energy nutrient contained in the food (in grams) by its correct physiological fuel value, and calculate their sum.

The dietary goals suggest that each day on average we should be getting no more than 30 percent of kcal from fat, 10–15 percent from protein, and at least 55 percent from carbohydrate. What percent of calories comes from each nutrient in the yogurt example, and how does this match the goals? Proceed to Step 2 in the figure: Divide the kcal contribution per nutrient by the total kcal content of the food portion, multiply by 100, and compare to the goals.

Nutrients provide energy to the body, and energy is a requirement for routine daily tasks as well as sports and athletic activities, which present even greater demands for energy. Are there any foods or supplements that can boost a person's energy and actually enhance athletic performance?

FIGURE 3.15 **Calculating the percentage of calories from each nutrient.**

Step 1. To determine calories (number of kcal) from each nutrient:
- 4 kcal/g of protein × 10g in yogurt = 40 kcal from protein
- 4 kcal/g of carbohydrate × 43g in yogurt = 172 kcal from carbohydrate
- 9 kcal/g of fat × 3g in yogurt = 27 kcal from fat

total kcal = 239 kcal

Step 2. To determine percentage of calories from each nutrient (using yogurt example):
- 40 kcal protein ÷ 239 kcal in the yogurt × 100 = 17% of yogurt kcal are protein
- 172 kcal carbohydrate ÷ 239 kcal in the yogurt × 100 = 72% of yogurt kcal are carbohydrate
- 27 kcal from fat ÷ 239 kcal in the yogurt × 100 = 11% of yogurt kcal are fat

3.6 ERGOGENIC SUBSTANCES

Ergogenic aids are substances that are promoted as agents that increase an individual's capacity for work and exercise. These substances run the gamut in terms of classification, ranging from amino acids like arginine to plant methylxanthines (caffeine), and from herbs such as ginseng to sodium bicarbonate (baking soda). See Table 3.14 for additional examples.

The claims made by the manufacturers of many of these substances are largely unproven. In certain cases, preliminary research indicates potential for some ergogenic aids, so, some may eventually be proven both safe and helpful. However, caution is advised in using even the most promising ones due to known side effects and as yet undetermined long-term effects.

TABLE 3.14 **Ergogenic Substances.** *These substances are promoted as being able to enhance muscle growth and athletic performance.*

Ergogenic Substances with Unproven Claims

Amino acids (e.g., arginine, ornithine, glycine) nonessential amino acids falsely promoted to increase muscle mass and strength by stimulating growth hormone and insulin. Individual amino acids do not significantly increase muscle mass or growth hormone. Weight lifting and endurance training do.

Anabolic steroids synthetic male hormones (related to testosterone) that stimulate growth of body tissues, with many adverse effects.

Bee pollen mixture of bee saliva, plant nectar, and pollen touted falsely to enhance athletic performance. May cause allergic reactions in people with a sensitivity to bee stings and honey allergies.

Carnitine a compound synthesized in the body from two amino acids (lysine and methionine) and required in fat metabolism. Falsely touted to increase the use of fatty acids and spare glycogen during exercise, delay fatigue, and decrease body fat. The body produces sufficient amounts on its own. No evidence that supplementation in healthy people improves energy or enhances fat loss.

Chromium picolinate Chromium is an esstential component of the glucose tolerance factor, which facilitates the action of insulin in the body. Picolinate is a natural derivative of the amino acid tryptophan. Falsely promoted to increase muscle mass, decrease body fat, enhance energy, and promote weight loss. Choose instead, a diet rich in whole, unprocessed foods.

Coenzyme Q10 a lipid made by the body and used by cells in energy metabolism; falsely touted to increase exercise performance and stamina in athletes; potential antioxidant role. May increase oxygen use and stamina in heart disease patients.

DHEA (Dehydroepiandrosterone) a precursor of the hormones, testosterone and estrogen; falsely promoted to increase production of testosterone, build muscle, burn fat, and delay the effects of aging. Long-term effects unknown, self-supplementation not recommended.

Ginseng a collective term used to describe several species of plants, belonging to the genus *Panax*, containing bioactive compounds in their roots. Falsely touted to enhance exercise endurance and boost energy. A lack of well-controlled research has yielded inconclusive evidence for the benefits of ginseng. The potential for adverse drug-herb or herb-herb interactions with ginseng exists.

Pyruvate a 3-carbon compound derived from the breakdown of glucose for energy in the body. Falsely promoted to increase fat burning and endurance. Side effects include intestinal gas and diarrhea, which could interfere with performance.

Ergogenic Substances with Some (Not All) Scientific Support for Claims

Creatine A nitrogen-containing substance made by the body which combines with phosphate to form the high-energy compound, creatine phosphate (CP). CP is stored and used by muscle for ATP production. Some (not all) studies show that creatine may increase CP content in muscles and improve short-term (<30 seconds) strenuous exercise performance (e.g., sprinting, weight lifting). Long-term effects unknown.

Caffeine a stimulant that increases blood levels of epinephrine; promoted for improved endurance and utilization of fatty acids during exercise. Consuming 2 to 3 cups of coffee (equal to 3 to 6 milligrams of caffeine per kilogram body weight) 1 hour before exercise may improve endurance performance. High caffeine consumption may cause dehydration, headache, nausea, muscle tremors, and fast heart rate.

HMB (beta-hydroxy-beta-methylbutyrate) a metabolite of the branched-chain amino acid leucine and promoted to increase muscle mass and strength by preventing muscle damage or speeding up muscle repair during resistance training. More research on long-term safety and effectiveness is needed.

Phosphate salt a salt with claims for improved endurance. Found to increase a substance in red blood cells (diphosphoglycerate) and enhance the cell's ability to deliver oxygen to muscle cells and reduce levels of disabling lactic acid in elite athletes. However, more research on safety and efficacy of phosphate loading is needed. Excess can cause loss of bone calcium.

Sodium bicarbonate baking soda is touted to buffer lactic acid in the body and thereby reduce pain and improve maximal-level anaerobic performance. May cause diarrhea in users due to the high sodium load; effects of repeated ingestions unknown.

SOURCE: For more information, see E. Coleman, *Eating for Endurance* (Palo Alto, CA: Bull Publishing Co., 1997); S.A. Sarubin, *The Health Professional's Guide to Popular Dietary Supplements* (Chicago, IL. The American Dietetic Association, 2000).

Key Points

- The human body requires approximately 50 nutrients, including water, carbohydrates, lipids, proteins, vitamins, and minerals, for optimal health.
- The Food Guide Pyramid and Dietary Guidelines were developed to enable Americans to achieve sound nutrition through proper food choices.
- Nutritional functions of water include regulating body temperature; serving as a carrier in the movement of nutrients during digestion, absorption, circulation; and serving as the medium for metabolic reactions.
- Nutritional functions of carbohydrates include supplying energy, maintaining blood glucose levels, and providing fiber to aid in gastrointestinal functioning.
- Nutritional functions of lipids include acting as an energy source; serving as a component of cell membranes; and synthesizing prostaglandins, hormones and bile.
- Nutritional functions of proteins include promoting nitrogen, acid-base, and fluid balance; acting as enzymes in digestion; serving as hormones, antibodies, and as components of body tissue; and providing energy.
- Nutritional functions of vitamins include acting as antioxidants, serving as coenzymes for enzymatic reactions, and disease prevention.
- Nutritional functions of minerals include assisting in bone growth, vision, blood clotting, maintaining electrolyte balance, and nerve functioning.
- The energy nutrients are carbohydrates, lipids, proteins, and alcohol, contributing 4, 9, 4, and 7 kcal/gram of energy, respectively.
- Fat replacers are carbohydrate-based, protein-based, or fat-based.
- Alternative sweeteners include the sugar alcohols and the artificial sweeteners.
- Food labels provide information on key nutrients and can be used to plan a healthy diet.
- Energy metabolism refers to the way the body uses carbohydrate, fat, and protein for energy.

Study Questions

1. Create a "real world" food pyramid based upon the following average daily serving consumption patterns. Servings: fats, oils, and sweets = 3.2; milk, yogurt, and cheese = 2.3; meat, poultry, fish, dry beans, eggs, and nuts = 2.8; vegetables = 2.4; fruits = 2.3; bread, cereal, rice, and pasta = 2.7. Compare this pyramid to the USDA Food Guide Pyramid—what can you conclude?

2. Explain the relationships between food protein, biological value, nitrogen excretion, essential amino acids, and protein synthesis.

3. Estimate the caloric value of a manufactured health food item containing 0.5 gram of linolenic acid, 25 grams of fructose, 15 milligrams of vitamin C, 3 grams of fish liver oil, 11 grams of corn syrup, and 2.5 grams of polyarginine (a polypeptide made up of repeating units of the amino acid arginine).

4. Agree or disagree with the following statements, justifying your answer.
   a. The most abundant lipid in the typical diet is cholesterol.
   b. The lipoprotein responsible for reverse cholesterol transport (away from body cells) is VLDL.
   c. The Food Guide Pyramid suggests 3 to 6 servings of fruit per day.
   d. The substance that emulsifies fat when it enters the small intestine is called intestinal lipase.
   e. A deficiency of energy or nutrients is correctly called malnutrition.
   f. The current recommendation calls for no more than 30% of kcal to come from fat in the diet. For a person consuming 2000 kcal per day, this would be 67 grams of fat.

5. Protein is one nutrient for which an RDA (Recommended Dietary Allowance) can be calculated. As an example of estimating the protein RDA, consider a 5-foot 4-inch 20-year-old female, with an actual weight of 145 pounds but a desirable body weight of 125 pounds. Calculate her protein RDA, according to the following steps:
   a. Convert the recommended weight from pounds to kilograms by dividing by 2.2 (since one kilogram is equivalent to 2.2 pounds).

b. Select the RDA factor of 0.8 grams of protein per kilogram of body weight for females of this age group (for males, the factor would be 0.9).
c. Multiply this factor times the recommended weight for this individual to obtain the RDA for protein.

Multiple Choice

6. Which is not a major mineral?
   a. calcium
   b. iron
   c. potassium
   d. sulfur
   e. sodium

7. If an olestra molecule could be enzymatically digested, approximately how many calories would it provide, assuming each of its six fatty acids was a long-chain fatty acid?
   a. 58
   b. 54
   c. 45
   d. 9
   e. 4

8. Which statement is true regarding food labels?
   a. Such labeling was established in 1973 by the USDA.
   b. Manufacturer name and address are optional.
   c. DRV covers protein, vitamins, and minerals.
   d. Ingredients are listed on labels in alphabetical order.
   e. "Nutrition Facts" contains information on fiber content.

9. Which statement about metabolic reactions is *false?*
   a. Breakdown of fats into fatty acids and glycerol is an example of a catabolic reaction.
   b. Peptide synthesis from amino acids represents an anabolic reaction.
   c. Conversion of maltose into glucose is an example of a catabolic reaction.
   d. Anabolic reactions are important during digestion of a meal.
   e. Conversion of peptides into amino acids is an example of a catabolic reaction.

10. Which sweetener statement is correct?
    a. Sucralose is a nutrient alternative sweetener.
    b. All alternative sweeteners are less sweet than sucrose.
    c. Aspartame provides 4 kcal/gram of energy.
    d. Acesulfame-K is a nutrient alternative sweetener.
    e. All current sugar substitutes are carbohydrates.

11. Which provides the most kcal per gram?
    a. olestra
    b. sucralose
    c. omega-3 fatty acid
    d. microparticulated protein
    e. pectin

12. Which organ produces a starch-digesting enzyme and releases it into the small intestine?
    a. pancreas
    b. liver
    c. gallbladder
    d. amylase
    e. large intestine (colon)

13. Many consumers purchase herbal, protein, vitamin, and other supplements that are sold at health food stores in an attempt to improve their nutritional and health status. Which of the following is true?
    a. Chromium picolinate increases muscle mass and causes weight loss.
    b. Carnitine improves athletic performance.
    c. Phosphate salts promote oxygen uptake in muscle.
    d. Ginseng boosts energy.
    e. Bee pollen enhances athletic performance.

14. Drinking water or milk can be a good source of the following minerals except
    a. fluoride.
    b. copper.
    c. calcium.
    d. phosphorus.
    e. iron.

15. Which is not a key function of water?
    a. dissolves mineral electrolytes
    b. component of extracellular fluid
    c. provides energy
    d. regulates body temperature
    e. carrier of nutrients

16. Which antioxidant vitamin helps to absorb iron?
    a. vitamin C
    b. thiamin
    c. vitamin A
    d. beta-carotene
    e. vitamin E

17. Which of the following nutrients are absorbed into the intestinal cells by simple diffusion?
    a. glycerol, glucose, and fatty acids
    b. small fatty acids
    c. cholesterol, glucose, and amino acids
    d. diglycerides, dipeptides, and disaccharides
    e. amino acids and glucose

18. Proteases can catabolize all of the following except:
    a. dipeptides
    b. diglycerides
    c. amino acids
    d. microparticulated protein
    e. tripeptides

19. A deficiency of energy or nutrients is called
    a. overnutrition
    b. overt wasting
    c. secondary deficiency
    d. undernutrition
    e. malnutrition

20. Which carbohydrate does not contribute as a source for blood glucose?
    a. glycogen
    b. fructose
    c. lactose
    d. starch
    e. cellulose

21. Insoluble fibers, including cellulose, are thought to
    a. indirectly lower cholesterol levels by binding bile and preventing its reabsorption.
    b. relieve constipation and other colonic disorders by increasing fecal bulk (mass and weight).
    c. reduce the risk of heart disease by destroying colonic bacteria.
    d. improve the lipid profile by entering the bloodstream and removing arterial cholesterol deposits.
    e. be present in oat bran.

22. Which statement is *false* ("in" means intake, "out" means output)?
    a. Nitrogen equilibrium is when N in = N out.
    b. Zero nitrogen balance is the same as nitrogen equilibrium.
    c. negative nitrogen balance is when N in < N out.
    d. negative nitrogen balance is when N out > N in.
    e. positive nitrogen balance is when N in < N out.

References

Boyle, M. A. 2001. *Personal Nutrition*, 4th ed. Wadsworth/Thomson Learning, Belmont, CA.

Brown, A. 2000. *Understanding Food: Principles and Preparation*. Wadsworth/Thomson Learning, Belmont, CA.

Hegarty, V. H. 1995. *Nutrition, Food, and the Environment*. Eagan Press. St. Paul, MN.

Position of the American Dietetic Association: Use of nutritive and nonnutritive sweeteners. 1998. *Journal of the American Dietetic Association* 98:580.

Sizer, F. and Whitney, E. N. 2000. *Nutrition: Concepts and Controversies*, 8th ed. Wadsworth/Thomson Learning, Belmont, CA.

Whitney, E. N., and Rolfes, S. R. 1999. *Understanding Nutrition*, 8th ed. Wadsworth/Thomson Learning, Belmont, CA.

Additional Resources

Alexander, R. J. 1997. *Sweeteners: Nutritive*. Eagan Press, St. Paul, MN.

Chemical & Engineering News. 1999. "Good" cholesterol's mode of action clarified. *CENEAR* 77 (18):39.

Gopalan, C. 1997. Dietetics and nutrition: impact of scientific advances and development. *Journal of the American Dietetic Association* 97:737.

Groff, J. L., Gropper, S. S., and Hunt, S. M. 1995. *Advanced Nutrition and Human Metabolism*, 2nd ed. West Publishing Company, Minneapolis/St. Paul, MN.

Lehninger, A. L., Nelson, D. L., and Cox, M. M. 1993. *Principles of Biochemistry*, 2nd ed. Worth Publishers, New York.

Mahan, L. K., and Escott-Stump, S. 1996. *Food, Nutrition, and Diet Therapy*, 9th ed. W. B. Saunders Company, Philadelphia, PA.

Read, M. H. 1998. Dietary fat practices: age, gender, nutrition knowledge. *Topics in Clinical Nutrition* 13:53.

Stauffer, C. E. 1996. *Fats and Oils*. American Association of Cereal Chemists, Eagan Press, St. Paul, MN.

Vecchia, C. L. 1998. Vegetable consumption and the risk of chronic disease. *Epidemiology* 9:208.

Woodward, B. 1998. Protein, calories, and immune defenses. *Nutrition Reviews* 56:S84.

Web Sites

http://www.health.gov/dietaryguidelines/
USDA Dietary Guidelines for Americans. Includes a one-page summary and a complete text of the Dietary Guidelines.

http://www.pueblo.gsa.gov/cic_text/food/food-pyramid/main.htm
USDA. A booklet that introduces consumers to The Food Guide Pyramid.

http://www.healthfinder.gov
U.S. Office of Disease Prevention and Health Promotion. Use this site to search the U.S. government health information banks for "nutrition."

http://ific.org
International Food Information Council. This site includes information on nutrition and food safety and "Food Insight" on-line magazine.

www.fda.gov/fdac/features/895_vegdiet.html
U.S. Food and Drug Administration. This page contains information on vegetarian diets.

www.eatright.org
American Dietetic Association. This comprehensive site has information on nutrition, health and lifestyle, vegetarian diets, and weight loss.

http://www.fda.gov/fdac/features/1999/699_sugar.html
FDA Consumer magazine provides information on artificial sweeteners.

http://www.fda.gov/fdac/features1696_fat.html
FDA Consumer magazine describes a host of fat substitutes.

Challenge!

CAN A HIGH-PROTEIN, HIGH-FAT DIET WORK?

For reasons that include the increase in sedentary lifestyle and the tendency to eat high-fat fast foods, weight gain has become a problem for millions of Americans. In an effort to stop weight gain and achieve weight loss, people turn to fad diets that promise quick and dramatic results even though these diet plans often run counter to the Dietary Guidelines and goals and the recommendations of the Food Guide Pyramid. For example, the high-protein, high-fat diet put forth by Dr. Robert C. Atkins has received much attention from the media and from celebrities. Can such a diet work?

Weight Loss Through Ketosis

The Atkins diet encourages liberal ingestion of high-fat and high-protein foods and restricts carbohydrate foods. People who follow a high-protein, high-fat, but low-carbohydrate diet do experience weight loss—for two reasons.

Weight loss on the Atkins diet is due to (1) water loss and (2) reduction in food intake and calories. When carbohydrate is restricted in the diet, hormonal changes occur. Such changes promote *diuresis*, or excretion of water from the body. Much of the early weight reduction with this diet is water loss through diuresis and increased urination. This is why people get off to a good start with the Atkins diet, and they feel the diet is really working for them and they are losing weight.

The second component of the diet that explains its success is another consequence of low carbohydrate intake. When a person restricts carbohydrate intake to low levels, the liver makes *ketone bodies*, which results in a situation called *ketosis*. This is precisely what happens during starvation—so, a high-protein, high-fat diet somehow fools the body into thinking it is starving. The body of a starving individual will use the ketone bodies as fuel for the brain and for general energy requirements. There can be fat loss with this diet, although the conversion of body protein and adipose (fat) tissue to ketone bodies clearly occurs as well.

But the key psycho-physiological effect of ketosis is that it suppresses the appetite. Even though people can eat all they want on this diet, they don't want as much because ketosis suppresses their appetite. So, the high-protein, high-fat composition of the diet allows the ketosis to be sustained. As a result, appetite falls, as does calorie intake, so weight loss occurs.

However . . .

There are two problems with this diet. The first is that there is no evidence that this type of diet is any more successful in keeping weight off in the long term than other fad diets. And people cannot stay on the diet because the consequences of ketosis are not all positive. One of the long-term effects of a body remaining in a state of ketosis is upset in the body's acid-base balance, which can lead to a serious condition known as *acidosis*. In acidosis the pH of the blood becomes dangerously low and the increased acidity affects the structure and functioning of proteins, specifically the metabolic regulators like enzymes and hormones. As a result, acidosis can be fatal.

The second problem relates to general health. The high-fat diet that people consume with this weight-loss program is not one that is high in mono- and polyunsaturated fat, but rather high in saturated fat: bacon and eggs, for instance. This type of fat intake creates problems with blood lipids and lipoproteins, and can increase the risk of heart disease. For example, LDL levels tend to be higher when the saturated fat intake goes up, and the other blood lipids can be adversely affected. This type of diet is potentially damaging to cardiovascular and long-term health.

The Atkins diet also minimizes intake of carbohydrate-rich foods such as bread, grain, pasta, fruit, and vegetables. This intake pattern is the opposite of the current recommendations of the USDA's Food Guide Pyramid. As a result, the Atkins diet is depleted in many of the nutrients and the carbohydrate-rich plant foods that provide fiber, which is important for gastrointestinal and cardiovascular health.

Plant foods also provide essential vitamins, minerals, and key phytochemicals that may prevent cancer and other chronic diseases. Phytochemicals do not occur to a significant extent in high-protein, high-fat foods. A dieter could say, "but I am including *some* fiber-rich foods in my diet along with the high-fat and high-protein foods." However, the bottom line is: Some is nowhere near enough.

What Is a Safe and Effective Way to Lose Weight?

Generally speaking, a good weight-loss diet does not restrict any particular food or food group, but is balanced in terms of foods and low in total calories. Carbohydrates should form about 60 percent of the caloric intake, protein around 15 percent, and fats about 25–30 percent. The kinds of fats should also be balanced between polyunsaturated, monounsaturated, and saturated. Protein should be derived from both animal foods and plant foods, although it is certainly possible to consume a healthy vegetarian diet, if properly planned. In addition, water intake should be increased to the target of eight glasses per day.

A balanced, calorie-restricted diet aided by full water intake will work to produce weight loss when the individual also engages in increased aerobic physical activity. The weight loss (which will be not only water weight but also significant body fat) will be gradual, but safe, at the rate of a pound or two per week. The notion of coupling sound nutrition to exercise is of course nothing new, and is in fact plain common sense. In the Atkins diet and all other fad diets, the promise sounds better than the reality.

A successful weight-loss diet or program is one that not only safely produces weight loss, but also *can be followed easily for life*. A healthy dietary and activity pattern that can be followed for life includes balance, variety, optimum water intake, regular aerobic exercise, and caloric regulation. The long-term effects of such a program remain positive and desirable for most people.

Conclusion

Yes, the Atkins diet can yield rapid weight loss, but it is not a healthy diet and cannot be recommended over a sound diet based on the consumption of a variety of foods as presented by the Food Guide Pyramid. A sound diet emphasizes grains, cereals, breads, pastas, whole fruits, and vegetables. When coupled to calorie reduction, increased water intake, and adequate aerobic exercise, a safe and effective long-term weight-loss program is the result.

Challenge Questions

1. Access the web site "Ask the Dietitian" at *http://www.dietitian.com/carbos.html*. Read through the information on carbohydrates and glycemic index to provide a response to this statement: "I am on a diet that says that weight can be controlled simply by limiting the consumption of foods having a high glycemic index (or glycemic effect). Consuming carrots, potatoes, and even low-fat snack foods like rice cakes causes a rapid rise in blood sugar and are making fat in my body."

2. List the specific advantages and disadvantages of the Atkins high-protein high-fat diet from the viewpoint of a typical consumer and from an overall health perspective. How would you advise a friend who believes that too high an intake of carbohydrate induces weight gain? Suggestion: Access the following web site for further information and insight: *http://www.heartinfo.com/nutrition/atkins103097.htm*.

3. Access the FDA web site that alerts consumers to fraud: http://www.fda.gov/fdac/features/1999/699_fraud.html. Read the "Tip Offs to Rip-Offs" concerning (a) over-the-counter (OTC) transdermal weight-loss patches, and (b) an unapproved weight-loss product marketed as an alternative to a prescription drug combination. In a short paragraph, identify the key points to consider when evaluating weight-loss products and diets, and apply any of them that you can to the Atkins diet.

CHAPTER 4

Food Chemistry I: Functional Groups and Properties, Water, and Acids

4.1 The Nature of Matter

4.2 Chemical Reactions in Foods

4.3 Functional Groups

4.4 The Chemical and Functional Properties of Water

4.5 The Chemical and Functional Properties of Food Acids

4.6 Food Acidity

Challenge!
Food Systems

CHAPTER OBJECTIVES

After completing this chapter, you will be able to:
- Describe the use of chemical symbols, formulas, and equations.
- Explain the types and importance of chemical bonds that occur in foods.
- List the fundamental classes of chemical reactions in foods.
- Explain the significance of food enzymatic reactions.
- Define oxidation, reduction, oxidizing agent, and reducing agent.
- List the major functional groups occurring in food molecules.
- Describe the chemical and functional properties of water.
- Describe the chemical and functional properties of food acids.
- Explain food acidity in terms of pH and titratable acidity.
- Distinguish features of food systems such as emulsions, foams, gels, and solutions.

Chemistry plays a central role in food science. For instance, the chemistry of ingredients is critical in food formulation and product development. Food flavors, colors, and textures depend upon the functional properties of key food molecules. Understanding the effects of temperature, oxygen, and light on the quality of processed foods requires knowledge of how food molecules respond to these factors. This chapter examines the chemical nature of matter, chemical bonding, fundamental chemical reactions, the functional groups of food molecules, and the functional properties of water and acid molecules in food. The chapter closes with a Challenge section on food systems.

4.1 THE NATURE OF MATTER

Foods are made of matter. Matter in turn is composed of pure substances, as either *elements* or *compounds*. **Elements** are the simplest type of pure substance that have mass and cannot be broken down into something else (without splitting atoms). These fundamental substances make up all other, more complex substances. Elements are showcased in the periodic table of the elements.

Examples of elements commonly found in foods are carbon, hydrogen, oxygen, magnesium, and sodium.

Compounds are two or more elements chemically bonded together in definite proportions by weight. For example, water is a compound composed of 11.11 percent hydrogen and 88.89 percent oxygen by weight. The properties of a compound are entirely different from the properties of each element of which it is composed.

In every sample of a compound, the elements that make it up occur in the same proportions. Even though compounds have unique identities, different from their constituent elements, they can be broken down into those elements under certain conditions. Simple examples of compounds are water, sugar, and salt.

Compounds are represented by formulas made up of elemental symbols. The important compounds in food science are called *organic compounds*. These compounds are the organic biomolecules that contain the element carbon, as well as hydrogen, oxygen, and sometimes sulfur, nitrogen, and phosphorus.

Matter can also be classified as a *mixture*. In a **mixture**, two or more substances are combined physically rather than chemically. For this reason, the separate substances tend to maintain their original properties. Three examples of mixtures are a solution, a dispersion, and an emulsion, which can be considered as types of food systems.

Chemical Symbols, Formulas, and Equations

Symbols. Chemical symbols are used to represent the elements of the periodic table. Scientists use this chemical shorthand in describing reactions between food components. **Symbols** are one- or two-letter abbreviations representing elements by themselves (e.g., H for hydrogen, Na for sodium) or elements as part of a compound or molecule (e.g., the C, H, and O atoms in CH_3COOH, acetic acid). The symbols for elements important in food science are shown in the partial periodic table in Figure 4.1.

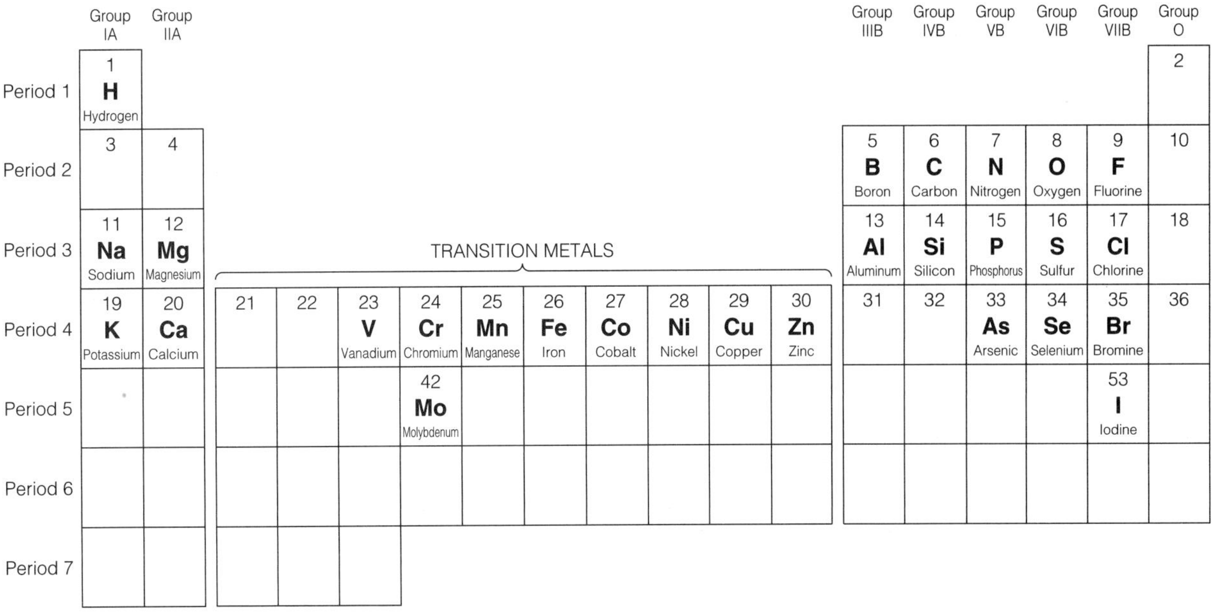

	Group IA	Group IIA											Group IIIB	Group IVB	Group VB	Group VIB	Group VIIB	Group O
Period 1	1 **H** Hydrogen																	2
Period 2	3	4											5 **B** Boron	6 **C** Carbon	7 **N** Nitrogen	8 **O** Oxygen	9 **F** Fluorine	10
Period 3	11 **Na** Sodium	12 **Mg** Magnesium				TRANSITION METALS							13 **Al** Aluminum	14 **Si** Silicon	15 **P** Phosphorus	16 **S** Sulfur	17 **Cl** Chlorine	18
Period 4	19 **K** Potassium	20 **Ca** Calcium	21	22	23 **V** Vanadium	24 **Cr** Chromium	25 **Mn** Manganese	26 **Fe** Iron	27 **Co** Cobalt	28 **Ni** Nickel	29 **Cu** Copper	30 **Zn** Zinc	31	32	33 **As** Arsenic	34 **Se** Selenium	35 **Br** Bromine	36
Period 5						42 **Mo** Molybdenum											53 **I** Iodine	
Period 6																		
Period 7																		

FIGURE 4.1 Elements from the periodic table important in food science. *The elements naturally prevalent in living matter that are essential for life include H, C, N, O, P, and S. Important dietary minerals are Cl, Na, Mg, K, Ca, Mn, Fe, Co, Ni, Cu, Zn, and I. The elements B, F, Al, Si, V, Cr, As, Se, Br, and Mo have specific cellular functions and may prove to be essential.*

Formulas. All of the elements present in a compound make up the **formula** of the compound. Chemical symbols are combined to define chemical formulas. Consider that hydrochloric acid, whose formula is HCl, is represented by two, not three, symbols: hydrogen and chlorine. Subscripts are used to indicate the proportion of elements. The formula for carbon dioxide, CO_2, indicates that there are two oxygen atoms for each carbon atom.

Equations. The chemist's version of a sentence is the **equation**. An equation is the written description of a chemical reaction, using chemical symbols and formulas. All equations place the *reactants* (the substances that react together, on either side of the plus sign) on the left side of a yield arrow, and the *product(s)* to the right of the yield arrow. All equations contain specific reactants and products, for example, the breakdown of glucose:

$$\underbrace{C_6H_{12}O_6 + 6\,O_2}_{\text{Reactants}} \rightarrow \underbrace{6\,CO_2 + 6\,H_2O}_{\text{Products}}$$

Electron Orbits and Chemical Bonds

In all chemical reactions, reactant atoms or molecules approach each other close enough so that they interact through some sort of chemical bonding event. What results is a new substance (the product) with different properties from the reactants.

The forces that hold atoms and molecules together are called **chemical bonds.** The formation of chemical bonds requires *energy*. When chemical bonds form, only the electrons of the reactant atoms are involved in the chemical bonding event, not the protons or neutrons. This concept is so important that it required a theory to explain how electrons are the key to chemical bonding.

The Danish physicist Niels Bohr proposed the theory that negatively charged electrons are in motion in orbits around the fixed positively charged atomic nucleus. The opposite charges result in attraction. In describing the electronic configuration of electrons in atoms, the atom is viewed as having a series of energy layers (the electron orbits). Each layer holds only a certain number of electrons.

The breakdown of glucose shows one glucose molecule and six oxygen molecules reacting to produce six carbon dioxide and six water molecules.

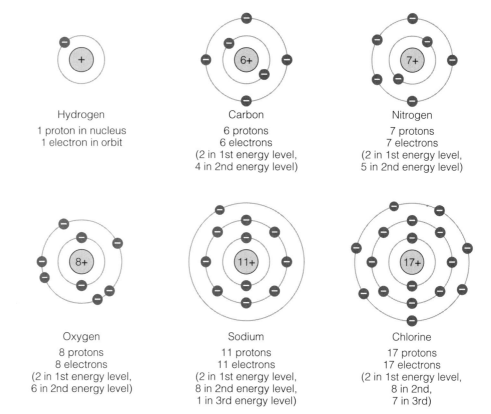

FIGURE 4.2 Electrons distributed in atomic structure via energy layers (orbits). *Hydrogen has one electron and orbit, while a more complex element like sodium requires 3 energy levels for its 11 electrons.*

FIGURE 4.3 Propionic acid structure. *You can match the structural formula $CH_3—CH_2—COOH$ with the structure. Three hydrogen atoms are bonded to an end carbon atom; two hydrogen atoms are bonded to a central carbon atom; and one hydrogen atom is bonded to an oxygen atom, which is bonded to a different end carbon atom by a single bond, while another oxygen atom is bonded to the same carbon atom by a double bond. All of these bonds connected in just this way contribute stability to the propionic acid molecule.*

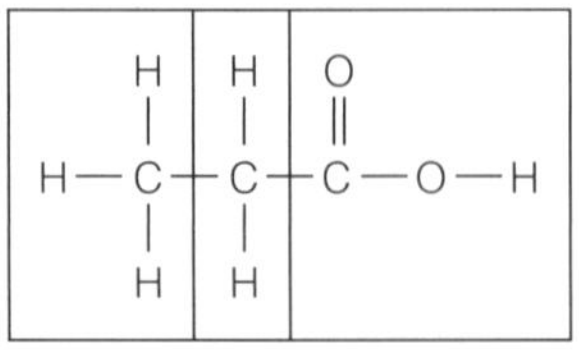

Farther from the nucleus, the layers get larger and can hold more electrons, of increasing motion. The layers are also called *energy levels*. The first energy level is the one closest to the nucleus. It has the lowest energy. Electrons in this level have the least energy of motion because they are the most strongly attracted to the nucleus. Figure 4.2 shows the electrons in the atoms of hydrogen, carbon, nitrogen, oxygen, sodium, and chlorine.

To find out how many electrons are possible in each energy level, scientists use a simple relationship, $X = 2n^2$. This says that X is the maximum number of electrons in energy level n. Using this relationship, in the first energy level ($n = 1$), there can be a maximum of 2 electrons; the second energy level ($n = 2$) holds a maximum of 8; the third holds up to 18, and the fourth can hold 32 electrons.

A chemical rule states that electrons fill levels in sequence, one level at a time. Most atoms become *chemically stable* when they have 8 electrons in their outermost orbital (or **valence shell**). Electrons can change orbitals, but only when they absorb or release energy, such as in the formation of a chemical bond. Chemical bonds form because the atoms seek a full complement of electrons (8) to fill the valence shells, making them stable.

Chemical Bonds in Foods

During chemical bonding, reactant substances are combined in a specific ratio as given by the formula of the product. In the example of *propionic acid*, which is an additive used to keep foods from spoiling, the chemical formula can be written $C_3H_6O_2$. However, to understand the bonding in propionic acid, the formula can be written CH_3CH_2COOH, or $CH_3—CH_2—COOH$. The latter is called a *structural formula*, which leads to the structure of propionic acid shown in Figure 4.3.

Covalent Bonds. Many compounds, like propionic acid, are produced through **covalent bond** formation of their H and O atoms. A covalent bond means *sharing* one or more pairs of electrons so that each atom can fill its valence shell. See Figure 4.4 for a description of covalent bonding in water molecules.

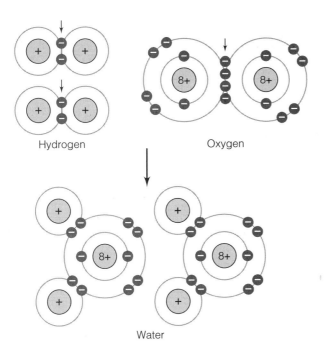

FIGURE 4.4 Covalent bonding in water molecules. *In water (H_2O) hydrogen and oxygen share electrons so that hydrogen fills its first energy level and oxygen fills its second. The most common single and double covalent bonds in food molecules include carbon to carbon (C—C and C=C), carbon to hydrogen (C—H), and carbon to oxygen (C—O and C=O) bonds. Carbon to nitrogen (C—N), nitrogen to hydrogen (N—H), phosphorous to oxygen (P—O and P=O), sulfur to sulfur (S—S), sulfur to hydrogen (S—H), and sulfur to oxygen (S—O and S=O) covalent bonds also occur.*

A *single covalent bond* is formed when one pair of electrons is shared, as in the case of water. Double and even triple covalent bonds are possible in food molecules. A *double bond* occurs in the case of the leavening gas CO_2 (bonds written out as O=C=O). In this case, two pairs of electrons are shared between one carbon and two oxygen atoms.

Food substances such as fats, proteins, vitamins, or carbohydrates involve covalently bonded carbon atoms that form the "backbones" of these food molecules.

Ionic Bonds. Some elements do not have the tendency to share electrons. Instead, they undergo **ionic bonding,** the filling of the valence shells through the *transfer* of electrons between reactants. For instance, two atoms achieve a full valence shell when one donates an appropriate number of electrons to the other—see Figure 4.5. The one that donates the negatively charged electrons becomes positively charged (since the number of protons exceeds the number of electrons for that atom) and is called a *cation*. A few examples of cations in food products and ingredients are Na^+ (sodium ion), K^+ (potassium ion), Ca^{+2} (calcium ion), Mg^{+2} (magnesium ion), NH_4^+ (ammonium ion), and Fe^{+2} (ferrous iron).

The atom that accepts the negatively charged electrons becomes negatively charged (since the number of electrons exceeds the number of protons for that atom), and is called an *anion*. Examples of anions in food products and ingredients are Cl^- (chloride), I^- (iodide), S^{-2} (sulfide), SO_3^{-2} (sulfite), HCO_3^- (bicarbonate), and NO_2^- (nitrite).

Hydrogen Bonds. Compounds containing oxygen or nitrogen bonded to hydrogen, such as H_2O and NH_3, interact by a weak force, called the hydrogen bond. A **hydrogen bond** is a kind of covalent bond, but there is *unequal* rather than equal sharing of electrons between the bonding atoms. This occurs because the relatively large nucleus of the nitrogen or oxygen attracts the electron from hydrogen more strongly than does the hydrogen atom's own small nucleus with only one proton (Figure 4.6).

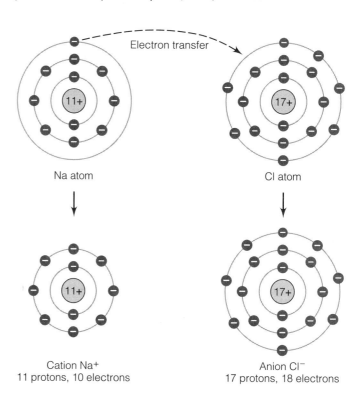

FIGURE 4.5 Electron transfer and the formation of cations and anions.

Another distinction between hydrogen bonds and other covalent bonds is that hydrogen bonds are *intermolecular bonds*—that is, bonds between two different molecules. This is very different from the bonding between a hydrogen atom and an oxygen atom *within the same* molecule, a situation called **intramolecular bonding.**

4.2 CHEMICAL REACTIONS IN FOODS

Chemical reactions can be classified as composition reactions or decomposition reactions. In a **composition reaction,** two or more substances are combined into a single product; the general type is A + B → AB. The important macromolecules found in foods are produced this way in nature, and sometimes during processing. As examples, amino acids combine to form polypeptides, and sugars combine to form polysaccharides.

The opposite of a composition reaction is a **decomposition reaction:** AB → A + B. Decomposition describes how food macromolecules come apart during processing conditions or by enzyme action. As an example, the decomposition of the disaccharide maltose generates two glucose molecules.

To be more complete, our discussion needs to include the notion that reactions involving food molecules are classifiable into two general types: those that involve enzymes and those that do not. (See Table 4.1—these reactions are discussed in detail in the next two sections.) Composition reactions and decomposition reactions may be enzymatic or non-enzymatic.

Enzymatic Reactions

Enzymes occur only in living systems, such as plant and animal tissue; they are present in the microorganisms that are found in food. Enzymes can cause quality changes in food color, texture, flavor, and odor during storage and use. Depending on the specific food and reaction, the outcome is desirable or undesirable. The softening of texture and the browning of fruits and vegetables are undesirable effects of enzyme activity. On the other hand, the food industry takes advantage of desirable enzymatic reactions carried on by bacteria to produce a variety of fermented foods, such as cheeses, certain meats, pickles, sauerkraut, teas, and yogurt.

FIGURE 4.6 Hydrogen bond between two water molecules.
Hydrogen bonding plays a crucial role in dissolving food ingredients and is a key intermolecular attraction in foods.

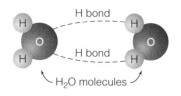

TABLE 4.1 Chemical Reactions of Food Molecules.

Chemical Reaction	Enzymatic	Nonenzymatic
Addition	no	yes
Condensation	yes	yes
Hydration	no	yes
Hydrolysis	yes	yes
Oxidation-reduction	yes	yes

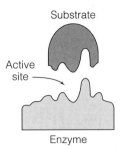

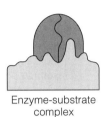

Enzymes are specialized protein molecules—polypeptides composed of amino acids (see Chapter 3). Enzymes are *biological catalysts*, which means that they cause chemical reactions to speed up. Enzymes catalyze reactions specific for particular substrates. For instance, the enzyme that breaks down starch into glucose will not work on protein.

The region on the surface of an enzyme where catalytic activity occurs is the **active site** (Figure 4.7). It is the location where the substrate joins with the enzyme to form what is called the enzyme-substrate complex, the ES complex, the transition state, or the reaction intermediate. The reactivity of this site is governed by the functional groups present there. Functional groups are arrangements of a few atoms that create particular properties of molecules.

The active site can also be thought of as a pocket in the enzyme's structure. It is a small, three-dimensional region that contains amino acids that bind noncovalently with the substrate. As an example, the active site of papain, an enzyme present in papaya, contains adjacent histidine and cysteine residues that form noncovalent bonds with functional groups of protein substrates.

Activation Energy and How an Enzyme Works. Two biochemists, Michaelis and Menten, first proposed an equation to express the manner in which enzymes (E) and substrates (S) combine in reversible steps to form a joined enzyme-substrate complex (ES), which can then either revert back to the free enzyme and free substrate, or generate the new product (P) and regenerate the free enzyme. The famous Michaelis-Menten Equation is given by

$$E + S \leftrightarrow ES \leftrightarrow E + P$$

Whether an enzymatic reaction will progress to completion, meaning to form product, depends on if it is energetically favorable. **Activation energy** or energy of activation is the amount of energy needed to convert substrate molecules from the ground or baseline energy state to the ES complex. Under proper reaction conditions, it represents the greatest barrier to enzyme product formation. The reason that enzymes speed the conversion of substrates into products is because they do something special—they lower the activation energy. Enzymes work best under certain conditions of temperature, pH, and amount of substrate present. These factors also influence whether or not a particular enzymatic reaction will generate product, and the time required to do so.

When a food scientist performs an assay on an enzyme to determine just how fast it functions (the study of *enzyme kinetics*), the amount of substrate must be kept as high as possible to create the condition known as a **zero-order reaction**. In a zero-order reaction the progress of the reaction depends solely upon enzyme concentration. Under zero-order reaction kinetics, the concentration of the ES complex is maintained at a constant level, and the amount of product formed is a linear function of time.

First-, second-, and third-order reactions are enzymatic reactions governed by the presence of one, two, or three unreacted substrates that are converted into products proportional to enzyme concentration. Thus, the rate of these enzymatic reactions is slower than in zero-order reactions.

Enzymatic Hydrolysis. In **enzymatic hydrolysis** reactions, an enzyme breaks large food molecules apart into smaller fragments. Examples of hydrolytic enzymes are carbohydrases (sucrase, lactase, maltase, amylase), lipases, and proteases. Certain of these enzymes function in food fermentation reactions carried out by microorganisms. The

FIGURE 4.7 The active site of the ES complex. *At the active site, the substrate is recognized and oriented spatially to facilitate catalysis and form of product. A highly reactive transition state of the substrate is steadied into the active site as it undergoes transformation to product. Noncovalent interactions in the active site include hydrogen bonding between amino acids of the enzyme and substrate.*

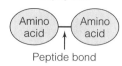

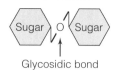

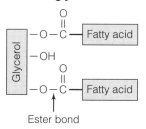

FIGURE 4.8 Synthesis of a dipeptide, a disaccharide, and a diglyceride—examples of condensation reactions. *In food polymers, special names are given to the types of bonds between the linked units. For carbohydrates, the bonds between sugar units are called* **glycosidic bonds.** *In lipids, fatty acids join to glycerol through* **ester bonds.** *In the case of proteins, the bonds between condensing amino acids are referred to as* **peptide bonds.** *Enzymes frequently play a role in such syntheses.*

enzymatic fermentation of sugar by yeast in beer creates hydrolysis, followed by oxidation and reduction reactions to produce carbon dioxide and alcohol (ethanol) as fermentation products.

$$\text{sucrose} \xrightarrow{\text{hydrolysis}} \text{glucose} \xrightarrow{\text{oxidation}} \text{pyruvate} \xrightarrow{\text{reduction}} \text{acetaldehyde} \xrightarrow{\text{reduction}} \text{carbon dioxide} + \text{ethanol}$$

During fermentation, the original food molecules are changed and are converted into fermentation products. The most common substrates for fermentation are carbohydrates. Carbohydrate fermentation yields a variety of products, including lactic acid, acetic acid, ethyl alcohol (ethanol), and carbon dioxide.

Enzymatic Oxidation-Reduction. In enzymatic oxidation-reduction reactions, the enzyme causes changes in the chemical structures of food molecules. **Enzymatic oxidation** reactions occur when oxygen is added to, or hydrogen or electrons are removed from, food molecules in the presence of an active enzyme. **Enzymatic reduction** refers to the gaining of one or more electrons, gaining hydrogen, or losing oxygen from the structure of a food molecule in the presence of an active enzyme.

The browning of fruits and vegetables is an example of an enzymatic oxidation reaction. Polyphenol oxidase (PPO) is the enzyme responsible for the formation of brown oxidized pigments in apple and banana slices exposed to air. The original pigment molecules react with oxygen and are transformed into brown melanin pigment molecules. In plant and animal cells, oxidation of one substance will not occur without the simultaneous reduction of another. Oxidation and reduction are called *coupled reactions* for this reason.

Enzymatic Polymerization. In **condensation,** separate reactant molecules can be joined through the action of enzymes. Condensation reactions can result in the generation of small molecules like dipeptides and disaccharides, or larger polymers (see Figure 4.8). A **polymer** is a high molecular weight molecule created by the repetitive reaction of hundreds or thousands of low molecular weight units.

Natural food substances that are examples of polymers include polysaccharides and proteins. When many glucose molecules join via condensation, starch is the result. The condensation of multiple amino acids produces polypeptides. Glycerides are formed when fatty acid molecules condense with glycerol. **Polymerization** is a type of reaction in which many molecular units (called monomers) are joined together end to end. Chemists refer to such products as addition polymers and condensation polymers, based upon the mechanism by which the units are joined.

Nonenzymatic Reactions

Chemical reactions in foods that do not depend on enzymes include addition, oxidation-reduction, hydrolysis, and condensation reactions.

Addition. **Addition** reactions take place in organic molecules that possess double or triple bonds between carbon atoms. In contrast to a single bond, a carbon-to-carbon double or triple bond is highly reactive. In foods, double bonds are found in unsaturated fatty acids. The addition of hydrogen to a fat creates an addition reaction, in which the double bond is broken and hydrogen is incorporated into the fatty acid structure. This addition reaction is known in the food industry as the hydrogenation of fat, a processing step that alters a fat's melting characteristics. Hydrogenation is used to make vegetable oils hard (solid) at room temperature; for example, margarine is made through hydrogenation.

Oxidation and Reduction. In oxidation-reduction reactions, electrons are transferred between substances. These reactions are significant because they can affect the color, quality, and acceptability of foods. In specific reactions with *carbon,* **oxidation** can be considered as the addition of oxygen, and **reduction** as the gain of hydrogen. Adding oxygen oxidizes carbon atoms, while adding hydrogen reduces carbon atoms.

General Condensation Reaction to Form an Ester:

organic acid + alcohol ⟶ ester + water

Example

$CH_3COOH + CH_3(CH_2)_4OH \rightarrow CH_3COO(CH_2)_4CH_3 + H_2O$

acetic acid + pentanol ⟶ pentyl acetate ester + water

FIGURE 4.9 Condensation reaction. *Artificial banana flavoring can be produced by mixing acetic acid with pentanol, causing the formation of the ester pentyl acetate via condensation.*

Reducing agents and oxidizing agents are used in the food industry. Stated simply, a **reducing agent** is a substance that causes another substance to become reduced, while the reducing agent itself becomes oxidized in a reaction. An **oxidizing agent** is a substance that causes another substance to become oxidized, while the oxidizing agent itself becomes reduced in a reaction. Food antioxidants such as ascorbic acid function as reducing agents; dough conditioners such as calcium peroxide or flour-bleaching agents such as chlorine gas function as oxidizing agents.

Condensation and Hydrolysis. Like oxidation and reduction reactions, condensation and hydrolysis reactions can be considered as opposites. Condensation and hydrolysis reactions represent composition and decomposition. In **condensation,** separate reactant molecules are linked together by special chemical bonds. The reactants lose hydrogen and oxygen atoms, which combine through a side reaction to form water—see Figure 4.9. Some food esters, including artificial flavors, are produced through condensation.

Nonenzymatic hydrolysis reactions refer to the breakdown of food molecules due to extremes of temperature or pH rather than by enzymes. Hydrolysis is the reverse process of condensation. In **hydrolysis,** a water molecule enters the region of the functional group of a larger molecule and splits it off. The OH group of the water molecule attaches to one of the newly split-off molecule pieces, and the hydrogen attaches to the other.

The breakage of the disaccharide maltose into glucose units by heat offers an example of nonenzymatic hydrolysis. Heat energy and acid are effective because they break chemical bonds in molecules.

4.3 FUNCTIONAL GROUPS

Functional groups are arrangements of just a few atoms that create particular properties of molecules. As such, they represent structural features as well as reactive sites. Many reactions of food molecules involve functional groups in some characteristic way (Table 4.2).

Alcohol Group —OH

An alcohol (OH) group is not the same as a hydroxide ion (OH^-). The OH group is called a hydroxyl group, and it does not ionize. Food alcohol compounds contain hydroxyl functional groups. An example of an important food alcohol is ethanol, C_2H_5OH. This substance can be produced various ways, such as by hydrolyzing the starch in potatoes or by fermentating the sugars in molasses. Another food alcohol, glycerol ($CH_2OH-CHOH-CH_2OH$), is a slightly sweet, water-soluble alcohol derived from animal fats and vegetable oils.

TABLE 4.2 Functional groups found in food molecules.

Name of Group	Functional Group	Examples (typical ending)
Alcohol group	R—OH	ethanol, glycerol (-ol)
Aldehyde group	R—C(=O)—H	furfural, diacetyl (-al)
Ketone group	R—C(=O)—R'	acetone (-one)
Carboxylic acid group	R—C(=O)—OH	citric acid, lactic acid, acetic acid
Amino group	R—NH$_2$	all amino acids (-amine)
Ester group	R—C(=O)—O—R'	fatty acid such as oleic acid
Phosphate group	R—O—P(OH)(=O)—OH	trisodium phosphate, sodium phosphate
Methyl group	R—CH$_3$	methionine, methanol
Sulfhydryl group	R—SH	amino acid cysteine

Aldehyde Group.   H—C(=O) [aldehyde group] / R [rest of molecule]

In an organic molecule called an **aldehyde,** oxygen is double-bonded to carbon. Formaldehyde (HCHO) is an odorous, water-soluble gas that is used as a germicide. Other aldehydes are cinnamic aldehyde, a sweet substance found in cinnamon spice; furfural found in coffee; and diacetyl, which has a buttery or butterscotch flavor.

Amino Group   —NH$_2$

A compound that contains an amino (NH$_2$) functional group is called an **amine.** Amino acids and proteins contain this functional group. Amines are derived from ammonia (NH$_3$) and, like ammonia, amines are basic substances, gaining H$^+$ (protons) in chemical reactions. Some food amines are produced by bacterial action. Examples of naturally occurring food amines are histamine, tyramine, dopamine, and serotonin, which are generated by the decarboxylation (loss of COO), and sometimes hydroxylation (addition of OH), of amino acid precursor molecules.

Carboxylic Acid Group   O=C—OH

The carboxylic acid group (or simply, carboxyl or acid group) can be regarded as an organic acid since it contains carbon in its structure. The COOH functional group ionizes to produce COO$^-$ plus a proton (H$^+$). Students sometimes confuse carboxylic acid groups with aldehyde groups because they both contain carbon-to-oxygen double bonds.

Such carbon atoms are termed *carbonyl carbon*. However, in an aldehyde, the other bond to C is always a single bond with H, while in the acid, it is a single bond to OH. Examples of food carboxylic acids are citric acid in citrus fruits, acetic acid (CH_3COOH) in vinegar, and lactic acid (CH_3 — CHOH — COOH) in milk.

The carbonyl group, represented as a covalent C=O bond, is part of a carboxylic acid functional group. An organic compound termed a carboxylic acid contains the OH (hydroxyl) group attached to the carbonyl group, as COOH.

$$-\overset{\overset{O}{\|}}{C}-OH$$

Ester Group $\quad -\overset{\overset{O}{\|}}{C}-O-C-$

Esters are organic compounds having a carbonyl-oxgen-carbon system. Animal fats such as lard and tallow and vegetable oils such as corn oil and peanut oil are esters of fatty acids attached to glycerol. Esters are present in fruits and give characteristic flavors and aromas. Isoamyl acetate exhibits the characteristic flavor and aroma of banana. Esters can be produced by combining an acid with an alcohol in a reaction called *dehydration synthesis*, a type of condensation. In the food industry, synthetic flavors are made this way. The condensation of formic acid with ethanol results in the esterification of the acid, producing ethyl formate, which has a characteristic rum aroma.

Ketone Group $\quad -C-\overset{\overset{O}{\|}}{C}-C-$

Ketones are organic compounds in which an interior carbon atom is double-bonded to an oxygen atom. The most frequently cited example of a ketone is acetone. Many students know of it as nail polish remover. Acetone is a combustible water-soluble liquid with a sweetish odor, and it is a strong solvent. Certain odorous ketones are produced during the cooking of meats and the brewing of beer.

Methyl Group $\quad -CH_3$

Methyl groups are present in a variety of organic molecules, which are said to be "methylated" with CH_3. Methyl group transfer reactions are important in the biosynthesis of certain bioactive compounds. The amino acid methionine functions as a methyl group donor. Methyl alcohol, or methanol, contains a methyl group bound to an OH group. In foods, plant pectins and gums contain methyl groups as methyl esters, which, as —COO—CH_3 groups, contribute to functional properties such as viscosity and gelation. The sweetener aspartame is a methyl ester of a dipeptide.

Phosphate Group $\quad -O-\overset{\overset{OH}{|}}{\underset{\underset{O}{\|}}{P}}-OH$

Phosphates are salts of phosphoric acid, H_3PO_4, an inorganic acid, which makes it a noncarboxylic acid. Phosphoric acid resembles a phosphate group except it has a double bond between the phosphorus and one of the oxygen atoms, and the remaining oxygen atoms are bound to hydrogen. Phosphoric acid, used as an acidulant, is a tribasic acid that ionizes to $H_2PO_4^-$, HPO_4^{-2}, and PO_4^{-3}.

The types of phosphates used in food processing include orthophosphates and polyphosphates. They are sometimes added to meat to improve texture, juiciness, and

TABLE 4.3 Chemical ionic groups found in food molecules.

Name of Ion	General Formula
Sulfide ion	S^{-2}
Sulfite ion	SO_3^{-2}
Sulfate ion	SO_4^{-2}
Ammonium ion	NH_4^{+}
Ferrous ion	Fe^{+2}
Ferric ion	Fe^{+3}
Bicarbonate ion	HCO_3^{-}
Nitrite ion	NO_2^{-}
Nitrate ion	NO_3^{-}

water-holding capacity. Trisodium phosphate is an antimicrobial used in poultry dip. Sodium phosphate is an emulsifying salt commonly used in the production of process cheese.

Sulfhydryl Group —SH

Sulfur is found in the same group of the periodic table as oxygen. The —SH group is called the **sulfhydryl** group and is similar to an alcohol group. Compounds containing this group are called **thiols**. Thiols are easily oxidized and yield disulfides:

$$2\ R-S-H \rightarrow R-S-S-R\ \text{(a disulfide)}$$

$$HOOC-\underset{H}{\overset{NH_2}{C}}-CH_2-SH$$

Cysteine

$$HOOC-\underset{H}{\overset{NH_2}{C}}-CH_2-S-S-CH_2-\underset{H}{\overset{NH_2}{C}}-COOH$$

Cystine

The amino acid cysteine contains a sulfhydryl group. The reduced state of sulfhydryl groups enables them to be reactive with oxygen. Oxidation of the SH group of two cysteine molecules joins them by a disulfide bond (S—S). The resulting dipeptide is called cystine. Sulfhydryl groups are involved in the development of gluten structure in bread dough.

Ionic Groups

In addition to serving as functional groups, a variety of atoms exist as **ions** that carry either negative or positive charges, depending on whether or not they have gained or lost electrons. Ions play crucial roles in foods, influencing food quality. Table 4.3 shows some of the ionic groups found in food molecules.

4.4 THE CHEMICAL AND FUNCTIONAL PROPERTIES OF WATER

Functional properties are the physical and chemical properties of food molecules that affect their behavior in foods during formulation, processing, and storage. These include sensory and mechanical properties of foods, such as flavor, texture, effects on water, and on the physical condition (e.g., gel vs. liquid) of the final product. Functional properties are determined by the functional groups of food molecules. Many of the functional properties of food molecules depend upon weak forces, especially the hydrogen bond.

The functional properties of water in foods include acting as a diluent and carrier of hydrophilic food ingredients, providing a medium for chemical and enzymatic reactions, and dispersing and solvent action. Water serves as a fat replacer and zero-calorie ingredient, is a component of gels and emulsions, acts as a medium for heat transfer, functions as a plasticizer, accounts for food moisture, and is a reactant or product in chemical reactions, such as condensation and hydrolysis.

Water Molecule Structure

In one water molecule, H_2O, the two hydrogen atoms are bonded to one oxygen atom. Each hydrogen atom of a water molecule shares an electron pair with the oxygen atom (look back at Figure 4.4). A charge separation (dipole) results, in which the

Functional properties of water in foods:

Component of colloidal dispersions
Fat-replacer ingredient
Heat transfer
Medium for chemical reactions
Medium for microbial growth
Plasticizer (a substance that, when added to a food system, makes it softer)
Product moisture
Solvent

oxygen atom displays a partial negative charge while the hydrogen atoms are partially positive (Figure 4.10).

The dipolar nature of water affects its physical characteristics like boiling point, freezing point, and vapor pressure. It also results in an electrostatic attraction between adjacent H_2O molecules (hydrogen bonding). The tetrahedral arrangement of the orbitals around the oxygen atom in a water molecule allows each molecule to form hydrogen bonds with as many as four neighboring water molecules.

Solvation and Dispersing Action

The fact that food molecules can form hydrogen bonds with water means they can be dissolved or dispersed. This important concept is termed *solubility*. Compounds that hydrogen bond easily to water to form solutions or colloidal dispersions are called *hydrophilic compounds*. These are generally charged or polar molecules themselves.

Many biomolecules, such as minerals, salts, vitamins, sugars, complex carbohydrates, amino acids, and proteins exist as polar (charged) substances. Water dissolves such substances by hydrating them. **Hydration** is the process by which water molecules surround and interact with solutes by acting as a solvent—see Figure 4.11. This is why hydration is also considered a solvation event. In this manner, water acts as a carrier for hydrophilic substances, as well as a diluent of food ingredients.

Water also disperses amphiphilic molecules. *Amphiphilic molecules*, such as proteins, certain vitamins, phospholipids, and sterols, contain both hydrophilic and hydrophobic regions in their structures. In water these form **micelles**, clusters of molecules in which the hydrophobic groups are directed away from the water while the polar (charged) groups are exposed on the external surface—see Figure 4.12. The nonpolar hydrophobic groups form a stable inner core due to forces called hydrophobic interactions. Thus, micelles are stabilized structures of amphiphilic molecules.

These interactions just described among water molecules and between water and food molecules are called **noncovalent interactions.** These are very important in food chemistry and include hydrogen bonding (intermolecular H bonds), ionic interactions (ions in water), and hydrophobic interactions (micelle structure).

Water Activity and Moisture

Water exists in one of several forms in foods. Chemically, each form is the same (H_2O), but differences exist in the physical and chemical conditions in which water can exist. These differences are significant from several perspectives, including food processing, food safety, and sensory evaluation.

The presence of water in foods is described as the moisture content or as the water activity of the food. The two terms, though, are not synonymous. Moisture refers to the absolute amount of water present in a food, while water activity has to do with the form in which the water exists in the food, such as free or chemically bound water. For this reason these terms not only indicate techniques or methods of analysis in determining the water content of foods, but can also indicate the form of the water being measured.

Moisture is the amount of water present in a food, as a component, relative to all the solid constituents, such as proteins, carbohydrates, and any nonwater liquid (i.e., oils). Most water in foods is called **free water.** Free water is lightly entrapped and therefore easily pressed from food matter. When this is done, the water can be seen and felt. Free water acts as a dispersing agent and solvent, and can be removed by drying foods.

Adsorbed water, or structural water, is a second type. This water associates in layers via intermolecular hydrogen bonds around hydrophilic food molecules. **Bound water,** sometimes called the water of hydration, is a third form of water in food. It exists in a tight chemically bound situation, such as within a crystalline structure, via water-ion and water-dipole interactions. Bound water does not exhibit the typical properties of water, failing to freeze at 0°C and failing to act as a solvent.

Now we can define water activity. **Water activity** (a_w or A_w) is a measure of the availability of water molecules to enter into microbial, enzymatic, or chemical reactions.

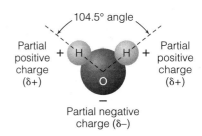

FIGURE 4.10 Structure of the water molecule showing intramolecular bond angle and charge separation. *The geometry of the water molecule is dictated by the shapes of the outer electron orbitals of the oxygen atom. The arrangement is nonlinear, having an H—O—H bond angle of 104.5 degrees. Since the oxygen nucleus with 8 protons attracts electrons more strongly than the hydrogen nucleus (only a single proton), the sharing of electrons in the intramolecular covalent bonds is unequal. In this dipole effect, the O atom bears a partial negative charge ($\delta-$), and each H atom a partial positive charge ($\delta+$).*

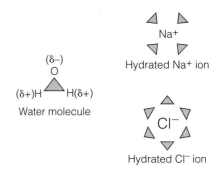

FIGURE 4.11 Water hydration of sodium and chloride ions. *When a salt such as sodium chloride becomes hydrated, the electrostatic attraction of the Na^+ and Cl^- ions for one another weakens, reducing their tendency to stay in their own sodium chloride structure. Instead, these ions interact with water molecules and go into solution as solutes, with the negative end of the water molecule attracts to the Na^+ ion and the positive end of the water molecule attracts to the Cl^- ion. Food compounds that have charged functional groups, such as COO^- (ionized carboxylic acids), NH_3^+ (protonated amines), and PO_4^- (phosphate esters), are water soluble for the same reason.*

TABLE 4.4 Moisture content and water activity of several foods.

Food	Moisture %	Water Activity
Ice at 0°C	100	1.00
Fresh meat	70	0.985
Bread	40	0.96
Flour	14.5	0.72
Ice at −50°C	100	0.62
Raisins	27	0.60
Macaroni	10	0.45
Potato chips	1.5	0.08

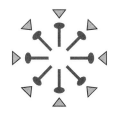

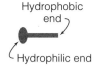

Amphiphilic molecule

FIGURE 4.12 Formation of a micelle showing ordered water molecules surrounding amphiphilic molecules.

This availability determines the shelf life of a food. Regarding the forms of water, bound water is inversely related to water activity: As the percentage of bound water in a food increases, the water activity decreases.

Water activity is calculated as the ratio of the water vapor pressure of the substance divided by the vapor pressure of pure water at the same temperature:

$$a_w = \frac{P}{P_0}$$

where P is the vapor pressure of the food and P_0 is the vapor pressure of pure water at the same temperature. The water activity of pure water is 1.0 according to this equation. All foods, since they contain some nonvolatile substances (substances that will not change to the gaseous state at ordinary temperatures and lower the vapor pressure of the water present) will have a_w values less than this. In simpler terms, water activity is a measure of relative humidity. By multiplying a_w by 100, the relative humidity (RH) of the atmosphere in equilibrium with the food (RH (%) or ERH) is obtained:

$$\text{RH (\%)} = 100 \times a_w$$

Table 4.4 gives the moisture content and water activity of several foods. Interestingly, some foods are stable at low moisture content, whereas others are stable at relatively high moisture content. For instance, peanut oil deteriorates at a moisture content above 0.6 percent, whereas potato starch is stable at 20 percent moisture.

How are measurements of food stability made? Water sorption isotherms (or MSI, **moisture sorption isotherms**) are graphs of data that interrelate the water (moisture) content of a food with its water activity at a constant temperature. A sorption isotherm indicates the water activity at which a food is stable and allows predictions of the effect of changes in moisture content on a_w, and therefore on storage stability—see Figure 4.13.

Water activity is known to be a temperature-dependent phenomenon. Therefore, moisture sorption isotherms must also display temperature dependence. Hence, at any given food moisture content, water activity will increase with an increase in temperature. Water sorption isotherm plots show this, and they are used to determine the rate and extent of drying, the optimum frozen storage temperature, and the moisture barrier properties required in food packaging materials.

Water as a Component of Emulsions

An emulsion is a type of colloidal dispersion, a system containing two liquids or phases that normally do not mix: a dispersed phase and a continuous phase. Water, as the aqueous component of an emulsion, can function in either a dispersed or a continuous phase. The water phase of an emulsion is hydrophilic in character while the fat phase is lipophilic (or hydrophobic). The presence of the water phase is therefore a prerequisite for an emulsion.

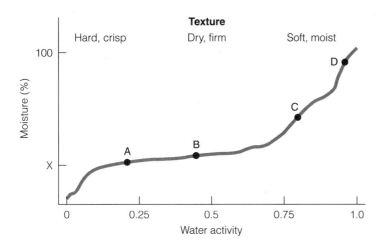

FIGURE 4.13 Moisture sorption isotherm (MSI). *The MSI plot shows how water activity and moisture differ. For example, a food product might have a relatively low constant moisture level (X) and yet increase significantly in water activity (points A to B). During storage of a dry food in a humid environment, as more moisture uptake occurs (points B to C to D), the proportion of bound water decreases, and the amount of free water increases. This changes a hard, crisp product to one that is soft and moist. For a product like a cracker, this indicates quality loss in terms of texture and potential for spoilage due to the increase in moisture and water activity.*

Water and Heat Transfer

Water acts as an important vehicle for heat transfer in foods during food processing operations and in food preparation. Water molecules always possess kinetic energy, as long as their temperature is above 0° Kelvin (−273°C). With the addition of heat (thermal) energy, the kinetic energy of water molecules increases. The temperature increase of water as it is heated is proportional to its kinetic energy increase. Water is able to act as a conductor of thermal energy to food molecules, a process called *heat transfer*.

Water as an Ingredient

Incorporating water as a component of processed foods is common practice, including frozen desserts that have been formulated to be low-fat or fat-free. However, increasing the amount of water in a food can have quality repercussions because water can act as a solvent, change state with temperature (freeze or thaw in response to temperature), and exhibit motion within a food system. In frozen foods, stabilizing the movement of water is desirable from a quality standpoint. Freeze-thaw cycles in stored foods can result in the production of concentrated and diluted portions of a previously homogeneous food product when freeze-thaw stability is poor.

Water as a Plasticizer

Water acts as a **plasticizer,** especially in low moisture and frozen foods. A plasticizer, when added to a polymer food system, lowers what is called the *glass transition temperature* (T_g). The glass transition temperature refers to the temperature at which a change in the physicochemical state and the mobility of the water and polymer molecule constituents of a food occurs, and it will be discussed in more detail in Chapter 13.

Water activity and the glass transition temperature show a somewhat steady relationship, the decrease in T_g with increasing water activity being linear over a wide range of a_w values. A plasticizer acts as a food system softener, increasing food polymer molecular volume as well as mobility. For example, consider starch in water. Increasing the water

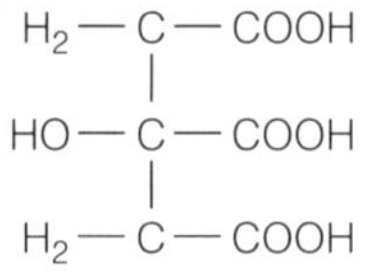

Citric acid (from lemons) $C_6H_8O_7$:
H₂—C—COOH
|
HO—C—COOH
|
H₂—C—COOH

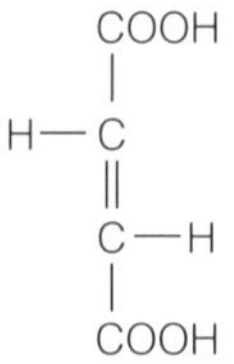

Fumaric acid (from mushrooms) $C_4H_4O_4$:
COOH
|
H—C
‖
C—H
|
COOH

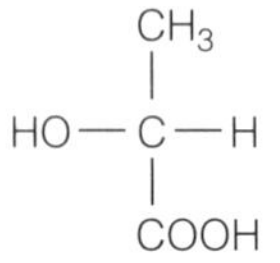

Lactic acid (from yogurt) $C_3H_6O_3$:
CH₃
|
HO—C—H
|
COOH

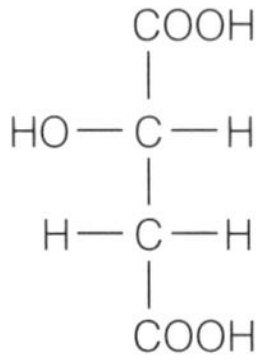

Malic acid (from apples) $C_4H_6O_5$:
COOH
|
HO—C—H
|
H—C—H
|
COOH

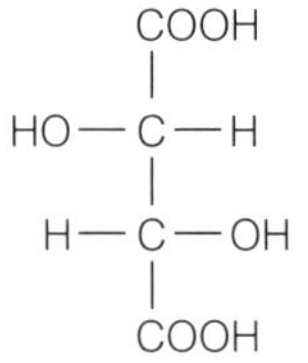

Tartaric acid (from grapes) $C_4H_6O_6$:
COOH
|
HO—C—H
|
H—C—OH
|
COOH

FIGURE 4.14 Chemical structures of important food acids.

◆ **Functional properties of food acids:**
  Antimicrobial agent
  Buffering system component
  Chelating agent
  Dough softening agent
  Hygroscopicity agent
  Leavening system component
  Gelation promoter
  Sequestrant/antioxidant synergist
  Sourness agent

content of a starch-plus-water system expands the volume and increases the freedom of motion of starch molecules.

4.5 THE CHEMICAL AND FUNCTIONAL PROPERTIES OF FOOD ACIDS

Acids have long been known to be important in foods. Early societies knew that a sour substance was produced during the fermentation of cider mixed with honey—the mixture was about 10 percent acetic acid. Acetic acid contains the carboxylic acid group and is thus an organic acid.

Food acids come in many varieties. Protein is composed of amino acids, which are responsible for the protein's shape, functionality, and nutritional quality. Fats are composed of fatty acids; for example, butyric acid is a small carboxylic acid four carbons in length—it has a foul smell and is present in rancid butter.

Tartaric acid functions in the leavening systems of many baked products. The crisp, tart flavor of citrus fruits like oranges, grapefruit, and lemons originates from citric acid. Phosphoric acid serves as an ingredient in many cola drinks.

Food Acid Structure

A typical food acid is a carboxylic or organic acid, containing the carboxylic acid group (COOH) attached to the remainder of the molecule. Food acids differ in where the group is located, as well as the number and arrangement of the other carbon, hydrogen, and oxygen atoms in the structure—see Figure 4.14. These differences influence physical and chemical properties. Acids lacking the carboxylic acid group, like phosphoric acid, are inorganic acids.

The functional properties of food acids range from influencing flavor to acting as antimicrobial agents. Certain food acids are added to sweetened beverages to extend and intensify sweet flavor. In dairy and baked products, acids impart a desirable sour flavor. Some foods or ingredients that must remain free-flowing have acids added to them because acids exhibit low **hygroscopicity,** which means a low attraction for moisture. Without the benefit of the acid, clumping of ingredients due to moisture pickup would result.

The functions of food acids are directly related to their molecular size and structure. The number of carboxylic acid groups in their structures creates differences in functional properties. Consider the similarities and differences between malic acid $C_4H_6O_5$ and fumaric acid $C_4H_4O_4$ (Figure 4.13). Although both are C_4 organic acids, they do not behave identically and offer distinct functional properties. Fumaric acid exists in a more "reduced state" (lack of oxygen atoms) at the central carbons in its structure (HC=CH) compared with malic acid (HCOH — HCH), which in part accounts for the difference.

Fumaric acid is an alkene, possessing a carbon-to-carbon double bond, and it is a less polar molecule than malic acid. As a result, fumaric acid is much less soluble in water than malic acid. Nor are these two acids identical in strength.

Acid Strength

Food acids characteristically donate (lose) protons. Fumaric and malic are termed *weak* acids, as are all food acids. A *weak acid* is one that is mainly in the form of —COOH but a small amount has H^+ separated, or dissociated, to form $COO^- + H^+$. *Strong acids* are acids that have large amounts of dissociated ions. Even a small dissociation is critical, though, because of its effect on pH, which in turn affects food properties like sweetness, sourness, flavor, mouthfeel, and appearance.

Chemists describe a weak acid as one having a small dissociation or **ionization constant,** K_a. The K_a is given for the ionization reaction HA ⇌ $H^+ + A^-$, where HA is the un-ionized acid, and H^+ (proton) and A^- (the conjugate base or anion of the original acid) are ionized species derived from the HA acid.

Fumaric acid has two acid groups in its structure—thus it has two protons to lose and two K_a values. Focusing on the ionization constant for the first carboxylic acid group:

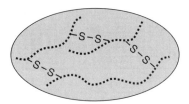
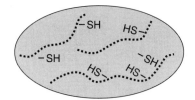

Without fumaric acid: disulfide bonds present in flour

Add fumaric acid: disulfide bonds break and form sulfhydryl groups, relaxing the dough

FIGURE 4.15 Fumaric acid effect on disulfide bonds in tortilla dough. *Fumaric acids acts as a reducing agent by adding hydrogen to disulfide bonds, which establishes reduced sulfhydryl groups. This prevents crosslinking of cysteine amino acids and the firming such crosslinks would cause. The result: a soft, workable dough. (Adapted from Woo, 1998.)*

$K_1 = 1 \times 10^{-3}$. This is a larger than for malic acid ($K_1 = 4 \times 10^{-4}$). Since the strength of an acid is directly proportional to its K_a value, fumaric is a stronger acid than malic.

The pK_a of an acid is another measure of its strength. The rule to remember: The lower the pK_a, the stronger the acid (because mathematically pK_a is the inverse of K_a). Thus, we would expect the pK_a of malic acid to be greater than that of fumaric acid, which is exactly the case.

Fumaric Acid and Dough Softening

When added to products like flour doughs, certain acids, like fumaric, can soften them. Wheat flour contains protein, and flour proteins contain the amino acid cysteine. Cysteine amino acids possess sulfhydryl groups in the reduced —SH form.

However, under typical oxidizing conditions of flour mixing, baking, and storage, the sulfhydryl groups become oxidized, losing hydrogen. The remaining sulfur atoms on each cysteine can join, forming **disulfide bonds** (—S—S—) between cysteine amino acids in flour dough proteins. These bonds or linkages represent a tightening of the dough structure. Such a firm dough is not easy to manipulate and process. However, if fumaric acid is added to the dough formulation, it breaks disulfide bonds, and a softer, more easily manipulated dough results.

What is the reason for this? The carboxylic acid groups in fumaric acid establish what are called "reducing conditions" in the dough, by donating H^+. Under such conditions, the hydrogen adds to the sulfur atoms present in the disulfide bonds, converting them back to —SH. Therefore, fumaric acid acts as a reducing agent, by adding hydrogen to disulfide bonds, which restores the reduced sulfhydryl groups—see Figure 4.15.

Salts of Organic Acids

Organic salts are compounds formed from organic acids in which the hydrogen atom of the acid group, COOH, is replaced by a metal ion such as sodium, calcium, or potassium (Figure 4.16). Food chemists recognize conditions that favor production of organic salts

FIGURE 4.16 Formation of an organic salt, sodium acetate, and its ionization.

$$CH_3-COOH + NaOH \rightarrow CH_3-\overset{\overset{O}{\|}}{C}-O^- Na^+ + H_2O$$

Acetic acid   Sodium hydroxide   Sodium acetate   Water

TABLE 4.5 Base cations and acid anions or organic salts. *Salts are ionic crystals composed of base cations and acid anions. Knowing this allows one to propose salt names. For example, using potassium as the base cation: potassium bromide, potassium iodide, potassium nitrate, and potassium sulfate.*

Base cations	Acid anions
Calcium (Ca^{++})	acetate
Hydrogen (H^+)	bicarbonate
Potassium (K^+)	bromide
Magnesium (Mg^{++})	iodide
Sodium (Na^+)	nitrate
Ammonium (NH_4^+)	phosphate
	sulfate
	sulfite

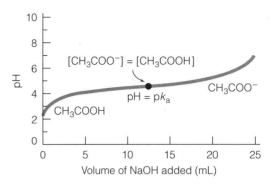

FIGURE 4.17 Titration curve for acetic acid, CH_3OOH, with the buffer zone occurring 0.5 pH units above and below the pK_a value. *If enough base is added to cause the pH to rise above pH 5.25, then the buffering action of the acetate will be overcome, with small additions of base causing large increases in pH.*

and their potential to affect food quality. Like their parent acid compounds, organic salts are able to undergo ionization.

When organic salts ionize, they produce ions other than hydrogen ions or hydroxide ions. For this reason, they are technically not behaving as acids or bases. You can think of a salt as containing the cation of a base (e.g., Na^+ is the cation of the base sodium hydroxide, NaOH) and the anion of an acid (e.g., acetate, or CH_3COO^-, is the anion of acetic acid). Table 4.5 lists some base cation and acids anion that can be combined to form organic salts.

Buffers

A **buffer** is a solution of a weak acid and its salt at a pH where the solution has the ability to maintain that pH when quantities of base are added. The milk acid, lactic acid, and its anionic conjugate, lactate, form a typical *buffer system*. The salt of lactic acid in milk is calcium lactate. This substance helps to resist changes in pH when milk is acidified.

The buffering action of any acid/salt system is limited to a pH range extending one-half pH unit on either side of the pK_a of an acid. Figure 4.17 shows this in a titration curve, which is a graph showing the change in pH as a base is added. (Titration is discussed in the next section.) As an example, acetic acid has a pK_a of 4.75. The buffering action of the acetic/acetate system extends from a pH of 4.25 to 5.25, reaching a maximum at its pK_a, which is a pH of 4.75.

Leavening

Leavening refers to the production of gas by yeast fermentation, by the reaction of an acid with baking soda in batter and dough products, or by the heating of salts. During fermentation, carbohydrates are converted to carbon dioxide gas and ethanol. The CO_2 gas diffuses into small air bubbles produced during the mixing of dough ingredients, which function as gas cell nuclei. See Figure 4.18.

The key to the leavening effect, which causes the "rise" in the product, is the production of gases that create expansion in the product. A variety of gases, alone or in combination, create leavening action or the leavening effect. Carbon dioxide gas is one such leavening gas, along with water vapor, air, ammonia, and ethanol.

Leavening acids are useful participants in the food leavening systems used for baking. In leavening systems, these acids generate hydrogen ions that facilitate the release of carbon dioxide from baking soda (sodium bicarbonate). The gas release causes the expansion of a baking dough or batter product, due to the increased pressure inside the gas nuclei.

⚐ Examples of leavening acids:

Leavening acids tend to be slow acting, except for MCP, tartaric acid, and its derivatives such as cream of tartar, which are fast acting.
Cream of tartar
Dicalcium phosphate dihydrate (DCPD)
Glucono delta lactone (GDL)
Monocalcium phosphate (MCP)
Sodium acid pyrophosphate (SAPP)
Sodium aluminum phosphate (SALP)
Sodium aluminum sulfate (SAS)
Tartaric acid

Yeast fermentation of sugars generates ethanol:

$$C_6H_{12}O_6 \dashrightarrow 2\,CO_2 + 2\,C_2H_5OH \uparrow$$

glucose   carbon dioxide   ethanol

Salts can decompose to produce ammonia, water vapor, and carbon dioxide:

$$NH_4HCO_3 \xdashrightarrow{heat} NH_3 \uparrow + H_2O \uparrow + CO_2 \uparrow$$

ammonium chloride salt   ammonia   water   carbon dioxide

The reaction of acids or acidic salts (HX) with sodium bicarbonate produces water vapor and carbon dioxide:

$$HX + NaHCO_3 \xdashrightarrow{H_2O + heat} NaX + H_2CO_3 \dashrightarrow \underbrace{H_2O \uparrow + CO_2 \uparrow}_{\text{leavening gases}}$$

acid salt   sodium bicarbonate   sodium salt   carbonic acid   steam   carbon dioxide

FIGURE 4.18 Production of leavening gases. *Leavening gases are produced via fermentation of carbohydrate, the heating of salts, and the reaction of leavening acid with sodium bicarbonate.*

Two critical properties of leavening acids are neutralizing value (NV) and dough reaction rate. NV is the amount of sodium bicarbonate that can be neutralized by 100 parts by weight of leavening acid. Dough reaction rate refers to the speed of the reactivity of a leavening acid in a dough, as the amount of carbon dioxide released.

Baking soda, sodium bicarbonate ($NaHCO_3$), is used in conjunction with leavening acids to produce the leavening gas CO_2. Baking soda is an alkaline substance. In the reaction, sodium bicarbonate ionizes to produce bicarbonate ions (HCO_3^-). These are converted into carbonic acid (H_2CO_3). The carbonic acid dissociates to produce the leavening gas carbon dioxide:

$$H_2CO_3 \rightarrow H_2O + CO_2 \uparrow$$

Baking powder is used alone without adding leavening acid, because it is a mixture of baking soda and acid (such as cream of tartar, $KHC_4H_4O_6$, potassium hydrogen tartrate). When hydrated in cold liquid, the acid releases CO_2 from the soda. The reaction may be fast or slow, depending on how quickly carbon dioxide is generated.

A special type of baking powder is called "double-acting," for example, sodium aluminum sulfate (SAS) plus monocalcium phosphate monohydrate (MCP). Unlike a single-acting baking powder that reacts quickly with baking soda to produce carbon dioxide as soon as water is added, this double-acting powder (called SAS-phosphate) reacts twice. The first reaction releases carbon dioxide in the dough through reaction of MCP (acting as the acid) in water with baking soda. For the second reaction to take place, SAS must be converted into an acid in order to release carbon dioxide from baking soda. The SAS heated in the presence of water is converted into sulfuric acid, which reacts with baking soda to release carbon dioxide during the baking step.

TABLE 4.6 Average pH values of selected foods. *Foods having pH below 3 are considered very acidic, while those having pH values ranging from 4 to 6 are considered weakly acidic. (Adapted from Handbook of Chemistry and Physics, CRC Press.)*

Food	Average pH
Banana	4.6
Carrot	5.1
Cherry	3.6
Corn	6.3
Egg	7.8
Lemon	2.3
Milk	6.5
Orange	3.5
Potato	5.8
Wheat flour	6.0

4.6 FOOD ACIDITY

The concept of acidity is not only related to the sensory perception of sourness, but to a chemical effect called the pH of a substance. Foods and beverages differ in pH because of their content of acids, which produce hydrogen ions (see Table 4.6). These ions can be detected by a hydrogen ion sensitive electrode in a device called a pH meter.

pH and the pH Scale

The hydrogen ion concentration, expressed on a logarithmic scale, of the free or dissociated hydrogen ions in a product is termed **pH**. To calculate pH, the following relationship is used: $pH = -\log[H^+]$

What this means mathematically is that pH is equivalent to the negative logarithm (log) of the hydrogen ion concentration. A log is an exponent. To calculate pH, a basic familiarity with exponents is helpful. For example, the log (or exponent) of 10^{-3} is -3. To determine the log value of a nonexponent, such as the number 4.42, a log table or a calculator is required.

The relative number of H^+ ions and OH^- ions present in a solution is measured on the pH scale. This scale reflects the concentrations of these ions in solution. The pH scale extends from a value of 0 to an upper limit of 14 units. Foods with pH values much lower than 7 are *acidic*, those with pH values higher than 7 are basic or *alkaline*, and those with a pH of 7 are *neutral*.

Low values indicate high hydrogen ion concentrations (concentration is represented by brackets, in chemical notation), while high values indicate low $[H^+]$. Strong acids (e.g., stomach acid, hydrochloric acid) have very low pH, while very strong bases (e.g., ammonia, a waste product of protein metabolism) have high pH.

Titratable Acidity

In addition to its pH, a food's acidity can be described in another way. **Titratable acidity** is a measure of the total acidity in a sample, both as free hydrogen ions (dissociated) and as hydrogen ions still bound to acids (undissociated).

The techniques needed to measure pH and titratable acidity differ. Titratable acidity is measured by careful additions of a base of known concentration (typically sodium hydroxide, NaOH) to the sample until a predetermined end point is reached. The end point may be an indicator color change at a particular pH, like the color change at pH 8.1 for the indicator dye phenolphthalein.

In the dairy industry, measuring titratable acidity monitors the progress of fermentation. Fermentation, the breakdown of carbohydrates into sugars and acid, is essential for cheese and fermented milk quality. Titratable acidity is used to measure the amount of base required to neutralize the components of a given quantity of milk and milk products, and is expressed as a percentage of the dominant (or principal) acid, which is lactic acid. A maximum allowable titratable acidity for milk is 0.15 percent lactic acid.

How is this lactic acid in milk formed? Fresh, raw milk contains practically no lactic acid. Lactic acid ($CH_3-CHOH-COOH$) can be produced from the lactose sugar present in milk by bacterial fermentation. If bacteria are present in milk, there can be an undesirable souring of the taste due to the fermentation reaction:

$$\text{Lactose hydrolysis} \longrightarrow \text{glucose} \xrightarrow{\text{bacterial fermentation}} \text{lactic acid}$$

Controlling microbial activity in milk is the most critical function in dairy manufacturing (Table 4.7). Since fresh, raw milk is not free from bacteria and other microorganisms, it must be pasteurized—a heat treatment to kill microorganisms. Through pasteurization, milk is kept free of microbes, and the lactic acid they would otherwise generate.

TABLE 4.7 Milk souring: titratable acidity and pH values of milk. *As the pH of milk decreases due to lactic acid production, several changes occur including production of acid (sour) flavor and odor, the coagulation (clotting) of the milk protein casein, and an increase in titratable acidity.*

Titrate Acidity	pH	Comment
0.14–0.18	6.5–6.7	Normal fresh milk
0.25–0.27	6.0–6.2	Acid odor detectable
0.26–0.30	5.6–6.0	Acid flavor detectable
0.50–0.60	4.4–4.8	Coagulation of casein

TABLE 4.8 The principal acid composition of several fruits.

Fruit	Principal Acid	Typical Acid Percentage
Apple	malic	0.27–1.02
Grape	tartaric/malic (3:2)	0.84–1.16
Orange	citric	0.68–1.20

In contrast to the situation in milk, certain fruits and vegetables may contain two acids rather than one. Knowing the stage of maturity of fruit is important in the fruit industry, and determining titratable acidity provides this information because the relative predominance of the acids shifts during ripening and maturity. For instance, in grapes, malic acid predominates prior to maturity, whereas tartaric is the major acid in the ripe fruit (Table 4.8).

pH and Acid Foods

Foods are termed acid foods, acidified foods, low-acid foods, and fermented foods according to the following definitions. An **acid food** is one that has a natural pH of 4.6 or below. *Natural pH* means the pH prior to processing

Acidified foods are low-acid foods to which acids are added. Acidified foods have water activity values exceeding 0.85 and a final equilibrium pH of 4.6 or below. By contrast, **low-acid foods** are foods having an equilibrium pH greater than 4.6 and a water activity (a_w) greater than 0.85. *Equilibrium pH* is the condition achieved when the solid and liquid parts of the product have the same pH. Pickles and pickled foods are examples of acidified foods.

Pickles are also an example of a fermented food. **Fermented foods** are low-acid foods subjected to the action of certain microorganisms. The microorganisms produce acid during their growth and reduce the pH of the food to 4.6 or below. Partially fermented foods requiring addition of acid to reduce the pH to 4.6 or less are also considered acidified foods.

Key Points

- Chemical symbols, formulas, and equations are useful in describing chemical reactions in foods.
- Chemical bonds include the electron sharing of covalent bonds, the electron transfer of ionic bonds, and the unequal electron sharing of hydrogen bonds.
- Chemical reactions in foods affect food color, texture, aroma, and flavor quality.
- Food chemical reactions are classified as enzymatic if an enzyme speeds up or catalyzes the reaction or nonenzymatic if the reaction takes place without an enzyme. Some reactions (condensation, hydrolysis, oxidation-reduction) can be either.
- Chemical reactions in foods include addition, condensation, hydrolysis, oxidation-reduction, and polymerization.
- Functional groups are arrangements of just a few atoms that create particular properties of molecules.
- Functional groups include acid, alcohol, aldehyde, ester, ketone, methyl, phosphate, and sulfhydryl.
- Functional properties are the physical and chemical properties of food molecules that affect their behavior in foods, as determined by the functional groups present.
- Water functions in emulsions, acts as a plasticizer, and promotes heat transfer in foods.
- Food acids function in dough softening, act as chemical leavening agents, and contribute flavor.
- Food acidity can be expressed in terms of pH and titratable acidity.
- Acid foods, acidified foods, and low-acid foods, are distinguished by their natural pH or equilibrium pH.

Study Questions

True/False

1. T/F: According to the reaction $H_2CO_3 + NaOH \leftrightarrow NaHCO_3 + H_2O$, baking soda is the salt of carbonic acid (H_2CO_3) in which one atom of the metal element sodium has replaced one hydrogen.

2. T/F: In chemical bonding, there can be a sharing or a transfer of protons between reactants.

3. T/F: In functioning as a dough softener, fumaric acid acts as an oxidizing agent.

Multiple Choice

4. The neutralization of an acid by a base is most closely related to
   a. titration.
   b. polymerization.
   c. condensation.
   d. oxidation.
   e. reduction.

5. Which one is *not* a food acid?
   a. propionic
   b. ethyl acetate
   c. lactic
   d. phosphoric
   e. cream of tartar

6. Which of the following statements is correct?
   a. Acetic acid is an acid that can donate two hydrogens.
   b. Tartaric acid can donate three hydrogens.
   c. Fumaric acid is an acid that can donate two hydrogens.
   d. Citric acid is a C4 organic acid.
   e. Lactic acid is an alkene.

7. Which statement regarding carbon dioxide is *false?*
   a. It is an important leavening gas in baked products.
   b. Its chemical formula is CO_2.
   c. An atom of carbon dioxide contains two oxygens covalently bonded to carbon.
   d. Carbon dioxide does not act as an acid or a base.
   e. It is produced during fermentation reactions.

8. Which element from the periodic table is not yet considered an essential dietary mineral?
   a. Na
   b. Br
   c. Cl
   d. Ca
   e. Zn

9. CH_3COOH is the chemical formula for
   a. acetic acid.
   b. propionic acid.
   c. lactic acid.
   d. malic acid.
   e. fumaric acid.

10. The carbon-to-carbon bonds in food molecules such as carbohydrates, proteins, fats, and vitamins are examples of
    a. covalent bonds.
    b. intermolecular bonds.
    c. ionic bonds.
    d. hydrogen bonds.
    e. electron transfer bonds.

11. When amino acids bond to form peptides, this is an example of a(n) _____ reaction.
    a. hydrolysis
    b. composition
    c. decomposition
    d. oxidation
    e. reduction

12. What is required for an enzymatic reaction to go to completion?
    a. presence of substrate
    b. presence of enzyme
    c. decrease in activation energy
    d. formaton of ES complex
    e. all of these

13. When oxygen is added to, or hydrogens or electrons are removed from, food molecules in an enzymatic reaction,
    a. enzymatic reduction occurs.
    b. hydration occurs.
    c. enzymatic oxidation occurs.
    d. solvation occurs.
    e. polymerization occurs.

14. Which food molecule is most prone to addition reactions?
    a. hydrogen
    b. unsaturated fatty acid
    c. protein
    d. any having a C=H double bond
    e. carbohydrate

15. Ascorbic acid functions as _____ because it causes other substances to become reduced while it becomes oxidized in chemical reactions.
    a. oxidizing agent
    b. reducing agent
    c. antioxidant
    d. both (a) and (b)
    e. both (b) and (c)

16. What is *false* regarding functional groups?
    a. They are arrangements of between 6–12 atoms in a single structure.
    b. Esters, acids, and ketones contain C=O.
    c. The alcohol group is OH.
    d. The carboxylic acid group is COOH.
    e. Functional groups are responsible for functional properties.

17. Which is *not* a functional property of water?
    a. heat transfer
    b. solvent action
    c. fat replacer
    d. hygroscopicity
    e. plasticizer

18. What is *incorrect* regarding water activity?
    a. MSI plots show food water activity values versus moisture content.
    b. It is a measure of relative humidity.
    c. The greater the free water content of a food, the higher the water activity.
    d. The maximum value for water activity is 1.0.
    e. In general, as water activity increases, food stability increases.

19. Which is not a functional property of food acids?
    a. sourness
    b. promote gelation
    c. leavening system component
    d. gain protons
    e. antimicrobial

20. The following are all true regarding the ionization of fumaric acid, $C_4H_4O_4$, except:
    a. can result in the loss of two protons
    b. produces OH^- (hydroxide) ions
    c. produces the conjugate base $C_4H_3O_4^-$
    d. produces the conjugate base $C_4H_2O_4^-$
    e. is a dissociation reaction

21. In acting as a(n) _____ in dough, fumaric acid promotes the _____ of disulfide bonds.
    a. reducing agent; breakage
    b. oxidizing agent; breakage
    c. antioxidant; formation
    d. oxidizing agent; formation
    e. reducing agent; formation

22. What is the correct statement regarding baking soda and baking powder?
    a. Baking powder is sodium bicarbonate.
    b. SAS is an example of a baking soda.
    c. Baking powder does not require added acid in order to produce leavening.
    d. Baking soda reacts with baking powder to produce oxygen plus water.
    e. Baking soda is a leavening acid.

Critical Thinking

23. Draw the reaction showing the oxidation of a sulfhydryl group.

24. Draw the hydrogen bonds between ethanol and water, glucose and water, and a dipeptide (glycine-glycine) and water. Comment on the reason these substances do or do not exhibit solubility.

25. Why are ionic compounds like NaCl and covalent molecules like H_2O polar?

26. Can water (H_2O), acetic acid ($HC_2H_3O_2$), ammonia (NH_3), and hydrogen chloride (HCl) mix?

27. Which do you think is a better indicator of the stability of a food to deterioration (its shelf life): water activity or moisture content? Why?

28. People who make their own jellies, jams, and preserves say that the addition of lemon juice, before cooking the fruit, makes the final product sweeter. How can this be, since the lemon juice is sour?

29. Sodium aluminum sulfate (SAS) has the chemical formula $Na_2Al_2(SO_4)_4$. It reacts with six water molecules to produce two molecules of aluminum hydroxide $Al(OH)_3$, one molecule of sodium sulfate Na_2SO_4, and three molecules of sulfuric acid, H_2SO_4 in the presence of heat. Write out the complete chemical equation that shows this reaction.

30. What would it mean chemically if a milk sample had a titratable acidity of 0.40?

31. Adenosine triphosphate (ATP) is the important molecule in bioenergetics. Its formula is $C_{11}H_{18}O_{13}N_5P_3$. Chlorophyll, the key molecule in photosynthesis, has the formula $C_{55}H_{68}O_5N_4Mg$. Vitamin B_{12}, which is essential for health, has the formula $C_{63}H_{90}O_{14}N_{14}PCo$. What similarities and differences exist between these molecules based upon your examination of their formulas?

32. A compound that yields ions other than H^+ ions or OH^- ions is called a salt. Are the following reactants (in bold) acids, bases, or salts?

    $Na_2CO_3 \rightarrow 2\,Na^+ + CO_3^{-2}$

    $H_2PO_4^- \rightarrow H^+ + PO_4^{-2}$

    $Ca(OH)_2 + 2\,H^+ \rightarrow Ca^{2+} + 2\,H_2O$

33. The chemical form produced when a proton (H^+) leaves an acid is known as the *conjugate base* of that acid. For the reaction $HA + B^- \rightarrow A^- + HG$, HA is the acid, B^- is the base, A^- is the conjugate base, HB is the conjugate acid. Predominant acids ionize to produce hydrogen ions plus the conjugate base of the acid. The chemical names and formulas present in four titration curve buffer zones are given below. Identify the most likely sample source (e.g., milk) for each acid, and the name of the conjugate base, corresponding to the four acids listed below.

    a. **lactic acid**, $HC_3H_5O_3$
       source ? _____
       conjugate base? _____
    b. **citric acid**, $H_3C_6H_5O_7$
       source ? _____
       conjugate base? _____
    c. **tartaric acid**, $H_2C_4H_4O_6$
       source ? _____
       conjugate base? _____
    d. **acetic acid**, $HC_2H_3O_2$
       source ? _____
       conjugate base? _____

References

Best, D. 1992. New perspectives on water's role in formulation. *Prepared Foods* 161(9):59.

Coultate, T. P. 1996. *Food: The Chemistry of Its Components*. Royal Society of Chemistry, Cambridge, UK.

DeMan, J. 1999. *Principles of Food Chemistry*. Aspen Publishers, Gaithersburg, MD.

Dziezak, J. D. 1990. Acidulants: ingredients that do more than meet the acid test. *Food Technology* 44(1): 76.

Fennema, O. R. 1996. *Food Chemistry*. Marcel Dekker, New York.

Miller, D. 1998. *Food Chemistry: A Laboratory Manual*. John Wiley & Sons, New York.

Mitchell, J. R. 1998. Water and food macromolecules. In: *Functional Properties of Food Macromolecules*. Aspen Publishers, Gaithersburg, MD.

Nielsen, S. 1998. *Food Analysis*. Aspen Publishers, Gaithersburg, MD.

O'Donnell, C. D. 1992. How to smooth sourness. *Prepared Foods* 161(5):107.

Woo, A. 1998. Use of organic acids in confectionery. *Manufacturing Confectioner* 78(8):63.

Additional Resources

Baum, S. J. and Hill, J. W. 1993. *Introduction to Organic and Biological Chemistry*. Macmillan Publishing Company, New York.

Brown, A. 2000. *Understanding Food: Principles and Preparation*. Wadsworth/ Thomson Learning, Belmont, CA.

Chen, P. S. 1979. *Inorganic, Organic, and Biological Chemistry*. Barnes & Noble Books, New York.

Hill, J. W., Feigel, D. M., and Baum, S. J. 1993. *Chemistry and Life: An Introduction to General, Organic, and Biological Chemistry*. Macmillan Publishing Company, New York.

Lehninger, A. L., Nelson, D. L., and Cox, M. M. 1993. *Principles of Biochemistry*. Worth Publishers, New York.

McWilliams, M. 1997. *Foods: Experimental Perspectives*. Prentice Hall, Upper Saddle River, NJ.

Mehas, K. Y., and Rodgers, S. L. 1997. *Food Science: The Biochemistry of Food and Nutrition*. Glencoe/McGraw-Hill, New York.

Oulette, R. J. 1994. *Organic Chemistry: A Brief Introduction*. Macmillan Publishing Company, New York.

Roos, Y. H. 1998. Role of water in phase-transition phenomena in foods. In: *Phase/State Transitions in Foods: Chemical, Structural, and Rheological Changes*. R. W. Hartel and M. A. Rao, eds. Marcel Dekker, New York.

Sackheim, G. I. 1991. *Introduction to Chemistry for Biology Students*. Benjamin/Cummings Publishing Company, Redwood City, CA.

Web Sites

http://www.ift.org/divisions/food_chem/
IFT Food Chemistry Division. This site has the composition and properties of food and food ingredients, and the chemical, physical, and biological changes in food constituents as a result of preparation, processing, and storage.

FOOD SYSTEMS

Food is matter—it's both as simple as that and as complicated as that. Matter can be divided into two general categories: homogeneous matter and heterogeneous matter. *Homogeneous* matter is of uniform composition throughout. In foods that are homogeneous, individual components cannot be visually discerned. Solutions, such as some beverages, are examples of *homogeneous matter*, in which one substance is dissolved into another. In heterogeneous matter, individual components can be visually discerned, and may be distributed unevenly.

Foods are mixtures composed of solid, liquid, and gas phases that are dispersed. The chemical properties, physical properties, and the behavior of food systems during processing and storage depend upon their phase makeup.

A **food system** is a dispersion containing two phases: a continuous phase and a dispersed phase. In a typical food system, solids, liquids, and gases are dispersed within other solids, liquids, and gases. For example, carbonated soda is a liquid (water) that contains dissolved solids (sugars and flavorings) and dissolved carbon dioxide gas. Another example is margarine, a water-in-oil type of emulsion in which water droplets constitute the dispersed phase within a continuous oil phase.

Food System Stability and Texture

The types and relative stability to separation of the phases within food systems are of utmost importance to food scientists. The functional properties of food macromolecules and ingredients in such systems affect their *physical stability*. Key ingredients, called emulsifiers and stabilizers, are often added to food systems to maintain phase stability. For a product like a frankfurter (a meat emulsion food system), such ingredients prevent the separation of the fat from the product during storage.

Texture is another quality characteristic of major importance influenced by the functional properties of food macromolecules and ingredients within food systems. In solid foods like butter and cheese, and semisolid foods like bread, mayonnaise, and pudding, ingredients and food molecules interact at the molecular level. Interacting forces involving food and ingredient molecules establish important chemical and physical equilibria, and result in a food's texture.

Key to understanding food systems is the notion of particle size. One nanometer (symbol nm) is 10^{-9} meter, or 0.000000001 meter, and one micrometer (or micron, symbol μ) is 10^{-6} meter, or 0.000001 meter. The three basic dispersion types based upon particle size are *solutions* (particles less than 1 nm in diameter), *colloidal dispersions* (particles 10-100 nm in diameter), and *suspensions* (large particles that remain suspended in the continuous phase).

Solutions

Solutions are homogeneous mixtures in which one substance (the *solute*) is dissolved in another (the *solvent*). In theory, solvents can be solids, liquids, or gases, but in foods, they are almost always liquid (water). Solutes may be solids (salt, sugar, vitamin C, or any solid particles small enough to dissolve), liquids (ethanol in alcoholic beverages), or gases (carbon dioxide in a carbonated soda). A solution is a special example of a mixture displaying both uniformity and transparency characteristics. Common food solutions are coffee, tea, and various cold beverages.

One of the key characteristics of a solution is the solubility of its solute particles. **Solubility** is the maximum amount of solute that dissolves in a specified volume of solvent at a specified temperature. Temperature has a significant effect on solubility, and the effect depends on whether the system consists of a gas solute or a solid solute. For example, an opened can of soda (carbonated with CO_2 gas) goes flat quicker if it is left out at room temperature compared with refrigeration temperature. This happens because the solubility of the CO_2 gas *decreases* at higher temperature, allowing it to escape from the soda, which as a result loses its "fizz."

On the other hand, the solubility of most solids *increases* when temperature is raised. A glass of iced tea to which sugar is added will reach a point of saturation, beyond which additional sugar will not dissolve, but will instead settle at the bottom. Heating this supersaturated iced tea solution will dissolve the excess sugar, in effect converting the supersaturated solution into an unsaturated one. The process of dissolving specific amounts of sugar through heating and then cooling to a solid is in fact the basis of crystalline candymaking.

Food Colloids

Food colloids are surface active ingredients such as fatty acids, glycerides, phospholipids, polysaccharides, and proteins. In terms of size, a colloid is a particle that is too large to dissolve and become the dispersed phase of a true solution. Instead, colloids become components of colloidal dispersions. Colloid particles may exist as charged particles (ions), as molecules, or as clusters of these, called *aggregates*. The particles in a colloidal dispersion do not settle out as they would in a suspension.

Colloidal dispersions include emulsions, foams, gels, and sols—see Table 4.9. Examples of food colloidal dispersions include any solid protein dispersions in liquid, such as the milk protein casein, the egg white protein albumen, and the protein in gelatin. Nonprotein substances that are colloidally dispersed in foods include the carbohydrate pectins and the minerals calcium and magnesium in milk.

Emulsions

An **emulsion** is a colloidal dispersion of two liquids, usually oil and water, that are immiscible (not mixable). If oil is dispersed in water, the emulsion is called an *oil-in-water emulsion (O/W)*. If water is dispersed in oil, the emulsion is called a *water-in-oil emulsion (W/O)*. Emulsions are opaque to look at, because the particle size in the dispersed phase is large enough to deflect rays of visible light. Many food products, including homogenized milk, cream, ice cream, butter, mayonnaise and even some meats (e.g., frankfurters before they are cased and processed) are emulsions.

Butter is a W/O emulsion that contains two liquid phases that are allowed to solidify. In the process of churning, most of the water is excluded from the final product. By law, butter must contain 80 percent butterfat by weight; approximately 16 percent by weight is water. Margarine, which uses vegetable oil instead of butterfat, is a W/O emul-

TABLE 4.9 Food colloidal systems. *The systems differ according to the state of matter in the dispersed and continuous phases. G = gas, L = liquid, S = solid. (Adapted from McWilliams, 1997.)*

Colloidal Dispersion	Dispersed/Continuous Phase	Example
Emulsion	L/L	salad dressing
Foam	G/L	egg white foam; whipped topping
Suspensoid	G/S	marshmallow
Gel	L/S	cooked egg custard
Sol	S/L	gravy

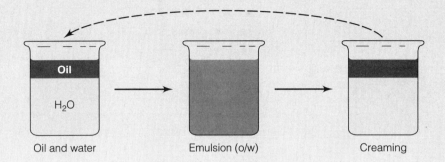

FIGURE 4.19 Creaming. *If an O/W emulsion is allowed to stand, oil droplets coalesce (combine) and rise to the top (creaming).*

sion also. Some imitation butter products are more than 50 percent water, and since they are not made from animal fat, contain no cholesterol or saturated fat.

Raw, unprocessed milk is an unstable (O/W) emulsion because it contains large fat globules that clump together and rise to layer on the top of the water phase. This effect is known as **creaming** (Figure 4.19). Milk from the supermarket is *homogenized*, a process that keeps the emulsion stable and prevents creaming. As a finished product, ice cream is an emulsion, yet its key ingredient (cream) is obtained by the creaming process—allowing the fat in milk to separate from the watery portion.

Emulsion Stability: Interactions Are Key. It is important in food science to develop an understanding of the interactions that take place in food systems, such as emulsions. Interactions include solvent:solvent (solvent-to-solvent), solute:solute (solute-to-solute), and solute:solvent (solute-to-solvent) molecular interactions. Depending upon which predominates, such interactions can either promote stability or destabilize a food system. In practice, many foods represent highly complex systems. Therefore, a basic oil-in-water emulsion example will be simplest to examine in terms of the key stabilizing interactions.

Although water and oil droplets do not mix in an O/W emulsion, it is incorrect to assume that they repel each other the way opposite magnet poles do. Individual oil molecules are actually attracted to water molecules by a force that is much greater than the attraction of two oil molecules to each other. You can observe this: A single drop of oil on a clean surface of water spreads out to form a thin monolayer, rather than balling up with only a small surface area in contact with the water. The spreading is a consequence of the attractions between the oil and water molecules, which are stronger than the oil:oil attraction that would produce a thick oil phase on top of the water.

However, the net cost of energy in putting all the oil molecules into a water solution is too large, because the force of water-to-water attractions is greater than water-to-oil attractions. The result is merging of oil droplets until separation of phases is achieved. Thus, emulsions are thermodynamically unstable food systems. However, with addition of surface active molecules (surfactants or emulsifiers) or thickening agents, the phases can be stabilized. This stabilization is very important in foods such as cream, egg yolk, gravy, ice cream, mayonnaise, milk, salad dressings, and sauces.

Emulsifiers. Emulsifiers such as monoglycerides, phospholipids, and sorbitan monostearates are **amphiphilic molecules,** which means they contain hydrophilic and hydrophobic regions in their structure. (Emulsifiers work by attaching to the surface of droplets in an emulsion.) The spatial orientation of the emulsifier molecules as they surround dispersed droplets matches the hydrophilic and hydrophobic portions to the respective emulsion phases—see Figure 4.20.

Thickening agents, although they help keep the phases in an emulsion from separating, are not the same as emulsifiers. Examples of thickening agents are polysaccharides, such as food gums and starches. Such "stabilizer" ingredients act to increase the viscosity of the continuous phase of an emulsion, which enhances emulsion stability by inhibiting droplet merging and phase separation.

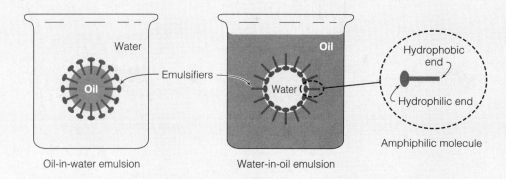

FIGURE 4.20 Emulsifiers—amphiphilic substances that form monomolecular layers between immiscible oil and water phases. *In an emulsion, the presence of two phases creates a surface tension at the interface between them. Surface tension refers to forces that cause two dissimilar liquids to separate from each other. Emulsifiers have the ability to reduce surface tension as a result of being amphiphilic. At oil-water interfaces they adopt an orientation in which the hydrophilic part is directed to the water while the hydrophobic part is directed to the oil. This minimizes the contact area between -phobic and -philic phases and reduces surface tension.*

Foams

Another example of a colloidal dispersion is a **foam**. What is distinctive about a foam is that the dispersed phase is a *gas*, within a liquid continuous phase. Foams are not permanently stable and are termed *metastable*. Over time, a foam achieves equilibrium under the force of gravity through the process of drainage of liquid from the foam. Surfactants (surface active agents) and proteins inhibit drainage by different mechanisms.

In whipped cream, air is incorporated (whipped) into cream, and the milk protein in the system acts as an emulsifier to trap the tiny air "cells." The high viscosity (thickness) and low surface tension of the liquid phase of this system favors a stable foam formation.

An example of a more dry food foam (termed a *suspensoid*, a gas phase dispersed in a semisolid) is aerated confectionery, like marshmallows. In such a product, a multifunctional ingredient such as gelatin is used. In addition to creating and setting the foam structure due to its lowering of surface tension and its gelling abililty, gelatin also maintains foam stability by increasing viscosity and prevents crystallization of sucrose. These effects all combine to provide a smooth texture and stable shelf life for aerated confectionery products.

Gels and Sols

Another special type of colloidal dispersion is a **gel:** a two-phase system in which a liquid is dispersed in a solid. Colloidal gels form when colloid molecules or particles associate in a liquid such that the solvent becomes immobile.

Many food gels are composed of concentrated polymer solutions that form three-dimensional networks. Gels are quite able to imbibe water that diffuses into polymer networks, where it becomes entrapped. Although the polymers themselves cannot diffuse out from a strong gel, if the gel breaks, a situation common to weak gels, polymers can move out into solution.

The formation of many types of food gels is usually based upon the functionality of either a polysaccharide or protein gelling agent. Polysaccharides such as gums, pectin, and starches are examples of plant-derived gelling agents. Legume proteins, such as those in soybeans, can be coagulated by heat, acid, or enzymes prior to setting into a tofu gel. Salt-soluble meat proteins (myofibrillar proteins) are able to form gels. Other animal proteins such as gelatin, as well as nonmeat proteins including casein in cheese and egg proteins, have the ability to form gels.

In a sense, the opposite of a gel is a **sol:** a solid dispersed in a liquid. Gravy is an example of a sol, which although a somewhat viscous food system, does not form a gel and is pourable. On the other hand, a starch suspension in water becomes a sol when heated due to the process of gelatinization. Upon cooling, a gelatinized starch sol converts into a gel. A starch gel therefore is a thickened sol that is no longer pourable because it has developed a solid continuous phase.

Challenge Questions

1. Give an example of one protein and two nonprotein substances that can form colloidal dispersions.
2. Describe the differences between a solution, an emulsion, a mixture, and a colloidal dispersion.
3. Discuss the specific type of liquid-in-liquid colloidal dispersion that is represented by (a) raw unprocessed milk, and (b) margarine.
4. Explain what is meant by the term *amphiphile*.
5. Provide examples of how two key quality characteristics are influenced by the functional properties of food molecules or ingredients within specific food systems.
6. Suppose you are developing a blended olive oil plus fish oil salad dressing in vinegar. What specific forces must be overcome by an emulsifier and/or stabilizer to promote emulsion stability?

CHAPTER 5

Food Chemistry II: Carbohydrates, Lipids, Proteins

5.1 Food Carbohydrates

5.2 Food Lipids

5.3 Food Proteins

Challenge!
Milk Protein Chemistry

CHAPTER OBJECTIVES

After completing this chapter, you will be able to:
- Identify important food sugars, the chemical reactions they participate in, and their functional properties.
- Describe the structure and functional properties of food polysaccharides including pectin, starch, and vegetable gums.
- Distinguish between the classes of lipid molecules and the chemical differences of fatty acids.
- State the important functional properties of food lipids, including aeration, crystallization, heat transfer, and mouthfeel.
- Describe the structure of food proteins, and list their functional properties.
- Explain the relationship between isoelectric point and protein functionality.
- Describe the composition of the casein micelle and the functional role of the alpha-, beta-, and kappa-casein polypeptides in the micelle.

Why does freshly baked bread smell so good? Why do fats get rancid? The answer: chemistry. The chemistry of food carbohydrate, lipid, and protein molecules affects their behavior in food systems during processing and storage, and chemistry contributes to food flavor, color, and texture. These macronutrient molecules interact with each other, and with acids, oxygen, vitamins, minerals, and water to create both desirable and undesirable effects. Understanding the functional properties of food carbohydrates, lipids, and proteins is critical in order to effectively formulate new food products, improve existing ones, and control processing costs. In addition, some value-added products rely on the unique functionalities of carbohydrate, lipid, and protein ingredients.

5.1 FOOD CARBOHYDRATES

Simple and complex carbohydrates were discussed in Section 3.2 in relation to nutrition. In this chapter we focus on their chemistry and the consequences of that chemistry.

The Structures of Sugars

All carbohydrates contain the elements carbon, hydrogen, and oxygen, and the basic building block of carbohydrates is the "simple" sugar. Some of the simple sugars such as fructose (fruit sugar) and glucose are single carbohydrate units (monosaccharides). Others, such as sucrose (table sugar) and lactose (milk sugar) are composed of two carbohydrate units joined together, forming a disaccharide. A simple sugar is also termed an **organic alcohol,** since it is a molecule that contains carbon atoms attached to —OH (alcohol) groups.

Remember from Chapter 3 that carbohydrates are classified on the basis of the number of sugar units they possess. In foods, carbohydrates exist as monosaccharides, disaccharides, oligosaccharides, or polysaccharides. Monosaccharides are single sugar molecules that are not chemically bonded to other sugar molecules.

Monosaccharides. Monosaccharides that contain 3 carbon atoms in their structure are called trioses; those with 5 carbon atoms are called pentoses, while those that contain 6 are called hexoses. The important monosaccharides found in foods are the hexoses glucose, fructose, and galactose. These share the same chemical formula, $C_6H_{12}O_6$. See Figure 5.1

Although the hexoses may have identical chemical formulas, slight differences in the location of the functional groups cause differences in functional properties, including sweetness and solubility. Fructose, which exists as a five-membered ring structure, is sweeter and more soluble than glucose, which exists as a six-membered ring structure.

Glucose is the most common monosaccharide in foods. Grapes, berries, and oranges contain 20 to 30 percent glucose (sometimes called dextrose). Glucose is considered an aldose because its carbonyl group (—C=O) at carbon atom 1 is in the form of an

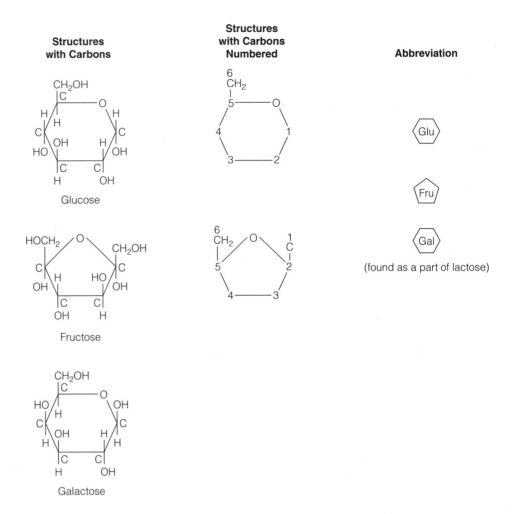

FIGURE 5.1 Structures of simple sugar monosaccharides glucose, fructose, and galactose.

aldehyde. Fructose is considered a ketose because its carbonyl group at carbon atom 2 is in the form of a ketone. Galactose is the only other food monosaccharide and occurs in the disaccharide lactose.

Monosaccharides exist in more than one structural form, as so-called straight chain or Fischer projection molecules and ring or cyclic Haworth projection form. In solution, the ring form predominates because the functional groups at carbon 5 (alcohol group) and carbon 1 (aldehyde group) undergo an intramolecular *cyclization* reaction which joins C1 and C5 to form a closed ring (as shown in Figure 5.1). Ring formation eliminates the presence of the carbonyl group, leaving instead an OH group.

Disaccharides. Monosaccharides are the basic units around which larger carbohydrates are built. Two monosaccharides joined together form a disaccharide, and the bond in the structure of a disaccharide is a *glycosidic bond*. The three important food disaccharides are sucrose, lactose, and maltose (Figure 5.2).

Sucrose, composed of glucose and fructose, is common table sugar and is widely used in food preparation in its crystalline form. In sucrose, glucose and fructose are linked between the aldehyde group of the carbon 1 atom of glucose and the ketone group of carbon 2 atom of fructose. Lactose, composed of glucose and galactose, is found only in milk and dairy products. Maltose is composed of two glucose units.

The Functional Properties of Sugars

Sugar molecules contain two important, reactive functional groups: the $-C=O$ carbonyl group and the $-OH$ alcohol group. The alcohol group is important for solubility and sweetness, while the carbonyl group is important for reducing activity and the

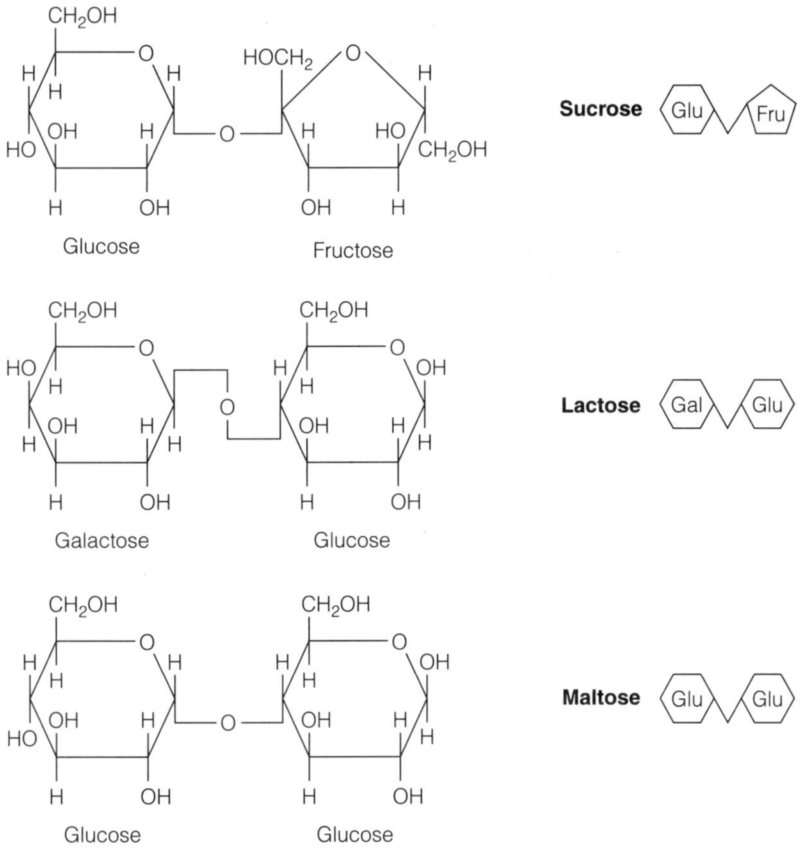

FIGURE 5.2 Structures of sugar disaccharides sucrose, lactose, and maltose.

Maillard browning reaction, which can cause color and flavor changes. Table 5.1 lists the functional properties of sugars that are discussed in this section.

The alcohol group is present in multiple locations on sugar molecules. For instance, glucose has five OH groups. Alcohol groups readily form hydrogen bonds with water molecules, making glucose molecules very water soluble. Other sugars are either more or less soluble than glucose, due to factors not readily attributable to the location or numbers of alcohol groups alone. For example, molecular size and molecular weight affect a sugar molecule's affinity for water, as does its crystallization behavior (i.e., the crystal lattice form that the solid sugar molecules exist in).

Reducing Sugars. Sugars that contain the aldehyde or ketone carbonyl group are called **reducing sugars**. By definition, reducing sugars react with other substances through oxidation-reduction chemistry to produce a reduced substance plus the oxidized sugar molecule. Thus, reducing sugars act as reducing agents. All monosaccharides and certain disaccharides are reducing sugars.

The term dextrose equivalent (DE) relates to solubility, reducing action, viscosity, and many other properties. **Dextrose equivalent** is a measure of the percentage of glycosidic bonds hydrolyzed in disaccharides and polysaccharides, which indicates the level of reducing sugar present. Pure dextrose has a DE value of 100, while starch, a glucose polymer, has a value of zero. In general, the higher the DE, the more soluble and the greater the reducing ability of a sugar.

Browning. Two important browning reactions can occur in sugars: Maillard browning and caramelization. Reducing sugars react with the amino acids, through the **Maillard reaction,** to produce brown color pigments in foods. **Maillard browning** is the browning of foods as a result of the Maillard reaction, and the brown pigments that form are called *melanoidins*. Unwanted brown colors and off-odors can develop during extended storage of such foods as baked products, dried egg whites, instant mashed potatoes, and certain food ingredients. However, not everything about the Maillard reaction is negative.

TABLE 5.1 Functional properties of sugars.

Browning
Crystallization
Humectancy/hygroscopicity
   (moisture properties)
Inversion
Sweetness
Texturizing

The pleasant aroma and browned surfaces of baked bread and other baked goods, the development of chocolate flavor, and the production of flavors in roast beef and other cooked meats is due to this reaction.

The Maillard reaction, known as nonenzymatic browning because enzymes are not part of the reaction, is a complex sequence of chemical reactions. For the sequence to begin, a sugar molecule having a free carbonyl group (C=O) and an amino group must be present in the reaction mixture. Sucrose, a nonreducing sugar, does not participate, unless it is hydrolyzed to glucose and fructose.

The Maillard reaction can be viewed as a sequence of three chemical reactions: condensation, rearrangement, and polymerization.

1. Condensation:

$$\text{reducing sugar} + \text{amino group} \leftrightarrow \text{glycosylamine}$$

During the initial condensation, it is important to picture the location of the reactive site on the sugar molecule. For an aldose sugar (glucose) it is C1; for a ketose sugar (fructose) it is C2.

2. Rearrangement:

$$\text{glycosylamine} \leftrightarrow \text{Amadori compounds (colorless)} \rightarrow \text{pyrazines}$$

Intermediate reactions with the potential for continued rearrangements are more complex in scope. In general, color and flavor development begins at this stage. Degradation of Amadori compounds favors formation of *furfurals*, cyclic aldehydes in the form of a 5-membered ether ring. Perhaps the most desired compounds produced during intermediate reactions are pyrazines. These have a pleasant aroma and are generated, for instance, in roasting coffee and baking bread. Controlled processing can be employed to make sure that the Maillard reaction does not advance to the stage where negative food efffects occur. This fact is exploited in, for example, the roasting of coffee beans and cacao beans (in cocoa manufacture). In addition, the Maillard reaction can be used to great advantage in developing process and reaction flavors (Chapter 6).

3. Polymerization:

$$\text{colorless intermediate compounds} \leftrightarrow \text{brown melanoidins}$$

The late-stage polymerization reactions generate the large molecular weight melanoidins. The polymerizations are irreversible and eventually result in significant darkening of food. Unpleasant taste and unpleasant aroma also develop, and there is excessive product moisture loss.

Figure 5.3 summarizes this sequence of events in the Maillard reaction. The Maillard reaction is influenced by chemical and physical factors. Amino acids having "extra" amino groups in their side chain structures, such as lysine, are better substrates in the initial Maillard reaction than others. Maillard browning is also accelerated by lower molecular weight sugars (reactivity: pentoses > hexoses > reducing disaccharides). Free amino acids are more reactive than peptides. Low product moisture (15% maximum) and pH in the 5–8 range also favor the reaction.

Caramelization is the formation of brown caramel pigments as a result of applying heat energy to sugars. The temperature required is approximately 200°C. Protein material, including enzymes, is not required for caramelization. Many of the flavors and colors in sugar-containing foods that are a result of caramelization are considered highly desirable. *Caramelen*, $C_{24}H_{36}O_{18}$, is the name of a caramel pigment that can undergo fragmentation and dehydration to form acids and other compounds that impart flavors.

The terminology related to the word *caramel* can be confusing because no single compound is identified as *caramel*. Caramel coloring refers to a brown color additive used in the food industry, while caramels refers to sauces or candies made from carbohydrates that have been allowed to caramelize. The color used in cola beverages is created by caramelizing sucrose in the presence of ammonium bisulfite.

> *The chemical reactivity of sugars: (1) reducing action and (2) oxidation and reduction affect functional properties like browning and sweetness.*

1. Condensation:
reducing sugars + amine ⟶ **glycosylamine**

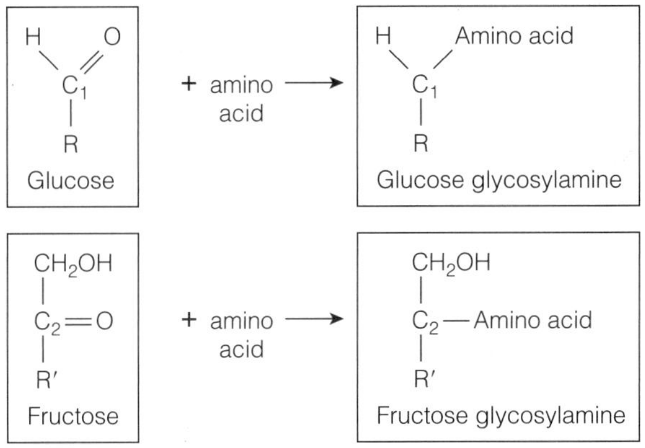

Glucose + amino acid ⟶ Glucose glycosylamine

Fructose + amino acid ⟶ Fructose glycosylamine

R and R′ are the rest of each sugar's structure

2. Rearrangement:
glycosylamine ⟶ colorless **intermediate compounds**

3. Polymerization:
colorless intermediate compounds ⟶ brown **melanoidin polymers**

FIGURE 5.3 The Maillard Reaction—the reaction of sugars with amino acids or proteins. *The presence of a reactive carbonyl functional group ($C=O$) from reducing sugars and the amino group from amino acids is a prerequisite for the reaction. Notice how the location of the carbonyl on glucose differs from fructose. Factors that enhance the reaction, in addition to substrates, include increasing the temperature, increasing the pH, and lowering the water activity of the food system.*

Crystallization. Sugars can exist in both soluble (as syrup) and crystalline states, which is one of their important characteristics. The formation of a crystalline structure, or **crystallization**, implies organized three-dimensional arrays of unit cells into a solid form. Crystallization depends on factors such as the moisture, temperature, and concentration of the sugar in a food system. A **crystal** is a solid made up of units in a repeating pattern. When sugars are purified from the initial plant material in commercial manufacture, crystallization is the key step. Sugars crystallize in two stages. The first stage is transfer of the sugar molecule to the surface of a crystal, and the second is the incorporation of the sugar into the crystalline structure.

In foods that contain sugars dispersed throughout the product in the soluble state, it is often an undesirable event when those sugars later recrystallize. An example is the formation of large lactose crystals in dairy products such as ice cream or sweetened condensed milk, causing the texture to be less smooth and more gritty, a process termed *graining.*

However, in the manufacture of a hard candy, crystallization is critical to the finished product quality. Sugars melt upon heating and recrystallize when cooled into a different crystal form and size than initially present. The best crystal size for hard candies is a small, fine crystal rather than a large, coarse crystal. Small crystal size is obtained by applying the proper time, temperature, and sugar concentration in the manufacturing procedure.

In hard candies, the sugar source also affects the grain of the finished product. Sugar crystals from sucrose are too large for producing good quality hard candy—the resulting product is not smooth or uniform. Glucose and fructose form smaller crystals and are suitable for candymaking. The food industry makes use of various agents, such as acids, to maintain the small, desirable crystal size. (See Table 5.2.)

Humectancy. The term **humectant** refers to a substance that has an affinity for moisture. (This is the same as saying that humectant substances are hygroscopic, since hygroscopicity refers to the ability to attract and retain moisture.) Carbohydrates in general, and sugars in particular, are used as humectant ingredients in the food industry. As such, they are effective in influencing the state of water in food systems. For instance, fructose, which is much more hygroscopic than sucrose, acts to reduce the water activity in a food system.

Humectants are able to hydrogen bond with water molecules, making water less available for microbial growth. Sugars play this role in food preservation. On the other

TABLE 5.2 Additives used in the manufacture of hard candy and the physicochemical basis of their functionality.

Agent	Basis of Functionality
Corn syrup	High in glucose content
Butter	Fat imparts smooth texture
Egg white	Inhibits large crystal formation
Cream of tartar	Promotes hydrolysis of sucrose
Vinegar	Promotes hydrolysis of sucrose

TABLE 5.3 Relative sweetness of sugars on a weight basis.
Fructose > Sucrose > Glucose > Maltose > Galactose > Lactose

hand, from the processing standpoint, if a sugar is to be added to an ingredient base that must remain free-flowing during manufacture, sucrose, because it is less hygroscopic, would be a better choice than fructose.

Inversion. The hydrolysis of sucrose to its component monosaccharides is carried out if a sweeter product is desired than when sucrose is present alone in a product. The mixture of the two monosaccharide endproducts, called **invert sugar,** is typically created in food products through the deliberate application of the enzyme invertase.

Oxidation and Reduction. Oxidation of the R—COH (aldehyde) group in sugars causes a loss of sweetness and coverts the aldehyde to an acid (HO—C=O) group:

$$\text{glucose (R—COH)} + \text{oxygen} \rightarrow \text{glucuronic acid (R—COOH)}$$

Reduction of the —C=O (carbonyl) group of reducing sugars causes formation of sugar alcohols, and these are moderately sweet:

$$\text{glucose} + \text{hydrogen} \rightarrow \text{sorbitol}$$
$$\text{fructose} + \text{hydrogen} \rightarrow \text{mannitol}$$
$$\text{maltose} + \text{hydrogen} \rightarrow \text{maltitol}$$

The use of sugar alcohols as alternative sweeteners was discussed in Section 3.4.

Sweetness and Texturizing. Sweeteners (referring specifically to simple sugars) contribute to sweet taste. However, each monosaccharide and disaccharide differs in sweetness. The relative sweetness of sugars is always in comparison to sucrose. For example, fructose is the most sweet while lactose is the least (Table 5.3).

Sugars also function to control the amount of water available as free, adsorbed, and bound water. This in turn affects food texture, shelf life, and microbial growth. Sugars bind to water by hydrogen bonds to their OH groups.

Competition between sugars and other substances in foods for water, and their direct interaction with molecules like starch, affect food texture. Starch gelatinization is delayed due to sugars. The effect is to reduce the viscosity and gel strength of starch-thickened mixtures like puddings. Sugars also act as texture tenderizers. In cakes, the presence of a critical amount of sugar creates a desirable soft texture and full volume.

Texture is discussed in detail in Chapter 6, and the Chapter 6 Challenge discusses the chemistry of sweeteners and sweetness.

Polysaccharides and Their Functional Properties

Complex carbohydrates are so named because of the size and complexity of their structures. These large carbohydrate structures are composed of joined sugar units and come in two sizes: oligosaccharides and polysaccharides. Complex carbohydrates of 10 or fewer (typically nonreducing) sugar units are called **oligosaccharides.** Two oligosaccharides, raffinose and stachyose, are present in dried beans, soybeans, peas, and lentils. These sugars, though poorly digested and absorbed by humans, are fermented in the large intestine by bacteria, which produce gases, and can result in flatulence.

Polysaccharide molecules are usually at least 40 or more sugar units in size, potentially comprising hundreds or more. Such molecules, for example, the starch fractions amylose and amylopectin, often contain reducing ends as well as nonreducing areas in their structures. This structural feature is important because nonreducing ends of carbohydrates are susceptible to enzyme attack. The use of enzymes in the food industry includes modifying starches to enhance their functionality in foods.

TABLE 5.4 Summary of functional properties of polysaccharides.

Functional Property	Polysaccharides
Bulking	Inulin, maltodextrins
Emulsion stabilizer	Inulin, vegetable gums
Fat replacer	Beta-glucans, cellulose, dextrins, maltodextrins, vegetable gums
Gelation	Inulin, pectins, starches, vegetable gums
Thickening/viscosity	Starches, vegetable gums
Water binding	Inulin, vegetable gums
Promotion of growth of probiotics	Fructooligosaccharides

Sugars can exist in two structural forms, called α and β. The forms differ on whether the OH group attached to the first carbon atom projects downward (α) or upward (β). When two monosaccharides are joined by a glycosidic bond, the linkage is α if the OH of the first cabon atom of the sugar unit is pointing down. This is the sugar unit whose first carbon atom is not part of the glycosidic bond. Maltose serves as an example:

α-maltose   β-maltose

Polysaccharides occur much more frequently than do oligosaccharides in raw plant foods and in manufactured foods employing them as ingredients. The important food polysaccharides include beta-glucans, cellulose, dextrins, fructooligosaccharides, maltodextrins, pectic substances, starch, and vegetable gums. Table 5.4 summarizes the functional properties of polysaccharides, and the substances are discussed below.

Beta-glucans. **Beta-glucans** are polysaccharides of glucose similar to cellulose, but less linear, occurring in oats, barley, and yeast. A variety of beta-glucan substances have been studied for their ability to lower cholesterol and activate the macrophage cells of the immune system. Beta-1,3-glucan is derived from the cell wall of common baker's yeast and is actually a complex of beta-1,3-glucan (about 70%) and other branched glucan linkages.

A water-soluble form of enzyme-treated oat flour containing beta-glucan soluble fiber, called Oatrim, can be used as a fat replacer and texturizing ingredient. This product offers reduced calories (1–4 kcal/gram) as a fat replacer in baked goods, fillings and frostings, frozen desserts, dairy beverages, cheese, salad dressings, processed meats, and confections.

Fibers, a diverse group, differ in structure. Most are polysaccharides—see Table 5.5.

Cellulose. The most abundant of all carbohydrate polymers, cellulose composes plant cell wall material. It consists of linear chains of glucose joined by β-1,4 glycosidic bonds. Although similar to amylose, the nearly insignificant structural difference accounts for the strength of cellulose fibers, making it impossible for human digestive enzymes to hydrolyze them.

A noncaloric microparticulate form of cellulose is available in the food industry that, when dispersed, forms a network of particles with mouthfeel and flow properties similar to fat. Therefore cellulose can replace some or all of the fat in foods such as dairy-type products, sauces, frozen desserts, and salad dressings.

Dextrins and Maltodextrins. **Dextrins** are polysaccharides derived from starch, linear arrays of glucose units bound by α-1,4 glycosidic linkages. They are produced commercially by hydrolyzing especially the amylose portion of starch. This process is termed pyroconversion or *dextrinization*. Since heat is used to carry out the reaction, the products are often referred to as pyrodextrins.

Dextrins can be used as 4 kcal/gram fat replacers that can replace all or some of the fat in a variety of products. Food sources for dextrins include tapioca, with

Important food polysaccharides:

Beta-glucan
Cellulose
Dextrins
Hemicellulose
Fructooligosaccharides (inulin)
Maltodextrins
Pectic substances
Starch
Vegetable gums (hydrocolloids)

TABLE 5.5 Categories of food fiber.

Category	Solubility	Fiber
Carbohydrate	Soluble	Beta-glucans
		Fructooligosaccharides (inulin)
		Gums
		Some hemicelluloses
		Pectin
Carbohydrate	Insoluble	Cellulose
		Some hemicelluloses
Noncarbohydrate	Insoluble	Lignan

applications including salad dressings, puddings, spreads, dairy-type products, and frozen desserts.

Dextrins should not be confused with **dextrans**. Dextrans are another example of a polysaccharide composed of glucose units. However, their bonding is α-1,6 glycosidic bonds, rather than the α-1,4 of dextrins. This polysaccharide is produced by certain bacteria and yeast and may be branched. Dextrans are classified as a food gum.

Maltodextrins are polysaccharide fragments derived from starch hydrolysis. In the food industry, corn starch is the typical source material used to generate maltodextrins. The large corn starch polymer is broken into smaller maltodextrin polymers, having a low degree of polymerization (DP).

Processed maltodextrins are gels or powders derived from carbohydrate sources such as corn, potato, tapioca, and wheat starches. Maltodextrins can be used as a fat replacer (4 kcal/gram), texture modifier, or bulking agent. Applications include baked goods, dairy products, salad dressings, spreads, sauces, frostings, fillings, processed meat, frozen desserts, extruded products, and beverages.

Maltodextrins are defined as having a dextrose equivalent of less than 20. The lower the DE, the less sweet the carbohydrate. At DE values of 20 and above, the substance is classified as corn syrup. *Corn syrup solids* are simply corn syrups dried to a crystalline form.

Fructooligosaccharides. **Fructooligosaccharides** (FOS) are naturally occurring sugars consisting of multiple units of sucrose joined to one, two, or three fructose molecules via glycosidic bond to the fructose portion of the sucrose molecule. Three types of these carbohydrates, based upon the number of fructose molecules in their structures, are produced commercially. Fructooligosaccharides are known as **prebiotics**. These substances promote the growth of **probiotics**, bacterial organisms believed to be beneficial to health. FOS are available in a wide variety of edible plants and foods such as banana, garlic, honey, barley, onion, wheat, tomato, rye, and brown sugar.

Fructooligosaccharides are selectively utilized as nutrients by the following bacteria: *Acidophilus, Bifidus,* and *Faecium*. FOS promote, stabilize, and enhance the proliferation of these beneficial bacteria into the gastrointestinal environment. Conversely, pathogenic bacteria including *Escherichia coli* and *Clostridium perfringens* have been shown to be unable to utilize FOS. Fructooligosaccharides may offer the following benefits to human gastrointestinal ecology: stimulate growth of *Acidophilus, Bifidus,* and *Faecium*; reduce fecal pH, toxic metabolites, serum cholesterol, and triglyceride levels; modify composition and rate of production of secondary bile acids; and reduce carbohydrate and lipid absorption, thereby normalizing blood glucose and serum.

Inulin: Dietary Fiber. The substance called **inulin** is a fructooligosaccharide that functions as a soluble dietary fiber. It occurs naturally in plants such as chicory root, onions, asparagus, and Jerusalem artichokes. It is composed of a chain of fructose units with a terminal glucose unit, with an average chain length of nine. Inulin is a β-2,1 fructan with the general formula Gf_n, where G is the glucosyl unit, f is fructose, and n is the number of units linked. Inulin is not digestible, although fermentation in the large intestine contributes approximately 1.5 kcal/gram. As a fructooligosaccharide, inulin is a prebiotic and stimulates the growth of beneficial bacteria.

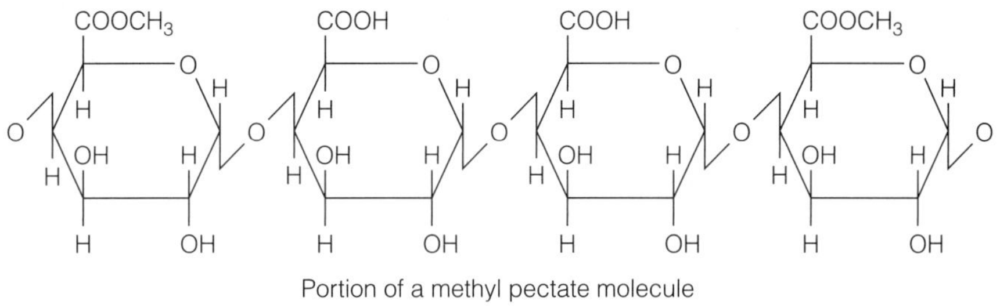

Portion of a methyl pectate molecule

FIGURE 5.4 The basic structure of pectic substances. *Methyl pectate is shown.*

When used as a food additive at low concentrations, inulin forms viscous solutions, while at concentrations above 30 percent, it can form a gel-like substance. In reduced-fat or nonfat systems, inulin provides a creamy mouthfeel through texture modification. It can be used to create a water-in-oil emulsion in which inulin binds the water and stabilizes the emulsion while providing creamy mouthfeel. As a reduced-calorie (about 1 kcal/gram) fat and sugar replacer, fiber and bulking agent, inulin has found application in yogurt, cheese, frozen desserts, baked goods, icings, fillings, whipped cream, dairy products, fiber supplements, and processed meats.

Pectic Substances. **Pectic substances** are high molecular weight polysaccharides found in plant cell wall middle lamellae. They are composed of galacturonic acid units, joined by α-1,4 glycosidic linkages. Some of the acid groups along the galacturonic acid chains become methylated (or "methoxylated") during fruit ripening. The methylated carboxylic acid groups on the galacturonic acid units are chemically considered to be esters of methanol. The term used for the proportion of methyl esters in a pectic substance is *degree of esterification*, (DE), not to be confused with dextrose equivalent (also abbreviated DE).

Although pectic substances are commonly referred to as pectin or methyl pectate (Figure 5.4), three chemically distinct substances have been identified:

- **Protopectin** refers to nonmethylated galacturonic acid polymers found in immature fruit.
- **Pectinic acid** is a methylated galacturonic acid polymer produced during ripening.
- **Pectic acid** is a short-chain demethylated derivative of pectinic acid associated with overripe fruit.

$$\text{Protopectin} \rightarrow \text{Pectin} \rightarrow \text{Pectic acid}$$

Immature fruit (no gel)   Ripe fruit (gels)   Overripe fruit (no gel)

In solution, pectin molecules become hydrated, with water "cages" surrounding the methyl esters. Hydrogen bonds occur between H and OH groups, and ion dipole interactions (e.g., between COO^- and alcohol groups or water molecules) are possible. Gelation of pectin is possible, but the conditions for gelation differ for what are termed LM (low methoxyl) pectin versus HM (high methoxyl) pectin. LM and HM pectin differ in degree of esterification. **LM pectin** is defined as having less than 50 percent of the carboxylic acids esterified with methanol, while **HM pectin** is defined as having more than 50 percent of the carboxylic acids esterified with methanol. HM pectin occurs naturally in fruits, while LM pectin is chemically modified pectin.

Pectin/Gelation. Because pectins are able to form colloidal dispersions, sols, and gels, they are widely used in the food industry. The inclusion of pectins in pie fillings to stiffen the texture is but one example. **Pectin gels** are systems containing a large volume of water within a three-dimensional solid network. Pectin and water interaction is considered similar to solute:solvent interaction. Pectins tend to remain dispersed in water due to attractive interactions between their carboxylic acid (COOH) and alcohol (OH) groups with water. For pectins to form gels, this tendency of pectin to remain dispersed in water

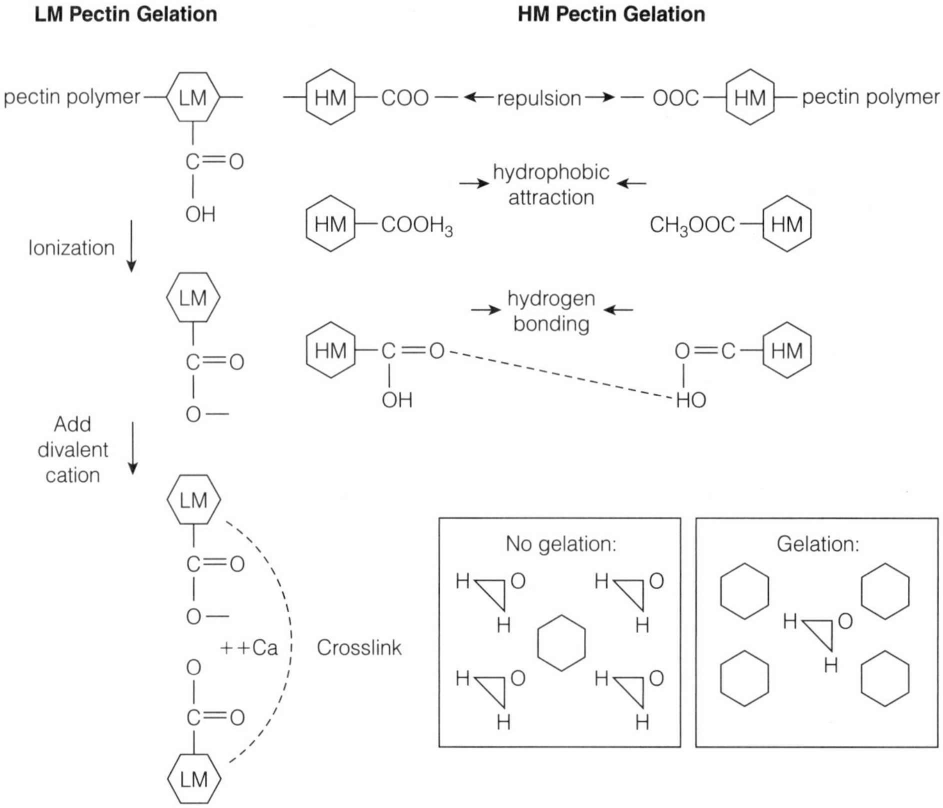

FIGURE 5.5 Comparison between LM and HM pectin gel formation.
There are two types of pectin based on methoxyl content, LM (low methoxyl) and HM (high methoxyl). Although each can form a gel, LM requires the presence of calcium ions, while HM requires sugar plus acid. In the gelation figures, water on the left prevents pectin:pectin interaction needed for gelation, whereas on right, with less H_2O available due to sucrose binding, some more pectin:pectin can occur.

must be overcome in favor of pectin:pectin associations. Conditions that promote interpolymer pectin associations and reduce polymer:water interaction are necessary.

The conditions required for HM pectin gelation are addition of acid and addition of sucrose. The addition of acid reduces the repulsive forces between polymers that keep them dispersed in water. Sugar acts to reduce pectin solubility by interacting with water itself, lowering the water activity of the system. As a result, sucrose addition decreases pectin polymer:water interactions, which allows for increased polymer:polymer association. This permits the pectin gel network to form. The decrease in pH assists intermolecular associations between adjoining pectin polymers in the network. Thus, HM pectins gel in the presence of sucrose, as long as a pH of 3.5 or less is achieved through addition of acid, such as citric acid. See Figure 5.5.

In the case of an LM pectin, the majority of carboxylic acid groups on the galacturonic acid polymers are not methylated. This allows for hydrogen bonding with water molecules through dipole-dipole interactions. In addition, the carboxylic acid groups are free to ionize to produce carboxylate groups (COO^-), which creates an abundance of negative charges and the possibility of ion-dipole interactions with water. The negative charges also repel LM pectin polymers apart. These reactions act as a barrier to gelation. However, when a metal such as calcium is added, carboxylate groups can crosslink with oppositely charged ions, such as Ca^{2+}, which pull neighboring LM pectin polymers together. Thus, low methoxyl pectin gels are set by the calcium ion concentration and not the soluble solids content of the system. See Figure 5.5.

Starch. Starch is a polysaccharide derived from plant sources such as corn, potato, rice, and wheat. A starch molecule is a polymer of 200 or more glucose units, having both branched (amylopectin) and unbranched (amylose) regions to its molecular structure (Figure 5.6). These starch polymers do not exist in the free state in plants. Instead, they occur as discrete, spherical aggregates called **starch granules** (Figure 5.7).

The amount of amylose and amylopectin in a starch is important because the behavior of heated starch in water, a common food processing step, depends on whether the source is relatively high in amylose or in amylopectin. Because starches often lose func-

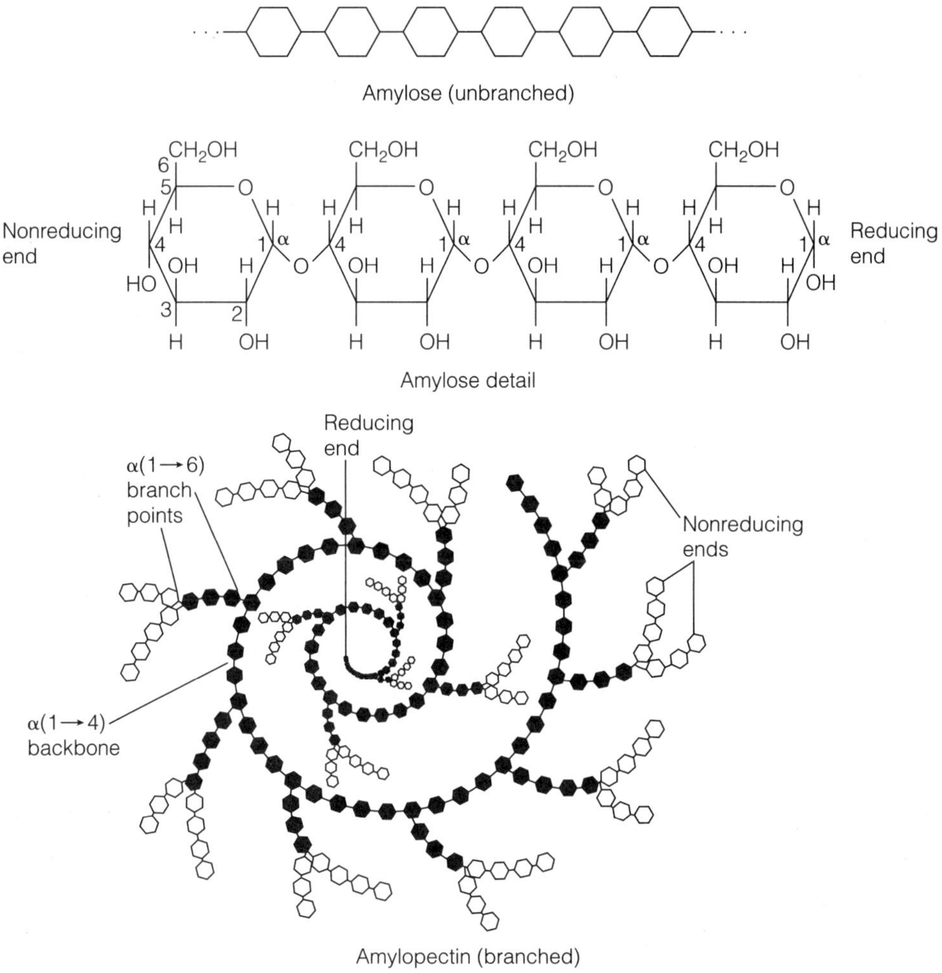

FIGURE 5.6 Amylose and amylopectin. *Starches are composed of two fractions, amylose and amylopectin, that occur as aggregates in structures called starch granules in plants.*

tionality when subjected to food processing conditions, certain heat or chemical modifications are employed. In most instances, modifications involve the alcohol groups on the starch polymer, or glycosidic bond cleavage. Modified starches function in film formation, exhibit freeze-thaw stability, pasting and gelling, enhanced solubility, and promote viscosity. An example of a chemically modified starch is **pregelatinized starch,** a type that increases product thickness with minimal thermal processing required.

Starch gelatinization. It is important to distinguish starch gelatinization from gelation, pasting, and retrogradation when discussing the functionality of starches. These distinct functional properties actually happen in sequence.

Starches do not form true solutions with water because starch molecules are too large. But when starch is heated in water, the bonds joining the starch fractions amylose and amylopectin are weakened or loosened, which allows water molecules to move in and form hydrogen bonds. This results in the gelatinization phenomenon. During **gelatinization,** heated starch granules absorb water and swell in size; it is an irreversible reaction. Since the molecular organization of starch granules is disrupted during gelatinization, there is an increase in entropy (randomness) as the starch crystallites melt from their organized crystal structures and become soluble.

Starch pasting and gel formation. A **starch paste** is a viscoelastic starch and water system that possesses both thick liquid-like (viscous) and solid-like (elastic) properties. The pasting process follows gelatinization and can be thought of as a continuation of it. Pasting encompasses three changes in starch: swelling, exudation, and disruption. In a starch paste, the majority of the starch granules have gelatinized, producing swollen and disrupted starch granules and exudative molecular matter. This material migrates into the

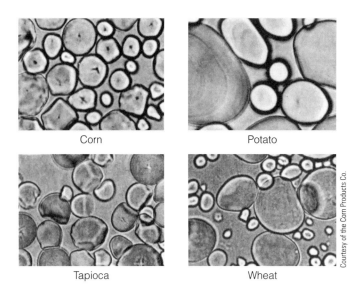

FIGURE 5.7 Starch granules—cytoplasmic plant structures storing starch polysaccharides. *Microscopic images of starch granules from various sources are shown. Within starch granules, hydrogen bonding interactions between amylose and amylopectin polymers, and between the polymers and water molecules are possible due to the presence of OH groups on the polymers. The nature of these associations changes when starch is heated in the presence of water.*

intergranule matrix, which is the region between adjacent starch granules. Such changes contribute to a noticeable increase in the viscosity of the system.

The interesting phenomenon that follows gelatinization and pasting is termed *gelation*. With respect to starch, **gelation** refers to the formation of a gel from a cooled paste. A **starch gel** is a rigid, thickened starch and water mixture that has the properties of a solid. Starches with high amylose content form gels more easily than do starches high in amylopectin. Pastes derived from high amylopectin starches are considered to be nongelling but have a gummy, cohesive texture. They are useful as thickening agents due to their ability to create an increase in viscosity.

Starch retrogradation. After heating and cooling, the starch polymers amylose and amylopectin and the intergranule matrix starch fragments can reassociate into an ordered structure, which represents a loss of entropy to the system. This process is defined as **retrogradation** and is due to intermolecular hydrogen bond formation, between linear amylose molecules especially. Therefore, retrogradation is seen particularly in high-amylose starch gels. In such systems, texture changes, from a gelled texture to a more solidified and rubbery one, occur with time. An increased tendency to release water from the gel, called **syneresis,** also occurs. Products that show syneresis are considered not to be freeze-thaw stable.

Figure 5.8 summarizes the gelatinization → pasting → gelation → retrogradation sequence.

Vegetable Gums. **Vegetable gums** are plant hydrocolloids—substances derived from plants that distribute in water as colloidal dispersions. They are composed of long-chain polymers of various hexoses and pentoses. Sources of food hydrocolloids include exudate gums such as gum arabic, carageenan (a seaweed extract), guar gum (a plant seed gum), methyl cellulose (chemically modified cellulose), and xanthan gum, a substance produced by microorganisms.

Hydrocolloid polymers may be linear or branched, which affects their functionality. Branched molecules are not as able to interact with each other through physical surface contact, compared to linear polymers. However, their side groups become entangled, and they gel more easily than the more linear hydrocolloids.

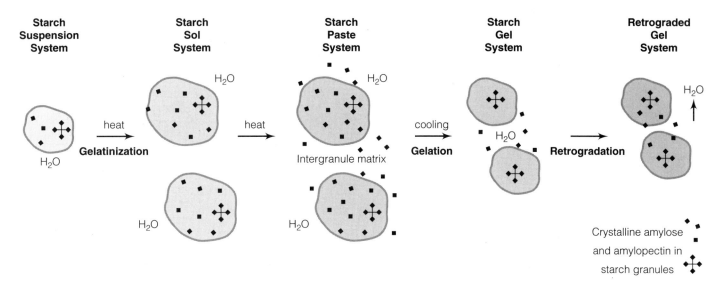

FIGURE 5.8 Relationships between starch gelatinization, pasting, gelling, and retrogradation. *The initial suspension of insoluble starch in water contains crystalline amylose and amylopectin starch polymers associating with each other via hydrogen bonding within starch granules. Heated starch undergoes gelatinization, losing its crystalline order through melting and interpolymer hydrogen bond breakage. The system is a sol of low viscosity, consisting of a liquid continuous phase and a solid dispersed phase. With continued heating, the system becomes a paste, and is less pourable due to an increase in viscosity as molecular material enters the intergranule matrix while the granules become swollen with water. As the gelatinized paste cools, an elastic, two-phase system forms consisting of a solid continuous phase of mainly amylose polymers holding a liquid dispersed phase, water, through intermolecular hydrogen bonds. This system, termed a starch gel, forms through the gelation process, in which water molecules associate with the starch polymers via intermolecular hydrogen bonds. However, in starch gels, gelatinized starch regions of high amylose content revert back to more crystalline structures upon storage, through retrogradation. Retrograded starch expels water from its three-dimensional structure via syneresis, which causes gel shrinkage and increased starch polymer intermolecular associations.*

Because they can be used at very low levels as food additives, gums are virtually noncaloric. This, coupled to their functionality, makes them appropriate fat replacers in certain food applications. High solubility, pH stability, and gelling ability are the desired characteristics of a food gum. As food additives, gums provide smooth texture, act as thickeners, and act as water binders. They are important in food processing as ingredients in reduced-fat foods, such as salad dressings and processed meats, to increase viscosity, and they also behave as emulsifiers.

5.2 FOOD LIPIDS

Lipids are organic substances that are relatively nonpolar. Lipids, as fats, oils, or waxes, are only slightly soluble in water but very soluble in organic solvents, owing to the presence of hydrocarbon chain structures. They are considered lipophilic or hydrophobic compounds and possess an interesting chemistry.

Structures and Types of Lipids

Fats and Oils. Fats and oils are chemically known as triacylglycerols. **Triacylglycerols**, or triglycerides, as they are often called, are triesters of glycerol and fatty acids— see Figure 5.9. Glycerol, a simple organic compound sometimes called glycerin, is a 3-carbon molecule containing three alcohol groups. Fatty acids are organic molecules

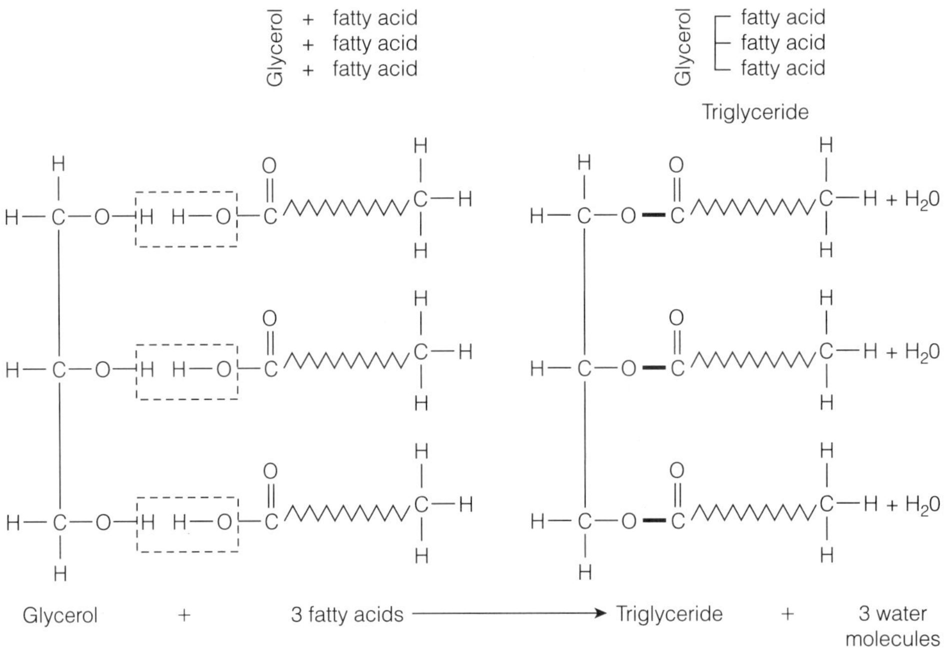

FIGURE 5.9 The chemical structure of a triglyceride. *The ester bonds appear in bold.*

that contain chains of carbon bound to hydrogen, plus an acid group (COOH) at one end and a methyl group (CH_3) at the other. Ester bonds hold fatty acids to glycerol, joining the OH groups of glycerol to the COOH groups of fatty acids, with the loss of water.

Saturated and unsaturated fats. Fats and oils are mixtures of fatty acids differing in chain length and degree of unsaturation. A fatty acid chain is termed **saturated** if it does not contain any carbon-to-carbon double bonds. In saturated fatty acids, each carbon atom (except the ones at each end) has two hydrogens attached. A fatty acid chain is termed **unsaturated** if it does contain carbon-to-carbon double bonds. Look back at Figure 3.8. Oils in particular contain a typically high proportion of unsaturated fatty acids. Unsaturated fatty acids can be monounsaturated or polyunsaturated. Monounsaturated fatty acids describe those having only one double bond in the carbon chain. An example of a common monounsaturated fatty acid is oleic acid. Polyunsaturated fatty acids are those having two or more double bonds. See Table 5.6.

TABLE 5.6 Examples of fatty acids, characterized according to saturation.
Unsaturated fatty acids can be either monounsaturated or polyunsaturated, depending upon how many C═C double bonds are present. A fatty acid is described by chemical convention in terms of both its chain length and number of double bonds. For example, C4:0 indicates a fatty acid containing four carbon atoms and zero double bonds. C18:2 is a polyunsaturated fatty acid having 18 carbon atoms and 2 double bonds.

Fatty Acid Name	Fatty Acid Type	Notation
Butyric	saturated	C4:0
Capric	saturated	C10:0
Palmitic	saturated	C16:0
Stearic	saturated	C18:0
Oleic	monounsaturated	C18:1
Linoleic	polyunsaturated	C18:2
Linolenic	polyunsaturated	C18:3
Eicosapentaenoic	polyunsaturated	C20:5
Behenic	saturated	C22:0

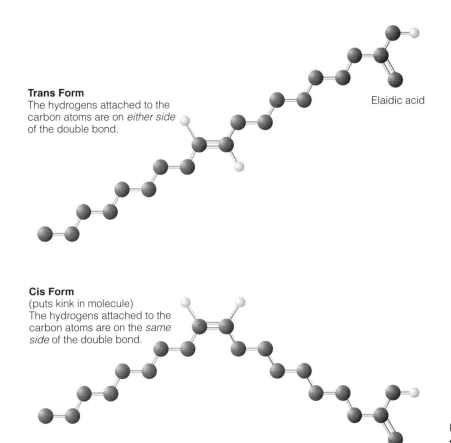

FIGURE 5.10 The cis and trans forms of a typical unsaturated fatty acid.

One special case of polyunsaturated fatty acids is the **omega (ω) fatty acid** discussed in Section 3.2. Structurally, an omega fatty acid has a certain number of carbon atoms between the terminal CH_3 (methyl) group and the last double bond (the one furthest from the COOH end) in the fatty acid. Look back at Figure 3.9 for examples of omega-3 and omega-6 fatty acids.

The chemical reactivity, with hydrogen or oxygen for example, of unsaturated fatty acids is determined by the position and number of double bonds. The higher the degree of unsaturation, the greater the reactivity—provided that the double bonds occur in a series with a single bond in between, categorizing them as **conjugated** double bonds. Also, when double bonds are separated by a methylene ($-CH_2-$) unit, greater reactivity results. In omega fatty acids, the *methylene interrupted* pattern exists. This is not the same as conjugated, because there are two, not one, carbon-to-carbon single bonds between double bonds:

$$C=C-C-C=C \} \text{ methylene interrupted}$$
$$C=C-C=C-C \} \text{ conjugated}$$

Cis and trans fats. Unsaturated fatty acids come in two configurations, defined by their structure at the double bonds. In the *cis* configuration, the hydrogen atoms bonding to the $C=C$ are located on the same side of the double bond—see Figure 5.10. Notice that the carbon chain segments on either side of the double bond bend toward each other in the cis configuration. In the *trans* configuration, the hydrogen atoms attached to the carbon atoms of the double bond are opposite each other. For the most part, unsaturated fatty acid double bonds exist in foods in the *cis* rather than the *trans* configuration. These structural differences affect properties such as melting point.

TABLE 5.7 Cis- and trans-fatty acids and melting point. *The MP of saturated fats is higher than for unsaturated fats, and the* trans *configuration results in a higher melting point than the* cis. *The trans form is therefore similar to a saturated fat, exhibiting the latter's straight structure and hardness at room temperature.*

Fatty Acid Name	Fatty Acid Type	Notation	MP (°C)
Stearic	saturated	C18:0	70
Oleic	*cis*-unsaturated	C18:1	19
Elaidic	*trans*-unsaturated	C18:1	43

Melting point. The *melting point* is the temperature at which a solid is converted into liquid. Fat molecules exist in crystalline forms, and the strength of the bonding forces between fatty acids in a fat crystal determine its melting point. Each crystalline form has its own characteristic melting point. Fatty acids exhibit unique melting points, and a pure substance such as a single type of fatty acid will exhibit a sharp melting point corresponding to a defined temperature. Triglycerides, with their array of 3 fatty acids, exhibit melting points based on which fatty acids are present. This determines whether a fat will be a liquid, a solid, plastic, or brittle at room temperature. A *plastic fat* is soft and pliable in consistency at room temperature. This is a consequence of containing a mixture of both solid and liquid triglycerides.

Melting point is determined by a variety of factors, including fatty acid chain length and degree of unsaturation. All else equal, short-chain fatty acids show lower melting points than long-chain, and saturated fatty acids have higher melting points than unsaturated ones. However, when a mixed fat, such as a triglyceride S-O-P (stearic-oleic-palmitic), is heated, the melting is gradual and over a range of temperatures rather than one distinct measurable value, reflecting the "impure" fatty acid composition. In addition, the cis configuration has a lower melting point than the trans configuration. So, for a C18 unsaturated fatty acid, the cis form will exist in the liquid form at a lower temperature than the trans form, which will remain as a solid fat longer (Table 5.7).

Flavor compounds. The statement that fat is important for flavor should be of no surprise to us. Food scientists learned from reduced-fat and fat-free food processing that when fat is removed from product formulations, much flavor is removed as well. In addition to fat serving as a carrier of flavor compounds, the contribution of fat molecules to flavor may be either direct, owing to their intrinsic flavor, or indirect, as a result of interaction with other substances.

For instance, the intense characteristic flavor of cod liver oil is recognizably different from that of a mild canola oil. If a salad dressing were made from each, the resulting flavors of the two dressings would be quite distinguishable. Another example is frying oils. In fried foods, the flavor of the oil is important because the oil becomes absorbed into the food. A food fried in corn oil exhibits a different flavor from one fried in sesame oil.

The interaction of fats or oils with other atoms, for instance oxygen, or molecules, such as proteins and carbohydrates, creates chemical products that impact food flavor. **Reversion flavor** is a mild off-flavor developed by refined oils that have become exposed to oxygen. Though undesirable, reversion flavors are less intense compared to those associated with oxidative rancidity; however, reversion flavors are still considered to lessen product quality.

Polar Lipids. Polar lipid molecules are found in the membranes of plant and animal tissue. Such molecules, though lipids, have a degree of water solubility owing to the presence of polar (potentially charged hydrophilic groups containing oxygen or nitrogen) atoms in their structure. In foods, **glycerophospholipids** are important examples of polar lipids. The structure of such a lipid shows why it is both fat (lipophilic) and water soluble (hydrophilic)—see Figure 5.11. Attached to the glycerol backbone of such a phospholipid, are two fatty acids, for instance oleic or palmitic (the lipophilic part), and a hydrophilic part that contains a phosphate group.

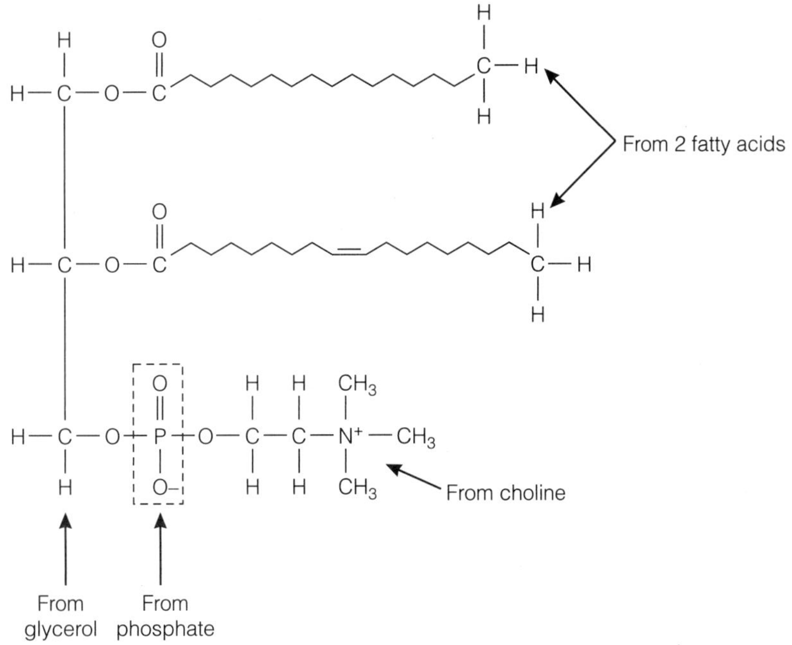

FIGURE 5.11 General structure of a phospholipid (lecithin).

Because of their amphiphilic structure, polar lipids can function in foods as emulsifiers. Lecithin (chemically termed phosphatidylcholine) is a phospholipid in which choline is the hydrophilic part of the polar lipid. Another example of a polar lipid is the sterol molecule, or "steroid alcohol," which is a lipid compound having a ring structure and a polar alcohol group. Cholesterol is the main sterol found in animal tissue, while beta-sitosterol is the predominant plant sterol.

Pigments. A variety of natural pigment molecules are lipids associated with food lipid matter. These may occur as esters of fatty acids or as crystal forms in liquid oil. The classes of pigments include the carotenoids, such as beta-carotene and lycopene, anthocyanins such as delphinidin, betalains (betacyanins) such as betanin, and chlorophylls. We will discuss them in Chapter 6.

Waxes. Waxes are esters of fatty acids and even-numbered long carbon chain alcohols, for example, stearyl alcohol $C_{18}H_{38}O$. They occur in nature as low melting point solids that coat plant leaves and fruits. Several waxes, including beeswax, candelilia wax, carnauba wax, and paraffin, are used in foods. In the food industry, waxes are used as protective coatings in some fruits and vegetables to increase their shelf life and flavor quality. This protection occurs because a fat coating provides a barrier to the migration of moisture into or out of a food. As an example, **carnauba wax,** an exudate from palm tree leaves, is a GRAS substance ("generally recognized as safe") that functions as a coating in chewing gum, sauces, fruits and vegetables, and confections.

Chemical Reactions of Lipids

The chemical reactions that lipids undergo include hydrogenation, hydrolysis, interesterification, and oxidation.

Fractionation. The process of **fractionation** splits an oil into its higher melting point components, such as stearic acid, and lower melting point components, such as oleic acid. Crystallization is accomplished, so that the crystal portion can be separated from the liquid portion. An example of the application of fractionation is the production of a good frying oil and a good plastic fat for shortening from the same starting material.

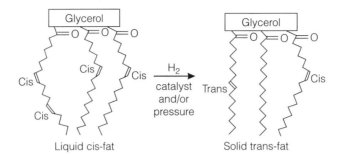

FIGURE 5.12 Hydrogenating vegetable oils can produce trans-fats.

Hydrogenation. **Hydrogenation** is the forced addition of hydrogen atoms to the unsaturated bonds in an unsaturated fat. This process raises the fat's melting point and is used in the food industry to "harden" liquid oils into semisolid fats. During hydrogenation, the fatty acids containing the most double bonds hydrogenate more quickly than the less saturated ones. Although raising the saturated fat level, the process can be controlled, as in the conversion of vegetable oils into butter, so that a high linolenic and oleic acid content is retained. Typically, not all of the fatty acid molecules in a sample react with hydrogen. A side effect of the process is the change of a percentage of the cis-unsaturated fatty acids to trans-unsaturated fatty acids (Figure 5.12).

Hydrogenation also decreases the tendency of a fat, such as a frying oil, to oxidize, since less unsaturation means less potential for chemical oxidation to occur. For instance, in commercial frying operations, it is common practice to use a frying oil such as corn or peanut oil that has been slightly hydrogenated because hydrogenation provides a longer usable "frying life" for the oil by making the oil's fatty acids less unsaturated and therefore more stable (less reactive with oxygen). These oils are also selected for use as frying oils because they have relatively high smoke points of about 440°F—higher, that is, than the fat frying temperature, which is 350°–400°F.

Hydrolysis. In the case of fats, **hydrolysis** is the reaction requiring heat plus the addition of water molecules to separate the fatty acids from the glycerol portion of a lipid (triglyceride) molecule. The glycerol can be further changed into a substance called *acrolein*, which produces odorous and irritating fumes in the smoke of an overheated fat.

$$\text{triglyceride} + 3H_2O + \xrightarrow{\text{Heat}} 3 \text{ fatty acids} + \text{glycerol} \xrightarrow{\text{Heat}} 3 \text{ fatty acids} + \text{acrolein} + 2H_2O$$

Hydrolytic rancidity results when stored fats become rancid by the hydrolysis reaction with water. Notice that free fatty acids are produced through hydrolysis. For each molecule of water that combines with the fat molecule, one free fatty acid is liberated. Heat usually causes hydrolytic rancidity to start, although hydrolytic rancidity can also be caused by the naturally present food enzyme lipase.

The size of the liberated fatty acids is important in determining rancidity. Shorter-chain fatty acids cause the objectionable flavors and odors associated with hydrolytic rancidity. If long-chain fatty acids are liberated, they do not contribute to off-flavor and odors. When butter is left out at room temperature for too long, hydrolytic rancidity causes the butter to smell and taste bad. What happened? The many short-chain fatty acids present in butterfat, especially C4 (butyric) and C6 (caproic), hydrolyze from glycerol, to cause unpleasant living at the breakfast table.

Interesterification. **Interesterification** is essentially the removal of fatty acids from glycerol and their subsequent rearrangement or recombination into numerous configurations, most of which differ from the original fat molecule. Industrially, interesterification can be applied to hard fats such as lard, to produce mixed glycerides that offer improved creaming. Since pure triglycerides contain 3 identical fatty acids, interesterification products of 2 different pure fatty acids are easy to predict in terms of their fatty acid profiles

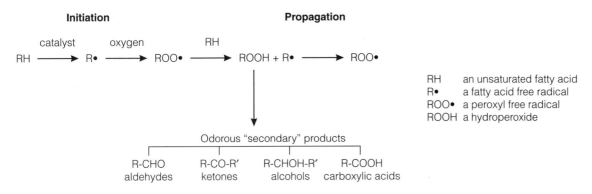

FIGURE 5.13 Simplified scheme of lipoxidation. *During the first step,* initiation, *free radicals are formed as loosely bound hydrogen atoms near the C═C bonds of a fatty acid (RH) are lost. The resulting fatty acid free radical (R•) reacts with oxygen (O$_2$) to produce an activated peroxide, or peroxyl free radical (ROO•). This radical reacts with a nearby unsaturated fatty acid (RH) to remove its H atom, which generates hydroperoxide (ROOH) from the peroxyl free radical and a new fatty acid free radical (R•) from the fatty acid. The newly formed fatty acid free radical reacts with oxygen, and the cycle is repeated, which is termed* propagation. *At some point, the hydroperoxides split apart through chemical decomposition into smaller short-chain organic molecules, including aldehydes and ketones. These contribute oxidized or rancid odors and flavors to the fat. Lipoxidation ends when all of the fat molecules have reacted, or when two unstable radical species react with each other, or when an antioxidant reacts with a radical, which stops the chain reaction. This is called* termination *of lipoxidation.*

and the proportions of each resulting triglyceride. Eight configurations of the fatty acids with glycerol are possible, each accounting for 12.5 percent (100/8) of the total.

For example, an interesterified mixture of equal quantities of triglycerides A and B, in which A is triolein (O—O—O) and B is tripalmitin (P—P—P), would result in the following quantities of possible triglyceride products: 12.5 percent each of O—O—O, O—P—O, P—O—P, and P—P—P, and 25 percent each of O—P—P and P—O—O. The last two account for 25 percent because O—P—P is equivalent to P—P—O (hence 12.5% twice = 25%), and P—O—O is equivalent to O—O—P (again, twice 12.5% = 25%).

Oxidation. In oxidation of food lipids, oxygen reacts with the double bonds of unsaturated fatty acids. The chemical result of this reaction is the production of small organic compounds, which in turn generate undesirable "rancid" odors in foods that contain oxidized fats or oils. The original fatty acids, along with their nutritional value, are lost due to oxidation.

What happens chemically during oxidation? The unsaturated fatty acid that reacts with oxygen forms a **hydroperoxide,** and a chain reaction results. Although not responsible for off-flavors and odors, the hydroperoxide decomposes further to yield the odorous aldehydes, ketones, acids, and alcohols, which cause rancidity. The quantity of hydroperoxide in a fat sample can be measured and is called the peroxide value (PV), which is used as a predictor of the level of oxidation in a fat.

The term **lipoxidation** is used to describe this chemical mechanism in which heat, light, or metals trigger a chain reaction within stored fats and oils that results in a fat becoming rancid. Lipoxidation is a sequence of steps, including initiation, propagation, and termination—see Figure 5.13 for a diagram of the overall mechanism.

To prevent or delay the onset of oxidative rancidity, fats and oils should be stored in cool areas away from light and metallic containers. In addition, using an antioxidant in the food product may be considered. Antioxidants are effective because they donate hydrogen atoms to the lipid fatty acid radicals, which regenerates the original fat molecules from the lipid radicals, stopping the chain reaction.

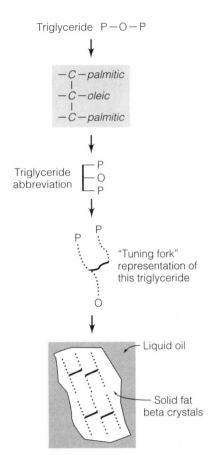

FIGURE 5.14 Fat beta crystals, showing the "tuning fork" packing of the component triglycerides. *The central portion of each triglyceride tuning fork represents the glycerol region, with the "arms" the fatty acids. Solid crystals pack into ordered structures dispersed within liquid oil.*

Important functional properties of food lipids:

Aeration
Crystallization
Emulsion phase
Flavor
Heat transfer
Mouthfeel
Plasticity
Tenderization

Polymerization. Once fatty acids have been hydrolzed from glycerol, they can link together to form very large molecules called fatty acid polymers. This kind of **polymerization** has consequences when it occurs in food fats and oils. For example, it is the cause of an increase in the viscosity of cooking oils. Polymerization occurs at the **smoke point** of a fat—the temperature at which overheating of a fat causes it to give off smoke.

The Functional Properties of Lipids

In addition to participating in chemical reactions, food fats and oils such as beef fat or tallow, pork fat or lard, milk fat, olive oil, and corn oil represent functional components of the foods we eat. Many people recognize that these substances contribute to flavor and calories. Fats, however, also contribute to food texture and the satiety or full, satisfied feeling occurring after consuming a meal. In addition, the ability of food lipids to exist as crystals, within liquid, semisolid, and solid fat forms affects the plasticity, structure, stability, and quality of many foods. Fats contribute to emulsion formation, the tenderness in pastries and baked products, the aeration of batters and doughs during mixing, fat-soluble flavor compounds, color compounds, and the mouthfeel of foods. Fats also serve as a medium of transferring heat to fried foods.

Aeration. In the production of cakes, cookies, and other baked goods that require a creaming step, fat as a plastic shortening is mixed with sugar. During the mixing, air bubbles are incorporated into the batter, physically held by the crystal molecule arrangements in the fat. This aerated batter is needed for the product to expand during baking to a desirable volume and height. The air bubbles become nuclei for gas expansion during baking, with the number of nuclei determined by the mixing. Heat causes steam (water vapor) to migrate to neighboring air bubbles, and the steam and the leavening gases expand during baking due to pressure and cause the bubbles to expand.

Crystallization. Food fats, whether they are solid, liquid, or something in-between at room temperature, contain very small solid fat crystals. These crystals are composed of the fat's triglycerides. Most triglyceride crystals have fatty acid chains that are inclined at a certain angle in their packing fit. The "crystal habit" of a triglyceride of uniform fatty acid composition, such as one that is uniformly stearic acid (S—S—S), is called the beta (β) crystal. Such crystals pack into three-dimensional arrays reminiscent of side-by-side tuning forks—see Figure 5.14.

Fat molecules pack into crystal lattice structures in a variety of different ways, a property of triglycerides termed **polymorphism.** The three predominant crystal forms are designated alpha (α), beta (β), and beta prime (β'), each having a unique packing organization. A given fat will tend to exist in one of these forms at a time. (However, cocoa butter, from which chocolate is made, must be considered superpolymorphic, in that it can exist in one of six crystal forms!) Each crystal form retains a characteristic melting point due to its unique crystal arrangement. Temperature changes experienced by a fat not only result in a change of state (example: a solid fat that is heated will melt and become an oil) but also a change in crystal form. The state of a fat, meaning its proportion of solid fat crystals to liquid oil at a particular temperature **(solid fat index),** the form of crystal packing arrangement, and the types of fatty acids in the triglyceride structures present in the crystal forms all affect food functional properties.

Emulsification. Emulsions are homogeneous dispersions of an oil and water phase. As an example, cake batter represents an emulsion. The fat phase of shortening plus air bubbles is created first during creaming with sugar. Air bubbles taken in during creaming are coated with fat crystals present in the liquid fat phase. The liquid or aqueous phase, which is added to the aerated fat after creaming contains flour particles, sugar, egg proteins, and other substances in water. These phases set up within the fresh batter system as a stable emulsion, an oil-in-water type. If fat were not present in the batter, the emulsion would not form.

TABLE 5.8 Fat replacers.

Substance	Characteristics	Uses
Caprenin	Replaces cocoa butter; composed of caprylic, capric, and behenic acids, and glycerine	confectionery
Emulsifiers	9 kcal/gram; used at low level to reduce calories	cake mixes, cookies, icings, and dairy products
Salatrim ("Benefat")	5 kcal/gram	baked goods, confections, and dairy products
EPG (esterified propoxylated glycerol)	Reduced-calorie fat replacer	replace meat for fats and oils in all typical commercial applications
Olestra (Olean)	Calorie-free ingredient made from sucrose and edible fats and oils; stable under high heat applications such as frying	salty snacks and crackers
Sorbestrin	Heat stable, liquid fat substitute composed of fatty acid esters of sorbitol (1.5 kcal/gram)	fried foods, salad dressing, mayonnaise, and baked goods

Flavor. As already mentioned, fats carry flavor compounds and also contribute to flavor directly owing to their own intrinsic flavor, or indirectly as a result of interaction with other substances. A special category of compounds called terpenes includes a variety of lipid flavor molecules from plants, called **terpenoids**. In citrus fruits, the terpenoids limnolene and citral contribute to flavor. Terpenoids can be isolated into their essential oil portions by a process known as steam distillation.

Heat Transfer. Frying fats (oils) transfer heat energy from the heat source to the surface of the frying food, which is immersed or in contact with the fat. As food makes contact with hot frying oil, moisture in the product escapes, and is evaporated off as steam into the atmosphere. A portion of the hot frying oil is absorbed by the food during frying as the moisture leaves, which promotes heat transfer inward. As a result, the fat content of a cake doughnut after frying can be four to five times the original fat content of the cake doughnut itself.

Mouthfeel. Fat is a lubricant in the mouth. As such, it helps to clear particles of food from the teeth, tongue, and gums during chewing. In addition, the creamy texture and smoothness of fat contribute to mouthfeel sensations. When fat is replaced in a food by a carbohydrate- or protein-based ingredient, the texture and mouthfeel of the resulting low-fat food can be altered. However, the use of fat-based fat substitutes has been accomplished, with not only calorie reduction, but retention of fat functional properties, such as mouthfeel. Several fat-based fat replacers are listed in Table 5.8.

Plasticity. The term **plasticity** refers to the physical property of a fat that describes its softness at a given temperature. A plastic fat will respond to an external force by deforming, as when squeezed or spread, but holds its shape on a flat surface. The rapid cooling of a melted fat results in a waxy solid made up of alpha crystals. These crystals are very unstable and change into clusters of needle-like beta prime crystals, which is the preferred form for plastic shortenings such as hydrogenated vegetable shortening. However, provided the opportunity, this crystal form will convert to stable beta crystals. This occurs through improper **tempering,** which means that during manufacture the product did not form to the correct crystal type because it was not held at the proper temperature for the prescribed time. Fats composed of beta crystals exhibit decreased plasticity and more graininess to the touch, as well as excessive oiliness.

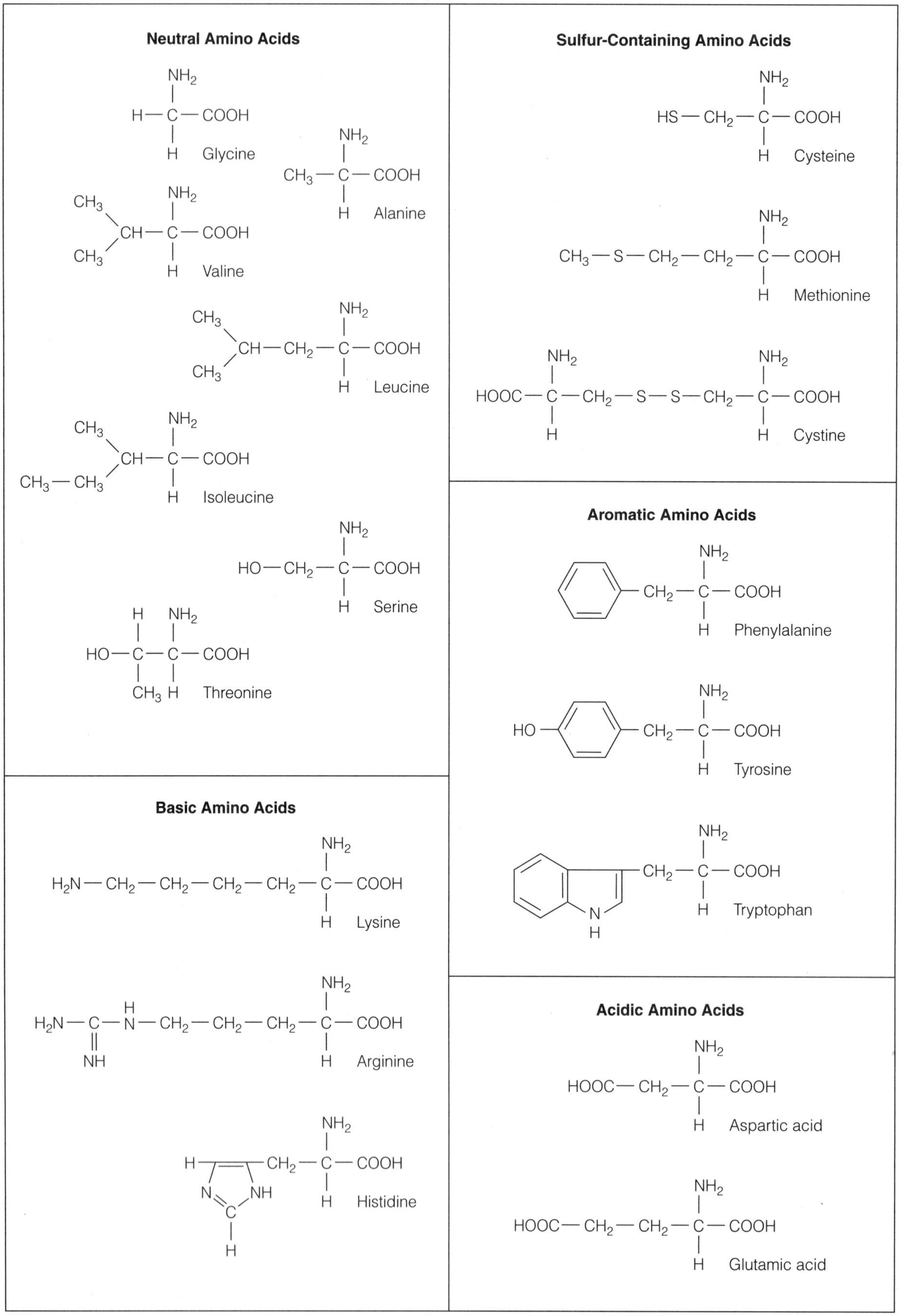

FIGURE 5.15 Classification of amino acids based upon their side chains.

Tenderization. In foods, fat acts as a tenderizer. The dramatic difference in tenderness between a cooked lean cut of meat (low fat content) and one that is marbled with fat (high fat content) illustrates this property. The lean meat seems dry and tough to chew, while the marbled cut seems moist, tender, and easier to chew.

The presence of fat in baked goods such as breads and rolls is important to structure. The solid portion of a shortening contributes to the overall structure of baked bread, by strengthening the loaf sides and preventing its inward collapse. In cakes, besides entrapping air bubbles in the batter during the mixing step, fats coat protein and starch molecules to inhibit gluten networks and starch structure, providing for the typical cake crumb texture. The result is a softer crumb. This process is referred to as a tenderizing effect, since it limits formation of a tough gluten protein dough structure.

5.3 FOOD PROTEINS

Proteins are polymers of amino acids and are referred to as polypeptides. Each amino acid shares a common structure, with an amine group (NH_2), an acid group (COOH), and a central carbon atom bonded to hydrogen and to a side chain (R). The distinct side chains or groups allow for classification into acidic, basic, neutral, aromatic, and sulfur-containing amino acids (Figure 5.15).

Proteins are composed of chains of amino acids joined by peptide bonds—see Figure 5.16. A peptide bond connects the acid end of one amino acid with the amino end of another. Two amino acids bonded together form a dipeptide—as in Figure 5.16. Most proteins are a few dozen to several hundred amino acids long and are called polypeptides.

As polypeptides, proteins can be described as either nonconjugated or conjugated. Most exist as *nonconjugated*, meaning they are not bound to other substances and contain only amino acids. *Conjugated proteins* are proteins combined with nonprotein substances such as carbohydrates and lipids into complex molecules. The glycoproteins are conjugated proteins that contain sugar molecules in their structure (example: the egg white proteins ovomucoid and ovalbumin contain sugar molecules and are thus glycoproteins). The low-density lipoproteins found in egg yolk are complexed with lipid. Another type of conjugated protein are phosphoproteins. These are proteins that contain phosphate groups in their structure. Caseins in milk are an example, due to the presence of phosphate in the casein micelle structure.

The Structure of Proteins

Within the protein itself, four specific levels of protrein structure have been identified: primary, secondary, tertiary, and quaternary. These levels provide a description of what the polypeptide is composed of and its shape. Primary (1°) is the linear sequence of amino

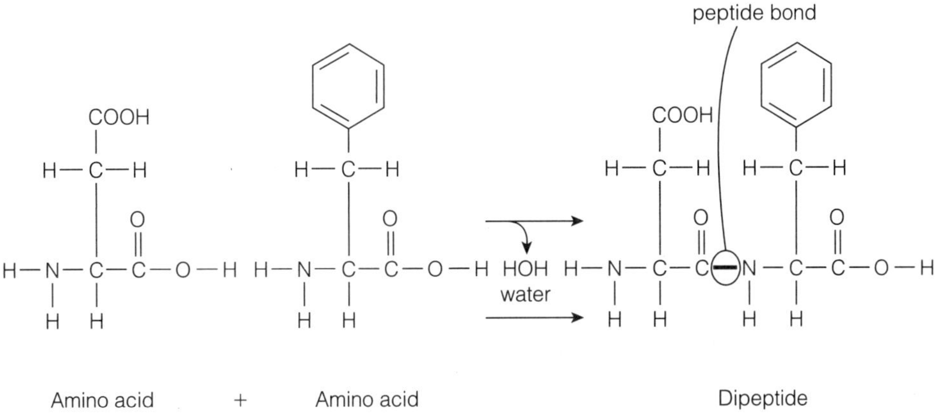

FIGURE 5.16 The formation of a peptide bond—a condensation reaction. *Amino acids link together by peptide bonds to form polypeptides.*

FIGURE 5.17 Sections of secondary protein structure. *The amino acids of a protein may assume the configuration of the alpha-helix or the beta-sheet. Notice the repeating sequence N—C—C—N—C—C. In the alpha-helix, notice how the helix passes behind and in front of an imaginary line.*

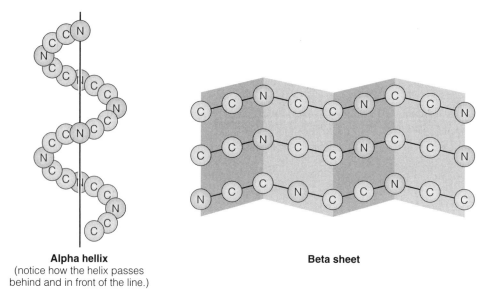

Alpha hellix
(notice how the helix passes behind and in front of the line.)

Beta sheet

acids in order within a polypeptide, such as glycine-glycine-proline, etc. Food chemists have a shorthand for this primary structure: N—C—C—N—C—C—, where N stands for the amine group (NH_2), C is the central carbon of the amino acid, and C stands for the acid group. The sequence N—C—C—N—C—C denotes a dipeptide.

Secondary (2°) refers to whether the amino acids together assume either an α-helix or β-sheet configuration within a polypeptide—see Figure 5.17. Tertiary (3°) describes the overall three-dimensional shape achieved by the folding of the entire protein molecule. This level of structure is stabilized by bonds such as hydrogen and disulfide between amino acids that happen to be in close proximity to one another due to the folded situation of the 3° structure. Some food proteins (e.g., caseins) are more "open" and less ordered than others.

Denaturation is the unfolding of such structure, usually due to acid or heat. In a denatured protein, the normal spatial arrangements, and importantly, the protein's functional properties, are altered or lost.

If a protein contains more than one polypeptide chain in its structure (such as in casein, collagen, myosin), its overall spatial structure is referred to as quaternary (4°).

What about protein shapes? Food proteins are typically either spherical like a ball or elongated like a twisted rope. The former situation describes many proteins that are soluble in water, for example, myoglobin. These are termed *globular* proteins. Elongated proteins are generally insoluble and are called *fibrous*, an example of an elongated protein is collagen.

Table 5.9 gives some examples of the three types of protein structure.

TABLE 5.9 Examples of food protein structures and shapes.

Food Proteins	Protein Structure	Protein Shape
Egg albumen	Globular	Spherical
Meat and legume globulins	Globular	Spherical
Collagen	Fibrous	Elongated
Elastin	Fibrous	Elongated
Glycoprotein: ovomucoid	Conjugated	Protein bound to carbohydrate and hemagglutinin
Lipoproteins: chylomicron	Conjugated	Protein bound to lipid LDL and VLDL
Metalloproteins: hemoglobin	Conjugated	Protein bound to metal ferritin, and myoglobin
Phosphoprotein: casein	Conjugated	Protein bound to phosphorus

The Chemical Reactions and Functional Properties of Proteins

Proteins in foods are responsible for specific colors, flavors, and textures. They can function as buffers, emulsifiers, enzymes, and fat replacers (Table 5.10). In addition, certain proteins can form gels and foams under proper conditions.

Buffering. Buffering refers to preventing a pH change by undergoing an ionization reaction. The buffering ability of proteins is a function of the amino acids that make up the protein structure. The carboxylic acid groups and the amino groups of amino acids can ionize and become charged:

$$COOH \rightarrow COO^- + H^+$$
$$NH_2 + H^+ \rightarrow NH_3^+$$

Therefore, carboxyl groups behave as acids, and amino groups behave as bases. Since proteins contain both amino and carboxyl groups, they can behave as both acids or bases and are called *amphoteric* substances. The degree of acidity or alkalinity affects food quality and depends on the pH of the medium surrounding the food protein.

At certain pH (in the acid range, having excess H^+ ions), the total number of + charges exceeds the total number of − charges on the surface of the protein molecule. In the acid pH range, the net charge on the protein would be positive, which helps it to be polar and soluble in water. At certain pH (in the alkaline range, having excess OH^- ions), the total number of − charges exceeds the total number of + charges on the surface of the protein molecule. In the alkaline pH range, the net charge on the protein would be negative, and helps it to be polar, and soluble in water.

Denaturation. Chemically, **denaturation** occurs as an unfolding of protein structure (due to H bonds breaking) without disrupting protein covalent bonds. Original properties of protein will change during denaturation (e.g., enzyme function will be stopped; solubility in water will decrease). Thermal processing, as in the roasting of meat, denatures the meat proteins actin, myosin, and myoglobin. Cooking an egg denatures egg white proteins including ovalbumin. Whipping egg whites denatures some of the protein and helps stabilize egg white foams.

Denaturation can be thought of as the relaxation of protein tertiary structure, with decreased solubility and altered functional properties. Denaturation behavior is an event unique to proteins (starch, lipids, and vitamins do not denature).

Coagulation is distinct from denaturation. Coagulation is the precipitation of protein as individual molecules aggregate (often as a result of energy input, such as heating, or acid treatment)—see Figure 5.18.

Emulsification. Proteins can stabilize emulsions by acting at the oil-water interface. This interface is interesting, because it represents a boundary between immiscible phases: oil and water. Since protein molecules contain both hydrophilic and hydrophobic characteristics, they can situate in between the two phases to stabilize them. This functional property is important in the formation of many common food products such as salad dressings, sauces, frankfurters, and sausages. Foods such as meat tissue, milk, eggs, and soy contain proteins that can be isolated and function as emulsifiers.

Enzymes. Enzymes are protein molecules that function to speed up chemical reactions without being used up in the process. They are used to produce products in the food industry that range from modified starches to protein hydrolysates. Several examples are given in Table 5.11. Modified polymer substrates have been found to offer enhanced funtionalities in many cases. As a specific example, the major sweetener used in soft drinks, high-fructose corn syrup (HFCS), is produced by a controlled enzymatic process that changes corn starch. Corn starch is converted into HFCS by several enzymes during three processing steps: (1) liquefaction via amylase to produce dextrins, (2) saccharification via fungal enzymes to produce glucose syrup, and (3) isomerization via glucose isomerase to produce a 42 percent fructose HFCS.

TABLE 5.10 Important reactions and functional properties of food proteins.

Reactions
   Denaturation
   Hydrolysis
Functional Properties
   Buffering
   Coagulation
   Emulsification
   Enzymes
   Fat reduction
   Foaming
   Gelation
   Solubility
   Water-holding capacity

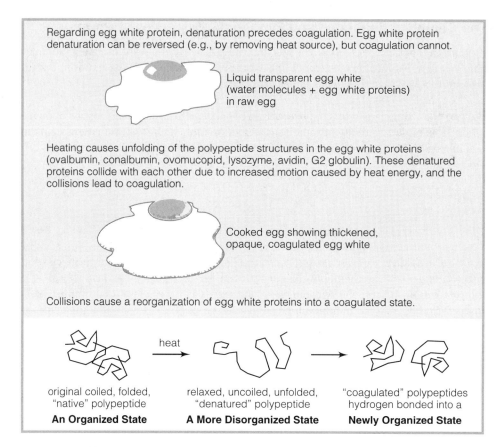

FIGURE 5.18 Denaturation and coagulation of egg white upon heating.

Factors affecting enzyme activity include temperature, pH, and moisture. Optimum temperature for enzyme function occurs at 30–40°C. Below 0°C, enzyme action slows down; above 45°C enzymes begin to denature and become inactivated. Optimum pH for enzyme activity is from 7 to 10; enzymes generally are inactive at pH < 6.2 and pH > 10.8. Water is important for all chemical reactions, including enzymatic reactions. It serves as a medium in which the reactants can be brought together. Dehydration of foods keeps enzymes and substrates apart, reducing enzyme activity. However, enzyme reactions are not eliminated even in dried fruits, which must be blanched before they are dried in order to stop polyphenol oxidase activity.

Fat Reduction. Microparticulated Protein (Simplesse®) is a reduced-calorie (1–2 kcal/gram) ingredient made from whey protein or milk and egg protein. It has been used in baked goods, coffee creamers, dairy products, salad dressing, sauces, and soups. Modified whey protein concentrate is another fat replacer. Controlled thermal denaturation of whey protein results in a functional protein with fat-like properties. Applications include baked goods, dairy products, frostings, and salad dressings.

Foaming. Foams are colloidal dispersions of gas in liquid (look back at the Challenge in Chapter 4). Proteins that are good foaming agents are egg, milk. and soy proteins. Food foams produced with proteins include ice cream, whipped toppings, and beer froths.

Gelation. Proteins can form a well-ordered gel matrix by balancing protein-protein and protein-solvent interactions in food products. These gel matrices can hold water, fat, and other food ingredients to produce various food products, including bread dough, comminuted meat products, gelatin desserts, tofu, and yogurt.

TABLE 5.11 Examples of several food enzymes. *Some enzymes (e.g., amylase, catalase, pectinase, protease, rennet) are used in the food industry to improve certain products or to create new ones. Other enzymes (e.g., chlorophyllase, lipase, PPO) are naturally present in foods and must be controlled or deactivated in order to prevent unwanted reactions from occurring in foods and beverages.*

Food Enzyme	Reaction
Amylase	Hydrolysis of starch
Chlorophyllase	Changes color of chlorophyll
Catalase	Decomposition of hydrogen peroxide (H_2O_2)
Glucose oxidase	Oxidation of glucose
Lactase	Hydrolysis of lactose
Lipase	Hydrolysis of triglycerides
Lipoxygenase	Oxidation of unsaturated lipid
Pectinase	Clarification of wines; modification of jam and jelly texture
Peptidase	Hydrolysis of proteins
Polyphenol oxidase (PPO)	Browning of fruits and vegetables
Proteases	Elimination of chill haze in cold meat packaging; meat tenderizers
Rennet	Coagulation of milk protein
Sucrase	Hydrolysis of sucrose

Hydrolysis. In protein hydrolysisis, protein + water + protease enzymes produce amino acids. On the other hand, nonenzymatic hydrolysis is the breaking apart of molecules due to extremes of heat or pH. This can occur to proteins during food processing operations as well as in the home during cooking, and will generate amino acids as breakdown products.

Solubility. Protein functional roles are related to the various characteristics of hydrated protein. Protein solubility is affected by pH and temperature. Highly soluble proteins are required to produce whipped products, protein films, and emulsions.

Water-Holding Capacity. The ability of a protein molecule to bind water has to do with the presence of hydrophilic and charged groups in its structure. As an example, meat can retain water during application of external forces such as cutting, heating, grinding, and pressing. This property is called the **water-holding capacity** (WHC) of protein. Several factors influence the water-holding capacity of proteins. The increase in water binding by meat proteins can be explained in terms of pH, salt, and temperature effects.

When the negative and positive charges on a protein equal each other, protein:protein interactions are at maximum. On the other hand, when the protein is not electrically neutral, these interactions lessen, allowing for greater water:protein associations. By increasing the salt concentration, more Na^+ and Cl^- ions are available to bind to the charged groups on protein fiber molecules. This reduces the protein fiber associations with each other in favor of increased protein fiber:water associations. As temperature increases to 80°C, water binding increases in proteins that form thermally induced gels, because gelation traps water inside a three-dimensional gel network in addition to creating gel surface binding of water molecules.

Key Points

- Reducing sugars such as the monosaccharides are reactive in foods and can contribute to both desirable and undesirable quality changes.
- Important functional properties of sugars include browning, crystallization, humectancy, inversion, sweetness, and texturizing.
- The chemical structures of starch and other polysaccharides are based upon glucose or other sugar units linked together.
- The Maillard reaction occurs in foods containing proteins and reducing sugars and can produce browning as well as flavor and aroma changes.
- Amylose and amylopectin, owing to their different structures, do not have the same functional uses in foods in terms of thickening and gelation.
- Heating starch in water leads to important physicochemical changes, including gelatinization, pasting, gelling, and retrogradation.
- Food lipids, as fats and oils, are composed of a variety of fatty acids in the form of triglycerides, contributing to food flavor and aroma.
- Fatty acids may contain double bonds (cis- or trans-unsaturated) or not (saturated), and these characteristics influence hardness, melting behavior, and tendency to oxidize.
- Fat hydrogenation is a process applied to liquid vegetable oils in order to solidify them.
- Lipid oxidation is a chemical reaction sequence that results in the formation of off-flavors and odors associated with rancidity.
- Food proteins are fibrous or globular, can be conjugated to other substances, and have primary, secondary, tertiary, and quaternary levels of structure.
- In solution, proteins are affected by pH changes, being unstable at their isoelectric point (isoelectric pH), and stable at other pH values, due to surface charge effects.
- Casein proteins are organized into submicelle and micelle structures.
- Functional differences exist between milk proteins casein and whey in terms of solubility, due to structural differences.

Study Questions

1. All of the following are monosaccharides except
   a. glucose.
   b. fructose.
   c. dextrose.
   d. maltose.
   e. galactose.

2. Which substance would *not* possess glycosidic bonds?
   a. galactose
   b. pectin
   c. lactose
   d. cellulose
   e. starch

3. Which choice does *not* belong with the others?
   a. Maillard reaction
   b. nonenzymatic browning
   c. caramelization
   d. condensation, rearrangement, polymerization
   e. reducing sugar + amino group → glycosylamine

4. In the commercial manufacture of sugar, which is the key step?
   a. caramelization
   b. gelatinization
   c. polymorphism
   d. graining
   e. crystallization

5. The reaction *glucose + hydrogen* → *sorbitol*
   a. produces a sugar alcohol.
   b. results in oxidation of the sugar substrate.
   c. results in reduction of the sugar substrate.
   d. Both (a) and (b) are correct.
   e. Both (a) and (c) are correct.

6. Which polysaccharide functions as both a prebiotic and ingredient for low-fat foods?
   a. beta-glucan
   b. inulin
   c. carrageenan
   d. methyl pectate
   e. retrograded starch

7. Branched hydrocolloid polysaccharides can form gels more easily than linear ones because
   a. gels form only in branched polysaccharides.
   b. branched polysaccharides are more soluble than linear polysaccharides.
   c. branched polysaccharides exhibit greater polymer:water interactions.
   d. branching brings reactive functional groups in close proximity through entanglement.
   e. linear polysaccharides exhibit greater polymer:polymer interactions.

8. Which describes the correct changes in a heated then cooled starch plus water system?
   a. starch suspension → sol → paste → gel
   b. starch paste → gel → paste → sol
   c. starch suspension → paste → sol → gel
   d. starch gel → paste → sol → gel
   e. starch sol → suspension → paste → gel

9. In the manufacture of hard candy based on sucrose, added acid would accomplish the following:
   a. hydrogenation of sucrose.
   b. decrease in sweetness.
   c. impart creamy texture.
   d. increase the content of nonreducing sugars.
   e. hydrolysis of sucrose.

10. All of the following are functional properties of polysaccharides *except*
    a. gelation.
    b. water binding.
    c. thickening.
    d. enhanced sweetness.
    e. fat replacement.

11. Which choice does not belong with the rest?
    a. oleic acid
    b. sixteen carbon atoms in length
    c. fatty acid
    d. C18:1
    e. monounsaturated

12. All of the following are lipid functional properties *except*
    a. aeration.
    b. heat transfer.
    c. gelation.
    d. flavor.
    e. plasticity.

13. Which is the *incorrect* statement regarding linoleic acid?
    a. It exists in nature in the cis form.
    b. It is a polyunsaturated fatty acid.
    c. It would have a higher melting point than stearic acid, C18:0.
    d. The location of the first double bond is between carbon 6 and 7 from methyl end.
    e. It is an omega-6 fatty acid.

14. The commercial _____ of oils is done to harden them and increase shelf life.
    a. oxidation
    b. hydrogenation
    c. hydrolysis
    d. interesterification
    e. lipoxidation

15. During lipoxidation, when a fatty acid free radical reacts with oxygen,
    a. hydroperoxide is produced.
    b. ROO• is produced.
    c. R• is produced.
    d. ROOH is produced.
    e. RH is produced.

16. What is *true* regarding S—S—S and S—O—P molecules?
    a. Both are monoglycerides.
    b. Both are pure substances.
    c. S—S—S is an unsaturated triglyceride while S—O—P is a saturated triglyceride.
    d. S—S—S would show a gradual melting curve while that of S—O—P would be sharp.
    e. their melting behaviors would differ.

17. In the enzymatic reaction: *PPO enzyme + fruit pigment → ES complex → PPO enzyme + brown pigment*, which substance is the oxidized molecule in the reaction, as written?
    a. PPO enzyme
    b. fruit pigment
    c. ES complex
    d. brown pigment
    e. all of these

18. Which statement regarding amino acids is *true*?
    a. Alanine, leucine, and serine are classified as neutral amino acids.
    b. Basic amino acids contain COOH groups in their side chains.
    c. The sulfur amino acids are methionine, histidine, and phenylalanine.
    d. The sequence of amino acids linked together by peptide bonds is referred to as tertiary structure.
    e. When two amino acids form a peptide bond, the atoms involved are N—C—C—N—C—C.

19. Which process does *not* take place in the cooking of eggs?
    a. denaturation, followed by coagulation
    b. denaturation of egg white proteins
    c. thickening of the egg white
    d. denaturation of egg yolk lipids
    e. reorganization of coagulated ovalbumin polypeptides

20. At the isoelectric pH of a protein,
    a. pH = pI
    b. pH < pI
    c. pI > pH
    d. solubility is at a maximum
    e. a protein is at its most stable state

Internet Search

At the PD Lab website (http://www.pdlab.com/pdproteinx.htm), read about the "salting in" and "salting out" *protein* effects. Then answer the following questions:

21. T or F: "Salting out" refers to reducing the solubility of a substance, such as protein, such that it may precipitate by the addition of added solute, for instance, salt.

22. T or F: Certain substances normally insoluble in water can be dissolved by the addition of salt; this is referred to as "salting out."

23. T or F: In a "salting in" scenario in which salt is added to protein in water, repulsion between proteins would be reduced, allowing them to approach and form weak bonds, and possibly to form a precipitate.

References

Alexander, R. J. 1998. *Sweeteners: Nutritive.* American Association of Cereal Chemists, Eagan Press, St. Paul, MN.

Best, D. 1988a. Archi-texturing foods with proteins. *Prepared Foods* 157(4):106.

Best, D. 1988b. Fats and oils crystallize formulation opportunities. *Prepared Foods* 157(9):10.

Blumenthal, M. M. 1991. A new look at the chemistry and physics of deep fat frying. *Food Technology* 45(2):78.

Carr, R. A. 1991. A development of deep frying fats. *Food Technology* 45(2):95.

Coultate, T. P. 1996. *Food: The Chemistry of Its Components.* Royal Society of Chemistry, Cambridge, UK.

Daniel, J. R., and Weaver, C. M. 2000. Carbohydrates: functional properties. In: *Food Chemistry: Principles and Applications.* Science Technology System, West Sacramento, CA.

Deis, R. C. 1996. New age fats and oils. *Food Product Design* 6(11): 27.

DeMan, J. 1999. *Principles of Food Chemistry.* Aspen Publishers, Gaithersburg, MD.

Duncan, S. E. 2000. Lipids: basic concepts. In: *Food Chemistry: Principles and Applications.* Science Technology System, West Sacramento, CA.

Fennema, O. R. 1996. *Food Chemistry.* Marcel Dekker, New York.

Fox, P. F. 1984. The milk protein system. In: *Food Science and Technology: Present Status and Future Direction,* J. V. McLouglin, and B. M. McKenne, eds. Boole Press, Dublin, Ireland.

Giese, J. 1994. Proteins as ingredients: types, functions, applications. *Food Technology* 48(10):50.

Hoseney, R. C. 1986. *Principles of Cereal Science and Technology.* American Association of Cereal Chemists, St. Paul, MN.

Kuntz, L. 1996. Bulking agents: bulking up while scaling down. *Food Product Design* 6(6): 27.

Mathewson, P. 1998. *Enzymes.* American Association of Cereal Chemists, Eagan Press, St. Paul, MN.

McWilliams, M. 1997. *Foods: Experimental Perspectives.* Prentice Hall, Upper Saddle River, NJ.

Miller, D. 1998. *Food Chemistry: A Laboratory Manual.* John Wiley & Sons, New York, NY.

Morris, V. J. 1998. Gelation of polysaccharides. In: *Functional Properties of Food Macromolecules.* Aspen Publishers, Gaithersburg, MD.

Schmidt, K. 2000. Lipids: functional properties. In: *Food Chemistry: Principles and Applications.* Science Technology System, West Sacramento, CA.

Smith, D. M., and Culbertson, J. D. 2000. Proteins: functional properties. In: *Food Chemistry: Principles and Applications.* Science Technology System, West Sacramento, CA.

Stauffer, C. E. 1996. *Fats & Oils.* American Association of Cereal Chemists, Eagan Press, St. Paul, MN.

Technical Committee of the Institute of Shortening and Edible Oils. 1988. *Food Fats and Oils.* Institute of Shortening and Edible Oils, Washington, DC.

Thomas, D. J., and Atwell, W.A. 1999. *Starches.* American Association of Cereal Chemists, Eagan Press, St. Paul, MN.

Tolstoguzov, V. B. 1998. Functional properties of protein-polysaccharide mixtures. In: *Functional Properties of Food Macromolecules.* Aspen Publishers, Gaithersburg, MD.

Wilhelm, C. 2000. Focusing on fat (again). *Food Quality* March/April 2000: 50.

Wilkes, A. P. 1992. Keeping it together: selecting and using emulsifiers. *Food Product Design* 2(5):88.

Additional Resources

Brown, A. 2000. *Understanding Food: Principles and Preparation.* Wadsworth/Thomson Learning, Belmont, CA.

Chandan, R. 1997. *Dairy-Based Ingredients.* American Association of Cereal Chemists, Eagan Press, St. Paul, MN.

Charley, H., and Weaver, C. 1998. *Foods: A Scientific Approach.* Prentice Hall, Upper Saddle River, NJ.

Corriher, S. O. 1997. *Cookwise: The Hows and Whys of Successful Cooking.* William Morrow and Company, New York.

Edwards, W. P. 2000. *The Science of Sugar Confectionery.* The Royal Society of Chemistry, Cambridge, UK.

Hartel, R. W. 1993. Controlling sugar crystallization in food products. *Food Technology* 47(11):99.

Hernandez, E., and Lusas, E.W. 1997. Trends in transesterification of cottonseed oil. *Food Technology* 51(5):72.

Kennedy, J. P. 1991. Structured lipids: fats of the future. *Food Technology* 45(11): 76.

Kevin, K. 1995. The next generation of fat replacers. *Food Processing* 56(7): 63.

McClements, D. J. 1999. *Food Emulsions: Principles, Practice, and Techniques.* CRC Press, Boca Raton, FL.

Platzman, A. 2000. Conjugated linoleic acid—miracle nutrient? *Food Product Design* 10(6):38.

Rasco, B., and Zhong, Q. 2000. Proteins: basic concepts. In: *Food Chemistry: Principles and Applications.* Science Technology System, West Sacramento, CA.

Weaver, C. 1996. *Food Chemistry Laboratory.* CRC Press, Boca Raton, FL.

Whitney, E. N., and Rolfes, S. R. 2002. *Understanding Nutrition,* 9th ed. Wadsworth/Thomson Learning, Belmont, CA.

Web Sites

http://www.ift.org/divisions/food_chem/
IFT Food Chemistry Division. The IFT site has the composition and properties of food and food ingredients, and the chemical, physical, and biological changes in food constituents as a result of preparation, processing, and storage.

http://www.pdlab.com/
PD Lab. This central site for product developers in the food, beverage, and pharmaceutical industries has food science, chemistry, references, popular consumer issues, suppliers, and a range of scientific information.

sugars	*http://www.pdlab.com/pdsugarsx.htm*
carbohydrates	*http://www.pdlab.com/pdcarbox.htm*
fats and oils	*http://www.pdlab.com/fatsx.htm*
protein	*http://www.pdlab.com/pdproteinx.htm*

MILK PROTEIN CHEMISTRY

An important food nutritionally, milk is an excellent source of high-quality protein. The chemistry and functional properties of milk proteins play essential roles in creating high-quality dairy products. For this Challenge, we will review some of the chemistry of milk proteins, especially as it relates to cheese.

The important milk proteins are caseins and whey proteins. Lactalbumin, lactoglobulin, and immunoglobulins are the major whey proteins. Caseins make up approximately 80 percent of all milk proteins and whey about 20 percent. Alpha $(\alpha)_{S1}$-casein, α_{S2}-casein, β-casein, κ-casein, and γ-casein are the major casein proteins. The gamma fraction derives from proteolytic enzyme breakdown in milk of β-casein. Each fraction exhibits chemical differences. Alpha-S1 contains eight special structures called phosphoserine units; alpha-S2 has about a dozen; beta has five phosphoserine units; and kappa, only one. Besides their biological and nutritional role, caseins and caseinates are important because of their structure, charge, and functional properties. Caseins become insoluble when milk is acidified to pH 4.6, whereas whey proteins remain colloidally dispersed.

Caseins exist in a disordered tertiary state as random coiled molecules. Whey proteins have a more ordered globular structure containing disulfide linkages. Like other globular proteins, they can be heat-denatured, which results in gel formation. Caseins are more resistant to heat denaturation because of their structure. However, they can form gels with addition of rennin, and precipitate in the presence of acid. Table 5.12 summarizes the major functional differences.

Casein and Isoelectric pH

A critical pH for proteins in solution called the **isoelectric point** (or isoelectric pH) occurs when the number of H^+ ions equals the number of OH^- ions: The total number of positive and negative charges on the surface of a protein molecule are equal. Thus, the net charge on the protein is zero. At the isoelectric pH (pI), the protein molecule is unstable and is at its most insoluble state. This instability causes the protein molecules in solution to form hydrogen bonds with each other (protein:protein interactions), producing protein clumps and precipitates that separate from water (Figure 5.19). This phenomenon is exploited in cheesemaking. The formation of a solid protein cheese curd is due to the clumping of casein at its isoelectric pH (a pH value below 5).

Casein Micelles

Caseins exist within micelles, each containing thousands of casein polypeptide molecules. **Casein micelles** are large colloidal particles composed of calcium phosphate complexed to casein. Within each micelle are aggregates of submicelles composed of alpha, beta, and kappa casein polypeptides. All of the casein fractions react with calcium and become precipitated except kappa. Calcium and phosphorus complex with the alpha and beta caseins, while kappa casein stabilizes the colloidal casein particles by surface binding to water (the hydophilic amino acids of kappa casein H-bond with water molecules).

Functional Property	Casein Proteins	Whey Proteins
Emulsification	Excellent	Good
Foaming	Good	Good
Gelation	No thermal gelation without added Ca^{++} or rennin	Thermal gelation at 70°C
Solubility	Very high (except pH < 4.6)	Soluble at all pH

TABLE 5.12 Major functional differences between casein and whey proteins in milk.

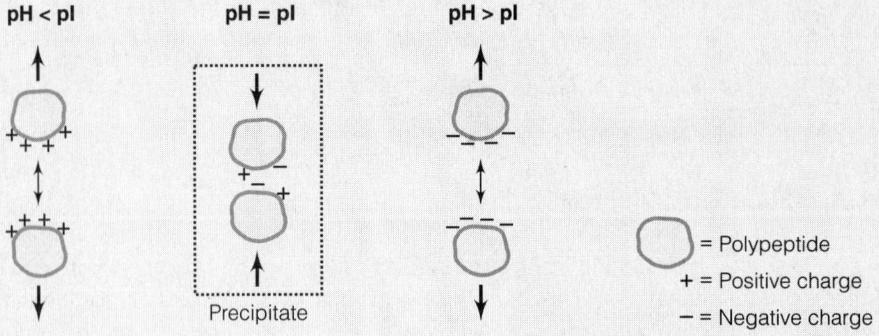

FIGURE 5.19 The effect of pH on the behavior of proteins in solution. *Above and below the isoelectric pH of a protein (e.g., casein), solubility is enhanced because the individual protein molecules carry net positive or negative charges, causing repulsion. Protein:protein interactions are minimized, while protein:water interactions are promoted. However, at the protein's isoelectric pH, which for casein is around pH 4.6, there is a net neutral charge, which, combined with hydrophobic attractions of certain amino acid side chains, contribute to the precipitation of protein molecules out of solution. (Adapted from Best, 1988a.)*

The phosphate groups of alpha and beta casein, but not kappa, react with Ca and P to form crosslinks between submicelles, or to form chains in which phosphate and also citrate are linked together with Ca. Kappa-casein can aggregate within each submicelle so that its hydrophilic, yet noncalcium binding, ends form a "cap" where no crosslinking occurs. This results in the kappa form dominating the micelle surface.

Casein micelles are stable in milk as colloids, unless rennin enzyme is applied, or a shift in pH down toward the pI is created. To make cheese from milk, casein micelles must be destabilized. Therefore, in cheese manufacture, the stabilizing effect of kappa-casein on the micelle surface must be eliminated. The enzyme rennin catalyzes the splitting of kappa-casein into the micelle remnant para-κ-casein, which is hydrophobic, and κ-casein macropeptide, which is lost to the whey liquid. In the presence of calcium, para-κ-casein becomes insoluble. Thus destabilized, the casein micelle remnants aggregate to form a gel, the cheese curd.

Challenge Questions

1. Which milk protein, casein or whey protein, would be a better choice as an ingredient in a protein formulation that needs to heat-set into a gel?
2. Whey protein, unlike casein, will not precipitate at its isoelectric point unless a prior heat treatment is applied. Why might heat make a difference?

Multiple Choice

3. The casein submicelle consists of all of the following except
   a. alpha-casein polypeptides.
   b. macropeptide.
   c. polar amino acids.
   d. beta-casein polypeptides.
   e. kappa-casein polypeptides.

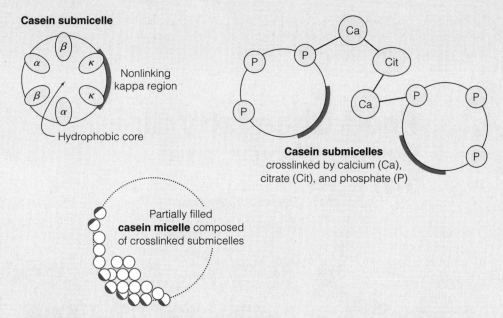

FIGURE 5.20 Casein submicelles and the casein micelle. *Recent models describe the micelles as aggregates of submicelles, each containing alpha-, beta-, and kappa-casein polypeptides. Their polypeptide chains fold into a spherical mass having at one end a predominance of hydrophilic amino acids, which are serine residues to which phosphate groups are esterified (in alpha-casein). At another end, kappa-casein does not crosslink, and its hydrophilic ends dominate the micelle surface.*

4. Casein submicelles contain crosslinks of
    a. phosphate.
    b. calcium.
    c. citrate.
    d. both (a) and (b).
    e. choices (a), (b), and (c).

5. The nonlinking regions of the submicelles and micelles
    a. contain alpha-casein polypeptides.
    b. form phosphate crosslinks.
    c. create important protein surface stability effects.
    d. form calcium crosslinks.
    e. are located at the hydrophobic core of the submicelles.

6. Kappa-casein macropeptide
    a. is a piece of the original casein micelle.
    b. is hydrophobic.
    c. is an enzyme that splits kappa-casein apart.
    d. helps whey proteins to aggregate in the cheese curd.
    e. forms crosslinks with calcium, citrate, and phosphate.

7. At the isoelectric pH of casein,
    a. the number of H^+ ions > the number of OH^- ions.
    b. the net charge on the protein is either negative or positive overall.
    c. protein:water associations are favored.
    d. casein solubility is increased.
    e. protein:protein associations are favored.

CHAPTER 6

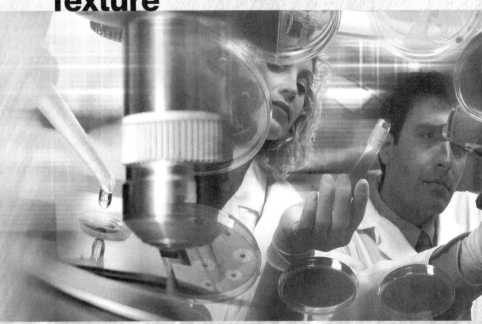

Food Chemistry III: Color, Flavor, and Texture

6.1 Food Color Chemistry

6.2 Food Flavor Chemistry

6.3 Food Texture

Challenge!
The Chemistry of Sweeteners and Sweetness

CHAPTER OBJECTIVES

After completing this chapter, you will be able to:
- Explain the chemical basis for red meat color.
- List the names and important chemical features of plant pigment molecules.
- List the certified food colorants and examples of those exempt from certification.
- Describe the four basic tastes and distinguish them from umami and astringency.
- State the significance of the Maillard and Strecker degradation reactions.
- Explain the purpose of encapsulation with respect to the flavor industry.
- Match specific texturizing ingredients with their functionalities.
- Explain the relationship between water activity and texture.
- Describe the chemical interactions that stabilize a gelatin gel system.
- Explain the chemical basis of sweetness according to the AH,B theory.

Color, flavor, and texture are three key factors that drive consumer acceptance of foods. When we encounter off-colors, off-flavors, or poor textures in foods, we do not buy or eat the foods. Understanding the chemistry of the molecules responsible for color, flavor, and texture as influenced by temperature, oxygen, and various elements and molecules has opened up a field of functional ingredient applications. Determining the effects of special functional ingredients including polysaccharides and proteins on preventing color, flavor, and texture changes has become a key area in food research and in product development.

What makes sugar sweet? In the Challenge section, the chemical basis of sweetness is explored.

6.1 FOOD COLOR CHEMISTRY

The appeal of color is a major influence on consumer perception and food acceptability and preference. Color can even affect the way consumers perceive flavor! In understanding the chemical basis of food color, the food technologist can address quality assurance issues regarding color stability, important in both unprocessed foods and processed foods during storage.

What Is Color?

Color describes a perception of a physical attribute of food arising from a collection of sensations. These sensations originate from the rods and cones in the retina of the eye. The rods are sensitive to lightness and darkness, while the cones are sensitive to red, green, or blue color. Light striking a food object is either absorbed, reflected, or transmitted. It is *reflected* light that determines the color of a food.

The electromagnetic spectrum of visible light is split into wavelengths ranging from 380 nm to 780 nm. The colors red, orange, yellow, green, blue, indigo, and violet are described by specific wavelengths in this range of visible light (Figure 6.1). If a food object reflects all of these colors, it is perceived as white, while one that absorbs all of them appears black. Broccoli is green because it absorbs white light except for the light rays that produce the sensation of green as perceived by the brain. The three classes of cones

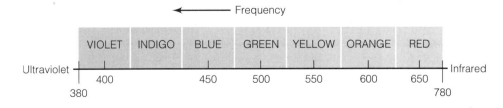

FIGURE 6.1 Color and visible light. *The wavelengths of color (in nm) derived from white light in the visible spectrum.*

TABLE 6.1 The five major groups of natural pigment molecules. *An understanding of the fundamental chemistry of these compounds provides the basis for the formulation and use of color additives, for predicting the effects of food processing on natural food color, and for understanding undesirable food color changes due to microorganisms, heat, oxygen, light, metal ions, and catalysts that enhance the rate of oxidation and reduction reactions.*

Pigment	Source	Solubility
Anthocyanins	Plants	Water-soluble
Betalains	Plants	Water-soluble
Carotenoids	Plants	Lipid-soluble
Chlorophylls	Plants	Lipid-soluble
Myoglobin	Animals	Water-soluble

overlap in their sensitivity to color, meaning that many shades of the color green, for instance, are possible, since green light stimulates not only green cones, but also red and blue ones.

The surface color of food objects can be characterized by three qualities: (1) **hue,** which is the actual color name (red, blue, and green), (2) saturation or **chroma** (the clarity and purity of the color), and (3) **intensity,** the range from lightness to darkness of color. In addition to color, food surfaces exhibit different qualities based upon light distribution. These surface appearance characteristics are (1) shiny, (2) glossy, (3) cloudy, and (4) translucent. The first two are a result of reflected light, while the last two are related to transmission of light through food objects.

Pigment Molecules

Color, in the physical sense, refers to a hue such as the red color of ground beef. The cause of the color has to do with the presence of special molecules in the beef called *pigment molecules*. Since color is one of the most important quality attributes of foods, we must understand the chemistry of pigment molecules responsible for food color. Pigment or **chromophoretic** compounds in foods constitute a structurally diverse group and possess extremely complex chemical and physical properties. Color compounds can be classified according to chemical structures, which will be discussed in the sections below.

There are five major groups of natural food pigment molecules. Four of these are distributed throughout the plant kingdom and the fifth is found in the animal kingdom (Table 6.1). The lipid-soluble chlorophylls and carotenoids and the water-soluble anthocyanins and betalains are found in plants. In animal tissues, meat color is due to the heme protein myoglobin. In some fish tissue such as salmon, the bright orange is due to carotenoid pigments, although these are not biosynthesized by the fish but derived from plant sources.

The Color Chemistry of Red Meat

Why is the color of fresh red meat bright red, but meat that is not fresh is brown? To understand the reason, we must examine the muscle pigment molecule myoglobin present in meats. **Myoglobin** is a single polypeptide, a globular protein containing the *globin* protein part and a prosthetic group called *heme*—see Figure 6.2a. Globin, as a polypeptide, is composed of amino acids. It is attached to the heme in both myoglobin and hemoglobin. Myoglobin is a special kind of protein that, like hemoglobin, has the ability to bind oxygen. It can bind oxygen because of its globin polypeptide tertiary structure and because of the heme, which is an iron porphyrin ring complex.

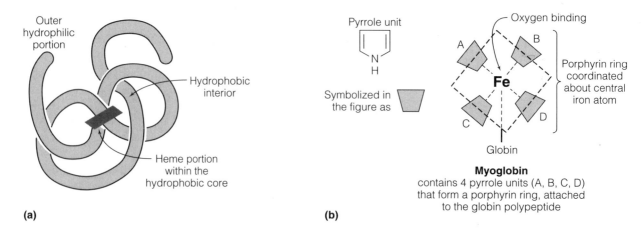

FIGURE 6.2 The structure of myoglobin. *(a) Myoglobin has a folded globular polypeptide structure. Everything except the heme portion is the globin. (b) The heme portion of myoglobin with the iron porphyrin ring complex.*

What exactly is a porphyrin ring? The porphyrin ring proper is composed of four *pyrrole units* (connected by methane bridges). The pyrrole units have nitrogen atoms in the corners. Iron (Fe) is present in the center of the porphyrin ring, linked by covalent bonds to the nitrogen atoms of the four pyrrole units. A fifth coordination site links the Fe atom to a nitrogen atom of the globin protein in myoglobin—see Figure 6.2b.

The iron atom has a sixth binding site and can bond to any atom (such as oxygen) able to donate a pair of electrons. In a living animal before slaughter, myoglobin serves to store the oxygen transferred to it by hemoglobin at this sixth site. The unoxygenated myoglobin (Mb) form exists in equilibrium with its oxygenated form (O_2Mb) in vivo:

$$Mb + O_2 \rightleftharpoons O_2Mb$$

Upon slaughter, the oxygenated myoglobin is no longer formed, so the shift in equilibrium is to the left toward all unoxygenated myoglobin. The iron in the unoxygenated myoglobin exists in the reduced state, as ferrous (Fe^{+2}) iron. Oxidation and reduction of the iron atom in myoglobin are linked to color changes in meat.

What happens when meat obtained from a slaughtered animal is cut and exposed to air? Myoglobin's functional property is to bind oxygen. After meat is cut and comes in contact with oxygen, the unoxygenated myoglobin's iron (ferrous iron, Fe^{+2}) loosely binds O_2 and converts the myoglobin to **oxymyoglobin,** creating a bright red color. The ferrous Fe^{+2} myoglobin binds oxygen at the sixth binding site, and the oxygen becomes part of the pigment complex. This process is called oxygenation, which is distinct from oxidation. If exposure to oxygen continues, oxidation does take place, and production of metmyoglobin results. **Metmyoglobin** is associated with aged meat exposed to air and produces a grayish or brown meat color. The iron in metmyoglobin changes from the reduced Fe^{+2} to the oxidized Fe^{+3} (ferric) state, which alters the color. Figure 6.3 shows these forms of myoglobin.

In nitrate-cured meats, myoglobin reacts with nitric oxide (NO) to produce **nitric oxide myoglobin,** or nitrosylmyoglobin, which is bright pink-red. When cured meats are cooked, they retain the pinkish or reddish color, due to the formation of nitrosohemochrome from denatured, coagulated nitrosylmyoglobin. Brown color development can occur during storage of cured meats, since oxygen and light exposure causes oxidation of ferrous iron to ferric iron.

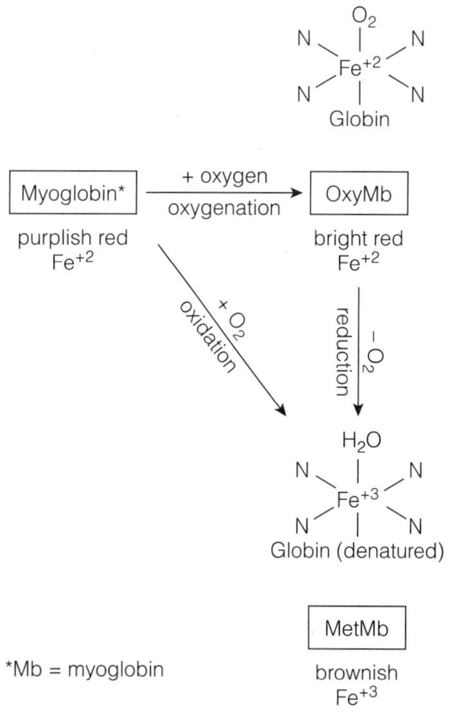

FIGURE 6.3 Chemical forms of myoglobin in meat.

*Mb = myoglobin

Myoglobin   —nitric oxide→   Nitrosylmyoglobin   —heat→
(purplish red) Fe^{+2} (red) Fe^{+2}

Denatured   —oxidation, light, air→   Denatured
Nitrosohemachrome   Nitrosohemochrome
(pink) Fe^{+2} (brown) Fe^{+3}

Other forms of myoglobin are possible, such as *choleglobin* and *sulfmyoglobin*, which result from bacterial action on contaminated meat to yield green colors. Oxidation of the porphyrin ring can produce a yellow or green color.

In addition to these chemical factors, meat color at the point of purchase depends upon factors such as the packaging material, the number of bacteria, the species of animal, the animal's age, the amount of myoglobin in the animal's tissue, and the particular muscle source.

The Color Chemistry of Fruits and Vegetables

The appealing color of fruits and vegetables is due to naturally occurring plant pigments that absorb and reflect light at certain characteristic wavelengths. All plant pigment molecules contain **conjugated double bonds,** which are alternating single and double carbon-to-carbon bonds: $C=C-C=C$. In addition, the chlorophylls contain metal-coordinated porphyrin rings. Color is a result of *resonance* within the ring structures in porphyrin rings (chlorophyll, myoglobin, and hemoglobin) as well as along conjugated carbon chains. *Resonance electrons* are spread across the atoms containing alternating single and double bonds, and their movement across carbon-to-carbon bonds gives color. Plant pigment molecules are classified into three groups based on structure: (I) the phenolic-based pigments (anthocyanins, anthoxanthins, and betalains), (II) the carotenoids and (III) the chlorophylls. See Table 6.2. The phenolic-based pigments contain at least one phenolic ring as a 6-membered aromatic ring, as exemplified by the benzene ring C_6H_6:

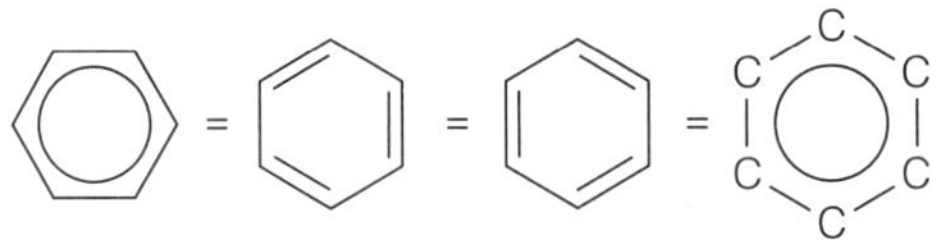

Anthocyanins. Anthocyanins are water-soluble flavonoid compounds that range in color from deep purple to orange-red. *Flavonoids* are chemically related phytochemicals and include the anthocyanins and anthoxanthins. Flavonoids contain two phenol rings and an intermediate ring of variable structure:

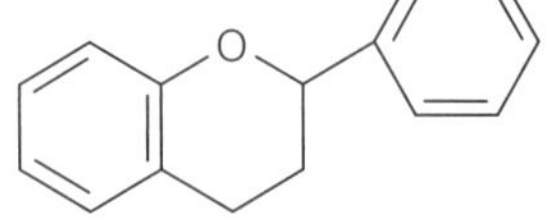

Anthocyanin pigment color is pH sensitive, being red in strong acid, colorless at pH 4, and blue at neutral pH. The chemical structure of anthocyanins is a combination of an aglycone molecule attached to a sugar molecule by a glycosidic bond—see Figure 6.4. An *anthocyanidin* is an anthocyanin molecule that has been broken apart into sugar and a free aglycone, which is the anthocyanidin.

Three prominent examples of anthocyanidins are pelargonidin, cyanidin, and delphinidin, which are red, red-blue, and blue pigments, respectively (Figure 6.5). When complexed to sugars, these would be called *anthocyanins* instead of anthocyanidins.

Anthoxanthins. Anthoxanthins are colorless or white pigments that can become yellow. They contribute only slightly to food color.

Betalains. The **betalains** represent a group of two types of water-soluble plant pigments: (1) betacyanins, which encompass about 50 violet-red pigments, and (2) beta-

6.1 Food Color Chemistry

TABLE 6.2 Chemical classification of plant pigments and food examples.

I. Phenolic-based Pigments	II. Carotenoids	III. Chlorophylls

Anthocyanins (red-purple)
   pelargonidin (red)
   cyanidin (red-blue)
   delphinidin (blue)
   Food examples: eggplant,
     radish, red cabbage,
     red potato
 	
Anthoxanthins (colorless, white,)
   but can become yellow)
   Food example: cauliflower (white,
     but becomes yellow at pH > 7.0)
 	
Betalains (purple-red/yellow)
   betacyanins (violet-red)
   betaxanthins (yellow)
   Food example: beets
	Carotenoids
	   carotenes
	     beta-carotene (yellow-orange)
	     lycopene (red-orange)
	     xanthophyll (yellow)
	   Food examples: carrots,
	     oranges, peaches, pineapples,
	     pink grapefruit, red and yellow peppers,
	     tomatoes, watermelon, winter squashes
		Chlorophylls (green
		   chlorophyll *a* (blue-green
		   chlorophyll *b* (green)
		   Food examples: broccoli,
		     green cabbage, kale,
		     lettuce, spinach

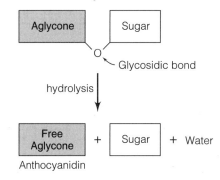

FIGURE 6.4 Relationships between anthocyanins, anthocyanidins, glucosides, and aglycones. *After hydrolysis of the sugar, the aglycone of the anthocyanin structure is referred to as an anthocyanidin.* Glucoside *is the generic term applied to an aglycone bonded to a sugar.*

xanthins, a series of about 20 yellow pigments. Betanin and betanidin are both examples of betacyanins, and vulgaxanthin is an example of a betaxanthin. Betanin is a glycoside found in beets, in which the sugar group (R) is glucose. Analogous to the anthocyanins/anthocyanidins, betanidin is the aglycone of betanin, in which the R group is hydrogen.

Betanin: glucose — O — R (R symbolizes the remainder of the betacyanin molecule)
Betanidin: H — O — R (R symbolizes the remainder of the betacyanin molecule)

Carotenoids. Carotenoids are a class of fat-soluble plant pigments that consists of carotenes and xanthophylls. Carotenes are hydrocarbons, while xanthophylls are oxygenated carotenoids containing alcohol, carbonyl, or other functional groups. Two examples of carotenes are beta-carotene (from carrots) and lycopene (from tomatoes). The substance lutein is an example of a xanthophyll. With the exception of lycopene, the carotenoids contain cyclic structural units called β-ionone rings, to which the functional groups are attached (Figure 6.6). These pigments contribute red, orange, and yellow as a result of resonance in isoprene units within the overall carotenoid structure. Carotenes are polymers of isoprene: $CH_2 = C — CH = CH_2$. They are easily oxidized at the double bond positions to alcohols, carbonyls, and other compounds. Such oxidation causes color loss in fruits and vegetables.

Chlorophylls. Chlorophylls are green lipid-soluble plant pigments that contain a porphyrin ring complexed to magnesium (Figure 6.7). In this regard, it is very similar to myoglobin. Phytol alcohol is attached by an ester linkage to one of the pyrrole groups, methyl alcohol to a second pyrrole group.

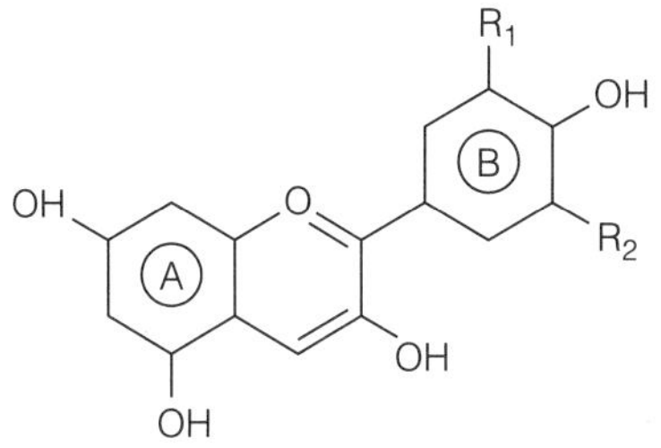

Anthocyanidin	Color	R_1	R_2
Pelargonidin	Red	H	H
Cyanidin	Red-Blue	OH	H
Delphinidin	Blue	OH	OH

FIGURE 6.5 Examples of three anthocyanidins and their color. *Pelargonidin, cyanidin, and delphinidin differ structurally by the number and position of hydroxyl groups on the B ring.*

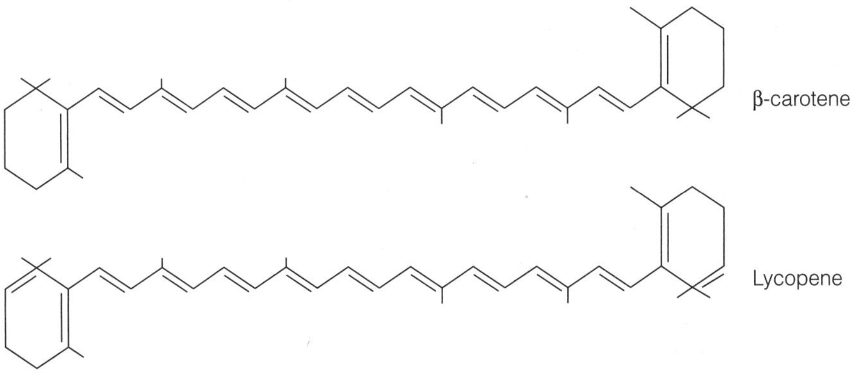

β-carotene

Lycopene

FIGURE 6.6 A comparison between the structures of beta-carotene and lycopene.

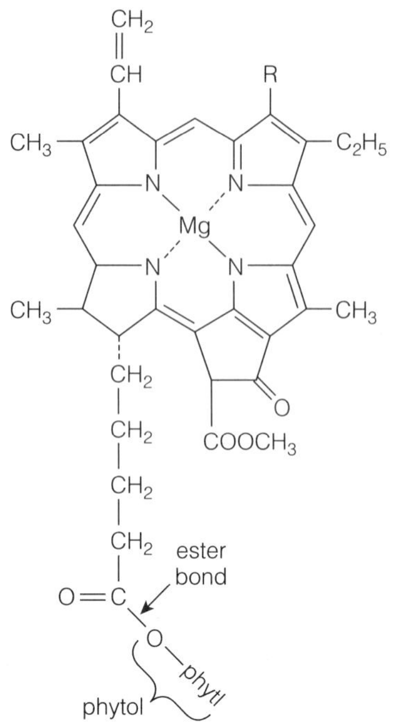

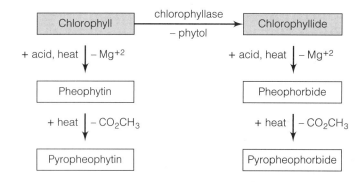

FIGURE 6.8 The effects of acid and heat treatment on chlorophyll.

FIGURE 6.7 The chemical structure of chlorophyll. *Magnesium is the central atom in the porphyrin ring. Phytl is a complex, long hydrocarbon chain, responsible for making chlorophyll insoluble in water.*

Plants contain two naturally occurring chlorophyll types: chlorophyll *a* and chlorophyll *b*. Vegetables may contain a number of chlorophyll *a* and *b* derivatives because of changes to these molecules during thermal processing or exposure to extreme pH environments. Such situations can alter their color. For example, chlorophylls *a* and *b* are degraded to produce a substance called **pheophytin.** In the process, acid and heat cause removal of the Mg^{+2} atom of chlorophyll, which is replaced in the pheophytin structure by hydrogen.

Reaction: Chlorophylls *a* and *b* + heat, acid → pheophytin *a* and *b*

An enzyme called *chlorophyllase* catalyzes the degradation of chlorophyll. This enzyme cleaves the phytol group, producing a compound called *chlorophyllide*. Stored, uncooked, green vegetables experience a limited amount of chlorophyllide production, which tints the color of cooking water green. Under certain conditions of heat and acid pH, vegetables may lose not only the phytol chain due to chlorophyllase activity, but also the magnesium atom. When this occurs, a chlorophyll derivative called *pheophorbide* results, which is a common cause of color change from green to gray-brown in canned green beans and brined cucumber pickles.

In general, the canning of green vegetables using low heat results in loss of bright green color and a change to dull olive green due to production of pheophytin. Sustained heating required during low-temperature, long-time processing can convert pheophorbides to *pyropheophorbides*, and pheophytins to *pyropheophytins* (Figure 6.8).

The Color Chemistry of Food Colorants

A **colorant** is a pigment used to impart color to a food or beverage. Around the year 1900, almost one hundred colorants were available for use in foods, even though their safety had not been tested or proven. These colorants were from two distinct categories of substances: (1) natural colorants or those naturally derived, such as annatto extract (containing two carotenoids: bixin and norbixin), turmeric (a dried herb), paprika oleoresins, cochineal extract, and caramel color; and (2) synthetic (artificial) colorants.

FD&C Colorants. Although the history of food colorants classes them as natural or synthetic, the FDA does not recognize any of today's colorants as "natural," even if they are directly obtained from nature. For regulatory purposes all color ingredients are additives. So, the two categories of food colorants today are: those certified as FD&C colorants and those that are exempt from certification. FD&C colorants are colorants certified as safe by the FDA for use in <u>f</u>oods, <u>d</u>rugs, and <u>c</u>osmetics. See Table 6.3.

The structures of the FD&C colorants contain phenolic rings having double bonds and various functional groups. For example, Allura Red contains three phenolic rings. One ring has the —OH group attached to it, while a second ring contains an —SO_3Na group. The third ring contains three different groups: one —SO_3Na group, one —OCH_3 group, and one —CH_3 group (Figure 6.9). The sodium salt form of sulfonic

TABLE 6.3 The nine certified food colorants approved for use. *Certification means that colorants that are manufactured must be sent to FDA for laboratory analysis before being shipped to customers for use. All FD&C colorants are water soluble.*

FD&C Colorant	Name
FD&C Blue No. 1	Brilliant Blue
FD&C Blue No. 2	Indigotine
FD&C Green No. 3	Fast Green
FD&C Yellow No. 5	Tartrazine
FD&C Yellow No. 6	Sunset Yellow
FD&C Red No. 3	Erythrosine
FD&C Red No. 40	Allura Red
Orange B	Orange B
Citrus Red No. 2	Citrus Red No. 2

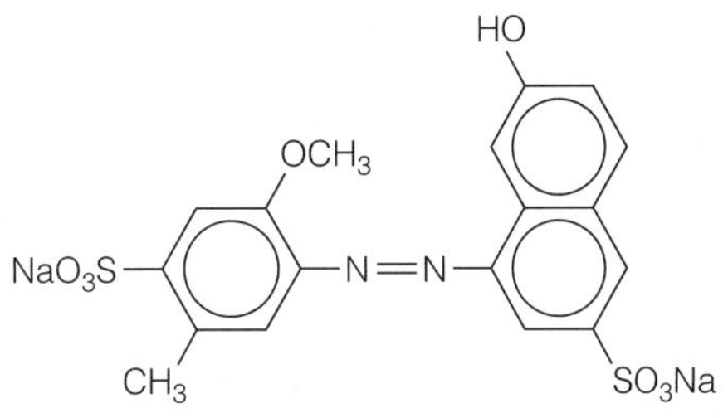

FIGURE 6.9 Chemical structure of FD&C Red No. 40, Allura Red.

acid (the —SO$_3$Na group) is important because its presence makes a colorant molecule more water soluble. This facilitates its incorporation into foods and beverages, and promotes its excretion from the body.

Food color suppliers manufacture what are called dyes and lakes. **Dyes** are water-soluble chemicals that are used to color entire food products, for example lollipops, throughout. A **lake** is an insoluble powder formed by precipitation of a water-soluble food colorant. The process involves the production of an aluminum, calcium, or magnesium salt of the colorant that reacts with aluminum oxide. Lakes are insoluble dispersions derived from dyes, and are used to color the surface of foods or fat-based products, including chocolates (Table 6.4).

Exempt Colorants. A wide variety of color substances are "exempt from certification" according to the FDA—see Table 6.5. These range from **annatto** (from the seeds of *Bixa orellana*, a tropical evergreen) to caramel, and from **cochineal** and grape skin extract to paprika (*Capsicum annum*).

Cochineal extract. The use of cochineal dates back to 5000 BC when it was used by Egyptian women to color their lips. What is it derived from? The dried bodies of cochineal insects, *Dactylopus coccus costa*, are treated with ethanol, and a red solution is produced. After the alcohol is evaporated, cochineal extract, high in carminic acid content, is used as the colorant.

Cochineal is classified as an anthraquinone, a type of polyphenol. The major cochineal pigment is carminic acid, a polyphenolic acid having three rings (polyphenol) and an acid (COOH) group. A quinone is a six-membered ring having double bonds to oxygen outside the ring:

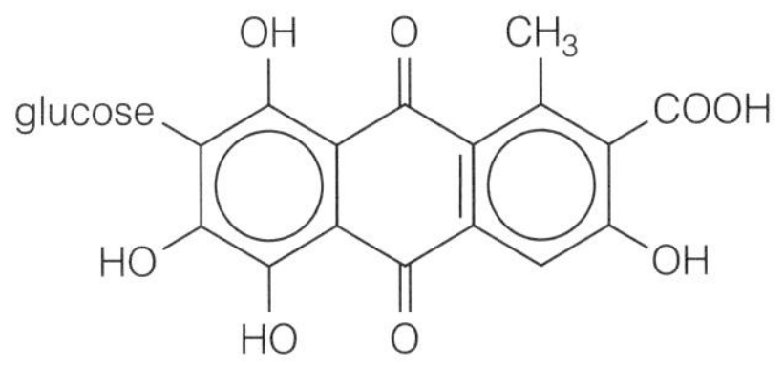

carminic acid

Caramel color. The preparation and use of caramel color dates back to nineteenth century Europe, when it was discovered that the carefully controlled heat treatment of sugars created brown "burnt sugar color." Although sauces or candies derived from this process are flavorful and are known as "caramels," caramel color is not a flavor but is simply a coloring agent used at low concentrations in foods and beverages.

According to the FDA, the following food grade carbohydrates can serve as reactants in the production of caramel color: dextrose (glucose), sucrose, invert sugar, lactose, malt syrup, molasses, and various starch hydrolysis products. What these starter materials all have in common is their glucose content, which is the key to producing brown color upon heating. The actual formation of caramel is through the nonenzymatic thermal process known as caramelization.

Owing to the complexity of chemical reactions during caramelization and the multiple products generated, it is inaccurate to consider any single compound as the caramel

TABLE 6.4 What is the difference between a dye and a lake?

Substance	Manufactured Form
FD&C Dye	
Water-soluble compound Produces color in solution	Dispersions, granules, pastes, powders
FD&C Lake	
Insoluble compound Combining dye with alumina Al(OH)$_3$ (dye content about 25%)	Dispersions for oil-based foods

TABLE 6.5 A selection of exempt color additives that have been approved for use in foods without the requirement of certification.

Color Additive	Category	Color	Food Use
Annatto extract	Carotenoid	Orange to purple	Dairy products
Canthaxanthin	Carotenoid	Yellow	Chicken products
Caramel	Caramelization	Brown	Beverages, confectionery
Cochineal extract	Anthraquinone	Red	Comminuted meats, jams
Ferrous gluconate	Iron	Black	Olives
Grape skin extract	Anthocyanin	Purplish-red	Beverages
Paprika oleoresin	Carotenoid	Orange	Prepared and frozen

TABLE 6.6 Classes, names, and uses of caramel color formulations.

Class	Charge	Name	Use
I	Negative	Plain caramel color	Spice blends, desserts
II	Negative	Caustic sulfite process caramel color	Liqueurs
III	Positive	Ammonia process caramel color	Baked goods, beer
IV	Positive	Sulfite ammonia process color	Soft drinks, soups

colorant. For this reason, the international food additive committee JECFA (Joint FAO/WHO Expert Committee on Food Additives) has grouped caramel formulations into four classes, depending on the food grade reactants used in their manufacture and the net ionic charge and presence of reactants—see Table 6.6.

6.2 FOOD FLAVOR CHEMISTRY

Flavor is a property of food material and the receptor mechanisms of the human body. Flavor involves both taste and aroma (Figure 6.10). The study of flavor includes the composition of food compounds producing taste or odor, and their interaction with receptors of the taste and smell sensory organs. Humans have four basic, or true, tastes: sweet, salty, sour, and bitter. The tongue possesses areas of taste sensitivity for these basic tastes and may possess two others, designated *umami* and *astringency*.

When tastant molecules bind to tongue receptor cells, they interact with specific proteins in these sense cells. This interaction creates a disturbance in the molecular geography of the surface, allowing an interchange of ions (a "chemical signal"). This reaction is followed by electrical depolarization in the sense cells, initiating a nerve impulse.

Chemical Structure and Taste

A first requirement for a substance to produce a taste is that it be water soluble. Thus sour, salty, bitter, and sweet substances contain hydrophilic functional groups. In general, all food acids are sour. Sodium chloride and other salts are salty, but the larger atoms provide an additional bitter taste. For example, potassium bromide is both salty and bitter. Bitterness is exhibited by alkaloids (nitrogen-containing organic compounds from plants) such

TABLE 6.7 Astringency and umami. *Astringency is a sensation of puckering in the mouth, and is believed to be a result of tannins or polyphenols reacting with proteins. Tannins are polyphenols of high molecular weight. Both polyphenols and tannins can form bonds to proteins due to the presence of many OH bonds on their structure. The 5′-nucleotides, also known as flavor potentiators, are associated with the* umami *taste response.* Umami *is described as a savory and delicious sensation.*

Food Compound	Taste Response
Polyphenols	astringent
Tannins	astringent
GMP (guanosine 5′-monophosphate)	umami
IMP (inosine 5′-monophosphate)	umami
MSG (monosodium glutamate)	umami
Ribotide® (co-crystalline mix of GMP and IMP)	umami

FIGURE 6.10 Flavor—the interaction of the basic tastes with aroma. *Flavor is not accomplished only with the tongue, but with the nose as well, as receptors in both the oral and nasal cavities are chemically stimulated by tastant molecules.*

as quinine, picric acid, and heavy metal salts. Umami is associated with 5′-nucleotides, MSG, IMP, and GMP. These can be produced commercially by sucrose fermentation. Sweetness is a property not only of sugars but also lead acetate, saccharin, aspartame, sugar alcohols, and other substances.

Table 6.7 lists food compounds that elicit the taste response of astringency and umami, which have been proposed as additional basic tastes.

Pungency. The sensation of "spicy heat" or "chemical heat" in the oral cavity is due to specific chemicals primarily from cruciferous vegetables and chili peppers. The sensation, identifiable as a warming or hot sensation in the mouth and lips, is termed **pungency**. Pungent molecules cause the effect, which varies in intensity and duration according to the active chemical component present—see Table 6.8. When pungent foods are consumed, endorphins are also released, creating a sensation of pleasure amidst the pain!

An example of pungent substances is the capsaicinoid family of molecules. **Capsaicinoids** are pungent alkaloid compounds that occur in chiles. Two important capsaicinoids are *capsaicin* and *dihydrocapsaicin*.

To systematize the potency ratings of pungent substances, the Scoville Organoleptic Test, which uses human subjects as tasters, has been developed. Low-heat and hotter chiles are quantified using the **Scoville heat unit** (Shu). Low-heat chiles give readings in the range of 200 to 2,500 Shu, while high-heat chiles reach 70,000 Shu. This is still low, however, compared to capsicum oleoresin, rated at up to 1,000,000 Shu. However, Scoville measures are highly variable due to poor reproducibility and panelist fatigue factors. An instrumental method of analysis using HPLC (high-pressure liquid chromatography)

TABLE 6.8 A selection of pungent compounds derived from chiles, ginger, pepper, and cruciferous vegetables (horseradish and mustard). *Compounds found in garlic and onion, while not properly termed pungent, are nevertheless responsible for potent flavors and aromas. In common with allyl isothiocyanate (AITC) formation from benign precursor glucosinolates (e.g., sinigrin) in cruciferous vegetables, the raw garlic and onion tissue is odorless. However, cutting the tissue liberates the enzyme alliinase, converting the innocuous cysteine substrate into allicin.*

Source	Species	Active Chemical Component
Chiles	*Capsicum* sp.	Capsaicin
Ginger	*Zingiber officinale*	Gingerol
Pepper	*Piper nigrum*	Piperine
Horseradish	*Amoracia lapathiofolia*	Isothiocyanates (derived from glucosinolates)
Mustard	*Brassica* sp.	Isothiocyanates (derived from glucosinolates)
Garlic	*Allium sativum*	S-alkyl cysteine sulfoxides
Onion	*Allium cepa*	S-alkyl cysteine sulfoxides

produces the most reliable potency rating results, according to the American Spice Trade Association. The HPLC method measures capsaicinoids in the more sensitive ppm (parts per million) range, which can be converted to Scoville heat units using the appropriate conversion factor.

Cooling Sensation. The opposite sensation to heat is the sensation of coolness in the mouth. This effect is familiar to those who chew spearmint or peppermint gum or have experienced those flavors in a dessert. Key substances responsible for the cooling effect are *menthol* and isomers of menthol.

Menthol, $C_{10}H_{20}O$, is a crystalline cyclic alcohol. It is the primary constituent of peppermint oil, which provides a fresh "minty" flavor. In addition to flavor, menthol compounds also generate a minty aroma. The combination of the two sensations sends a greater signal stimulation to the brain. Menthol is used as a flavoring in baked goods, chewing gum, confectionery, frozen dairy products, jams and jellies, nonalcoholic beverages, and soft candy.

Polyols, or sugar alcohols, are also able to produce the sensation of coolness in the mouth (in addition to providing sweetness with fewer calories—see Chapter 3). **Polyols** are polyhydric alcohol counterparts of the sugars maltose, mannose, sucrose, and xylose, and the corresponding polyols are maltitol, mannitol, sorbitol, and xylitol. Xylitol is found naturally in plums, raspberries, and strawberries. Sorbitol has been the most frequently used polyol because of its humectant properties. Used in products such as shredded coconut and marshmallows, sorbitol prevents these products from drying out. Lactitol and isomalt are two additional polyols used in the food industry.

Recall from Chapter 5 that a humectant substance attracts moisture.

Process and Reaction Flavors

Many foods have vastly different flavors in the raw state than in the cooked state, especially bread and meats. Both duration and temperature of thermal processing affect the final flavor of cooked and processed foods. Chemical reactions responsible for these changes are known as nonenzymatic browning reactions or "chemical (carbonyl-amine) browning." Browning reactions may involve caramelization of sugars or the Maillard reaction between naturally occurring reducing sugars and amino acids, amines, small peptides, and protein.

For example, consider bread. The delightful aroma of freshly baked bread is lost upon cooling and storage due primarily to volatility of aroma molecules. Bread flavor originates mainly from baking and fermentation processes. Crust formation and browning during baking are primary contributors to bread flavor. In the baking of bread, the Maillard reaction generates dicarbonyl compounds such as glyoxal (HOC—COH). These can react with amino acids like glycine (H_2N—CH_2—COOH) through a series of steps, referred to as the **Strecker degradation,** to produce a pyrazine. Pyrazines are similar to benzene rings except two of the carbons in the ring are replaced by nitrogen. Pyrazines contribute to the flavor and aroma of bread.

Pyrazine structure   Benzene structure

During *fermentation,* yeast chemically breaks down sugars in the dough. As a result, a number of alcohols are formed: ethanol, isoamyl alcohol, isobutyl alcohol, as well as the following acids: formic, capric, lactic, succinic, pyruvic, and hydrocinnamic. These products all contribute to bread flavor.

Meat also exhibits a specific cooked flavor profile. In meat, flavor development is related to the action of heat on precursor substances within the tissue. This is primarily due to the Maillard browning reactions, although fat molecules and sulfur-containing compounds (such as hydrogen sulfide and methyl mercaptan) also contribute to meat flavor. Basic meat flavor components include amino acids, peptides, sugars, and the nucleotide

TABLE 6.9 Miniglossary of flavor terms.

Flavor Term	Definition
Artificial flavor	Flavor produced through chemical synthesis
Essential oil	Oil obtained from plant matter by steam distillation, retaining the characteristic flavor; a concentrated natural flavor extractive
Extract	Flavor obtained through the use of alcohol as solvent
Natural flavor	Flavor obtained from natural source(s)
Oleoresin	Pure concentrated oil-soluble extractive of a natural spice or herb
Resin	Any naturally oily plant substance
Spice extractive	Main grouping of essential oils and oleoresins
Top note	The dominant, initial aromatic flavor of a substance
Volatile oil	Synonymous with essential oil

inosine 5′-monophosphate (IMP). Cooked beef aromatic compounds such as furan, pyrazine, pyridine, pyrrole, and thiophene may also impact meat flavor.

Producing Food Flavors. Not only are food flavorists busy studying the natural flavors present in whole foods, but in addition, they isolate and develop flavoring ingredients, called flavor compounds or flavor additives. These include artificial flavors, essential oils, extracts, natural flavors, oleoresins, process flavors, and reaction flavors—see Table 6.9.

The selection of foods that provide process flavors involves precursors that provide desirable aromas, such as roasted or boiled notes, when heated. **Process flavor substances** are substances obtained by heating a mixture of ingredients, not necessarily themselves having flavor properties, of which at least one contains nitrogen and another is a reducing sugar.

Commercially, **reaction technology** can be used to produce what are referred to as *reaction flavors*. **Reaction flavors** are flavors produced by chemical reactions taking place under controlled conditions. Typically, reactants such as sugars and proteins are combined at specified pH, a_w, and temperature until the desired product results. As an example, to produce a cheese process flavor, milk proteins, short- and long-chain fatty acids, amino acids, salt, water, and top notes are blended together. A **top note** is the predominant initial aroma or flavor characteristic for a substance. Process machinery heats the ingredient mixture under agitation for a specified time and temperature to ensure uniformity of flavor development.

Enzyme-Produced Flavors. A large number of flavor compounds can be produced by enzymology if the right enzyme system and substrate can be found. For meat flavor, a tenderizing enzyme (protease) is added to meat and the mixture is heated and held at an optimal temperature (approximately 140°F (60°C)) for the enzyme to function. Once the meat is liquefied, the temperature is raised to "kill" the enzyme (180°F; 82.2°C) and water is expelled from the system. The flavoring concentrate that results can be heated again to produce further "cooked" notes. In natural fermented products, enzymes are responsible for flavor formation in tea, coffee, chocolate, cheese, and plant products like onions, garlic, and mustard.

Enzymology and fermentation are often combined in the production of flavors. The enzymes are "harvested" from bacteria, yeasts, and molds that are grown industrially. For example, yeast-derived flavors and flavor enhancers have been manufactured from brewers and other yeasts. Different flavors can result when yeasts are grown on corn syrup vs. whey protein. Not only the choice of growth medium, but the strain of yeast, the fermentation conditions, and the form of processing can affect the flavor obtained from the yeast.

Yeasts that have been grown to produce flavors during their metabolic fermentations must be killed and hydrolyzed to obtain the various flavorful metabolites. Terms such as **autolyzed yeast** and autolyzed yeast extracts indicate that the yeasts have been centrifuged to remove cell wall material in order to concentrate such flavors and flavor precursor substances as free amino acids, peptides, monosaccharides, and Maillard reaction products generated during thermal processing. For instance, the processing can be designed to extract 5′-nucleotide-rich extracts from RNA, with the final product used as a flavor enhancer.

Flavor Enhancers

Although no body of data in the scientific literature distinguishes flavors from flavor enhancers, proposed FDA regulations do differentiate them. The FDA has proposed that a **flavoring** is a substance that has a flavor of its own at the level at which it is used in a food, while **flavor enhancers** and flavor potentiators do not themselves impart flavor, but rather intensify in some manner the flavors that are naturally present or are added to food.

Protein Hydrolysates.

A **protein hydrolysate** is a protein breakdown product obtained via enzyme or chemical (acid, alkali) action. Protein hydrolysates function in foods as both flavorings and flavor enhancers.

The purposes of protein hydrolysates include enhancing food nutritive value, enhancing protein functionality, and adding flavor. Proteins are hydrolyzed in a series of steps to produce proteoses, peptones, peptides, and amino acids, which differ in size and molecular weight. Whey protein, casein, and soy protein are typical sources used in the food industry for the production of protein hydrolysates. **Hydrolyzed vegetable proteins** (HVP) are derived from soybeans and are used for flavor, and as alternate sources of protein to meat and dairy products. However, they are not usually considered in the same category as protein hydrolysates.

Flavor Encapsulation.

Encapsulation is a technique applied to flavors to accomplish convenience, stabililty, and timed release. For example, encapsulated oleoresins and oils are dry powders to which thin, polymeric coatings have been applied. The coatings are often composed of maltodextrin, modified starches, or vegetable gums. They offer a temporary barrier to heat, low pH, oxygen, and moisture, making the flavor compound stable during storage. Encapsulated flavors often provide enhanced incorporation into dry mixes and superior dispersion characteristics in aqueous food systems.

Flavors, as well as vitamins and other substances, have been successfully encapsulated with specially designed fat systems that coat the encapsulated material. This type of system is an effective oxygen and moisture barrier, and offers delayed release of the flavor compound until the fat melts. For example, flavor oil droplets can be protected via polymer encapsulation, in which an insoluble wall of gelatin polymers envelops the flavors and is stable to heat (will not melt). The flavor is added to a specific product and is released only by the mechanical force of chewing. This approach offers advantages for certain baked foods, extruded foods, fried foods, and processed meats.

In summary, encapsulation technology protects flavors in a variety of potentially destructive conditions:

- **Thermal processing conditions:** baking, deep-fat frying, extrusion, retorting
- **Storage conditions:** low temperature, high temperature, oxidation, pH
- **Consumer conditions:** cooking, freezing, refrigeration, reheating

The Chemistry of Flavor Deterioration

Off-flavors, sometimes confused with "taints," refer to an unwanted flavor development in foods. A taint results from external contamination from the environment, whereas off-flavor results from internal chemical changes in foods during processing or storage. Off-flavors are different from the characteristic flavors normally associated with a particular product. Food scientists must recognize the possible sources of factors and reactions that cause off-flavors in foods. Table 6.10 lists several substances that are sources of off-flavors.

TABLE 6.10 Several substances responsible for specific off-flavors in foods.

Substance	Off-flavor	Food
Aldehydes	rancid	foods containing oleic and unsaturated fatty acids
Lipoxygenase*	"beany"	soy foods
Maillard intermediates	"cooked"	milk products
Phospholipid	WOF	meats
Putrescene	putrid, rotting flesh	meats
Trimethylamine	"fishy"	dairy products

*Lipoxygenase is not the odorous material. The reaction this enzyme catalyzes is the breakdown of unsaturated lipids in soybeans to produce short-chain fatty acids; the latter are the odorous substances.

One type of off-flavor that can develop in dairy products is a fishy flavor. This results from the formation of *trimethylamines* by the hydrolysis and oxidation of lecithin, a naturally occurring phospholipid in milk. Factors that promote development of this flavor include high acidity and the presence of prooxidants such as metals or metal salts of iron and copper. Although SH groups present in casein proteins can act as antioxidants to inhibit the reaction, it is thought that heat processing uncoils the milk proteins. In uncoiled protein, the sulfhydryl groups become exposed, and may form crosslinks or enter other reactions, rather than inhibit phospholipid oxidation.

Lipid food material is subject to two types of chemical reactions that lead to rancidity. **Hydrolytic rancidity** reactions produce off-flavors due to liberation of free fatty acids by water hydrolysis and enzyme action. Heat acts as a catalyst for this reaction, and lipase enzymes can as well. The rancid smell results from the liberation of specific, short-chain fatty acids from triglycerides, such as butyric acid from butterfat:

$$\text{triglyceride} + \text{lipase enzyme} \rightarrow \text{short-chain fatty acids (odorous)} + \text{glycerol}$$

Keeping fats, such as butter, cold in the refrigerator prevents hydrolytic rancidity. However, the generation of short-chain fatty acids during the fermentation periods in cheese and cocoa production contributes in a positive way to flavor developments.

Oxidative rancidity involves reactions between unsaturated fatty acids (UFA) and oxygen, producing hydroperoxides. The breakdown of hydroperoxides results in small molecular weight compounds like acids, alcohols, aldehydes, and ketones. These small compounds are the ones responsible for the off-flavors and odors. Heat or light energy initiate the process:

$$\text{UFA} + \text{heat, light, oxygen} \rightarrow \text{hydroperoxides} \rightarrow \text{alcohols, aldehydes, ketones (odorous)}$$

Cooked meat that is refrigerated and later reheated is subject to a type of oxidative rancidity called **warmed-over flavor** (WOF). WOF is an unpleasant, stale taste in reheated meats. In this reaction, iron acts as a catalyst. Meats that have high phospholipid content in the fatty portions (specifically, the lipid is *phosphatidyl ethanolamine*, which contains many unsaturated fatty acids in its structure) exhibit warmed-over flavor problems.

The unsaturated fatty acids of triglycerides are involved in producing WOF as well. Pork is more susceptible than beef or lamb in developing WOF, and turkey is more likely than chicken to develop it. A good question is "can WOF be minimized?" Reheating methods have been shown to influence WOF. Microwave reheating results in less WOF than does conventional oven reheating. Studies have shown that WOF can be inhibited by adding nitrite or antioxidants.

6.3 FOOD TEXTURE

The **texture** of a food is a quality parameter of major importance. It refers to the perception of food structure when a food item is held by the fingers, pushed by the tongue against the palate, or chewed by the teeth and sensed within the oral cavity (Table 6.11). Texture is determined by the microstructure of animal and plant tissue, and it is influ-

TABLE 6.11 **All foods possess texture; examples of low- and high-quality food textures.**

Poor Texture/ Quality	Good Texture/ Quality
chalky peanut butter	smooth, spreadable peanut butter
rubbery marshmallows	spongy, fluffy marshmallows
dry, stale bread	soft, fresh bread
gritty pudding	creamy pudding
limp celery	firm celery
soggy snack chips	crisp snack chips
tough, dry meat	tender, juicy meat

enced by the presence of texturizing ingredients. As a physical characteristic of food, texture can be evaluated by human sensory perception as well as by mechanical testing. Since changes in texture reflect changes in quality, food scientists must learn how to apply techniques that measure food texture.

Mouthfeel and food texture are closely related parameters. **Mouthfeel** encompasses the entire spectrum of a food's physicochemical characteristics inside the mouth, from the initial sensation inside the oral cavity to the first bite, through chewing, and the act of swallowing. Food texture changes as it is chewed, as structures are broken down and moisture is released.

Texture Classification

Objective texture measurement refers to the use of analytical equipment to determine (1) food microstructure and macrostructure, and (2) resistance to a cutting force that causes shear, or an applied force that causes deformation and/or flow behavior. *Rheology* is the study of the flow of matter in response to force. The rheological behavior of food tells the food scientist (1) how a fluid food like a beverage, oil, chocolate, sauce, syrup, or gravy will pour, (2) how a fluid will be sensed by the tongue, and (3) how long it will take for the fluid food to be cleared from the palate.

Fluid foods are classified as either Newtonian or non-Newtonian, depending on their flow behavior. These terms will be described in more detail in Chapter 13, Food Engineering. A food's viscosity measurement is a type of rheological assessment that provides information on the flow properties of foods. In turn, these are related to the physical state of a food: liquid, semisolid, and solid. Liquid or "fluid" foods exhibit greatly increased flow (low viscosity) compared to semisolid foods (higher viscosity).

Solid foods typically show no flow behavior, but deform (change shape) or shear (cut or break) in response to force. Table 6.12 describes some texture parameters for solid and semisolid foods.

The Influence of Chemical Forces in Water and Fat Systems on Texture

Even though texture is a physical property of food, chemical forces and interactions between food molecules underlie texture. Thus, food scientists have developed a base of ingredients that can be added to products to influence the overall texture. These "texturizing agents" can impart body by increasing viscosity, promoting gelation, increasing firmness through the binding of water or by causing the crosslinking of molecules, or through the stabilization of emulsions.

Water-Based Systems. Of key importance to food texture is the presence of water (moisture) and the manner in which it is chemically positioned. Water can exist in foods in one of three situations: (1) as free water, (2) as adsorbed water, and (3) as bound water. Free water becomes entrapped within a gel as a consequence of gelation. This is the case with water present in a protein gel, a pectin gel, or a vegetable gum gel.

Food water activity (a_w) can be viewed as a predictor of food texture. Food texture ranges from hard and crisp (lowest a_w) to dry and firm (intermediate a_w), to soft and moist (highest a_w). Hard textured foods are associated with bound water and relatively low moisture and a_w levels, whereas dry, firm foods, while retaining relatively low moisture, are associated with adsorbed water and exhibit a wide range of a_w levels. Soft, moist foods are associated with free water and the highest levels of moisture and water activity.

Fat-Based Systems. Whole foods or food systems, including fat-based ones such as cocoa butter in chocolate, are illustrative of texture. Cocoa butter is the lipid material from which chocolate is made. As a fat, it contributes a smooth, creamy consistency, a softness, and a pleasant mouthfeel. The chemistry behind such textural properties of fats is rather complex. In simple terms, the consistency of any fat is the result of an organized three-dimensional network of solid fat crystals embedded in liquid oil. The crystals are

TABLE 6.12 Texture of semisolid and solid foods as defined by mechanical characteristics.

Texture Parameters	Explanation
Cohesiveness and chewiness	Tender versus chewy versus tough
Fracturability	Force at which a solid food splits or "fractures"
Hardness	Soft versus firm versus hard
Springiness—elastic	When force is removed, item returns to original shape
Springiness—plastic	Force results in an impression that does not spring back

TABLE 6.13 Functional ingredients used as texturizing agents.

Substance	Category	Functionality
Sucrose	Simple carbohydrate	Firm texture, drying
Starch, maltodextrin	Polysaccharide	Moisture retention, texture bulking, fat-like mouthfeel
Gums (hydrocolloids)	Polysaccharide	Thickening and stabilizing
Pectins	Polysaccharide	Thickening, stabilizing, and gelling
Gelatin	Polypeptide	Stabilizing and gelling
Soy protein (isolates, concentrates)	Polypeptide	Influencing texture of soy-meat analogs; tenderizing baked products

composed of specific glycerides, such as triglycerides having saturated-unsaturated-saturated acyl groups (remember that *triglyceride* is synonymous with *triacylglycerol*).

The soft texture of fat depends upon which triglycerides are present within the solid fat crystals, and which acyl groups are present in the triglycerides. The size, shape, and stability of the fat crystals (cocoa butter is also polymorphic and can exist in a variety of crystal forms) affects the conversion of solid fat to liquid oil. This change occurs at the melting point, which is different for every food fat, and results in differences in hardness at any given temperature.

The Chemistry of Food Texturizing Agents

A wide variety of **texturizing agents** is available to the food industry. These promote viscosity, increase firmness, and cause gelation. Texturizing agents tend to be starches, non-starch polysaccharides, and proteins—see Table 6.13. When extruded foods are formulated with them, a diversity of texture effects can be achieved.

Viscosity and gelling are two functional properties of texturizing agents that affect food texture. Starches of varying amylose-to-amylopectin ratios, proteins (dairy, fish, meat, soy), and hydrocolloids all can form gels when their molecules arrange into stable three-dimensional networks. On the other hand, denatured proteins and non-gelling polysaccharides, such as cooked starch granules, cannot form strong covalent bonds required for gelation to occur. Instead, they contribute to increased product viscosity.

Polysaccharides. As a hydrocolloid, the polysaccharide carrageenan creates a weak gel in chocolate milk that breaks when the milk is poured. The effect is to convince the consumer that the chocolate milk is "richer" and creamier" in texture than ordinary milk. Polysaccharides can interact with each other in such a way as to enhance gelation. For example, gel rigidity can be greatly strengthened when combining modified starch with iota-carrageenan, compared to either component alone. The presence of cations, such as calcium in low methoxyl pectin, promote gelation. Synergism is also seen in protein:protein systems, for example, casein:gelatin mixtures.

Fat Replacers. From a functional standpoint, fat replacers can be classified as fat substitutes and fat mimetics. **Fat substitutes** have the same physical properties of fat (e.g., olestra) and replicate fat functions when used in foods. **Fat mimetics** do not possess all of the true fat physical properties, such as flavor and flavor release, but can imitate some of

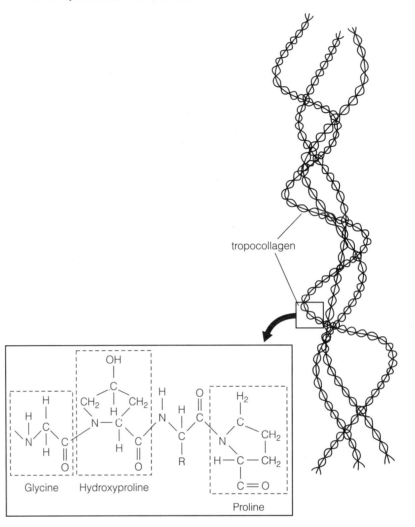

FIGURE 6.11 **The Structure of collagen.** *Collagen is a fibrous, structural protein of high tensile strength. Collagen consists of separate collagen fibers that are joined together by chemical crosslinking. Each collagen fiber consists of parallel arrays of tropocollagen proteins. These proteins are each considered a superhelix of three self-associating tropocollagen polypeptide monomers, composed of amino acids such as glycine, proline, and hydroxyproline.*

them, such as creaminess. Fat mimetics alter texture by controlling water in food systems. A typical fat mimetic is a combination of a polysaccharide and water. This combination produces gelling characteristics that help imitate fat in foods. Characteristics may vary from a firm, cuttable gel suitable for pie fillings to a softer, more tender gel, good for puddings.

Sugars. Sugars compete with protein and starch for water in food systems. This reduces the amount of water available for starch gelatinization and for protein and starch gelation. The end result may be an increase in viscosity without formation of a strong gel, a change that alters the product texture and storage stability. In addition to these sugar:water:starch and sugar:water:protein interactions, hydrocolloid:starch interactions affect texture.

Collagen and Gelatin. Gelatin, another texturizing agent, is a soluble protein derived from insoluble collagen present in animal connective tissue. To understand how gelatin is obtained from collagen, as well as its functionality, we must examine the molecular organization of collagen. The same chemical forces that stabilize collagen in connective tissue are active in the stabilization of gelatin gel systems.

What collagen and gelatin have in common is a molecule called *tropocollagen*. Each collagen fiber contains parallel arrays of tropocollagen, a polypeptide superhelix that actually consists of three distinct polypeptide chains—see Figure 6.11. These chains are tightly intertwined and are stabilized by interpolymer hydrogen bonding. The chemical bonding involves amino acid residues (R—C=O and R—NH), as well as OH groups bonded across polypeptide chains within each superhelix, via water bridges.

In a sense, gelatin is locked within collagen's three-dimensional structure, only to be liberated when collagen is subjected to acid, alkaline, or heat treatment. Heating, for instance, causes hydrolysis of the intertwined tropocollagen monomer-to-monomer bonds, which causes separation of the polypeptide monomer chains from each other. There is also some dissociation of adjacent collagen fibers due to the breaking of interfiber crosslinks. The resulting "free, single" tropocollagen strands are considered individual gelatin molecules.

The gelatin molecules so produced are isolated and concentrated into a powdered ingredient. When used as a food ingredient, gelatin powder exists as a crystalline glassy polymer. Upon addition of hot water, the gelatin disperses as random coil polypeptides. The water solubilizes the gelatin and forms a sol, in which water is the continuous phase and gelatin, the dispersed (Figure 6.12). Cooling increases the viscosity of the system. Eventually, polymer:polymer gelatin associations result in gelation, a three-dimensional network of gelatin polymers stabilized by noncovalent bonds at junction zones.

Within the gel, individual gelatin polymers exist mainly as random polypeptide coils in the amorphous (protein network) gel regions, where there may be hydrogen bonding to water at exposed polar groups along each polypeptide. The crystalline regions refer to the more highly ordered junction zones, where gelatin polymer:polymer interactions predominate and crystallites partly reform. Liquid water becomes entrapped in regions of higher entropy away from junction zones. Here, solvent:solvent interactions predominate, although the entrapped water exhibits restricted mobility. Figure 6.13 illustrates the three coexisting states.

The tenderness of meat is one of its most desirable sensory attributes. In fact, consumer studies show that it is the most important sensory attribute of beef. The use of weak acid additives like lemon juice (citric acid source) or vinegar (acetic acid source) is the traditional method for overcoming meat toughness. These promote swelling of collagen, and this swelling enhances chain dissociation through breakage of noncovalent bonds, including hydrogen bonds. Proteolytic enzymes of plant origin, such as *papain*, from papaya, are used to reduce both connective tissue and muscle fiber toughness because they enzymatically separate collagen constituents by hydrolyzing the chemical linkages joining them together.

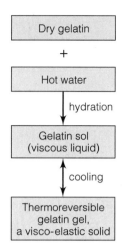

FIGURE 6.12 Gelation of gelatin and hot water. *Hydration of gelatin with hot water produces a sol, which becomes increasingly viscous as it cools. Gelatin results when junction zones form between self-associating gelatin polypeptides.*

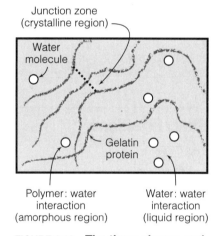

FIGURE 6.13 The three phases and interactions in a gelatin gel matrix. *Imagine three physicochemical states coexisting within the gelatin gel: (1) crystalline, (2) amorphous, and (3) liquid. These states maintain the gel's structure, without allowing structural collapse or liquid flow out of the gel.*

Key Points

- Food pigment molecules include anthocyanins (red-purple), betalains (violet-red and yellow), carotenoids (red, orange, yellow), chlorophyll (green), and myoglobin (purple-red).
- Red meat color comes from the protein myoglobin, whose chemical structure is a porphyrin ring bonded to a central iron atom. Oxidation of the iron causes color changes.
- Fruit and vegetable colors come primarily from compounds that contain conjugated double bonds (anthocyanins, carotenoids, chlorophylls). Chlorophylls also contain magnesium-coordinated porphyrin rings.
- The nine certified food colorants contain phenolic rings with double bonds and various functional groups.
- Flavor is a combination of sensations, resulting from chemical interactions between flavor molecules and receptors in the oral and nasal cavities.
- In addition to the basic sweet, sour, salty, and bitter tastes, umami (savory) and astringency (dryness and puckering of the mouth) are important.
- Process and reaction flavors due to the Maillard and Strecker reactions, contribute to flavor development.
- Enzymes are used commercially to produce flavorings and flavor enhancers.
- Off-flavors result from chemical reactions in foods, including trimethylamine formation in fish, lipid oxidation in oils, and as warmed over flavor in meats.
- Food texture is determined by its microstructure, as perceived by tactile, oral, and nasal senses.
- Analytical texture evaluation measures food structural elements and rheological characteristics such as viscosity.
- The presence of water in food systems in the free, adsorbed, and bound states influences food hardness and softness.
- Texturizing agents including starches and gelatin affect food texture through the processes of thickening and gelation.

Study Questions

True/False

1. T/F: Dyes and lakes contain conjugated chromophore systems in their molecular structures.
2. T/F: Flavor results from the interaction of the basic tastes with aroma and touch.
3. T/F: Process flavor precursor substances may or may not have flavor properties prior to heating.
4. T/F: Water activity is not a predictor of food texture.
5. T/F: Crystalline regions called junction zones are characteristic of gelatin gels.

Multiple Choice

6. Which statement regarding color perception is true?
   a. Rods are sensitive to red, green, and blue color.
   b. Reflected light that determines the color of a food.
   c. The surface hue refers to the degree of light or dark perception.
   d. Cones are sensitive to light and dark.
   e. The surface chroma refers to the actual color perceived.

7. In addition to color qualities, food surfaces exhibit all of the following appearance characteristics *except*
   a. translucency.   d. shine.
   b. cloudiness.   e. intensity.
   c. gloss.

8. Which molecule(s) contains metal-coordinated porphyrin structures?
   a. anthocyanin
   b. chlorophyll
   c. myoglobin
   d. chlorophyll and myoglobin
   e. anthocyanin, chlorophyll, and myoglobin

9. Which factors can change the quality of food color?
   a. oxygen   d. oxygen and metals
   b. heat   e. oxygen, heat, and metals
   c. metals

10. Which statement shows the *incorrect* color provided by the molecule?

a. chlorophyll = green
   b. cyanidin = red
   c. delphinidin = blue
   d. carotenoids = yellow, orange, and red
   e. betacyanin = violet-red

11. The certified colorant substance _____ contains three phenolic rings and is water soluble.
    a. Agent Orange
    b. lutein
    c. FD&C Red No. 40
    d. lycopene
    e. xanthophyll

12. Which pairing of food compound to flavor response is *incorrect*?
    a. quinine, astringent
    b. GMP, umami
    c. lactic acid, sour
    d. polyphenol, astringent
    e. potassium chloride, salty

13. Which food is pungent due to the presence of isothiocyanates?
    a. chiles
    b. mustard
    c. garlic
    d. onion
    e. pepper

14. Shu is the unit of measure for
    a. pungency.
    b. sweetness intensity.
    c. viscosity.
    d. sourness.
    e. cooling sensation.

15. Which in *not* an example of a polyol?
    a. xylitol
    b. isomalt
    c. maltitol
    d. mannose
    e. sorbitol

16. The Strecker Degradation
    a. results in the production of aromas.
    b. results in the production of flavors.
    c. can generate a pyrazine product.
    d. results in the production of aromas and results in the production of flavors.
    e. results in the production of aromas, results in the production of flavors, and can generate a pyrazine product.

17. The definition of _____ is a food's dominant initial aromatic flavor.
    a. oleoresin
    b. top note
    c. essence
    d. volatile
    e. extractive

18. Identify the *incorrect* pairing related to off-flavors.
    a. short-chain fatty acids, rancid
    b. trimethylamine, fishy
    c. phospholipid, WOF
    d. Maillard intermediates, sweet
    e. lipoxygenase, beany

19. Identify the *incorrect* pairing related to food texture.
    a. semisolid foods, low viscosity
    b. solid foods, no flow behavior
    c. gel, entrapped free water
    d. hard, crisp texture, lowest water activity
    e. fluid foods, low viscosity

20. Which two functional properties allow texturizing agents to affect food texture?
    a. precipitation and solubility
    b. microparticulation and encapsulation
    c. viscosity (thickening) and gelling
    d. sweetening and hydration
    e. gelatinization and hydrolysis

21. Fat mimetics
    a. possess identical flavor as fats.
    b. results in the production of flavors.
    c. do not exhibit creamy texture.
    d. possess identical flavor release as fats.
    e. act by controlling water in food systems.

22. Hydration of gelatin produces a _____ that becomes increasingly _____ as it cools.
    a. emulsion, stable
    b. sol, viscous
    c. gel, viscous
    d. sol, fluid
    e. gel, fluid

23. Allura Red is a colorant that contains the sodium salt form of sulfonic acid (the —SO$_3$Na group) as well as a methyl and methoxyl group. Which statement below is *correct*?
    a. The methyl and methoxyl groups are more hydrophilic than the —SO$_3$Na group.
    b. Allura Red is completely water insoluble.
    c. The —SO$_3$Na group is hydrophobic.
    d. Each of the three functional groups of Allura Red is hydrophobic.
    e. Allura Red is a water-soluble colorant due to the —SO$_3$Na group.

24. Sorbitol is identical in structure to glucose, except at C1. Sorbitol is CH$_2$OH at C1, whereas glucose is C=O at the C1 position. Which statement is *true*?
    a. Sorbitol is the reduced form of glucose.
    b. Sorbitol would take part in the Maillard reaction.
    c. Sorbitol provides more calories per gram than glucose.
    d. Sorbitol is not a polyol, but glucose is.
    e. Sorbitol is the oxidized form of glucose.

25. Examine the following structure and select the answer choice that best describes it:

    —CH$_2$—CH=CH—CH$_2$—CH=CH—

    a. glucoside
    b. short-chain fatty acid
    c. conjugated double bonds
    d. unsaturated fatty acid
    e. saturated fatty acid

26. Which statement is *incorrect*?
    a. The aglycone of an anthocyanin is called the anthocyanidin.
    b. Hydrolysis of a glucoside produces the aglycone, sugar, and water.
    c. Aglycone joins to a sugar by a glycosidic bond.
    d. Delphinidin is an example of an anthocyanidin.
    e. The aglycone of a betanidin is called betanin.

27. Which is *not* the correct pairing of a sugar and a polyol derived from it?
    a. mannose, mannitol
    b. sorbose, sorbitol
    c. maltose, maltitol
    d. xylose, xylitol
    e. lactose, lactitol

Critical Thinking

28. Suppose the following ingredients are used to produce an imitation bacon product. Identify the functions of as many of the ingredients as you can, speculating on which would provide color, flavor, and texture.

Caramel powder	3.8 g
Indigotine	0.1 g
HVP	35.8 g
Ribotide	0.5 g
Salt	83.3 g
Sucrose	54.5 g

References

Carden, L. A., and Baird, C. S. 2000. Flavors. In: *Food Chemistry: Principles and Applications*. Science Technology System, West Sacramento, CA.

Coultate, T. P. 1996. *Food: The Chemistry of Its Components*. Royal Society of Chemistry, Cambridge, UK.

DeMan, J. 1999. *Principles of Food Chemistry*. Aspen Publishers, Gaithersburg, MD.

Fennema, O. R. 1996. *Food Chemistry*. Marcel Dekker, New York.

Francis, F. J. 1999. *Colorants*. American Association of Cereal Chemists, Eagan Press, St. Paul, MN.

Giese, J. 1994. Modern alchemy: use of flavors in foods. *Food Technology* 48(2):106.

Giese, J. 1995. Measuring physical properties of foods. *Food Technology* 49(2):54.

Miller, D. 1998. Food Chemistry: A Laboratory Manual. John Wiley & Sons, New York.

Nielsen, S. 1998. *Food Analysis*. Aspen Publishers, Gaithersburg, MD.

O'Donnell, C. D. 1992. How to smooth sourness. *Prepared Foods* 161(5):107.

Shallenberger, R. S., and Acree, T. E. 1967. Molecular theory of sweet taste. *Nature* 216:480–482.

Wong, D. W. 1989. *Mechanism and Theory in Food Chemistry*. Van Nostrand Reinhold, New York.

Wrolstead, R. E. 2000. *Colorants*. In: *Food Chemistry: Principles and Applications*. Science Technology System, West Sacramento, CA.

Additional Resources

Best, D. 1988. Archi-texturing foods with proteins. *Prepared Foods* 157(9):10.

Brown, A. 2000. *Understanding Food: Principles and Preparation*. Wadsworth/Thomson Learning, Belmont, CA.

Charley, H., and Weaver, C. 1998. *Foods: A Scientific Approach*. Prentice Hall, Upper Saddle River, NJ.

Clark, A. H. 1998. Gelation of globular proteins. In: *Functional Properties of Food Macromolecules*. Aspen Publishers, Gaithersburg, MD.

DeRovira, D. 1996. The dynamic flavor profile method. *Food Technology* 50(2):55.

Dziezak, J. D. 1990. Chemical and biochemical mechanisms of sweetness. *Food Technology* 44(1):76.

Giese, J. 1994. Spices and seasoning blends: a taste for all seasons. *Food Technology* 48(4):88.

Hegenbart, S. 1992. Flavor enhancement: making the most of what's there. *Food Product Design* 2(11): 56.

Hegenbart, S. 1994. Learning and speaking the language of flavor. *Food Product Design* 4(8):33.

Kuntz, L. 1998. Colors au naturel. *Food Product Design* 8(3): 60.

Marsili, R. 1992. Food color: more than meets the eye. *Food Product Design* 6(10):66.

O'Donnell, C. D. 1997. Colorful experiences. *Prepared Foods* 166(6):32.

Pszczola, D. 1997. Feelin' good: ingredients that add texture. *Food Technology* 51(11):82.

Nagodawithana, T. 1995. Flavor enhancers: their probable mode of action. *Food Technology* 49(4):79.

Shallenberger, R. S. 1996. The AH,B glycophore and general taste chemistry. *Food Chem* 56(3):209-214.

Vaclavik, V. A. 1998. *Essentials of Food Science*. Chapman & Hall, New York.

Vieira, E. R. 1996. *Elementary Food Science*, 4th ed. Chapman & Hall, New York.

Web Sites

http://www.mapinternet.com
An extensive listing of food color, flavor, and texture ingredient suppliers.

http://www.ift.org/divisions/food_chem/
IFT Food Chemistry Division

http://www.pdlab.com/
PD Lab. A central site for product developers in the food, beverage and pharmaceutical industries.

http://www.pdlab.com/colorguide.htm
A Basic Guide to Food Color Concentrates

http://www.chilepepperinstitute.org
A presentation of chile pungency is found at The Chile Pepper Institute at New Mexico State University.

THE CHEMISTRY OF SWEETENERS AND SWEETNESS

Human beings presumably have always had a liking for foods that taste sweet. Early in history, sugar beets, sugarcane, honey, and syrups served as sources of sweet molecules—the simple sugars. Because sweet foods are highly desired, researchers have been working to define the molecular mechanisms involved in initial sweet taste reception and the biochemical and human physiological events that result. A chemical basis of sweetness has been proposed, and in the absence of complete understanding, let's look at the proposed theory.

Theory of Sweetness—The Overview

The tongue has traditionally been divided into four taste zones, which accommodate the roughly 10,000 taste receptors for the primary tastes, including sweetness. It has been proposed that food molecules responsible for these tastes can bind in some way to specific protein receptor molecules on the tongue for a period of time, based upon electrical charge complementarity. This is the basis of the chemical theory for sweetness, called the AH,B Sweetness Theory of Shallenberger.

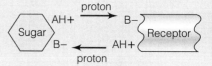

FIGURE 6.14 Sweetness theory. *Tastant sugar molecule having AH+ and B− sites is attracted to the complementary regions of the tongue protein receptor molecules through donation of a proton.*

The sweetness of sugars and also nonsugars has been proposed to be due to the presence of an AH,B structure of the tastant substance, in which A and B represent electronegative atoms (oxygen) while AH indicates hydrogen bonding potential. Only one AH,B unit per molecule is thought to be able to bind to receptors—see Figure 6.14. This is critical to the theory, because it explains why polymers of glucose (starch and starch derived polysaccharides) are not sweet.

The theory also states that the substances binding tightest to receptor sites are sweetest, and that these contain a third site in the AH,B system, called gamma. This produces a completed triangle, AH—B—γ. Gamma is thought to be a nonpolar or lipophilic site, which interacts with a similar site of the protein receptor.

Theory of Sweetness—The Details

The AH,B theory states that the tongue receptor sites are either electropositive or electronegative (AH+ or B−), and that sweet-tasting compounds also possess electropositive and electronegative portions to their molecules. An **electronegative** designation implies an element or molecule with a strong attraction for electrons, while **electropositive** indicates a weaker attraction for them. The AH+ is called the proton donor, and the B− is the proton acceptor, similar in concept to the idea of acids and bases.

According to the theory, a sweet molecule will align its AH+ region to the receptor's B− region and bind to it by donating a proton, while the sweet molecule's B− region will align with the receptor's AH+ region. This alignment can be considered as the formation of intermolecular hydrogen bonds as presented in the AH,B model. The chemical interaction between sweet sugar molecules and tongue receptor molecules is thought to initiate the sweet taste transduction event. In **sweet taste transduction,** a chemical stimulus, the tastant molecule, is converted into electrical impulse. The electrical impulse (depolarization) is sent to the brain and interpreted as a "sweet" sensation.

The sweet portion of the tastant is referred to as a **glycophore** and consists of the AH and B units together with a third unit called γ (gamma), to form a tripartite structure. The glycophore is a *dipolar compound*, which means that it has areas of opposing forces (attractive and repulsive) on opposite ends. This can be rationalized as different areas on the same molecule differing in the tendency to lose or gain a proton (H+). The proton donor is the glycophore AH group, while the proton acceptor is termed the glycophore B group.

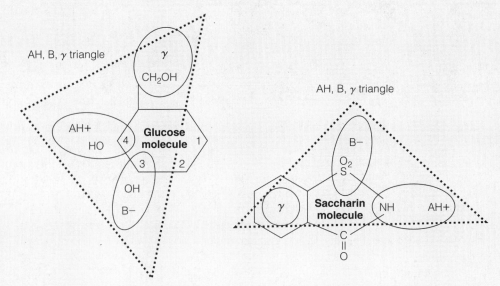

FIGURE 6.15 Identifying the AH and B regions of two sweet tastant molecules, glucose and saccharin.

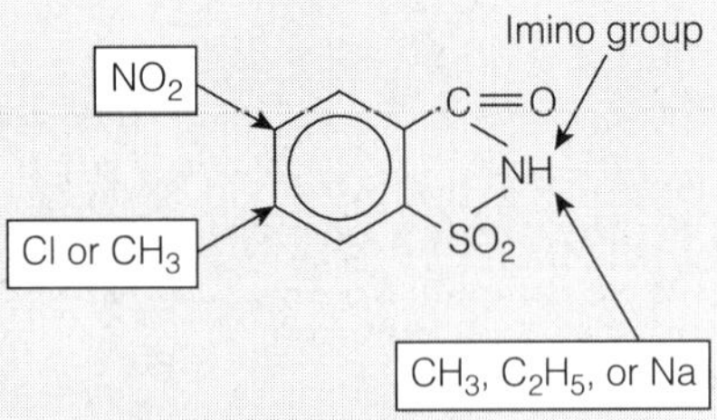

FIGURE 6.16 Saccharin and structural derivatives: effect on sweetness. *Introduction of a methyl or a chloride group in the* para *position reduces saccharin sweetness by one half. Introduction of a nitro group in the* meta *position makes the compound very bitter. Substitutions at the imino (NH) group by methyl (CH_3) or ethyl (C_2H_5) groups renders the molecule tasteless. However, adding a sodium group yields sodium saccharin, which is very sweet.*

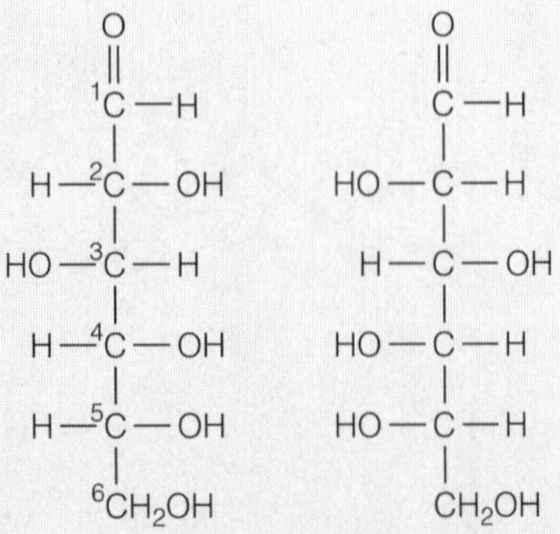

In this model, the glycophore binds to the appropriate protein receptor site of the tongue via intermolecular hydrogen bonding.

Hydrogen bonding involves an attraction between the proton of the H in the AH area of the glycophore and the unbonded electrons of the B area of the receptor. The receptor molecule's AH region behaves similarly with the B region of the glycophore. Thus the sweet taste is initiated by this hydrogen bonding between the AH,B glycophore and similar AH,B receptor units.

The γ unit is any hydrophobic site. A hydrophobic site usually means a nonpolar area of hydrocarbon: multiple —CH_2— groups, one or more methyl groups (—CH_3), or a carbon (benzene) ring structure, all of which have low affinity for water or forming any hydrogen bonds. The γ unit functions to direct and align the molecule as it approaches the receptor site.

The AH,B glycophore model can be applied to many carbohydrates with a *1,2 glycol structure*. Experimentally, the AH,B structure in many sweet compounds has been located. For instance, in glucose, the C_4—OH has the AH function, while C_3—O is the B unit—see Figure 6.15.

It is believed that the size of tastant molecules influences binding with AH,B sites. Larger molecules will cover fewer binding sites and decrease the perceived sweetness. The more sites that are bound, as with smaller molecules, the greater the perceived sweetness. The more concentrated the sweet food item in the vicinity of the binding sites, the more the AH,B sites will be saturated.

Other Influences on Sweetness

Another factor influencing sweetness is the chemical nature of the functional groups in a sweet molecule. From this standpoint, saccharin is interesting because, although saccharin is 500 times sweeter than sucrose, minor chemical changes in its functional groups affects its sweetness. For example, substitution compounds of saccharin exhibit varying sweetness or can be without sweet taste altogether—see Figure 6.16 for examples.

Forms of amino acids and sugars also play a role in sweet taste. Various stereoisomers (same chemical formula but different arrangements of atoms in the structure, designated L or D) of amino acids, for instance, influences their taste. Several amino acids are essentially without taste, whereas others (phenylalanine, leucine, valine, histidine, isoleucine, and tryptophan) are bitter in the L form and sweet in the D form. A similar dichotomy exists in sugars. Some researchers suggest that L-glucose is slightly salty and not sweet, compared to the sweet D-glucose form, the predominant form in nature. The problem of why L-isomers of sugars are not sweet may be related to the inability of the gamma region to align with the corresponding receptor site.

Challenge Questions

1. Which functional group(s) could represent the AH region of a sweet molecule?
   a. OH group
   b. NH group
   c. CH_3 group
   d. OH group and NH group
   e. OH group, NH group, and CH_3 group

2. Substituting the imino group on saccharin with methyl or ethyl groups eliminates the sweetness. Why might this be?
   a. A tasteless (nonsweet) stereoisomer would be formed.
   b. The molecule has been made too large by this substitution to be sweet.
   c. The AH has been replaced with hydrophobic groups.
   d. Hydrogen bonding at the receptor site would be enhanced.
   e. There is no logical reason.

3. What is *true* about D and L forms of molecules?
   a. They denote the directivity of specific OH groups.
   b. They usually exhibit the same level of sweetness.
   c. They denote the reactivity of specific COOH groups.
   d. They denote the reactivity of specific hydrophobic groups.
   e. They possess identical AH,B and gamma spatial orientations.

4. The AH, B and gamma regions on a sweet molecule describe
   a. the aglycone
   b. the chromophoretic regions
   c. the glucoside
   d. an anthocyanidin
   e. the glycophore

5. Look up the structure of acesulfame-K. Which statement is *incorrect*?
   a. The AH region is likely to be NH.
   b. It exhibits an AH,B glycophore structure.
   c. The CH_3 group likely aligns the molecule to the receptor site.
   d. The B region is likely to be SO_2.
   e. There is no gamma region on this molecule.

Internet Search

6. Perform a search on the Internet for the keywords "sweet taste transduction" or access the library for the November 1991 *Food Technology* article "Chemical and biochemical basis of sweetness" by G. G. Birch. List the basic steps involved in sweet taste transduction.

CHAPTER 7

Food Additives, Food Laws, and Dietary Supplements

7.1 What Is a Food Additive?

7.2 Food Laws and Regulations in the United States

7.3 The Enforcers of Food Laws

7.4 The Approval Process for Food Additives

7.5 The Nutrition Labeling and Education Act of 1990

7.6 The Dietary Supplement Health and Education Act of 1994

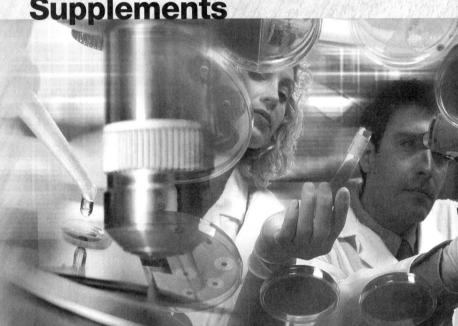

Challenge!
Regulation of Functional Foods, Bioengineered Foods, and Organic Foods

CHAPTER OBJECTIVES

After completing this chapter, you will be able to:
- State the legal definition of a food additive.
- Explain the purposes for the various types of food additives.
- Summarize the history of food law in the United States.
- Discuss milestones in food law during the last five decades of the twentieth century.
- Define the distinction between the FDA and the USDA.
- Describe the approval process for food additives.
- Explain the importance of the NLEA.
- Describe the impact of the DSHEA regulation.
- Explain how regulatory efforts have addressed functional foods, GM foods, and organic foods.

In the thirteenth century, the king of England proclaimed the "Assize of Bread," which was the first English food law. This early law prohibited the addition of foreign material such as ground peas or ground beans into bread flour (which had been the practice). In the United States, the state of Massachusetts enacted the first general food adulteration law in 1785. Within a hundred years after that, the federal government began to take an active part in protecting the public health from the fraud and potential dangers of the food supply, including food additives. People wonder about the necessity and the safety of additives and dietary supplements. This chapter deals specifically with the nature of additives, dietary supplements, and the regulations that govern their use in the United States. By and large, other countries have adopted similar approaches. Looking ahead, this chapter serves as a useful prelude to Chapters 11 and 12 on food safety and food toxicology.

7.1 WHAT IS A FOOD ADDITIVE?

In the broadest sense, a food additive is any substance added to food. A more useful definition is that a food additive is a chemical or other substance that becomes a part of a food product either intentionally or accidentally. Most food additives are **intentional additives,** meaning they were purposely added, and these include such substances as sugar, salt, corn syrup, baking soda, citric acid, and vegetable coloring. Intentional additives must receive approval from the Food and Drug Administration before they can be used in foods.

Indirect additives are contaminants—substances that accidentally get into a food product during production, processing, or packaging. Indirect additives are for the most part anticipated, with controls in place to restrict the occurrence of such accidental additives to a minimum level. Adulteration is the deliberate addition of cheap ingredients to a food to make it appear to be of high quality. Intentional food adulteration is illegal in the United States.

Although food regulatory agencies and food legislation will be the focus later in the chapter, examine this portion of the definition of food additive as found in the 1958 amendment to the Federal Food, Drug, and Cosmetic Act (FFDCA) to get a better idea of the legal definition of food additive.

> The term "food additive" means any substance the intended use of which results or may reasonably be expected to result, directly or indirectly, in its becoming a component or otherwise affecting the characteristics of any food (including any substance intended for use in producing, manufacturing, packing, processing, preparing, treating, packaging, transporting or holding food; and including any source of radiation intended for any such use). . . .

Notice that this definition refers to any substance used in the production, processing, treatment, packaging, transportation, or storage of food, and that it includes radiation, or the ionizing energy used to destroy microorganisms. So, even though we usually think of an additive as an ingredient, it can also refer in a specific case (food irradiation) to a processing step.

Examples of indirect additives:

Antibiotics
Dioxins—chemical pollutants, used in bleached paper food/drink containers
Dirt and dust
Hair
Hormones
Insects
Microwave packaging—active and passive packaging materials and chemical constituents that can migrate into food at high heat (500°F)

FIGURE 7.1 Uses of food additives. *Without additives, bread would quickly mold and salad dressing would separate into oil and water phases and become rancid.*

Despite their association with modern-day food production, food additives have been in use for centuries: salt to preserve meats and fish; herbs and spices as flavor enhancers; sugaring of meats and fruits; and pickling with vinegar. Vitamin and mineral additives have reduced serious nutritional deficiencies among Americans and help assure a constant availability of wholesome, appetizing, and affordable foods.

The Uses of Food Additives

In our society, having wholesome, nutritious, safe, and fresh foods available in abundance year-round is something we take for granted. Part of the reason we enjoy such quality has to do with the use of food additives. In an industrial rather than an agrarian society, additives are important to help to keep food wholesome and appealing while en route to market. During this journey, foods are subjected to a number of factors such as temperature changes, oxidation, and exposure to microbes that can change the original composition. Additives are key in maintaining the food qualities and characteristics consumers demand (Figure 7.1).

Some of the key uses for food additives are as follows:

1. *To maintain product consistency.* What makes salt flow freely? How can salad dressings and peanut butter stay smooth and not separate? Certain ingredients such as emulsifiers, stabilizers, thickeners, and anticaking agents help ensure consistent food texture and characteristics.
2. *To improve or maintain nutritional value.* Nutrients in food can either be lacking or lost during processing. Cereals, milk, margarine, and other foods can be enriched or fortified by additives such as vitamins A and D, ascorbic acid, niacin, iron, riboflavin, thiamin, and folic acid.
3. *To maintain palatability and wholesomeness.* Foods naturally lose flavor and freshness due to aging and exposure to natural elements such as oxygen, bacteria, and fungi. Preservatives such as butylated hydroxyanisole (BHA), butylated hydroxytoluene (BHT), ascorbic acid, and sodium nitrite help to slow product spoilage and rancidity while maintaining taste.
4. *To provide leavening or control acidity/alkalinity.* Leavening agents enable cakes, biscuits, and other baked goods to rise during baking. Certain additives modify the acidity and alkalinity of foods for proper flavor, taste, and color.
5. *To enhance flavor or impart desired color.* Many spices and natural and synthetic flavors enhance the taste of foods, while color additives enhance the appearance of certain foods to meet consumer expectations.

In addition to these particular functions of food additives, basic principles influence the use of food additives in the food industry. These principles or requirements guide the application of each additive:

- The safety of a food additive for human consumption must never be in doubt. For a new food additive, this is achieved through extensive testing and validation, at the expense of the manufacturer, to the satisfaction of the FDA.
- A food additive must function in food systems in accordance with its stated function under specific conditions of use. This is termed the additive's *efficacy*.
- A food additive must not significantly diminish the nutritional value of the food in which it is functioning, nor should it be used to compensate for improper manufacturing practices or inferior product characteristics in a way that would deceive the consumer.
- A food additive should be detectable by a defined method of analysis.

The Major Types of Food Additives

Food additives have been classified by the FDA into more than two dozen groupings of substances based upon their functionality. Certain additives can have multiple functions. For example citric acid, depending on the level added to foods, can act as an acidulant, a flavor enhancer, a preservative, and a sequestrant. Here are specific food additives, grouped according to function.

Anticaking and free-flowing agents are substances that keep ingredients in a powder form for ease of incorporation into formulations during product manufacture. Examples include silicates and talc.

Antimicrobial agents act to inhibit the growth of bacteria, yeasts, and molds and thus function as preservatives. Examples of antimicrobial agents are sodium benzoate, fatty acid salts such as calcium propionate, sodium nitrite and nitrates, sodium chloride, sulfur dioxide, sorbic acid, and oxidizing agents like chlorine, hydrogen peroxide, and iodine.

Antioxidants act to inhibit the oxidation of fats and pigments, which would otherwise result in product rancidity and altered color. Sample antioxidants are BHA, BHT, propyl gallate, ascorbic acid, and tocopherols.

Colorants, or food colors, are added to certain foods to offset color loss due to storage or processing of foods, or to correct for natural variations in food color. As discussed in Chapter 6, color additives are either certifiable or exempt from certification. The FDA approval process for certifiable additives, known as **color additive certification,** assures the safety, quality, consistency, and strength of a color additive prior to its use in foods. In the United States nine certified colors are currently approved for use in foods.

The artificial colorants include certified FDA *dyes* (water-soluble colorants available in powder, liquid, or paste form) and *lakes* (suspensions of organic colorants coated onto metallic salts). Table 6.3 listed the certified FD&C colorants. The exempt colorants are pigments obtained from plant, animal, or mineral sources (see Table 6.5), and they must meet certain legal criteria for specifications and purity.

During the 1970s, some scientists suggested that food additives including colorants might be linked to clinical symptoms such as childhood hyperactivity. However, controlled studies have failed to find evidence to support the notion that food additives cause hyperactivity or learning disabilities in children. A Consensus Development Panel of the National Institutes of Health (NIH) concluded in 1982 that no scientific evidence supported the claim that additives or colorings cause hyperactivity, and scientific studies conducted after 1982 continue to support this conclusion.

Curing agents for meats contain sodium nitrite, which helps retain the pink color of cured meats, as well as acting as a preservative.

Dough strengtheners are substances used to improve the machinability of bread dough during processing. These include emulsifiers such as SSL (sodium stearoyl lactylate), EMG (ethoxylated monoglyceride), and DATEM (diacetyl tartaric acid esters of mono- and diglycerides).

Emulsifiers keep fat globules dispersed in water or water droplets dispersed in fat. These effects are important in such diverse processed food systems as butter, frankfurters, cakes, salad dressings, and ice cream. Lecithins, monoglycerides, and diglycerides are examples of food emulsifiers. Distinct from emulsifiers are **emulsifying salts,** which function to enhance natural emulsifier activity in food systems such as process cheese. Sodium and potassium phosphates and citrates are examples of emulsifying salts.

Enzymes, biological catalysts that occur naturally in foods, are used by the food industry for use as beneficial food additives. Examples of enzymes include pectinase, used in jelly manufacture; glucose oxidase, which prevents nonenzymatic browning in powdered egg white and removes traces of oxygen in certain beverages; and invertase, which is used in the manufacture of chocolate-covered cherries.

Flavorings may be natural or synthetic and are added for flavor production or modification. Natural essential oils, the odorous components of plant extracts, are used as flavorings. Synthetic flavorings are often ester compounds, such as amyl acetate, which provides artificial banana flavoring. **Flavor enhancers** such as monosodium glutamate (MSG) and flavor potentiator substances identified chemically as $5'$-nucleotides are used to make foods taste more delicious ("umami" effect).

Humectants are substances that attract water within a food product, which may lower the product's water activity. The hygroscopic nature of the monosaccharide fructose makes it an excellent humectant for use in sweetened baked goods. Polyhydric alcohols or polyols such as glycerol (also known as glycerine), sorbitol, mannitol, and propylene glycol are also effective humectants.

Leavening agents such as baking powder are used to enhance the leavening effect, rise, or "oven spring" of dough in baked products. Baking powder is a combination of baking soda (sodium bicarbonate, a base) and potassium acid tartrate (cream of tartar, an acid). The chemical reaction requires the addition of water and produces carbon dioxide gas, which is responsible for the leavening effect.

Nutritional additives are included in foods such as breakfast cereals, baked goods, and drinks to boost nutrient intake and provide for a more balanced diet. Vitamin D is added to milk, and B vitamins and the mineral iron are added to baked products. Certain foods are *enriched*, while others are *fortified*, with nutrients. **Enrichment** denotes the addition of nutrients lost during processing in order to meet a specific standard for a food. Bread, flour, and rice are examples of enriched foods. A related term, **fortification,** means the addition of nutrients, either absent or present in insignificant amounts. Fortification provides nutrients that are often lacking in the diet in order to prevent or correct a particular nutrient deficiency. Iodine in salt and calcium and antioxidants in orange juice are examples.

Nonnutritive sweeteners are compounds that provide much greater sweetness intensity per amount when compared to sucrose. The very small quantity of nonnutritive sweetener required translates into negligible calorie and nutrient contribution in a food product. Examples are aspartame, acesulfame potassium (acesulfame-K), and saccharin.

Nutritive sweeteners are compounds that provide significant calories from carbohydrates in addition to a level of sweetness intensity. These include sucrose, fructose, maltose, lactose, polyhydric alcohols such as xylitol and sorbitol, as well as molasses and honey.

Oxidizing agents occur in food mainly as residuals (of chlorine or iodine, for example) from application as sanitizing agents of food processing equipment. In addition, hydrogen peroxide, another oxidizing agent, is used in the dairy industry as an antimicrobial, with all residual levels removed by the addition of the enzyme catalase. Oxidizers or oxidizing agents also act as bleaching agents to whiten food material such as flour. Benzoyl peroxide and sodium hypochlorite are used to bleach starch and flour.

pH control agents are acidulants, which lower food pH, and alkalis or alkaline compounds, which increase food pH. Acidulants include malic acid, tartaric acid, phosphoric acid, citric acid, and vinegar (contains acetic acid). In many cases an acidulant exerts multiple effects in addition to pH lowering, such as the enhancement of flavor and the inhibition of microorganisms. Alkaline compounds such as sodium hydroxide and potassium hydroxide are used to neutralize the excess acid developed in fermented foods in order to prevent undesirable flavor development. Sodium hydroxide is also used to modify the functionality of food starches. Sodium carbonate reduces the hardness of drinking water and formula water, while sodium bicarbonate is a leavening component.

Processing aids include not only acidulants and alkalis, but also buffers and phosphates. Buffers are added to help maintain a constant pH in a food, by balancing the hydrogen and hydroxide ions to protect its color, flavor, or some other pH-sensitive characteristic. Orthophosphates, citrate, citric acid, and sodium bicarbonate can function in foods as buffers. Other food phosphates, such as polyphosphates, act to increase the water-holding capacity of meats and to stabilize emulsions.

Sequestrants act to combine with metal elements, such as copper and iron, which are active in oxidation reactions. By forming complexes with them, sequestrants (also called chelating agents) inhibit the development of off-flavors and odors due to oxidation and can protect antioxidants to extend their effectiveness. Sequestrants are routinely added in metal canned food products, including beer. Common sequestrants include citric acid, polyphosphates, and EDTA (ethylenediamine tetra acetate).

Stabilizers and thickeners combine with water in foods to increase product viscosity, to form gels, and to prevent product crystallization. Starch, pectin, gums (arabic, carrageenan, guar), cellulose, gelatin are commonly used to thicken such diverse products as gravies, pie fillings, dairy products, cake frostings, and puddings.

Surface active agents or surfactants act as wetting agents, lubricants, dispersing agents, and emulsifiers, by affecting the surface tension of materials present in food systems. They are added to foods during processing to reduce stickiness, promote mixing, improve baking properties, and either destabilize foams or promote foaming. Examples of surface active agents are lecithin, monoglycerides, diglycerides, and polysorbate emulsifiers or surfactants, commercially known as Tweens.

Understanding the functions of food additives is only a part of what food scientists must know about them. Food law and regulation is another part. Knowing the rules about *rules* is critical to understanding the approval processes and the interaction among government, business, and consumers.

7.2 FOOD LAWS AND REGULATIONS IN THE UNITED STATES

Differentiating Laws and Regulations

The study of food law includes understanding the agencies that have regulatory authority, such as the FDA (see Section 7.3), which write regulations, check on compliance, and levy penalties against violators. Food law is also concerned with the history of specific acts and regulations related to foods. But what are regulations, and are they the same or different from acts, laws, ordinances, and statutes?

An *act* (starting life as a bill before Congress) becomes a law when the President of the United States signs it (e.g., Federal Food, Drug, and Cosmetic Act). Laws require adherence by all citizens—businesses and the public at large. Those in violation are guilty of having broken the law.

A *statute*, by legal definition, is not strictly a law, but rather a rule or administrative code issued by governmental agencies at all levels (municipal, county, state, and federal). However, since they are adopted under special authority and carry with them penalties for violations, statutes have the force of law.

A *code* is the legal term for a collection of statutes and rules. The **Code of Federal Regulations** (CFR) is a yearly codification of rules published in the Federal Register by agencies of the Department of Health and Human Services (HHS) and the Department of Agriculture. The CFR is an invaluable resource of extensive and authoritative information; it is divided into 50 titles; for instance, CFR 21 is assigned to the FDA branch of HHS.

The *Federal Register* is a legal newspaper published daily by the National Archives and Records Administration (NARA) in partnership with the Government Printing Office. It contains published and proposed rules and regulations. A new regulation does not become official until it is published in the Federal Register.

Regulations and *rules* have similar meaning, although they are not the same as statutes and laws. Rules may be considered external policy statements issued by agencies, such as the USDA, that ultimately become regulations. A regulation begins life as a *pro-*

TABLE 7.1 Definitions.

Act	Bill approved by both houses of Congress (Senate and House of Representatives) and signed into law by the President
Directive	Internally written instructions regarding policy and procedure within an agency (e.g., FDA)
Law	An act of Congress that has been signed by the President
Ordinance	Law enacted by local legislative process
Regulation	Administrative code or rule issued by governmental agencies; regulations are implemented to enforce statutes
Rule	Regulation that has been put forth in its final form
Statute	Federal or state written and enacted rules and administrative codes having the force of law

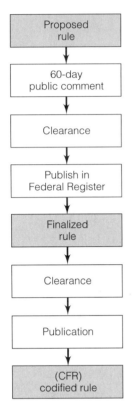

FIGURE 7.2 The life cycle of a new rule, from proposal through finalization and codification.

posed rule. Proposed rules are available in the Federal Register and are filed for public inspection and comment. Open hearings are held to obtain input from stakeholders, including consumer advocates, government officials, food industry representatives, and the public, for the purpose of possible rule modification. After a specified time period, opportunity for comment ends, and the finalized rule is published in the Federal Register and subsequently listed in the Code of Federal Regulations.

Proposed Rule. A proposed rule is written whenever a change is needed in an agency's regulations. The public is provided the opportunity to comment on proposed changes to regulations through the Administrative Procedures Act. For this reason, a rule is published in proposed form first, with (usually) a 60-day comment period. Most proposed rules are developed to implement legislative (law) requirements, but not all rules are based on laws. A proposed rule may be developed to implement legislative (law) requirements, to clarify a particular policy, or as a response to suggestions received by an agency.

After completing the appropriate clearance process based on the rule's designation by the Office of Management and Budget (OMB) and through appropriate signature, the proposed rule is published in the Federal Register and the public comment period begins.

Any rule, whether proposed or final, has two parts. The first part, the *preamble*, discusses the changes being made (or proposed) by the rule. The second part, the *regulatory text*, describes how the CFR is being changed by the rule. In a proposed rule, the preamble explains the proposed provisions for commenters. In the final rule, the preamble addresses the comments received on the proposed rule and how those comments changed the rule in its final form.

Finalized Rule. Once the 60-day comment period is completed, comments from the public are reviewed and categorized based on the relevance of their support of (or disagreement with) the changes proposed in the rule. After careful consideration, the final rule is then drafted to incorporate appropriate changes (Figure 7.2).

The final rule must once again be designated by the OMB and complete the clearance process appropriate to its designation. A rule that is in clearance is often called "a docket." Once signed and published, it carries an effective date, which shows the public the date the changes made by the rule go into effect. It may also carry an implementation date, which tells regulated entities the date by which they must begin acting under the new/revised regulations.

Codifying the Final Rule into the CFR. Once a year, the Office of the Federal Register codifies all changes made by interim and final rules during the year into a new revision of the CFR (generically referred to as "the regulations"). The time of year varies by title. For USDA, the rules are codified as of December 31 of each year and issued in a new edition of the CFR dated January 1 of the following year. Each year's new edition of the CFR, though dated January 1, is usually not available until several months later (around April or May).

The CFR is divided into 50 titles (USDA regulations are in Title 7). Each title is divided into chapters, each chapter representing a Government agency (FSIS is Chapter III of Title 7). A Chapter is further divided into Parts, each covering a broad subject area. The part is further divided into sections, which break the broad part topic into more specific topics. Each section is broken down into paragraphs outlining specific requirements where applicable (Table 7.2).

Early Events and Legislation

The beginnings of interest in laws for foods and food additives on a national scale can be traced to President Lincoln in 1862, with his appointment of Charles Wetherill, a chemist, to serve in the newly formed Department of Agriculture, or USDA. The Bureau of Chemistry in the USDA evolved into the Food and Drug Administration (FDA) in the Department of Health and Human Services. In 1879, Peter Collier, as Chief Chemist of the USDA, began an investigation into food adulteration practices in the United States.

TABLE 7.2 Divisions of the CFR. *The CFR is subdivided into titles, each one further divided into specific parts. Important titles related to food include Title 7 (Agriculture, subtitle USDA regulations), Title 9 (Animals and Animal products, Chapter III, FSIS), and Title 21 (Food and Drugs; Chapter I, FDA). For example, Title 9, Chapter III, Part 417, Section 417.1 presents the definition of HACCP by the Food Safety and Inspection Service (FSIS) mission area of the USDA.*

CFR divisions:
- Title: (subject of title)
  - Chapter (government agency)
    - Part (subject area)
      - Section (specific topic)
        - Paragraph (outlines specific requirements)

TABLE 7.3 Timeline of selected governmental legislation and regulation of the U.S. food supply.

Year	Legislation
1906	Pure Food and Drug Act (FDA)
1906	Federal Meat Inspection Act (USDA)
1914	Federal Trade Commission Act (FTC)
1935	Federal Alcohol Administrative Act (U.S. Dept. of the Treasury)
1938	Federal Food, Drug, and Cosmetic Act (FDA)
1944	Public Health Services Act (USDA)
1946	Agricultural Marketing Act (USDA)
1947	Federal Insecticide, Fungicide, and Rodenticide Act (EPA)
1954	Pesticide Residue Amendment to 1938 (EPA)
1956	Fish and Wildlife Act (U.S. Dept. of Commerce)
1957	Poultry Products Inspection Act (USDA)
1958	Food Additives Amendment to 1938
1960	Color Additives Amendment to 1938
1966	Fair Packaging and Labeling Act (FDA)
1967	Wholesome Meat Act (USDA)
1970	Egg Products Inspection Act (USDA)
1972	Federal Environmental Pesticide Control Act (EPA)
1974	Safe Drinking Water Act (EPA)
1977	Saccharin Study and Labeling Act
1990	Nutrition Labeling and Education Act (FDA)
1994	Dietary Supplement Health and Education Act
1996	Food Quality Protection Act
1996	Pathogen Reduction Hazard Analysis and Critical Control Points (HACCP) Systems Final Rule
1997	Pathogen Reduction Act
1997	Safe Food Act

In 1883, staff members under Dr. Harvey Wiley studied this issue with increasing intensity, and as a result Congress introduced more than 100 food and drug bills during the 25 years that followed.

A defining event held in Washington in 1898 was the **Pure Food Congress,** which was spearheaded by Wiley. This event focused national attention on the growing movement to enact federal legislation against misbranding and adulteration of foods. President Theodore Roosevelt involved his administration in this movement by signing into law the **Food and Drug Act of 1906.** This law prohibited interstate commerce in misbranded and adulterated foods, beverages, and drugs.

On the same day, June 30, 1906, the **Meat Inspection Act** was passed to regulate meat quality and safety. The motivational force behind this was the public outcry following the publication of *The Jungle* by Upton Sinclair. In the book, Sinclair described in shocking detail the abhorrently unsanitary conditions prevailing at the time in the meat industry. The use of poisonous preservatives and dyes in foods as well as the worthless claims for cure-all potions, medicines, and elixirs also helped promote passage of these 1906 laws.

Table 7.3 lists important legislation regarding the U.S. food supply. We will examine some of the specific food laws that have been enacted to protect U.S. consumers.

The 1938 FFDCA and Amendments

Federal Food, Drug, and Cosmetic Act of 1938. The basis of modern food law is the **Federal Food, Drug, and Cosmetic Act (FFDCA) of 1938,** which gave the FDA authority over food and food ingredients and defined requirements for truthful labeling of ingredients.

The FFDCA required a **standard of identity** for several special foods—a detailed listing of the type and quantity of ingredients and means of preparation. The standard defines what a product is, its legal name, and its required ingredients. Figure 7.3 shows part of the standard of identity for sour cream.

Standards of identity were written to provide guidelines to food manufacturers to ensure the sensory quality of those specific food products. Table 7.4 lists some of the foods for which standards of identity have been published. Two other standards required by the FDDCA of 1938 are *standards of minimum quality*, which define minimum standards of quality (product texture, color, and so forth), and *standards of container fill*, which define how full the food container must be and how it is measured.

Food Additives Amendment of 1958. The **Food Additives Amendment (FAA) to the FFDCA** of 1958 requires FDA approval for the use of an additive prior to its inclusion in food. It also requires the manufacturer to prove an additive's safety for the ways it would be used. The FAA exempted two groups of substances from the food additive regulation process. All substances that the FDA or the USDA had determined were safe for use in specific foods prior to the 1958 amendment were designated as **prior-**

TABLE 7.4 Some foods for which standards of identity exist. *In addition, FSIS has approximately 80 food standards of identity and composition that are codified as Federal regulations for meat and poultry.*

Milk and cream
Cheeses and related products
Frozen desserts
Bakery products
Cereal flours and related products
Macaroni and noodle products
Canned fruits
Canned fruit juices
Fruit butters, jellies, and jams or preserves
Fruit pies
Canned vegetables
Vegetable juices
Frozen vegetables
Eggs and egg products
Fish and shellfish
Cacao products (cocoa, chocolate)
Tree nut and peanut products
Margarine
Sweeteners and table syrups
Food dressings and flavorings (mayonnaise and salad dressing, vanilla flavoring)

---

§ 131.160   Sour Cream.

(a) **Description**. Sour cream results from the souring, by lactic acid producing bacteria, of pasteurized cream. Sour cream contains not less than 18 percent milkfat; except that when the food is characterized by the addition of nutritive sweeteners or bulky flavoring ingredients, the weight of the milkfat is not less than 18 percent of the remainder obtained by subtracting the weight of such optional ingredients from the weight of the food; but in no case does the food contain less than 14.4 percent milkfat. Sour cream has a titratable acidity of not less than 0.5 percent, calculated as lactic acid.

(b) Optional ingredients.

(1) Safe and suitable ingredients that improve texture, prevent syneresis, or extend the shelf life of the product.

(2) Sodium citrate in an amount not more than 0.1 percent may be added prior to culturing as a flavor precursor.

(3) Rennet.

(4) Safe and suitable nutritive sweeteners.

FIGURE 7.3   Excerpt from the standard of identity for sour cream (21 CFR 131.160). *The U.S. government has established standards of identity for a number of dairy foods. These standards define a food's minimum quality (e.g., minimum and maximum content requirements for various constituents such as milk fat), required and permitted ingredients (e.g., vitamins A and D), and processing requirements. Most dairy foods with a standard of identity conform to the FDA standard and regulations published in the Code of Federal Regulations.*

sanctioned substances. Examples of prior-sanctioned substances are sodium nitrite and potassium nitrite, which are additives used to preserve luncheon meats.

The second category of substances excluded from the food additive regulation process are "generally recognized as safe" or **GRAS substances.** GRAS substances are those whose use is generally recognized by experts as safe, based on their extensive history of use in food before 1958 or based on published scientific evidence. Salt, sugar, spices, vitamins, and monosodium glutamate are classified as GRAS substances, along with several hundred other substances.

Since 1958, the FDA and the USDA have continued to monitor all prior-sanctioned and GRAS substances in light of new scientific information. If new evidence suggests that a GRAS or prior-sanctioned substance may be unsafe, federal authorities can prohibit its use or require further studies to determine its safety. In 1960, Congress passed similar legislation governing color additives.

The Color Additives Amendments.   The 1960 **Color Additives Amendments** (CAA) to the FFDCA requires dyes used in foods, drugs, cosmetic, and certain medical devices to be approved by the FDA prior to their marketing. In contrast to other food additives, colors in use before the legislation were allowed continued use only if they underwent further testing to confirm their safety.

The Delaney Clause.   Both the Food Additives and Color Additives Amendments included a clause that prohibited the approval of an additive if it was found to cause cancer in humans or animals. This clause is often referred to as the **Delaney Clause,** named for its Congressional sponsor, James Delaney. The Delaney Clause stated:

> That no additive shall be deemed to be safe if it is found to induce cancer when ingested by man or animal, or if it is found, after tests which are appropriate for the evaluation of the safety of food additives, to induce cancer in man or animal . . .

The Delaney Clause established a "zero-cancer-risk" standard for food additives, including pesticide residues. Over time, this provision led to the cancellation of about eight additives: a veterinary drug, a veterinary feed additive, a flavoring agent, the sweetener saccharin, indirect food additives from packing materials, and several color additives. Because all these additives had substitutes available at the time except for saccharin, there was little noticeable impact on the food supply. The Delaney Clause is absolute in meaning—there are no gray areas of interpretation. It does not allow consideration of whether the cancer risk is "negligible" (extremely small) or whether other benefits of the product might outweigh the cancer risk.

The zero-tolerance aspect of the Delaney Clause became an issue at the Environmental Protection Agency (EPA), which was developing a realistic standard for pesticide safety just beyond zero tolerance. The problem: if residues of carcinogenic pesticides were found to concentrate in a processed food, the EPA could not set a tolerance or maximum legal safety limit for that pesticide/food combination according to the Delaney Clause.

However, since 1958, science had developed methods to detect minute quantities of residues. The EPA believed these advanced methods could be employed to help establish realistic pesticide safety limits, realizing also that no definition of safety was explicitly provided in the statute. "Safe" as it was interpreted according to the legislative history and by the courts, came to mean "a reasonable certainty of no harm." Factors used by the FDA to determine "safe" food included (1) probable consumption of the substance, (2) cumulative effects in the diet, taking into account chemically or pharmacologically related substances, and (3) safety factors that in the opinion of qualified scientific experts were generally recognized as appropriate.

In 1996, the Delaney Clause was essentially laid to rest, with the adoption of "negligible risk" (*de minimus*) in place of "zero tolerance" by the FDA. The agency now deems additives (as well as pesticides and contaminants) as safe if lifetime (70 year) use presents no more than a one-in-a-million risk of cancer to human beings.

The 1938 FFDCA, the 1958 FAA, the 1960 CAA, and the Delaney Clause compose only a small part of the food law history in the United States. This chronology of the legislative activity relating to foods, additives, and safety demonstrates the sequential development not only of food law but also of scientific knowledge, consumerism, and food industry practices.

Other Legislation and Significant Regulatory Actions

Processed Foods Innovations. During the 1940s through the 1960s, the processed food industry experienced a dramatic surge in growth, called the **chemogastric revolution.** Processed foods, for example, frozen orange juice, breaded shrimp, frozen desserts, and frozen "TV dinners," began to appear in stores. All of these foods used new chemical ingredients and packaging, and they set the stage for future legislation concerning additives and manufacturing practices. For example, the **Fair Packaging and Labeling Act** of 1966 was passed to counter problems with underweighing of products, particularly cereals. Each product label was required to identify the product, the name and place of business of manufacturer, the net quantity of contents, and the net quantity of a serving when the number of servings is represented.

Pesticides and Toxicants. In 1954 Congress passed the **Pesticide Residue Amendment** as Section 408 of the 1938 Act. This amendment split jurisdiction for pesticides between the USDA, which could register pesticides for specific uses, and the FDA, which was charged with setting tolerances for residues of pesticides in raw agricultural products. The burden of proof was shifted to the pesticide sponsors, whereby pesticide residues found in food would mean the food was adulterated, unless a sponsor received a specific tolerance or an exemption statement.

The discovery of the *Aspergillus* mold which produced the powerful toxin **aflatoxin** was made following the outbreak of "turkey x disease" in England. Its discovery marked the emergence of the field of study of mycotoxins. Aflatoxin grows naturally and is particularly associated with crops stressed by drought. More than 100 mold species that produce toxins have been found growing on peanuts, corn, rice, cottonseed meal, oats, hay, barley, sorghum, cassava, and millet. The FDA chose to set an **action level** for mycotox-

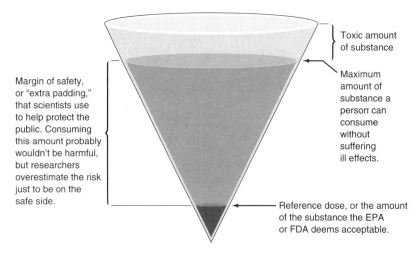

FIGURE 7.4 The Concept of Margin of Safety. *Due to agrarian practices and environmental conditions, pesticides and other potentially harmful substances cannot be totally absent from foods. However, the FDA and EPA establish margins of safety for such substances to protect the public. These set limits to what is legally permitted in foods and beverages, according to a safety factor.*

ins such as aflatoxin. An action level is a level for contamination of a food, below which no court enforcement action is necessary. Figure 7.4 illustrates the concept of margin of safety, which is used to set action levels.

Cyclamates. In 1969 **cyclamate,** an alternative sweetener used in beverages, was banned as a food ingredient because tests conducted at the New York Food Research Laboratory showed that a saccharin/cyclamate mixture caused cancer in experimental laboratory mice. Also in that year the **White House Conference on Food, Nutrition and Health** recommended systematic scientific review of generally recognized as safe (GRAS) substances. Citing the prohibition of cyclamates and a subsequent loss of public confidence in the continued food use of other previously approved substances, President Nixon directed the FDA, in his consumer message, to study the safety of GRAS food ingredients and to initiate a safety review of GRAS substances.

Subsequently the FDA compiled data and evaluated the scientific literature over several years on approximately 600 food ingredient substances in the GRAS review program. The FDA estimated there were approximately 10,000 food packaging materials and 400 functional food additives in addition to those already under review that remained to be reevaluated. The FDA published revised provisional listing regulations in the Federal Register, which required new chronic toxicity studies using more modern protocols on 31 color additives as a condition of their continued provisional listing.

Saccharin. In 1972, the sweetener **saccharin,** which had been included in the FDA's original published list of GRAS substances for addition to food, was removed from the GRAS list. It was reclassified as a substance approved for interim use within prescribed limitations, pending completion of further studies. The final rule on saccharin, published in 1973, prescribed limitations on the amount of saccharin allowed per serving of food and on the purposes for which saccharin could be used in food. The restrictions were intended to limit the amount of saccharin in the American diet and to confine saccharin to uses already prevalent at the time.

As a result of experimental data that showed saccharin to be a carcinogen in rats, the FDA proposed to ban saccharin under the general safety provisions of the Food Additives Amendment and the Delaney Clause. In response, Congress passed the **Saccharin Study and Labeling Act** on November 23, 1977. The law imposed an 18-month moratorium on the FDA's proposed saccharin ban and required that saccharin packets and foods con-

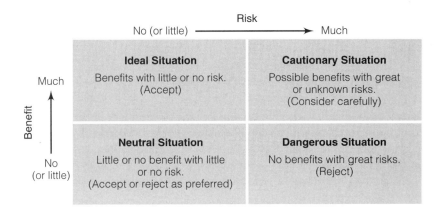

FIGURE 7.5 Risk-benefit relationships. *Certain substances that are known to provide a risk to human health also have been shown, at certain levels of use, to be of some benefit. It is possible to weigh the risks versus the benefits of consuming such substances (e.g., herbal medicines), or employing them as functional additives (e.g., colorants) by categorizing the possible outcomes as ideal, neutral, cautionary, or dangerous.*

taining it bear the warning: "Use of this product may be hazardous to your health. This product contains saccharin which has been determined to cause cancer in laboratory animals." FDA regulations required that this warning also appear in retail establishments that sell products with saccharin and on some vending machines as well.

Congress has since extended the moratorium on prohibiting saccharin a number of times during the last two decades of the twentieth century. On December 21, 2000, President Clinton gave saccharin a clean "bill" of health by signing legislation that removed the warning label. This action was prompted by conclusions reached by the NTP (National Toxicology Program), AMA (American Medical Association), ADA (American Dietetic Association), and other agencies worldwide as to the safety of saccharin, based upon knowledge accumulated since 1977.

Red Book.   In 1982 the FDA published its first **Red Book,** officially known as *Toxicological Principles for the Safety Assessment of Direct Food Additives and Color Additives Used in Food.* Also, the FDA issued the first regulation under its "constituents policy." The FDA announced that if an additive was not carcinogenic, but a contaminant of that additive was a known carcinogen, the additive itself could be approved if a *risk-benefit analysis* showed the risk to be negligible. Figure 7.5 shows a risk-benefit matrix; the actual measurement of risk must be determined for each case.

Nutritional Labels.   Under the **Nutrition Labeling and Education Act (NLEA),** the label on almost every food product in the United States was changed. For the first time, some health claims for foods were authorized (in four categories), the food ingredient and nutritional content panels on labels were standardized, serving sizes were standardized, and terms such as *low-fat* and *light* were standardized. A more detailed account of the NLEA is in Section 7.5.

Biotechnology.   The FDA published its original **biotechnology policy** in the Federal Register in 1992, in which the FDA stated that it would focus on the safety of a food and not the process by which the food was developed. Under this policy, the food products of biotechnology were to be treated no differently than ordinary food, except in cases in which the genetic engineering process added a new substance to the food that is not GRAS or that is significantly different in structure, function, or amount than substances found ordinarily in food. In connection with the biotechnology policy, the FDA granted approval of the animal drug **sometribove,** a recombinant bovine somatotropin (BST) product for increasing milk production in dairy cows. The approval process generated public debate and controversy. The FDA also approved the **Flavr Savr™ tomato,** the first whole food produced using genetic engineering.

7.3 THE ENFORCERS OF FOOD LAWS

Learning a bit about food policy within the United States government helps us appreciate the efforts of the organizations involved. In the United States, the federal government regulates foods and food additives through the FDA, with additional food legislation and regulation enforced by the USDA and other offices. Before focusing our attention on them, it is important to acknowledge a few of the international agencies concerned with foods.

INTERNATIONAL FOOD AGENCIES

Within the United Nations are three major food agencies: the Food and Agriculture Organization (FAO), the World Health Organization (WHO), and the **Codex Alimentarius Commission.** The latter is an international organization established in 1962 by the FAO and WHO. It is made up of over 90 countries. The Commission establishes international food quality standards through a thirteen-volume publication, the **Codex Alimentarius,** to facilitate world trade (Figure 7.6). The Joint FAO/WHO Expert Committee on Food Additives (JECFA) sets standards for the purity of food additives at the international level. These standards are set for the nations within the EEC (European Economic Community).

National Food Agencies

USDA: www.usda.gov

USDA—United States Department of Agriculture. The United States Department of Agriculture (USDA) is involved with food processing and marketing and employs specially trained inspectors to carry out its functions. Figure 7.7 shows the seven divisions of the USDA. USDA inspectors work within several branches and subbranches of the USDA with respect to food additives. One such branch is the **Food Safety and Inspection Service,** or **FSIS,** within the Food Safety area.

The FSIS administers the Federal Meat Inspection Act (FMIA), the Poultry Products Inspection Act (PPIA), and the Egg Products Inspection Act (EPIA), which set the national standard for meat, poultry, and egg inspection. FSIS is responsible for the safety, wholesomeness, and labeling of meat and poultry products. FSIS regulates all raw beef, pork, lamb, chicken, turkey, and processed meat and poultry (hams, sausage, soups, stews, pizzas, and frozen dinners) containing 2 percent or more cooked poultry and 3 percent or more red meat. FSIS inspectors check all meat and poultry sold in interstate and international commerce (as an import-export inspection system). They visually check animals before slaughter and carcasses after slaughter for disease and chemical residues. FSIS inspectors also check animal products during processing, handling, and packaging (Figure 7.8). FSIS sets various meat and poultry production standards, such as the use of food additives, packaging, labeling, as well as standards for facilities and equipment sanitation and cleanliness.

FDA: www.fda.gov

FDA—Food and Drug Administration. The Food and Drug Administration or FDA is part of the Department of Health and Human Services (HHS). The FDA regulates all foods except meat and poultry, which fall under USDA jurisdiction. It develops standards for composition, quality, nutrition, and safety, and enforces federal regulations on labeling, food and color additives, food sanitation, and food safety. The FDA interprets and enforces the Food, Drug, and Cosmetic Act of 1938 as presently amended, as well as the Fair Packaging and Labeling Act of 1966. The FDA regularly inspects food processing plants.

The FDA's role in conducting a recall is tempered by the fact that the FDA has no authority under the Federal Food, Drug, and Cosmetic Act to order a recall, although it can request a company to recall a product. Most recalls of products regulated by the FDA are carried out voluntarily by the manufacturers or distributors of the product. If a company will not comply, then FDA can seek a court order authorizing the federal government to seize the product. This cooperative relationship between FDA and its regulated industries has proven over the years to be the quickest and most reliable method to remove potentially dangerous products from the market.

Under FDA guidelines, companies are expected to notify FDA when recalls are started, to make progress reports to FDA on recalls, and to undertake recalls when asked to do so by the agency. The guidelines categorize all recalls into one of three classes according to the level of hazard involved:

- **Class I** recalls are for dangerous or defective products that predictably could cause serious health problems or death. An example is a food found to contain botulinum toxin.
- **Class II** recalls are for products that might cause a temporary health problem.
- **Class III** recalls are for products that are unlikely to cause any adverse health reaction, but that violate FDA regulations. An example might be food packages that contain less than the amount stated on the label.

FDA regulatory functions include the setting of rules to ensure consumer protection (Table 7.5), communicating the rules to industry and monitoring compliance, educating the public regarding consumer protection, performing specific scientific review and support to ensure adequate and up-to-date rules, and enforcing applicable laws and regulations. In addition, the FDA has produced directives called GMPs (Good Manufacturing Practices) to establish requirements for worker cleanliness, education, training, and supervision. The design and ease of cleaning and maintenance of buildings, facilities, and equipment also come under the umbrella of GMPs, as does the requirement for adequate record keeping to ensure quality control.

In 1991 a blue ribbon panel appointed by the Secretary of Health and Human Services, known as the Edwards Committee, evaluated the resources and responsibilities of FDA. In its final report, the Committee admonished the agency to "recognize that approval of useful and safe new products can be as important to the public health as preventing the marketing of harmful or ineffective products" and recommended that the agency continue to "develop a flexible range of regulatory pathways, all of which uphold current standards of safety and efficacy, but which reflect the fact that not all drugs, devices, and foods are alike." The Committee's report also predicted "current demands and constraints on the FDA are expected to remain throughout the 1990s. These include demands for improved performance, continued high quality standards, and an increasing number and complexity of submissions, despite insufficient resource relief."

Based upon the Edwards report, the importance of continual FDA monitoring and evaluation was obvious. Reorganization or budget expansion to improve FDA efficiency

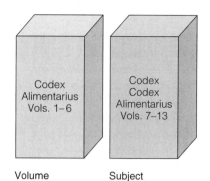

Volume	Subject
1A	General requirements
1B	General requirements (food hygiene)
2A	Pesticide residues in foods (general)
2B	Pesticide residues in foods (maximum residue limits)
3	Residues of veterinary drugs in foods
4	Foods for special dietary uses
5A	Processed and quick-frozen fruits and vegetables
5B	Fresh fruits and vegetables
6	Fruit juices
7	Cereals, pulses (legumes) and derived products, and vegetable proteins
8	Fats and oils and related products
9	Fish and fishery products
10	Meat and meat products, soups and broths
11	Sugars, cocoa products and chocolate, and miscellaneous
12	Milk and milk products
13	Methods of analysis and sampling

FIGURE 7.6 Content of the *Codex Alimentarius*.

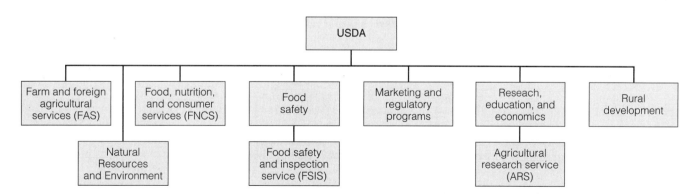

FIGURE 7.7 USDA Organizational Chart. *The USDA has seven mission areas. **ARS**, the Agricultural Research Service, is part of the Research, Education, and Economics Division of the USDA. **FSIS**, the Food Safety and Inspection Service, is the single agency under the Food Safety Division of the USDA.*

FIGURE 7.8 Meat and poultry inspection by the USDA. *USDA's FSIS has the authority to detain and seize product that is in violation of the Federal Meat or Poultry Products Inspection Acts. Although FSIS does not have mandatory recall authority, it recommends to establishments when a recall is necessary if public health is threatened. Historically, all establishments have complied when required. If an establishment were to refuse to recall a product, FSIS could invoke its detention and/or seizure authority and remove the product from commerce.*

and public service was to be considered. On November 21, 1997, the **FDA Modernization and Accountability Act of 1997** was enacted. The purpose of the act, which was submitted as a reform bill, was to reduce the bureaucracy inherent in the health claim approval process. Specifically, the new law amended section 403(r) of the FFDCA to authorize truthful, nonmisleading health claims. These claims must be based on the published authoritative statements of scientific bodies of the U.S. government, those with official responsibility for public health protection or research directly related to human nutrition, such as the CDC (Centers for Disease Control), the NAS (National Academy of Sciences) or the NIH (National Institutes of Health).

The act required manufacturers intending to submit health claims supported by authoritative governmental decrees to submit a premarket notice to the FDA at least 120 days before introduction of the food into interstate commerce. The premarket notice must concisely describe the claim itself plus the authoritative statement relied upon. Such a procedure eliminated the need for FDA preapproval of such claims. However, the FDA retains its full scope of enforcement powers to remedy misleading claims, including product seizure, injunction, and criminal penalties.

Other U.S. Food Regulatory Entities. In addition to the USDA and FDA, other government centers and offices function in some way with respect to food regulation, inspection, quality, or safety. Several are presented below, along with a statement of their

TABLE 7.5 FDA contamination levels. *The FDA acknowledges the possibility of insect and rodent pest contamination in food facilities and that not all indirect additives can be eliminated. Therefore it has established a policy that permits acceptable levels of "filth."*

Food	Acceptable Levels of Filth
Chocolate	Up to 4 rodent hairs/sample
Coffee beans	Up to 10% insect-infested
Fig paste	Up to 6 insect heads/100 g
Fish (fresh frozen)	Up to 5% "definite odor of composition"
Mushrooms (canned)	Up to 20 maggots/100 g (drained)
Peanut butter	Average of 30 insect fragments/100 g
Popcorn	Either one rodent pellet/sample or one rodent hair/two samples
Spinach	Either 50 aphids, thrips, or mites, or 8 leaf miners; or, in 24 pounds, two spinach worms or worm fragments
Tomato paste	Either 30 fly eggs or 15 eggs plus one larva

primary purpose. Beyond these federal offices, state and local government agencies cooperate with the federal agencies to develop uniform food safety standards and regulations.

APHIS—Animal and Plant Health Inspection Service. The Animal and Plant Health Inspection Service protects U.S. animal and plant resources from diseases and pests and operates an agricultural quarantine inspection program for imported agricultural products. APHIS is a mission area of the USDA (www.aphis.usda.gov).

CFSAN—Center for Food Safety and Applied Nutrition. The Center for Food Safety and Applied Nutrition is a branch of the FDA within the Department of Health and Human Services (www.vm.cfsan.fda.gov/list.html). Its mission is to promote food safety, protect consumers from fraud, promote sound nutrition, and facilitate innovation. CFSAN operates seven special offices for specific products and issues on policy guidance, programs, enforcement actions, and scientific capabilities: (1) Office of Cosmetics and Colors, (2) Office of Field Programs, (3) Office of Food Additive Safety, (4) Office of Nutritional Products, Labeling, and Dietary Supplements, (5) Office of Plant and Dairy Foods and Beverages, (6) Office of Seafood, and (7) Office of Scientific Analysis and Support.

EPA—Environmental Protection Agency. The U.S. Environmental Protection Agency (EPA) is a distinct agency not under the USDA nor FDA. It is responsible for a number of activities that contribute to food security within the United States, in areas such as food safety, water quality, and pesticide applicator training. EPA's primary contribution to food security is through its program to regulate the use of pesticides (www.epa.gov).

FCCC—Food Chemicals Codex Committee. The Food Chemicals Codex Committee, as a part of the National Academies of Sciences, sets standards (i.e., grades) for the identification and purity of food additives (www.iom.edu/iom/iomhome.nsf/Pages/Foods+Chemicals+Codex).

FTC—Federal Trade Commission. The Federal Trade Commission enforces consumer protection laws, such as new food advertising standards and the use of terminology in ads such as *low, high, lean, less, reduced, lite* (see Section 7.5) (www.ftc.gov).

NMFS—National Marine Fisheries Service. The National Marine Fisheries Service in the Department of Commerce is responsible for seafood quality and identification, fisheries management and development, habitat conservation, and aquaculture production. It also operates a voluntary inspection program for fish products (www.nmfs.noaa.gov).

7.4 THE APPROVAL PROCESS FOR FOOD ADDITIVES

Today, food and color additives are more strictly regulated than at any time in history. Food additive regulations require evidence that each additive substance is safe at its intended level of use before it can be used in foods. All additives are subject to ongoing safety review as scientific understanding and methods of testing continue to improve. Over the years improvements have been made in increasing the efficiency of the approval process and in ensuring the safety of all additives.

To market a new food or color additive, a manufacturer must first petition the FDA for its approval. Approximately 100 new food and color additives petitions are submitted to the FDA annually. Most of these petitions are for so-called indirect additives such as packaging materials. A food or color additive petition must provide convincing evidence that the proposed additive performs as it is intended. Animal studies using large doses of the additive for long periods are often necessary to show that the substance would not cause harmful effects at expected levels of human consumption. Studies of the additive in humans also may be submitted to the FDA.

In deciding whether an additive should be approved, the FDA considers the composition and properties of the substance, the amount likely to be consumed, its probable long-term effects, and various safety factors. The absolute safety of any substance can never be proven. For this reason, the FDA must determine if the additive is safe under the proposed conditions of use, based on the best scientific knowledge available. If an additive is approved, the FDA issues regulations that may include the types of foods in which it can be used, the maximum amounts to be used, and how it should be identified on food labels. As a means of follow-up evaluation, federal officials carefully monitor the extent of consumption of the new additive and results of any new research on its safety to assure its use continues to be within safe limits.

Additives proposed for use in meat and poultry products must receive authorization not only from the FDA but also by the USDA. A good example is food irradiation. Irradiation, although it is a process of applying ionizing energy to foods to preserve them and maintain freshness, is regulated as a food additive. See Table 7.6. The FDA alone regulates all irradiated foods except meats and poultry, which are jointly regulated by the USDA and FDA.

Testing Additive Safety

As chemicals, all food additives have the potential to cause harm, which is termed **toxicity**. At the approved low levels of use, food additives are efficacious (perform useful functions) rather than toxic. If the levels of a proposed additive required for efficacy are high enough to result in measurable toxicity, the FDA would rule that the additive is toxic and not grant approval for its use.

On what basis is toxicity determined? To answer the question, several types of toxicity studies must be described. Animal feeding studies, or the exposure of animal or human cell cultures *in vitro* to additives, are typically performed. Several toxic responses may develop in these systems: *Acute toxicity* is a rapid toxic response to a chemical additive; *subacute toxicity* is a less rapid but still a short-term response; and *chronic toxicity* indicates a long-term response to a low level of additive exposure.

Specific toxic effects investigated include teratogenic effects, mutagenic effects, and carcinogenic effects. A **teratogen** refers to a substance that causes abnormal fetal development and birth defects. A **mutagen** is a substance that causes a change (mutation) in the base sequence of a cell's DNA. A **carcinogen** causes cancer in a test animal. Furthermore, a mutagenic substance can induce tumors and other forms of cancer in an organism.

The Ames Test

A simple test called the **Ames test** was developed by researcher Bruce Ames at the University of California at Berkeley to identify the mutagenic potential of chemical substances. This is important since mutagenic substances often turn out to be carcinogenic. The test uses a bacterial strain of *Salmonella typhimurium* that has a defective (mutant) gene that prevents it from being able to make the amino acid histidine (His) from the ingredients in its culture medium. Normal *S. typhimurium* cells are able to grow in the absence of histidine because they can make their own histidine, but mutants cannot. When placed into culture medium lacking histidine, mutant bacterial cells (called His⁻ for histidine-requiring) die.

However, this type of mutation can be reversed, and the result is a fully functional organism, called a revertant, that is able to grow on medium lacking histidine. What enables the reversion is the presence of a mutagen. For instance, His⁻ cells grown in medium lacking histidine die, but if a carcinogen such as 2-aminofluorine is added, the mutagenic effect of this chemical causes bacteria to regain the ability to live in this medium.

Thus, if the addition of a test substance to colonies of His⁻ *Salmonella typhimurium* causes reversion, visualized as colonies growing on the culture dish, the test substance is a

TABLE 7.6 Food irradiation approvals by the FDA. *The FDA approved the first use of irradiation on a food product in 1963 when it allowed radiation-treated wheat and wheat flour to be marketed. In approving a use of radiation, FDA sets the maximum radiation exposure for the product, measured in units called kiloGray (kGy). The most recent approval for irradiation was for the irradiation of eggs in the shell in July 2000.*

Food Approved	Dose (kGy)	Application
Dry or dehydrated enzyme preparations	Up to 10	Eliminate microorganisms
Spices	Up to 30	Decontamination; eliminate insects and microorganisms
Fresh fruits and vegetables	Up to 1.0	Inhibit maturation/sprouting; control insects
*Poultry	1.5–3.0	Eliminate *Salmonella* and other pathogens
*Pork (for *Trichinella*)	0.3–1.0	Eliminate *Trichinella*
*Red meat (beef, lamb, pork, veal, venison)	Up to 4.5 (fresh) Up to 4.5 (frozen)	Eliminate *E. coli* 0157.H7 and other pathogenic and spoilage bacteria
Shellfish (clams, crab, oysters)	Petition in progress	Eliminate *Vibrio* and other pathogenic and spoilage bacteria
Shell eggs (consumer eggs)	3.0	Eliminate *Salmonella enteritidis* and other pathogens

*Also approved by the USDA.

mutagen. A nonmutagen would fail to cause reversion of the *Salmonella*, which would not be able to grow into colonies on the dish.

Aflatoxin (a peanut and grain mold toxin) and safrole (once used as a flavoring agent in root beer) test positive in the Ames test, while saccharin does not. While the test is rapid and inexpensive, bacterial cells are different from human cells. It is therefore typical to screen a substance quickly with the Ames test and then confirm mutagenicity and carcinogenicity using long-term studies in animals such as rats.

7.5 THE NUTRITION LABELING AND EDUCATION ACT (NLEA) OF 1990

The 1969 White House Conference on Food, Nutrition and Health led to major changes in various federal food policies, including the food stamp program and various FDA regulations, including those on "imitation" foods and nutrition labeling. Nutrition labeling was established by the FDA in 1973 to help consumers be aware of the nutrient content of foods. At that time, the focus of the label was directed at the micronutrient content of foods, in order to prevent deficiency diseases. However, in 1990 Congress passed the Nutrition Labeling and Education Act (NLEA), one of the rare examples of nutrition (as distinct from food manufacturing) legislation. The 1990 regulations require nutrition labeling for almost all foods for sale to consumers and regulated by the FDA. The FDA revised the layout and content of the nutritional label to focus on calorie and macronutrient content, which reflected consumers' dietary concerns regarding the role these factors play in chronic disease prevention.

The 1990 law did not go into effect until 1994, in order to allow the drafting and finalizing of regulations and to give food manufacturers time to perform the required analyses on their products and create the new labels. The USDA's FSIS developed a parallel set of regulations requiring nutrition labeling of most meats, meat products, poultry, and poultry products. This initiative was undertaken to give meat and poultry products equal standing in the marketplace.

In addition to the mandatory nutrition labeling requirements, guidelines for voluntary nutrition labeling of products such as raw fruit, vegetables, and fish have been established and are contained in the CFR in a specific section: 21 CFR 101.45. The FSIS guidelines for voluntary nutrition labeling of single-ingredient raw meat and poultry products can be found in 9 CFR 317.345 and 9 CFR 317.445.

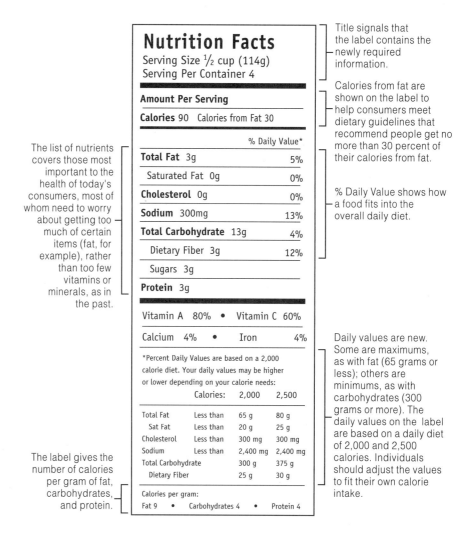

FIGURE 7.9 Example of a nutrition label.

The Nutrition Facts Label

As we discussed in Chapter 3, the standard "Nutrition Facts" label contains the following information: serving size, quantitative amount per serving of each nutrient except vitamins and minerals, amount of each nutrient (except sugars and protein) as a percent of the Daily Value for a 2000 calorie (kcal) diet, and a footnote with Daily Values for selected nutrients based on 2000 and 2500 kcal diets (representing average female and male calorie intakes). See Figure 7.9.

Daily Value is a new term on the revised 1990 nutrition label. In a way, this term has replaced the earlier terms RDA (Recommended Dietary Allowance) and USRDA. To understand Daily Value, you must know two other new terms: Reference Daily Intake (RDI) and Daily Reference Value (DRV). The term RDI replaces the USRDA and gives reference intakes of specific vitamins and minerals. The DRVs on the other hand give reference intake suggestions for nutrients for substances such as fat, cholesterol, fiber, protein, sodium, and carbohydrates. Table 7.7 lists the RDIs and DRVs.

The nutrient units listed for RDIs and DRVs vary from International Units (IU) and micrograms (μg) to milligrams (mg) and to grams (g), which might be a source of confusion to consumers. Therefore, the FDA decided to present the nutrient content as "% Daily Value" on the Nutrition Facts label rather than use the mix of units. A statement on the label indicates the Percent Daily Values are based upon 2,000 kcal diet and that a given consumer's need for each nutrient could be less or more, depending on his or her actual calorie requirement.

TABLE 7.7 RDIs for vitamins and minerals and DRVs for food components. *RDIs for vitamin A are listed as Retinol Activity Equivalents (RAE) to account for the different activities of retinol and provitamin A carotenoids. RDIs are also listed in International Units (IU) because food and some supplement labels list vitamin A content in International Units. The relationship between RAE, micrograms (mcg) and IU is that 1 RAE in micrograms = 3.3 IU.*

Vitamin	RDI	Mineral	RDI	Food Component	DRV
Vitamin A	5000 IU*	Calcium	1000 mg	Fat	65 g
Vitamin C	60 mg	Iron	18 mg	Saturated fatty acids	20 g
Vitamin D	400 IU	Phosphorus	1000 mg		
Vitamin E	30 IU	Iodine	150 μg	Cholesterol	300 mg
Vitamin K	80 μg	Magnesium	400 mg	Total carbohydrate	300 g
Thiamin	1.5 mg	Zinc	15 mg	Fiber	25 g
Riboflavin	1.7 mg	Selenium	70 μg	Sodium	2400 mg
Niacin	20 mg	Copper	2 mg	Potassium	3500 mg
Vitamin B_6	2 mg	Manganese	2 mg	Protein	50 g
Folate	400 μg	Chromium	120 μg		
Vitamin B_{12}	6 μg	Molybdenum	75 μg		
Biotin	300 μg	Chloride	3400 mg		
Pantothenic Acid	10 mg				

*1 μg vitamin A = 3.3 IU.

General Product Labeling

In addition to the nutrition information contained in the Nutrition Facts panel, the following information is presented:

- Product name and place of business
- Product net weight
- Product ingredient contents (in decreasing order of amount)
- Company name and address
- Product code (UPC bar code)
- Product dating if applicable
- Religious symbols if applicable
- Safe handling instructions if applicable (e.g., raw meats)
- Special warning instructions if applicable (e.g., aspartame)

Regarding product ingredient content, since NLEA, full ingredient labeling is required on "standardized foods," which previously were exempt. Ingredients are always listed in descending order of amount present (Figure 7.10). The ingredient list includes, when appropriate: FDA-certified color additives, such as FD&C Blue No. 1, by name; sources of protein hydrolysates, which are used in many foods as flavors and flavor enhancers; and declaration of caseinate as a milk derivative in the ingredient list of foods that claim to be nondairy, such as coffee whiteners. This requirement is to alert individuals who may be allergic to such additives and need to avoid them.

The NLEA also requires that beverages that claim to contain juice must declare the total percentage of juice on the information panel. In addition, the FDA's regulation establishes criteria for naming juice beverages. For example, when the label of a multi-juice beverage states one or more—but not all—of the juices present, and the predominantly named juice is present in minor amounts, the product's name must state that the beverage is flavored with that juice or declare the amount of the juice in a 5 percent range—for example, "raspberry-flavored juice blend" or "juice blend, 2 to 7% raspberry juice."

Nutrient Content Descriptors

The NLEA regulations indicate that only specific terms may be used to describe the level of a nutrient in a food and are defined by the FDA, such as *free, low, extra lean,* and *reduced*. Table 3.3 listed the criteria for use of such terms.

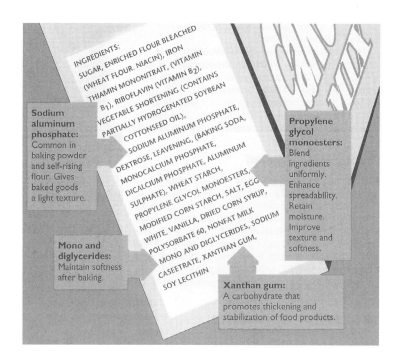

FIGURE 7.10 Typical ingredient list. *The ingredient list enables the consumer to identify food additives present in a product, in order of weight (most to least).*

Use of "Percent Fat Free." A product bearing this claim must be a low-fat or a fat-free product. In addition, the claim must accurately reflect the amount of fat present in 100 g of the food. Thus, if a food contains 2.5 g fat per 50 g, the claim must be "95 percent fat free."

Use of "Healthy." "Healthy" is a term describing a food low in fat and saturated fat and containing limited amounts of cholesterol and sodium. In addition, if it's a single-item food, it must provide at least 10 percent of one or more of vitamins A or C, iron, calcium, protein, or fiber.

Use of "Fresh." Although not mandated by NLEA, the FDA issued a regulation for the term *fresh*. The agency took this step because of concern over the term's possible misuse on some food labels. The regulation defines *fresh* when it is used to suggest that a food is raw or unprocessed. In this context, *fresh* can be used only on a food that is raw, has never been frozen or heated, and contains no preservatives (irradiation at low levels is allowed.) "Fresh frozen," "frozen fresh," and "freshly frozen" can be used for foods that are quickly frozen while still fresh. Blanching is allowed. Other uses of the term *fresh*, such as in "fresh milk" or "freshly baked bread," are not affected.

Baby Foods. The FDA is not allowing broad use of nutrient claims on infant and toddler foods. The terms *unsweetened* and *unsalted* are allowed on these foods, however, because they relate to taste and not to nutrient content.

Health Claims

As defined by the Nutrition Labeling and Education Act of 1990 (NLEA), a health claim is any claim on a food label or in labeling (including material that accompanies a food product) that expressly, or by implication, characterizes the relationship of any substance to a disease or health-related condition. Health claims can be presented in several ways: through third-party references (e.g., the National Cancer Institute); symbols (e.g., a heart); or appropriate vignettes or descriptions. The claims must be stated plainly so that consumers can understand the relationship between the nutrient and the disease and the nutrient's importance in relationship to a daily diet.

TABLE 7.8 Specific approved health claims.

1. Calcium and osteoporosis
2. Dietary lipids (fat) and cancer
3. Dietary saturated fat and cholesterol and risk of coronary heart disease
4. Dietary sugar alcohol and dental caries
5. Fiber-containing grain products, fruits, and vegetables and cancer
6. Folic acid and neural tube defects
7. Fruits and vegetables and cancer
8. Fruits, vegetables and grain products that contain fiber, particularly soluble fiber, and risk of coronary heart diease
9. Sodium and hypertension
10. Soluble fiber from certain foods and risk of coronary heart disease
11. Soy protein and risk of coronary heart disease
12. Stanols/sterols and risk of coronary heart disease

The list of approved health claims (nutrient and disease or condition) is given in Table 7.8. Three recently approved claims involve soluble fiber, soy protein, and plant sterols and stanols.

Soluble Fiber from Whole Oats and Coronary Heart Disease. The soluble oat fiber component beta-glucan, an example of a nondigestible form of dietary fiber, has been shown to reduce total cholesterol and LDL cholesterol. As a result, the FDA approved a health claim for rolled oats, oat bran, and whole oat flour (good sources of beta-glucan, >0.75 gram per serving). Such a claim might read "soluble fiber from foods such as oat bran, as part of a diet low in saturated fat and cholesterol, may reduce the risk of coronary heart disease (CHD)."

Soy Protein and Coronary Heart Disease. In 1999, the FDA authorized use of health claims about the role of soy protein in reducing the risk of CHD on labeling of foods containing soy protein. This approval is based on the FDA's conclusion that foods containing soy protein included in a diet low in saturated fat and cholesterol may reduce the risk of CHD by lowering blood cholesterol levels. Studies demonstrate that 25 grams of soy protein daily in the diet is needed to show a significant cholesterol-lowering effect. To qualify for this health claim, a food must contain at least 6.25 grams of soy protein per serving, the amount that is one-fourth of the effective level of 25 grams per day.

An example of a health claim about the relationship between diet and the reduced risk of heart disease is "Diets low in saturated fat and cholesterol that include 25 grams of soy protein a day may reduce the risk of heart disease. One serving of (name of food) provides ____ grams of soy protein."

Plant Sterol and Stanol Esters and Coronary Heart Disease. The FDA has authorized use of labeling health claims about the role of plant sterol or plant stanol esters in reducing the risk of CHD for foods containing these substances. The FDA had concluded that plant sterol esters and plant stanol esters may reduce the risk of CHD by lowering blood cholesterol levels.

Plant sterols are present in small quantities in many fruits, vegetables, nuts, seeds, cereals, legumes, and other plant sources. Plant stanols occur naturally in even smaller quantities from some of the same sources. For example, both plant sterols and stanols are found in vegetable oils. Foods that may qualify for the health claim based on plant sterol ester content include spreads and salad dressings. Among the foods that may qualify for claims based on plant stanol ester content are spreads, salad dressings, snack bars, and dietary supplements in softgel form. Studies show that 1.3 grams per day of plant sterol esters or 3.4 grams per day of plant stanol esters in the diet are needed to show a significant cholesterol-lowering effect. To qualify for this health claim, a food must contain at least 0.65 grams of plant sterol esters per serving or at least 1.7 grams of plant stanol esters per serving.

7.6 THE DIETARY SUPPLEMENT HEALTH AND EDUCATION ACT OF 1994

The **Dietary Supplement Health and Education Act (DSHEA)** of 1994 changed the definition and regulations for dietary supplements. Previously, the FFDCA had governed dietary supplements. The FDA traditionally considered dietary supplements to be composed only of essential nutrients, such as vitamins, minerals, and proteins. The Nutrition Labeling and Education Act of 1990 added "herbs, or similar nutritional substances," to the term "dietary supplement." Through the DSHEA, Congress expanded the meaning of the term "dietary supplements" beyond essential nutrients to include such substances as ginseng, garlic, fish oils, psyllium, enzymes, glandulars, and mixtures of these.

The DSHEA established a formal definition of **dietary supplement** using several criteria. It stated:

A dietary supplement:

- is a product (other than tobacco) that is intended to supplement the diet that bears or contains one or more of the following dietary ingredients: a vitamin, a mineral, an herb or other botanical, an amino acid, a dietary substance for use by man to supplement the diet by increasing the total daily intake, or a concentrate, metabolite, constituent, extract, or combinations of these ingredients.
- is intended for ingestion in pill, capsule, tablet, or liquid form.
- is not represented for use as a conventional food or as the sole item of a meal or diet.
- is labeled as a "dietary supplement."
- includes products such as an approved new drug, certified antibiotic, or licensed biologic that was marketed as a dietary supplement or food before approval, certification, or license (unless the Secretary of Health and Human Services waives this provision).

Examples of dietary supplements:

Amino acid pills
Fish oil pills
Flaxseed fortified bread
Garlic
Ginseng
Herbs and herbal remedies
L-carnitine
Melatonin
Nonnutrients (e.g., bee pollen, shark cartilage)
Nutrient-fortified formulas
Vitamin and mineral supplements

The DSHEA provided that retail outlets may make available third-party printed literature to help inform consumers about any health-related benefits of dietary supplements. These provisions stipulated that the information must not be false or misleading; cannot promote a specific supplement brand; must be displayed with other similar materials to present a balanced view; must be displayed separate from supplements; and may not have other information attached (such as product promotional literature).

Additionally, the DSHEA provided for the use of various types of statements on the label of dietary supplements: health claims and a new category of claims—structure/function. Structure/function claims regarding the use of a dietary supplement cannot state it can be used to diagnose, prevent, mitigate, treat, or cure a specific disease. The only way to make such a claim is to seek approval under the new drug provisions of the FFDCA. For example, a product may not carry the claim "cures cancer" or "treats arthritis." Appropriate health claims authorized by the FDA, such as the claim that calcium may reduce the risk of osteoporosis, may be made in supplement labeling if the product qualifies to bear the claim.

Under DSHEA, supplement manufacturers can make statements about classical nutrient deficiency diseases, provided these statements disclose the prevalence of the disease in the United States. In addition, manufacturers may describe the supplement's effects on "structure or function" of the body or the "well-being" achieved by consuming the dietary ingredient. However, to use these claims, manufacturers must substantiate that the statements are truthful and not misleading. The product label must display the disclaimer: "This statement has not been evaluated by the Food and Drug Administration. This product is not intended to diagnose, treat, cure, or prevent any disease." Unlike health claims, nutritional support statements do not need to be approved by the FDA before manufacturers market products displaying such statements. However, the agency must be notified no later than 30 days after a product that bears the claim is first marketed.

Safety Testing of Dietary Supplements

While vitamin and mineral supplements are regulated much like foods, the other categories of dietary supplements, such as amino acids, nonnutrients, and herbs, are largely unregulated. They are not considered to be drugs, even though they are often purchased because they are believed to be effective treatments for certain disease conditions. Since they are not viewed as drugs by the FDA, they are not subjected to the same vigorous safety testing as drugs. Such products are considered safe until demonstrated to be hazardous according to the FDA.

Dietary Supplement or Drug?

In 1999 a federal district court ruled against the FDA and in favor of the manufacturer, Pharmmanex Inc., regarding a product named *Cholestin*. Cholestin is a cholesterol-lowering product containing Chinese red yeast milled rice, and as such, has nutritive value apart from any pharmacological effects. It had been marketed in 1996 as a dietary supplement but ran afoul of the FDA.

The FDA, in 1997, notified the manufacturer that it considered Cholestin an unapproved drug rather than a dietary supplement because it contained mevinolin, a natural chemical identical to lavostatin (lavostatin is an active ingredient in Mevacor, a prescription cholesterol-lowering drug). One year later, the FDA banned the importation of the red yeast rice, effectively making the sale of Cholestin illegal in the United States. The manufacturer responded by filing a federal lawsuit against the FDA, contending it had overstepped its regulatory bounds as outlined under DSHEA.

In ruling in favor of the manufacturer, the court provided an interpretation of the FFDCA as amended by DSHEA. It based its decision on the intended use of Cholestin as a dietary supplement (in capsule form), arguing that it was a product meant to supplement the diet to help consumers maintain a healthy cholesterol level. In other words, the fact that Cholestin contained an active ingredient that can be found in a pharmaceutical product was not at issue. The court felt that Cholestin was adequately described by the definition of a dietary supplement.

The lesson is that product ingredients alone do not necessarily classify the product as a dietary supplement, a food, or a drug. Even with its active drug component, Cholestin has been ruled a dietary supplement. In addition, intended use is critical to a product's classification according to the FFDCA. The marketing of a product, in terms of promotional material and so forth, can also affect its status. The example of Cholestin provides the food and pharmaceutical industries insight into the issue of product classification in light of DSHEA regulations.

Key Points

- Food additives have been used for many years to preserve, flavor, blend, thicken, and color foods, and have played an important role in reducing serious nutritional deficiencies among Americans.
- Food additives help maintain product consistency, improve nutritional value, maintain palatability, provide leavening, and enhance flavor or color.
- Federal regulations require evidence that each food additive is safe at its intended level of use before it can be added to foods.
- All additives are subject to ongoing safety review as scientific understanding and methods of testing continue to improve the regulation of additives.
- Food and color additives are more strictly regulated now than at any time in history.
- NLEA mandated a change in nutritional labeling, standardized key label features and use of terms, and authorized health claims for specific foods and ingredients.
- DSHEA expanded the definition of the term *dietary supplement* beyond essential nutrients to include such substances as ginseng, garlic, fish oils, and herbs, and safety testing is not required before marketing these substances.

Study Questions

1. The important difference between intentional and indirect additives is
   a. indirect additives are more easily controlled.
   b. intentional additives are undesirable.
   c. indirect additives are less costly.
   d. intentional additives are less costly.
   e. indirect additives are undesirable.

2. Which processing step is regulated as a food additive?
   a. canning
   b. irradiation
   c. extruding
   d. freezing
   e. pasteurization

3. Which is *not* a function of food additives?
   a. reduce cost of formulation
   b. maintain product smoothness
   c. maintain nutritional value
   d. acidity control
   e. flavor enhancement

4. All are guiding principles for additive use except
   a. performance.
   b. detection.
   c. efficacy.
   d. deception.
   e. safety.

5. Which statement concerning GRAS is true?
   a. GRAS means "generally regarded as safe."
   b. GRAS substances are exempt from the FAA.
   c. The FDA has no interest in GRAS testing.
   d. Even if new evidence shows a GRAS substance to be unsafe, it remains GRAS.
   e. GRAS substances are in violation of the Delaney Clause.

6. In a product like bread, why is calcium propionate used?
   a. antioxidant
   b. emulsifier
   c. antibiotic
   d. dough strengthener
   e. antimicrobial agent

7. Sorbitol
   a. is a polyol.
   b. provides sweetness.
   c. acts as a humectant.
   d. is a polyol and provides sweetness.
   e. is a polyol, provides sweetness, and acts as a humectant.

8. Which additive would be effective in canned foods to inhibit oxidation reactions?
   a. lecithin
   b. benzoyl peroxide
   c. EDTA
   d. tartrazine
   e. silicates

9. Which substance enhances the natural emulsifier activity present in process cheese?
   a. enzymes
   b. emulsifying salts
   c. humectants
   d. oxidizing agent
   e. reducing agent

10. All are examples of stabilizers or thickeners *except*
    a. pectin.
    b. starch.
    c. guar gum.
    d. gelatin.
    e. glucose.

11. Which *event* focused national attention on food misbranding and adulteration?
    a. Pure Food Congress
    b. 1906 Food and Drugs Act
    c. 1966 Fair Packaging and Labeling Act
    d. 1938 FFDCA
    e. White House Conference on Food, Nutrition and Health

12. As legally interpreted, "reasonable certainty of no harm" defines
    a. unlikely hazard.
    b. safe.
    c. pseudotoxin.
    d. nonnutrient.
    e. approved pesticides.

13. The FDA published its biotechnology policy in the
    a. Federal Register.
    b. NLEA.
    c. DSHEA.
    d. Red Book.
    e. Pesticide Residue Amendment.

14. Which risk-benefit relationship indicates the best scenario from the health and safety standpoint?
    a. favorable
    b. cautionary
    c. ideal
    d. neutral
    e. dangerous

15. Which is *not* a U.S. food-related agency?
    a. NMFS
    b. FCCC
    c. CFSAN
    d. AMS
    e. FAO

16. In deciding to approve a food additive, this agency considers its composition and properties, the amount likely to be consumed, and long-term effects.
    a. FAA
    b. FCC
    c. FDA
    d. EPA
    e. Congress

17. This test was developed to determine the mutagenic potential of a substance.
    a. CFR test
    b. Ames test
    c. Pesticide Residue Test
    d. GRAS test
    e. FFDCA test

18. Specific terms such as *free*, *low*, and *reduced* used on a food label are subject to _____.
    a. NLEA
    b. DSHEA
    c. standards of identity
    d. health claim approval
    e. CFR

19. The substance that is considered a drug rather than a dietary supplement due to the presence of mevinolin is
    a. Cholestin
    b. Sometribove
    c. Lavostatin
    d. Cyclamate
    e. Chemogastrin

20. Which legislation changed the definition and regulations for dietary supplements?
    a. DEHA
    b. NLEA
    c. FFDCA of 1958
    d. DSHEA
    e. FFDCA of 1938

21. Which fact about citric acid does *not* describe its multifunctionality?
    a. It contains six carbon atoms and has three carboxylic acid groups in its structure.
    b. It acts as an acidulant and sequestrant.
    c. It can act as a flavor enhancer and metal chelator.
    d. It can be used as a pH control agent and contribute sourness.
    e. It donates H^+ ions and forms complexes with metals.

22. The main difference between *enrichment* and *fortification* is
    a. enrichment denotes addition of nutrients but fortification does not.
    b. fortification provides nutrients lost during processing.
    c. fortification provides missing nutrients.
    d. enrichment provides nutrients lacking in the diet.
    e. fortification denotes addition of nutrients but enrichment does not.

23. The main difference between a mutagen and a carcinogen is
    a. mutagens cause birth defects.
    b. mutagens are never linked to cancer.
    c. mutagens cause cancer.
    d. carcinogens cause birth defects.
    e. mutagens cause mutations in cellular DNA.

24. The only health claim related to fruits and vegetables regards
    a. stanol and sterol esters and CHD.
    b. polyols and dental caries.
    c. soluble fiber and CHD.
    d. soy protein and CHD.
    e. calcium and osteoporosis.

25. Which statement best demonstrates the connection between Reference Daily Intake (RDI), Daily Reference Value (DRV), and Daily Value?
    a. The units for DRV, RDI, and Daily Value are standardized in milligrams.
    b. DRV and RDI values are given as "% Daily Value" on the Nutrition Facts label.
    c. RDIs and DRVs cover macro- and micronutrient Daily Value intake needs, respectively.
    d. The units for DRV, RDI, and Daily Value are standardized in milligrams, and DRV and RDI values are given as "% Daily Value" on the Nutrition Facts label.
    e. DRV and RDI values are given as "% Daily Value" on the Nutrition Facts label, and RDIs and DRVs cover macro- and micronutrient Daily Value intake needs, respectively.

References

Blair, S. N. and coworkers. 2000. Incremental reduction of serum total cholesterol and low-density lipoprotein cholesterol with the addition of plant stanol ester-containing spread to statin therapy. *The American Journal of Cardiology* 86:46.

Boyle, M. A. 2001. *Personal Nutrition*, 4th ed. Wadsworth/Thomson Learning, Belmont, CA.

Brown, A. 2000. *Understanding Food: Principles and Preparation*. Wadsworth/Thomson Learning, Belmont, CA.

Gilmore, R. 2000. GMOs: Progress at a cost. *Food Technology* 54(6):146.

Hall, R. L. 1978. Food additives and their regulation. In: *Agricultural and Food Chemistry: Past, Present, and Future*, pp. 222–233, R. Teranishi, ed. AVI Publishing Company, Westport, CT.

IFT. 2000. Genetically modified organisms (GMOs). *Food Technology* 54(1):42.

Mermelstein, N. 1994. Nutrition labeling regulatory update. *Food Technology* 48(7):62.

Murray, J. 2000. Functional foods: Rocky regulatory road ahead. *Food Product Design* 10(4):15.

Nesheim, M.C. 1998. Regulation of dietary supplements. *Nutrition Today* 33:62.

Pape, S. 2000. New frontiers for a new millennium. *Prepared Foods* 171(1):19.

Termini, R.B. 1999. More than ingredients, intended use. *Food Quality* (6)5:18.

Additional Resources

Chapman, N. 1999. What's the big deal over 'little' organic? *Prepared Foods* 170(11):26.

Charley, H., and Weaver, C. 1998. *Foods: A Scientific Approach*. Prentice Hall, Inc. Upper Saddle River, NJ.

Giese, J. 1995. Vitamin and mineral fortification of food. *Food Technology* 49(5):110.

Kracov, D.A. 1998. Making sense of organics. *Food Processing* 59(2):49.

McWilliams, M. 1997. *Foods: Experimental Perspectives*. Prentice Hall, Upper Saddle River, NJ.

Nielsen, S. S. 1999. *Food Analysis*. 2nd ed. Aspen Publishers, Gaithersburg, MD.

Potter, N. H., and Hotchkiss, J. H. 1995. *Food Science*, 5th ed. Chapman & Hall, New York.

Sizer, F., and Whitney, E. N. 2003. *Nutrition: Concepts and Controversies*, 9th ed. Wadsworth/Thomson Learning, Belmont, CA.

Vaclavik, V. A. 1998. *Essentials of Food Science*. Chapman & Hall, New York.

Whitney, E. N., and Rolfes, S. R. 2002. *Understanding Nutrition*, 9th ed. Wadsworth/Thomson Learning, Belmont, CA.

Web Sites

http://www.access.gpo.gov/nara/cfr/cfr-table-search.html
Code of Federal Regulations (National Archives and Records Administration)

http:/www.fda.gov
Food and Drug Administration

http://vm.cfsan.fda.gov/list.html
FDA's Center for Food Safety and Applied Nutrition

http://vm.cfsan.fda.gov/label.html
FDA's Food Labeling and Nutrition

http://www.ams.usda.gov/nop
National Organic Program

http://www.usda.gov
U.S. Department of Agriculture

http://www/usda.gov/fsis
USDA's Food Safety and Inspection Service

http://www.nal.usda.gov/fnic/foodcomp
Nutrient Data Laboratory (USDA Nutrient Database for Standard Reference, Release 14)

Challenge!

REGULATION OF FUNCTIONAL FOODS, BIOENGINEERED FOODS, AND ORGANIC FOODS

The start of the twenty-first century was an active time in the area of food and agricultural policy regulations. Three areas in particular received much attention: functional foods, GM foods, and organic foods.

Functional Foods

Development of functional food products continues as consumer demand for these products increases. Regulations governing this growing market have drawn increasing attention from both critics and proponents. The problem in regulation is that the FFDCA does not provide a definition for functional foods. The terms *functional foods*, *nutraceuticals*, *pharmafoods*, and *designer foods* are all marketing terms. You might think the FDA is confused as to how it might regulate such products, if at all.

The Center for Science in the Public Interest (CSPI), a consumer advocacy group that focuses on food and nutrition issues, has urged the FDA to issue regulations governing the use of label claims on functional foods. On the other hand, the National Food Processors Association (NFPA) believes the FDA already has complete enforcement authority to ensure that label claims are scientifically supported and do not mislead consumers. The organization thinks that functional foods need to be regulated, simply, as foods. In this interpretation, the FDA already has regulatory authority to oversee the safety of functional foods and to ensure that claims on these products do not mislead consumers. However, as it stands, functional foods could be regulated as drugs, dietary supplements, food ingredients, or medical foods. What probably matters most in the regulatory arena is the intended use of the food.

What is the current FDA approach to functional foods? The FDA asks a number of questions, including:

- What is the public perception of the product?
- Is it sold in a conventional market or in a limited, controlled market?
- Is the manufacturer trying to skirt existing regulations?
- Is the product really a drug masquerading as a food?
- Does it pose a hazard?

In the absence of clear regulations, the FDA will need to look at each functional food on a case-by-case basis.

Case A: Benecol. Consider an action by the FDA that temporarily stopped the sale of a functional food. The agency said that McNeil Pharmaceuticals had to prove that its Benecol cholesterol-lowering margarine was safe before it was marketed. In response, McNeil said Benecol was not a new food, but rather a dietary supplement and therefore did not require FDA approval. The FDA then cited a law saying dietary supplements may not masquerade as foods.

Benecol, made from an ingredient in trees, looks and tastes like regular margarine and would be sold in stores next to butter. Benecol contains plant stanol ester, a nutrient the makers claim will help consumers better manage their cholesterol. Specifically, Benecol's active ingredient blocks the absorption of LDL, or "bad," cholesterol.

The FDA ultimately reviewed product testing conducted by McNeil and stated it had no further concerns about the product's safety. The FDA said McNeil would not be violating federal labeling rules by claiming the product contained an ingredient that "helps promote healthy cholesterol levels." However, the FDA also suggested that Mc-

Stanol esters are a new food ingredient derived from naturally occurring substances in plants. Plant stanols (sterol-like material) are combined with canola oil to form stanol esters, and are whipped into a spread. Stanol esters block absorption of cholesterol from the digestive tract, lowering LDL cholesterol levels.

Neil continue to monitor, through sound scientific studies, the safety of consumer dietary exposure to plant stanol esters.

Case B: Vitamin O. The Federal Trade Commission warned that health claims and ads for the dietary supplement "Vitamin O" were "blatantly false" and that the substance contained little more than saltwater. The commission filed a preliminary injunction against the manufacturer to stop the spread of the company's health claims. Ads allegedly claimed the supplement could cure or prevent ailments such as cancer, heart disease, and lung disease.

The manufacturer said it had sold nearly a million bottles of "Vitamin O" at the cost of about $25 each, including shipping. It claimed that customers benefit from the "stabilized monoatomic oxygen" in the product, but not that the product cured anything. However, scientists say the placebo effect (belief in the product alone rather than any true biological activity) is at work with "Vitamin O." Two companies paid a large fine to settle Federal Trade Commission charges for "Vitamin O." As part of the May 2000 settlement, the defendants were prohibited from representing that "Vitamin O" or any food, drug, or dietary supplement they market is effective against any life-threatening disease, or has any other health benefits, unless they possess competent and reliable scientific evidence to support the representation. The FTC also warned buyers to beware when purchasing dietary supplements.

Clearly, the issue of functional foods regulations is complicated. As more new foods are developed, the FDA and other government agencies will investigate ways to promote both consumer safety and commerce regarding functional foods.

Bioengineered Foods

We will discuss biotechnology in detail in Chapter 14; here we will look at regulation of foods produced with this technology. "Genetically modified organism" describes the use of genetic engineering to alter the genetic makeup of animals, microorganisms, and plants. In genetic engineering, in vitro enzyme manipulations of DNA, rather than conventional breeding techniques, are used to improve or add desired traits. Crops, foods, or food products developed using genetically modified organisms (GMOs) are sometimes termed genetically modified foods (GM foods). We will use the term *bioengineered foods.*

All foods developed through genetic engineering are subject to the original Federal Food, Drug, and Cosmetic Act in terms of their safety. More recently, in May 1992, the FDA published its "Statement of Policy: Foods Derived from New Plant Varieties." The policy states that the FDA has no basis for concluding that bioengineered (genetically engineered) foods differ from other foods in any meaningful way, or that they present a greater safety concern. FDA labeling regulations are stringent if the use of biotechnology alters the food materially or introduces a known allergen into the product. But since the FDA said that bioengineered foods would not be subject to regulation unless the biotechnology process added a new substance, many bioengineered foods are not labeled as such. This policy and the nonlabeling of all bioengineered foods clashed head-on with consumer and environmental groups. Questions raised focused on the safety of biotechnology, the safety of foods derived from GMOs, the environmental impact of GMOs, the potential to create "super pests," the danger to the Monarch butterfly, labeling issues of bioengineered foods, and the lack of good faith regulatory action in light of the Starlink corn fiasco in fall 2000 (see Chapter 14). At the time of this writing, consumer confidence in and acceptance of bioengineered foods was low.

Pre-market Review. At present, the FDA does not require food companies to consult when developing new foods or food products through genetic engineering, but it does encourage this consultation and has received cooperation from the food industry. While not currently mandatory, developers of genetically engineered foods do consult with the FDA prior to marketing their products. This is termed *pre-market review.* The process includes a science-based product safety assessment. In May 2000, the FDA announced plans to strengthen this process to ensure all genetically engineered foods are as safe as their nonengineered counterparts.

Subsequently, a court has ruled in favor of FDA policies. The U.S. District Court for the District of Columbia on September 29, 2000 agreed with the FDA view that genetically engineered foods as a class do not require pre-market review and approval of a food additive petition, and that special labeling for genetically engineered foods is not required solely because of consumer demand or because of the process used to develop these foods. Mandatory labeling of foods developed through genetic engineering is currently not required.

The FDA is moving in the direction of tighter regulation of foods developed through genetic engineering. In January 2001 the FDA issued a proposed rule and a draft guidance document concerning such foods. If finalized as a regulation, it would require food developers to notify the FDA at least 120 days in advance of their intent to market any food or animal feed developed through biotechnology and to provide information to demonstrate that the product is as safe as its conventional counterpart. In a separate action, the FDA issued a draft guidance document that provides direction to manufacturers wanting to label their food products as being made with or without ingredients developed through biotechnology.

Labeling Bioengineered Foods. The 2001 draft guidance for the labeling of foods developed through genetic engineering will help manufacturers ensure that their labeling is truthful and not misleading. The FDA considers the terms "derived through biotechnology" and "bioengineered" as acceptable on food labels. The following terms: "GM free," "GMO," and "modified" are not acceptable. The FDA's plans on mandatory consultation and labeling guidance will benefit from consumer input at public sessions, with the finalized rulings to be developed over the course of a few years.

The FDA realizes that certain manufacturers may want to develop informative label statements in terms of bioengineered foods or foods that contain ingredients produced from bioengineered foods. The following are FDA-worded examples of some statements that might be used, and are intended to provide guidance as to how similar statements can be made without being misleading.

"Genetically engineered"
or "This product contains cornmeal made from corn that was produced using biotechnology."

The information that the food was bioengineered is optional, and this kind of simple statement is not likely to be misleading. Here's another example:

"This product contains high oleic acid soybean oil from soybeans developed using biotechnology to decrease the amount of saturated fat."

This example includes both required and optional information. When a food differs from its traditional counterpart such that the common or usual name no longer adequately describes the new food, the name must be changed to describe the difference. Because this soybean oil contains more oleic acid than traditional soybean oil, the term "soybean oil" no longer adequately describes the nature of the food. A phrase like "high oleic acid" is required to appear as part of the name of the food to describe its basic nature. The statement that the soybeans were developed using biotechnology is optional, as is the statement that the reason for the change in the soybeans was to reduce saturated fat.

In the next example, the change in texture is a difference that *may* have to be described on the label. If the texture improvement makes a significant difference in the finished product, it would require disclosure of the difference for the consumer. However, the statement must not be misleading. The phrase "to improve texture" could be misleading if the texture difference is not noticeable to the consumer:

"These tomatoes were genetically engineered to improve texture."

FIGURE 7.11 The USDA organic seal can be used on the labels of foods that are certified to be organically grown.

Organic Foods

The USDA has established the first set of quality and production standards for organic foods, developed under the Organic Food Production Act of 1990. Under these regulations, *organic* refers to food production practices that avoid most synthetic pesticides and fertilizers, genetically modified crops, antibiotics in livestock production, irradiation, and using sewage sludge as fertilizer. Foods using the "USDA Organic" labeling will be certified as grown and processed under specified conditions.

The **Agricultural Marketing Service** (AMS) of the USDA published final regulations on procedures for organic food production as the National Organic Program final rule in the Federal Register (December 21, 2000). This requires all but the smallest organic operations to be certified by a USDA accredited agent and provides requirements for organic food production. One requirement is that products or ingredients identified as organic may not be produced using biotechnology methods.

Thus, language has been developed for standards and labeling that now attaches specific meaning to organic foods. These are a step forward in terms of informing the public about the meaning of the term *organic* regarding a food's origin and production.

The final standard also outlines labeling requirements for organic products, certification and record-keeping requirements, and accreditation requirements for producers of organic foods. Three interesting provisos in the ruling were:

- to increase the minimum percentage of organic ingredients in products labeled "Made with Organic Ingredients" from 50 percent to 70 percent
- to utilize the EPA's 5 percent pesticide residue tolerance as a compliance threshold
- to allow wine containing sulfites to be labeled "Made with Organic Grapes"

In addition, the new standards included a "commercial availability provision" which required that "organic" products—which are defined as having at least 95 percent organic ingredients—have their remaining ingredients also sourced from organically certified sources.

Organic Certification and Label. The USDA cautioned that organic labels were designed to be "marketing tools," not government statements or certifications about food safety or nutritional quality. The National Organic Program (NOP) went into effect on October 21, 2002. To display the USDA Organic seal (Figure 7.11), an organically grown product must have been certified by an accredited USDA organic certifying agent and been determined to meet the NOP requirements.

The national organic standards provide for adequate segregation of the food throughout distribution to assure that nonorganic foods do not become mixed with organic foods. The FDA position is that practices and record keeping that substantiate the "certified organic" statement would be sufficient to substantiate a claim that a food was not produced using bioengineering.

Challenge Questions

1. Vitamin O
   a. was marketed as a dietary supplement.
   b. stabilizes oxygen in the air we breathe.
   c. is mainly monomolecular oxygen in liquid form.
   d. was marketed as a dietary supplement and stabilizes oxygen in the air we breathe.
   e. does not provide a placebo effect.

2. GMOs
   a. are bacteria that have been genetically modified.
   b. can be labeled as "genetically engineered."
   c. represent an application of biotechnology.

d. are bacteria that have been genetically modified and can be labeled as "genetically engineered."
   e. can be labeled as "genetically engineered" and represent an application of biotechnology.

3. Consider this label statement on a bioengineered food product: "This product contains high oleic acid soybean oil from soybeans developed using biotechnology to decrease the amount of saturated fat." Which is true?
   a. "Contains high oleic acid soybean oil" is an optional label statement.
   b. "From soybeans developed using biotechnology" is an optional label statement.
   c. "Contains soybean oil from soybeans" is a required label statement.
   d. "To decrease the amount of saturated fat" is a required label statement.
   e. "Contains high oleic acid soybean oil" is a required label statement.

4. Organic foods
   a. may not contain organic fertilizers.
   b. have certification and record-keeping requirements.
   c. can be genetically modified.
   d. must display the radura symbol if irradiated.
   e. are regulated by the National Organic Program (NOP).

5. The FDA
   a. required mandatory labeling of organic foods starting September 2000.
   b. required food companies to participate in consultation with FDA as part of a mandatory pre-market approval process beginning in May 1992.
   c. provides guidance to food companies regarding bioengineered foods.
   d. views bioengineered foods as less safe than conventional counterparts.
   e. declared in May 1992 that all genetically engineered foods required pre-market approval.

CHAPTER 8

Understanding Food Processing and Preservation: Animal Products

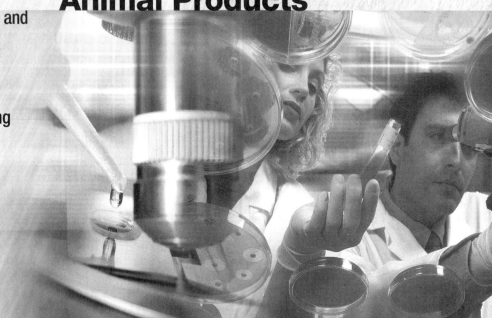

8.1 Food Processing—From Field and Farm to Consumers

8.2 What Is Heat Transfer?

8.3 Food Preservation—Preventing Food Spoilage

8.4 Dairy Products Processing

8.5 Egg Processing

8.6 Meat Processing

Challenge!
Food Irradiation

CHAPTER OBJECTIVES

After completing this chapter, you will be able to:
- Explain reasons why foods are processed, including maintaining their freshness, nutritional value, and to extend shelf life.
- List the unit operations, and discuss how they serve as underlying principles to guide the processing of the wide variety of foods.
- Describe the importance of heat transfer and how it occurs through conductive, convective, and radiant energy transfer mechanisms.
- Provide examples of chemical antimicrobial agents, such as acidulants, short-chain fatty acids, and sulfur dioxide.
- Distinguish pasteurization and blanching as examples of mild heat processes from sterilization, a more severe heat treatment.
- Define D value (decimal reduction time), the 12D concept, and TDT (thermal death time), and explain how each provides different information relative to thermal processing and food safety.
- Provide a general flowchart to indicate the steps required to process milk, yogurt, ice cream, and cheese.
- Describe how egg processing involves separation, mixing, pasteurization, and drying operations.
- Explain the steps in meat, poultry, and fish processing that are directly related to preservation (e.g., drying, thermal treatment) and those that are not (comminution).
- Explain irradiation processing and its potential to improve food safety by destroying pathogenic microorganisms.

Chapter 2, Food Categories and Composition, introduced the important consumer foods, with an emphasis on food examples and the nutrients they contain. In this chapter and the next, the features of food processing will be related to these food commodities. An understanding of the principles of food processing and the special processing methods that are used to preserve foods is essential for anyone preparing for a career in the food industry. Food processing has become the largest industrial activity of many nations. The automated processing and manufacture of packaged foods has provided consumers with an unprecedented variety and abundance of food choices. Before beginning our examination of food processing and preservation, let us distinguish between the two.

Food processing refers to the conversion of raw animal and plant tissue into forms that are convenient and practical to consume. Food processing is accomplished through mechanical action, heating, extrusion, and other manipulations. **Food preservation,** on the other hand, is the use of specific thermal and nonthermal processing techniques to minimize the number of spoilage microorganims in foods, making them safe and giving them an extended shelf life. These techniques include canning, refrigeration, freezing, dehydration, high pressure, irradiation, and the use of certain food additives that inhibit the growth of microorganisms.

8.1 FOOD PROCESSING—FROM FIELD AND FARM TO CONSUMERS

Food processing includes all the operations by which raw foodstuffs are made suitable for consumption or storage. The concept of processing encompasses the basic preparation of foods, the alteration of a food product into another form, and preservation and packaging techniques. Food processing innovations have resulted in new consumer food products, such as concentrated fruit juices, freeze-dried coffee, and instant foods. For our purposes, we can say that foods are processed for one of two reasons: (1) to preserve them so that they remain fresh, wholesome, nutritious, safe, and free from the effects of spoilage for a certain length of time, and (2) to manufacture specific desirable food products that exhibit a certain shelf life.

Factors used in food processing to inhibit or destroy microorganisms

Heating
Freezing
pH control
a_w
Antimicrobial chemicals

All raw foods are perishable commodities. From the time of harvest of a plant food or the slaughter of an animal to be used for food, raw plant and animal tissue undergoes deterioration. One of the prime causes of spoilage is the amount of **biologically active water** in the tissue. Tissues with high biologically active water content (such as leafy vegetables or red meat) will deteriorate in only a matter of days, whereas dry seeds, which contain mostly structural rather than biologically active water, can be stored for years.

The major causes of food spoilage are microbial growth, enzymatic reactions, and chemical changes, especially oxidation. These processes occur most rapidly at high water content as well as at optimal conditions of temperature, pH, and other environmental conditions. The principles of food preservation are based upon manipulation of these environmental conditions. For example, microorganisms require an optimum temperature for growth; if they are forced to exist in a colder environment, their metabolic processes (including growth, reproduction, and production of toxins) will slow down. In addition, most microorganisms can be killed when forced to endure high processing temperatures. In practical applications, often a combination of factors including pH, temperature, and chemical perservatives are employed to inhibit or destroy microorganisms in foods.

The Unit Operations of Food Processing

The number of operating steps required in the processing and manufacture of the multitude of food products available is huge. Exact procedures vary from company to company and from country to country due to, among other things, technology availability. Food scientists would surely be frustrated in attempting to apply sound and practical approaches to food processing if it were not for the underlying principles of unit operations or the system of individual activities—the unit operations, offering a systematic approach to the processing of food.

Unit operations are the broad categories of common food processing operations in practice in the food industry. There are about a dozen categories of unit operations. Most cover a wide variety of products and applications. For instance, the heat exchange process (or simply, heating) is used in each of these product operations: pasteurizing milk, baking biscuits, roasting peanuts, and commercially sterilizing canned tuna.

Table 8.1 lists the unit operations in the food industry. A primary unit operation can be defined by many different subactivities or actions. For instance, the unit operation of mixing encompasses all of the following actions:

agitating	emulsifying
beating	homogenizing
blending	whipping

Let's look at each of the primary unit operations in Table 8.1. The first, **materials handling,** refers to the manner in which raw commodities such as crops and animals or animal by-products are harvested and transported to a food processing facility. For example, eggs are a by-product of hens, and material handling of eggs includes gathering the eggs and transporting them to an egg processing facility.

The unit operation of **separating** isolates a desirable part of a food raw material from another part. A solid food part can be removed from another solid part, or a liquid is taken from a solid food. Examples include removing the skin of potatoes in preparation for frozen french fry production (solid from solid separation) and pressing the juice from oranges to make frozen concentrated orange juice (liquid from solid separation).

Cleaning and sanitizing deal with food items themselves as well as surfaces and equipment that come in contact with food. **Cleaning** is straightforward—it implies the removal of surface dirt, debris, and associated bacteria by washing with water and detergent, when practical. Debris often means chemical residue or organic material on food that can serve as nutrients for microorganisms. Methods used to remove unwanted microorganisms from foods besides cleaning include centrifugation, filtration, and trimming. Milk, fruit juices, and syrups can be centrifuged to remove spores, bacilli, yeasts, and molds due to their mass. Liquid foods and beverages respond well to filtration, while

TABLE 8.1 Unit operations in the food industry.

• Materials handling	• Pumping	• Drying
• Separating	• Mixing	• Forming
• Cleaning	• Heat exchange	• Packaging
• Disintegrating	• Evaporation	• Nonthermal methods

TABLE 8.2 Types of cleaning compounds used in the food processing industry.

Type	Advantages	Disadvantages
Strong alkaline (e.g., NaOH)	Strong dissolving power, corrosive and caustic, inexpensive	Little effect on mineral deposits, damage to skin and tissue
Heavy-duty alkaline	Good dissolving power, CIP (cleaning-in-place) application, fat removal	Low to noncorrosive, removes skin oils
Strong acid (e.g., HCl)	Strong dissolving power, corrosive and caustic, mineral removal, encrusted debris removal	Possible toxic gases upon heating
Wetting agents	Promote suds formation, emulsifying action, noncorrosive	
Solvent cleaners	Dirt-dissolving power, degreasing action	

fruits, vegetables, and animal carcasses are washed. At times foods such as vegetables, fruits, and hard cheeses are trimmed to remove areas of contamination.

Cleaners are amphiphilic compounds having hydrophilic and hydrophobic structure. They interact with water and dirt/debris by suspending these particles in solution, with the cleaner's hydrophilic portion soluble in water and the hydrophobic portion soluble in the debris. Food items are cleaned only, whereas work surfaces and equipment are cleaned and also sanitized. The most effective approach when cleaning and sanitizing an area involves a pre-rinse, application of the detergent cleaner solution, post-rinse, acid rinse, and sanitizing application and rinse. Table 8.2 lists the advantages and disadvantages of some typical cleaning compounds used in the food industry.

Sanitizers are chemical compounds that are bacteriostatic and bactericidal agents. Bacteriostatic action refers to inhibiting the growth of microorganisms, while a bactericide destroys microorganisms. Sanitizers exhibit varying degrees of effectiveness against gram-negative bacteria (*Pseudomonas*, *E. coli*, *Campylobacter*, *Salmonella*, *Yersinia*) and gram-positives (*Staphylococcus*, *Streptococcus*, *Clostridium*, *Listeria*), bacterial spores, fungi and yeasts, and viruses. In general, spores are more resistant than vegetative cells, and gram-negatives are more susceptible to control methods than gram-positives. To be effective, the surfaces must be free of dirt and debris when sanitizers are applied.

Sanitizers come in a variety of forms, including those that are chlorine based (HOCl, hypochlorous acid), **iodophores** (iodine-based sanitizers), QAC or QUATS (quaternary ammonium compounds, which are actually detergent-sanitizers with both cleaning and sanitizing properties), and acid anionic surfactants. Each type of chemical sanitizer has advantages and disadvantages, so the food sanitation worker must know the proper application. A comprehensive table of sanitizers is presented in Chapter 11, Food Safety.

Disintegrating is the unit operation that refers to particle size reduction of foods. An example is cutting meat into small pieces, which is referred to as **comminution.** The grating of cheese and the dicing of fruits and vegetables are further examples of disintegrating.

Pumping is a mechanical method of moving food material from point A to point B during processing. Materials that are pumpable include semisolid foods such as yogurt and chunky soups, and liquid foods such as syrups, beverages, and liquid or dispersed ingredients.

Mixing refers to the blending of food ingredients to create a food product. We have already seen that mixing encompasses such actions as agitating, beating, blending, emul-

Gram-negative bacteria have a thin cell wall, an outer membrane, and stain pink in a gram stain. Gram-positive bacteria have a thick cell wall, no outer membrane, and stain blue-violet. See Chapter 10.

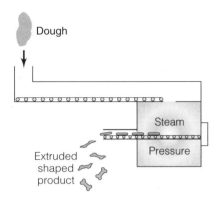

FIGURE 8.1 Forming. *Forming is an important unit operation in extrusion cooking of dough to produce an extruded product with a specific shape, such as a breakfast cereal or pet food.*

sifying, homogenizing, and whipping. Emulsifying a mixture of oil, water, and flavor ingredients creates stable salad dressings. Homogenizing is a process applied to milk, while whipping air into cream creates soft spread butter. Materials that are mixed together may be solid with solid, liquid with solid, liquid with liquid, and gas with liquid. Liquid soup broths are mixed with solid ingredients such as vegetables, pasta, flavorings, and meats to create canned soup products. The carbonation of water is an example of mixing a gas with a liquid.

Heat exchange is the application or removal of heat from a food. The roasting of coffee beans and the freeze-drying of coffee are both examples of heat exchange. Heat exchange is a processing approach and qualifies as a unit operation. Heat exchange is *macro*, whereas *heat transfer* is at the micro level; heat transfer occurs between molecules and between phases in food systems.

Evaporation is the removal of moisture from a food to concentrate its solids content. This unit operation is usually accomplished through controlled heating and is used, for example, in the manufacture of evaporated milk, in the processing of sugarcane into white crystalline sugar, and in the concentration of fruit juices for freezing.

Drying is a more extensive approach to moisture removal, in which product moisture is reduced to a mere few percent. Liquid foods such as milk and egg whites, as well as whole meat and fruit and vegetable pieces, can be dried. Spray drying is a method of choice for many liquids, whereas tunnel drying or vacuum freeze-drying is more useful for raw food pieces.

Foods are often formed into specific shapes during processing—this unit operation is known as **forming**. Breakfast cereals, pet foods, confectionery, and pasta products are examples of foods that are shaped upon extrusion—see Figure 8.1. Sugar confectionery, such as boiled sweets and chocolate candies, are often poured into and hardened within molds, to create recognizable shapes.

Packaging of foods protects them from the environment and offers convenience for retailers and consumers. Of primary concern are prevention of contamination by microbes, and protection from oxygen, moisture, light, unwanted flavor and odor transfer, and pest access. Containers traditionally used include glass bottles; tin, aluminum, or other metallic cans; and paper (cardboard). More recent packaging materials include waxed cardboard, plastic, and flexible plastic or metallic pouches. Outer packaging is typically some type of plastic. We will examine packaging in detail in Chapter 13, Food Engineering.

Newer strategies for food preservation, other than typical chemical, drying, and heat exchange methods, include nonthermal processing approaches. These strategies will be discussed more fully in Section 8.3.

The Basic Principles of Food Processing

There are six basic principles of food processing to achieve preservation: moisture removal, heat treatment, low-temperature treatment, acidity control, traditional nonthermal processing, and innovative nonthermal processing. The goal of each is to reduce or remove conditions that allow spoilage microorganisms to grow.

Moisture Removal. Since microorganisms need water to survive, removal of biologically active water through drying or dehydration stops their growth. It also reduces the rates of enzyme activity and chemical reactions. **Moisture removal** is achieved through drying, dehydration, evaporative concentration, and intermediate moisture processing. Intermediate moisture processing is used to create *intermediate moisture foods* (IMF) characterized by moisture contents of 15–50 percent, and a_w values of 0.6–0.85.

The preservation of foods by drying is probably the oldest food preservation process. The most prevalent drying methods used today are **sun drying** to produce dried fruits and nuts over a period of days or weeks; **drum drying** to dry potatoes into flakes, or to create fruit and vegetable juices and cereals; and **spray drying** to dry milk, eggs, syrups, and instant coffee. **Freeze-drying** or *lyophilization*, which first freezes the product and

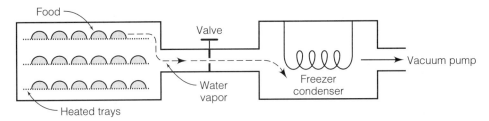

FIGURE 8.2 Freeze-drying. *Foods are processed by low-pressure chilling under a vacuum, allowing moisture to form ice crystals slowly, which are then evaporated into the gas phase without going through the liquid phase.*

then evaporates off moisture by application of a vacuum, is a popular approach applicable to many foods—see Figure 8.2.

Although dried foods will have reduced moisture content, the growth of microorganisms in the food actually depends on **water activity,** not moisture. For any given moisture content, one food may support the growth of bacteria, yeasts, or molds, while another food of equal moisture content will not, because the water activity is lower. The lower the water activity of a food, the less likely microorganisms can survive and grow in that food. Molds can grow at lower water activities (0.60) than yeasts, and yeasts grow at lower water activities (0.70) than spoilage bacteria (0.90). Water activity can be decreased in a food by adding soluble components, such as sugars and salts. One pathogenic bacterium that can grow at low water activity is *Staphylococcus aureus*, which grows at a level of 0.84 and higher.

Heat Treatment. The second basic principle of food processing to achieve preservation is **heat treatment.** Forms of heat treatment used in food processing include pasteurization, blanching, baking, canning to achieve commercial sterility, extrusion cooking, and microwave cooking.

The origin of modern heat processing techniques was developed more than 200 years ago. French scientist Nicolas Appert discovered that foods could be preserved if they were placed in sealed jars containing boiling water; a cycle of boiling created a vacuum inside the jars (see Section 1.2). Later, using sealed metal cans, it was discovered that heating food in the can by steam under pressure created a higher temperature than with just steam alone and reduced the treatment time. This idea is the principle behind today's canning and retort processing methods. The important heat treatment (or *thermal processing*) methods used are sterilization, pasteurization, blanching, and retort canning.

Sterilization refers to the complete destruction of microorganisms. This usually requires 121°C (250°F) of wet heat for 15 minutes or its equivalent. This means *every particle of the food being treated* must reach and maintain this temperature for 15 minutes. In a can, there may be a relatively slow transfer of heat depending on several factors, including the amount of food, its composition, and proportion of solid to liquid. The time required to achieve true sterility could be several hours. However, this length of exposure to high heat would cause undesirable quality changes in the food components. Fortunately, many foods need not be completely sterile to be safe and have a reasonable shelf life. These foods are treated to achieve commercial sterility rather than to achieve total sterility.

The term **commercially sterile** means the degree of sterilization at which all pathogenic and toxin-producing organisms are destroyed, as well as any spoilage organisms. Commercially sterile foods may contain a small number of heat-resistant bacterial spores, but these will not multiply under usual conditions within the food. However, if they are isolated and given favorable conditions outside the food, they may become viable. Most canned and bottled food products are commercially sterile and have a shelf life of 2 years or more.

Pasteurization is not the same type of heat treatment as sterilization—it involves a low-order heat treatment (approximately 80–90°C; 176–194°F) below the boiling point of water. Pasteurization has two primary objectives. In the case of milk and liquid eggs, pasteurization is used to specifically destroy organisms known to occur in those foods that could affect public health. The second more general objective is to extend food product shelf life from a microbial and enzymatic point of view. Pasteurized products still contain many living organisms capable of growth, thus limiting their storage life compared with commercially sterile products. Many pasteurized foods must be stored under refrigeration for this reason (to inhibit the growth of surviving organisms).

Flash pasteurization is a high-temperature, short-time (HTST) treatment in which pourable products, such as juices, are heated for 3–15 seconds to a temperature that destroys pathogenic microorganisms. The product is subsequently cooled and packaged. Flash pasteurization is a very rapid form of aseptic processing. As such, it reduces the thermal stress on the product and maintains product freshness, nutrient content, and flavor. Most drink boxes and pouches use this pasteurization method as it allows extended, unrefrigerated storage while providing a safe product.

Blanching is a type of heat treatment applied to fruits and vegetables that is specially intended to inactivate natural food enzymes. This practice is common when such products are to be stored frozen, because the freezing process alone will not completely stop enzyme activity. The blanching of sliced potatoes for french fries prior to freezing them for sale to fast food operations or retailers is a good example of a blanched vegetable. Green beans to be canned are blanched prior to canning.

Heating is applied to the **canning** of foods as well. The specific time and temperature of heating is defined by the TDT, *thermal death time*, which identifies the parameters required to destroy the spores (described as an inactive, dormant, or seed-like "resting" phase) of microorganisms such as *Clostridium botulinum*. This bacterium is pathogenic, producing disease in humans by making a deadly toxin. In this case, the toxin is released into the food prior to consumption, causing food intoxication (see Chapter 11, Food Safety). The concept of TDT is discussed more fully in Section 8.3, thermal processing.

Low-Temperature Treatment. The third basic principle of food processing to achieve preservation is **low-temperature treatment,** which refers to cold storage: refrigeration and freezing. In freezing, the goal is to cool food as quickly as possible from the danger zone for microbial growth (5–60°C; 40–140°F). Fish, meats, poultry, and dairy products should be held at temperatures above, but as near to (0°C) (32°F) as possible, to maintain good eating qualities and to prevent spoilage due to excessive enzyme activity and/or microbial action. Foods can be preserved by freezing them because freezing solidifies the water, which lowers the water activity, making the water unavailable for bacteria and unavailable to serve as a medium or solvent for chemical deterioration reactions.

The water in foods does not start to freeze exactly at 0°C (32°F) because it contains dissolved substances, which lower the freezing point (freezing point depression). During freezing, water molecules in foods begin to form ordered ice crystal structures, which pack together and restrict molecular freedom of movement. All detectable water in food is converted to ice when the temperature of the food reaches −60°C (−76°F).

Direct expansion refrigeration is a common method that pumps a gaseous refrigerant through a coil. The refrigerant expands as it moves through the coil, and air moving over the coils cools and then cools the food product. Three general types of refrigerant gases are in use: liquid nitrogen, carbon dioxide, and ammonia. The first two are considered disposable refrigerants, while ammonia is a re-useable refrigerant.

A word about **IQF foods.** A rapid freezing technique called Individual Quick Frozen (IQF) process uses carbon dioxide (CO_2) cryogenic equipment to quickly cool cooked foods from 71°C (160°F) to −23°C (−10°F) in 10 minutes. This process enables products to pass through the dangerous temperature zone very quickly, leaving very low and safe microbial counts. IQF products are safer when precooked and stored frozen. Some food products that have been IQF processed include ground beef, chunky chili, chopped beef, chicken a la king, and sausage. The vegetable industry utilizes fluidized bed freezers to create IQF products as well.

TABLE 8.3 Examples of acidulants used in the food processing industry. *Acetic acid is present in vinegar, while the acid salt sodium acetate can be used as an additive to decrease a food's pH. Citric acid is present in citrus fruits; malic acid is a component of apples; and all of these acids also can enhance food flavor. Sodium benzoate and calcium propionate are acid additives that are effective antimicrobial agents. One type of inorganic acid, namely phosphoric, is used extensively in foods and beverages.*

Acid	Comment
Acetic	Provides flavor, decreases pH
	Sodium acetate is salt form present in vinegar.
Benzoic	As sodium benzoate, effective antimicrobial agent
	Occurs naturally in cranberries
Citric	Provides flavor, decreases pH, acts as chelating and sequestering agent
	Occurs naturally in citrus fruits
Lactic	Provides tartness
Malic	Provides flavor
	Occurs naturally in apples
Phosphoric	Provides flavor and tartness in beverages
	Enhances juiciness in meats (as phosphate)
Propionic	As calcium propionate, effective antimicrobial agent
	Produced in some cheeses
Tartaric	Present in baking powder as potassium tartrate salt
	Occurs naturally in grapes

Acidity Control. The fourth basic principle of food processing to achieve preservation is acidity control, which refers to controlling the pH of a food through, for instance, the use of acidulants. The acidification of many foods sufficient to kill microorganisms would make them unpalatable, so a combination of acidification, heat treatment, and refrigeration storage is usually employed. Recall from Chapter 4 that acidulants are present naturally in certain foods or can be added during formulation or processing—see Table 8.3.

High-acid foods are those that are at pH < 4.6, such as fruits, certain vegetable-fruits like tomatoes, and products derived from them. High-acid foods are naturally low in pH due to the presence of citric, malic, or tartaric acid. Through fermentation, lactic acid or other organic acids may be produced in a food, which creates a preservative effect. On the other hand, acidified foods have acids specifically added to them. When canned, acidic foods do not need to be heated to high temperatures to achieve commercial sterility (the target temperature is below 100°C). The reason is that the acid is an environmental stress that renders the microorganisms and their spores more easily destroyed by heat, since it denatures protein molecules and weakens cell membranes. And spore-forming bacteria generally do not grow in foods having pH values of 4.5 and less.

Traditional Nonthermal Processing. The fifth basic principle of food processing to achieve preservation is through traditional nonthermal means such as appropriate packaging and chemical additives with antimicrobial, antioxidative, or other key functional properties.

Antimicrobial chemical preservatives. These chemical additives are termed **preservatives.** Strictly speaking, a preservative substance would function to maintain or preserve a food product's freshness. Antimicrobial preservatives include the food acidulant additives just presented, including the short-chain fatty acids: acetic, sorbic, and propionic. These as well as the nonacids sucrose and sodium chloride are effective.

Other substances present in spices and essential oils, such as eugenol (cloves), allicin (garlic), cinnamic aldehyde (cinnamon), allyl isothiocyanate (mustard), and thymol (sage and oregano) have also demonstrated antimicrobial activities. Sulfur dioxide is a multi-

TABLE 8.4 Types and properties of antimicrobial substances used in the food processing industry. *Salts of acids, such as sodium acetate, calcium citrate, sodium benzoate, calcium propionate, potassium sorbate, and sodium nitrite all have antimicrobial properties, as does sodium chloride and sucrose. Sulfur dioxide (SO_2) can be derived from a variety of sulfur additives such as bisulfite, and can create a multitude of effects in foods.*

Antimicrobial Compound	Effective Against	Some Food Applications
acetic acid salt (sodium acetate)	bacteria, molds	bread; as vinegar in pickled products and mayonnaise
benzoic acid salt (sodium benzoate)	molds and yeasts	ketchup, jams, syrups, orange juice products, syrups
Na and Ca propionate	bacteria, molds	bread, cake, cheese foods
potassium sorbate	bacteria, molds	breads
salt (sodium chloride)	bacteria, yeast, mold	baked products, canned foods, meats
sodium nitrite	*Clostridium*	cured meat products
sodium benzoate	molds and yeasts	condiments, fruit juices
sugar (sucrose)	bacteria, yeast, mold	baked products, fruit preserves, meats
sulfite, sulfur dioxide (SO_2)	bacteria, yeast, mold	dried fruit, lemon juice, molasses, wines

➤ BHA: butylated hydroxyanisole
BHT: butylated hydroxytoluene
TBHQ: tertiary-butylated hydroquinone (an antioxidant used to stabilize edible oils and fats to oxidation)

functional substance—besides antimicrobial activity, it posseses antibrowning properties with respect to both Maillard and enzymatic browning. Table 8.4 lists some of the antimicrobial compounds.

Antioxidant compounds also are important in food preservation, since oxygen can cause foods to lose freshness and quality. These compounds include phenolic substances, some of which serve double duty as effective antimicrobial compounds, such as BHT, BHA, TBHQ, vanillin, and propyl gallate. Two other popular antioxidants are alpha-tocopherol (vitamin E) and ascorbyl palmitate, a form of vitamin C.

Packaging. Packaging, another form of traditional nonthermal processing, is more fully covered in Chapter 13. However, as a means to preserve foods, it deserves mention now. A main purpose for packaging is to offer protection from biological, chemical, and physical factors that, if permitted to contact a food, would hasten its deterioration. Packaging in a modified gas atmosphere (called MAP, **modified atmosphere packaging**) is a popular way to prevent oxidation reactions in foods. If the barrier properties are carefully selected, a packaging material maintains a modified atmosphere inside the package and extends a food's shelf life. Packaging materials such as polyvinyl chloride, polyvinylidene chloride, and polypropylene offer low moisture permeability, perfect for storing dehydrated foods.

Packaging materials with low gas permeability are used for fat-containing foods in order to diminish oxidation reactions. Because fresh fruits and vegetables respire, they require packaging materials, such as polyethylene, that have high permeability to gases. Another approach to delaying the deterioration or senescence of fresh fruits and vegetables during storage is by depleting the surrounding atmosphere of oxygen and enriching it with carbon dioxide. This can be accomplished by sealing the fruit or vegetable in an airtight package and allowing it, through normal respiration, to use up the oxygen in the package and to thus create its own modified atmosphere high in CO_2 content.

Another modified atmosphere strategy is to remove oxygen from the headspace of packaged goods such as single serving baked goods, instant coffee, fat-containing snack foods, and dry milk by pulling a vacuum inside the package, and adding N_2 (nitrogen gas) or carbon dioxide. Vacuum packaging of retail cuts of fresh meat is practiced both en route to the retail outlet and for use at the retail level. This type of modified atmosphere package prevents oxidative changes in myoglobin, thus maintaining a fresh purple-red color to the meat and preventing the growth of aerobic spoilage organisms. Such packages may also be flushed with carbon dioxide or nitrogen as a means of maintaining quality and freshness.

So-called **smart packages** meet the special needs of certain foods, like oxygen-sensitive baked products formulated with fat. For example, packages made with oxygen-absorbing materials such as iron remove oxygen from the inside of the package to protect such foods from oxidation. A different class of smart package, packaging containing antimicrobial agents embedded in the packaging, has recently been developed. In addition, precut salad greens packaged in bags specially designed to permit gas exchanges result in extended freshness.

Nonthermal Processing Innovations. The sixth basic principle of food processing to achieve preservation is through innovative **nonthermal processing,** including irradiation, high pressure, and pulses of light and electric fields. These are discussed more fully in Section 8.3, and irradiation is highlighted in the Challenge section.

8.2 WHAT IS HEAT TRANSFER?

Heat transfer is required to destroy microorganisms in many processed foods. The nature of heat transfer will be revisited in Chapter 13, Food Engineering; our focus here will be to understand it in terms of processing. **Heat transfer** refers to the manner in which heat energy is transferred from a heat source to food particles in a container, such as a can, bottle, or pouch. Transfer is typically a result of three processes: conduction, convection, and radiant energy. In conduction, heat transfer occurs when heat moves through a material due to molecular motion. In convection, it is due to the movement of a heated fluid from hot regions to cold. In radiant heating, it occurs when heat is transferred directly between objects without an intervening medium.

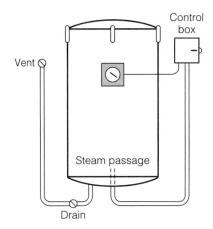

FIGURE 8.3 Retort. *The food processing industry uses a retort chamber to create temperatures in excess of 100°C inside cans being heated within the retort. Retorts may be oriented either vertically or horizontally, and include controls for steam pressure heating, venting, draining, and recording the process temperature.*

Heat Transfer in a Retort Canner

In a retort canner heat energy is transferred from the heating source to the food inside each container of food (can) by conduction and convection. **Retort processing** is the procedure used to heat sealed cans in order to destroy bacteria and spores. Retort processing utilizes a chamber with steam valve jets that allow steam to enter the chamber for precise temperature control. Cans are placed inside this large retort chamber on stackable pallets, and steam is injected in. The steam transfers heat into each can placed inside the retort. By injecting steam under pressure, temperatures exceed the boiling point of water (100°C; 212°F) inside each can within the retort chamber (Figure 8.3).

Consider this question: Which would burn your hand more: a 400°F (204°C) oven or 212°F (100°C) steam? The steam heated to 212°F. The reason has to do with heat transfer. Air offers poor transfer of heat compared to steam (water vapor). For this reason, retort units must be vented properly so that all the air is removed before a thermal process is begun. Air removal allows the correct processing temperature to be reached.

Heat Transfer Within a Can

To understand heat transfer in a can, we must review conduction and convection. As mentioned earlier, **conduction** is heat that is transferred between food molecules inside the can via molecular collisions. **Convection** occurs when heat is transferred through a liquid according to density differences, and it is unique to liquid foods or foods packed in liquid. To determine if a food has been heated sufficiently within a can, the temperature of the last area within the can to heat up is measured. Where is this location?

When a conduction heat process is begun, this **cold point** is toward the center of the can because heat would have been conducted equally inward from all sides—See Figure 8.4. When a conduction plus convection heat process is begun, the cold point is below the center of the can due to the movement of fluid from the bottom (colder, more dense) to the top (warmer, less dense) as it heats up. Did you catch the importance of measuring the cold point? Since the cold point is the last area within a can to heat up, it is measured to determine if heat has penetrated a container of food to completely heat all the food par-

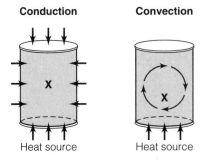

FIGURE 8.4 Heat transfer in a can via conduction and convection, the cold point (X). *In a conduction heat process, the coldest point is toward the center of the can since heat is conducted inward equally from all sides. In liquids, convection heating causes the cold point to be below the center.*

ticles to the desired temperature. The time it takes the cold point to heat up and reach the required temperature determines the overall processing time and temperature parameters for a particular product.

Vacuum Canning

Canned foods are packed under vacuum—that is, all air is removed from inside the can. This condition is necessary for several reasons. First, removal of the air prevents the cans from swelling and bursting open if they are stored at reduced pressure. In areas of high altitude, cans previously sealed at sea level with air inside would experience expansion due to the expansion of gas in the can, and could rupture due to pressure increases. Similarly, cans stored in hot climates would swell as air inside would expand in volume.

Second, canning under vacuum removes a source of oxygen that is potentially damaging to foods by producing off-color and off-flavor. In the canning process, foods are filled or packed into cans and then automatically sealed by can-sealing machinery. The can is open as the food enters, then a lid is dropped over the can with a protruding edge. The action of rotating rollers first introduces a crimp or bend in the edge of the lid, and finally a seal is made with the help of a plastic gasket located in the outer rim of the cover as the lid is pressed on. Foods packaged into cans in this manner enjoy the benefit of a **hermetic seal,** which is airtight, meaning that no gases are able to enter or exit from the can.

8.3 FOOD PRESERVATION—PREVENTING FOOD SPOILAGE

Foods are biological materials derived from once living matter. They are subject to spoilage and decomposition, which means they are perishable commodities. Spoilage refers to the loss of food quality as a result of specific biological, chemical, and physical changes. This quality decline is observable as a deterioration in food appearance, flavor, odor, and texture. Besides being unpalatable, spoiled foods may exhibit altered nutritional properties but are usually not dangerous in terms of their safety.

Microorganisms such as yeasts, molds, and bacteria are responsible for the biological changes in spoiled foods. Most of the time, these organisms travel through the air and make contact with and contaminate food surfaces. Bacteria are often introduced to foods through unsanitary handling, equipment, or environmental conditions. **Biological changes** in foods stem from the action of microorganisms fermenting carbohydrates into smaller organic molecules such as acids and alcohols and into gases such as hydrogen sulfide and carbon dioxide. For example, these biological changes happen in opened containers of milk. **Chemical changes** in foods refer to the action of microbial enzymes called proteases and lipases on food protein and lipid molecules, respectively. Figure 8.5 shows how proteases degrade a globular protein into its constituent peptides, and peptidases break the peptides down into amino acids. Such chemical changes occur in meats that are unrefrigerated and in cheese during ripening. The products of these chemical changes result in significant changes in food texture, flavor, and odor. **Physical changes** in foods include loss of moisture due to evaporation and separation of phases. Loss of moisture contributes to bread staling, while phase separation (oily layer) occurs in freshly ground (vs. formulated and processed) peanut butter.

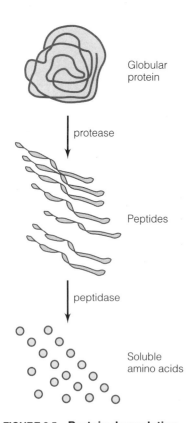

FIGURE 8.5 Protein degradation enzymes. *Enzymes that hydrolyze proteins are called proteinases (proteases) and peptidases. Contamination of foods with microorganisms, which have such enzymes, results in both physical and chemical changes in food.*

Thermal Processing for Food Preservation

The goal of preservation methods such as thermal processing is to delay food perishability and inhibit food spoilage. Thermal processing targets both spoilage and pathogenic organisms and seeks to destroy them and their spores. Both wet and dry heat can be applied to kill microbial populations in foods. The criterion for microbial cell death is failure to grow and reproduce. Simply causing damage to cells is insufficient; the cells must be demonstrably killed. Various testing methods are used to establish cell death, including

TABLE 8.5   D values showing heat resistance of various foodborne microorganisms.

Organism	Heat Resistance
Bacillus cereus	D_{100} = 2–8 minutes
Campylobacter jejuni	D_{55} = 60 seconds
Clostridium botulinum (type A, proteolytic strain)	D_{121} = 0.21 seconds
C. botulinum toxin	Destroyed in 5 min at 85°C (185°F)
Escherichia coli	$D_{71.7}$ = 1 second
Listeria monocytogenes	$D_{71.7}$ = 3 seconds
Staphylococcus aureus	$D_{71.7}$ = 4.1 seconds

Source: Adapted from Garbutt, 1997.

lack of typical chemical changes produced by microbial metabolic activity, such as gas production, lack of formation of colonies on culture plates, and lack of turbidity in broth cultures.

Although heat can and does kill microorganisms, some species require more severe heating, meaning they exhibit what is called heat resistance. **Heat resistance** is the ability of an organism to survive thermal processing of a particular time and temperature combination that destroys non–heat-resistant organisms. To cause the thermal death of organisms in processing, specific time-temperature combinations are required according to their heat resistance and according to the physical makeup (density, ratio of solid to liquid present) of the food. From the processing standpoint, not only is destruction of microbes essential, but this must be achieved without adversely affecting the sensory quality of the heat processed food, such as canned meats.

D Values. It is a simple matter to experimentally determine the time and temperature combinations required to kill any particular microorganism. Heating a species-specific bacterial population (called a pure culture) to specific temperatures, removing samples at timed intervals, and plating the sample to look for survival and growth is a standard approach.

When the number of survivors (log value) is plotted on the *y*-axis of a graph, and the heating time is plotted on the *x*-axis, a straight line survival curve is obtained when semilog (semilogarithmic) graph paper is used. On semilog paper, the *x*-axis is linear, and the *y*-axis is logarithmic—divided into equal parts that represent log cycles, such as 10^1, 10^2, 10^3, etc. See Figure 8.6. When the graph passes through one log cycle (for instance, in decreasing in survivors from 10^3 organisms to 10^2 organisms), it means that 90 percent of the bacterial population has been killed, and only 10 percent has survived. The time required for a bacterial population to pass through one log cycle, in which 90 percent of the organisms have been killed, is referred to as the **decimal reduction time,** or **D value.**

D values vary among organisms owing to their variation in heat resistance. Although D values represent time, the processing temperature involved in the kill is indicated as a subscript number, for example, D_{121}. This D value, D_{121}, indicates the time required to kill 90 percent of a bacterial population at 121°C (250°F). The number of organisms present in a sample does not affect the D value—it will be the same whether there are 10 organisms or 10 million organisms present. Table 8.5 gives the D values for several microorganisms.

D values offer the basis for calculating process times in the food industry. The microorganism that is used to measure the effectiveness of a heat processing time-temperature combination is a nonpathogenic bacterium, *Bacillus stearothermophilus*. This organism was chosen because of its great heat resistance. It is even more heat resistant than the spores of the pathogen *Clostridium botulinum*, which is the most heat-resistant pathogen. If *B. stearothermophilus* is killed by a particular set of processing parameters, then there is confidence that all other spoilage as well as pathogenic organisms that may have been in the food have also succumbed to the heat treatment.

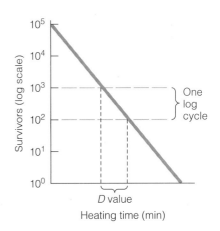

FIGURE 8.6 Bacterial survivor curve, identifying the D value.

12D Concept. To be certain that a can of a particular food item is safe, the **12D concept** is applied. To establish a generous safety margin for thermally processed foods, foods are processed for a time-temperature combination that results in not one log cycle reduction, but twelve log cycle reductions. With this 12D treatment, the food is commercially sterile (except thermophilic organisms). Said another way, the probability is that one out of 10^{12} cans would have one spore of *Clostridium botulinum* surviving (in one trillion cans!). This is so unlikely that clearly 12D processing is an effective thermal treatment.

In practical terms, unprocessed canned foods like potato soup, fruit cocktail, or beef stew rarely contain over a billion organisms per can, which is 10^9 organisms. Even so, the standard measure used in thermal death processing is the length of time required to kill 60 billion spores of *Clostridium botulinum* in a can. This has been determined as equivalent to heating all parts of the food in the can (remember the cold point) to 121°C (250°F) for 2.5 minutes. Since heating does not instantaneously make all areas of the food within a can equally hot, a lag "come up" time is always taken into account in calculating safe processing times. In actual practice, since it is known that certain thermophilic organisms are even more resistant to this type of heat treatment than *C. botulinum*, foods are heated to the equivalent of 7 *minutes* at 121°C (250°F), which includes the "come up" time to reach 121°C (250°F). Thus, in applying the 12D concept, the chances of *Clostridium botulinum* spores surviving are so small that they can be ignored.

TDT—Thermal Death Time. Let's look at the **thermal death time (TDT)** graph in Figure 8.7. A TDT graph (on semilog paper) plots heating time in minutes on the *y*-axis and temperature (°F or °C) on the *x*-axis. Microorganisms are destroyed by heat in direct proportion to their numbers in a sample, such as canned food. This is not a contradiction to a statement we made earlier, that the number of organisms present in a sample does not affect the D value. What it means is that under constant thermal conditions (at the same temperature) the same *percentage* of organisms (or spores of *Clostridium botulinum*) are killed per time interval.

It also means that at any given thermal processing time (say, 30 minutes) the greater the numbers of spores present, the longer the heating time required to kill the spores. For example, if applying 100°C of heat for 2 minutes destroys 90 percent of a population of microorganisms, then after 4 minutes 90 percent of the survivors of the initial treatment will be destroyed (90% of 10%=9%). This represents a 2D heat treatment, which means 90 + 9, or 99 percent of the population has been killed. Can you calculate the percentage of the microorganism population that would remain after a 12D heat treatment?

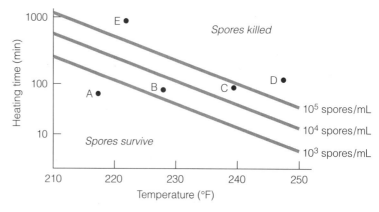

FIGURE 8.7 TDT (thermal death time) plot, showing combination of time and temperature to kill various spore concentrations. *Three cans contain various numbers of C. botulinum spores: 10^5 spores/ mL, 10^4 spores/ mL and 10^3 spores/ mL. As spore concentration increases, the processing conditions required to deactivate them become more severe. (See Study Question 29 for questions regarding points A, B, C, D, and E.*

Ohmic Heating. Traditional thermal processes are not the only approaches currently being used to destroy foodborne microbes. **Ohmic heating** or resistance heating is an efficient alternative to traditional methods, in which a food product is subjected to an alternating current. The current is applied to opposing electrodes with the food sample in between (Figure 8.8). To be effective, the product is situated within a conducting ionic solution, such as salt brine. Although a significant amount of heat can be generated in this fashion, food product particles do not undergo typical heat damage associated with surface to interior temperature gradients because all of the food particles are heated in a nearly simultaneous manner. As a result, even a low-acid food that contains large particles of meat suspended in liquid, such as beef stew, can be rendered commercially sterile without suffering thermal damage due to overprocessing.

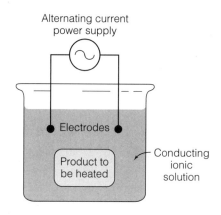

FIGURE 8.8 Ohmic heating. *Foods are processed by subjecting them to an alternating current within a conducting ionic solution.*

Traditional Nonthermal Processing for Food Preservation

Nonthermal processing refers to methods of food preservation without the use of heat processing. The two categories of nonthermal processing are traditional and innovative. The use of chemical additives that have antimicrobial properties and the design and application of food packaging that offers a barrier to moisture and gases are considered traditional nonthermal processing approaches. We discussed these in Section 8.1.

Innovative Nonthermal Methods of Food Preservation

Pulsed electric fields, oscillating magnetic fields, and high-pressure processing are among three technologies being investigated to effectively inactivate microorganisms and enzymes in food products while retaining nutrients and freshlike qualities. Combined methods, known as hurdle technology, is another area of special attention in which conventional methods such as pH and water activity control are applied with various emerging nonthermal techniques. Such technologies are designed to be safe both for the environment and for those using the equipment involved with each operation. Before we examine each nonthermal method, let's discuss hurdle technology a bit.

Hurdle Technology. Food safety can be achieved through the use of a single preservation method, known as a *hurdle*. This requires a forceful approach (i.e., a high level of a particular hurdle) that may adversely affect a food's taste, texture, and acceptance. A **hurdle** then is simply a stress placed on a microorganism that it must overcome in order to survive, grow, and reproduce in food. In terms of food preservation, hurdles are equivalent to the factors affecting microbial growth: water activity, pH, temperature, pressure, and chemical antimicrobials. Food scientists have also demonstrated that a combination of more than one hurdle, at lower levels than when used singly, can be effective in food preservation.

Hurdle technology creates a combination of suboptimal growth conditions in which each hurdle factor alone is insufficient to prevent the growth of spoilage and pathogenic organisms, but hurdles used in combination provide effective control. Figure 8.9 demonstrates the concept—as an organism attempts to "jump" over each hurdle, it becomes less able to overcome the next, and the cumulative effect of the hurdle combination prevents the growth and ultimate survival of the organism. Lower levels of acids or chem-

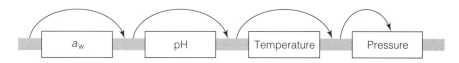

FIGURE 8.9 Hurdle technology. *Low-level hurdles are combined to prevent microbial growth in food (a_w = water activity).*

ical preservative can be used in combination with temperature and water activity controls, which maintains food taste, texture, and acceptability.

PEF—Pulsed electric fields. Microbial inactivation by high-intensity pulsed electric fields (PEF) is a new technology for food preservation that results in minimal influence on food temperature and quality. PEF is suitable for such products as fruit juices and eggs. A patented process called CoolPure™ has been developed by a California company (PurePulse Technologies). PEF involves the application of a short burst of high voltage to a fluid food placed between two electrodes, and it has the potential to improve the economy and efficiency of energy usage and to provide microbiologically safe, nutritious, and freshlike quality foods.

Fresh and concentrated apple juice, skim milk, liquid whole eggs, and pea soup have been successfully preserved in the continuous PEF treatment system. Sensory evaluations, shelf life determination, and chemical and physical analyses are being conducted to coincide with three selected packaging methods. Microscopic examinations can confirm the damage to microorganisms without concurrent food structure destruction.

OMF—Oscillating Magnetic Fields. Preservation of foods with high-energy oscillating magnetic fields (OMF) inactivates microorganisms and denatures enzymes in fresh or prepared foods. The advantages OMF potentially provides are an aseptic treatment of solid and liquid foods inside a flexible package, generating minimal heat within the food, and reducing energy requirements. Possible applications include roast beef, milk, and juices.

HPP—High-Pressure Processing. High-pressure processing (HPP) is also referred to as Pascalization, named after French mathematician and scientist Blaise Pascal, who pioneered the study of atmospheric pressure in the mid-1600s. Microorganisms (but not spores) are destroyed by high pressure, and at a uniform pressure throughout the food that enables complete preservation, HPP does not affect quality. HPP is not time/mass dependent, so processing time is minimal. Possible applications include surimi, eggs, rice, milk, and juices. HPP is extremely promising because it can improve the rheological and functional properties of foods to yield a fully nutritious and flavorful product.

PLT—Pulsed Light Technology. Pulsed light technology (PLT) utilizes brief bursts of high-intensity light energy to cause selective damage to the cell membranes of bacterial cells without the disruption of food tissue. It was approved in 1996 by the FDA as an irradiation processing method. A patented process called PureBright®, has been developed by a California company (PurePulse Technologies). The intensity of PureBright is 20,000 times that of sunlight. At this level, food surfaces and packaging materials are decontaminated, and certain enzymes may even be deactivated. Not only vegetative bacteria, but yeasts, molds, bacterial spores, and viruses are said to be destroyed. At the same time, no significant nutrient reduction is reported.

PLT uses intense and short duration pulses of broad-spectrum white light, including wavelengths in the ultraviolet to the near infrared region. Food material is exposed to a few flashes applied in a fraction of a second to provide a high level of microbial inactivation.

Pulsed light is produced using engineering technologies that multiply power many fold. Accumulating electrical energy in an energy storage capacitor over relatively long times (a fraction of a second) and releasing this storage energy to do work in a much shorter time (millionths or thousandths of a second) magnifies the power applied. The result is very high power during the duty cycle, with the expenditure of only moderate power consumption (Dunn and others, 1995). The technology is applicable mainly in sterilizing or reducing the microbial population on the surface of packaging materials or food surfaces. Since packaging material used in aseptic processing has been traditionally sterilized with hydrogen peroxide, which may leave undesirable residues in the food or package, light pulses can be used to reduce or eliminate the need for chemical disinfec-

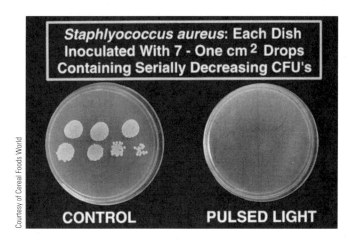

FIGURE 8.10 **The effect of pulsed light on *Staphylococcus aureus* survival.** *Each dish was inoculated with seven drops of colony-forming units (CFUs) of* S. aureus, *each drop decreasing (from top left to lower right) in concentration of the organism. In the control dish at left, the seven CFUs all grew, even at low concentrations. But the pulsed light* processing destroyed all* S. aureus *at all concentration levels.*

*2 Flashes, 0.75 3/cm² per Flash

tants and preservatives. Pulsed light may also be used to extend the shelf life or improve the quality of produce.

The lethality of the light pulses results from induction of photochemical and photothermal reactions. The shorter wavelengths in the ultraviolet range of 200-320 nm are more effective than the longer wavelengths due to their higher energy levels. The conjugated carbon-to-carbon double-bond systems in proteins and nucleic acids of the organisms absorb the light, leading to damage that causes cell death. Inactivation of DNA occurs by several mechanisms, including chemical modifications and structural cleavage. Damage to proteins, membranes, and other cellular material probably occurs concurrently with the nucleic acid destruction. Figure 8.10 shows the lethal effect of PLT on seven bacterial colonies compared to a control. It has been observed that the motility of *E. coli* ceases immediately after exposure to pulsed light.

8.4 DAIRY PRODUCTS PROCESSING

Chapter 2 introduced the various types of milks and dairy products. We have also examined ice cream as an emulsion and considered the isoelectric point of casein in cheese processing. In this section, we will focus on specific processing steps in the manufacture of milk, cheese, and yogurt.

Milk, Cheese, Ice Cream, and Yogurt

Milk from the dairy farm to be processed is first clarified cold. It is centrifuged at slow speed to separate out dirt and sediment but not the cream. It is then pumped into a storage tank where the milk is sampled for butterfat content and adjusted to the desired fat level. This is accomplished by addition of either skim milk (to decrease) or butterfat (to increase) the fat level to meet regulatory standards. Figure 8.11 outlines the steps in milk processing.

Milk is fortified with vitamin D, at the rate of 400 IU per quart, and 2,000 IU of Vitamin A. It undergoes a mild thermal processing called **pasteurization.** The standard method for pasteurization of milk uses HTST processing. HTST (high-temperature, short-time) processing corresponds to one of three time and temperature combinations. Milk is heated quickly to either: 72°C (161°F) and held 15 seconds, 88°C (191°F) for 1 second, or 90°C (194°F) for 0.5 seconds.

Milk is marketed in a number of different forms to appeal to the varied needs of the consumer. Federal standards of identity have been established for these milk products that enter interstate commerce:

whole milk
low-fat milk
skim milk
fat-free milk
acidophilus milk
buttermilk

evaporated milk
sweetened
    condensed milk
UHT milk
dry milk

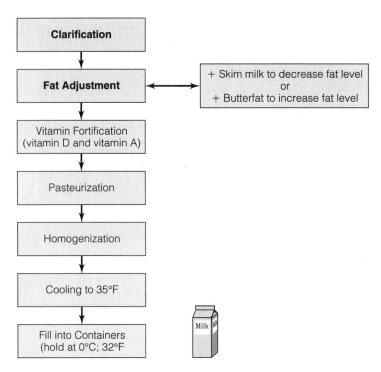

FIGURE 8.11 Sequence of steps in milk processing.

While still hot, the milk is homogenized to prevent cream separation. **Homogenization** decreases the size of fat globules dispersed in milk to prevent them from clustering and forming a cream line on the surface. The process pumps milk under pressure through very small openings in a machine called a homogenizer. The fat in the milk is immediately coated with protein films, which act as emulsifiers to keep the small, homogenized fat particles from coalescing and separating from the milk.

Processed milk is rapidly cooled to 1.6°C (35°F), mechanically filled into waxed, plastic-coated cardboard or semirigid plastic containers and held at 0°C (32°F), and shipped to retail outlets.

Cheese

Cheese is a concentrated dairy food that is allowed to cure or ripen in order to develop full flavor. In cheese processing, milk is coagulated into a curd through the use of enzymes (e.g., chymosin or rennin) or acid. After coagulation of casein, the major milk protein, the curd is cut, heated, and pressed to remove whey, the liquid serum part of the milk.

Figure 8.12 outlines cheese processing and discusses the importance of ripening.

The *Penicillium roqueforti* and *P. glaucum* types of mold (which are injected into the curd prior to ripening) are responsible for the mottled blue-green color of roquefort, gorgonzola, and stilton cheeses. Molds generate their own enzymes, which penetrate and soften the cheese, and generate flavors. Gas-producing microorganisms create the familiar holes in Swiss cheese. These organisms grow and produce carbon doixide gas during early stages of ripening, when the cheese is soft and elastic. Putrefactive bacteria give Limburger cheese its characteristic odor. Milder cheeses (such as brick and monterey jack) are ripened for shorter periods of time than strong cheeses (e.g., blue cheese may be aged 3-9 months). Characteristics of flavor, aroma, texture, and appearance are therefore developed in each particular cheese.

Ice Cream

Ice cream processing, introduced in Chapter 2, requires a sequence of steps that includes blending of the mix ingredients, pasteurization (heating to 71°C, 109°F for 30 minutes), homogenization, aging the mix, freezing, packaging, and hardening (Figure 8.13). Pre-

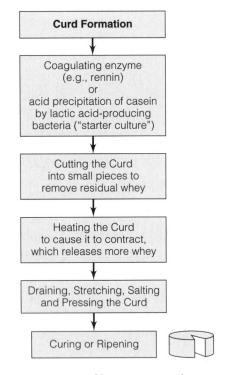

FIGURE 8.12 Cheese processing and the importance of ripening.
Ripening allows cheese to develop the desired characteristics of its own particular type. Some specific ripening changes are: formation of lactic acid from lactose by lactic acid-producing bacteria; digestion of protein by protease enzymes into end products (amino acids and peptides); development of molds in mold-ripened cheese; gas formation; development of characteristic flavors, as a consequence of fat decomposition by lipase enzymes.

mium ice cream contains about 12 percent butterfat (cream). Injection of air (called overrun) into the product lightens the texture. The presence of fat in ice cream requires the addition of emulsifiers to the mix. Aging involves crystallization of this lipid matter in the ice cream prior to freezing. Aging also promotes the ability of fat to coat air cells and enhance product firmness. The mix is frozen to cause ice crystals to form. The freezing point of ice cream is related to the sweetener used, which is typically sucrose or corn syrup. Freezing generally causes about half of the available water in the product to convert into the frozen state. This enables product to flow easily into containers during packaging. Following packaging, hardening of product at low temperature (−34°C, −29°F) causes most of the product's water to go into the frozen state. Manufactured ice cream needs to be hardened to be transported for commercial distribution. Quality ice cream exhibits appropriate color, flavor, creamy texture, body, and melting characteristics.

Of additional importance to ice cream quality are controls that prevent excessive ice crystal formation and growth in the product during manufacture and storage, which would otherwise produce a gritty texture. The substances often used to control ice crystal development are carrageenan, locust bean gum, and gelatin, which act as stabilizers during storage. Another potential quality problem has to do with the crystallization of the sugar naturally present in the product: lactose. During storage, and depending on the product solids content and temperature, lactose can crystallize and separate from what is called the unfrozen matrix of the ice cream. This change can be detected by the consumer as a sandy texture. However, with the processing equipment and stabilizer technology currently available, sandiness is not the problem in ice cream it once was.

Yogurt

Yogurt is a fermented, coagulated milk product. Yogurt is also called a cultured milk product because to make yogurt, manufacturers add bacterial cultures to the starting material—milk. The action of the cultures coagulates the milk proteins, creates a sour flavor, and preserves the product.

The yogurt processing steps are outlined in Figure 8.14. Milk is pasteurized (85°C, 185°F) for 30 minutes and homogenized. Cultures of *Lactobacillus delbrueckii spp bulgaricus* and *Streptococcus salivarius spp thermophilus* are added, and the mixture is subjected to controlled incubation and fermentation, at approximately 43°C (109°F). During this time, an interesting interaction takes place. The two organisms stimulate each other's growth. The *Strep* organism initiates fermentation, producing formic acid and carbon dioxide. The pH drop due to acid in turn stimulates growth of the *Lactobacillus* organism, and its production of small peptides and amino acids. These act to stimulate growth of the *Streptococcus* organism, and the generation of lactic acid. In this way, the two bacteria together create a faster fermentation that produces more acid and flavor in the yogurt than if the organisms were used separately. The final pH of the product ranges from 3.7 to 4.3, which exerts a preservative effect and imparts tartness. Yogurts are manufactured to be either stirred or preset. In addition to plain unflavored, added flavors commonly include fruits, coffee, vanilla, or chocolate.

8.5 EGG PROCESSING

Eggs have been recognized as a good overall source of nutrients and especially of high biological value protein. Although the cholesterol content of eggs (slightly over 200 mg per egg on average) decreased consumption during the 1980s and 1990s, recent statements by the American Heart Association have contributed to increased consumer confidence in eggs. For the food processor, eggs offer particular challenges because of their breakable physical nature, as well as their tendency to decline in quality over time. Maintenance of egg quality before and during processing are critical. Figure 8.15 shows egg quality classes.

At the preprocessing stage, eggs are placed in liquid CO_2 to allow for rapid chilling compared to standard chilling prior to cold storage. Reduced shell cracks and decreased

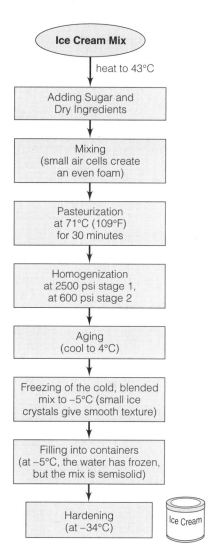

FIGURE 8.13 Steps in making ice cream. *Ice cream contains milkfat (as butterfat), milk solids nonfat (MSNF), sugar, stabilizer, emulsifier, flavoring, water, and air.*

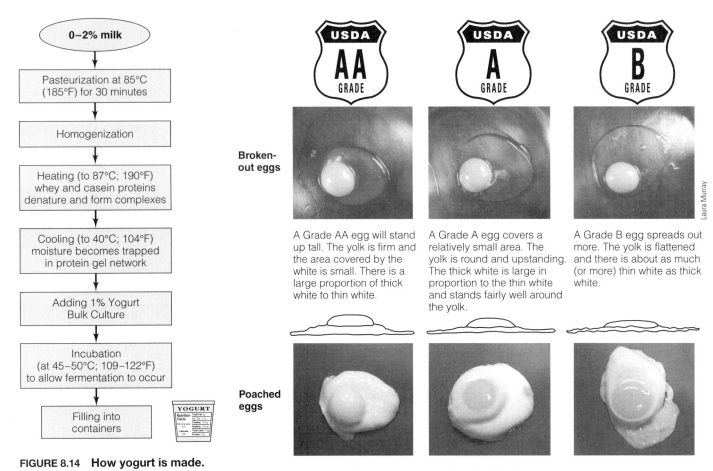

FIGURE 8.14 How yogurt is made.

FIGURE 8.15 Egg quality classes according to USDA grade.

growth of *Salmonella* are associated with this treatment. Egg processing varies according to two distinct product directions: (1) the whole egg in the shell, called *shell eggs*, and (2) the separated white and yolk products. We will focus our discussion on the further processing of the latter—see Figure 8.16.

Egg Product Processing

At an egg processing plant, eggs to be further processed are weighed, tested for freshness, washed, disinfected, and candled before going to the breaking machine. **Candling** is the method of examining eggs in front of a bright light to check for staleness, blood clots, embryo development, and quality grade. Candled eggs are next broken in preparation for further processing. **Breaking** refers to the process of shell breaking to separate the egg white from the egg yolk. During breaking, special machinery carefully separates the white and yolk in a manner that reduces the likelihood of rupturing the egg yolks. If the yolk membrane breaks, white and egg yolk will mix, and the functionality of the egg white will be reduced. However, eggs with broken yolk membranes can be further processed as whole egg products.

The separation process actually creates three products: (1) the egg yolk, (2) the egg white, and (3) whole egg (mixed yolk plus white). Each product is processed by a separator machine to filter out chalazae, yolk membrane, or small shell pieces. The purified product is then stored in refrigerated tanks, where the product is standardized for color, dry matter, and fat content. Standardizing the fat content of whole egg product is accomplished by adding yolk or egg white. After standardization, the three base products are ready for further processing.

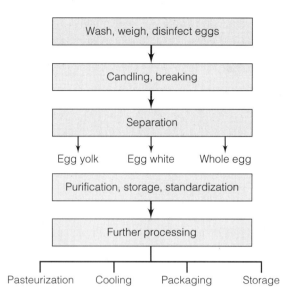

FIGURE 8.16 Processing steps for white and yolk products.

The liquid products are pumped through stainless steel tubes to the mixing department. At this point, sugar or salt can be added according to the nature of the desired product. Adding salt and sugar also extends product shelf life.

Following the mixing step comes pasteurization to destroy pathogenic bacteria, and it occurs in a flow-through pasteurizer. Egg yolks and whole eggs are heated to a maximum 65 to 67°C (149–153°F), while egg white is heated to a maximum of 55°C (131°F) for a period of time that depends on the amount of sugar or salt added. Care is taken to prevent the protein from coagulating. Following pasteurization, the products are quickly cooled to maintain nutrient quality. The liquid egg products are then packaged in cans or cartons within a plastic inner bag and stored under refrigeration.

Freezing and drying steps are applied to liquid egg products to improve their shelf life. The freezing takes place at −35°C (−95°F), and frozen storage is achieved at a temperature of −18°C (−64.4°F). Drying can be either spray, foam, tray, plate, or freeze-drying. We will focus on spray drying and plate drying. In **spray drying** a fine stream of the liquid product is dried very quickly in midair by heat. In spray drying, the egg white is first treated to prevent the browning of the egg white powder when heated. Browning is a result of small amounts of glucose in egg white, and it is prevented by yeast fermentation or by addition of glucose oxidase enzyme. In this way, glucose, which would otherwise contribute to the undesirable brown products of the Maillard reaction, is rendered unreactive and the desired white color is maintained.

Spray drying is practiced with all three base products while plate drying is used only for drying egg white. The process of **plate drying** produces a crystal layer of egg white that is ground into granules. During plate drying, concentrated egg white is sprayed onto metal plates after fermentation. The egg white is dried at 45°C to 50°C (113–122°F) for a minimum of 48 hours. The resulting egg white crystals have an extremely high foam-forming capacity. The foam is very stable and contains less bacteria than egg white powder. Crystal egg white is primarily used in the manufacture of candy bars and candies.

The various egg products can be made into powders, making them convenient to use in the baking food industry and in product development. Powdered egg white is produced by pumping the liquid egg white through a hyperfiltration system. This concentrated egg white is made into powder by means of nozzle pulverization and high temperature (150°C; 302°F) treatment. Whole egg and egg yolk can be made into powder without concentrating the liquid products. The various dried products are very similar to each other in dry matter content: Whole egg and egg white powder contain about 95 percent, and egg yolk powder about 93 percent to 94 percent. The egg powder is stored in plastic lined boxes for 14 days at a temperature of 70°C (158°F) in a post-pasteurization "hotroom" to eliminate microbial contamination.

TABLE 8.6 Examples of products generated from egg processing.

Egg Category	Product
Liquid egg	Egg yolks
	Egg whites
	Whole eggs
Frozen egg	Sugared egg yolks
	Salted egg yolks
	Whole eggs and yolks with corn syrup
Dried egg	Spray-dried egg white solids
	Flake albumen
	Free-flowing whole egg solids
	Egg yolk solids
Specialty egg	Frozen omelets
	Frozen quiche mix
	Frozen scrambled egg mixes
	Egg substitutes

Egg products represent processed and convenience forms of eggs obtained from breaking and processing shell eggs. These products are classified as refrigerated liquid egg products, frozen egg products, dried egg products, and specialty egg products—see Table 8.6.

Egg Substitute

The final egg product to be discussed is **egg substitute.** This product is based on egg white (albumen) and maintains a high nutritional value without a whole egg's cholesterol and fat content (because the cholesterol is in the yolk). A typical product contains 99 percent egg white and less than 1 percent vegetable gums such as xanthan and guar, beta-carotene for color, salt, natural flavor, and assorted vitamins and minerals. As such, it is essentially fat-free, with one-quarter cup of the liquid product equivalent to a whole egg and providing zero fat grams compared to five in a whole egg. Touted as a "healthy real egg product," egg substitute is processed via pasteurization and must be stored under refrigeration.

8.6 MEAT PROCESSING

Meat can be defined as edible animal flesh, including processed or manufactured products derived from such tissue. This definition of meat would then include not only beef, but also chicken and other poultry, and such animal flesh as pork, fish, lobster, lamb and, even alligator.

Animal tissue is typically processed into a number of familiar products, such as cured bacon, frankfurters, fish sticks, ham, and luncheon meat. To create these products, the processing steps must take into account composition and safety issues. Muscle tissue is the main component of most of these products. The conversion of muscle tissue into meat involves removing connective tissue, fat (adipose) tissue, and associated blood vessels by processing. The physical and chemical properties of muscle and these other tissues influence the quality of processed meat.

Meat Quality

The tenderness of meat is its most important quality attribute, along with flavor and juiciness. It is influenced by a number of factors including the amount of adipose and connective tissue present. **Connective tissue** is composed of a watery dispersion of stromal protein matrix. Two kinds of connective stromal proteins are common in meats: elastin and collagen. Recall from Chapter 2 that collagen molecules are fibrous, elongated proteins that contribute to meat toughness. Collagen is broken down and denatured during the cooking method of **braising,** forming a gelatinous substance that makes meat more tender. Collagen becomes more resistant to breakdown and denaturation with age, explaining the greater toughness of meat from old animals.

Another factor that influences meat tenderness is the **grain** of the meat. The basic unit of muscle tissue is the muscle cell or muscle fiber, which consists of muscle fibrils or myofibrils. These myofibrils contain contractile proteins in a specialized array or organization of thick myosin and thin actin filaments (look back at Figure 2.10). Visible muscle is therefore a collection of muscle fibers, held together by connective tissue into bundles. Grain is related to the muscle bundle size present in meat and is termed either fine or coarse. Fine-grained meats are more tender and have smaller bundles, whereas coarse-grained meats are tougher and have larger bundles.

Rigor Mortis. Related to meat toughness is the concept of rigor, or rigor mortis. **Rigor mortis** is a transient postmortem physical and biochemical event that takes place in animal muscle tissue. It causes a loss of extensibility of the tissue, which makes meat very tough. In rigor mortis, the actin and myosin components of the myofibrils form permanent crosslinkages. These chemical bonds lock the muscle fibers in place, rather than

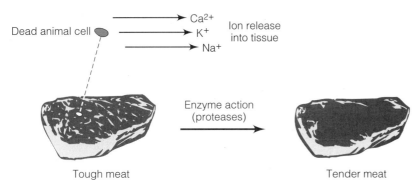

FIGURE 8.17 Events associated with rigor mortis and its resolution. *Rigor mortis (rigor) refers to stiffening of muscle as a result of a final contraction as an animal dies. Following an animal's death, specific enzymes responsible for maintaining the ionic calcium, potassium, and sodium balance within and outside muscle cells cease to function. As a result, membrane-bound enzymes release these ions, which diffuse through the tissue and contribute to toughening. Resolution of rigor mortis is the natural softening that results from changes in muscle ultrastructure mediated by enzymes called neutral proteases. As a result of the activity of these enzymes, myofibrillar proteins are broken down. Other enzymes called matrix metalloproteases are active during postmortem aging and are believed to degrade connective tissue, contributing to postrigor tenderness.*

allowing for expansion and contraction. In a living animal, energy would be available to break such bonds. After death, myosin stays locked onto actin, even if the muscle is trying to relax. Thus, rigor mortis and toughness develop (Figure 8.17).

Cold Shortening. Cold shortening is a quality problem that results from the rapid chilling of carcasses immediately after slaughter, before the glycogen in the muscle has been converted to lactic acid. Specifically, cold shortening occurs when the muscle is chilled to less than 15°C (60°F) before the completion of rigor mortis. With glycogen still present as an energy source, the cold temperature induces an irreversible contraction of the muscle in which the actin and myosin filaments shorten. Cold shortening causes meat to be as much as five times tougher than normal. This condition occurs in lean carcasses lacking adequate fat covering as insulation from the cold. The application of a high-voltage current (electrical stimulation) to postmortem carcasses minimizes this defect by forcing muscle contractions that use up muscle glycogen.

Thaw Rigor. Thaw rigor, similar to cold shortening, is a condition that results when meat is frozen before it enters rigor mortis. When thaw rigor meat is thawed, the leftover glycogen allows for muscle contraction and the meat becomes extremely tough.

PSE. Pale, soft, and exudative (PSE) meat is the reault of a rapid postmortem pH decline in pork while the muscle temperature is too high. The combination of low pH and high temperature adversely affects muscle proteins, reducing their ability to hold water. This condition is manifested by mushy, dripping meat texture and a pale surface appearance. PSE appears to be a stress-related and inheritable condition. Pigs having a genetic condition known as porcine stress syndrome (PSS) are more likely to yield PSE meat.

Other specific quality changes in meats that can occur include undesirable color changes (oxidation of myoglobin), the production of slime layer (bacterial polysaccharide synthesis), and fatting out (separation of fat in certain processed meats).

Meat Preservation and Processing

Processed meat refers to a whole muscle product that has been transformed into a manufactured product by chemical, enzymatic, or mechanical treatment. Processed meats are derived from animals slaughtered and portioned into primal (available at wholesale) or

TABLE 8.7 Comparison of processed meats grouped according to major meat ingredient. *Additives and processing techniques differ for each product.*

Beef	Pork/Ham	Beef and Pork	Veal and Pork	Liver
Beef bologna	Blood sausage	Club bologna	Bockwurst	Braunschwieger
Beef salami	Bratwurst	Cervelat	Bratwurst	Liverwurst (pork)
Pastrami	Capacolla	Frankfurters*	Veal loaf	
	Chorizo	Honey loaf	Wiesswurst	
	Frizzies	Hot dog*		
	Ham	Knackwurst		
	Ham bologna	Luncheon meat		
	Linguica	Mettwurst		
	Lola/Lolita	Mortadella		
	Luncheon meat	Olive loaf		
	Lyons	Peppered loaf		
	New England-style sausage	Pimento loaf		
	Old-fashioned loaf	Salami		
	Pork sausage	Smokies		
	Proscuitto	Weiner*		
	Salsiccia	Vienna sausage*		
	Scrapple			
	Thuringer			

*Terms used interchangeably.

⚐ **Food processing for preservation: meat preservation methods.**

Canning
Chemical additives
Cold storage
Curing
Drying
Fermentation
Heating
Irradiation
Packaging

⚐ **Classification of common red meats.**

Meat type	Animal Source
Beef	Cattle (bovine)
Lamb	Young sheep (ovine)
Mutton	Sheep older than two years (ovine)
Pork	Hogs (porcine)
Veal	Calves (bovine)
Venison	Deer

subprimal (available at retail) cuts, as well as so-called variety meats, trimmings, and by-products. These are all under the inspection jurisdiction of the USDA's FSIS (Food Safety Inspection Service). Processed meat exhibits altered appearance, flavor, texture, and shelf life compared to whole meat. Table 8.7 groups some processed meats by their major ingredient.

Meat is a a perishable commodity. The need for meat preservation and the trend toward urbanization and industrialization are the main reasons for the development of the processed meat industry. So, is there any difference between meat processing and meat preservation? The two are tightly related. Think of it this way: Preservation methods such as the use of chemical additives, freezing, and thermal processing help keep meat from spoiling, help retain nutrients, and contribute to a safe and wholesome condition (see list in margin). Such preservation methods constitute what we consider food processing. However, not all processing steps are directly related to preservation. For example, when sausage meat is stuffed into casing material, this operation is a processing rather than a preservation step. Comminution is another example of a processing practice that is not related to preservation.

Red Meat Processing

Most of the processing steps described here also apply to fish and poultry processing. These products are typically processed to preserve them due to their perishability.

Canning. A common method of meat preservation is canning. Canning involves sealing meat in a container and then heating it to destroy all microorganisms capable of food spoilage. Canned meat is therefore a commercially sterile product. Canning requires the processing of canned product in a retort cooker (at 83 kilopascals or 12 psi) to achieve the destruction of bacterial spores at temperatures exceeding the boiling point of water. Three minutes at 121°C (250°F) for small consumer-sized canned products (4 to 8 oz.) has been found to be an effective time and temperature combination and is commonly used. Under normal conditions, canned products can safely be stored at room temperature indefinitely.

A special variation called **aseptic canning** involves the separate sterilization of food and canning container. The commercially sterile meat is continuously introduced (at

TABLE 8.8 Several common processed meat ingredients and their functions.

Ingredient	Functions
Nitrate (NO_3)	Nitrite is the active form; antioxidant that slows rancidity
Nitrite (NO_2)	Inhibits spoilage and pathogenic organisms; contributes to flavor; prevents warmed-over-flavor; stabilizes color
Phosphates	Increase juiciness; inhibit rancidity; retain moisture; solubilize proteins
Salt	Cure ingredient; extracts myofibrillar proteins; provides flavor; tenderizes
Spices	Contribute specific flavors; inhibit bacterial growth
Sugar	Binds to water; facilitates browning reactions; provides sweetness to counteract salt; acts as substrate for fermentation
Water	Provides calorie reduction (fat replacer); carries and distributes dry ingredients; contributes to juiciness and tenderness
Erythorbate	Acts as antioxidant

149°C; 300°F) into sterile containers via an aseptic "hot fill" operation. The containers are cooled, sealed in a sterile environment to ensure their safety and shelf stability.

Canned meat products do not require refrigeration until they are opened. They offer convenience and availability because they are fully cooked and can be consumed directly from the can. The thermal process does result in slight flavor and texture changes.

Examples of canned meats:

Beef stew   Spam
Meatballs   Deviled ham
Corned beef   Vienna sausages

Chemical Additives. Ingredients added to processed meat include antioxidants (ascorbic acid and erythorbic acid) that function to retain processed meat color, nitrate (NO_3) and nitrite (NO_2), phosphates, salt, spices, sugar, and water (Table 8.8). Nitrates are converted into nitrites by bacteria during meat curing. Slow cured meats use nitrate, whereas nitrite is used directly in other cured meats (usually at levels of 50 ppm or less). Nitrite is converted into **nitric oxide** (NO), which acts as a reductant, and accelerates color development in cured meats.

At certain concentrations, salt increases the tenderness of meat. The presence of salt is one of the reasons that cured meats such as ham are more tender than uncured meats. Salt apparently exerts its influence on tenderness by softening the connective tissue protein (collagen) into a more tender form. Although once used at high enough levels in salted meats to effect preservation, at current usage levels, salt no longer functions as a preservative.

A number of vegetable enzymes such as bromelin (from pineapple), ficin (from fig), and papain (from papaya) are used to tenderize meat. They act to break down collagen and elastin connective tissue.

Cold Storage. Cold storage refers to both refrigeration and frozen storage. The reduction in temperature associated with cold storage is the most important factor influencing bacterial growth in cold stored meats. **Refrigerated storage** is the most common method of meat preservation. The typical refrigerated storage life for fresh meats is 5 to 7 days. Pathogenic bacteria are inhibited at temperatures below 3°C (38°F).

Freezer storage (−18°C, 0°F) offers an excellent method of meat preservation. Although freezing does not kill spoilage and pathogenic microorganisms, it stops their growth and reproduction. It also slows down deteriorative chemical reactions. How frozen meats are packaged is also important. Packaging must allow for minimum air contact to prevent moisture loss during storage. The length of time meats are held in frozen storage also determines their quality. Freezing meats allows them to be stored without significant quality loss for much longer periods compared to refrigeration. For instance, beef can be stored for 6 to 12 months, lamb for 6 to 9 months, pork for 6 months, and sausage products for 2 months.

With the availability of mechanical refrigeration, the process of freezing is widely practiced commercially. Rapid or "flash freezing" is especially effective with certain types of meat tissue, like fish. Although beef and venison benefit from an aging process with respect to tenderness, other meats are frozen quickly after slaughter. Freezing is accomplished by cold air blown by fans (blast freezing) or by placing meat in contact with refrigerated surfaces (contact freezing).

The rate of freezing affects meat quality. Slow freezing creates large ice crystals in the meat tissue that ruptures cell membranes. When such meat is thawed, much of the original moisture found in the meat is lost as purge. **Purge** refers to juices that flow from the meat, as water, soluble material, and extracellular fluid. For this reason cryogenic freezing, the use of supercold substances like liquid nitrogen to freeze meat tissue is often used commercially to maximize product quality.

However, when meat is equilibrated to room temperature following frozen storage, thawing often causes detrimental quality changes. In contrast to freezing, thawing should be a slow process. To minimize moisture loss, meats are best thawed in the refrigerator with packaging left intact. Meat thawed under warm water subjects the meat's outer layers to danger zone temperatures (5–60°C; 40–140°F) for long periods of time before it is completely thawed. This situation provides a conducive environment for microbial growth and increases the risk of food poisoning.

Comminution. The term **comminution** refers to meat particle size reduction. It is not a preservation method but is the unit operation by which muscle tissue is chopped, diced, emulsified, ground, and transformed into minute particles for incorporation into a sausage. *Sausage* is understood to include bologna, bratwurst, "breakfast" pork sausage, frankfurters, Italian sausage, liverwurst, luncheon meat "loaf form" varieties, Polish sausage, salami, and Vienna sausage.

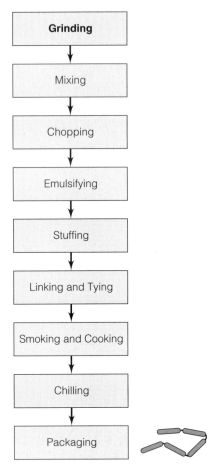

FIGURE 8.18 Steps involved in sausage processing.

Figure 8.18 diagrams the steps involved in sausage processing. The manufacture of sausage is a continuous sequence, beginning with grinding and ending with packaging. Not all types of sausage require all the steps shown in Figure 8.18. For instance, not all dry sausages require smoking. The emulsification step takes place following fine chopping in the presence of salt to produce a meat batter. In the meat batter emulsion, lean meat particles and fat particles within a complex system containing cellular components, proteins, seasonings and spices, and water. The batter is stuffed into casings to generate a formed shape like that of a weiner.

Curing. **Curing** refers to the addition of salt, sugar, and sodium nitrite to meats for the purposes of color development, flavor enhancement, preservation, and safety. In dry curing, of hams and pork sausages, processors add the dry cure ingredients to meat without the addition of water. Moisture from the meat combines with the cure to form a brine, which diffuses into the processed meat. Another method of adding dry cure ingredients is through the use of tumblers, which act like cement mixers to mechanically force ingredient absorption into the meat. This tumbling forces the myofibrillar protein myosin out of the tissue. Extracted myosin acts like a "glue" to allow separate small pieces of meat to adhere together.

When cure ingredients are dissolved in water to create a brine, this is called pickle curing. Occasionally phosphate and ascorbic acid are added to the pickle solution. The most common pickle curing method is a process called **needle stitch pumping,** in which a single needle with multiple openings or multiple needles inject brine directly into meat tissue.

Drying. Drying, a common method of meat preservation, removes moisture so that microorganisms cannot grow. Ancient practices probably included sun drying meats to achieve preservation. Dried meats exhibit an extended shelf life and can be stored at room temperature. Examples of dried meats include dry sausages, freeze-dried meats, and jerky products. **Freeze-drying** requires low-pressure chilling under a vacuum. The process allows for moisture in the meat to form ice crystals slowly. These are evaporated into the gas phase via sublimation, without passing through the liquid phase. Industrial freeze-drying is carried out at about −45°C (−50°F). The frozen items are transferred to cold condenser plates, leaving behind the dried and cold product. Freeze-dried foods require rehydration (reconstitution) prior to consumption. As long as oxygen and water are excluded by the packaging of freeze-dried products, quality can remain acceptable for decades.

Fermentation. A traditional form of food preservation used in the meat industry is fermentation. When fermentative bacteria are added to meat, they produce acids. These acids contribute to flavor and also decrease the pH of the meat, inhibiting the growth of unwanted microorganisms. Examples of fermented meats include summer sausage and bologna. See Chapter 10, Food Microbiology and Fermentation for more information.

Irradiation. Irradiation, or radurization, is a pasteurization method accomplished by exposing meat to low and medium doses of radiation generated by electron accelerators or by exposure to gamma sources (the Challenge section provides more details). Irradiated meats are typically indistinguishable from unirradiated meats, but they have significantly lower microbial contamination. Irradiation can be performed on packaged meat. Once opened to the atmosphere, the antimicrobial effect of irradiation is lost and the meat is subject to spoilage. The storage life of irradiated, refrigerated meat is greatly extended.

Restructuring. Flaked or ground beef or pork can be reformed into loaves or portions resembling steaks through the process of restructuring. Smoked sliced beef and boneless hams are examples of restructured meat.

Smoking. The practice of smoking meat was originally intended as a preservation method. It has evolved into a method for imparting desirable smoked appearance and flavor characteristics to ham, turkey, and other meats. Smoking is accomplished through the exposure of meat to the natural smoke from burning hickory wood, or the application of the key wood smoke ingredients in liquid form.

Vacuum Packaging. Many spoilage organisms common to meats are aerobic. For this reason meats can be vacuum-packaged to extend the storage life under refrigerated conditions to approximately 100 days. Also, since oxygen can cause off-flavors and odors when in contact with unsaturated meat lipids, vacuum packaging is useful to inhibit the development of rancidity in meat.

Fish Processing

Although fish are cold-blooded vertebrates, fish muscle tissue is biochemically similar to that of mammals. It is lower in connective tissue and contains certain associated problems like endogenous enzymes, free amino acids, and small peptides associated with quality loss. Fish lipids contain high levels of polyunsaturated fatty acids, making them unstable and highly subject to rancidity. Although shellfish usually have high levels of cholesterol, finfish are similar to red meat products in this respect.

Regarding preservation and spoilage of fish, most spoilage is due to bacterial action. One spoilage characteristic found in fish and not in other muscle foods is **trimethylamine** formation, which we examined in Chapter 2. This substance is responsible for the "fishy" smell associated with spoiling fish and is often used as an index of fish quality. Rapid chilling of fish immediately after catching is the most important part of preservation. For freezing to be effective, freshly caught fish must be kept cold and then frozen to −30°C (−22°F) within 2 hours of capture. Since fish caught at sea are often held for several days at only 0°C (32°F) until reaching the market, they are not truly "fresh."

Fish is marketed up to 40 percent as fresh, 30 percent as frozen and canned, 15 percent as cured, and 10–15 percent of fish is sold as comminuted or sectioned and formed products. Figure 8.19 diagrams the steps in canning finfish. **Cured fish** is of three kinds: salted, smoked, and pickled. Salted fish is processed in either dry salt or a brine solution (e.g., dry salted cod). Smoked fish is mildly salted fish (e.g., kippers, which are herring cut

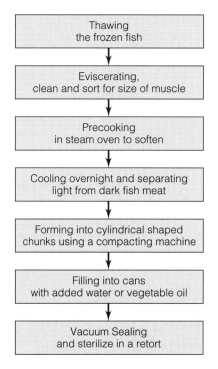

FIGURE 8.19 Steps in the canning of finfish, such as herring, mackerel, salmon, sardines, and tuna. Evisceration means removal of the internal organs (viscera).

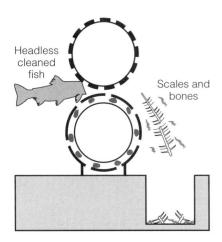

FIGURE 8.20 Specialized deboning equipment for obtaining minced fish. *This process is much more efficient than filleting. The equipment contains a belt-driven perforated drum through which fish flesh is forced under pressure. Headless, eviscerated fish pass between a rotated crusher roll and a perforated cylinder or drum. The minced flesh collects within the drum, while fish waste (scales, bones) is collected at the output stage of the device.*

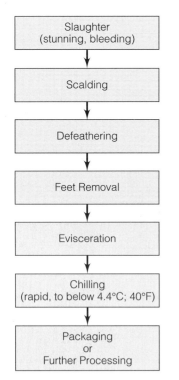

FIGURE 8.21 Steps involved in poultry processing. *(Evisceration means removal of the viscera, or internal organs, sometimes called "entrails").*

lengthwise in half) that receives a smoke treatment. Pickled fish is typically herring cured in a brine containing spices and vinegar.

We will briefly examine two types of processed fish products: fish sticks and minced fish, such as surimi. In the fish industry, fish stick processing consists of creating rectangular three-dimensional sticks from frozen fish meat. The sticks are individually covered with a breaded batter, placed into a continuous fryer, cooked, and then cooled, packaged, blast frozen, and shipped to retail markets.

Surimi is a popular form of processed fish flesh, or mincemeat, to which sugar or sorbitol is added as a *cryoprotectant* to extend shelf life. To produce surimi, a sequence of multiple washings is essential. After harvest, fish heads are removed and the body cavities are thoroughly cleaned and washed. Processing occurs by passing fish through sophisticated deboning equipment that yields the white and sticky mincemeat—see Figure 8.20. The extreme pressure of the process separates the choicest meat from the rest of the carcass. The consistency of the end product, which ranges from large flakes to a smooth paste, depends on the amount of pressure and shear action present. Processors apply heat to the minced meat to make chunk or composite-molded products. The meat can be formed and colored to resemble a variety of seafood meats, like crab meat.

Poultry Processing

Poultry refers to fowl, that is, birds (chicken, duck, turkey), and Figure 8.21 outlines the processing steps for whole birds. The poultry carcasses are then sold at retail or processed into products by the poultry processing industry. Most chicken is marketed fresh. Fresh chicken products are normally packaged in an oxygen-impermeable film. The limited shelf life for poultry is due to microbial spoilage and oxidation of unsaturated fats (autoxidation). Controlled atmosphere storage can be used to extend the shelf life to 20 days. Over the past couple of decades, consumption of poultry in the United States, especially chicken, has increased significantly.

Poultry muscle tissue has different characteristics than red meat, and poultry processors must keep these in mind during processing and cooking. A key distinction is the higher white muscle fiber content in poultry, which is related to a reduced myoglobin pigment content in poultry. As a result, in curing, there is less cure color development in poultry. Commercial cooking of chicken is low-temperature, long-time (LTLT) to minimize quality loss and to maintain palatability. The USDA requires an internal cook temperature of 71°C (160°F) for uncured poultry and 68°C (154°F) for cured poultry.

Processed chicken products include cured whole birds or parts. Mechanically deboned chicken (MDC) can be processed into chicken loaves or rolls by binding meat pieces together in a manner similar to that used to make comminuted red meat loaves. Incorporation of chicken or turkey meat into comminuted meat products is permitted by the USDA. The agency allows up to 15 percent poultry meat content in comminuted red meat products like frankfurters and bologna.

> MDC, which has a fresh storage life of up to 5 days and a frozen storage life of 10–12 weeks, has many uses:
>
> canned chicken soup
> chicken breast patties
> chicken nuggets
> frozen meals
> luncheon meats
> chicken pot pies
> chicken rolls
> chicken roasts
> chicken tenders

Key Points

- Unit operations, including materials handling, cleaning, separating, and mixing, are categories of common food processing operations in practice in the food industry, which correspond to a broad variety of products and applications.
- The principles of food processing to achieve preservation are based upon controlling microorganisms and include moisture removal, heating, cooling, acidity control, traditional nonthermal processing, and innovative thermal processing.
- Heat transfer refers to movement of heat energy from a heat source to food particles.
- Thermal processing techniques include blanching, canning, and pasteurization while traditional nonthermal processing methods rely upon preservatives and packaging.
- Thermal methods rely on D values, TDT plots, and the 12D concept to calculate processing times and temperatures to eliminate microbial survivors in foods.
- Innovative nonthermal processing methods are oscillating magnetic fields, pulsed electric fields, pulsed light, high-pressure processing, and irradiation.
- The standard method for milk pasteurization uses high-temperature, short-time processing, for instance, 161°F for 15 seconds.
- Egg processing methods produce liquid, frozen, dried, and specialty egg products.
- Meat quality depends upon its fat, muscle, and connective tissue composition and upon changes that affect color and tenderness, such as rigor mortis and oxidation.
- Processed meats can be formulated with salt, phosphates, water, and other ingredients, and can be comminuted and restructured into novel products.

Study Questions

Multiple Choice

1. The conversion of raw animal and plant tissue into forms that are convenient and practical to consume is called
   a. food preservation.
   b. disintegration.
   c. food processing.
   d. food engineering.
   e. unit operations.

2. Commercially sterile foods
   a. contain living pathogens.
   b. can never spoil.
   c. are heated to the equivalent of 121°F.
   d. have a 2-year shelf life.
   e. need refrigeration.

3. Which is *true* regarding sterilization and blanching?
   a. They are used to produce commercially sterile foods.
   b. Sterilization is a less severe heat treatment.
   c. Meats are typically blanched and vegetables are sterilized.
   d. They are both forms of thermal processing.
   e. Freezing done prior to blanching and sterilizing reduces pathogens.

4. The main use of iodophores is in
   a. chemical preservation.
   b. emulsification.
   c. surimi processing.
   d. acidity control.
   e. sanitizing.

5. Agitating, emulsifying, and blending
   a. are separate unit operations.
   b. are sanitation methods.
   c. are methods to extend shelf life.
   d. eliminate microbes.
   e. represent the same unit operation.

6. The pressing of peanut oil from peanuts is an example of the _____ unit operation.
   a. forming
   b. sanitizing
   c. disintegrating
   d. separating
   e. pressing

7. Foods are processed because
   a. foods may contain biologically active water.
   b. raw foods are perishable.
   c. foods are contaminated.
   d. both (a) and (b).
   e. choices (a), (b), and (c).

8. Technology differences between countries
   a. would impact the unit operations of their respective food industries.
   b. would change the effect of thermal processing on microbes.
   c. would change the stability of foods with respect to pH.
   d. would have no effect on their ability to feed their respective populations.
   e. would not cause greater needs for preservation in some countries over others.

9. In a TDT plot, as spore concentration increases,
   a. kill time and temperature increases.
   b. kill time and temperature decreases.
   c. kill time and temperature does not change.
   d. survival decreases.
   e. survival decreases as temperature decreases.

10. Nitric oxide acts as a _____ to accelerate color development in cured meats.
    a. cryoprotectant
    b. reductant
    c. antioxidant
    d. bactericide
    e. promotor

11. Identify the *incorrect* pairing:
    a. liquid egg, egg whites
    b. frozen egg, sugared egg yolks
    c. specialty egg, egg yolk solids
    d. dried egg, salted egg yolks
    e. specialty egg, egg substitute

12. Which condition is common to meats that are frozen before rigor mortis is established?
    a. cold shortening
    b. thaw rigor
    c. prerigor
    d. PSE
    e. BSE

13. Comminuted meat
    a. plays a role in sausage processing.
    b. is a preservation technique.
    c. may contain chicken meat.
    d. both (a) and (b).
    e. both (a) and (c).

14. Identify the correct pairing:
    a. acetic acid, cheese
    b. nitrite, vegetables
    c. sulfur dioxide, meats
    d. propionate, bread
    e. benzoate, jams

15. The procedure used to heat sealed cans to destroy bacterial spores is called
    a. rendering.
    b. retort processing.
    c. convection.
    d. hermetic sealing.
    e. decimal reduction.

16. Which is the *incorrect* pairing?
    a. biological change, evaporation
    b. biological change, presence of microbes
    c. chemical change, lipid molecule oxidation
    d. physical change, phase separation
    e. physical change, moisture loss

17. The ability of an organism to survive thermal processing is properly termed
    a. heat tolerance.
    b. decimal reduction.
    c. heat resistance.
    d. TDT.
    e. D value.

18. Adding _____ International Units of vitamin D to milk is a form of _____ .
    a. 400, enrichment
    b. 1,000, fortification
    c. 400, fortification
    d. 1,000, enrichment

19. Which processing step is *not* also a preservation step?
    a. Add calcium propionate.
    b. Stuff sausage meat into casing.
    c. Apply 121°C for 7 minutes.
    d. Curing.
    e. Apply 1 kGy of irradiation to strawberries.

20. Which statement is *false*?
    a. N gas can replace oxygen in a MAP.
    b. IQF processing often utilizes carbon dioxide cryogenic equipment.
    c. Flash pasteurization is an example of a HTST treatment.
    d. High acid foods have pH values > 4.6.
    e. THBQ is a phenolic antioxidant and antimicrobial.

21. Freeze-dried meats have virtually all of their moisture removed (as ice) via sublimation (vaporization of solids), while retaining much of their original shape.
    a. This is a true statement as written.
    b. The statement is false because sublimation does not occur as stated.
    c. The statement is false because they gain moisture during the freeze-drying process.
    d. The statement is false because freeze-dried meats do not retain their original shape.
    e. The statement is false because meats cannot be freeze-dried.

22. Traditional nonthermal processing methods
    a. are used for food preservation.
    b. include OMF, which applies a uniform high pressure to kill microbes.
    c. include HPP, which applies a magnetic field to kill microbes.
    d. include FEP, which applies a burst of high voltage to kill microbes.
    e. may inactivate microorganisms, and probably enzymes as well.

23. Dissociated hydrolysis products of braised meat are primarily
    a. glycogen molecules.
    b. gelatin molecules.
    c. intact collagen molecules.
    d. insoluble proteins.
    e. free amino acids.

24. Certain manufactured cheese spreads are shelf stable (no spoilage) at room temperature—why?
    a. They probably contain antimicrobial additives.
    b. The temperature, pH, and water activity of the cheese provide a hurdle effect.
    c. The product was pasteurized and sealed in an airtight package.
    d. choices (a) and (c)
    e. choices (a), (b), and (c)

25. D values written as D_{121} indicate
    a. the specific temperature required to kill 90 percent of a population of organisms.
    b. that thermophilic spores are present at 121°C.
    c. the specific time required to kill 90 percent of a population of organisms heated to 121°C.
    d. the specific temperature (121°C) required to kill 100 percent of a population of organisms.
    e. the specific time (121 seconds) required to kill 90 percent of a population of organisms.

Short Essay

26. Grinding is a common meat processing step. Provide a detailed explanation of the possible physical and microbiological consequences of meat grinding.

27. Explain what sodium ascorbate and sodium erythorbate are, and how they might be useful in meat processing.

28. Are cold storage and freezing of meats the same thing?

29. Study the TDT graph in Figure 8.7, which shows time and temperature combinations required to kill three spore concentrations. Which spore concentrations are destroyed and which survive at each of the points A, B, C, D, and E?

References

Andrews, L. S., Ahmedna, M., Grodner, R. M., Liuzzo, J. A., Murano, P. S., Murano, E. A., Rao, R. M., Shane, S., and Wilson, P. W. 1998. Food preservation using ionizing radiation. *Review of Environmental Contamination and Toxicology.* 154:1.

Brody, A. 1996. Integrating aseptic and modified atmosphere packaging to fulfill a vision of tomorrow. *Food Technology* 50(4):56.

Cooksey, K. 2000. New technologies: portable packaging. *Food Product Design* 10(6):137.

Dunn, J., Ott, T., and Clark, W. 1995. Pulsed-light treatment of food and packaging. *Food Technology* 48 (9): 95.

Garbutt, J. 1997. *Essentials of Food Microbiology.* Arnold, London.

Gibson, R. M. 1995. Fermentation. In: *Physico-Chemical Aspects of Food Processing.* S. T. Beckett, ed. Chapman & Hall. London.

Hegenbart, S. 1992. Shelf-stable products: technology with "keeping quality." *Food Product Design* 2(2):26.

Holdsworth, J. E., and Haylock, S. J. 1995. Dairy products. In: *Physico-Chemical Aspects of Food Processing,* S. T. Beckett, ed. Chapman & Hall. London.

James, S. J. 1995. Freezing and cooking of meat and fish. In: *Physico-Chemical Aspects of Food Processing,* S. T. Beckett, ed. Chapman & Hall. London.

Kuntz, L. A. 1994. In the can: a look at retorting. *Food Product Design* 4(8):106.

López-González, V., Murano, P. S., Brennan, R.E., and Murano, E.A. 2000. Sensory evaluation of ground beef patties irradiated by gamma rays versus electron beam under various packaging conditions. *Journal of Food Quality* 23:195.

Marshall, R.T., and Arbuckle, W. S. 1996. *Ice Cream.* Chapman & Hall, New York.

Mermelstein, N. 1998. Interest in pulsed electric field processing increases. *Food Technology* 52(1):81.

Meyer, R. S., Cooper, K. L., Knorr, D., and Lelieveld, H. L. M. 2000. High pressure sterilization of food. *Food Technology* 54(11):67.

Murano, E. A., Murano, P. S., Brennan, R. E., Shenoy, K., and Moreira, R. G. 1999. Application of high hydrostatic pressure to eliminate *Listeria monocytogenes* from fresh pork sausage. *Journal of Food Protection* 62:480.

Additional Resources

Charley, H., and Weaver, C. 1998. *Foods: A Scientific Approach.* Prentice Hall, Upper Saddle River, NJ.

Edwards, M. C. 1995. Change in cell structure. In: *Physico-Chemical Aspects of Food Processing,* S. T. Beckett, ed. Chapman & Hall. London.

Eeles, M. F. 1995. Sugar confectionery. In: *Physico-Chemical Aspects of Food Processing,* S. T. Beckett, ed. Chapman & Hall. London.

Ennen, S. 2001. Irradiation plant gets USDA approval. *Food Processing* 62(3) 90-92.

Guy, R. C. E. 1995. Breakfast cereals and snackfoods. In: *Physico-Chemical Aspects of Food Processing,* S. T. Beckett, ed. Chapman & Hall. London.

Guy, R. C. E. 1995. Cereal processing: The baking of bread, cakes and pastries, and pasta production. In: *Physico-Chemical Aspects of Food Processing,* S. T. Beckett, ed. Chapman & Hall. London.

Herrington, T. M., and Vernier, F. C. 1995. Vapour pressure and water activity. In: *Physico-Chemical Aspects of Food Processing,* S. T. Beckett, ed. Chapman & Hall. London.

Kindle, L. 2000. refrigeration systems. *Food Processing* 61(9):116.

Lechowich, R. 1993. Food safety implications of high hydrostatic pressure as a food processing method. *Food Technology* 47(6):170.

McWilliams, M. 1997. *Foods: Experimental Perspectives.* Prentice-Hall, Upper Saddle River, NJ.

Murano, E. A. 1995. Irradiation of fresh meats. *Food Technology* 49:52.

Murano, P. S., Murano, E. A., and Olson, D. G. 1998. Irradiated ground beef: sensory and quality changes during storage under various packaging conditions. *Journal of Food Science* 63:548.

Potter, N. M. and Hotchkiss, J. H. 1995. *Food Science*. Chapman & Hall. New York.

Street, C. A. 1995. Multi-component foods. In: *Physico-Chemical Aspects of Food Processing*, S. T. Beckett, ed. Chapman & Hall. London.

Vaclavik, V. A. 1998. *Essentials of Food Science*. Chapman & Hall, New York.

Vieira, E. R. 1996. *Elementary Food Science*, 4th ed. Chapman & Hall, New York.

Zind, T. 1999. The coming of carbon dioxide. *Food Processing* 60(2):81.

Web Sites

www.ift.org/divisions/nonthermal/
IFT Nonthermal Processing Division

http://www.foodprocessing.com/
Food Processing Magazine

horticulture.tamu.edu/newsletters/food/tfpwb599.p65.html
Texas Food Processor, Texas Agricultural Extension Service

www.fpmsa.org/
Food Processing Machinery and Supplies Association

www.preparedfoods.com
Prepared Foods Magazine

www.gov.ns.ca/nsaf/elibrary/archivelives/poultry/laypull/further.htm
Nova Scotia Agriculture and Fisheries: Further Egg Processing

www.foodonline.com/
Food Online: Virtual community for the food processing industry

www.highpressure.org.uk/
High pressure food processing (UK): industry and academia technology transfer

www.commerce.ca.gov/california/economy/profiles/food.html
The California Food Processing Industry

www.agnic.org/cc/d_q100.html
Agriculture Network Information Center. This site has general descriptions of food processing industries, equipment and processing techniques of food and drink manufacture.

Challenge!

FOOD IRRADIATION

The focus of food preservation is to maintain the freshness and nutrient quality of foods by preventing spoilage caused by the microorganisms that contaminate food. Food irradiation is the application of ionizing energy to foods in order to preserve them. A side benefit is that pathogenic organisms are also destroyed, which means no possibility of foodborne illness from irradiated foods.

In the early 1950s, as part of President Eisenhower's "Atoms for Peace" program, the Atomic Energy Commission (now part of the U.S. Department of Energy) examined the promise of irradiation. This research, and studies performed in other countries, established that the most important benefit from irradiation was this control of pathogenic microorganisms and that a practical and effective dose depended on the particular food application. In 1980, an expert committee of scientists formed by the World Health Organization, the Food and Agriculture Organization, and the International Atomic Energy Agency designated foods irradiated up to 10 kGy as safe. More significant is a statement issued in 1997, based upon a decade and a half of additional studies, in which this committee declared that foods irradiated at any dose are safe.

To consider the practical application of this technology, two very important questions must be addressed: (1) Is it safe to produce and safe to consume irradiated foods? (2) Does the quality of food change appreciably when it is treated with irradiation? Before pursuing answers, it's a good idea to learn some basic definitions.

What Is Irradiation?

We must distinguish between several similar sounding terms: radiation, irradiation, and food irradiation. **Radiation** is the emission and propagation of energy through matter or space by electromagnetic disturbances called photons. These photons have associated with them units of energy measured in electron volts. **Irradiation** is the process of applying radiation to matter.

Food irradiation refers to the application of radiation, as ionizing energy, to foods. It is a nonthermal processing method that offers a means of preservation that destroys microorganisms without heating the food material. For this reason it has been referred to as a "cold pasteurization" process. Foods can be irradiated within their packaging and remain protected against postprocessing contamination until opened by users. Pasteurization doses do not render the product sterile but do extend shelf life, without "cooking" the food.

Ionizing energy is simply a form of energy in the electromagnetic spectrum, for example, an Xray generates ionizing energy. The electromagnetic spectrum is an organized scale of electromagnetic radiation, which includes gamma radiation, radio waves, visible light (Figure 6.1), and microwaves based upon frequency, wavelength, and energy value.

Sources of Radiation Used in Food Irradiation. Food irradiation involves the use of either high-speed electrons from an *e-beam* electron accelerator or high-energy gamma ionizing energy such as that given off by radioisotopes Co-60 (radioactive cobalt) or Cs-137 (radioactive cesium). Cesium-137 irradiators, like those for cobalt-60 and e-beam, are intended to treat fully processed and packaged foods.

Electron beam (e-beam) technology uses high-energy electrons to destroy harmful microorganisms within seconds by penetrating products in their final shipping packaging. Pathogen reduction through electron beam technology is limited by the thickness and density of the treated product. With e-beam technology, ground beef can be effectively treated at a depth of at most 4 inches in its final retail package.

Factors that influence the application of e-beam or gamma for particular foods include product characteristics and process factors. Process parameters of importance include source of ionizing energy, the dose and dose rate, and packaging environment, while product characteristics include product temperature, density, thickness, and composition. For the same level of energy, gamma rays can penetrate approximately 3 feet of material having the density of water, which is greater penetrating power than high-speed electrons. But e-beam technology is fast and efficient, taking only 2 minutes of processing for what may take 20 minutes with gamma or Xrays.

Dose and Dose Rate. Ionizing energy processes create enough of an absorbed dose to destroy microorganisms. The unit of absorbed dose in food is given in kilogray (kGy) units, equal to 1000 grays. One gray is equivalent to the absorption of 1 joule of energy by 1 kilogram of food. The dose of irradiation is divided into three levels: (1) **radicidation** is less than 1.0 kGy and is called a low dose, (2) **radurization** is a dose of 1 to 10 kGy and is called a medium dose, and (3) **radapperization** is a dose above 10 kGy, such as 20 to 30 kGy, which can sterilize a food product, and is called a high dose. The foods consumed in the space program by astronauts are radapperized, but consumer foods have only been approved for low- and medium-dose levels.

When ionizing energy makes contact with food, it penetrates the surface and a portion of its radiation energy is absorbed by the food. This absorbed dose increases with exposure time, as more ionizing energy collides with food matter. The idea of **dose rate** is the dose accumulated per unit of time. Gamma sources provide slower dose rates of approximately 1,000 Gy per hour, while e-beam accelerators provide a fast dose rate of about 1,000 Gy per second. Although scientists have speculated that dose rate (hence radiation source) affects the microbial and sensory aspects of foods differently, this has not been unequivocally demonstrated.

As in thermal processing, the irradiation of food microorganisms produces a D value—equivalent to the dose of radiation required to kill 90 percent of the population of microorganisms in a food. The water activity of a food has a considerable influence on food irradiation efficiency. Microorganisms present in drier (lower a_w) foods are more resistant to irradiation than those occurring in high a_w foods. For example, the D value for *Salmonella typhimurium* in fresh poultry meat is about 0.5 kGy, while in frozen poultry the D value increases to 1.7 kGy.

Does the Food Become Radioactive?

The doses of energy absorbed by irradiation-processed foods do not make them radioactive. To become radioactive, food would need to be exposed to a minimum of 15 MeV of energy. This is a lot of energy. The energy possessed by an electron is called an **electron volt** (eV). One eV is the amount of kinetic energy gained by an electron as it accelerates through an electric potential difference of 1 volt. It is usually more convenient to use a larger unit such as megaelectron volt (MeV) when discussing food applications, and 1 MeV is equal to one million electron volts. Energy output Co, Cs, and e-beam generators is carefully regulated, with the maximum energy outputs of 5 or 10 MeV being too low to induce radioactivity in foods.

Although many consumers know that food irradiation cannot make foods radioactive, they may be unaware that all foods are naturally radioactive. Physicists have calculated the amount of naturally occurring radioactivity in the unirradiated foods consumed daily to be on average 100 bequerel units of radioactivity. We are therefore exposed to a certain low background level of radioactivity, because about 0.01 percent of the natural elements found in foods (such as carbon, phosphorus, potassium, and sulfur) exist in their radioactive isotope forms in whole foods.

Perhaps even more surprising—a person consuming a diet of only irradiated foods would be exposed to less background radioactivity than a person consuming fresh, unirradiated foods. This apparent contradiction occurs because irradiated foods can be stored for long periods of time, while fresh foods cannot. During the longer storage of an irradiated food, the natural radioisotopes have time to undergo decay, at a rate dependent upon their specific half-life, into unradioactive elements. Therefore, when the consumer finally eats the previously stored irradiated food item, it contains less of its pre-storage natural background radioactivity (and zero radioactivity due to irradiation processing). Over time, the body tissue of such an individual would incorporate less and less natural radioactivity, and would have a lower background radioactivity associated with it.

What Does Irradiation Do to Microorganisms?

Irradiation has both direct and indirect effects on biological materials. The direct effects are due to the collision of the photons of radiation with the atoms in the molecule of living matter. The indirect effects are due to the formation of free radicals (unstable molecules carrying an extra electron) during the radiolysis (radiation-induced splitting) of water molecules.

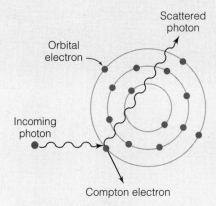

FIGURE 8.22 The Compton Effect. *Incoming high-energy photons create the Compton effect when they interact with orbital electrons of matter, such as those of microorganisms or food molecules.*

Direct Effects. Ionizing radiation kills microorganisms by damaging the biomolecules of their cells. Incoming photons of radiation hit electrons in the atoms of the microorganisms or food molecules. In the collision, some of the incoming photon's energy is transferred to the electron, changing the photon's direction ("scattering"), which is the *Compton effect*. In addition, and particularly significant, the electron, (called the Compton electron) is ejected from the matter and is free to collide with neighboring electrons. Electron ejection in atoms and molecules breaks chemical bonds, since bonding is the transfer or sharing of orbital electrons. It is believed that only one out of every six billion chemical bonds in bacteria or food molecules are broken by irradiation. Although this is too few to cause appreciable changes in foods, in a molecule of bacterial DNA, even a low magnitude of damage such as this interrupts normal cell metabolism and division. (See Figure 8.22.)

In particular, the radiation energy causes nicks or breaks in the chemical bonds that hold the key organic biomolecules together. These key biomolecules are the nucleic acids (DNA, RNA) and cell proteins (enzymes and cell membrane proteins) of the microorganisms. DNA is a double-stranded molecule of nucleotides. The base pair of the nucleotides are the rungs of the DNA helix. The intramolecular (interatomic) radiation damage can break the bonds connecting the base pairs, impairing cellular biochemistry and metabolism, and cell death results. See Figure 8.23.

Pathogenic bacteria that are able to produce resistant endospores in foods such as poultry, meats, and seafood can be eliminated by radiation doses of 3 to 10 kGy. If the dose of radiation is too low, then damaged DNA can be repaired by specialized enzymes. If oxygen is present during irradiation, the bacteria are more readily damaged. Doses in the range of 0.2 to 0.36 kGy are required to stop the reproduction of *Trichinella spiralis* (the parasitic worm that causes trichinosis) in pork, although much higher doses are necessary to eliminate it from the meat.

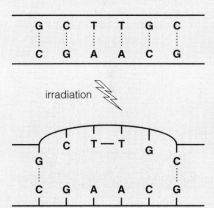

FIGURE 8.23 Direct effects of ionizing energy. *Targets are the nucleic acids and cell proteins of microorganisms. The base pairs shown include G-C (guanine-cytosine), T-A (thymine-adenine).*

Indirect Effects. Irradiation also exerts indirect effects on microorganisms that destroy them. Radiation energy absorbed into a food does not make it radioactive. What does happen, though, involves the radiolysis of water molecules to produce free radicals (see Figure 8.24). The term **free radical** refers to an atom or molecule having an unpaired electron. These free radicals are highly reactive and quickly form stable products—they can combine with one another or with oxygen molecules to produce powerful oxidizing agents that can damage bacterial cell components. These unstable free radicals quickly react with bacterial cell membranes to change or damage their structure, resulting in bacterial death. Long-term testing in laboratory animals has confirmed that consuming free radicals does not create a toxicological or otherwise harmful burden. Unirradiated bread toast actually contains more free radicals than some very dry irradiated food products, like spices.

Pathogenic microorganisms can be destroyed by the combined direct and indirect effects of irradiation. Thus, irradiated foods are safe from pathogens and represent value-added products.

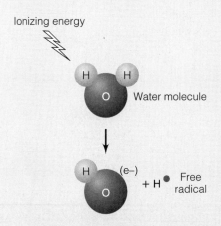

FIGURE 8.24 Indirect effects of ionizing energy. *The production of free radicals from water molecules in foods*

How Does Irradiation Affect Food Quality?

Having just mentioned the production of free radicals and their effect on microorganisms, a good question is: Do they also affect food molecules? The answer is yes, but since only a small fraction (one out of every six billion) of a food's molecules are affected by free radicals, irradiated foods remain wholesome and of high quality. However, let's examine the changes to food molecules when they are affected by free radicals. The food molecules of importance are water, lipids, proteins, carbohydrates, and vitamins. Radiation energy generates a degradative reaction when it interacts with food molecules, and this is called radiolysis. The products of radiolysis are called *radiolytic products*. Researchers have studied the effects of various doses of irradiation upon both food macronutrients (water, lipids, carbohydrates, and protein) and micronutrients (vitamins and minerals).

Food water. The radiolysis of water molecules produces hydroxyl radicals, a highly reactive species that interacts with the organic molecules present in foods. The products of these interactions cause many of the characteristics associated with the spoilage of food, such as off-flavors and off-odors. Strategies used to minimize negative effects include (1) apply lowest effective irradiation dose, (2) irradiate at low temperature, (3) choose appropriate packaging in terms of moisture and oxygen barrier properties.

Food lipids. In the absence of oxygen, radiolysis of lipids leads to cleavage of the interatomic bonds, producing compounds such as carbon dioxide, alkanes, alkenes, and aldehydes. In the presence of oxygen, lipids are highly vulnerable to oxidation by free radicals, a process that yields peroxides, carbonyl compounds, and alcohols. The consequent rancidity, resulting from the irradiation of foods containing a high unsaturated fatty acid content, negatively impacts their sensory quality. To minimize such harmful effects, foods are vacuum-packaged and held at very low temperatures during irradiation. Low temperature is effective because it drastically slows molecular motion and therefore inhibits the progress of chemical reactions. Other strategies include formulating with lipid-soluble antioxidants and using the lowest irradiation dose possible.

Food proteins. Proteins are not significantly degraded at the low doses of radiation employed in the food industry. For this reason irradiation does not inactivate enzymes involved in food deterioration, as most enzymes survive doses of up to 10 kGy. The biological value of protein in irradiated foods remains high. Availability of essential amino acids is not compromised.

Food carbohydrates. On the other hand, large carbohydrate molecules (as food polysaccharides) are broken down by irradiation. This depolymerization reduces the gelling and other functional properties of long-chain polysaccharide molecules such as starches and gums. However, in most foods some protection against these deleterious effects is provided by other food constituents. Also, the effect of irradiation on the nutritional and energy role of simple sugars is negligible.

Food vitamins. The vitamins have varying degrees of sensitivity to irradiation. Vitamins A, C, E, and B_1 (thiamin) have been shown to be sensitive to irradiation, especially at higher doses. This sensitivity is most apparent in foods packaged under air (aerobically). Therefore, the nutritional losses for a food product can be high if air is not excluded during irradiation processing at higher dose levels. It is good to recall, though, that conventional heat processing causes significant vitamin loss too.

The bottom line with respect to the effect of food irradiation on food nutrients is that irradiation causes changes to food molecules, particularly at high levels (above 10 kGy). At sterilization levels (on the order of 50 kGy) not only are nutrient losses a problem, but undesirable texture, flavor, and color changes occur. A few foods, like tomatoes, do not respond well to the process even at approved doses because of excessive softening. But the vast majority of foods for which the process has been approved hold up very well nutritionally and functionally when certain precautions, such as using the lowest possible dose, avoiding the presence of oxygen, and irradiating in the frozen state if possible, are taken.

Does Irradiation Create Unique Radiolytic Products?

The answer to this question turns out to be rather simple: No. **Radiolytic products** are unstable atoms or molecules derived from substances naturally present in foods treated by ionizing energy. A free radical is one such type of radiolytic product. When the technology was new, consumer activist groups were concerned that "unique" radiolytic products might be formed as a result of irradiation processing and that these might pose specific health threats, such as cancer.

However, no unique radiolytic products have been found in irradiated foods, although scientists have been searching for them for years using sophisticated, sensitive analytical equipment. Radiolytic products are produced by irradiation, but these products are not unique to this form of processing. They are the same chemical species produced in foods by other conventional processing methods, and their occurrence in irradiated foods, in concentrations on the order of parts per million, has been shown to be less than those formed in conventionally heat-processed foods, such as canned foods, toasted bread, and deep-fat-fried chicken.

One point more: the issue of being able to detect whether a food has received irradiation treatment. Several experimental methods, such as those based on ESR (electron spin resonance), detection of DNA damage in killed microorganisms, and detection of radiolytic products have been developed. However, the food industry is still lacking a definitive approach that could offer a simple, rapid test to unequivocally demonstrate that detection of a specific endpoint determinant in a food was caused by the irradiation process itself and not some other factor.

Are Irradiation Facilities Safe?

Facilities worldwide generate the kind of energy required to irradiate foods, including gamma (as either Co-60 or Cs-137) and electron beam (e-beam) facilities. The majority of such facilities are the Co-60 type, and both food and nonfood (especially medical and surgical equipment) products can be irradiated. When anthrax-tainted letters were sent through the mails, all of the mail at those postal sorting and processing centers was irradiated to kill any remaining anthrax.

The cobalt for a gamma facility is transported in lead-shielded containers specially designed not to damage or break despite any accident en route. This has resulted in an excellent safety record for close to half a century, with no incidents of radiation hazard in North America despite the transport of over 1 million shipments of Co-60.

Risk to employees at an irradiation plant is addressed and prevented by the adoption of in-house safety control measures. These include adequate shielding of the irradiation source (in the case of cobalt, a pool of water is used to house the radioactive cobalt, because water, at a depth of 9 or more feet, offers an effective shield), and the installation and operation of a standard failsafe system design. Laws and regulations have been enacted since 1960 to establish limits and standards for facilities and procedures with respect to irradiating foods. Facilities are subject to regular inspections, audits, and reviews to ensure compliance.

Can there be a Chernobyl-type disaster at a food irradiation plant? No—there is zero chance of that kind of nuclear meltdown scenario because the food irradiation source cannot generate neutrons, the specific subatomic particle required to create the chain reaction for a meltdown to occur. Disposal of radioactive waste is also not an issue, since no radioactive material is produced at a food irradiation facility. There is therefore zero net accumulation of radioactivity in the form of waste.

FIGURE 8.25 The Radura Symbol.
The FDA requires the symbol on the packaging of irradiated foods.

FIGURE 8.26 A comparison of unirradiated strawberries and those irradiated at 1.0 kGy. *In approving a food application of radiation, such as for fruits, the FDA sets the maximum irradiation dose the product can be exposed to, measured in kiloGray (kGy) units.*

Regulation of Irradiated Foods

Although irradiation is a nonthermal process, it is considered a food additive by the FDA and regulated as such. As part of the approval for any irradiated food, the FDA requires that the foods include labeling with either the statement "treated by ionizing energy" or "treated by irradiation" and the international symbol for irradiation, the **radura** (Figure 8.25). Irradiation labeling requirements apply only to foods sold in stores to consumers, for example, irradiated fresh strawberries. When used as minor ingredients in other foods, however, the label of the other food does not need to describe these ingredients as having been irradiated. For example, if irradiated spices are used in soup, as minor ingredients, the soup does not need to have an irradiation label. Irradiation labeling also does not apply to restaurant foods.

The FDA approved the first use of irradiation on a food product in 1963 when it allowed radiation-treated wheat and wheat flour to be marketed. In approving a use of radiation, FDA sets the maximum radiation dose the product can be exposed to, measured in kilograys (kGy) (See Figure 8.26). The FDA's 1997 red meat approval (jointly approved by the USDA in December 1999) and 2000 shell egg approval added two more product categories to the list of foods the agency has approved for irradiation since 1963. These include poultry, fresh fruits and vegetables, dry spices, seasonings, and enzymes—see Table 8.9.

The Future for Irradiated Foods

Recent studies estimating the costs of foodborne illnesses in terms of personal suffering, time away from work, and mortality have created a renewed interest in irradiation technology. Health experts say that in addition to reducing *E. coli* O157:H7 contamination, irradiation can help control the pathogens *Salmonella* and *Campylobacter*, two chief causes of foodborne illness. The Centers for Disease Control and Prevention estimates that *Salmonella*— a pathogen found in poultry, eggs, meat, and milk—causes illness in as many as 4 million and kills 1,000 people per year nationwide. *Campylobacter*, a pathogen found mostly in poultry, is responsible for 6 million illnesses and 75 deaths per year in the United States. A May 1997 presidential report, "Food Safety from Farm to Table," estimates that millions of Americans are stricken by foodborne illness each year and some 9,000, mostly the very young and elderly, die as a result.

Widespread use of irradiation in the food processing industry however has not occurred, despite awareness of these statistics and the FDA approvals. Food processors are concerned about consumer acceptance of irradiated food, about the economics of processing food with existing commercial irradiation systems, and about the logistics of shipping time and costs.

USDA's Agricultural Research Service and private industry have a cooperative research and development agreement to further test and evaluate a commercial size Cs-137

TABLE 8.9 All approved uses of radiation on foods to date, the radiation dose allowed, and the purpose for irradiating the foods.

Food Approved For	Dose (kGy)	Application
Dry or dehydrated enzyme preparations	up to 10	Eliminate microorganisms
Spices	up to 30	Decontamination; eliminate insects and microorganisms
Fresh fruits and vegetables	up to 1.0	Inhibit maturation/sprouting; control insects
Poultry	1.5 – 3.0	Eliminate *Salmonella* and other pathogens
Pork (for *Trichinella*)	0.3 – 1.0	Eliminate *Trichinella*
Red meat beef, lamb, pork, veal, venison	up to 4.5 kGy (fresh) up to 4.7 kGy (frozen)	Eliminate *E. coli* 0157.H7 and other pathogenic and spoilage bacteria
Shellfish (clams, crab, oysters)	petition in progress	Eliminate *Vibrio* and other pathogenic and spoilage bacteria
Shell eggs (consumer eggs)	3.0 kGy	Eliminate *Salmonella enteritidis* and other pathogens

irradiator to be installed at a USDA research facility. More than 40 other facilities nationwide primarily handle sterilization of medical supplies, though these plants may be approved to irradiate food products. In 2001, a facility in Illinois was granted approval for the irradiation of meat and poultry, which potentially will expand the application of this technology to consumer meat and poultry products.

The future of irradiated foods ultimately rests in the hands of the consumer. If consumers express a desire to be able to purchase irradiated foods, producers will listen. Helping to prevent the discomfort, illness, and even deaths of children and others who consume foods such as hamburgers contaminated with *E.coli* is a strong argument for irradiation. And the manner in which irradiated food labeling is presented has been shown to impact consumer attitudes. For example, in a 1998 survey, four of five people ($n = 1,003$) indicated that they would be "somewhat likely" or "very likely" to buy food products for themselves or their children if labeled "irradiated to kill harmful bacteria."

The safety and quality aspects of irradiated food must continue to be communicated clearly and honestly to consumers. Such consumers then become educated regarding the pros and cons of irradiated foods. Their freedom to choose, and the opportunity to choose, should be championed by the food processing industry.

Challenge Questions

1. In the United States, irradiation is considered to be an additive; which specific regulation is it subject to?
2. Explain the difference between dose and dose rate.
3. Suppose you work at a poultry processing plant, and the boss wants to irradiate the chicken meat, which he knows is on average 30 percent contaminated, to eliminate *Salmonella*. He tells you "let's use as low a dose as possible; I know that approval is for 1.5-3.0 kGy, but just because its approved, we don't have to go that high. Give it 1.0 kGy of treatment." The plant produces both fresh (refrigerated) and frozen chicken meat and chicken meat products. What should your response be?
4. Why do you think that the irradiation of shellfish, particularly oysters, would be effective in preventing unwanted foodborne disease?

Multiple Choice

5. The process of applying radiation to matter is called
   a. irradiation.
   b. ionizing energy.
   c. Compton effect.
   d. food irradiation.
   e. radiation.

6. A medium irradiation dose is
   a. radurization.
   b. radicidation.
   c. between 1 and 10 kGy.
   d. both (a) and (c).
   e. both (a) and (b).

7. The irradiation D value for *Clostridium botulinum* spores is 3.5, while for *E. coli* the D value is 0.2; therefore what is the conclusion?
   a. Irradiation is ineffective against these organisms.
   b. 90 percent of *C.botulinum* spores are killed with 3.5 kGy of irradiation dose.
   c. *E. coli* is more resistant to irradiation processing.
   d. When adjusted for temperature, both microorganisms have the same D value.
   e. Both must be gram-positive organisms.

8. Which is *incorrect?*
   a. Irradiation can cause vitamin loss in foods.
   b. Key targets of ionizing energy in microorganisms are DNA, RNA, enzymes, and other proteins.
   c. Irradiated foods do not become radioactive.
   d. Irradiation at approved levels does not cause foods to heat up.
   e. Irradiation creates unique radiolytic products in foods not seen with any other processing treatments.

9. How could irradiation best be incorporated into the hurdle concept?
   a. Combine minimal low-dose irradiation and mild thermal processing.
   b. Apply a 20 kGy dose of irradiation and then refrigerate the product.
   c. Apply a low dose in conjunction with a pH control agent or chemical antimicrobial.
   d. Apply a 20 kGy dose of irradiation and then refrigerate the product and apply a low dose in conjunction with a pH control agent or chemical antimicrobial.
   e. Combine minimal low-dose irradiation and mild thermal processing and apply a low dose in conjunction with a pH control agent or chemical antimicrobial.

CHAPTER 9

Understanding Fat, Sugar, Beverage, and Plant Product Processing

9.1 Processing of Fats and Oils

9.2 Sugar Processing

9.3 Beverage Processing

9.4 Processing of Cereal Grains

9.5 Fruit and Vegetable Processing

9.6 Soybean Processing

9.7 Chocolate Processing

Challenge!
Enzymes in Food Processing—
The Protein Hydrolysates

CHAPTER OBJECTIVES

After completing this chapter, you will be able to:
- Provide several examples of tests that assess fat and oil quality.
- Explain the difference between interesterification and fractionation.
- List the sequence of steps required to produce refined sugar from sugarcane.
- Differentiate wet from dry milling.
- Discuss the key processing aspects of bread, pasta, and snack food.
- Explain what is meant by minimally processed fruits and vegetables.
- Discuss the effects of pickling, canning, dehydration, and freezing on fruits and vegetables.
- Outline the basic approaches to produce soy isolates and concentrates.
- Describe the method in which cocoa butter is converted into chocolate.
- Assess the potential for protein hydrolysates to act as functional ingredients.

Did you know it takes more than a hundred hours of slow mixing to make a chocolate bar? In this chapter we continue the focus on processing of specific foods, particularly those of plant origin. We apply the principles of food processing and the unit operations to fats and oils, sugars, beverages, cereal grains, fruits and vegetables, soybeans, and chocolate. The chapter closes with a Challenge section on the use of enzymes in food processing, specifically the manufacture of protein hydrolysates. Taken together, Chapters 8 and 9 give a broad introduction to the topic of food processing and preservation. Many of the facts and concepts in these chapters will be applied later, particularly in Chapters 11, 12, and 15.

9.1 PROCESSING OF FATS AND OILS

In general usage, lipids that have a relatively high melting point and are solid at room temperature are called fats, while those that have lower melting points and are liquid at room temperature are called oils. All food lipids are mixtures of triglycerides, usually with two or three different fatty acids attached to each triglyceride molecule. Because each fatty acid has its own melting point, these triglycerides melt over a range of temperatures rather than at one specific temperature (e.g., a melting range from 120–130°F (49–54°C), vs. a 128°F (53°C) melting point).

At the outset, lipid (fat) *processing* must be distinguished from *refining*. Processing of fats means the removal or extraction of a fat or oil from its natural source, such as a kernel of corn, the carcass of a hog, or a cottonseed. **Refining** refers to the removal of impurities from the extracted fat or oil. When food lipid matter is obtained via pressing and rendering as discussed below, it is referred to as an edible **crude oil.** Crude oils are subjected to a number of commercial refining processes designed to remove compounds that may contribute especially to flavor and color instability (see list in margin). Oxidative stability of oils is generally improved by refining, as are color and flavor. Toxicants such as aflatoxins in peanut oil and gossypol in cottonseed oil are also removed in the refining process. The term *RBD oil* refers to crude oil that has undergone three essential treatments: refining, bleaching, and deodorizing.

A typical solvent extraction system to obtain oil from oilseed kernels consists of the following operations: (1) cleaning to remove dirt and debris, (2) grinding the kernels, (3) steaming (tempering or cooking) of the seed "meat," (4) flaking the small pieces between flaking rolls, (5) extracting the oil with solvent, (6) separating the meal, or *marc*, from the oil-solvent solution (*miscella*), and (7) removing the solvent from both the miscella and the marc. The oil is further refined to improve its stability, while the marc is then desolventized and toasted, and then either ground or pelletized, for use as a food ingredient (e.g., fat-free soy flour) or for use in animal feeds.

A few basic production methods are employed to obtain and refine fats and oils from animal, marine, and plant sources. The following text is a list of common lipid processing techniques; it is not a sequence of processing steps. As you read each description, consider how these fit in as unit operations.

Compounds removed through oil refining which would otherwise contribute to undesirable flavors, odors, and product instability:

aldehydes
free fatty acids
gums
lecithin
ketones
soaps

- **Rendering** is the heating of fatty meat scraps in water, which allows the fat to melt and rise to the surface to be separated from the tissue and water.
- **Pressing** is the mechanical squeezing of oil from oilseeds. Seeds are first cooked slightly to break down the cell structure and to melt the fat for easier release of oil. Seeds may also be ground or cracked, and the expelled oil may be further cleaned by filtering or centrifugation.
- **Solvent extraction** is the separation of oil from cracked seeds using a nontoxic fat solvent such as hexane. Hexane is percolated through seeds in a countercurrent solvent flow process and dissolves the oil. After the oil is extracted, the solvent is distilled off and recovered for reuse. More oil can usually be obtained this way than by pressing alone. However, the practice of solvent extraction with organic solvents is not favored by some health-conscious groups preferring "natural" cold-pressed oils. Nor are extraction procedures that require washing steps desirable, since these introduce wastewater into the processing equation. Wastewater treatment is important to avoid harming the environment, but it can represent a major activity and expense for the food processor.
- **Deodorization** is the application of steam heat in a vacuum chamber to strip away certain odor-causing low molecular weight compounds from oils. This technique is applied not only to fish oils but also to crude canola, safflower, and soy oils.
- **Degumming** is the first refining step for oils such as soybean oil and others high in phospholipids. The process mixes 3 to 5 percent water with the oil at 50–60°C (122–140°F), and separates the hydrated phospholipids by centrifugation.
- **Neutralization** removes free fatty acids from a fat. Caustic soda (a strong alkali solution) is mixed with a heated fat and the mixture is allowed to stand until the aqueous phase settles. This aqueous phase called "foots" or "soapstock" in the industry is separated and used for making soap. The free fatty acids are bound to alkali in the soap. Residual soapstock is removed from the oil by adding water and centrifuging.
- **Bleaching** refers to the removal of colored substances from oil. The process uses diatomaceous earth clays mixed with oil samples at 90°C (194°F). The colored materials (carotenoids, chlorophylls, etc.) adsorb onto clays and activated charcoal. The bleaching medium is removed from the oil by filtration.
- **Hydrogenization** is the process to saturate double bonds and make an oil more solid and more resistant to oxidative rancidity. A by-product of hydrogenation is the production of trans double-bonded fatty acids. Interestingly, linoleic acid, which is a dietary essential fatty acid, becomes biologically inactive when converted to the trans form.
- **Winterizing** refers to a refrigeration treatment of oils for a specific purpose. Without such treatment, some triglycerides composed of saturated long-chain fatty acids with high melting points would crystallize (separate as solids) from the sample. Such crystals are undesirable and cause haze in a product meant to be a transparent, free-flowing liquid. To prevent haze in refrigerated salad oils, corn oils, and others, the process of winterizing subjects the oils to between 0 and 2°C (32–36°F). The crystals that form are then removed from the oils.
- **Plasticizing** refers to softening a hard fat, which changes the fat's consistency. The state of crystallization influences the consistency and functional properties of the more solid fats of the sample. Agitation, heating, and cooling rates all affect the crystallization rate (slow vs. fast) and crystal forms exhibited by fats from smallest to largest (alpha, beta, and others). Rapid cooling favors formation of small alpha crystals (important in candies and frozen desserts). Slow cooling of melted fat favors formation of large, coarse crystals. If solid butter melts and is then cooled, it will have a different crystal structure than the original butter because the crystals will be few in number, large, and of the coarse beta crystal form. These can be reconverted to smaller, fine-grained crystals if it is melted again, and rapidly cooled with agitation.

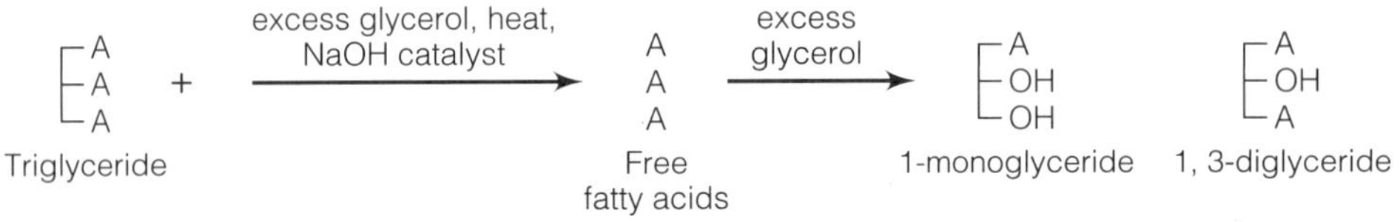

$$\begin{bmatrix} A \\ A \\ A \end{bmatrix} \xrightarrow{\text{excess glycerol, heat, NaOH catalyst}} \begin{matrix} A \\ A \\ A \end{matrix} \xrightarrow{\text{excess glycerol}} \begin{bmatrix} A \\ OH \\ OH \end{bmatrix} \begin{bmatrix} A \\ OH \\ A \end{bmatrix}$$

Triglyceride → Free fatty acids → 1-monoglyceride, 1,3-diglyceride

FIGURE 9.1 The preparation of mono- and diglycerides. *The designation A represents the fatty acid attached to glycerol at specific positions of the glyceride structures.*

- *Mono- and diglyceride preparation* refers to isolating these triglyceride derivatives for use as emulsifiers. Glycerol containing only one or two fatty acid residues rather than three (as with triglycerides) can be prepared from a batch of triglycerides to which excess free glycerol is added. When this mix is heated to 200°C (392°F) in the presence of catalyst, some of the fatty acid molecules will separate (dissociate) from the glycerol molecule and react with the extra glycerol to produce both mono- and diglycerides (see Figure 9.1). Mono- and diglycerides are both hydrophilic and hydrophobic due to their free OH groups on the glycerol and the presence of fatty acid residues, respectively. Thus, they are excellent emulsifiers, since they are soluble in water and oil.

Processing of Specific Fats

In this section, we briefly examine the processing of a highly saturated fat of animal origin, milkfat, to produce a dairy spread product via fat fractionation. We also consider the meaning and application of a technique called interesterification. In Section 9.7, we'll cover the processing of a unique confectioner's fat—cocoa butter—from which chocolate is made.

Milkfat is a mixture of glycerides found in milk, milk products, and butter. It can be processed via **fractionation** to yield butter fractions of varying melting points. Traditional butter right out of the refrigerator will not spread until it warms nearer to room temperature, which is inconvenient. Fat fractionation can be used to produce a lower melting point butterfat that offers improved "spreadability" right out of the refrigerator.

What is the fat fractionation process? During a dry crystallization process, milkfat is heated and melted, and then cooled under controlled conditions. This process generates solid and liquid phases that are easily separated by vacuum or pressure filtration. Fat fractions can include a high melting point fraction that is very hard at room temperature, a traditional butterfat, a lower melting point fraction that is softer than butter, and a very low melting point fraction that is liquid at room temperature (Figure 9.2). Blending these fractions in the right proportions yields a product that is spreadable in the desired temperature range.

Wouldn't it be useful to be able to modify the fatty acid distribution within a fat to produce a more plastic fat? That is the idea behind interesterification. **Interesterification** refers to the removal of fatty acids from the glycerol portion of food triglycerides, such as those occurring in lard, and their subsequent recombination into numerous configurations. These new configurations contain a random distribution of the fatty acids within a glyceride molecule (Figure 9.3). The term esterification is a reminder that fatty acids are attached to glycerol via ester bonds, which break and reform during an interesterification reaction. All of the fat molecules in a melted and esterified sample participate in the reaction.

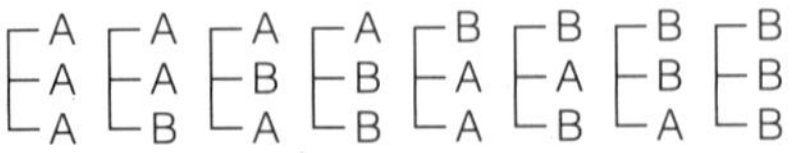

The reason to process fats via interesterification is to create a new fatty acid distribution of glyceride molecules, which in turn affects the melting point and crystallization behavior of the product. Through interesterification, a coarse-crystal, grainy-textured fat like lard can be changed into one with a lower melting point, finer crystal size, more plastic, and smoother texture and mouthfeel. In addition, the process has been used to produce fats with reduced calories, by changing the fatty acid composition of the triglycerides, allowing shorter-chain fatty acids to react with glycerol. This produces triglycerides with short- and medium-chain fatty acids that provide about 5 rather than 9 calories per gram (see discussion of fat-based fat replacers Salatrim and caprenin in Chapter 3).

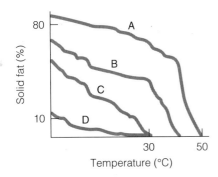

FIGURE 9.2 Comparison between melting behavior and solid fat content of butter (B) and several fractions (A, C, D), displaying different melting point ranges. *These are a function of the percent solid fat content. When the solid fat percentage is high, the product melts at a higher temperature than when the solid fat percentage in the sample is low. Melting behavior affects the ease of spreading butter fractions onto bread.*

$$\begin{bmatrix} A \\ A \\ A \end{bmatrix} \begin{bmatrix} A \\ A \\ B \end{bmatrix} \begin{bmatrix} A \\ B \\ A \end{bmatrix} \begin{bmatrix} A \\ B \\ B \end{bmatrix} \begin{bmatrix} B \\ A \\ A \end{bmatrix} \begin{bmatrix} B \\ A \\ B \end{bmatrix} \begin{bmatrix} B \\ B \\ A \end{bmatrix} \begin{bmatrix} B \\ B \\ B \end{bmatrix}$$

FIGURE 9.3 Random distribution of two different fatty acids, designated A and B, within a single triglyceride molecule.

TABLE 9.1 Tests applied to processed fats.

Test	Purpose
Iodine value	Predicts level of fat unsaturation
Peroxide value	Measures peroxide content due to lipid oxidation
Acid value	Measures free fatty acid content due to lipid hydrolysis
Saponification value	Predicts free fatty acid molecular size
Smoke point	Measures an oil's heat stability

Chemical and Physical Testing Of Fats

Processed fats must be tested to gain information regarding specific food applications, to measure deterioration (e.g., rancidity), to check performance vs. purchase expectations, and to examine physical properties. Table 9.1 gives the tests applied to processed fats and their purposes, and discussion of these tests follows.

Iodine value is used to measure the degree of unsaturation of a fat or oil. The principle of the test is that iodine binds to fat. Iodine value is defined as the number of grams of iodine that are absorbed by a 100-gram quantity of a fat.

Peroxide value or **PV** is a means of measuring lipid oxidation, and it too uses iodine. As we learned in Chapter 5, unsaturated fatty acids are subject to lipid oxidation, a chemical process that results in peroxide formation. The principle of the PV test is based on the reaction of potassium iodide with the oxidized portion of unsaturated fatty acids. Potassium iodide releases iodine, and the amount of iodine liberated is directly proportional to the amount of lipid oxidation as measured by standard titrometric solutions or spectrophotometrically. When little or no iodine is liberated and detected, little or no lipid oxidation has taken place.

Acid value has to do with hydrolytic rather than oxidative rancidity and refers to the splitting of glycerides into component parts: glycerol and free fatty acids, caused by an enzyme (lipase) rather than by oxygen. **Acid value** is a measure of the number of free fatty acids present in a fat. Because free fatty acids are normally chemically bound in the fat molecule, their presence indicates hydrolytic rancidity has occurred. In performing acid value analysis, a base (KOH) is added to 1 gram of the test fat or oil, and the amount of base it takes to neutralize the acid pH of the fat is called the acid value.

Saponification value is a test that gives the average molecular weight of the fatty acids in a fat. This value is important because molecular weight influences physical firmness. It also offers an index to the potential to produce odor, with short-chain or low molecular weight fatty acids being most odorous. Saponification value is the number of milligrams of KOH required to convert 1 gram of fat into soap. Saponification value is high for a fat that contains many low molecular weight fatty acids. Think of 1 gram of fat containing only short-chain (hence, low molecular weight) fatty acids—it would contain more fatty acids than would 1 gram of fat containing the heavier long-chain fatty acids.

Smoke point is the temperature at which smoke emanates continuously from the surface of a lipid heated under standard conditions. If heated to higher temperatures, the fat or oil sample will flash, and still further, burn. Knowing these specifics is critical when selecting fats for deep-fat frying.

9.2 SUGAR PROCESSING

The most common sugar is sucrose, a crystalline sweetener used in foods and beverages. As a food processing term, "sugar" usually refers to sucrose, a disaccharide found in almost all plants. Sucrose is found at highest concentrations in sugarcane (*Saccharum officinarum*) and sugar beets (*Beta vulgaris*). The former is a tall grass growing in subtropical and tropical climates, while the latter is a root crop growing in temperate zones at high latitudes. Table 9.2 gives their sucrose composition.

Cane sugar refers to the sucrose product obtained from sugarcane and is generally produced in two stages because the sucrose decomposes after harvesting the cane. The

TABLE 9.2 Sucrose composition by weight of major plant sources.

Plant Source	Percent Sucrose
Sugarcane	7–18
Sugar beet	8–22

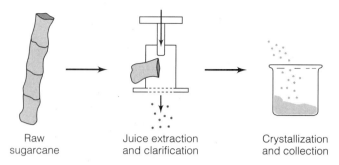

FIGURE 9.4 Overview of sugarcane processing.

manufacture of raw cane sugar occurs in the cane-growing countries, while refining into food-grade products takes place in the sugar-consuming countries. Sugar beets, which can be stored without sucrose decomposition, are processed into white sugar in one stage. Sugarcane processing usually consists of the following steps: extraction of the cane juice by mill or diffusion, clarification of the juice, concentration of the juice to produce syrup by evaporation, crystallization of sugar from the syrup, and separation and drying of the crystals, as refined sugar (Figure 9.4). The by-products of cane sugar and beet sugar production include fiber and molasses, which is the residual concentrated syrup from which no more sugar can economically be removed.

Extraction

To extract the cane juice, cane is first chopped into chips to expose the tissue and open the cell structure, which allows for the efficient extraction of the juice. The crushed cane proceeds through a series of roll mills, where it is forced against a countercurrent of water known as water of maceration. Streams of cane juice mix with maceration water and combine into a dilute juice. Extraction of sugar by a process called countercurrent diffusion results in an average rate of 93 percent extraction, compared to 85–90 percent by milling. However, disposal of the large amounts of water used by diffusers is a costly environmental problem and requires operation of water treatment systems.

Neutralization and Clarification

The mixed juice from the extraction mills or diffuser is purified by addition of calcium hydroxide (lime) and heat (about 99–104°C, 210–220°F). These treatments inactivate enzymes in the juice and raise the pH from a natural acid level of 5.0–6.5 to a neutral pH. Controlling pH during sugar manufacture is important because sucrose inverts, or hydrolyzes, to its component monosaccharides glucose and fructose below pH 7.0. Once neutralized, the juice is pumped into a continuous clarification vessel, which is a large heated tank in which clear juice flows off the upper part while process residue settles below.

Concentration and Crystallization

The clarified juice is pumped to a series of devices called evaporators, in which steam is used to concentrate the juice into evaporator syrup, consisting of 55–59 percent sucrose and 60–65 percent by weight total solids. The concentrated syrup from the evaporators is further evaporated under vacuum to achieve supersaturation. With the addition of seed crystal, this "mother liquor" yields a solid precipitate of about 50 percent by weight crystalline sugar. Crystallization is accomplished through a series of steps. The first yields what is called "A" sugar, and leaves behind a residual mother liquor known as A molasses. The A molasses is concentrated to yield a B variety, and the low-grade B molasses is concentrated to yield C sugar and final molasses, known as blackstrap. Blackstrap molasses contains approximately 25 percent sucrose and 20 percent invert (a mixture of glucose and fructose).

Separation and Drying

The next step in sugar processing is crystal separation and drying. In this stage, crystals and mother liquor are separated via centrifugation. In the process, the mother liquor is spun off the crystals, and a fine jet of water is sprayed on the sugar to reduce the syrup coating on each crystal. This washing process is performed in an effort to produce high-purity raw sugar. The overall recovery of sugar from cane juice averages between 70 and 80 percent.

The washed sugar is next allowed to dry and cool and is placed in bulk storage. This "raw" sugar has a pale brown to golden yellow color and a sucrose content of 97–99 percent plus a moisture content of 0.5 percent. Raw sugar, although sold in the United States largely in the health food industry, is typically further refined to produce white granular sugar or white powdered sugar.

Sugar Refining

Sugar refining refers to the production of high-quality sugars from remelted raw cane sugars. Sugar refining is conducted in the consuming regions at large refineries, which produce a range of products including white sugar cubes, powdered and granulated white sugar, and light and dark brown sugars. At these refineries, the raw sugar is washed ("affined"), dissolved ("melted"), clarified, decolorized, and crystallized. The finished sugar products are then dried, packaged, and stored.

9.3 BEVERAGE PROCESSING

The variety of beverages identified in Chapter 2 included water, sports drinks, soft drinks, fruit juice, iced tea, iced coffee, cocoa drinks, and alcoholic beverages. We focus in this section on the processing of bottled water and soft drinks. Fruit juice manufacture is covered in Section 9.5, and the processing of milks was covered in Chapter 8.

Water Beverages

Drinking (or potable) **water** is water that is intended for human consumption and sealed in bottles or other containers with no added ingredients, except optional antimicrobial agents (Figure 9.5). Drinking water can be used as an ingredient in beverages like diluted juices or flavored bottled waters. **Bottled water** includes natural mineral waters, carbonated waters, and sweetened, flavored waters that are typically carbonated. Table 9.3 defines the major types.

FIGURE 9.5 Examples of manufactured bottled waters.

TABLE 9.3 Differences between various forms of manufactured drinking water.

Distilled water is a purified water that has been produced by a process of distillation and meets the definition of "purified water" in the United States Pharmacopeia. Distillation converts liquid water into steam, with any impurities collected as the steam is condensed to make pure water.

Mineral water is bottled water containing not less than 250 parts per million total dissolved solids. It is distinguished from other types of bottled water by its constant level and relative proportions of mineral and trace elements at the point of emergence from an underground water source.

Purified water is bottled water produced by distillation, deionization, reverse osmosis (reverse osmosis is a process by which water is reduced to a nonmineral state by passing through a plastic membrane under pressure, which separates the water from other elements) or other suitable process and that meets the definition of "purified water" in the most recent edition of the United States Pharmacopeia.

Sparkling water is bottled water that, after treatment and possible replacement of carbon dioxide, contains the same amount of carbon dioxide that it had at emergence from the source, such as an underground spring. Manufacturers are permitted to add carbonation to previously noncarbonated bottled water products and label such water appropriately (e.g., "sparkling spring water").

The primary aims of bottled water processing are (1) to ensure a safe product and (2) to preserve the properties ascribed to the water, such as mineral content or flavor, up to the point of consumption. At the bottling plant, careful sanitation is employed to prevent the occurrence of microorganisms in the bottles. Ozone also can be employed to sanitize the water.

Ozone is an unstable, colorless gas that acts as a powerful oxidizer and a potent germicide. Ozone is used in the bottled water industry because it controls the growth of bacteria in water. Figure 9.6 explains how ozone kills the bacteria. Ozone is a preferred sanitizer because it does not leave a residual taste, which is the case with chlorine. Ozone also has a much higher disinfection potential than chlorine. The sanitized water is filled into bottles that can be made of glass, polyethylene terephthalate (PET), or HDPE (high-density polyethylene). Although plastics such as PET and HDPE are commonly used because of their low cost and light weight, a drawback is that they are not biodegradable.

The manufacture of flavored bottled water is straightforward, involving the unit operations of mixing and blending to achieve a homogeneous solution of the flavor substance in the water. Typically, natural or artificial fruit flavors are used. Flavors are present individually, or in combination, according to the product goal. Since flavor ingredients represent a potential source of contamination, a means of temperature adjustment can be used to apply a heat treatment to ensure microbial destruction in the product.

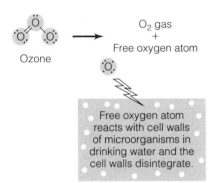

FIGURE 9.6 How ozone (O_3) works as a sanitizer in beverages. *The variables determining the effectiveness of ozone in killing bacteria are contact time and residual ozone concentration achieved in the product water. Ozone consists of three atoms of oxygen. Once generated, it takes only a brief time for it to break apart and return to its natural form of oxygen as two atoms. As this chemical change occurs, the free atom of oxygen seeks out any foreign particles in the water (such as bacterial cell walls) and binds to them, disintegrating bacteria or organic matter and protecting the water from waterborne contamination.*

Soft Drink Beverages

Did you know that the term *soft drink* was originated to distinguish nonalcoholic beverages from hard liquor? **Soft drinks** are nonalcoholic carbonated or noncarbonated beverages, usually containing a sweetening agent, edible acids, and natural or artificial flavors. Soft drinks include cola beverages, fruit-flavored drinks, ginger ale, and root beer—Table 9.4 distinguishes the true soft drinks.

The first attempts to manufacture carbonated soft drinks were the result of a desire to duplicate the naturally effervescent, mineral-rich waters that flowed from the springs at well-known European spas. Early experimenters believed that the effervescence was the source of the reputed healthful properties of the waters, and they therefore concentrated on this gaseous nature. By the late 1700s reports in London were published regarding this phenomenon by scientists including Joseph Priestley, who engaged in experiments with carbonation. Joseph Campbell, an Englishman, first patented carbonated water in London in the 1790s. Flavoring was added, and "soda pop" was invented. The pop was the sound made when the stopper or cork to the container was removed.

The building of factories and bottling plants in cities throughout Europe and in the United States followed, making the first bottled soda waters available to consumers in the turn of the nineteenth century. Some advertisements included claims that soft drinks "pepped up" tired people, calmed the nervous system, or made sad people feel happy. Some pharmacies sold soft drinks that contained such (at the time legal) drugs as heroine, codeine, and cocaine. These are no longer permitted in beverages, but many soft drinks still contain the stimulant caffeine.

A brief review of the origin of a popular soft drink is instructive. John Pemberton invented a caramel-colored syrup in 1886; when diluted and carbonated, this syrup became known as Coca-Cola. The reason for the product name: It originally contained cocaine from the coca leaf and was rich in caffeine from the kola nut. This premiere flavored soft drink was first patented in 1893.

Although modern readers of this book are familiar with slogans characterizing soft drink beverages as "the real thing" or "thirst quenching," Pemberton's bookkeeper originally marketed Coca-Cola as the "tonic for all ailments." Since that time, many flavored soft drinks have been marketed. In 1984, in response to the public demand for more healthful and less fattening foods, soft drink manufacturers began formulating with natural juices. Vitamin-enriched soft drinks and sugar-, caffeine-, and sodium-free soft drinks also became popular in the late twentieth century.

For the production of soft drinks, water quality is critical, since poor quality water affects the color, odor, flavor, and clarity of the finished product. For this reason, water treatment is performed in many soft drink bottling plants. In some, the water treatment equipment consists simply of a sand filter to remove minute solid particles and an activated carbon purifier to remove color, chlorine, and any other tastes or odors that may be present. In most plants, however, water is treated by process known as **superchlorination** and coagulation. In this process, the water is exposed to a high concentration of chlorine and to a flocculant that removes microorganisms. The water is then passed through a sand filter and activated carbon purifier.

Carbonation is a key step that dissolves carbon dioxide in soft drinks during processing. The characteristic effervescence (fizz and sparkle) of soft drinks is the result. **Carbonation** refers to the saturation of water with carbon dioxide under pressure in which the CO_2 gas dissolved in the water becomes carbonic acid. Carbon dioxide is supplied to the soft drink manufacturer either in liquid or solid form. The solid form, *dry ice*, is simply carbon dioxide that has been compressed and frozen so that it is more economical to transport. As the pressure is released, the carbon dioxide changes into a gas, which can be dissolved in a liquid such as water. To achieve carbonation, water or the finished beverage mixture is chilled and cascaded in thin layers over a series of plates in an enclosure containing carbon dioxide gas under pressure. The amount of gas the water will absorb increases as the pressure is increased and the temperature is decreased.

Because they contain carbonated water, or *soda water*, soft drinks are also called soda pop, soda, or pop. In the manufacture of soft drinks, special attention is given the purity and uniformity of ingredients: water, carbon dioxide, sugar or sugar substitute, acids, flavoring, and optional coloring. The sugar is dissolved or diluted with processed water, then combined with flavoring substances. Food-grade acidulants, principally citric acid, are added to give the mixture tartness. Natural flavors are derived from fruits, nuts, berries, and other plant sources. Natural or artificial coloring may also be added, and sometimes preservatives are used to protect the beverage from spoilage.

Although many beverages are sweetened with sucrose, high-fructose corn syrup, or fruit juice concentrates, synthetic sweeteners are used in the production of low-calorie beverages. These are usually not carbohydrate-based substances. They include acesulfame-K, aspartame, saccharin, and sucralose (see Section 3.4). These sweeteners are characteristically many times sweeter, gram for gram, than sucrose—see Table 9.5. Aspartame provides 4 kcal per gram, but since it is intensely sweet, only a small amount is used, so that its caloric contribution in a standard beverage serving is negligible. Alternative sweeteners may impart a degree of bitterness as well as sweetness to a beverage.

Finished soft drinks are produced by diluting the flavoring syrup with carbonated water. One of two methods can be used. In one method, a mixture of noncarbonated water and flavoring syrup is combined with highly carbonated water and then bottled. In the second method, syrup is measured directly into bottles and the bottles are filled with

TABLE 9.4 Examples of true soft drinks.

True soft drinks	Non soft drinks
cola beverages	coffee
fruit-flavored drinks	tea
ginger ale	milk
root beer	cocoa
soda water	undiluted fruit and vegetable juice
seltzer water	
tonic water	

TABLE 9.5 Nonnutritive sweeteners used in beverages. *The term* nonnutritive *indicates that although the substance imparts sweet taste, it does so with little or no caloric or nutritional value.*

Sweetener	Sweetness (sucrose = 1)	Taste Characteristics	Uses	Acceptable Daily Intake (mg/kg of body weight)
Acesulfame-K (Sunette)	130–200	Rapid onset, persistent side-tastes at high concentrations	Approved for use in table-use sweeteners, dry beverage mixes, and chewing gum	15
Aspartame (Nutrasweet)	180	Clean, similar to sucrose, no bitter aftertaste	Approved for use in table-use sweeteners, dry beverage mixes, chewing gum, beverages, confections, fruit spreads, toppings, and fillings	50
Saccharin	200–700	Slow onset, persistent aftertaste, bitter at high concentrations	Approved for use in soft drinks, fruit juice drinks, and other beverages; table-use sweeteners; processed fruits; chewing gum and confections; gelatin desserts; salad dressings; and baked goods	2.5
Sucralose (Splenda)	600	Can withstand high temperatures without losing flavor	Approved for use in soft drinks, baked goods, chewing gums, and table-use sweeteners	15

*The FDA has calculated an Acceptable Daily Intake (ADI) for nonnutritive sweeteners as the level at which a substance is safe to consume were it to be consumed every day for a person's entire life.

carbonated water injected under high pressure. Such pressurization automatically mixes the syrup and carbonated water. In either case, the sugar content is reduced from about 60 percent in the syrup to about 10 percent in the finished beverage.

The blending of syrups, mixing with carbonated water, and filling of containers is carried out by automated machinery. Bottles are passed along on an assembly line, and automatic fillers fill from 30 to more than 1,000 containers per minute. The bottles are capped by another machine on the assembly line, inspected, then packed in cartons or cases ready for shipping. Carbonated soft drinks are packaged for sale in a variety of containers, including glass bottles, tin or aluminum cans, and plastic bottles. Since they are not under pressure, noncarbonated soft drinks may be packaged not only in bottles and cans but also in treated cardboard cartons.

Special Beverage Categories

Noncarbonated soft drinks are produced with much the same ingredients and techniques as those for carbonated soft drinks. However, because they are not protected from spoilage by carbonation, noncarbonated soft drinks are usually pasteurized, which can be done in bulk or by continuous flash pasteurization either prior to filling or in the bottle. **Powdered soft drinks** are made by blending flavoring material with such ingredients as dry acids, gums, and artificial color. If a sweetener has been included, the addition of water is all that is required to prepare the drink. **Nutraceutical beverages** are drinks formulated with special functional ingredients that promote some aspect of health or reduce the risk of certain diseases. Table 9.6 lists just a few examples of these ingredients.

9.4 PROCESSING OF CEREAL GRAINS

Cereal grain processing refers to the conversion of cereal grains into food products or ingredients (e.g., flour). Cereal grains are technically classified as dry fruits, derived from the seeds of plants in the grass family—Table 9.7 compares wheat grain to some of the "wet" fruits. The term **cereal** technically refers to any grain used for food, and **grain** refers to a small hard seed produced by plants that are grasses. Barley, buckwheat, corn (maize), millet, oats, rice, rye, sorghum, and wheat are grass plants that produce the common cereal grains. What is known by the term "cereal" in common usage is usually ready-to-eat breakfast cereal made from these grains. However, breakfast cereals are only one type of processed grain product. What different methods are used to convert grain into ingredients for the production of breakfast cereal, bread, pasta, and snack chips?

TABLE 9.6 Partial listing of functional ingredients that can be incorporated into water, juice, or milk-based formulations to produce nutraceutical beverages. *In the manufacture of these beverages, it is important to manage certain factors that affect the quality of the finished product. These include ingredient stability to moisture and oxidation prior to use, ingredient purity, the solubility and bioavailability of the ingredient, and the flavor, color, and mouthfeel of the beverage product.*

Functional Ingredient	Comment
Galactomannan	Polysaccharide food gum that can absorb fat and bind sugars, thereby aiding weight loss and helping in the maintenance of diabetes
Mineral salts	Including calcium lactate and other salts of iron, magnesium; associated health benefits such as reducing the risk of osteoporosis, anemia and fatigue
Probiotics	Microorganisms that colonize the intestine and inhibit the growth of harmful pathogens; associated with health benefits such as reducing the risk of cancer
Protein hydrolysate	Associated health benefits such as speeding postexercise muscle recovery
Soy isoflavone	Associated health benefits such as reducing the risk of cardiovascular disease and cancer

TABLE 9.7 Comparing nutrient content of fruits and vegetables to a cereal grain, such as wheat. *Note the shift to carbohydrate and protein and away from moisture.*

Food type	Carbohydrate (%)	Protein (%)	Fat (%)	Water (%)
Wheat grain	73	12.6	1.5	12
Potato	20	2	0.1	78
Carrot	9	1.1	0.2	89
Lettuce	2.8	1.3	0.2	95
Banana	24	1.3	0.4	74
Apple	13	0.3	0.4	86

Cereal processing is a complex operation because grains are typically not consumed in the raw, unprocessed state. The key process is *milling*—grinding of grain into a form that is easily incorporated into foods or cooked.

Milling procedures that crush grain kernels are based upon the structure of the kernels. Milling methods are dry and wet, depending on the particular grain in question and the product desired. For instance, wheat is dry milled (crushed with grinding stones) to separate the endosperm from the outer coverings (bran) and the germ in the processing of flour. On the other hand, much corn is milled by a wet process to obtain corn syrup, fructose, and maltodextrins. Corn oil is obtained from the germ portion of the kernel, not from the starchy portion of the kernel, in a process that does not require wet milling.

Certain cereal grains, rice being a prime example, are polished to remove most of the bran and germ and isolate the endosperm layer. The starchy endosperm portion of the grain forms the raw material of cereal flours, dry milled products such as cornstarch, and even whole milled grains such as rice and barley.

Wheat Milling

The main classification for wheat grains distinguishes between the hard and soft types. In wheat milling the terms *hard* and *soft* wheat describe the firmness of the kernels and relate to the strength of the gluten developed when doughs are made from milled flours. Flours developing **strong gluten** have a higher protein content, with an elastic gluten suited to breadmaking. Flours developing **weak gluten** are low in protein, their weak, fragile gluten producing a softer, more fragile dough for cakes and biscuits.

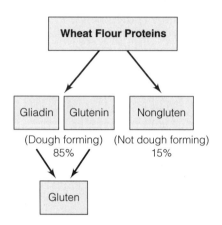

FIGURE 9.7 Dough characteristics are due to gluten, a combination of two proteins: gliadin and glutenin.

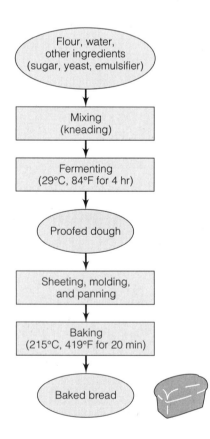

FIGURE 9.8 Breadmaking: a series of basic steps.

What about pasta? Doughs are made not only from strong and weak gluten flours, but also from semolina. Generally, semolina, rather than flour, is used to make pasta. **Semolina** is the result of milling certain hard ("durum") grains. Doughs made from it are strong but do not exhibit the same level of elasticity as strong gluten flour doughs (see the discussion on pasta processing).

Wheat flour processing consists of a complicated series of grinding rollers, sieves, and purifiers. Improvements in milling techniques have resulted in flour produced by the gradual particle size reduction process. The first step in grinding is performed between steel cylinders with grooved surfaces.

The initial grinding is a process called first break, in which corrugated rolls break wheat into coarse particles. This "broken" wheat is sifted through successive screens of increasing fineness, and directed through a series of air currents called purifiers that separate bran and separate particles called *middlings*. Reducing rolls decrease the size of the middlings into flour. Bleaching and aging of the flour is done to produce a uniform white color and to allow for proper performance in the manufacture of baked goods.

Breadmaking

Bread is the product of baking doughs from a mixture of flour, water, salt, yeast, and other ingredients. The flour-to-water ratio in bread is about 3:1. The purest flour, selected from the purest flour streams released in the mill, is often called **patent flour.** It has very low mineral content, is about 12 percent protein, and is devoid of bran specks and other impurities; it is suited to most breadmaking operations.

The basic breadmaking process involves mixing of ingredients until the flour is converted into stiff dough, followed by baking the dough into a loaf. The goal of the breadmaking processes is to produce dough that will rise easily and have eating properties desired by the consumer. Dough must be extensible enough for it to relax and to expand while it is rising. Quality dough is extensible—it will stretch out when pulled. It also must be elastic, having the strength to hold the gases produced during fermentation, and stable enough to hold its shape and cell structure. This viscoelastic character of the dough depends upon adequate gluten formation. Gluten is formed from two proteins present in flour: *gliadin*, important for its stickiness, and *glutenin*, important for its elasticity (Figure 9.7). These form the gluten network or matrix when mixed with water and create the special extensible and elastic properties of dough.

Figure 9.8 outlines the steps and products in breadmaking. The gluten network of dough is a complex system that requires proper mixing in order to form. Mixing evenly distributes the ingredients and allows the development of the gluten network. Natural flour lipids (glycolipids) bound to carbohydrate associate with gluten proteins in the dough matrix. During mixing, flour proteins form gluten as they and the starch molecules in the flour become hydrated. Together, after baking, they form the final product crumb structure. Initial heating causes gluten proteins to stretch and expand as the gas cells in the dough increase in volume (oven spring), and then the protein denatures and starch undergoes gelatinization.

Overmixing produces dough that is very extensible but with reduced elastic properties. Undermixing may create small, unmixed dough areas that will remain unrisen in the bread. After mixing, the dough is allowed to rise (ferment). During fermentation, dough changes from a dense mass lacking extensibility into smooth, extensible dough with good gas-holding properties. Yeast cells grow, the gluten protein pieces form networks, and alcohol and carbon dioxide are formed from the breakdown of starch and sugar. The carbon dioxide entrapped within gas cells in the dough causes the dough to rise. During fermentation each yeast cell forms a nucleus around which carbon dioxide bubbles form. Thousands of these bubbles form cells inside the dough, each surrounded by a thin film of gluten. The increase in dough size occurs as these cells fill with gas and expand.

After the first rise, the dough is *sheeted.* Sheeting dough means manipulating it between two rollers to develop the dough for breadmaking. Dough sheets rather than balls are generated. Sheeting is a gentler process than mixing, using much less energy, and can be the preferred method for dough development for certain products.

Breadmaking plants use an action roller system for sheeting, in which a pair of rotating rollers shuttle back and forth over two separate belt conveyors. The sheeting mechanism is designed to sheet the dough without giving excessive work to the dough structure. Once sheeted, dough can be cut and molded for bread baking.

The dough is panned and returned to the fermentation cabinet. After about 55 minutes, the dough has proofed to a fixed height. This is the second rise. After the final proofing sequence is completed, the dough is transferred to the oven for baking. The baking process transforms unpalatable raw dough into an aerated and delicious baked product. The physical changes involved in this conversion are quite complex. As the intense oven heat penetrates the dough the gas cells increase in size, which creates what is called "oven spring." Much of the carbon dioxide produced by the yeast is present in the dough. At baking temperatures, this carbon dioxide is converted to a gas, and moves into existing gas cells. This change expands the cells further and reduces the solubility of the gases. Heat also increases the rate of fermentation until the dough reaches the temperature at which the yeast dies, about 46°C (115°F). Alcohol produced during fermentation evaporates during the baking process.

At about 60°C (140°F) starch granules swell, and in the presence of water released from the gluten, the outer wall of the starch granule cell ruptures and the starch inside forms a thick gel-like paste, which helps form the structure of the dough. The gluten strands surrounding the individual gas cells are transformed into the semirigid structure commonly associated with the breadcrumb.

The natural enzymes present in the dough become inactive at different temperatures during baking. Alpha-amylase, the enzyme that converts starch into sugars, remains active until the dough reaches about 75°C (167°F). Extra sugars produced by this enzyme are available to sweeten the breadcrumb and produce the attractive brown crust color due to browning reactions during baking.

Some interesting events happen during baking. At a crust surface temperature in excess of 100°C (212°F), moisture is driven off. The crust retains a crisp, browned characteristic as opposed to the soft, white, aerated crumb. The two areas of the baked bread (crust and crumb) have different compositions due to the moisture difference. Importantly, the browning reactions experienced by the crust provide unique flavor differences, which are due to the temperature difference. The crust temperature reaches over 200°C (392°F) while the internal temperature of the crumb achieves only about 98°C (208°F). The crumb environment holds saturated steam, whereas the crust layer is in contact with the open air environment, allowing for evaporation from the crust surface.

Breakfast Cereal

Before examining how breakfast cereals are manufactured, here's a bit of history. This type of product was developed in association with a sanitarium, in part by one of its patients! Mr. J. H. Kellogg, an associate with the Battle Creek Sanitarium in Michigan, developed a healthful meal devoid of animal tissue by creating biscuits made from a dough mixture of wheatmeal, oatmeal, and cornmeal. The dough was baked until it was dry and had browned, and the product was ground and packed. A patient at the sanitarium named C. W. Post had the idea that such a product could be mass marketed and started his own breakfast cereal business. A brother of J. H. Kellogg named W. K. Kellogg also began manufacturing cereal products in flaked, granular, shredded, and puffed forms, and the Post-Kellogg breakfast cereal processing industry rivalry began.

Some breakfast cereals, such as rolled oatmeal, require cooking, while others are packaged ready-to-eat. Ready-to-eat cereals are typically consumed with milk or dry out of the box. The types of ready-to-eat breakfast cereals include flaked, extruded, puffed, shredded, and toasted varieties. Although wheat and rice flakes are available, the most popular flaked breakfast cereals are made from corn, cooked under steam pressure. Special machinery is used to separate the individual kernels so that they can be flaked and toasted. Figure 9.9 outlines the processes and intermediate products in making corn flakes.

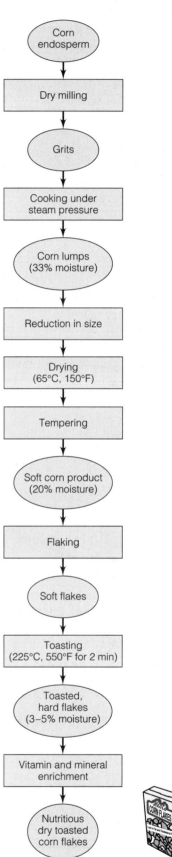

FIGURE 9.9 Manufacture of a flaked breakfast cereal. *This example begins with milled corn to isolate the endosperm. Wheat can also be processed into flakes.*

Pasta Processing

Pasta products are known as cooked and dried **alimentary pastes.** They include macaroni, egg noodles, spaghetti, and lasagna noodles. Semolina, not flour, is the proper form of cereal used to make pasta. Semolina is coarsely ground durum wheat, a type of high-protein hard wheat. Assorted macaroni products are made by combining a specific form of semolina from durum wheat with water. Alimentary pastes made with the addition of eggs create a richer egg noodle variety of pasta. The use of hard durum semolina contributes quality to alimentary paste products. The gluten that develops in the semolina is strong compared to bread flour gluten but not as elastic as that required for breadmaking.

Figure 9.10 shows the steps in making spaghetti. The procedure for most pasta products consists of adding water to semolina to produce a plastic homogeneous alimentary paste (referred to as a dough mass). Certain kinds of pasta, such as macaroni, are extruded as round, hollow dough. Noodles are instead extruded as flattened dough strips, while spaghetti dough is rounded but not hollow. This mixture is extruded through special dies, under pressure, producing the desired size and shape, and is then dried. A dry extruded product of about 12 percent moisture, is desirable. At this low moisture content, mold and yeast growth is inhibited during production and shelf stability is provided up to the point of cooking. Special packaging has been developed to market freshly made, undried, high-moisture uncooked pasta held at refrigeration temperatures.

Snack Foods

Developed in the 1950s, the potato chip was the original and most popular savory snack food. However, highly successful snack foods made from corn were subsequently developed, including corn puffs and corn chips. The general approach in processing corn into a corn chip involves the creation of dough, cutting or forming into chip shapes, cooking (deep-fat frying or baking), and seasoning for flavor—see Figure 9.11.

9.5 FRUIT AND VEGETABLE PROCESSING

The special characteristics of fruits and vegetables that make them desirable foods must be maintained as much as possible during processing and storage. In this section, the typical postharvest handling procedures and preprocessing measures applied to fruits and vegetables will be reviewed. And we will look at the specific processing steps for several common fruit and vegetable products. Although some differences between fruits and vegetables affect their processing and storage, many of the comments follow apply equally to fruits and vegetables.

A *vegetable* is a plant or plant part that is served either raw or cooked as part of the main course of a meal. A grouping of vegetables based upon plant part is shown. A *vegetable-fruit* is sometimes referred to as the fruit part of a plant that is not sweet and is usually served with the main course of a meal (cucumbers, squash, and tomato). See Table 9.8.

A **fruit** represents the edible, sweet-tasting fleshy seed-bearing ovary of flowering plants. Anyone studying an apple tree realizes that apples begin as flowers. After pollination, the ovary and nearby tissues enlarge and take on a more characteristic apple size and shape as the seeds develop inside. A fleshy fruit like an apple is the enlarged ovary of a flower, requiring fertilization or pollination to stimulate the rapid cell division that leads to differentiation and the formation of the fruit structure (Figure 9.12). Plant hormones that originate from the embryonic seeds are believed to cause rapid cell expansion as the developing fruit increases in size and weight, orchestrating its development into a mature, ripened fruit.

The texture of plant matter like fruits and vegetables depends upon its composition. Developing fruit cells possess small vacuoles in their tissue and begin to accumulate carbohydrate and lipid matter rather than protein. Cell walls of young plants are relatively thin and are composed largely of cellulose held together by pectin. As plants age, the walls

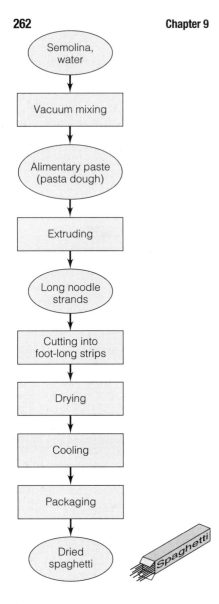

FIGURE 9.10 The manufacture of spaghetti pasta. *Although pasta processing utilizes hard wheat and ingredients similar to bread, it is an unleavened product. Pasta doughs are converted into dried shapes that must be cooked in boiling water before being eaten.*

TABLE 9.8 Vegetables—plants and plant parts.

Leaves	Seeds	Roots	Tubers
lettuce	corn	carrot	potato
spinach	peas	radish	ginger

Bulbs	Flowers	"Veg-fruits"	Stems/Shoots
onion	broccoli	tomato	celery
garlic	cauliflower	squash	asparagus

thicken, and contain higher amounts of tough, fibrous carbohydrate materials called hemicellulose and lignin. In contrast to cellulose, these carbohydrates do not soften significantly when cooked.

Most fresh fruits and vegetables are high in water, low in protein, and low in fat. They are good sources of both digestible carbohydrates (sugars and starches) and indigestible carbohydrates (fiber, as cellulose and pectic materials). As a group, fruits are generally good sources of specific vitamins and minerals, natural sugars, organic acids, flavor compounds, cellulose, pectin, fiber, and certain physiologically active compounds called phytochemicals. The bright colors and sweet flavors of fruits contribute to their popularity with consumers. These characteristics are related to the ripeness of a fruit.

Much of the appeal of fruits and vegetables lies in their colors, which are due to the color pigment molecules found mainly in the chloroplasts and chromoplasts within plant cells. Fruit and vegetable pigments are classified into several major groups, including the anthocyanins, carotenoids, and chlorophylls (see Section 6.1). These pigments, and hence the fruit or vegetable color, can be altered to varying degrees during processing.

Most fruits are harvested as close as possible to the time they are to be consumed as fresh produce. **Harvesting** is the collecting of fruits and vegetables at the specific time of peak quality of color, texture and flavor. The correct time to harvest a particular fruit varies by species and intended use (fresh vs. stored vs. processed), governed by whether the product will be sold and consumed quickly or is to be stored for days, weeks, months, or even a year. The apple and banana are examples that, although harvested while immature, still ripen satisfactorily.

The time of harvest can also correspond to the stage of maturity that is associated with optimum eating quality or processing characteristic (e.g. canning, pickling). Harvest time does not always correspond to ripeness. Some fruits and vegetables may be harvested in the immature state, at a time just before full ripeness, or when fully ripe, depending on the species and intended use. **Climacteric fruits** and vegetables continue to ripen after harvest, even demonstrating an increased respiration rate, with maximum respiration achieved just before full ripening. The best time to harvest these fruits is just prior to full ripening, when the tissue is yet firm (e.g., apple, banana). **Nonclimactic fruits** and vegetables respire at the same rate or less than before harvest, exhibiting their maximum respiration rate before being harvested. The best time to harvest nonclimacteric crops is at the fully ripened stage (e.g., grape, strawberry). See Table 9.9.

Ripening

Maturity represents the stage of development when the fruit is picked, at or just before the ripened stage. **Ripeness** is the optimum or peak condition of flavor, color, and texture for a particular fruit. Some fruits are picked when they are mature but not yet ripe. For instance, cherries, if picked when fully ripe, would be damaged because of their softness. Some fruits continue to ripen after they are picked and may become overripe if picked at peak ripeness.

Ripening refers to the transformation of a fruit from an immature stage of development that has undesirable eating attributes to one that has palatable eating quality and desirable color, flavor, texture, firmness, and aroma. Ripening is due to the natural internal physiological processes in a fruit that are under enzymatic and hormonal control. These are manifested in physical as well as chemical changes in the fruit. When storing fruits after harvest, recognizing their perishable nature is of extreme importance.

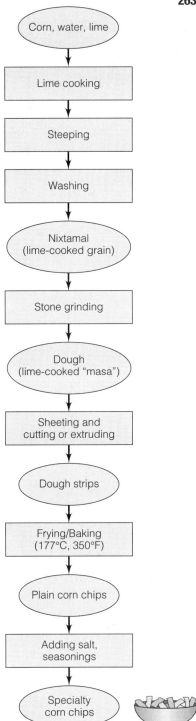

FIGURE 9.11 The manufacture of corn chips. *Lime cooking is used to soften and remove the husk material from the corn kernels in preparation of a pregelatinized dough.*

Apple flower

Pollinated apple ovary

Immature apple fruit

Mature apple fruit on tree

FIGURE 9.12 From flower to fruit. *Apples begin as flowers and develop gradually after pollination into fully-formed fruits. In mid-summer, apples are green and hard, with a sour taste. At maturity, the cell walls change and the texture softens. The ratio of acid to sugar decreases, and the flavor becomes characteristic of a sweet apple.*

TABLE 9.9 Classification of edible fruits according to respiratory patterns.

Firm = Climacteric (produce ethylene)
apple, apricot, avocado, banana, mango, melon, papaya, passion fruit, peach, pear, plum, tomato
Ripe = Nonclimacteric (ethylene sensitive)
cherry, cucumber, fig, grape, grapefruit, lemon, olive, orange, pineapple, strawberry

Respiration plays an essential role in postharvest fruit and vegetable ripening and quality. Respiration refers to the biological oxidation of organic molecules to produce energy plus carbon dioxide and water. In addition to carrying on respiration, fruits undergo moisture loss through pores in the tissue, a process called **transpiration**. Both respiration and transpiration contribute to fruit perishability. These processes continue in harvested fruits, as long as oxygen, substrate molecules, and moisture are present.

Fruit development can generally be divided into three major phases: growth, maturation, and senescence. The growth phase encompasses cell division and enlargement, causing an increase in size. Maturation, usually reached at the end of growth phase, is exhibited as an increase in sweetness and characteristic flavor and color development. **Senescence** is the phase associated with deteriorative processes, leading to aging and death of tissue, and quality decline. At the stage of senescence, a fruit is considered overripe and inedible—see Figure 9.13.

Because ripening leads to tissue breakdown, fruits represent a highly perishable commodity. As such, fruits vary in their postharvest longevity. For a comparison, strawberries possess a postharvest longevity of 7 to 10 days, while apples or lemons can be stored successfully for months. The goal of many fruit and vegetable processors is to prepare freshly harvested product prior to processing or storage to enhance postharvest longevity.

Processing

The most typical preprocessing unit operations are trimming, washing, and blanching (Table 9.10). **Trimming** refers to detaching superfluous plant parts from fruits and vegetables that are being processed into specific products. As an example, peas are shelled (removed from the pods, which are discarded) prior to canning or freezing.

Washing involves the use of water as a soaking medium or direct spray (pressurized water). The addition of detergents and a disinfectant like chlorine, with agitation, helps the soaking process remove dirt and debris, while at the same time reducing microbial contamination. Since cleaning can remove the natural outer waxy surface from fruits and vegetables, it is standard practice to reapply a waxy coating after the fruit or vegetable is washed and adequately dried if the fruit or vegetable is to be stored or distributed fresh. The waxing coating offers protection from desiccation during storage, extends shelf life, and improves appearance in many instances when polished before shipment.

FIGURE 9.13 Postharvest changes in fleshy fruit and vegetable tissue. *One of the products of respiration, carbon dioxide, decreases the respiration rate of stored fruits and vegetables. Ethylene, a gaseous hormone produced by some fruits and vegetables during ripening, accelerates the ripening process. As fruits and vegetables ripen, their cell walls become softer, starch converts into sugars, and sugars can convert back into starch. Pigments such as the carotenoids—orange-yellow—and anthocyanins—red, are produced. Overripeness (senescence) is associated with a decrease in attractive color and loss of edible appeal.*

TABLE 9.10 Three common operations in the early processing of fruit and vegetable tissue.

Washing: After harvesting, vegetables are washed to remove field soil surface microorganisms, and chemical/pesticide residues; the wash water usually contains detergents and sanitizers.

Skin removal (peeling): Hot water or steam is used to soften and loosen the skin, which can be removed by pressurized jets of water.

Cutting/trimming: Many vegetables require cutting or trimming: asparagus spears are cut to a precise length; olives are pitted by aligning them in small cups and then pushing a plunger through them.

Further preprocessing unit operations include the rapid application of mild heat process, known as **blanching**, used to inactivate the enzymes responsible for browning (polyphenol oxidase) and tissue softening (pectinase) in fruits and vegetables. Another important enzyme is peroxidase, a heat-resistant enzyme. If inactivated, then the other significant enzymes will have been inactivated too. Food processors know the amount of heat treatment required to destroy peroxidase in different vegetables, and sensitive chemical tests have been developed to detect its destruction.

Blanching also separates unwanted peel and removes odors, flavors, and slime from produce. Blanching as a unit operation for most commercial products is a short time heating in water at temperatures below 100°C (212°F). Because fruits and vegetables differ in size, shape, heat conductivity, as well as the natural levels of their enzymes, blanching treatments have to be established on an experimental basis for each. As a rule, the larger the food form, the longer it takes for heat to reach the center. See Table 9.11.

Packaging

Typical packaging systems for fresh fruit involve a simple plastic breathable bag or overwrap. In an effort to improve upon this basic packaging, specific modified atmosphere packaging (MAP) has been developed. In MAP, the barrier properties of the material are carefully selected according to the respiration characteristics of the fruit or vegetable. The goal is to allow an exchange of gases and moisture that produces the optimal storage environment so that freshlike attributes are maintained. This technology produces what are called **minimally processed vegetables.**

Minimally processed vegetables normally do not contain any preservatives and have not gone through any heat or chemical treatment. The most common minimally processed vegetable products are packaged cut lettuces sold as salad and stir-fry mixes. Such produce undergoes additional preparation steps of washing, sorting, grading, cutting, and packaging into retail packages. Vacuum packing and modified atmosphere packaging are employed to decrease the respiration rate and therefore the senescence of cut vegetables. In most cases, air is replaced by an atmosphere high in carbon dioxide or nitrogen and low in oxygen. This modified atmosphere acts to extend the fresh shelf life of these products.

TABLE 9.11 Time and temperature blanching parameters for some vegetables.

Blanching is followed by a cooling operation. Cooling of vegetables (to under 37°C, 99°F) in water after blanching avoids excessive softening of the tissues. Cooling in water can be achieved by sprays or by immersion. The temperature of the cooling water has to be as low as possible to avoid loss of water-soluble nutrients. Natural cooling is not recommended because the cooling rate is too slow and generates significant losses in vitamin C content.

Vegetable	Temperature (°C)	Time (min)
Carrots	90	3–5
Green beans	90–95	2–5
Peas	85–90	2–7
Peppers	90	3

Storage

Managing the temperatures under which fruits and vegetables are stored is critical to maintaining quality. Improper temperatures can hasten the deterioration of fruits, while certain temperature treatments can prolong their quality. Changes that take place during storage may impact color, flavor, and aroma. Softening of the flesh and manifestations of spoilage (by bacteria and fungi) especially during senescence are often observed in stored fruits and vegetables. Warm storage temperatures promote microbial growth and chemical changes. The rate of chemical reactions in fruit generally doubles for every increase of 10°C. Therefore, one wise strategy to prolong freshness and quality is to reduce the storage temperature.

Two types of cooling used in the food industry for incoming fruits and vegetables are *hydrocooling*, the immersion into cold water, and *vacuum cooling*, the practice of moistening food items and then placing them under vacuum to induce evaporative cooling.

Temperatures that are too cold can damage fruits and vegetables. **Chilling injury,** which is characterized by pitting and browning of the surface and the flesh, is due to subjecting a fruit to a severe cooling treatment. Assuming that fruits are properly cooled, they can be processed immediately, or put into storage for later processing or for later sale as fresh produce. Lemon, avocado, banana, tomato, eggplant, mango, and other tropical fruits display chilling injury when kept in prolonged cold storage. As a result, these fruits can fail to ripen properly. Chill-injury–sensitive fruits and vegetables should not be stored below 37°F (3°C).

However, proper application of cold storage is quite useful (see Table 9.12). Fruit life can be extended by both refrigeration and controlled atmosphere (CA) storage. In the latter, oxygen is held at about 3 percent and carbon dioxide at 5 percent, while temperature is cooled to a level best suited to the particular fruit. Air is composed roughly of 21 percent oxygen, 78 percent nitrogen, and less than 1 percent carbon dioxide, so the proportion of gases in CA storage is significantly different from atmospheric levels. This type of storage environment delays senescence and further quality loss in fruits.

A further strategy of CA storage to extend fruit longevity through appropriate storage is to promote the removal of ethylene gas. This "ripening hormone," which speeds the rate of ripening, is produced by fruits such as apples, avocados, bananas, cantaloupes, honeydew melons, peaches, pears, tomatoes, and watermelons. Its production by one fruit can affect the ripening of another stored in proximity to it. Adequate ventilation during storage helps remove ethylene and minimizes its effects.

A final storage method is **hypobaric storage**—low-pressure storage—which affects production of ethylene. Pressures as low as 80 and 40 mm of mercury and temperatures of 5°C (40°F) reduce ethylene production and respiration rates, to maintain quality in fruits for up to several months.

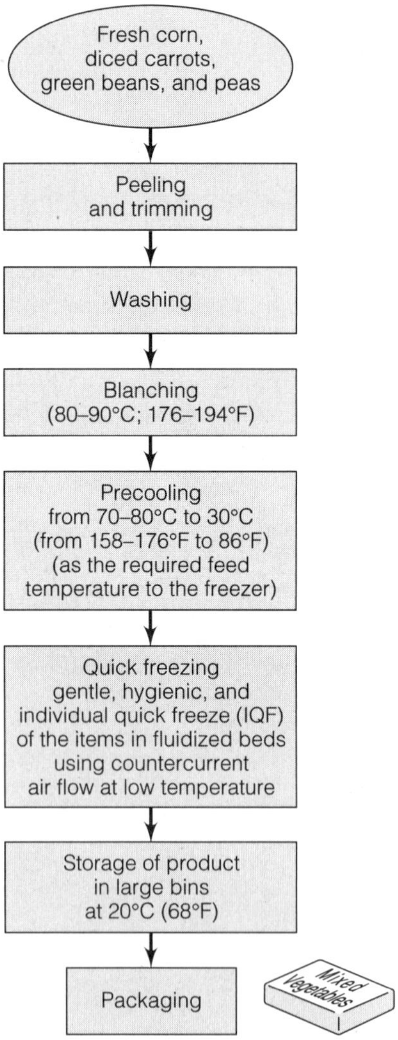

Freezing

Freezing fruits and vegetables is a common industry practice. Freezing is a means of stabilizing blueberries, peaches, raspberries, and strawberries. The process also works very well for vegetables of all kinds, including beans, carrots, corn, onions, and peas. Product must be held well below the freezing point of water—typically at a minimum of −23°C (−10°F)—to achieve extended storage life. The more rapid the freezing, the more improved the texture upon thawing.

Frozen fruits and vegetables offer high quality and nutritive value. The high quality of frozen foods is mainly due to the development of a technology known as the individually quick frozen method. IQF does not allow large ice crystals to form in fruit and vegetable cells during rapid freezing. And since each piece is individually frozen, particles do not stick together. Figure 9.14 shows the processing steps in the manufacture of frozen mixed vegetables. The basic primary processing and cleaning steps are coupled to final freezing of a mixutre of fresh corn, diced carrots, green beans, and peas.

Various other freezing techniques have been employed to preserve vegetables. These include belt-tunnel freezing, blast freezing, fluidized-bed freezing, and plate freez-

FIGURE 9.14 The processing steps involved in the manufacture of frozen mixed vegetables.

Fresh corn, diced carrots, green beans, and peas
↓
Peeling and trimming
↓
Washing
↓
Blanching (80–90°C; 176–194°F)
↓
Precooling from 70–80°C to 30°C (from 158–176°F to 86°F) (as the required feed temperature to the freezer)
↓
Quick freezing gentle, hygienic, and individual quick freeze (IQF) of the items in fluidized beds using countercurrent air flow at low temperature
↓
Storage of product in large bins at 20°C (68°F)
↓
Packaging

TABLE 9.12 Three refrigerant gases used in cooling operations in the food industry.
Direct expansion refrigeration pumps a gas refrigerant into a coil, and it expands as it moves across the coil, cooling the air around the product. Secondary cooling uses a liquid rather than a gas, such as glycol or brine. These are pumped into a coil, which becomes very cold and cools the area around the food.

Refrigerant	Application
Ammonia (reusable)	High volume, long-term storage
Carbon dioxide	Batch cooling; less rapid cooling
Liquid nitrogen	Batch cooling, very rapid cooling

ing. The choice of method depends on the kind of vegetable to be frozen, among other things. Most vegetables frozen commercially are intended for direct consumer use or for further processing into soups, prepared meals, or specialty items.

Manufacture of Fruit Juice

Fruit juices are beverages derived from whole fruits: apples, cranberries, grapes, grapefruits, lemons, oranges, pears, and more exotic varieties. Such beverages can exhibit different characteristics, ranging from naturally cloudy products to fully clarified juices. The processing of fruit juice involves four key steps: washing, extraction, clarification, and preservation via pasteurization—see Figure 9.15.

Fruits and vegetables are washed prior to processing into juice. In apple juice processing, a suitably rigorous preliminary wash is required to ensure the removal of mycotoxin-producing microorganisms.

Juice extraction involves a pureeing system that produces a mash of ground apple, followed by a pressing operation, which forces juice from the mash. In the case of some juices, enzymes are used to liquefy the mash. The clarity of the resulting juice from either method may be low, in which case the addition of pectinase may be required. The juice is monitored for pectin content using a qualitative pectin check, which consists of combining juice with ethanol and checking for gel formation (a positive indication that pectin is still present).

Filtration systems consisting of diatomaceous earth (DE) or ceramic membranes are employed in juice filtration systems. The juice that results is devoid of particulates and is of excellent clarity and transparency. Such juice can be pasteurized and bottled as a beverage, or made into a frozen or room temperature concentrate. Frozen concentrates are produced by passing the juice through an evaporator. This brings the level of dissolved soluble solids to 45 percent by weight. Reconstitution of the juice by the consumer requires a three-to-one water-to-concentrate dilution, creating a soluble solids level of approximately 12 percent in the finished product.

In January 2001 the FDA issued a ruling designed to improve the safety of fruit and vegetable juice and juice products. The rule was prompted by a number of foodborne illness outbreaks and consumer illnesses associated with juice products, including a 1996 *E. coli* O157:H7 outbreak associated with apple juice products and two citrus juice outbreaks attributed to *Salmonella* spp. in 1999 and 2000. Under the rule, juice processors must use Hazard Analysis and Critical Control Point (HACCP) principles for juice processing. Implementation of a HACCP system increases the protection of consumers from illness-causing microbes and other hazards in juices. Clarified juices are now heat processed (pasteurized) for preservation and safety purposes. The juice is heated to 88°C (190°F) and then bottled as a shelf-stable product.

Canning

Large quantities of vegetable and fruit products are canned, and foods packaged in materials other than metal cans are considered "canned" by food processing standards if the food undergoes the canning preservation process (being placed in an airtight container

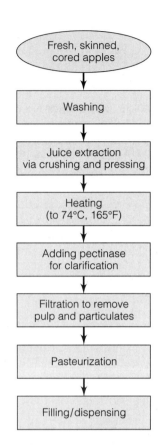

FIGURE 9.15 The processing steps involved in the manufacture of apple juice.

FIGURE 9.16 Applesauce is a high-acid commercially sterile hot-fill food. *Hot filling of foods having pH values of 4.0 or less, such as certain fruit juices and applesauce, can be considered commercially sterile without further thermal processing. Spore-forming bacteria cannot germinate at pH values of 4.0 or less. Such foods are heated to a temperature range of 90–96°C (195–205°F), filled into containers heated to the same temperature, hermetically closed, inverted for 5 minutes at the heat temperature, and cooled down to about 80°C (175°F). The 5-minute exposure to the heat temperature is sufficient to kill all non–spore-forming bacteria, yeasts, and molds that might be present in the food or the container.*

and heated to destroy microorganisms). A typical process flow for a canning operation covers basic food process unit operations performed in sequence: harvesting; receiving; washing; grading; heat blanching; peeling and coring; can filling; removal of air under vacuum; sealing/closing, retorting/heat treatment; cooling; labeling; and packing.

For high-acid fruit products the most typical thermal process is canning, in which fruit or fruit products are hot-filled or heated in a hermetically sealed container. A **hermetically sealed container** is any package, regardless of its composition (i.e., metal, glass, plastic, polyethylene-lined cardboard), that is capable of maintaining the commercial sterility of its contents without refrigeration after processing. The process temperature is generally about 93°C (200°F). Canning is of great importance to the peach, pear, and pineapple industries—these fruits require only pasteurization due to their acid content. Canned mixed fruit, called fruit cocktail, has been a consumer favorite for years.

In the canning of vegetables, cut pieces (which are blanched) are packed into cans and put through severe heat treatment to ensure the destruction of bacterial spores. The containers are sealed while hot to create a vacuum inside when they are cooled to room temperature. Properly processed canned vegetables are commercially sterile and can be stored at room temperature for years. Thermal processing for canned foods requires the equivalent of 121°C (250°F) of heat be reached in all parts of the can for 7 minutes. Processed cans are cooled to room temperature, labeled, and packaged for either storage or immediate distribution.

Due to the severe heat treatment, some canned vegetables exhibit inferior quality and lower nutritive value than fresh and frozen products. The nutrient most susceptible to destruction in canning is vitamin C. For high-quality products, HTST (high-temperature, short-time) processing is used: 10–15 seconds of heat at a temperature range of 70–75°C (158–167°F). In addition, **aseptic canning** is practiced in the processing of certain fruits and vegetables. In the process, presterilized containers are filled with a sterilized and cooled product. The product is sealed into a sterile atmosphere with a sterile container cover. The procedure avoids the slow heat penetration inherent in traditional in-the-can heating and results in products of superior quality.

Pickling

Fermentation and pickling of vegetables are examples of chemical preservation. The common denominator in both fermented and pickled vegetables is that acid is used to preserve the products. In the case of pickling, the product is usually preserved with added acetic acid (vinegar), but may be preserved by the addition of salt, sugar, or alcohol. Figure 9.17 shows the processing steps with a salt solution. High sugar content also acts as a fruit preservative by reducing water activity so that microorganisms cannot grow. The most common pickled food item is, not surprisingly, the pickle, which is a cucumber that has been acidified with vinegar and flavored with salt and other spices or seasonings. Other pickled vegetables are banana peppers, beets, carrots, cauliflower, and green beans.

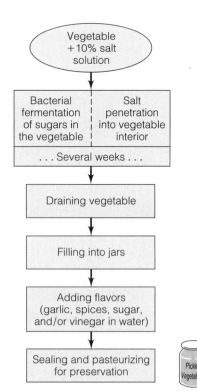

FIGURE 9.17 The process of producing a pickled vegetable, as a 10 percent salt solution plus vegetable to be pickled.

In fermentation the action of bacteria on carbohydrates produces acids in fermented vegetables (see Section 10.5). Pickled vegetables include cucumbers, green tomatoes, onions, radishes, and cabbages. Pickled (traditional dill) cucumbers produced by fermentation instead of a pickling process yield what are called brined pickles. Fermentation to produce a pickled vegetable is a slow process, whereas pickling is a rapid one.

Dehydration

Dehydration is among the oldest and most common forms of food preservation. In dehydration, moisture is driven off by the application of heat, resulting in a stable food product that has a moisture content below that at which microorganisms can grow. Dehydration offers a number of important advantages in the processing of fruits and vegetables. Dehydrated or dried fruit has a virtually unlimited shelf life when held under proper storage conditions, and it offers a concentrated source of flavorful nutrients. Fruits that are typically dried include apples, apricots, dates, figs, grapes, pineapples, and plums.

There are three basic systems for dehydration: sun drying, which has been applied for raisin production, hot air dehydration, and freeze-drying. Specialized equipment is required to accomplish drum drying, freeze-drying, spray drying, tunnel drying, and vacuum drying. Although dehydration offers a convenient product form, it can require careful inactivation of enzymes, often accomplished by blanching. To control enzymatic browning, a fruit is often treated prior to dehydration with a form of sulfur, such as sodium sulfite or sodium bisulfite. The sulfite is multifunctional, serving as a preservative, an antimicrobial agent, and an aid in heat transfer. Sulfur dioxide can be added as an antioxidant and preservative. It can inhibit both enzymatic and nonenzymatic browning.

Manufacture of Instant Mashed Potatoes. Not all dried foods are fruits. A familiar dehydrated vegetable is instant mashed potatoes. The process of converting raw potatoes into potato flakes involves many stages—see Figure 9.18. Producing good-quality potato flakes is the result of controlling the operating parameters of all the equipment involved in the production process. Once produced, the flakes are reconstituted with water to produce the "instant" mashed potato. In contrast to mashed potatoes made with freshly cooked potatoes, instant mashed potato product exhibits a pasty texture and different flavor, which is caused by the rupture of potato starch granules during processing.

Hot-Break/Cold-Break Processing

The wide range of textures of fruits and vegetables, both fresh and processed, is due in great measure to the integrity of specific cellular components of these plants and the changes that occur to them. Plant tissue generally contains much water. **Turgor** is the structural rigidity of plant cells due to their being filled with water. The state of turgor, or water content, is the most important factor in determining the texture of fruits and vegetables. When plant tissue is damaged or killed due to freezing, heat processing, or storage, denaturation of proteins within plant cell membranes occurs, which causes the water to move out of the plant cell. This causes a loss of osmotic balance, and a wilted texture results.

Recall that pectin and related molecules (termed pectic substances) are complex polymers of sugar acids. Found in the middle lamella of the plant cell, they are referred to as cement-like because they function to help hold plant cells together. Pectic substances may be either water soluble or insoluble, and affect fruit and vegetable texture—both whole foods and processed ones.

The texture of tomato products is determined by the processing used—hot-break processing or cold-break processing. Selecting the type of processing depends on the product thickness desired. Think of the wide variation in product thickness of these tomato products: tomato juice, tomato ketchup, tomato paste, tomato sauce, tomato soup, and salsa.

A key enzyme in tomatoes is **pectinase** (pectin methyl esterase). When the peel of the tomato is broken, and air makes contact with the interior of the fruit, oxygen from the air activates the enzymatic reaction, which "destroys" the pectin of the tomato. Without

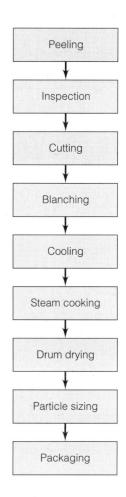

FIGURE 9.18 Instant mashed potato processing. *Drying is a key processing variable in the production of instant mashed potatoes. The product is dried to approximately 6 percent moisture in a fluidized-bed drier, and to produce the flakes, a steam-heated drum drier is used. The cycles of cooking, cooling, drying, and rehydration prior to consumption creates texture and flavor that is different from freshly prepared mashed potatoes.*

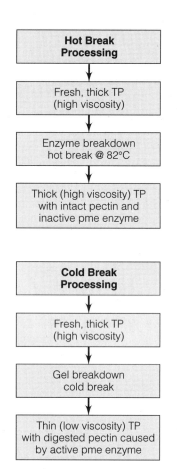

FIGURE 9.19 Hot-break and cold-break tomato processing. *Hot-break processing inactivates pectinase enzyme (pectin methyl esterase), resulting in high tomato paste (TP) viscosity. In cold-break processing, active pectinase lowers the viscosity of the product.*

pectin, it is not possible to obtain the level of viscosity required in certain tomato-derived products.

The objective of the **hot-break process** is to inactivate this enzymatic process. Heat treatment of about 82°C (180°F) destroys the enzyme activity of pectinase (Figure 9.19) and results in a product rich in pectin, of high viscosity, and with a very low syneresis (the separation of the liquid from the solids in the product, caused by the failure to inactivate the pectin-degrading enzymes). This type of product, of high consistency, is particularly suitable as a semifinished product for the production of sauces, ketchup, tomato paste, and other tomato by-products.

The **cold-break process** uses no heating, which allows pectinase to lower the viscosity of the product. This results in a slightly better color and flavor and delivers a product with low pectin, lower viscosity, and a higher syneresis. It is especially suitable for products such as juice, soups, and all products where good flavor and color are primary issues.

During tomato processing, the cell structure partially collapses. Calcium chloride can be used as a processing aid to maintain structure. It acts as a firming agent by bonding with soluble pectic substances (pectin substrate) in the tomato to form a calcium-pectate gel. Calcium pectates are water insoluble and help maintain structural rigidity, even during heat processing. These linkages help maintain a certain viscosity in the final product.

The manufacture of tomato paste, illustrates these processing approaches (Figure 9.20). By definition, tomato paste must contain at least 22 percent tomato solids, typically ranging from 24–31 percent. The processing parameters used to manufacture tomato paste result in either a hot-break or a cold-break paste. Freshly chopped tomatoes allowed to stand (cold-break paste) would yield a gradually thinner (lower viscosity) product than one that undergoes hot-break processing. The cold-break process is used to produce a product having high tomato solids content but a thin consistency. In a tomato juice, for example, a cold-break paste would give more tomato flavor and color without the unwanted viscosity.

9.6 SOYBEAN PROCESSING

Soybean (*Glycine max*) is a grain that is not a cereal but a leguminous seed that is grown widely around the world. Originating in Asia, it has become an important crop in the United States, which became the major world producer by late in the twentieth century. Nutritionally inferior, the wild soybean has been domesticated and crossed to produce a soybean rich in oil (approximately 20 percent) and protein (30–48 percent). Products derived from this crop are extensively utilized in both food and nonfood applications.

One method of soybean processing utilizes an expansion/expelling method. The expansion/expelling method of processing has gained popularity due to the quality of the by-products produced, plus the freedom from environmental hazards associated with solvent extraction. The expander and press equipment combined offer a continuous operation, meaning that product passes through at a steady rate (as opposed to a batch operation).

In the expansion/expelling method, raw soybeans are fed through a series of augers, screeners, and controlled rate feeders into the expanders. The beans experience extreme temperature and pressure conditions in the expanders, and the oil cells of the bean are ruptured as the product exits the expander. The high temperature (150–177°C; 300–350°F) cooks the meal and oil, yielding a high-quality food grade product, and moisture is released as steam. The hot meal slurry from the expanders is fed into a continuous oil expeller where the meal is squeezed under pressure and the free oil is expelled. Two key products are separated and recovered from this process: soybean meal and soybean oil.

The *soybean meal* exits the press as both a dry powder and chunks, and is a high-quality product free of the pathogens and metabolic inhibitors such as trypsin inhibitors associated with the raw soybeans. The meal contains about 7 percent oil, 45 percent pro-

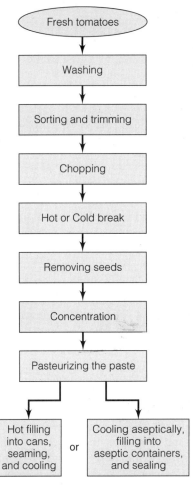

FIGURE 9.20 The processing steps involved in the manufacture of tomato paste.

TABLE 9.13   Processed soy food products.

Soy Milk: Soy milk is produced and used in the fresh state in China and as a condensed milk in Japan. In both of these preparations, certain antinutritive factors (antitrypsin and soyin) are largely removed. In the Western world most soy products are treated chemically or by heat to remove these antinutritive factors along with the unpopular "beany" taste due to lipoxygenase activity.

Soy Flour: Soybean is milled to produce soy flour. The flour is often used in a proportion of less than 1 percent in bakery operations. It stiffens dough and helps to maintain crumb softness.

Soy Protein Isolate (also called Isolated Soy Protein): A recent development is the isolation of the soybean proteins for use as emulsifiers and binders in meat products and substitutes. Enzyme-modified proteins provide useful egg-albumen supplement for whipped products.

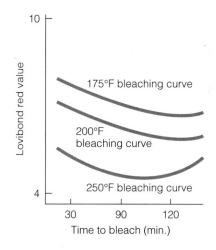

FIGURE 9.21 Bleaching of soybean oil: effect of time and temperature. *Bleaching lessens oil color intensity and also removes residues generated in refining steps. The Lovibond rating system assigns color values to processed oils. Minimum Lovibond red values are acheived sooner at higher temperature. The maximum Lovibond red value for processed oils is 3.5 units, although values up to 6.0 can be sold at reduced price.*

tein, and 94 percent dry matter. The oil content is considerably higher than solvent extracted meal.

The expelled *soybean oil* flows from the press and is collected in a basin, screened, and recycled back into the press. A semiclean oil results, which is pumped through a filter press to the final storage tank. With removal of gum residue (bleaching—see Figure 9.21), the oil is considered "first refined" and is typically highly unsaturated. It has to be completely filtered prior to shipping.

Foods derived from soybeans are recognized as potentially significant dietary sources of protein. These "soyfoods" include tofu and soy milk. With the increased interest in vegetarianism and an expanded awareness of the nutritional and functional properties of soy protein and other soy nutrients, soyfoods such as soy yogurt, soy ice cream, soy cheese, soy meat analogs such as sausages and burgers, and soy flour pancakes are all available. Descriptions of three basic soy products are given in Table 9.13. The interest in nutraceutical foods and ingredients has prompted the marketing of foods containing significant levels of soy isoflavones, such as genistein, thought to promote cardiovascular health.

A low-cost extrusion process can be employed to produce texturized soy protein products that offer excellent protein quality and low fat from soy flour. *Protein texturization* refers to a process that creates a fiberlike structure in a proteinaceous material. This process is the basis for producing texturized vegetable protein or **texturized soy protein** from soybeans. Texturized soy protein is extruded soy flour that can serve as an ingredi-

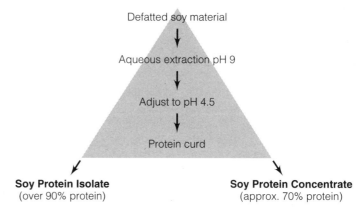

FIGURE 9.22 Distinction between soy protein isolates and soy protein concentrates. *Washing, centrifugation, and pH adjustment are performed to separate carbohydrates from the protein fraction.*

ent in food products. Such protein products are able to withstand hydration and cooking to produce meat extenders and analogs of various sizes, shapes, and colors.

Soy protein products include defatted soy flakes, soy meal, soy flour and grits, soy isolates, soy concentrates, texturized soy proteins, full-fat soy flour. Soy ingredients are also important in the sports beverage and food industry. Two common types are **soy protein isolates** and **soy protein concentrates**—see Figure 9.22.

Soybeans have been genetically engineered to increase nutritional value. One of the strategies employed in modifying soybeans is to increase the oleic acid content. Such bioengineered soybeans are referred to as high oleic acid soybeans. Oleic acid (C18:1) is a monounsaturated fatty acid that contains one double bond in the eighteen-carbon chain. Nutritionally, this fatty acid is beneficial to health and is preferred in the diet over polyunsaturated fatty acids. By using modern technology three lines of high oleic acid soybeans have been produced. The high oleic acid soybeans were generated by the suppression of a soybean gene involved in the metabolic pathway responsible for fatty acid biosynthesis. The oleic acid content of the genetically modified soybean is about 80 percent while the content in non–genetically-modified soybean is about 20 percent.

Soybean oil has poor oxidative stability due to naturally high levels of polyunsaturated fatty acids (such as linoleic acid). High oleic acid soybean oil is considered to have superior properties to that of standard soybean oil because of its reduced levels of the oxidatively unstable polyunsaturated fatty acids. Thus, high oleic acid soybean oil can be used for a number of food applications, including deep-fat frying, without the need for additional processing, such as chemical hydrogenation. High oleic acid soybean oil also offers improved nutritional properties compared to conventional soybean oil or partially hydrogenated soybean oil because of the increased levels of monounsaturated fatty acids and because its replacement of other oils may lead to a decreased consumption of saturated fatty acids.

High oleic acid soybean oil is expected to be used in the formulation of crackers and breakfast cereals and in frying applications in the food industry and food services and might replace heat-stable fats and oils, such as hydrogenated soybean oil or palm olein/vegetable oil blends. A consequent by-product of the soybean processing will be soybean meal and soy isolates. These products could appear in a wide range of processed foods such as improved soy milks already in the market and processed meats.

9.7 CHOCOLATE PROCESSING

Chocolate is a food derived from cacao beans from the tree *Theobroma cacao*. Chocolates are a form of confectionery in which the original chocolate material is often blended with sweeteners and other ingredients. This original material is called **cocoa butter,** which is the form of fat found in the cacao beans. At a certain point in the processing of cacao beans, they are referred to as cocoa beans.

The processing of cacao beans to produce chocolate is shown in Figure 9.23. Cacao beans are harvested, cleaned, fermented for 5–7 days to develop initial flavor compounds, dried to remove much of the moisture, and shelled to produce *nibs*. Following this processing, the dried nibs are roasted to develop flavor, in part due to the Maillard reaction and the removal of volatile acids. After roasting, with the outer shells removed, the cocoa beans are ground. Grinding generates enough heat to liquefy the fat in the beans. This results in a separation of liquid called the chocolate liquor, which is 55 percent fat (as cocoa butter) and a mixture of carbohydrates, proteins, tannins, ash, about 2 percent moisture, plus the compounds theobromine and caffeine (Figure 9.24).

Upon solidification, the chocolate liquor forms bitter (unsweetened) baking chocolate. Pressing the chocolate liquor at 600 psi extracts the cocoa butter fat. What is left behind as a dry cake can be ground and sifted to produce cocoa powder for baking and drink applications.

The processing of cocoa butter for chocolate manufacture includes mixing, refining, conching, tempering, and possibly certain product manipulations such as molding and enrobing. To produce milk chocolate using a traditional method, special mixers called melangeurs mix chocolate liquor with milk solids or unsweetened condensed milk and the proper amount of cocoa butter, to achieve a 25 percent fat level. Drying the mixture produces a chocolate crumb, which is pulverized through refining rollers to reduce the particle size and form a smooth chocolate paste. The paste is *conched*, which is a slow mixing process (100 or more hours) with heat added to increase product thickness and smoothness. The ingredient lecithin may be added at this point to improve flow behavior and to stabilize ingredients. Conching contributes important flavor and texture changes to the product.

Tempering of the conched product refers to the manner in which temperature of the product is manipulated to achieve the development as the desired stable crystal structure in finished chocolate. This step is key in the further processing of cocoa butter for chocolate manufacture. Properly manipulating the properties of this fat in terms of its melting behavior is critical. Cocoa butter is an unusual fat in that it can be solidified into six crystal forms depending upon conditions of tempering—see Figure 9.25. This phenomenon is referred to as **polymorphism.**

The crystal varieties (polymorphs) of chocolate form in a definite sequence. The less stable ones form first and transform into the more stable forms. The less stable forms (e.g., alpha) have "loose" molecular arrangements, which results in lower melting points. The more stable polymorphs have tighter (more dense) arrangements of the triglycerides that make up the crystal lattice, resulting in a higher melting point. Through proper tempering, all of the crystal forms are present, but the proper sequence of heating and cooling selects for the stable beta (β) form at the expense of the others. A 2–3 day cooling step to complete the crystallization ends chocolate processing, allowing packaging of this consumer favorite to begin.

A word about bloom in chocolate. *Bloom* is a visual defect that can appear on the surface of chocolate that removes the glossy appearance. It occurs when chocolate is stored at high temperature (above 27°C, 80°F) or experiences widely fluctuating temperatures, causing cocoa butter to crystallize on the surface as whitish streaks. Chocolate bloom develops naturally with time, but it can be brought on prematurely. It can form if the manufacturing process fails to include a tempering step in which the temperature is carefully raised and lowered to ensure that fat crystals grow in the correct form, size, shape, and number.

Although the exact mechanism of bloom formation remains elusive, many scientists feel that it involves fat crystal transformation from the β crystal form to form VI (Figure 9.25). Form VI crystals are more stable than the β form, so all chocolate should eventually bloom, unless preventive measures are taken. Richard Hartel of the Department of Food Science at the University of Wisconsin has a theory to explain how visual fat bloom develops even in properly tempered chocolates. A first requirement is that liquid fat must

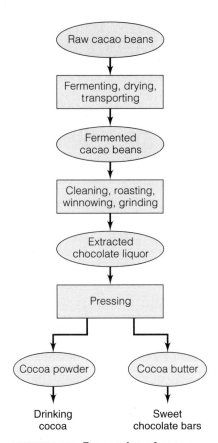

FIGURE 9.23 Processing of cacao beans in chocolate manufacture. *Only after roasting are the beans referred to as* cocoa beans.

Theobromine
(3, 7-dimethylxanthine)

Caffeine
(1, 3, 7-trimethylxanthine)

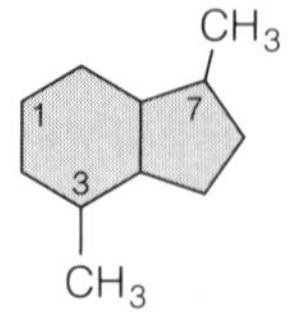

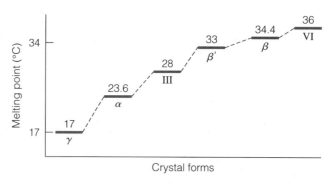

FIGURE 9.24 Relationship between theobromine and caffeine. *These two substances occur naturally in chocolate. (The structures are only generally drawn, without double bonds, to highlight the positioning of the methyl groups.)*

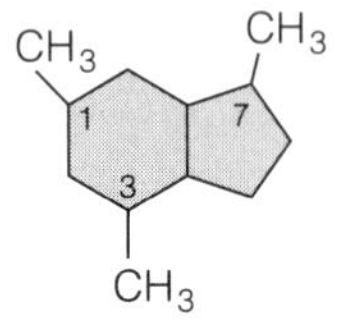

FIGURE 9.25 The art and science of chocolate manufacture. *With proper tempering, controlled crystallization of the cocoa butter is achieved. Heating melts the lower melting point crystals, and maintains the stable beta crystals upon cooling, which produce chocolate with proper performance characteristics (snap, gloss, and texture). During tempering, the melted chocolate is cooled to the point where some beta crystals begin to form. These function as "seed crystals." The cooled chocolate is then warmed to a temperature just below the melting point of the beta form. This melts all of the other crystalline forms due to their lower melting points, and selects for the stable beta crystal form. Temperatures are kept below 36°C to preclude formation of form VI crystals, which are not desirable in high-quality chocolate products and may be responsible for chocolate "bloom."*

be able to get to the surface of the chocolate. Temperature fluctuations would cause the fat crystals to melt and then recrystallize. High melting point crystals carried in any liquid fat reaching the surface could recrystallize as the "seeds" of bloom development.

Are there ways to inhibit or prevent bloom? Milkfat is commonly used to inhibit fat bloom, and skimmed milk powder exceeds whole milk in preventing bloom formation. By adding high melting point milkfat fractions to their chocolate mix, Hartel and his research team have observed a delay in polymorph transition from the beta form to form VI. Although the precise mechanism is unclear, denser crystal structures that form could potentially stop liquid fat from moving to the surface and recrystallizing. It has also been suggested that certain minor lipids, if present, might slow down the transformation of crystals from the beta form to form VI. Since milk fat can decrease the rate of fat crystallization, the chocolate would shrink less upon cooling, decreasing the number of microscopic cracks, and reducing the likelihood of liquid fat reaching the surface and causing bloom.

Key Points

- Fats and oils are refined to purify and prepare them for further processing.
- Interesterification of oils and fractionation of fats are methods used to create new fats and select for specific functionalities.
- Sugarcane processing consists of extraction, clarification, concentration, crystallization, and separation and drying into refined sugar.
- The processing of bottled waters such as natural mineral waters, carbonated waters, and sweetened, flavored waters includes a sanitizing processing step.
- A variety of funtional ingredients can be incorporated into water, juice, or milk-based formulations to produce nutraceutical beverages.
- Cereal grain processing converts cereal grains into food products, including bread, breakfast cereal, pasta, and snack foods.
- Preprocessing of fruits and vegetables includes trimming, washing, and blanching in order to minimize microbial contamination and deactivate enzymes.
- Preservation of fruits and vegetables is accomplished by chilling, dehydration, and picklilng.
- Hot-break and cold-break tomato processing affects the pectin methyl esterase enzyme, which in turn influences tomato paste viscosity.
- Soybeans are processed into a variety of products, including soy oil, nutraceutical ingredients, defatted soy flakes, soy meal, soy flour and grits, soy isolates, soy concentrates, texturized soy proteins, and full-fat soy flour.
- High oleic acid content soybeans have been genetically modified to increase nutritional value and extend shelf life.
- Tempering is a key step in the further processing of cocoa butter for chocolate manufacture that affects its crystal forms and melting behavior.

Study Questions

1. The sucrose product obtained from sugarcane is
   a. molasses.
   b. honey.
   c. cane sugar.
   d. sugar beet.
   e. sugar root.

2. Why does the addition of sugar to a dough mix delay gluten development?
   a. Sugar binds to glutenin.
   b. An emulsion is formed instead of a dough.
   c. Sugar promotes starch gelatinization.
   d. Yeast consumes the sugar.
   e. Sugar competes for water.

3. Which of the following is the correct sequence for sugarcane processing?
   a. extraction, separation, concentration, clarification, crystallization
   b. extraction, concentration, clarification, crystallization, separation
   c. extraction, concentration, crystallization, separation, clarification
   d. extraction, clarification, concentration, crystallization, separation
   e. extraction, separation, clarification, concentration, crystallization

4. What reaction leads to the production of flavor precursors in raw cacao beans?
   a. respiration
   b. fermentation
   c. flaking
   d. compounding
   e. photosynthesis

5. The main goal of sugarcane processing is
   a. crystallization
   b. compounding
   c. tempering
   d. fermentation
   e. evaporation

6. Which statement about cereal grains is false?
   a. Cereal grain processing converts grains into foods.
   b. Grains refers to the hard plant seeds.
   c. Rice is polished to remove bran and germ layers.
   d. Dry milling is used to produce corn oil from corn.
   e. Most grains are consumed in the processed state.

7. What is true regarding hot filling fluid foods like applesauce and most fruit juices?
   a. Hot filling requires heating to 200°C.
   b. High-acid foods are rendered commercially sterile without further thermal processing.
   c. The FDA requires that all hot-filled fluid foods and beverages be labeled "aseptic."

d. Hot filling causes spores to germinate and promotes growth of non–spore-forming bacteria.
e. Hot filling achieves 100 percent sterility (true sterility).

8. The ratio of _____ to _____ in bread dough is about 3:1.
   a. flour; water
   b. gluten: water
   c. water: gluten
   d. glutenin; starch
   e. water; flour

9. In this process, animal tissue is heated in steam to melt and separate out the fat.
   a. refining
   b. degumming
   c. winterizing
   d. rendering
   e. bleaching

10. Overmixing of dough ingredients
    a. enhances dough extensibility.
    b. dehydrates the starch.
    c. causes underdevelopment of gluten.
    d. leads to the best quality bread.
    e. enhances dough elasticity.

11. Bananas are climacteric fruits; therefore,
    a. they demonstrate maximum respiration rate before harvest.
    b. the best time to harvest them is before they are fully ripe.
    c. they do not continue to ripen after harvest.
    d. minimum respiration rate is achieved after harvest, just before ripening.
    e. the best time to harvest them is after they are fully ripe.

12. What technique using potassium iodide is used to measure oxidative rancidity?
    a. ADV
    b. PV
    c. HMO
    d. GMO
    e. nRT

13. The oxidative breakdown of organic biomolecules in fruits and vegetables
    a. produces $CH_2 = CH_2$.
    b. is called senescence.
    c. does not have anything to do with respiration.
    d. does not play a role in ripening or quality.
    e. requires the input of O_2 and releases carbon dioxide, water, and energy.

14. Which statement is true?
    a. Cocoa butter is composed of chocolate liquor and other substances.
    b. Chocolate liquor is composed of cocoa butter and other substances.
    c. Chocolate liquor is obtained by pressing cocoa butter.
    d. Cocoa butter is correctly referred to as unsweetened baking chocolate.
    e. Grinding the cacao bean nib produces roasted flavor notes.

15. Purified water meets the definition of
    a. United States Department of Water Systems Management.
    b. Bottled Water Association of America.
    c. United States Pharmacopeia.
    d. Pure Food and Water Act.
    e. Environmental Protection Agency.

16. What is the role of ozone?
    a. to sanitize fruit juice
    b. to increase the pH of a processed beverage
    c. to provide carbonation to soft drinks and seltzer water
    d. to add flavor to soft drinks
    e. to sanitize bottled water

17. Which term does not belong?
    a. gluten
    b. glutamine
    c. glutelin
    d. glutenin
    e. wheat flour protein

18. Which sequence is the correct one for breadmaking?
    a. mixing, sheeting, panning, fermentation
    b. mixing, fermenting, proofing, baking
    c. mixing, proofing, fermentation, baking
    d. molding, kneading, proofing, panning
    e. kneading, rising, gelatinizing, pasting

19. Which is *not* a processing step most associated with soft drinks?
    a. superchlorination
    b. carbonation
    c. sweetening
    d. flavoring
    e. fractionation

20. Which soy product has been produced through bioengineering?
    a. soy protein isolate
    b. soy protein concentrate
    c. high oleic acid soybean oil
    d. texturized soy protein
    e. high oleic acid soybean protein

21. In cold-break tomato processing
    a. heating step is used to make a more viscous product.
    b. heating is not employed.
    c. pme is deactivated.
    d. enzymes are thermally activated.
    e. heating step is used to make a thinner product.

22. A Lovibond value greater than 6.0 for a soybean oil most probably means
    a. oil color is very light.
    b. bleaching time is too brief.
    c. oil has been heavily blanched.
    d. oil is rancid.
    e. oil has been heavily bleached.

23. Suppose a team of food scientists are creating a new cooking oil for the space program, one that needs to have a melting point of 150°F (66°C). The cooking oil used previously had a melting point of 133°F (56°C). Which of the following statements then makes the most sense?
    a. The new fat will likely be unsafe.
    b. They should consider starting with a fully saturated fat.
    c. They should hydrogenate the original oil.
    d. They should force water molecules into the original oil to raise its boiling point.
    e. They should force hydroperoxide radicals onto the original oil.

24. Raw sugar from cane juice
    a. results from crystal separation and drying.
    b. has a moisture content of 5.0 percent.
    c. is pale golden to brown in color.
    d. results from crystal separation and drying; and is pale golden to brown in color.
    e. results from crystal separation and drying, has a moisture content of 5.0 percent; and is pale golden to brown in color.

25. What chemical reaction do you think would prove useful in the synthesis of chocolate flavor from sugars and proteins?
    a. Maillard
    b. compounding
    c. tempering
    d. fermentation
    e. melangeur

26. What is meant by tempering?
    a. fats like cocoa butter that exist in more than one crystal form
    b. selecting the α (alpha) crystal form of cocoa butter during melting and cooling
    c. selecting the β′ (beta prime) crystal form of cocoa butter during melting and cooling
    d. temperature manipulation so that cocoa butter will solidify into the desired, stable (β) crystal form
    e. keeping the temperature high enough that all of the crystal forms remain melted

27. To successfully operate a corn oil processing plant, what would be *most* essential?
    a. knowing how to press the corn oil from the kernel
    b. knowledge regarding the wet milling of corn
    c. knowing how to separate the oil from the endosperm
    d. knowledge regarding the dry milling of corn
    e. knowledge regarding the dry milling of wheat

References

Bamforth, C. W. 1995. Beer and cider. In: *Physico-Chemical Aspects of Food Processing*. S. T. Beckett, ed. Chapman & Hall. London.

Beckett, S. T. 1995. Chocolate confectionery. In: *Physico-Chemical Aspects of Food Processing*. S. T. Beckett, ed. Chapman & Hall. London.

Eeles, M. F. 1995. Sugar confectionery. In: *Physico-Chemical Aspects of Food Processing*. S. T. Beckett, ed. Chapman & Hall. London.

Guy, R. C. E. 1995. Cereal processing: the baking of bread, cakes and pastries, and pasta production. In: *Physico-Chemical Aspects of Food Processing*. S. T. Beckett, ed. Chapman & Hall. London.

Hartel, R. W. 1998. Phase transitions in chocolate and coatings. In: *Phase/State Transitions in Foods: Chemical, Structural, and Rheological Changes*. M. A. Rao and R. W. Hartel, eds., Marcel Dekker. New York.

Hoseney, R. C. 1986. *Principles of Cereal Science and Technology*. American Association of Cereal Chemists, St. Paul, MN.

Jones, H. F., and Beckett, S. T. 1995. Fruits and vegetables. In: *Physico-Chemical Aspects of Food Processing*. S. T. Beckett, ed. Chapman & Hall. London.

Lahl, W. J., and Braun, S. D. 1994. Enzymatic production of protein hydrolysates for food use. *Food Technololgy* 48(10): 68–71.

Minifie, B. W. 1989. *Chocolate, Cocoa, and Confectionery: Science and Technology*, 3rd ed. Van Nostrand Reinhold, New York.

Pedersen, B. 1994. Removing bitterness from protein hydrolysates. *Food Technololgy* 48(10): 96–98.

Additional Resources

Bennion, M. 1995. *Introductory Foods*. Prentice Hall. Upper Saddle River, NJ.

Bergenstahl, B. 1995. Emulsions. In: *Physico-Chemical Aspects of Food Processing*. S. T. Beckett, ed. Chapman & Hall. London.

Bowler, P., Loh, V. Y., and Marsh, R. A. 1995. Preserves and jellies. In: *Physico-Chemical Aspects of Food Processing*. S. T. Beckett, ed. Chapman & Hall. London.

Charley, H., and Weaver, C. 1998. *Foods: A Scientific Approach*. Prentice Hall, Upper Saddle River, NJ.

Gibson, R. M. 1995. Fermentation. In: *Physico-Chemical Aspects of Food Processing*. S. T. Beckett, ed. Chapman & Hall. London.

McWilliams, M. 1997. *Foods: Experimental Perspectives*. Prentice Hall, Upper Saddle River, NJ.

Potter, N. M., and Hotchkiss, J.H. 1995. *Food Science*. Chapman & Hall, New York.

Talbot, G. 1995. Fat eutectics and crystallization. In: *Physico-Chemical Aspects of Food Processing.* S. T. Beckett, ed. Chapman & Hall. London.

Vaclavik, V. A. 1998. *Essentials of Food Science.* Chapman & Hall, New York.

Vieira, E. R. 1996. *Elementary Food Science,* 4th ed. Chapman & Hall, New York.

Web Sites

www.scisoc.org/aacc
American Association of Cereal Chemists (AACC), an international organization of cereal science and other professionals.

www.agr.gc.ca/food/profiles/fruitveg/fruitveg_e.html
The Canadian Fruit and Vegetable Processing Industry (Food Bureau of Agriculture and Agri-Food Canada).

www.agr.ca/food/profiles/profiles_e.html
The Canadian Food and Beverage Sector Profiles

www.feodora.de/english/herstellung/kakaoverarbeitung.html
Cocoa processing

www.chocolateandcocoa.org/Research/researchindex.htm
American Cocoa Research Institute, cocoa research

www.monitorsugar.com/htmtext/btintro.htm
The Monitor Sugar Processing Facility.

www.uniqema.com/food/confectionery.htm
Confectionery products.

www.nutraceuticalalliance.com/research_oils.htm
Nutraceutical Alliance—Research: Oils as Functional Foods

ENZYMES IN FOOD PROCESSING—THE PROTEIN HYDROLYSATES

Enzymes have found multiple uses in the food processing industry, as processing aids and in the production of novel food ingredients. Proteins are polymers of amino acids joined by peptide linkages. When acted upon by enzymes called proteases, the peptide bonds break, and fragmentation of the polypeptide occurs via hydrolysis. The products of enzymatic hydrolysis of proteins are called **protein hydrolysates.** As such, protein hydrolysates represent modified proteins generated by enzyme technology.

These products vary in size and as a result offer different functionalities. Food processors have developed methods to produce protein hydrolysates for specific reasons. Protein hydrolysates are produced by reacting substrate protein material with protease enzymes. The specific type of enzyme used influences the functional nature of the product. Exogenous enzymes are a type not present in the food material itself, but obtained from other sources, such as bacteria. Exogenous protease enzymes may exhibit *exoenzyme* or *endoenzyme* activity. These create breaks in polypeptides in different regions, resulting in production of hydrolysates (Figure 9.26).

The sequence of steps to manufacture a protein hydrolysate is shown in Figure 9.27. To properly hydrolyze the substrate material, processors use a combination of *exo-* and *endopeptidase* enzymes. As an example, the substance **pancreatin,** which is a mixture of two proteases (chymotrypsin and trypsin) exhibits both exo- and endopeptidase activity. This enzyme mixture can act specifically to break arginine-lysine peptide bonds. Other proteases such as bromelain, ficin, papain, or bacterial proteases (e.g., subtilisin) offer varying degrees of selectivity or nonselectivity regarding their site of action along a polypeptide chain.

In general, the protein hydrolysates commercially produced fall into three categories: (1) flavor compounds or flavor enhancers, (2) nutritional additives, and (3) bioactive nutraceutical substances. The best food protein sources used as substrates to produce these compounds are soybean proteins and milk proteins—the caseins and whey. Work has also been carried out with egg, fish, and gelatin proteins.

Soybeans are unique among plant products because of their relatively high natural protein content (almost 40 percent) and because they are good sources of essential amino acids including glutamic acid, lysine, and tryptophan. Soy protein hydrolysates include hydrolyzed vegetable protein, used for making meat analog products (soy burgers), and hydrolyzed soy protein, used to produce alternative milk products (soy milk).

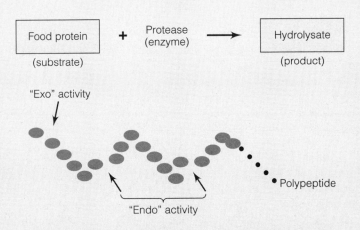

FIGURE 9.26 Action of exogenous and endogenous ("endo") enzyme activity to produce protein hydrolysate. *Proteins are polymers of amino acids joined by peptide bonds. Protease enzymes break the peptide bonds at outer and inner polymer regions to produce protein hydrolysates of various sizes. Protease enzymes used in creating protein hydrolysates are selected based upon their specificity, ability to hydrolyze the substrate to the extent desired, as well as their cost and availability.*

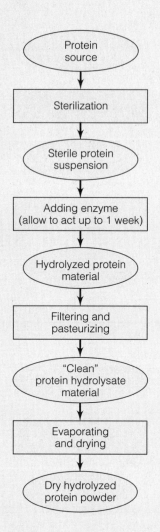

FIGURE 9.27 Flowchart for the manufacture of protein hydrolysate ingredient.

Enzymatic hydrolysis of protein material, be it soy, milk protein, or wheat gluten, produces hydrolysates of various sizes and molecular weights. Let's examine a parameter called the **AN/TN ratio**. By definition, this ratio is a measure of the relative amount of amino nitrogen (AN) present in a hydrolysate relative to the total amount of nitrogen (TN) in the intact protein substrate. As the molecular weight of a hydrolysate decreases, the AN/TN ratio increases (Table 9.14). In practice, although this ratio is a fraction, it is multiplied by 100 to give a number between 1 and 100. Therefore, if a measured AN/TN ratio is 0.60, in common usage processors talk about it as being 60 (which is 0.60 times 100). A protein hydrolysate with an AN/TN ratio of 70 is a smaller molecule than one with a ratio of 50.

The AN/TN ratio is an indicator of the hydrolysate's size and potential function properties. Since hydrolysates are produced by enzymatic catalysis, factors affecting enzyme kinetics will similarly affect the reaction rate and can be used to control the reaction. These factors include temperature, pH, and reaction time. Since enzymes themselves are proteins, they are inactivated by high temperature and low pH, which causes the unfolding of their structure (denaturation). Practice has shown that hydrolysate production is most efficient at temperatures in the range 90–120°F (32–49°C) and at neutral pH.

Through proper selection of enzymes for a particular substrate and by influencing the hydrolysis reaction by selecting the appropriate processing controls, processors can control and predict the size, amino acid content, and therefore functionality of resulting hydrolysates. Some of the important hydrolysate properties include emulsifying ability, foaming ability, gelling action, moisture binding, nutritional value, solubility, and whipping characteristics.

Processors must also consider the potential for microbial contamination of the hydrolysate during processing and storage and then take appropriate steps to eliminate this possibility. After all, the hydrolysate offers a rich source of nutrients, which are likely at a pH and temperature conducive to bacterial survival (including the pathogens) and growth. Processors must also standardize the nutrient content and availability, flavor, and storage stability of their hydrolysates.

Certain peptide fragments found in hydrolysates contribute to bitter flavor, especially peptides that contain hydrophobic amino acids, meaning amino acids that have hydrophobic side chains. Food processors have several approaches available to handle bitterness, including masking the bitterness by adding polyphosphate compounds, using further enzyme treatment to convert the bitter peptides to nonbitter "plastein" products, and passing the hydrolysate through a special adsorbant material to capture and separate out the bitter peptides. Table 9.15 shows which amino acids are most likely to be bitter.

Nutraceutical research on protein hydrolysates has produced substances such as **glutamine peptide** (GP). GP is produced from wheat gluten, and is reported to boost the immune system. It is purported to promote other physiological benefits and is marketed to the sports nutrition industry as well as in clinical settings.

Other claims for GP include improvement in postexercise muscle recovery, providing glycogen replenishment by supporting gluconeogenesis, diminishing oxidative stress in the elderly, enhancing recovery from stress and fatigue, and providing an easily

TABLE 9.14 Relationship between hydrolysate size (as MW, molecular weight) and AN/TN ratio.

HYPOTHETICAL PROTEIN HYDROLYSATE CHARACTERISTICS		
Hydrolysate	MW	AN/TN
Intact protein	30,000	<0.10
Proteose	7,000	0.10
Peptone	2,000	0.25
Peptides	500	0.40
Peptides and free amino acids	260	0.60

assimilated source of dietary glutamine for gastrointestinally impaired and immunocompromised patients. Potential food uses for GP includes not only its production from wheat via enzymatic hydrolysis, but also the manufacture of nutrition bars, beverages, powdered drink mixes, and liquids packaged in retort pouches containing this nutraceutical protein hydrolysate ingredient.

Challenge Questions

1. What is *not* true regarding GP?
   a. GP is a protein hydrolysate.
   b. GP is derived from enzyme-treated wheat gluten.
   c. GP is a larger polypeptide than gluten polypeptides.
   d. GP is claimed to enhance health.
   e. Amylase treatment of wheat would not be effective in producing GP.

2. Fragmentation of polypeptides
   a. eliminates protein functionality.
   b. enhances protein functionality.
   c. inhibits production of hydrolysates.
   d. is caused by exo- but not endopeptidase action.
   e. is a random event that cannot be controlled.

3. The AN/TN ratio
   a. is a measure of the relative amount of amino nitrogen (AN) present in a hydrolysate.
   b. decreases as the molecular weight of a hydrolysate decreases.
   c. is a measure of the total amount of nitrogen (TN) present in an intact protein substrate.
   d. increases as the molecular weight of a hydrolysate decreases.
   e. is a measure of the total amount of nitrogen (TN) present in a hydrolysate.

4. Why is hydrolysate production most efficient at temperatures in the range of 90–120°F and at neutral pH?
   a. Hydrolysates are unstable in acid.
   b. High pH stabilizes hydrolysates.
   c. Temperatures below 90°F cause denaturation.
   d. A pH of 7 destabilizes the hydrolysates.
   e. Hydrolysates lose functionality at 100°F and pH of 6.8.

5. Which hypothetical protein hydrolysate below would be most bitter? (Arg = arginine, Asn = asparagine, Asp = aspartic acid, Gly = glycine, Leu = leucine, Met = methionine, Val = valine).
   a. Gly-Gly-Asp
   b. Asp-Met-Asp
   c. Leu-Val-Phe
   d. Glu-Met-Asp
   e. Asp-Glu-Arg

TABLE 9.15 Relative bitterness as related to hydrophobicities of various amino acids. *Hydrophilic amino acids and hydrolysates from hydrophilic proteins (e.g., gelatin) tend to be nonbitter, while hydrophobic amino acids and hydrolysates from hydrophobics proteins tend to be bitter. The chemical nature of the amino acid R group determines hydrophobicity.*

Amino acid	Bitterness value
Asparagine	Lowest
Glycine	
Alanine	
Arginine	
Methionine	Moderate
Lysine	
Phenylalanine	Highest
Tyrosine	
Isoleucine	
Tryptophan	

CHAPTER 10

Food Microbiology and Fermentation

10.1 What Are Microorganisms?

10.2 Factors Affecting Microbial Growth

10.3 Foodborne Microorganisms

10.4 Food Spoilage by Microorganisms

10.5 Microbial Fermentation

Challenge!
Microbial Sampling to Verify Food Quality

CHAPTER OBJECTIVES

After completing this chapter, you will be able to:
- List the four types of foodborne microorganisms.
- Explain the six factors that affect microbial growth, including temperature, pH, and water activity.
- Discuss the sources that contribute to the microbial flora of foods.
- Describe the microorganisms associated with meats, seafood, fruits and vegetables, and dairy products.
- Explain how food spoilage occurs.
- Describe the microbial fermentation of milk products, meat products, fruit and vegetables, and cereal grains.
- Discuss how microbial sampling can be used to verify food quality.

Food microbiology is the study of microorganisms in foods: bacteria, yeasts, molds, protozoa, and viruses. Biology and chemistry are basic to this topic, as scientists study the effects that microorganisms have on food, both beneficial and detrimental. Disease-causing microorganisms are studied within the context of food safety. Those that cause food spoilage are the concern of food quality. Yet others are useful because they can produce substances that are beneficial in food processing—important in fermented food product development. In this chapter we examine the types of microorganisms found in foods, the factors that affect their survival and growth in foods, where they are found, and how they contaminate food. We then concentrate on microorganisms that cause spoilage in foods and those used in food production. The information in this chapter serves as a good foundation for and link to Chapter 11, Food Safety, which looks closely at the microorganisms that cause foodborne illness.

10.1 WHAT ARE MICROORGANISMS?

Simply stated, **microorganisms** are living entities that are too small to be seen with the naked eye. These comprise bacteria, viruses, protozoa, and fungi such as yeasts and molds. The primary function of microorganisms is self-perpetuation. Many microorganisms utilize organic matter (carbohydrates, proteins, lipids, etc.) in nature to form inorganic compounds (nitrates, sulfates, etc.). Parasites and viruses depend instead on a living host to obtain most nutrients and to carry out metabolic reactions required for growth.

All living organisms are classified as either **procaryotes** or **eucaryotes.** Procaryotes (from the Greek meaning "before nucleus") are organisms with no nucleus in their cells—the bacteria. The cells of eucaryotes (from the Greek meaning "true nucleus") have a nucleus, and this group includes the fungi, protozoa, plants, and animals. Viruses are neither because they are noncellular. However, they are considered a life form.

Like all organisms, microorganisms have scientific names consisting of a genus name and a species name, such as the bacterium *Escherichia coli.* Typically, organisms belonging to the same genus share one or more prominent *phenotypic*, or apparent, characteristics. Organisms belonging to the same species share many phenotypic characteristics as well as being genetically very similar, with at least a 70 percent similarity in their nucleic acid material (e.g., RNA or DNA).

Bacteria are unicellular organisms, measuring about 1 micron in length. They are found just about everywhere in nature, including soil, air, water, and the intestinal tract and mucous membranes of animals and humans. They are divided into *gram-positive* and *gram-negative* cells, according to whether they can retain crystal violet in the cell membrane during a staining procedure known as **gram staining.** The gram stain was developed by Dr. Christian Gram, a Dutch physician, in 1888. Gram-negative bacteria have a thin cell wall and an outer membrane, while gram-positive bacteria have a thick cell wall and no outer membrane.

Bacteria are also classified according to their shape: the spherical coccus, the rod-shaped bacillus, and the cell with twists—the spirillum (see Figure 10.1). An example of

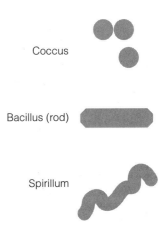

FIGURE 10.1 Typical shapes of bacterial cells.

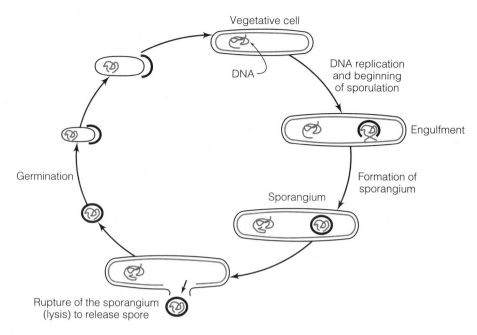

FIGURE 10.2 Life cycle of the bacterial spore. *The sporangium is the newly formed spore inside the remnant of the dead vegetative cell. Only the spore is alive, but dormant.*

the coccus type is *Staphylococcus aureus*, one of the leading causes of foodborne illness. Coccus-shaped organisms can also occur in the *diplococcus* arrangement, where two cells remain attached after cell division. Rod-shaped organisms include most disease-causing bacteria, such as *Salmonella enteritidis*, *E. coli* O157:H7, and *Yersinia enterocolitica*. Organisms such as *Campylobacter*, the leading cause of diarrhea in the world, has a curved-rod shape, a variation of the spirillum.

Some bacteria are able to develop into **spores** by coating their membrane and cell wall with extra layers of material in a process called *sporulation*—see Figure 10.2. Spore development often occurs as a response to unfavorable growth conditions, such as lack of nutrients or lack of water. In this state, bacteria are able to significantly increase their survival to processing treatments such as heating, drying, and irradiation. However, they are in a pseudodormant condition, unable to grow or divide. Once conditions become favorable again, or if exposed to a short heat treatment, the spores germinate into **vegetative cells,** resuming growth and metabolic activities.

The fungi include two types of microorganisms important to food microbiology (Figure 10.3). **Molds** are multi- or unicellular, found in decaying organic matter. They grow in the form of tangled mass called the *mycelium*, which spreads readily and may cover several inches within 3 days. The mycelium is composed of filaments called *hyphae*, which are basically nucleated tubes containing a cytoplasm and several nuclei. Molds of importance in foods multiply by spores known as *conidia*, which sit on top of the hyphae, giving mold a dusty look. Some molds produce toxins, antibiotics, and even enzymes that are useful in food production.

Yeasts are the other type of fungi, unicellular in structure. They can grow over a wide range of conditions. Most yeasts do not grow in the form of mycelia, but rather as single cells that are spherical or oval-shaped. This allows for greater surface area to volume of medium, easily distributing the cells and allowing for faster growth than molds.

Protozoa are single-celled eucaryotes that are classified by morphology, locomotion, and life cycle. The protozoa of interest to food scientists are parasites. They do not grow in food but require at least one animal host to carry out their life cycle. Most are *phagotrophic*, able to ingest particulate food. Parasitic protozoa such as *Giardia lamblia* are found in the intestinal tract of both wild and domestic animals. They exist in the form of a cyst, which germinates upon ingestion with the help of stomach acid and proteases from the pancreas. One cyst yields several **trophozoites,** motile parasites that penetrate the small intestine of the infected animal. If the cyst does not germinate, it can pass through the body and be excreted in the feces.

Viruses are obligate parasites. For the most part, they are host-specific, with animal viruses not able to infect plants and plant viruses not able to infect animals. Viruses are very simple in structure, composed of a proteinaceous capsule in which nucleic acid resides (Figure 10.4). The ones associated with food are typically RNA-containing viruses. They attach to host cells by receptors, and then either inject their nuclear material into the host or become engulfed by the host. Once inside the host, the virus nucleic acid is replicated using the host's enzymes, and virus particles are synthesized. The mature virus is released, with a *virulent virus* causing the host cell to lyse or rupture. Most food viruses are considered **temperate viruses,** which insert their nucleic acid into the host's DNA, leaving the host cell intact. The host may eventually die but lysis is not necessary.

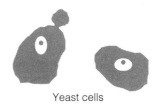
Yeast cells

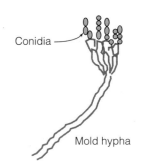
Conidia
Mold hypha

FIGURE 10.3 Typical yeast and mold structures.

10.2 FACTORS AFFECTING MICROBIAL GROWTH

The environment has a profound effect on the survival and growth of microorganisms. We will discuss the availability of nutrients, the amount of available water and oxygen, pH, the temperature of the medium, and the presence of inhibitory substances as factors affecting the ability of microbial cells to function in foods.

Nutrient Availability

Most nonparasitic organisms can be classified as chemotrophs or phototrophs. **Chemotrophic** organisms require chemicals for metabolism. Chemotrophs can be subdivided into **lithotrophic** organisms, if they require inorganic compounds such as minerals, or **organotrophic** organisms, if they require organic compounds such as carbohydrates. **Phototrophic** organisms are those that require energy in the form of light to live. Foodborne microorganisms are chemotrophic. The metabolic reactions used to digest these compounds will be discussed later in this chapter.

The nutrient needs of microorganisms depend on the organism and on other factors such as temperature—microorganisms may become more exacting in their requirements during refrigeration. Most foods contain sufficient nutrients to support the growth of nonparasitic organisms. The organisms that predominate in a food are the ones that can utilize the food components optimally. For instance, *pectinolytic* organisms occur in pectin-containing foods such as vegetables, while organisms that metabolize lactose are found in milk. Some organisms that cannot grow by themselves in a particular food may be able to do so if other organisms are present. For example, bacteria able to digest food components with special enzymes into low molecular weight compounds can serve to provide the organisms that lack these enzymes with nutrients that they require.

Water Activity

Bacteria are complex microorganisms, with requirements that are often more strict than those of other organisms. They need quite a bit of water to survive, as measured in terms of **water activity,** or a_w, the amount of water available for microbial growth. It is water that is unbound, not involved in chemical interactions with food components. Most bacteria require a minimum a_w of 0.90, with a value of 1.0 being the maximum possible. This minimum compares with 0.80 for yeasts and 0.70 for molds. There are exceptions, with the bacterium *Staphylococcus aureus* able to survive in environments where the a_w is 0.83, and the yeast *Saccharomyces cerevisiae* requiring at least 0.90.

At low water activities, microorganisms die because water inside the cell diffuses out in an effort to balance the osmotic pressure. In other words, since the concentration of solutes is greater outside the cell than inside, migration of water takes place in order to equalize the concentration. This migration results in cell death due to dehydration shrinkage.

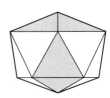

FIGURE 10.4 Structure of food-borne virus. *The capsule consists of ribonucleic acid (RNA) enclosed in a protein capsule.*

TABLE 10.1 Water activity (a_w) and pH requirements of microorganisms.

Microorganism	Minimum a_w	Minimum pH	Optimum pH	Maximum pH
Bacteria*	0.90	4.5	6.5–7.5	9.0
Yeasts	0.80	3.0	4.0–6.5	8.5
Molds	0.70	2.0	4.5–6.8	11.0

* Some, such as lactic acid bacteria, can tolerate very low pH conditions.

Acidity/Alkalinity

The pH of the medium in which microorganisms live must be in a certain range. For instance, many bacteria cannot survive at pH values less than 4.5, with a preference for values near neutrality. Deviations from these values can and do occur in foods due to interactions called food effects, which will be discussed in a moment. Yeasts can live at pH 3.0, with a maximum of 8.5. This broad range enables yeasts to survive in environments not suitable for most bacteria. Molds can tolerate even more extreme pH values than yeasts, with tolerance to pH between 2.0 to 11.0. Table 10.1 summarizes water activity and pH requirements for bacteria, yeasts, and molds.

The pH level alters a microorganism's ability to transport molecules in or out of the cell through the cell membrane. In an acidic environment of low pH, protons (H^+) saturate the membrane, making it difficult for cations to move in or out. At high pH, hydroxide (OH^-) ions saturate the membrane, preventing the movement of anions in or out of the cell. In addition, proteins, which are very sensitive to pH changes, are denatured and precipitate out of solution, disrupting the metabolic activity of the cell.

Oxygen

The amount of oxygen in the environment is also a crucial survival/growth factor for microorganisms. In discussing the oxygen content of a medium, we need to understand that we are really talking about the oxidation-reduction potential of that medium. This **redox potential** depends on the ratio of total oxidizing (electron-accepting) molecules to the total reducing (electron-donating) molecules in the medium. An **oxidized** environment means that the molecules have a relatively high affinity for electrons, while a **reduced** environment means that the molecules have a low affinity for electrons. To better illustrate this point, consider the reduction reaction of pyruvate by bacteria, as shown in Figure 10.5. In this reaction, pyruvate (an oxidized molecule) readily accepts the electrons from the protonated form of nicotinamide adenine dinucleotide (NADH + H^+), which causes it to become reduced into lactic acid. At the same time, NADH + H^+, a reduced molecule, is oxidized to NAD^-.

Molds are **aerobes,** requiring oxidized conditions, thus oxygen, to be present. Some bacteria, notably the ones that cause food spoilage, are also *strict aerobes.* Most bacteria that cause disease, such as *Salmonella,* are **facultative anaerobes.** This means that they prefer an aerobic environment but have the capability of growing even if oxygen is not present (i.e., a reduced environment). Other bacteria are classified as **microaerophilic**—they require some oxygen to be present but cannot tolerate the levels

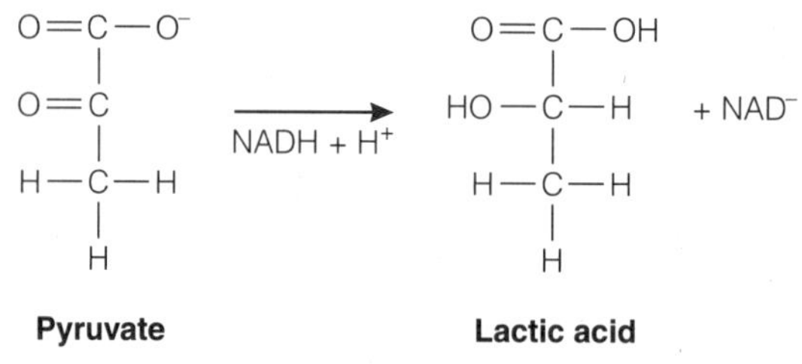

$$\text{Pyruvate} \xrightarrow{NADH + H^+} \text{Lactic acid} + NAD^-$$

FIGURE 10.5 Reduction of pyruvate by bacteria.

TABLE 10.2 Classification of microorganisms according to temperature requirements.

Microorganism	GROWTH TEMPERATURE		
	Minimum (°C)	Optimum (°C)	Maximum (°C)
Psychrophiles	−10	10–15	18–20
Psychrotrophs	−5	20–30	35–40
Mesophiles	5–10	30–37	45
Thermotrophs	10	42–46	50
Thermophiles	25–45	50–80	60–85

present in aerobic environments, usually 21 percent O_2. *Campylobacter* is a microaerophilic organism, requiring 6 percent oxygen. Finally, some microorganisms are strict **anaerobes** that cannot tolerate any oxygen, requiring the environment to be completely reduced.

Aerobic microorganisms die due to lack of oxygen because of their inability to produce **adenosine triphosphate** (ATP), the necessary molecule for cell-building reactions. Anaerobic microorganisms die due to too much oxygen because of their inability to remove toxic oxygen-derived radicals, such as $O_2\bullet^-$ (**superoxide radical**), from the cell. These radicals can disturb the metabolism of the cell by replacing essential molecules in chemical reactions.

Temperature

Refrigeration techniques are a modern development, but restricting the growth and survival of microorganisms in foods by temperature control has been practiced for centuries. Biblical accounts of food being stored and preserved in cool caves during the time of the Roman siege on Masada date from the first century AD. Today we classify microorganisms into five categories, according to their ability to tolerate specific temperatures—see Table 10.2.

To understand Table 10.2 we need to define some important terms regarding temperature and microbial survival and reproduction. The **maximum growth temperature** is the temperature that causes inactivation of an organism's enzymes and structure damage to the extent that these outbalance the enhanced ability to synthesize new cell material. **Optimum growth temperature** is the temperature that corresponds to the shortest generation time (time it takes for the cells to divide), usually a matter of minutes. **Minimum growth temperature** is the temperature corresponding to the longest generation time, usually exceeding one thousand minutes.

Psychrophiles are organisms that prefer low temperatures, whereas **psychrotrophs** prefer higher temperatures, but can grow at low temperatures. Similarly, **thermotrophs** merely tolerate high temperatures, whereas **thermophiles** prefer them. **Mesophiles,** a classification to which most disease-causing microorganisms belong, cannot tolerate extremes of temperature, preferring the levels found in the tissues of humans and animals.

Cell death due to low temperatures occurs because of a slowing down of reaction rates during metabolism and because of a decrease in cell membrane fluidity, which slows down transport of nutrients into the cell. Death due to high temperatures occurs because of inactivation of enzymes, as well as denaturation of cell structural components.

Food Effects

The nature of the food in which microorganisms live can affect how and if they can cope with conditions that may not be suitable for survival and growth. For instance, foods with the same water activity but different amounts of antimicrobial substances, such as phosphates, may have different levels of microorganisms. Foods that are high in protein can exert a **buffering effect,** such that microorganisms are able to live and grow in the food even if the pH is below the minimum levels necessary for survival.

Certain components and characteristics of foods can prevent a change in the redox potential of the food in spite of the oxygen content of the atmosphere in which it is packaged. This is called the **poising effect,** and it depends on the presence of reducing compounds such as some sugars, the ability of the food tissue such as fruit and vegetables to use oxygen, and the pH of the food.

The more alkali the pH, the more negative is the redox potential, thus the more reduced and anaerobic is the food. In addition, the food's access to the atmosphere immediately in contact with it can affect how oxidized or reduced it is. For example, foods "at rest" (do not require active preparation) compared with those that are mechanically agitated can differ in redox potential. This difference affects the oxygenation of the product and can select for specific microorganism survivors: anaerobic versus aerobic.

Microorganisms can also survive and even grow at temperatures designed to inhibit them. Such survival may be due to the insulating or shielding effect that a high-protein and high-fat food can have on cells. However, some microorganisms can inherently survive extremes of temperatures through special abilities. Some bacteria are able to switch their metabolism to synthesize more unsaturated fatty acids, as well as more short-chain fatty acids, which may allow for survival at lower temperatures. These compounds act to strengthen the cell membrane, protecting it from damage caused by ice crystals during freezing. Other organisms can synthesize enzymes that are adapted to function at relatively low temperatures.

Using the Hurdle Concept

The strategy known as the hurdle concept was introduced in Section 8.3—see Figure 8.9. Nonlethal levels of the various factors discussed above can be used in combination to inhibit or reduce microorganisms in foods. For example, combining the absence of oxygen, such as in vacuum packaging, with refrigeration can inhibit the growth of aerobic bacteria commonly involved in spoilage of fresh meats. In this example, lack of oxygen and low-temperature storage act as two hurdles to impede bacterial growth. The hurdle concept is more effective when several hurdles are applied.

Examples of some products in which the hurdle concept is applied for food preservation are fermented dairy products, where a low pH due to the presence of lactic acid is combined with refrigeration. Foods that are partially or fully cooked, such as frankfurters, also contain inhibitory substances that prevent growth by any organism that may have survived the heat treatment. Pepperoni is an example in which more than two hurdles are applied: cooking, addition of curing agents, and lowering of water activity result in almost complete inhibition and elimination of contaminating microorganisms, resulting in a shelf-stable product.

10.3 FOODBORNE MICROORGANISMS

Sources of Microorganisms

Microorganisms can be found just about anywhere in nature, including air, water, and soil. Gusts of wind pick organisms up from the soil, and they become airborne. Splashing of water containing microorganisms against surfaces forms **aerosols,** causing microbial cells to become airborne. The air near Earth's surface is more contaminated than air at higher altitudes, due in part to a higher surface temperature and in part to being closer to sources of microorganisms such as soil and water. And the air over land is more contaminated than the air over oceans. With respect to season, the summer months are highest in terms of microbial content of air.

Our planet is a reservoir of water and moisture. A certain amount of this water becomes contaminated by contact with soil as well as with air carrying microorganisms. Even the most pristine lake has approximately 10^5 to 10^6 microorganisms per milliliter. Deep lakes become thermally stratified during the summer, so that the surface of the water is warmer than the rest. It is not surprising, then, that the surface is where most microbial activity takes place.

What is the effect of water depth? The more shallow the body of water, the more contamination it has because oxygen from the atmosphere is readily available for microbial growth. A few generalizations regarding bacteria and water: Bacteria survive in lakes and oceans by degrading the excess organic material excreted by algae, as well as the remains of dead animals, such as fish (Figure 10.6). Marine bacteria tolerate and even require salt, making the sea environment, which contains 3.5 percent salt, a suitable environment. Water can also become contaminated with microorganisms living inside the intestinal tract of aquatic animals.

Soils harbor many types of microorganisms in a reduced state of activity because the soil offers a relatively poor environment. Lack of nutrient availability, dry conditions, and low temperature make soil a somewhat hostile environment for microbial growth. For this reason, most bacteria found in soil are spore-formers because spore production gives them a survival advantage when environmental conditions fall below optimum. Spores also offer resistance to mild or moderate food preservation and processing conditions.

The microbial flora of foods consists of microorganisms originating from the following sources:

- microorganisms associated with the raw food
- microorganisms acquired during handling and processing of food
- microorganisms that survived the preservation and storage treatments applied to the food.

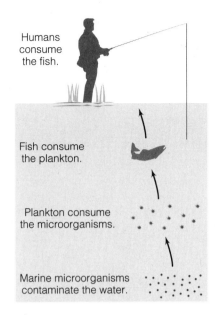

FIGURE 10.6 Contamination of water with microorganisms. *Microorganisms in water contaminate sea life, thereby providing a sequence of passage of microorganisms from water to plankton to fish to humans.*

Raw foods become contaminated through contact with air, water, or soil, or through contact with already contaminated surfaces. Foods of plant origin, for example, are in direct contact with air, water (both rainwater and irrigation water), and soil during production on the farm. Animals are also in direct contact with these sources. Animals may acquire microorganisms through consumption of contaminated feed and by contact with insects and other pests. Animals can harbor microbial contaminants in their intestinal tract, their skin, hair, and mucous membranes, making it easy for the spread of microorganisms through contact with other animals and their feces. Similarly, plants can also become contaminated with organisms associated with animals through contact with manure used as fertilizer.

During handling and processing of raw foods, contamination can come from the foodhandler, as humans also harbor microorganisms on their skin, hair, mucous membranes, intestines, and on their clothing. Utensils and equipment used in the harvesting and processing of foods can carry contaminants, which can be transmitted to other surfaces as well as directly to food. Finally, depending upon the severity of preservation and storage time and temperature treatments, some microbes may survive.

Types of Microorganisms Found in Food

Of all the kinds of microorganisms found in food, three types are most relevant: (1) those that spoil food, (2) those that are pathogenic, or cause disease, and (3) those that are useful for food production. Although the following discussion will include all three types, more thorough information on pathogenic organisms can be found in Chapter 11, Food Safety.

Muscle Foods. Fresh meat cuts usually have a microbial load of approximately 10^3 total organisms per gram. The organisms consist primarily of bacteria, yeasts, and molds, although viruses and other parasites can also be present. The interior of the meat is free of contaminants, or **sterile,** unless the animal has had a prior infection. Contamination of fresh cuts occurs from contact of the muscle tissue of the animal carcass with the contaminated hide of the animal or with the contents of the intestinal tract during slaughter. The use of knives, saws, and other equipment during fabrication can add contaminants to the surface of the meat.

If the meat is to be ground, the grinding operation increases the surface area of the meat and distributes any contaminants that were on the surface to all parts of the product.

Thus, ground products can contain as many as 100,000 (10^5) total organisms per gram of food.

The organisms usually present in fresh beef and pork are psychrotrophic aerobes that cause spoilage, such as *Pseudomonas* and *Acinetobacter*. Foodborne pathogens like *Salmonella* can also be present. **Mechanically deboned** meat, such as poultry, can have contaminants added when the special machinery removes the bones. A small amount of bone powder added to the product actually raises the pH slightly, which favors the growth of bacteria.

Most fresh meat is packaged under vacuum prior to shipping. The residual oxygen that remains in a vacuum package is depleted by aerobic bacteria and molds, causing an increase in the concentration of carbon dioxide on the meat's surface. This reduces the environment inside the package even more, rendering it anaerobic.

The population of microorganisms on the meat shifts from primarily gram-negative bacteria, such as *Pseudomonas*, to mostly gram-positives. Most of these organisms are **lactic acid bacteria**, able to ferment sugars under anaerobic conditions to form lactic acid. Examples are *Lactobacillus, Lactococcus, Leuconostoc, Streptococcus*, and *Micrococcus*. In addition, *Brochothrix thermosphacta* and other facultative anaerobic spoilage organisms are present and can grow under these conditions. These organisms cause a decrease in the pH of the product during growth, which is somewhat inhibitory to foodborne pathogens.

In the case of seafood, *Pseudomonas* is found in fish obtained from cold waters, as are *Acinetobacter, Moraxella, Shewanella*, and *Flavobacterium* species. In contrast, *Micrococcus* and *Bacillus* species are more commonly found in seafood from warmer, more tropical waters. Bacteria are usually present on the outside slime layer of fish, on the gills, and inside the intestines. The microflora of these products is influenced by the water quality from which they are harvested. As far as pathogens are concerned, it is not uncommon to find *Vibrio* and *Aeromonas* species present in seafood.

Harvesting contributes largely to the contamination of these products, with handling and storage on ships affecting their microbial quality. For example, the practice of cooling the fish by dumping ice on top of the catch can actually add microorganisms if the ice itself is contaminated. In addition, if ice is placed on a large load of fish, only the ones at the upper surface become refrigerated, with those at the middle and bottom of the pile not being cooled at all. This can lead to the proliferation of microorganisms, including pathogens.

Table 10.3 lists the microorganisms that contaminate muscle tissue, fruits and vegetables, and dairy products. Table 10.4 lists the conditions where specific microorganisms are found.

Fruits and Vegetables. Since fruits have a low pH, they do not support the growth of bacteria very well. However, acid-tolerant types such as *Lactobacillus* and *Leuconostoc*, as well as yeasts and molds, can be found in these products. As with fresh meats, vegetables contain predominantly lactic acid bacteria. Spore-forming organisms such as *Bacillus* and *Clostridium* species may also be found because of the relatively high content of these organisms in soil.

Contamination of fruits and vegetables occurs primarily during harvesting. However, the quality of the water used for irrigation, as well as the type of fertilizer used, probably influences the initial deposition of contaminants. If the water is not chlorinated or filtered, it can add microorganisms to the surface of the vegetables, including human pathogens. In the case of organically fertilized produce, bacteria can contaminate the crops if animal feces are used as the fertilizer. However, if this material is **composted,** a procedure based on fermentation and heat application, pathogenic organisms can be eliminated before it is used for fertilizer. During harvesting, contamination can take place from dirty equipment as well as from field workers. Microorganisms can also be added at the processing plant if sanitation procedures are not in place.

Dairy. Lactic acid bacteria such as *Leuconostoc, Lactobacillus,* and *Micrococcus* are often found in unpasteurized milk, along with pathogens such as *Mycobacterium, Coxiella,* and *Listeria* species. Obligate aerobes like *Pseudomonas, Flavobacterium,* and *Alcaligenes* are also

TABLE 10.3 Examples of microorganisms found in various foods.

Food	Organism
Muscle tissue (red meat, poultry, seafood)	*Acinetobacter* *Bacillus* *Pseudomonas* *Salmonella*
Fruits and vegetables	*Bacillus* *Lactobacillus* *Leuconostoc* Yeasts and molds
Dairy products	*Coxiella* *Lactobacillus* *Leuconostoc* *Listeria* *Micrococcus* *Mycobacterium* *Pseudomonas*

TABLE 10.4 Microorganisms responsible for spoilage of meat and poultry.

Microorganism	Conditions Where Found
Pseudomonas	Aerobically packaged meat and poultry stored under refrigeration
Acinetobacter, Moraxella	Aerobically packaged meat and poultry stored under refrigeration, present in larger numbers in high-pH meat like pork and lamb
Brochothrix thermosphacta	Modified-atmosphere and vacuum-packaged meat and poultry at pH 5.8 or above
Shewanella putrefaciens	Modified-atmosphere and vacuum-packaged meat and poultry at pH 6.0 or above
Clostridium laramie	Vacuum-packaged meat and poultry stored under refrigeration
Lactic acid bacteria	Modified-atmosphere and vacuum-packaged meat and poultry at pH 5.3 or above
Yeasts and molds	Aerobically packaged meat and poultry at a_w from 0.85 to 0.93

found. These organisms are psychrotrophic, able to grow at the typical storage temperature for milk (3 to 7°C). Hay, grass, and other types of animal feed, as well as water and soil are sources of contamination of raw milk on the farm.

The udder area of the cow supports the growth of psychrotrophic bacteria coming from these sources. Equipment used in milking can become contaminated with organisms on the teat area of the animal and can subsequently serve as sources of contamination of raw milk. Pasteurization eliminates most microbial contaminants. However, milk can become contaminated with psychrotrophic bacteria from air and equipment during the packaging operations after processing.

10.4 FOOD SPOILAGE BY MICROORGANISMS

Microorganisms utilize the carbohydrates and proteins in foods as energy sources for cell growth and reproduction. In accomplishing these tasks, bacteria, yeasts, and molds cause food to spoil or degrade.

Metabolizing or Producing Carbohydrates

During fruit and vegetable spoilage, carbohydrates such as polysaccharides, monosaccharides, and disaccharides are metabolized by microorganisms. **Pectin,** the polysaccharide found in many fruits, is made of galacturonic acid molecules linked by chemical bonds known as α-1,4 linkages. Organisms containing the enzyme **pectin esterase** or "pectinase" can split these bonds, resulting in fruit and vegetable rot. The bacteria *Bacillus polymyxa* and *Erwinia carotovora* cause soft rot in celery and carrots by this mechanism, and the mold *Penicillium citricum* causes soft rot in citrus fruits.

Cellulose, another polysaccharide commonly found in vegetables, is composed of glucose molecules held together by β-1,4 linkages. **Cellulase** enzyme produced by *Bacillus* species, as well as by the molds *Aspergillus* and *Penicillium* break down this molecule, resulting in softening of vegetables.

Microorganisms can also metabolize the monosaccharides and disaccharides found in dairy products. The lactic acid bacterium *Lactobacillus* utilizes lactose to produce lactic acid, and the yeast *Saccharomyces cerevisiae* metabolizes various simple sugars to produce ethanol. Milk spoiled by formation of lactic acid has a characteristic sour odor and taste, as well as thick consistency. Ethanol, although a product of microbial spoilage, can also be considered desirable, for instance, in the production of wine and beer.

In addition to metabolizing carbohydrates, microorganisms can cause spoilage of foods containing sugars by *producing* carbohydrates that alter the texture and flavor of those foods. *Leuconostoc mesenteroides* and *Bacillus mesentericus* are bacteria capable of synthesizing **dextrans,** glucose units connected by α-1,6 linkages. Similarly, *Bacillus* *megaterium* can produce **levans,** fructose units linked by β-2,6 bonds. Both dextrans and levans are responsible for the "ropy" consistency of spoiled fruit juices and make up the slime layer that forms on fruits and vegetables.

Metabolizing Proteins

Microbial spoilage can also involve proteins. Meat, a protein-rich food, is spoiled by a variety of microorganisms, some of which degrade the sugars in the tissue, while others metabolize protein. Aerobic bacteria like *Pseudomonas, Shewanella, Brochothrix thermosphacta,* and lactic acid bacteria like *Lactococcus* and *Lactobacillus* produce lactic acid by metabolizing glucose. This causes a souring of odor and taste. In addition, levans and dextrans are produced, which alter texture. In the case of proteins, *Clostridium laramie,* an anaerobic bacterium, catabolizes protein into amino acids, which impart souring odors and flavors to meats.

Several anaerobic and facultative anaerobic bacteria utilize amino acids in their metabolic reactions, producing foul-smelling compounds. Examples are the conversion of the amino acid lysine into **cadaverine** by the enzyme lysine decarboxylase, and conversion of ornithine into **putrescine** via ornithine decarboxylase. Also, the amino acids alanine and tryptophan can be deaminated to form pyruvate and ammonia, or indole and ammonia, respectively, which impart a strong odor to meat. Similarly, **hydrogen sulfide,** a malodorous compound, is formed by the breakdown of the amino acid cysteine by anaerobic and facultative anaerobic bacteria.

Mold Growth

In addition to bacterial spoilage, meats are subject to mold growth, with changes in meat color being the primary result. Examples of meat color changes are black spots from growth of *Cladosporium herbarum,* white spots from *Sporotrichum,* and fuzzy growth from *Thamnidium.*

10.5 MICROBIAL FERMENTATION

Microorganisms, like all living things, metabolize nutrients in order to grow. By metabolizing nutrients, microbial cells produce adenosine triphosphate (ATP) and other substances. If metabolism is carried out under conditions where oxygen is absent, it is called **fermentation.** See Figure 10.7—some of the products of fermentation change the character of the food they are in.

One of the ways in which microorganisms metabolize nutrients is called *glycolysis.* Glycolysis generates ATP through what is called substrate-level phosphorylation of sugars, with pyruvate as an end product. It is important in processing that several microorganisms are able to further metabolize pyruvate to produce organic compounds that are useful in the production of fermented foods. The groups of organisms most frequently used for the production of fermented foods are the lactic acid bacteria. They require amino acids, B vitamins, nitrogenous bases (e.g., purines and pyrimidines), and an optimal pH (ranging from 4 to 4.5 pH) for growth.

Lactic acid bacteria can be classified as either homofermentors or heterofermentors. **Homofermentors** are organisms that produce one single compound, such as lactic

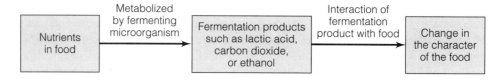

FIGURE 10.7 Fermentation, a metabolic process carried out under anaerobic conditions.

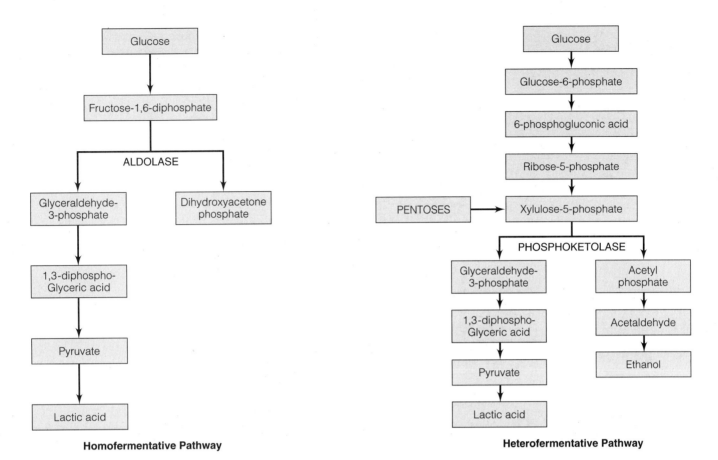

FIGURE 10.8 Homofermentative vs. heterofermentative metabolic pathways. *In glycolysis, homofermentors utilize glucose to produce ATP and one other substance (lactic acid in this case). Heterofermentors utilize glucose to produce ATP and more than one compound (lactic acid and ethanol in the case at right).*

acid, as a result of carrying out fermentation reactions. **Heterofermentors** produce more than one compound. Figure 10.8 shows the series of reactions in homofermentative and heterofermentative metabolic pathways.

Lactic acid lowers the food pH, imparting unique flavors. Production of additional compounds by heterofermentative organisms can result in desirable flavor changes. For instance, diacetyl, recognized as a buttery flavor, is produced by the reaction of pyruvate with lipoic acid. Two other products, formic acid and acetaldehyde, can be formed by the decarboxylation of pyruvate. Lactic acid bacteria are used as **starter cultures** to start the fermentation process to produce cheese, butter, cultured buttermilk, cottage cheese, yogurt, sausage, and fermented vegetable products, among many others. Table 10.5 lists the lactic acid bacteria used as starter cultures.

Yeasts are used in the food industry for fermentation. For example, *Saccharomyces cerevisiae* is commonly used in breadmaking because it generates leavening gas (CO_2) by metabolizing carbohydrate. But yeast in bread is only one of many food fermentation examples. We will now discuss several foods that can be produced by microbial fermentation.

The Fermentation of Milk

In the process of fermenting milk, lactic acid bacteria are used to lower the pH of the milk through the production of lactic acid from lactose. This causes a gel to form when the pH reaches 5.2, with precipitation of proteins taking place at pH 4.6. The temperature of fermentation is usually kept at 20°C (68°F) so that the lactic acid is produced slowly, which

TABLE 10.5 Lactic acid bacteria used in food fermentations. *Starter culture is usually added as a freeze-dried powder. Starter cultures may consist of a single strain or a mix of bacteria.*

Microorganism	Type of Fermentation
Carnobacterium piscicola	Heterofermentor
Enterococcus	Homofermentor

Lactobacilluss
   Group I—*Thermobacterium*
     L. acidophilus
     L. delbrueckii subspecies *bulgaricus*
   Group II—*Streptobacterium*
     L. plantarum
   Group III—*Betabacterium*
     L. fermentum
	Homofermentors
	
	 Heterofermentors
	
	 Heterofermentors

Lactococcus lactis
   Subspecies *lactis*
   Subspecies *cremoris*
   Subspecies *diacetylactis*
	
	 Homofermentor
	 Homofermentor
	 Heterofermentor

Leuconostoc cremoris	Heterofermentor
Pediococcus acidilactici	Homofermentor
Propionibacterium shermanii	Homofermentor
Streptococcus thermophilus	Heterofermentor
Vagococcus	Homofermentor

achieves a uniform curd (Figure 10.9). Fermentation products can affect the texture and flavor of the product, depending on the starter organisms used.

Besides fermentation by lactic acid bacteria, coagulation of milk proteins can be accomplished by adding the enzyme **chymosin** (rennet). This reacts with **casein,** the major protein in milk, forming a gel of calcium phosphate paracaseinate. This type of fermentation is called **rennet coagulation.** Another method used to coagulate casein is the addition of both lactic acid bacteria and rennet, which usually results in a softer coagulate than when rennet is used alone.

Cheddar cheese is made from pasteurized whole milk, to which *Lactococcus lactis* ss. *cremoris* or *Lactococcus lactis* ss. *lactis*, is added. A coloring agent such as **annatto** is added and the mixture is incubated at 30°C (86°F) for approximately 30 minutes to allow the proteins in milk to coagulate. The product is then cooked at 39°C (101°F), and the **whey** is drained. The cheese is pressed for 16 hours, and more whey is drained. The product is dried for 5 days at 10°C (50°F), covered with wax, and ripened at 4.4–7°C (40–44°F) for up to 12 months.

Cottage cheese is made from skim milk, which has been pasteurized. The starter culture used is similar to that for cheddar cheese. The difference is that incubation is carried out at 30°C (86°F) for about 4 to 5 hours to allow for the slow curdling of the milk. The **curd** is cut into cubes and cooked at 54°C (129°F). The whey is then drained and cream and salt are added. This results in a soft cheese texture.

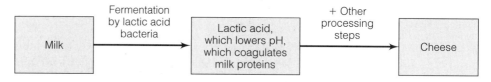

FIGURE 10.9 The fermentation of milk.

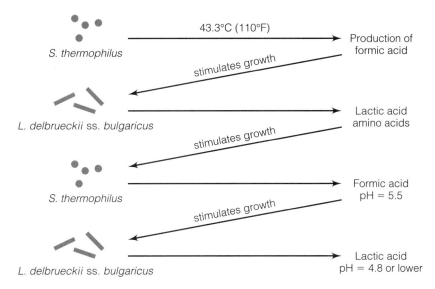

FIGURE 10.10 Series of steps in the fermentation of milk to produce yogurt.

Yogurt is one of the most interesting products made by microbial fermentation, from the standpoint of the interaction of the organisms involved in the reactions. Typically, whole milk is heated at 85°C (185°F) for 30 minutes, and cooled to 43°C (109°F). The starter cultures consist of *Lactobacillus delbrueckii* ss. *bulgaricus* (a homofermentor) and *Streptococcus thermophilus* (a heterofermentor) in a 1:1 ratio. The inoculated milk is incubated at 43.3°C (110°F) for about 6 hours until the pH drops to 4.3–4.5.

During this 6-hour period, a **symbiotic** relationship develops between the two organisms. First, *S. thermophilus* grows fast, producing formic acid and other flavor compounds. The presence of formic acid stimulates the growth of the *Lactobacillus* and the production of lactic acid. Although considered a homofermentor, this organism can produce minor amounts of other flavor compounds as well. Proteases secreted by the organism result in the release of amino acids from the milk proteins. This stimulates the growth of *Streptococcus* (which lacks proteases), producing more lactic acid until the pH drops to 5.5. *Lactobacillus* then resumes its growth, causing the pH to drop further down to 4.3–4.5 (Figure 10.10). The fermentation process is then slowed down by cooling the yogurt to 4.4°C (40°F) until the pH drops to 4.2, at which time growth stops.

The Fermentation of Meat

In the fermentation of meat, cultures to be used must be approved by the U.S. Department of Agriculture. These cultures are usually salt-tolerant, able to grow in 2 to 3.5 percent salt. The end product of choice is lactic acid, which helps in flavor and color development, and in texture. The approved cultures are:

Pediococcus cerevisiae   *Lactobacillus plantarum*
Pediococcus acidilactici   *Micrococcus varians*
Pediococcus pentosaceus

The *Pediococcus* organisms are available commercially as a product called **Lactacel**. These organisms grow rapidly at 32 to 49°C (90–120°F), which is the temperature commonly used for production of summer sausage (Figure 10.11). *L. plantarum* is used for dry sausage manufacture, and it is able to inhibit the pathogen *Staphylococcus aureus*. *M. varians* is able to use the nitrates in meat and reduce them to nitrites. In Europe, a nonpathogenic strain of the organism *S. aureus* can be used as a starter culture for meat fermentation because it helps develop a redder color on the product, but it is not permitted in the United States.

In commercial settings, frozen cultures are used. These cannot be refrozen, thus they come in small quantities to minimize waste, in the range of 10^6 cells per gram. They

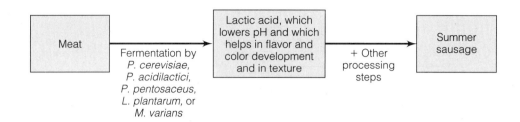

FIGURE 10.11 **The fermentation of meat.**

are added to the product as the last ingredient (just after curing and seasoning). The pH of the meat must drop to 5.3 or less to comply with USDA requirements. The pH can be controlled by varying the amount of sugar added, the temperature of fermentation, and the time. Minimizing the time it takes to reach the target pH is important in order to inhibit pathogenic organisms that can tolerate the curing agents.

Alternatives to starter cultures consist of **chemical acidulants.** Glucono delta lactone (GDL) is one of these. It can be added at a level of 0.5 percent in products such as sausage to achieve desirable results. GDL must be added in a controlled manner to limit any adverse effects in the product, such as protein denaturation, an overly soft texture, and flavor changes due to the fast pH drop.

Semi-dry sausage, containing 50 percent moisture, is prepared by adding 3 percent salt, sugar, seasonings, and nitrite to cubed beef. *P. cerevisiae* is then added and fermentation is carried out at 4°C (39.2°F) until the pH drops to 5.3 (approximately 10 days). The product is then cooked to an internal temperature of 68°C (154°F). Since the a_w is 0.9 to 0.94, refrigeration is necessary, unless the product is also smoked. Examples are summer sausage and bologna.

Fermenting Fruit and Vegetable Products

The starter culture is typically composed of the normal mixed flora of the raw vegetable. In the case of sauerkraut, for example, the cabbage is cut and 2.5 percent salt is added at 18°C (64°F), which inhibits the growth of gram-negative organisms. By closing the vat containing the cabbage, brine forms as the fermentation reaction begins. The typical cultures used consist of:

Leuconostoc mesenteroides *Lactobacillus brevis* *Pediococcus cerevisiae*

Lactic acid and carbon dioxide are the main products of fermentation. The lactic acid lowers the pH, and the carbon dioxide creates and maintains anaerobic conditions (Figure 10.12). This prevents oxidation of ascorbic acid in the cabbage, which would darken it.

Fruit can be fermented to make cider, where yeasts instead of bacteria are used in a mild fermentation (Figure 10.12). Apples are ground into a pulp, followed by the pressing of the pulp to release the juice. The juice is strained to remove large pulp pieces and stored with the yeast culture to allow sedimentation of particles at 4.0°C (39.2°F) for several days. The now clarified, mildly fermented juice can be pasteurized.

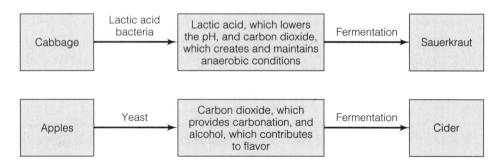

FIGURE 10.12 **Fermentation to produce sauerkraut and apple cider.**

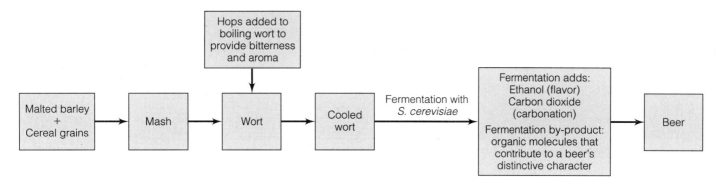

FIGURE 10.13 The production of beer involves fermentation with *Saccharomyces cerevisiae*.

Fermenting Cereal Grains

Beer is a popular alcoholic beverage made with barley, rice, and corn as the raw materials. These grains supply carbohydrates to the yeast *Saccharomyces cerevisiae*. **Hops** are plant flowers that contain essential oils that contribute bitterness compounds for flavor, and tannins for color. Barley is germinated and dried, a process that produces **barley malt**. Germination activates the enzymes needed to break down the starch in the barley malt, releasing individual sugars. This is essential so that the yeast can metabolize the individual sugar molecules. Cereal adjuncts like corn and rice are added to supply additional carbohydrates. The end products of the fermentation are ethanol and carbon dioxide.

The steps involved in beer making vary, according to the style of beer, but some steps are basic. The malted barley and the cereal grains are cooked together to form a mash. During this period, starches in the grains gelatinize, making them more susceptible to enzyme attack. The cooking is carried out at 38°C (100°F) and increased slowly to 77°C (170°F) so as not to inactivate the enzymes. The liquid that results after mashing is the **wort**.

Hops are added to the wort, and they are brewed by boiling for 2.5 hours. This step sterilizes the wort, inactivates enzymes, precipitates proteins that would affect turbidity, and extracts flavor from the hops. The cooled wort is inoculated with *Saccharomyces cerevisiae*, and the fermentation is carried out at 10°C (50°F) for approximately 9 days (Figure 10.13). During this time, the pH drops to 4.0. The pH can be adjusted to make different beer products. For example, **ale** is made at pH = 3.8 (takes 5 to 7 days), while **lager** is made at pH = 4.2 (takes 7 to 12 days). Finally, the beer is cooled to 0°C (32°F), filtered to remove the yeast, and stored to mellow the flavor. Beer can then be pasteurized at 60°C (140°F) for 15 to 20 minutes, or filtered once more before filling into cans.

Saccharomyces can also be used in breadmaking, with bacteria added for special flavor development. Sourdough bread is made by adding *S. cerevisiae* and *Candida krusei*, another yeast; both produce carbon dioxide. Bacteria such as *Lactobacillus plantarum* and *Lactobacillus acidophilus* are added for flavor.

Key Points

- Microorganisms such as bacteria, yeasts, molds, protozoa, and viruses can be found in many places, including air, soil, water, and the intestinal tract of humans and animals.
- Factors that affect microbial growth include nutrient availability, a_w, pH, oxygen content, temperature, and the nature of the food they live in.
- Microorganisms found in foods originate in the raw product and are acquired through handling, processing, and storage.
- Carbohydrates are spoiled by lactic acid bacteria, which break them down into simple sugars and metabolize them to produce malodorous compounds and lower the pH.
- Proteins are spoiled aerobically by *Pseudomonas*, *Acinetobacter*, and molds such as *Thamnidium*. Lactic acid bacteria spoil proteins under anaerobic conditions to produce off-odor compounds from decarboxylation and deamination of amino acids.
- Fermentation is the biochemical process by which individual sugar molecules are metabolized by microorganisms to produce various by-products, including lactic acid.
- Lactic acid bacteria are often used in fermentation reactions to lower pH. Homofermentors produce one compound, while heterofermentors produce more than one compound.
- Milk is fermented by bacteria to produce cheese and yogurt, among other products; meat is fermented to produce sausages; and vegetables like cabbage are fermented to produce fermented foods such as sauerkraut.
- Yeasts are used to ferment grain products such as barley to produce beer by the formation of ethanol from glucose, as well as to produce bread by the formation of carbon dioxide.
- Two-class and three-class attribute sampling plans are used by the food industry to evaluate the microbial quality of foods, as well as to predict shelf life of products (in Challenge section).

Study Questions

1. How does pH affect microbial growth?
2. What is the poising effect?
3. What are lactic acid bacteria, and where are they found?
4. What reactions take place in the formation of off-odor compounds from lysine and ornithine?
5. How do molds spoil foods?
6. The graph below depicts two events over time. The left y-axis (and the solid line in the graph) shows the log of the growth of microorganisms in broth in colony-forming units per milliliter (CFU/mL). The right y-axis (and the dotted line) graph the oxidation-reduction potential of the medium, measured in millivolts (mV). Based on the environmental conditions and the pattern of growth, what conclusions can you draw regarding the types of microorganisms that are likely to be present?
7. A friend of yours is trying to make cheese from milk proteins and water only. What steps and ingredients would be needed to successfully do it?

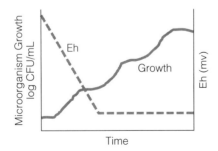

Multiple Choice

8. Bacteria are classified as
   a. eucaryotes.
   b. procaryotes.
   c. Protista.
   d. algae.
   e. parasites.

9. When bacterial spores germinate, the resulting organism is called
   a. oocyte.
   b. spore.
   c. vegetative cell.
   d. sporocyte.
   e. spirilla.

10. Bacteria that require organic compounds for survival are known as
    a. lithotrophs.
    b. chemolithotrophs.
    c. lithochemotrophs.
    d. organotrophs.
    e. lithoorthotrophs.

11. Most bacteria cannot grow when the water activity is
    a. above 0.7.
    b. below 0.7.
    c. 0.95.
    d. above 1.0.
    e. below 0.95.

12. Bacteria that require 6% of oxygen are called
    a. microaerophiles.
    b. macroaerophiles.
    c. aerobic.
    d. oxotrophs.
    e. microanaerobes.

13. Psychrotrophs are organisms that
    a. tolerate cold temperatures.
    b. prefer cold temperatures.
    c. cannot survive at cold temperatures.
    d. prefer hot temperatures.
    e. tolerate hot temperatures.

14. Mesophilic organisms can grow at
    a. refrigeration temperature.
    b. freezing temperature.
    c. room temperature.
    d. very high temperature.
    e. the boiling point of water.

15. The principle behind the application of several treatments to prevent bacterial growth is
    a. The Peter Principle.
    b. The Hurdle Effect.
    c. The Inhibitory Effect.
    d. The Antimicrobial Principle.
    e. The Inhibitory Principle.

16. One of the main products of bacterial fermentation is
    a. hydrochloric acid.
    b. lactic acid.
    c. malic acid.
    d. erucic acid.
    e. ascorbic acid.

17. Spoilage microorganisms cause the shelf life of foods to
    a. decrease.
    b. increase.
    c. remain unchanged.
    d. reverse itself.
    e. decrease then increase.

18. Bacterial fermentation is an important part of the process known as
    a. composting.
    b. breathing.
    c. bubbling.
    d. aerobiosis.
    e. comminuting.

19. Dextrans are produced by
    a. lactic acid bacteria.
    b. pathogenic bacteria.
    c. dextroses.
    d. yeasts.
    e. mold.

20. An example of a mold that spoils meats is
    a. *Thamnidium*.
    b. *Conidium*.
    c. *Putridium*.
    d. *Salmonella*.
    e. *Saccharomyces*.

21. One of the products of meat spoilage is
    a. spoiline.
    b. cadaverine.
    c. oxygen.
    d. pectin.
    e. starch.

22. Homofermentors are organisms that
    a. produce several products during fermentation.
    b. produce only one product during fermentation.
    c. produce no products during fermentation.
    d. produce oxygen during fermentation.
    e. produce cadaverine.

23. Fermentation causes the pH of foods to
    a. increase.
    b. decrease.
    c. remain unchanged.
    d. increase until a certain point.
    e. increase then decrease.

24. Starter cultures are used for
    a. their curative properties.
    b. fermentation of foods.
    c. dehydration of foods.
    d. decontamination of foods.
    e. deterioration of foods.

25. A commercial starter culture used in meat fermentation is
    a. Rennet.
    b. Lactacel.
    c. Fermentum.
    d. Lactaid.
    e. Lactalbumin.

26. A yeast used in the making of bread is
    a. *Saccharomyces*.
    b. *Actinomyces*.
    c. *Bradyomyces*.
    d. *Salmonella*.
    e. *Streptococcus*.

References

Banwart, G. J. 1989. *Basic Food Microbiology*, 6th ed. Aspen, Gaithersburg, MD.

Castillo, A. L., Lucia, L., Goodson, K. J., et al. 1998. Comparison of water wash, trimming, and combined hot water and lactic acid treatments for reducing bacteria of fecal origin on beef carcasses. *Journal of Food Protection* 61: 823.

Jay, M. J. 2000. *Modern Food Microbiology*. Van Nostrand Reinhold, New York.

Mountney, G. J., and Gould, W. A., 1988. *Practical Food Microbiology and Technology*. Van Nostrand Reinhold, New York.

Additional Resources

Doyle, M. P., Beuchat, L. R., and Montville, T. J., eds. 1997. *Food Microbiology Fundamentals and Frontiers*. ASM Press, Washington, DC.

Geise, J. 1998. New developments in microbial testing. *Food Technology* 52(1):79.

Perry, J. J., and Staley, J. T. 1997. *Microbiology: Dynamics and Diversity*. Saunders College Publishing, Fort Worth, TX.

Ray, B. 1996. *Fundamental Food Microbiology*. CRC Press, Boca Raton, FL.

Viljoen, B. C., Goernaras, I., Lamprecht, A., et al. 1998. Yeast populations associated with processed poultry. *Food Microbiology* 15: 113.

Web Sites

http://www.asmusa.org
American Society for Microbiology. This web site for the largest microbiology association in the United States has information on general, clinical, and industrial microbiology topics.

http://www.sfam.org.uk/
The Society for Applied Microbiology (sfam), the oldest microbiological society in the UK, is dedicated to the advancement of the study of microbiology. The Society has six Interest Groups: Bioengineering; Educational Development; Environmental; Food Safety and Technology; Infection, Prevention and Treatment; and Molecular Biology.

http://helios.bto.ed.ac.uk/bto/microbes/microbes.htm
The University of Edinburgh: "The Microbial World." This site has profiles of microorganisms and educational materials covering bacteria, yeasts, molds, and fungi.

http://www.cellsalive.com/index.htm
Cells Alive! Spectacular images of microorganisms, including parasites.

http://www.prenhall.com/brock/
Prentice Hall Publishers "Biology of Microorganisms Textbook Online." An online study guide for the textbook *Biology of Microorganisms* by Brock.

MICROBIAL SAMPLING TO VERIFY FOOD QUALITY

One of the most critical tasks carried out by industry is that of evaluating the quality of foods being produced. **Quality** involves not only the physical and chemical parameters of a food but also its microbial character. Determining the number of microorganisms helps to verify that a product meets industry guidelines or self-imposed limits and helps predict the shelf life of foods. A necessary tool for the gathering of information on the microbial quality of foods is the concept of microbial sampling.

For microbial enumeration to have any meaning, sufficient sampling of a particular food must be performed. The goal is to achieve a high degree of confidence that the samples represent the entire production lot. Samples should be taken randomly to eliminate bias from the samples. Only with such **representative sampling** can accurate conclusions about the microbial quality of the lot be drawn.

Figure 10.14 shows characteristic curves of Percent Defective and Probability of Acceptance for two sample sizes. Part (a) is based on fewer samples ($n = 5$) compared to part (b) ($n = 10$). Thus, the fewer number of samples, the greater the probability at any given percent defective lot that the lot will be accepted—for instance, for criterion of acceptance $C = 3$, the probability of acceptance of a 50 percent defective lot is 0.8 for the $n = 5$ sample. For the same C, the probability of acceptance of a 50 percent defective lot is only 0.18 when 10 samples are taken. So, it is better to take more samples to decrease the probability of accepting a defective lot of product/food.

Two sampling plans have been devised to conduct sampling for microbial testing: two-class attribute sampling and three-class attribute sampling. In **two-class attribute sampling,** each sample unit is classified as either acceptable or unacceptable. First, a specific number of samples are taken. If a certain number of those samples exceeds a certain limit in the number of microorganisms, then the lot is rejected. For example, a certain food company requires that its supplier provide it with raw meat with a total microbial load not to exceed 10^2 organisms per gram. Let's say that the number of samples to be taken is 5 and that the most defective units the company will accept is 2. If more than 2 samples are found to exceed 10^2 cells/g, the lot must be rejected.

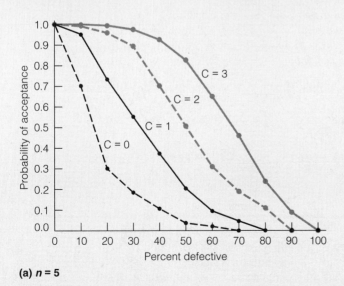
(a) $n = 5$

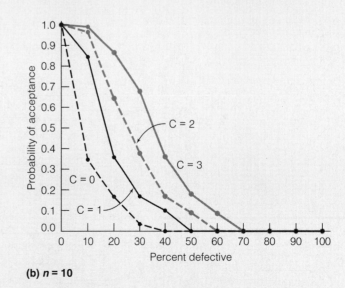
(b) $n = 10$

FIGURE 10.14 Characteristic curves for different sample sizes (n) and different criteria of acceptance (C) for a two-class attribute sampling plan.

In **three-class attribute sampling**, samples with microbial counts less than a certain level are deemed acceptable. Samples with counts between this value and another, higher value are marginally acceptable; samples with counts greater than the higher value are unacceptable. For example, suppose 10 units from a food lot are sampled, specifying that total bacterial counts greater than 10^2 cells/g but lower than 10^3 cells/g represent marginal acceptability. If 3 of the samples have levels between 10^2 and 10^3 cells/g, or if any have levels greater than 10^3 cells/g, the lot must be rejected.

As the number of samples taken (n) increases, the probability of accurately concluding that the lot should be rejected increases. Put another way, the more samples taken, the higher the confidence of being correct when a lot is rejected based on the information provided by the samples. In addition, as the number of samples exceeding the limit increases, the lower the confidence of being correct when a lot is rejected.

A key consideration in attribute sampling is that it should only be done when the distribution of the defect is **homogenous** and random. Such is the case in a production lot where each unit has an equal chance of having the same level of microbial contaminants. An example is ice cream or some other mixed product that is prepared in a batch, and then subdivided into individual units at the time of packaging. Any microbial contaminants are distributed evenly throughout the batch, with all units derived from the batch receiving the same level of microorganisms during mixing. For this reason, attribute sampling plans are independent of lot size. As long as the number of microorganisms is evenly distributed within the lot, the sample size does not matter.

There are many cases in which the distribution of the defect is random. An example is the production of broiler chickens, where each bird has a different level of microbial contamination. In this situation, the number of samples needed to reach an accurate conclusion about the lot must be large. Even so, if the results indicate that the lot has not exceeded the specifications of the sampling plan, we must not assume that the lot should be accepted. This is because the degree of confidence in our results is low when the distribution of the defect is not homogenous. However, if the results indicate that the lot has exceeded the specifications of the sampling plan, it can be assumed that the lot should not be accepted.

Challenge Questions

1. A new food company hires you as Quality Assurance Supervisor. The company produces fresh, nonpasteurized celery juice sold at health food stores nationwide. Your boss tells you that no more than 3 samples out of 10 can exceed 10^2 total microorganisms per gram.
   a. What type of sampling plan would you develop?
   b. If you find 2 samples exceeding 10^2 cells/g, will you reject the lot?
   c. Will you deem the lot as safe as a lot in which you find 0 samples to exceed 10^2 cells/g?

2. In a two-class attribute sampling plan, each sample is classified as:
   a. positive or negative
   b. marginal or acceptable
   c. acceptable or unacceptable
   d. marginal or unacceptable
   e. significant or insignificant

3. In sampling plans, as the number of samples taken increases, the probability of concluding that the lot should be rejected:
   a. decreases
   b. increases
   c. stays the same
   d. is irrelevant
   e. cannot be predicted

4. Attribute sampling is only valid when the defect distribution is:
   a. homogeneous
   b. heterogeneous
   c. ramdom
   d. predictable
   e. concrete

CHAPTER 11

Food Safety

11.1 What Is Foodborne Illness?

11.2 Types of Biological Hazards in Food

11.3 The Most Common Biological Hazards in Food

11.4 What Is Mad Cow Disease?

11.5 Preventing Foodborne Illness

11.6 HACCP—A Preventive Approach

Challenge!
Risk Assessment for Biological Hazards

CHAPTER OBJECTIVES

After completing this chapter, you will be able to:
- Describe what is meant by foodborne illness and the associated hazards.
- Explain how biological hazards cause disease.
- Identify the most common biological hazards responsible for foodborne disease.
- Describe the pathway of infection of several microorganisms and parasites.
- Identify commonly used sanitizers in the food industry.
- Identify the major contributing factors to foodborne illness.
- Explain what is meant by mad cow disease.
- Describe how a HACCP plan is structured.
- Critique risk assessment calculations associated with food biological hazards.

Food safety has gained national and international attention, mainly from a series of foodborne illness outbreaks caused by consumption of food contaminated with pathogenic, or disease-causing, bacteria. Most notably, cases attributed to consumption of undercooked hamburgers in Washington state and California in 1991 contaminated with the organism *Escherichia coli* O157:H7 focused attention on food safety by government regulators, the food industry, scientists, and consumers. In response, the government has increased funding for food safety research and revamped its inspection system for beef, poultry, and seafood to include preventive methods to control foodborne hazards.

The attacks on the World Trade Center in New York and the Pentagon in Washington, DC on September 11, 2001 have raised concerns over the possibility of terrorist plans to contaminate the food supply of the United States and elsewhere. The concerns regarding the use of biological warfare, or bioterrorism, intensified immediately following September 11, when *Bacillus anthracis* spores, the causative agent of anthrax, were spread through the U.S. mail.

It is important to differentiate between the terms *food safety* and *food security*. The former deals with disease-producing or toxic hazards that are accidentally introduced into the food supply, while the latter refers to the planned contamination of food with these substances as a result of malicious and criminal intent. The need for prevention is common to both.

The prevention of intentional "food supply terrorism" comes under the umbrella term **food biosecurity**. It is a concern not only of public health officials, but of law enforcement as well. Food companies utilize the HACCP Hazard Analysis Critical Control Points System to provide food safety. It is believed that the same principles can be applied to prevent a terrorist's food contamination plot. However, it is important to discern that protection against intentional harm involves practices such as employee hiring screens, not currently regulated by government food inspection authorities.

In this chapter, we examine the causes of foodborne illness, as well as agents and strategies available to minimize their occurrence. We will see the role that bacteria, molds, viruses, and parasites play in foodborne illness and identify specific sanitizers effective as inhibitors of particular biological hazards. The purpose and importance of HACCP in promoting food safety will be addressed, and we will discuss what CCPs, GMPs, and SSOPs are and their relationship to HACCP.

11.1 WHAT IS FOODBORNE ILLNESS?

Foodborne illness is defined as any illness resulting from ingestion of food. Several clues indicate that an illness is foodborne. For instance, if the agent causing the disease is detected in a sample of the food the victim has eaten, if there is a cluster of cases among persons who have nothing else in common except that they ate the same food, if the intestinal tract is affected, or if the symptoms are like those suffered in foodborne illness, we conclude that the illness in question may be foodborne. The hazards associated with foodborne illness are of three types: biological, chemical, and physical.

Biological hazards include bacteria, molds, viruses, and parasites such as protozoa, flatworms, and roundworms (Figure 11.1). **Chemical hazards** include chemical sub-

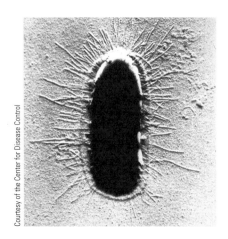

FIGURE 11.1 *Salmonella,* a biological hazard to humans, causing a food infection.

TABLE 11.1   Three types of hazards associated with foodborne illness.

Hazard	Examples
Biological Hazards	Bacteria (*Listeria monocytogenes, Salmonella*, etc.)
	Parasites (*Cryptosporidium, Taenia*, etc.)
	Viruses (hepatitis A, Norwalk, etc.)
	Toxins (mycotoxins)
Chemical Hazards	Cleaners and sanitizers (soap, chlorine, etc.)
	Lubricants (grease, oil, etc.)
	Antibiotics
	Hormones
Physical Hazards	Glass
	Woodchips
	Plastic
	Metal

stances that occur naturally in foods, such as plant toxins, and those that are added to food, such as antibiotics. **Physical hazards** include bone, metal, plastic, and any other foreign matter that can cause damage to the consumer if ingested. Examples of these hazards are given in Table 11.1. In this chapter, we will concentrate on biological hazards to foods. Chemical hazards will be discussed further in Chapter 12, Food Toxicology.

11.2 TYPES OF BIOLOGICAL HAZARDS IN FOOD

Bacterial Causes

Bacteria cause disease in humans according to the following classification: **infection, intoxication,** and **intoxification.**

Foodborne Infection.   Infectious bacteria are those that invade the intestinal tract. By consuming the food, the individual ingests the bacteria, which may be as many as 10,000 cells/g. Upon ingestion, the organisms make their way into the intestinal tract, whereupon they colonize it. During colonization, the epithelial cells lining the intestine are damaged, disrupting the uptake of solutes into the body by these cells. The result is an imbalance in osmotic pressure, causing water to be secreted from the tissues back into the intestinal tract. This excess of water is responsible for the loosening of stool, or diarrhea.

At the same time, the damage caused to the intestinal lining sends a message to the brain that triggers the vomiting response. In addition, if the organism in question is gram-negative, a fever also ensues, because of the pyrogenic effect of outer membrane components of these bacteria. Examples of such infective microorganisms are *Salmonella, Yersinia,* and *Shigella*. The onset of disease is usually from 12 hours to 2 days, depending on how many cells were ingested.

Table 11.2 presents the bacteria that cause food infections and some of the characteristics of the disease they cause.

Foodborne Intoxication.   Bacteria that cause intoxications are those that produce toxin in the food during growth. Foods can be contaminated with toxin-producing organisms like *Staphylococcus aureus* or *Clostridium botulinum*. If these foods are stored improperly in such a way as to allow growth and toxin production by these organisms, foodborne intoxication can result. The onset of disease is very fast, with mere hours elapsing from the time the food is consumed to the time that symptoms are first experienced, because the toxins are rapidly absorbed through the intestinal tract, reaching the target organs very rapidly.

Table 11.3 (p. 308) presents these bacteria. Note the presence of *E. coli*. There are three varieties of *E. coli* (see Table 11.4, p. 308), and the enterotoxigenic variety causes

TABLE 11.2 Bacteria that cause food infections.

Bacterium	Latency Period (duration)	Principal Symptoms	Typical Foods	Mode of Contamination	Prevention of Disease
Salmonella species (salmonellosis)	12–36 hr (2–7 days)	Diarrhea, abdominal pain, chills, fever, vomiting, dehydration	Raw, undercooked eggs; raw milk, meat, and poultry	Infected food-source animals; human feces	Cook eggs, meat, and poultry thoroughly; pasteurize milk; irradiate chickens
Camplyobacter jejuni (campylobacteriosis)	2–5 days (2–10 days)	Diarrhea, abdominal pain, fever, nausea, vomiting	Infected food-source animals	Chicken, raw milk	Cook chicken thoroughly; avoid cross-contamination; irradiate chickens; pasteurize milk
Escherichia coli (enteroinvasive)	at least 18 hr (uncertain)	Cramps, diarrhea, fever, dysentery	Raw foods	Human fecal contamination, direct or via water	Cook foods thoroughly; general sanitation
Listeria monocytogenes (listeriosis)	3–70 days	Meningoencephalitis; stillbirths; septicemia or meningitis in newborns	Raw milk, cheese, and vegetables	Soil or infected animals, directly or via manure	Pasteurization of milk; cooking
Yersinia enterocolitica (yersiniosis)	3–7 days (2–3 weeks)	Diarrhea, pains, mimicking appendicitis, fever, vomiting	Raw or undercooked pork and beef; tofu packed in spring water	Infected animals especially swine; contaminated water	Cook meats thoroughly; chlorinate water
Vibrio parahaemolyticus	12–24 hr (4–7 days)	Diarrhea, cramps; sometimes nausea, vomiting, fever; headache	Fish and seafoods	Marine coastal environment	Cook fish and seafoods thoroughly
Vibrio vulnificus	In persons with high serum iron (1 day)	Chills, fever, prostration, often death	Raw oysters and clams	Marine coastal environment	Cook shellfish thoroughly
Shigella species (shigellosis)	12–48 hr (4–7 days)	Diarrhea, fever, nausea; sometimes vomiting, cramps	Raw foods	Human fecal contamination, direct or via water	General sanitation; cook foods thoroughly

food intoxication. (*Entero* means "within the intestine," and *toxigenic* means "produces (generates) toxin.")

Foodborne Intoxification. Intoxification is caused by ingestion of bacteria that, once inside the small intestine, begin to produce toxin. The organism *E. coli* O157:H7 is a good example of this type of pathogen, able to produce toxins after damaging the lining of the intestine. These toxins are absorbed by the body, reaching the kidneys where they cause substantial damage to the convoluted tubules. In addition to serotype O157:H7, other enterohemorrhagic *E. coli* serotypes are O145:H-, O26:H11, O104:H21, and O111:NM (Table 11.5, p. 309).

The mechanism of intoxification is shown step-by-step in Figure 11.2 with the bacterium *Campylobacter jejuni*.

Mycotoxins from Molds

Some molds associated with food are able to produce highly toxic substances known as **mycotoxins.** These compounds have no apparent usefulness to the mold, yet it is possible that it may use them as a way to use up amino acids, acetate, and pyruvate to prevent overaccumulation of these compounds. Whatever the purpose, mycotoxins range in toxicity from causing typical foodborne illness symptoms, such as vomiting, to more serious conditions such as gangrene, bone marrow destruction, renal disorders, and liver cancer.

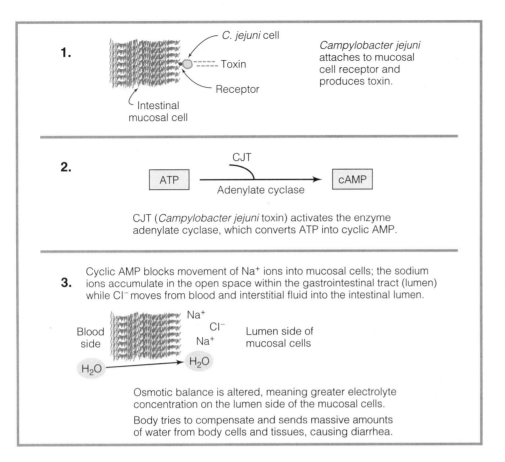

FIGURE 11.2 **Mechanism of foodborne intoxification.** *Certain organisms are capable of producing infection-like foodborne illness as well as intoxifications. Case in point:* Campylobacter jejuni, *a gram-negative rod.* C. jejuni *causes infection when fewer than 1,000 cells are ingested. In addition, it produces a soluble protein "exotoxin" into its environment. Since in the host this is the gastrointestinal tract, the toxin is referred to as an* enterotoxin, *and the organism is enterotoxigenic.*

Virus Transmission

Viruses do not grow in foods, yet food can serve as a carrier for these parasites from which disease can develop. Seldom does the victim die, mainly due to production of antiviral antibodies by the host's immune system, although this occurs very slowly. The incubation period for foodborne viral illness is usually several weeks, during which time the viral particles invade host cells and replicate. Transmission of the virus can be through the fecal-oral route.

For example, a food handler suffering from hepatitis A doesn't wash his/her hands after using the restroom. The hands become contaminated with fecal material that contains viral particles. Touching foods that are to be eaten without any further preparation, such as sandwiches and salads, can transfer the virus to the unsuspecting consumer (Figure 11.3).

Ingestion of Parasites

Parasites, specifically the protozoa, are typically harbored in the intestinal tract of animals. That is the primary source of contamination for humans. When the parasite is shed in the feces of an animal, it is in the form of a cyst. The cysts can contaminate the surface of meat, and thus be ingested through consumption of undercooked meat. Once inside the body, the cysts germinate into growing cells that can persist inside the body for life (Figure 11.4). The onset of the disease is typically at least one week.

A second type of parasite, the flatworm, exists in the form of eggs in the soil. Cattle can ingest the eggs as they forage. Once inside cattle, the eggs will release embryos, which

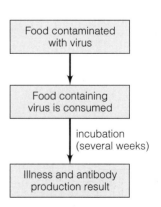

FIGURE 11.3 **Mechanism of virus transmission.**

TABLE 11.3 Bacteria that cause food intoxication.

Bacterium	Latency Period (duration)	Principal Symptoms	Typical Foods	Mode of Contamination	Prevention of Disease
Staphylococcus aureus (staphylococcal food poisoning)	½–8 hr (6–48 hr)	Nausea, vomiting, diarrhea, cramps	Ham, meat, poultry products, cream-filled pastries, whipped butter, cheese	Handlers with colds, sore throats, or infected cuts; food slicers	Thorough heating and rapid cooling of foods
Escherichia coli (enterotoxigenic)	10–72 hr (3–5 days)	Profuse watery diarrhea; sometimes cramps, vomiting	Raw foods	Human fecal contamination, direct or via water	Cook foods thoroughly; general sanitation
Clostridium perfringens	8–24 hr (12–24 hr)	Diarrhea, cramps, rarely nausea and vomiting	Cooked meat and poultry	Soil, raw foods	Thorough heating and rapid cooling of foods
Clostridium botulinum (botulism)	12–36 hr (months)	Fatigue, weakness, double vision, slurred speech, respiratory failure, sometimes death	Types A&B: vegetables, fruits; meat, fish, and poultry products; condiments; Type E: fish and fish products	Types A&B: soil or dust; Type E: water and sediments	Thorough heating and rapid cooling of foods
Clostridium botulinum (botulism, infant infection)	Unknown	Constipation, weakness, respiratory failure, sometimes death	Honey, soil	Ingested spores from soil or dust; or honey colonize intestine	Do not feed honey to infants—will not prevent all
Bacillus cereus (diarrhea)	6–15 hr (12–24 hr)	Diarrhea, cramps, occasional vomiting	Meat products, soups, sauces, vegetables	From soil or dust	Thorough heating and rapid cooling of foods
Bacillus cereus (emetic)	½–6 hr (5–24 hr)	Nausea, vomiting, sometimes diarrhea and cramps	Cooked rice and pasta	From soil or dust	Thorough heating and rapid cooling of foods

TABLE 11.4 The varieties of the *E. coli* organism and the illnesses produced.

E. coli variety	Disease	Principal Symptoms
Enteroinvasive	Infection	Cramps and diarrhea with fever
Enterotoxigenic	Intoxication	Watery diarrhea with no fever
Enterohemorrhagic	Intoxification	Hemorrhagic colitis (profuse bloody diarrhea)

will eventually penetrate the intestinal tract of the animal. The embryos will travel to other tissues, such as muscle, and there will develop into larvae. Consumption of meat from these animals will result in ingestion of the larvae by humans, causing disease.

11.3 THE MOST COMMON BIOLOGICAL HAZARDS IN FOOD

Bacteria—The Main Culprits

In their summary report published in 1997, the Centers for Disease Control reported that in 1992, the leading causes of foodborne illness were all bacterial in nature, with more than 26 percent of the cases being caused by *Salmonella* alone.

Outbreaks of foodborne illness attributable to bacteria usually occur for one of three reasons: product is not cooked properly in order to destroy the hazard; product is not stored at the appropriate temperature in order to prevent bacterial growth; or product was contaminated with a bacterial agent and not treated further before consumption.

TABLE 11.5 Foodborne bacteria that cause intoxification.

Bacterium	Latency Period	Principal Symptoms	Typical Foods	Mode of Contamination	Prevention of Disease
Campylobacter jejuni	2–5 days (2–10 days)	Diarrhea, abdominal pain, chills, fever, vomiting, diarrhea	Infected food-source animals	Chicken, raw milk	Cook chicken thoroughly; avoid cross-contamination; irradiate chicken; pasteurize milk
Escherichia coli (enterohemorrhagic: serotypes O157:H7, O145:H-, O26:H11, O104:H21, O111:NM)	12–60 hr (2–9 days)	Watery, bloody diarrhea	Raw or undercooked beef, raw milk	Infected cattle	Cook beef thoroughly; irradiate beef
Vibrio cholerae (cholera)	2–3 days hours to days	Profuse, watery stools; sometimes vomiting; dehydration; often fatal if untreated	Raw or undercooked seafood	Human feces in marine environment	Cook seafood thoroughly; general sanitation

Salmonella. *Salmonella* is often quoted as the leading cause of foodborne illness in the United States, with several significant outbreaks by *S. Typhimurium* over the years. In 1984, low-fat milk used in making dessert served to conventioneers in Chicago was contaminated with the organism. Approximately 19,000 became ill, and many had to be hospitalized.

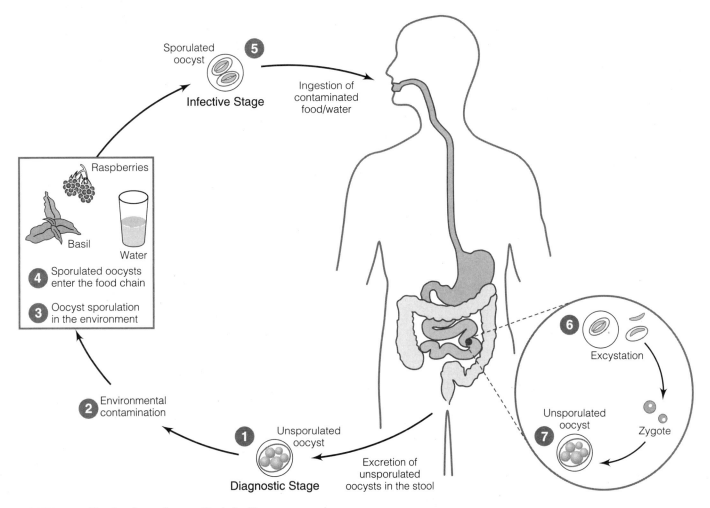

FIGURE 11.4 Mechanism of parasitic infection.

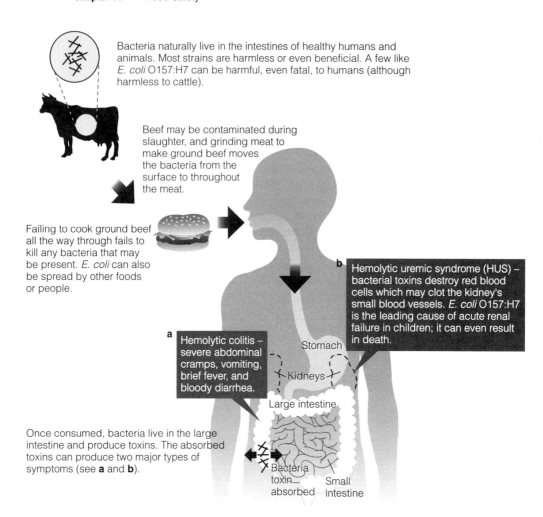

FIGURE 11.5 Mechanism of infection due to *Escherichia coli* serotype O157:H7.

S. enteritidis is another important pathogen in this genus, responsible for over 8,000 cases and 40 deaths attributed to consumption of contaminated undercooked eggs. As a result, the Food and Drug Administration (FDA) defined shell eggs as a "potentially hazardous food" in 1990. Food has not been the only vehicle of infection, with reptiles being a known reservoir for the organism. From 1970 to 1971, more than 280,000 cases of **salmonellosis** involved pet turtles and 300 other animals, prompting the FDA to ban interstate shipment of turtles in 1975.

***E. coli* O157:H7.** Perhaps the best-known outbreaks of foodborne illness in the last few years have involved the organism *Escherichia coli* serotype O157:H7 (pronounced and written as the letter O *not* the number 0). It is found in cattle, with dairy cows currently being considered the primary reservoir for the organism. In 1993, there were 477 cases of **hemorrhagic colitis** in Washington state and California from consumption of undercooked, contaminated hamburgers sold by a particular fast-food restaurant chain.

This disease manifests itself by profusely bloody diarrhea. In 6 percent of the cases, the victims developed a complication known as **hemolytic uremic syndrome,** which affects the kidneys. As a result, four children died, prompting the Food Safety and Inspection Service (FSIS) of the United States Department of Agriculture (USDA) to revamp its meat inspection system. Figure 11.5 shows the mechanism of infection.

Listeria monocytogenes. In 1983, 41 people in the Maritime Provinces of Canada became ill after consumption of cabbage obtained from a field that had been fertilized with animal manure. The farmer who grew the cabbage owned a flock of sheep in which two sheep had died of **listeriosis.** He fertilized the field with manure from his flock, essentially contaminating it with the organism. There were 9 stillbirths caused by consumption of the cabbage by pregnant women. In addition, of 23 babies that were born afterward, 6 died shortly after birth from complications such as meningitis. Since *Listeria monocytogenes* is psychrotrophic, growth occurred even though the cabbage had been refrigerated.

The most notable outbreak of listeriosis, however, occurred in 1985 in California, where 142 people came down with the disease after consuming Jalisco-style cheese made from milk that may not have been properly pasteurized. There were 48 deaths, including several stillbirths.

Yersinia enterocolitica. The organism *Yersinia enterocolitica* is found in rivers, streams, and lakes due to the contact between these bodies of water and wild animals that harbor the organism. The largest outbreak of **yersiniosis** occurred in 1976 in New York state, during a visit by school children to a dairy. Consumption of contaminated chocolate milk led to an outbreak where 220 children became ill. The abdominal pain suffered was misdiagnosed as appendicitis in 36 children, with 16 of them actually having their appendix removed. Contaminated chocolate syrup was the suspected source.

In 1982, a multi-state outbreak of yersiniosis took place, and 1,000 people became ill after consuming contaminated milk. It was determined that the crates used to transport milk cartons were also used to carry outdated milk to a pig farm. The crates were contaminated with mud and manure and brought back to the dairy. Bacteria probably accumulated on the crates, contaminating the outside of milk cartons sold to consumers, who then introduced them into the product during opening.

Clostridium botulinum. **Botulism** is a paralytic illness, caused by consumption of products contaminated with the organism *Clostridium botulinum*. This organism produces the most powerful toxin known, and it affects the nervous system. Although botulism cases are relatively rare in the United States, with only 34 cases reported in 1994, the illness is so severe that it is worth mentioning. Most outbreaks result from consumption of foods improperly canned in the home, such as green beans. Pressure cooking is required to destroy *C. botulinum* spores, which can be present in foods (such as vegetables) that come in contact with soil, where the organism is found.

Vibrio cholerae. *Vibrio cholerae* is a foodborne pathogen that has caused a veritable epidemic of **cholera** in Latin America in the last few years. Thousands of people have died since 1991, with hundreds of thousands becoming ill. These outbreaks have been caused by contact with waters contaminated with the organism. Other species, such as *V. parahaemolyticus* are not as important as *V. cholerae* on a worldwide basis. However, it is considered the leading cause of foodborne illness in Japan, where raw seafood is frequently consumed.

Molds—Ergotism and Aleukia

Mycotoxins associated with foodborne illness are produced primarily by molds belonging to the group Deuteromycetes (Imperfect fungi). The mold *Claviceps purpurea* is found in rye, wheat, barley, and oats, with toxin being produced under conditions that are cool and damp. It causes **gangrenous ergotism,** in which a burning sensation of the feet and hands develops into a loss of circulation, often resulting in the necessity of limb amputation.

This mold also causes **convulsive ergotism,** with the toxin being chemically similar to the drug LSD. Not surprisingly, hallucinations and convulsive seizures are the main symptoms. In the United States, grains are considered **ergoty** if they contain more than 0.3 percent of *Claviceps* mold by weight. The mold is usually removed during milling, but grains destined as cattle feed may be contaminated.

Fusarium mold species are responsible for several types of illness, depending on the mycotoxin produced. *F. sporotrichioides* produces trichothecene toxin, which causes **alimentary toxic aleukia.** The mold is found in wheat, oat, barley, and rye. Toxin produc-

TABLE 11.6 Viruses implicated in cases of foodborne illness.

Virus	Onset (duration)	Principal Symptoms	Typical Foods	Mode of Contamination	Prevention of Disease
Hepatitis A virus (hepatitis A)	10–50 days (2 weeks to 6 months)	Fever, weakness, nausea discomfort; often jaundice	Raw or undercooked shellfish; sandwiches, salads, etc.	Human fecal contamination, via water or direct	Cook shellfish thoroughly; general sanitation
Norwalk-like viruses (viral gastroenteritis)	1–2 hr (1–2 days)	Nausea, vomiting, diarrhea, pains, headache, mild fever	Raw or undercooked shellfish; sandwiches, salads, etc.	Human fecal contamination, via water or direct	Cook shellfish thoroughly; general sanitation
Rotaviruses (viral gastroenteritis)	1–3 days (4–6 days)	Diarrhea, especially in infants and young children	Raw or mishandled foods	Probably human fecal contamination	General sanitation

tion is stimulated at very low temperatures (-1 to $-10°C$; 30.2 to 14°F), and ordinary baking does not affect it. The first stage of the disease consists of a burning sensation in the mouth only a few hours after consumption. This progresses to the esophagus and stomach, which become inflamed, resulting in vomiting, diarrhea, and cramps. The second stage consists of a cessation of symptoms for about two months during which the person feels well. During this time, however, bone marrow is being destroyed and the patient develops leukemia, anemia, and secondary bacterial infections. Hemorrhagic patches appear subcutaneously. In the third stage, necrosis of the skin and muscles takes place, along with bronchial pneumonia, hemorrhages in the lungs, stomach, and intestines, leading to death in as many as 80 percent of the patients.

F. moniliforme is found in feed grains like corn. The toxin is produced at near freezing temperatures and is triggered by temperature cycling from low to moderate to low. The toxin is thought to cause esophageal cancer.

Aspergillus flavus is another mold that produces a mycotoxin of significance to human health. Wheat, flour, peanuts, and soybeans carry this mold. If consumption of the toxin is high, it can result in death due to liver necrosis. In 1974, 397 persons were affected, with 108 deaths in India due to consumption of corn containing 0.25 to 15 mg/kg of aflatoxin type B1. Sublethal doses cause chronic toxicity, while low doses are thought to promote liver cancer.

Viruses in Foods

Hepatitis type A ranks as the sixth leading cause of foodborne illness in the United States. It is transmitted by food, associated with shellfish. It causes profound malaise, headache, anorexia, nausea, abdominal discomfort, and may be followed by jaundice. It takes about 4 weeks to develop, with the virus being shed in the feces during this time. Usually, transmission is by the fecal-oral route, with foods subject to fecal contamination due to a foodhandler, or to contact with sewage. Table 11.6 summarizes the viruses that cause foodborne illness.

Norwalk and Norwalk-like viruses were first recognized in a school outbreak in Norwalk, Ohio in 1968, with water suspected as the source. The symptoms consist of gastroenteritis with vomiting and diarrhea. The typical Norwalk pattern is an incubation of 1 to 2 hours, followed by illness that lasts 1 to 2 days. The virus is also shed in the feces during the symptomatic period.

Some viruses resemble the Norwalk but are antigenically different. This means that antibodies that react with one virus type do not react with the other. These viruses cause the same symptoms as Norwalk viruses.

Viruses are easily inactivated by heat, thus thorough cooking is sufficient to render the food safe from these agents. Most viral foodborne illness occurs from eating food contaminated after cooking or to lightly cooked shellfish.

TABLE 11.7 Common parasites implicated in cases of foodborne illness.

Parasite	Onset (duration)	Principal Symptoms	Typical Foods	Mode of Contamination	Prevention of Disease
PROTOZOA					
Giardia lamblia (giardiasis)	5–25 days (varies)	Diarrhea with greasy stools, cramps, bloat	Mishandled foods	Cysts in human and animal feces, directly or via water	General sanitation; thorough cooking
Cryptosporidium parvum (Cryptosporidiosis)	2–14 days (2–3 weeks)	Diarrhea; sometimes fever, nausea, and vomiting	Mishandled foods	Oocysts in human feces	General sanitation; thorough cooking
Cyclospora	2–14 days (2–3 weeks)	Diarrhea; loss of appetite; fatigue; weight loss	Raw fruits and vegetables that may be irrigated with contaminated water	Cysts in feces	General sanitation, thorough cooking
Toxoplasma gondi (toxoplasmosis)	10–23 days (varies)	Resembles mononucleosis; fetal abnormality or death	Raw or undercooked meats; raw milk; mishandled foods	Cysts in pork or mutton, rarely beef; oocysts in cat feces	Cook meat thoroughly; pasteurize milk; general sanitation
ROUNDWORMS					
Trichinella spiralis (trichinosis)	8–15 days (weeks, months)	Muscle pain, swollen eyelids, fever; sometimes death	Raw or undercooked pork or meat of carnivorous animals (e.g., bears)	Larvae encysted in animal's muscles	Thorough cooking of meat; freezing pork at 5°F (−15°C) for 30 days; irradiation
FLATWORMS (TAPEWORMS)					
Beef tapeworm (*Taenia saginata*)	10–14 weeks (20–30 years)	Worm segments in stool; sometimes digestive disturbances	Raw or undercooked beef	"Cysticerol" in beef muscle	Cook beef thoroughly or freeze below 23°F (−5°C)
Pork tapeworm (*Taenia solium*)	8 weeks–10 years (20–30 years)	Worm segments in stool, sometimes "cysticercosis" of muscles, organs, heart, or brain	Raw or undercooked pork; any food mishandled by a *T. solium* carrier	"Cysticerol" in pork muscle; any food—human feces with *T. solium* eggs	Cook pork thoroughly or freeze below 23°F (−5°C), general sanitation

Parasites—Protozoa and Worms

Table 11.7 presents the most common parasites implicated in foodborne illness.

Protozoa. *Giardia lamblia* was first discovered by Antonin Van Leeuwenhoek as he examined his own feces under a microscope he designed and built. *Giardia* occurs in the form of a pear-shaped cyst, which germinates upon ingestion. One cyst yields two **trophozoites,** which have eight flagella that help propel them by a falling-leaf type of motility. The trophozoites penetrate the intestinal wall but not deeply, causing cramps, nausea, weight loss, severe diarrhea, vomiting, and flatulence. Its onset is one to two weeks, and it can last for up to 3 months. It is highly contagious, with contaminated water and contaminated meat being the prime sources of infection.

Toxoplasma gondii is commonly harbored in house cats. Oocysts in fecal material are ingested, which pass to the intestine where they release eight sporozoites. These pass through the intestinal wall into the circulatory system, able to multiply rapidly in many parts of the body. At this point they are referred to as *tachyzoites* (*tachy* means fast), which form clusters surrounded by a protective wall. They are then called *bradyzoites* (*brady* means slow), since they do not grow yet persist in the body for life. Figure 11.6 shows this sequence of infection.

When immunity is suppressed, these bradyzoites break and release the tachyzoites, which multiply rapidly bringing another acute infection. These are easily transmitted to

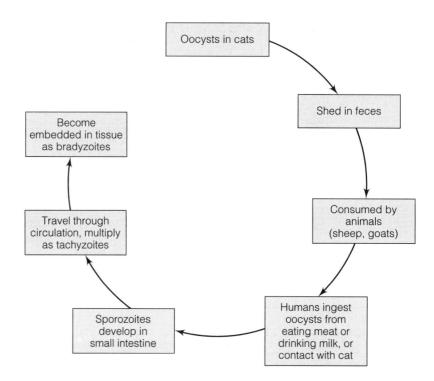

FIGURE 11.6 Pathway of Infection by *Toxoplasma gondii*.

the fetus through the placenta, thus are a danger to unborn children. Most people experience no symptoms, with some exhibiting rash, headaches, fever, muscle aches and pains, and swelling of the lymph nodes, all of which mimic infectious mononucleosis. In fact, 50 percent of Americans have antibodies to *T. gondii*, suggesting that infection with this organism is common. The cysts are easily destroyed by heating and freezing.

Cryptosporidium and *Cyclospora* are similar organisms, both of which exist as oocysts and are found in untreated water. Oocysts are resistant to chlorine used in municipal waters, thus these organisms have been implicated in outbreaks related to swimming activities. Once ingested, the oocysts give rise to four sporozoites in the intestinal or respiratory tract, depending on whether the water was ingested or inhaled. The sporozoites attach superficially to these surfaces, developing a feeder organelle that helps the parasite ingest nutrients from the host.

The sporozoite transforms into a trophozoite and then to a schizont. The schizont releases several merozoites, which penetrate other epithelial cells of the intestine. Merozoites undergo cell division and give rise to eight oocysts, some of which have a thin wall and some have a thick wall. The thin-walled oocysts give rise to an autoinfection, since they penetrate the intestinal wall and repeat the cycle described above. The thick-walled oocysts are shed in the feces or sputum, and infect other people who ingest material contaminated with the feces. The symptoms develop within 2 to 14 days, with voluminous watery diarrhea, anorexia, weight loss, dehydration, and abdominal discomfort lasting up to 3 weeks.

Flatworms. Flatworms are another type of parasite that can be transmitted via foods to cause disease in humans. Of the flatworms, the tapeworms *Taenia saginata* (beef) and *Taenia solium* (pork) are the best known. Cattle and swine are intermediate hosts for these parasites (host in which the parasite develops and matures), with humans being the **definitive hosts** (host in which the parasite multiplies). These worms do not have vascular, respiratory, or digestive systems—using those of the host to obtain nourishment.

Eggs in the soil are ingested by cattle or swine, depending on the tapeworm. Ingestion of eggs can also occur by **coprophagy,** where feces infected with the eggs are directly eaten by the animal. Eggs are called *oncospheres*, and these release embryos inside the intestinal tract of the host. The embryos penetrate the intestinal wall and are carried to other tissues such as muscle, tongue, and heart by the circulation. In these tissues, they form larvae called "cysticerci" over a period of 3 months. In this stage, the larvae are called *Cysticercus bovis* for the beef tapeworm, and *Cysticercus cellulosae* for the swine tapeworm.

Humans ingest the cysticerci in undercooked beef and pork. The adult form of the tapeworm develops inside the human intestine, where it can live for up to 25 years and grow to 4-6 meters. Usually there are no symptoms, but in 3 percent of the cases, cysticerci form in the eye, which may lead to vision problems. In 5 percent of the cases, cysticerci form in muscles, causing massive muscle enlargement in the patient. In 60 percent of the cases, cysticerci form in the central nervous system, causing neurological symptoms such as epilepsy. This takes place on average 5 years after ingestion of the eggs.

Roundworms—Trichinella. Roundworms are another parasite that can be transmitted through foods. The most commonly known is *Trichinella spiralis*, the organism responsible for **trichinosis**. It has the same intermediate and definitive host, which can be animals or humans. The adult form lives in the intestine of mammals like swine, dogs, bears, and marine animals. Females can lay about 1,500 eggs. The eggs contaminating the soil or feces can be consumed by animals.

Once inside the intestinal tract, the eggs develop into larvae, which burrow through the intestinal wall, lodging in certain muscles like the eye, tongue, diaphragm, and biceps. The larvae then grow in the muscle, and 6 to 18 months later develop into **cysts** by curling up and becoming enclosed in a calcified wall. No further development takes place, with cysts remaining viable inside the muscle for up to 10 years.

Consumption of muscle tissue, or meat, contaminated with cysts results in germination of the cysts inside the body. These lay eggs and the cycle of infection is repeated, except this time it is inside a human host. Severe muscle pain results, with facial swelling within 15 days. Diarrhea, vomiting, and fever can result in some cases. Trichinosis can lead to death due to heart failure. The cysts are easily destroyed by cooking at 77°C (170.6°F).

11.4 WHAT IS MAD COW DISEASE?

What has been termed "mad cow disease" is actually one of several transmissible spongiform encephalopathies (TSEs) specifically called bovine spongiform encephalopathy (BSE). TSEs are a class of disease, in which the brain of the victim is riddled with holes, taking on a spongelike consistency. Neurological symptoms such as convulsions, and eventually death, result. The big question is: Can the disease be transmitted from animals to humans via meat consumption? In attempting to answer this question, to which at present there is only speculation, the several types of TSEs must be clarified.

Transmissible Spongiform Encephalopathies (TSEs)

Scrapie in sheep and goats results in loss of coordination and intense itching (scraping) in these animals. It has been around for over 200 years, with an incubation period of about 4 years before symptoms develop. *Kuru* in humans is also known as "the laughing death." It is found only in New Guinea and is acquired through ritual cannibalism. The natives believe in honoring the dead by eating their brains. This practice has stopped and so has the incidence of kuru in that country.

Creutzfeldt-Jakob Disease (CJD) in humans causes dementia. Most cases result in death within 1 year, with some victims taking about 10 years to die. Roughly 85 percent of CJD cases are "sporadic" and occur in people over 60 years of age. Some cases (about 15%) are inherited, and a smaller percentage (5%) occur through the treatment of other medical problems ("iatrogenic"), such as in conducting cornea transplants. In these cases, the patient is believed to be infected directly through contaminated surgical instruments. The total number of CJD cases per year in the United Kingdom before 1995 was only 30.

New variant CJD (nvCJD) in humans is the mystery disease that has emerged since 1995, affecting teenagers and young adults. It is different from typical CJD, since it does not fit the "sporadic," "inherited," or "iatrogenic" profiles described above. The cause is

Known transmissible spongiform encephalopathies (TSEs):

Creutzfeldt-Jakob Disease (CJD)
Kuru
Mad cow disease (BSE)
New variant CJD (nvCJD)
Scrapie

unknown, but some scientists suspect that victims have somehow acquired the disease through consumption of beef from BSE-affected cows.

BSE, or mad cow disease, causes loss of coordination, convulsions, and apprehension in cattle. It was first observed in Great Britain in a farm in Kent in 1985, and later identified by a vet lab in 1986, with 59 percent of dairy herds and 15 percent of beef herds affected. The origins of BSE were traced to feeding the cows a feed containing bone meal obtained from dead (diseased) sheep that were perhaps suffering from scrapie. As a result, in 1989, animal-derived feed supplements were banned in the United Kingdom. The question is: Can nvCJD cases in humans be due to consumption of beef from "mad" cows?

Causes of TSEs

A very good question to raise at this point is: what causes TSEs, including nvCJD? In the 1990s, scientists discovered that a small protein, known as a **prion**, or PrP, is the cause of TSEs, both in animals and in humans. It is believed that the prion travels through the spinal cord and reaches the brain, where it causes damage. An important fact is that the gene that codes for PrP resides in the chromosome of many animals and humans normally, yet without causing sickness. Perhaps the reason is that PrP can occur in two forms: a *normal conformer* and an altered shape, called a *rogue conformer*.

Once produced in the body through heredity or acquired through infection, PrPsc causes disease by attaching to the normal PrPc and causing it to unfold and flip configuration from an alpha helix to assume the shape of the PrPsc (beta sheet). A beta sheet can be thought of as a less tightly organized secondary structure of amino acids. Somehow, this "new" PrPsc causes other normal PrPc molecules to change configuration to the beta sheet, resulting in more PrPsc being formed in a sort of chain reaction.

This conversion of PrPc into PrPsc occurs inside neurons and somehow causes the destruction of these cells, resulting in holes in the brain. The conversion of normal into rogue prion is an amazing event, almost as if the rogue prion were "self-replicating" itself. All the more amazing since this is not a living organism like a virus or bacterium, but merely protein material.

Table 11.8 lists the common features of the TSEs.

Is PrPsc Transmitted to Humans Through Consumed Beef?

Figure 11.7 shows a possible sequence of transmission of mad cow disease across two species barriers from sheep to cattle to humans. What does the evidence say?

Arguments Against. Here are the arguments and evidence against the notion of rogue PrPsc being transmitted to humans through eating beef. First, it is believed that the more the amino acid sequence of the disease PrPsc resembles that of the host's normal PrPc, the more likely it is that the host will acquire the disease if exposed to the disease prion. Sheep prions differ from cattle prions by only 7 amino acids, while cattle and humans differ by more than 30 amino acids. Given the dissimilarity between human and cattle prions, it has been suggested that this proves that transmission of mad cow disease to humans is impossible. The difference in number of amino acids between the species is just too great.

Second, it has been shown in the laboratory that transgenic mice containing human normal PrPc produce nvCJD when injected with human disease PrPsc, but not when injected with cattle disease PrPsc. Third, the distribution of CJD and nvCJD in the world does not coincide with the incidence of scrapie in sheep or of BSE in cattle. In other words, countries that have a high incidence of scrapie or mad cow disease do not have a high incidence of CJD or nvCJD, and countries that have a low incidence of mad cow disease do not have a low incidence of nvCJD. This is the opposite of what common sense would predict.

Arguments For. There is also argument and evidence that suggests BSE causes nvCJD. First, Macaque monkeys injected with the BSE prion die from nvCJD symptoms.

TABLE 11.8 Common features of TSEs.

Prolonged incubation period (months to years)
Causes a progressive and debilitating neurological condition
Always fatal
Brain tissue from infected animals and humans shows scrapie associated fibrils (SAF)
Pathological changes to the central nervous system include vacuolation and astrocytosis (misshapen cells)
Disease agent does not elicit a detectable immune system response in host
No live animal diagnostic test available
Highly resistant to processing, including antibiotic and antiviral agents, heat, irradiation, and ultraviolet light

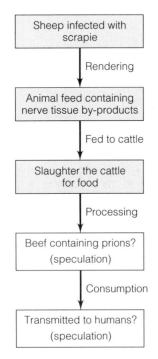

FIGURE 11.7 Speculation (unproven) regarding a possible mode of transmission of mad cow disease form infected sheep through cattle to beef consumption by humans.

Although injected rather than fed prions, this indicates a possible link between mad cow disease and nvCJD. Second, high numbers of BSE cases in cattle have been shown to correlate in timing to the incidence of the 31 nvCJD cases in the United Kingdom. Third, tests have shown a similarity between BSE and nvCJD prions structurally, although not with the sporadic type of CJD.

Why Great Britain? People have wondered why Great Britain experienced the mad cow and nvCJD problem more than any other country. Possible reasons are that the British produce more sheep than the United States, thus they have many more scrapie cases. Second, quite a bit of the feed in the United Kingdom used to contain animal by-products. Third, but most significant, the British eliminated the use of steam-heating (suggested steam treatment parameters of 140°C (284°F) for 30 minutes at 3.6 bars of pressure) of rendered products. This alone may have resulted in survival of the scrapie prion, which may have then been easily transmitted to cattle through the feed.

The incidence of BSE in Europe has of course had a devastating impact on the European beef and sheep industries. The cost was estimated at $2.75 billion for the year 2001, with a reduction of 375,000 tons of export capacity for the European Union. In February 2001, France, Belgium, Germany, Italy, and Spain uncovered more than a dozen BSE cases. Later that year, Japan reported its first case of mad cow disease, resulting in the destruction of about 5,000 animals by the Japanese government.

So far, the United States has avoided BSE contamination. However, approximately 1200 cattle in Texas were quarantined in the year 2001 because of a violation of the FDA's 1997 prohibition on using ruminant material in feed for other ruminant animals. This occurred when a very low level of prohibited material (domestic in origin) was found in the cattle feed. As a result, experts feel that the key to safety is a responsible U.S. beef industry. A risk assessment study was conducted by Harvard University, and the results reported in early 2002 (www.aphis.usda.gov/oa/bse). The scientists at Harvard developed a mathematical model that would predict the incidence of mad cow disease in the U.S. The report affirms that even if one takes into consideration a worst-case scenario in which the feed ban is not perfectly followed, "the United States is highly resistant to BSE." These are comforting words but it is clear that complete compliance with the feed ban is the key to keeping the U.S. free of BSE. The industry must work to tighten import control, achieve total compliance with the FDA's feed ban, and implement a strong surveillance program of high-risk animals in order to protect itself and, most importantly, consumers.

11.5 PREVENTING FOODBORNE ILLNESS

Of the factors involved in foodborne illness, improper holding or storage temperatures ranks as number one, with foods consumed in restaurants causing the vast majority of outbreaks (see Table 11.9). In addition, poor personal hygiene and improper cooking temperature are responsible for a significant number of outbreaks. Thus, in considering the ways in which foodborne illness caused by biological hazards can be prevented, we can summarize them into three categories:

1. preventing or minimizing contamination
2. preventing or minimizing growth of the hazard
3. eliminating or reducing the hazard

The following discussion will examine each of these strategies to improve food safety.

Preventing Food Contamination

To accomplish this goal, we must first realize that contamination of raw foods with pathogenic organisms is almost inevitable. Microorganisms and parasites are ubiquitous, making it very difficult to avoid contact with fruits, vegetables, or food animals on the farm, or

TABLE 11.9 Factors contributing to outbreaks of foodborne illness in 1992 in the United States.

Contributing Factor	Number of Outbreaks
Improper holding temperature	149
Poor personal hygiene of food handler	70
Inadequate cooking	70
Contaminated equipment and utensils	44
Cross-contamination of cooked and raw foods	23
Food obtained from unsafe sources	17

SOURCE: Centers for Disease Control.

TABLE 11.10 Sanitizers commonly used by the food industry.

Sanitizer	Characteristics	Advantages	Disadvantages	Common Use
Chlorine	Causes oxidation of molecules; most effective at low pH	Kills gram-positive, and gram-negative bacteria and spores; cheapest sanitizer; nontoxic	Deteriorates in the presence of organic matter; corrosive to metals; poor penetration	200 ppm to treat water and to sanitize processing equipment
Iodine	Disrupts cell permeability; works best at low pH (mixed with phosphoric acid)	Nonirritating to skin so can be used for handwashing; effective against cells and somewhat against spores; prevents accumulation of minerals	Stains plastic surfaces; if minerals are present it does not remove them well; costs more than chlorine; vaporizes at 50°C (122°F)	25 ppm as a handwash and to sanitize processing equipment
Quaternary Ammonium Compounds	Inhibit enzymes; disrupt membrane permeability; most effective at high pH	Good penetration of porous surfaces; natural wetting agent (surfactant) so has detergent ability; forms bacteriostatic film; stable in presence of organic matter and high heat; noncorrosive; nonirritating	Not effective against most gram-negative bacteria; film forms on equipment	500 ppm to sanitize floors; 200 ppm to sanitize walls
Organic Acids (lactic, acetic, propionic)	Affect metabolic processes of the cell; disrupt membrane permeability	Toxicologically safe; not affected by minerals such as found in hard water; stable at high heat and in presence of organic matter	Corrosive to metal; costly	130 ppm to sanitize stainless steel surfaces

with seafood in contaminated waters. Thus, a more realistic goal is to state that contamination with biological hazards should be prevented as much as possible, or minimized.

The food industry depends on good manufacturing practices, or **GMPs,** and sanitation procedures to accomplish this goal. As part of GMPs, foods should be produced in such a way as to minimize contact with contaminated soil, water, or air. **Cross-contamination** occurs when microorganisms are transferred from one food to another. This is a critical control point that must be minimized through careful processing and sanitation procedures.

During processing, foods should be handled in a sanitary manner, making sure that all food-contact surfaces, whether equipment, utensils, or hands, are thoroughly cleaned to remove organic matter and prevent cross-contamination. Once this is done, a **sanitizer** is applied just prior to processing to kill any biological hazards such as bacteria that may remain. Table 11.10 summarizes the sanitizers used in the food industry.

Preventing Proliferation of Foodborne Microorganisms

Table 11.11 lists the strategies to control microorganisms in food. Temperature is a good method to inhibit microbial growth—foods should be stored either below 40°F or above 140°F to maintain their safety. The range of temperatures 40–140°F (4.4–60°C) is known as the **danger zone** (Figure 11.8), because most disease-causing microorganisms proliferate at these temperatures if maintained for at least two hours. Thus, an increase in mi-

TABLE 11.11 Strategies to control microorganisms in foods.

Antimicrobial agents
Canning
Drying
Freezing
MAP (modified atmosphere packaging)
Nonthermal processing (pulsed electric field, pressure, ionizing energy, etc.)
Refrigeration

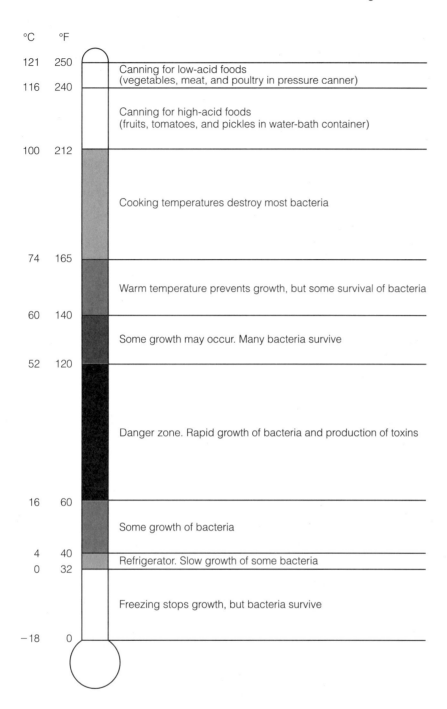

FIGURE 11.8 The food temperature danger zone.

crobial numbers can result, as well as production of toxin by toxigenic organisms, rendering the food unsafe. When foods are cooked and are to be stored, the temperature should be lowered to below 4°C (about 40°F) as quickly as possible to minimize the time that the food spends within the danger zone of temperatures. This can be accomplished by storing hot foods in small portions.

The best-known method to prevent proliferation of a biological hazard in foods is refrigeration. Lowering the temperature of the medium slows down metabolic reactions of the organism, slowing down its growth. Most pathogenic organisms cannot grow below 5°C (41°F), and even those that can are often not able to synthesize toxins.

The food industry relies heavily on refrigeration to prevent the growth of biological hazards. For example, most manipulations involved in fabrication of meats are carried out in processing rooms where the temperature is kept at 4 to 7°C (39.2 to 44.6°F). In addition to refrigeration, freezing can be used to prevent microbial growth. It should be noted that freezing kills only about 10 percent of bacteria, thus it is used primarily to prevent growth.

Another method is the use of preservatives that do not kill the organisms but simply prevent them from growing. This effect is termed **bacteriostatic** because bacteria are not able to multiply even though they remain viable. For example, sulfites, which are used in fruits to delay enzymatic browning, are bacteriostatic to lactic acid bacteria. Sodium nitrite is used in the curing of meats to stabilize the red color and contribute to flavor development. However, it also inhibits germination of *Clostridium botulinum* spores and the subsequent production of neurotoxin.

Drying can be used to inhibit germination of spores of pathogenic bacteria in a number of food products. However, if those products are rehydrated, the opportunity for germination and toxin production return. Smoking is another method, used primarily in cured and fermented products. In applying heat to a fuel such as wood to create smoke, the smoking process dehydrates the food, cooks it, and deposits smoke components, some of which are bacteriostatic. This process mainly causes reduction of biological hazards rather than inhibition of growth, thus it will be discussed in further detail in the next section.

Eliminating or Reducing Biological Hazards

One of the most widely used methods for reducing the number of, or completely eliminating biological hazards is heat. It can be used in conjunction with other methods such as drying and smoking to enhance its **bactericidal** ability (killing of bacterial cells). Heat inactivates enzymes due to protein denaturation, which essentially renders the biological hazards unable to carry out their metabolic processes. It also damages structural cell components, resulting in death. The effectiveness of this treatment on eliminating or reducing hazards depends on a few factors: (1) type of heat treatment, (2) type of food, and (3) type of biological hazard.

Regarding the treatment, the higher the temperature and the longer the time of heat application, the more effective the treatment. However, if the food is dry, the transfer of heat may not be efficient enough to reach the contaminating organisms. In addition, if the food has a high fat content, this can act as an insulator to protect microorganisms during heating. Finally, some organisms are better equipped to survive high temperature conditions. For instance, bacterial spores can survive at least ten times longer in conditions that would eliminate vegetative cells and other biological hazards. Also, gram-positive bacteria are much more resistant to heat than gram-negative cells due to their thicker cell wall.

Smoking can be applied at the same time as heat to eliminate biological hazards. Some of the components of smoke are phenolic compounds, which act as bacteriostatic and bactericidal agents. Alcohols such as methanol are also produced and deposited on food during smoking, which have known antimicrobial properties. In addition, organic acids such as formic and acetic acids are given off, which serve to inhibit and kill microbial contaminants in smoked foods.

Food irradiation is considered a recent technology for the elimination of pathogenic organisms in foods, although its application as a food preservation method has been studied since the 1950s. The Chapter 8 Challenge showed how ionizing radiation kills organisms such as parasites, bacteria, and molds by ionizing the DNA molecule. The radiation breaks the chemical bonds holding the molecule together, resulting in an inhibition of the ability of the organism to synthesize proteins or to replicate.

11.6 HACCP—A PREVENTIVE APPROACH

In 1959, the National Aeronautics and Space Administration (NASA) asked the Pillsbury Company, in collaboration with the U.S. Army Natick Laboratories and several food science experts, to develop a system that would ensure the safety of foods to be used in the space program. Thus the **Hazard Analysis Critical Control Points** system, or **HACCP** (ha-sip), was born.

The system was designed to identify the hazards that could occur to foods during their preparation, to identify the steps in production of the food that would have to be controlled in order to produce the safest product possible, and to monitor that these steps were actually being controlled. HACCP was so successful that the canning industry began applying its principles. It is designed as a *preventive system*, because through controlling the points in the process that can reduce or eliminate hazards, one can *prevent* the production of unsafe product as much as possible.

In 1985, the National Academy of Sciences (NAS) began a study to determine whether criteria could be established for specific foods in terms of the levels of allowable microbial contaminants. Instead of establishing microbial criteria, the NAS published a report endorsing and recommending that HACCP be adopted by the food industry. In 1988, the National Advisory Committee for the Microbiological Criteria for Foods was established, composed of experts in food safety and microbiology. This committee published a document in 1989 outlining seven principles for the application of HACCP to food processing. This document was revised in 1992 and 1997, attempting to harmonize the principles with those adopted by the international body Codex Alimentarius.

Principles of a HACCP System

According to the August 14, 1997 revision, the HACCP system consists of the following seven principles:

1. Conduct a hazard analysis.
2. Determine critical control points.
3. Establish critical limits.
4. Establish monitoring procedures.
5. Establish corrective actions.
6. Establish verification procedures.
7. Establish record-keeping and documentation procedures.

The first step toward implementing HACCP is to assemble a HACCP team, which should be composed of individuals representing every aspect within the organization: product development, production, sanitation, sales, management, etc. The team should be interdisciplinary to ensure that all issues are considered in developing a HACCP plan. Obviously, management must be committed and provide adequate resources to the HACCP team to succeed in their task.

The second step involves a description of the food being produced and its distribution, developing a flow diagram of the production steps in making that product. Once these preliminary steps are complete, the HACCP team follows the seven principles to develop a HACCP plan.

According to Principle 1 and using the flow diagram, each step of production is analyzed to determine whether a significant hazard is introduced, increased, or controlled in that step. A **hazard** is defined as any biological, chemical, or physical entity that can harm the consumer. A **significant hazard** is one that has a high probability of occurrence, given that GMPs are being followed, or that it is of sufficient severity to warrant its control even if the likelihood of its presence is low. Preventive measures are also identified at each of the steps.

Following Principle 2, **critical control points** (CCPs) are identified. CCPs are those points or steps during the process for which control is essential in order to produce the safest food possible. Next, **critical limits** for each CCP are established, according to Principle 3. Ideally, these are numerical values that, if exceeded or not met, will result in the loss of control of the particular CCP.

Each CCP also needs to be monitored, per Principle 4. **Monitoring** consists of conducting procedures that enable the determination of whether a critical limit is being maintained or not. Ideally, monitoring is carried out automatically and continuously, as occurs when electronic thermographs are used to measure temperature.

Corrective actions must also be established according to Principle 5, in case the monitoring procedure reveals the violation of a critical limit. These actions consist of specific outlined steps that dictate what is to be done to correct the cause of the violation, and what is to be done with the product that was produced while the critical limit was violated.

Principle 6 consists of **verification,** where procedures are outlined that will help determine whether control of a CCP is being maintained. These procedures can include calibration of thermometers used in monitoring and microbial analysis of the product. Finally, Principle 7 states that **records** must be kept of all procedures, including the hazard analysis, justification for decision, identification of CCPs within the flow diagram, monitoring records, corrective action records, verification records, and other documentation.

GMPs and SOPs

Before HACCP is even considered, and for it to be successful, certain programs must already be in place. GMPs outlining how employees are trained, building requirements, processing requirements, pest management requirements, and sanitation requirements of a food production or processing plant must be followed. Plants must have written **standard operating procedures,** or SOPs, which outline the step-by-step procedures they are to follow to comply with GMP requirements.

An example of an SOP is the step-by-step instructions for cooking a certain product. Another example is the step-by-step procedure for cleaning a certain type of equipment. This latter SOP is known as an SSOP, or **sanitation standard operating procedure,** because it describes an activity involved in the sanitation of a food processing plant.

If GMPs and SOPs are not in place, establishing a HACCP plan is doomed to failure. Consider the example of a beef slaughter plant in which the employees never wash their hands, the knives used for trimming the carcasses are never sanitized, the floors are never cleaned, all of which constitute violations to GMPs. Every step during production of such carcasses, from the removal of the hide to the splitting of the carcass, would need to be considered a critical control point. The washing of knives, the washing of hands, etc. would all be steps that would have to be controlled in order to produce the safest product possible.

On the other hand, if GMPs were adhered to in the first place, the HACCP plan would not have to include operations dealing with sanitation and employee hygiene as critical control points, allowing for *true* critical control points to be identified. The bottom line is that if every step that is carried out in a plant is identified as critical, then the definition of "critical" loses its meaning and importance. Thus, identification of critical control points must be done after a GMP program is well established, so that only those steps along the process that are truly essential to producing the safest product possible are identified.

Key Points

- Foodborne illness is caused by biological, chemical, and physical hazards.
- Biological hazards include bacteria, molds, parasites, and viruses.
- Leading bacterial foodborne pathogens include *Salmonella, Escherichia coli* O157:H7, *Campylobacter jejuni,* and *Listeria monocytogenes.*
- Examples of foodborne mycotoxins are those produced by *Claviceps, Aspergillus,* and *Fusarium* molds.
- Typical foodborne parasites are the protozoa *Giardia,* the flatworm *Taenia,* and the roundworm *Trichinella.*
- Foodborne viruses include the hepatitis A virus and the Norwalk virus.
- Foodborne illness is typically caused by introducing contaminants into foods, by failing to store foods at the appropriate temperature, or by failing to process foods properly.
- Biological hazards in foods can be controlled by inhibiting their growth through methods such as refrigeration and the use of bacteriostatic agents, or can be reduced or eliminated by processing methods such as heating, smoking, antimicrobial agents, or irradiation.
- Although not a proven foodborne illness, mad cow disease has had a devastating impact on the world's beef industry, and research is needed to establish a connection between contaminated beef consumption and illness in humans.
- The Hazard Analysis Critical Control Point system is a preventive method that can be used to control foodborne hazards during production and processing of foods.

Study Questions

1. Define food safety, and list the three types of hazards found in foods.

2. Describe how intoxications caused by bacterial agents occur, and provide an example of an organism that causes this type of illness.

3. Define the terms *bacteriostatic* and *bacteriocidal*, and provide an example of a food processing strategy for each.

4. Explain why HACCP without SOPs and GMPs is futile.

Multiple Choice

5. One of the three types of foodborne hazards is
   a. temperature.
   b. biological.
   c. time of cooking.
   d. raw meat.
   e. food handler.

6. Toxins produced by molds are called
   a. verotoxins.
   b. mycotoxins.
   c. toxicotoxins.
   d. bot toxins.
   e. marine toxins.

7. A host organism in which a parasite multiplies is the
   a. intermediate host.
   b. multiplicative host.
   c. determinative host.
   d. definitive host.
   e. primary host.

8. The mold that produces aflatoxins is called
   a. *Claviceps.*
   b. *Aspergillus.*
   c. *Thamnidium.*
   d. *Clostridium.*
   e. *Deuteromyces.*

9. The serotype of *Salmonella* most often associated with egg-related outbreaks is
   a. Enteritidis.
   b. Typhimurium.
   c. Multocida.
   d. Poona.
   e. Montevideo.

10. The international body involved in the definition of HACCP is
    a. International Association of Sanitarians.
    b. Codex Alimentarius.
    c. Codex Sanitarius.
    d. International Association for Food Safety.
    e. Food and Drug Administration.

11. *Clostridium botulinum* is
    a. a spoilage organism.
    b. a mold.
    c. capable of producing mycotoxins.
    d. the cause of hemolytic uremic syndrome.
    e. a sporeformer.

12. The danger zone of temperatures lies between
    a. 40 and 140°C.
    b. 0 and 100°C.
    c. 40 and 140°F.
    d. 0 and 45°C.
    e. 4 and 140°C.

13. The letters GMP stand for
    a. good manufacturing practices.
    b. good manufacturing procedures.
    c. good modes of practice.
    d. good modular procedures.
    e. good mentoring practices.

14. The letters SOP stand for
    a. standard orders of practice.
    b. sanitation operating practices.
    c. standard operating procedures.
    d. standard operational practices.
    e. sanitation operational principles.

15. The genus responsible for botulism is
    a. *Cyclospora*.
    b. *Cryptosporidium*.
    c. *Salmonella*.
    d. *Vibrio*.
    e. *Clostridium*.

16. Cholera is caused by an organism belonging to the genus
    a. *Clostridium*.
    b. *Vibrio*.
    c. *Salmonella*.
    d. *Cyclospora*.
    e. *Escherichia*.

17. The presence of antibiotic residues in food matter
    a. is an example of a biological hazard.
    b. is an example of a chemical hazard.
    c. is an example of a physical hazard.
    d. causes a food infection.
    e. is an example of a mycotoxin.

18. What is PrP protein?
    a. a polypeptide composed of 231 amino acids
    b. a prion
    c. a protein that is proposed to have two forms: normal and rogue
    d. choices (a) and (b) are correct
    e. choices (a), (b), and (c) are correct

19. The virus most often linked to foodhandlers is
    a. hepatitis A.
    b. hepatitis C.
    c. Norwalk.
    d. *Salmonella*.
    e. *Claviceps*.

20. Consumption of *Salmonella* cells results in
    a. intoxication.
    b. intoxification.
    c. exfoliation.
    d. infection.
    e. intrusion.

21. Consumption of toxin results in
    a. intoxification.
    b. infection.
    c. intoxication.
    d. imagination.
    e. intrusion.

22. *Taenia saginata* is an example of a
    a. bacterium.
    b. virus.
    c. mold.
    d. yeast.
    e. flatworm.

23. The leading cause of foodborne illness in the United States is
    a. molds.
    b. bacteria.
    c. yeast.
    d. viruses.
    e. parasites.

24. Contact between foodhandlers and turtles has resulted in outbreaks of
    a. cysticercosis.
    b. hepatitis.
    c. *E. coli* poisoning.
    d. salmonellosis.
    e. mold.

25. Which statement best describes the situation in an animal suffering from a TSE?
    a. The rogue form of the prion is decreased in the diseased animal.
    b. Its rogue prion protein shapes have more alpha helix, which indicates less organization.
    c. The rogue prion secondary structures are more organized than in the normal prion.
    d. The rogue prion secondary structures are less organized than in the normal prion.
    e. The rogue prion secondary structures are indistinguishable from those in the normal prion.

26. Imagine that John has had a drink of water contaminated with *Cyclospora* oocysts. However, the drinking water was chlorinated to the standard level. Which of the following is true?
    a. John will not contract a *Cyclospora* infection.
    b. The oocysts are destroyed by the chlorine.
    c. Sporozoites will attach to tissue in John's respiratory tract.
    d. Both John will not contract a *Cyclospora* infection and the oocysts are destroyed by the chlorine.
    e. John's intestinal tract will become infected.

27. How does CJT *directly* contribute to illness?
    a. The exotoxin destroys cells of the intestinal lining.
    b. The exotoxin activates an enzyme that results in osmotic imbalance.
    c. The exotoxin attaches to the mucosal cells, setting up an irritation.
    d. The exotoxin releases toxic levels of electrolytes into the blood.
    e. The intestinal lumen becomes blocked.

28. Enterohemorrhagic *E. coli* O157:H7
    a. is spread from the interior of meat tissue to the surface by grinding operations.
    b. can result from ingestion of cross-contaminated foods.
    c. can never be associated with fruit and vegetable consumption.
    d. creates a hemolytic uremic syndrome that is responsible for intestinal blockage.
    e. will be destroyed in ground beef cooked to an internal temperature of 140°F.

29. All of the following tend to support the idea that consuming BSE beef will make a person ill with mad cow disease, except
    a. BSE, scrapie, and CJD are all examples of transmissible spongiform encephalopathies.

b. human and cattle prions differ by more than 30 amino acids.
c. monkeys injected with BSE prion die from nvCJD symptoms.
d. the conversion of prions is the cause of all TSEs, in animals and humans.
e. rogue PrPsc is suggested as being transmissible to humans through eating contaminated beef.

Internet Search

Access the IFT website at *http://www.ift.org/education/food_industry/lesson1.shtml#milkprocessing* to answer the following questions.

30. Why is milk pasteurized?
    a. because it is virtually a sterile product
    b. because it is a nutritious food
    c. to destroy the microorganisms that can cause souring
    d. to sterilize it
    e. because postmilking handling creates opportunities for contamination

31. How do quality assurance workers test milk to insure optimum safety?
    a. perform nutrient analyses
    b. sterilize the milk
    c. perform SPC
    d. pasteurize milk samples
    e. perform CIP

32. Which is the *true* statement?
    a. Grade A pasteurized milk from an individual producer will contain 10^4 bacteria per mL or less, according to the U.S.P.H.S.
    b. Low SPC counts in milk samples do not guarantee absence of toxins.
    c. The primary goal of having microbiological standards for milk is to extend its shelf life.
    d. The organisms in cows that cause Q fever and tuberculosis are not transmissible to humans.
    e. The objective of pasteurization is to destroy pathogenic organisms in milk.

33. Quality assurance personnel perform all of the following except
    a. monitor compositional standards of finished milk.
    b. follow governmental regulations.
    c. perform in-process sample testing.
    d. operate homogenizer.
    e. keep records and write reports.

34. In performing a standard plate count,
    a. the goal is to quantify the number of viable aerobic organisms in a sample.
    b. plates are incubated for 72 hours.
    c. microorganisms are reported as number of colonies per mL in the diluted samples.
    d. samples are diluted onto agar plates to obtain TNTC counts.
    e. culture plates are labeled from 10^1 to 10^5 as the dilution factor.

Critical Thinking

35. Four hours after attending a wedding reception, one of the bridesmaids becomes ill with diarrhea and vomiting. When asked what she ate at the evening dinner, she recalls having consumed cured sausages, which had been left at room temperature since that morning. Given the rapidity of the development of symptoms, the type of product involved, and the symptoms themselves, what foodborne bacterial pathogen do you think was responsible?

36. A short-order cook sets out to prepare lunch for the restaurant patrons. He begins early in the morning by frying up some onions, which he stores in warm oil in a covered dish. He also slices up tomatoes and pickles. Right before lunchtime, he begins to fry the ground beef patties he will use in serving hamburgers. A few days after eating at the restaurant, a number of patrons begin to complain of having difficulty with their vision, as well as muscle weakness and fatigue. What foodborne bacterial pathogen do you think was responsible?

37. One of the instructors leading a workshop on food safety decides to have lunch at a nearby hotel restaurant. She orders a chicken dish from the buffet, which is served to her lukewarm. The next day, she begins to suffer from diarrhea, and by the following morning, she is suffering from vomiting, more diarrhea, fever, and even chills. What foodborne bacterial pathogen do you think was responsible?

38. You are conducting a hazard analysis as part of developing a HACCP plan for a company that produces refrigerated entrees. What specific bacterial hazards would you identify on fresh meat and on fresh vegetables at the receiving step of the process (i.e., as provided to you by the farmer)?

39. You open a new candy store and meet with your vendor. She introduces you to a line of German-made candy containing beef-based gelatin for improved texture, called "Mamba." Should you have any concerns selling the product at retail?

References

Ennen, S. 2001. The battle for cattle sanity. *Food Processing* 62(3):14.

Giese, J. 1994. Antimicrobials: assuring food safety. *Food Technology* 48(6):102.

Murano, E. A. 2000. Food safety in the 21st century. *Journal of Food Distribution Research* 31(1): 64.

Pierson, M. D., and Corlett, D. A. 1992. *HACCP: Principles and Applications.* Van Nostrand Reinhold, New York.

Pierson, M. D. and Stern, N. J. 1986. *Foodborne Microorganisms and Their Toxins: Developing Methodology.* Marcel Dekker, New York.

Smulders, F. J. M., ed. 1987. *Elimination of Pathogenic Organisms from Meat and Poultry.* Elsevier Science Publishers, New York.

Additional Resources

Andrews, L. S., Ahmedna, M., Grodner, R. M., Liuzzo, J. A., Murano, P. S., Murano, E. A., Rao, R. M., Shane, S., and Wilson, P. W. 1998. Food preservation using ionizing radiation. *Reviews of Environmental Contamination and Toxicology* 154:1.

Brown, N. E., Murano, E. A., and Kotinek, S. 1996. Evaluation of microbial hazards of pork products in institutional foodservice settings—Part I. *Dairy Food and Environmental Sanitation* (3)16:14.

Cliver, D. O., ed. *Foodborne Diseases.* 1990. Academic Press, New York.

Doyle, M. P., ed. 1989. *Foodborne Bacterial Pathogens.* Marcel Dekker, New York.

Hui, Y. H., Gorham, J. R., Murrell, K. D., and Cliver, D. O. 1994. *Foodborne Disease Handbook.* Marcel Dekker, New York.

López-González, V., Murano, P. S., Brennan, R. E., and Murano, E. A. 1999. Influence of various commerical packaging conditions on survival of *E. coli* O157:H7 to irradiation by electron beam vs. gamma rays. *Journal of Food Protection* 62:10.

Mountney, G. J., and Gould, W. A. 1988. *Practical Food Microbiology and Technology.* Van Nostrand Reinhold, New York.

Murano, E. A., and Pierson, M. D. 1991. Effect of heat shock and growth atmosphere on the heat resistance of *Escherichia coli* O157:H7. *Journal of Food Protection* 55(3):171–175.

Web Sites

http://www.cic.info@pueblo.gsa.gov
Federal Consumer Information Center. This site includes food safety information; click on Food.

http://vm.cfsan.fda.gov/list.html/
FDA Center for Food Safety and Applied Nutrition. This site provides information on foodborne pathogens and research activities in food safety.

http://www.nal.usda.gov/fnic/foodborne/foodborn.htm
USDA's Foodborne Illness Education Center.

http://www.cdc.gov/foodnet
FoodNet Active Surveillance System for Foodborne Diseases. This CDC site provides information and statistics on outbreaks of foodborne illness.

http://www.haccpalliance.org
International HACCP Alliance. This web site provides information on Hazard Analysis Critical Control Point system, access to generic plans for meat products, and regulatory information.

http://www.open.gov.uk
Updates on mad cow disease.

www.who.int/fsf
World Health Organization Food Safety Programme. This site provides global information on foodborne illness, food irradiation, and related topics.

http://www.ift.org/education/food_industry/lesson1.shtml#milkprocessing
IFT web site provides a learning exercise in food safety and quality assurance dealing with milk processing.

Challenge!

RISK ASSESSMENT FOR BIOLOGICAL HAZARDS

Risk assessment refers to the determination of the risk posed by an entity or situation to a population. The term was used in the 1960s to determine the risk that carcinogens (cancer-causing agents) in foods would pose to consumers. Animal bioassays were developed to identify compounds that were carcinogenic. Analytical methods were developed to determine the concentration of the chemical that could cause adverse health effects on animals. This concentration was defined as the **dose-response** of the particular chemical.

In spite of the advances made in risk assessment of chemical hazards, the same has not been true regarding biological hazards such as bacteria, primarily because a dose-response is difficult to assess with biological systems. Variabilities in the expression of virulence characteristics by organisms, differences among strains of the same organism, and variability in the immune response of the test animal make the quantification of dose-response unpredictable and difficult to repeat. Thus, to carry out risk assessment of biological hazards, scientists have developed the concept of **probability of risk.** In this method, the probability, or likelihood, of an adverse event occurring due to consumption of a biological hazard is determined. In addition, the magnitude, or severity, of the risk, even if unlikely to happen, is taken into consideration.

To carry out a risk assessment of biological hazards, we apply the following steps:

1. Identify and characterize the hazard (is it bacterial, viral, etc.?)
2. Assess the dose required (number of microorganisms) necessary to produce illness.
3. Assess the exposure of the population to the particular biological hazard (prevalence of the organism in a particular food).

In conducting a risk assessment, several mathematical models of probability can be used. The most common is the Poisson distribution, expressed as:

$$P = 1 - e^{-RN}$$

where P = probability (in percentage) of infection with a foodborne pathogen, also known as the risk; R = a constant specific to a particular pathogen, based on its growth characteristics, also known as the probability of getting sick if one cell of the pathogen is consumed; and N = the number of cells of the pathogen present in a particular food.

To obtain the value of N, data on the levels of a particular foodborne pathogen in a particular food can be collected through surveillance studies of that food. From epidemiological data obtained from health agencies, P can then be determined, the percentage of individuals in a population that have developed the particular foodborne illness caused by the pathogen. Using these values, we can determine R for a particular organism:

$$R = \frac{-[\ln(1-P)]}{N}$$

Having determined R, we can use it to calculate the risk of infection, or P, for other situations where the prevalence of the particular organism is different (as in the case of a different food). Now that you know about risk assessment, consider the following scenario:

In the hypothetical country of Illnessia, 16 million (16×10^6) people live, all of whom love to eat smoked fish. Each person eats 300 servings of smoked fish per year, with each serving consisting of 50 grams of smoked fish. Of those servings, 1 percent are contaminated with *Listeria monocytogenes*. The typical level of this organism in Illnessian smoked fish is 10,000 (1×10^4) cells per gram. There are 480 cases of listeriosis per year in Illnessia. From this information, see if you can answer the Challenge questions that follow.

Challenge Questions

1. How many servings of smoked fish, eaten by the Illnessian population, are contaminated with *L. monocytogenes*?
2. What is the probability of developing listeriosis by eating one serving of smoked fish? In other words, how many cases of listeriosis occur in Illnessia per contaminated serving of smoked fish? (the answer to this question is the value of *P*).
3. How many cells are ingested in one serving? (the answer to this question is the value of *N*).
4. Given *P* and *N*, calculate the value of *R* for *L. monocytogenes*, or the probability that ingesting a single cell will produce an active case of listeriosis in Illnessia.

Multiple Choice

5. The concept of dose-response is not easily applied to
   a. chemical systems.
   b. vitamins.
   c. physical systems.
   d. biological systems.
   e. foods.

6. The term "R" for a particular organism is based on its
   a. toxicity.
   b. ability to mate.
   c. ability to grow.
   d. dose-response.
   e. level in a particular food.

7. To assess the exposure of a population to a hazard, _____ should be determined.
   a. the toxicity of the organism
   b. the ability of the organism to grow
   c. the percentage of the population affected
   d. the probability of infection
   e. the prevalence of the organism in a food

CHAPTER 12

Food Toxicology

12.1 What Is a Food Toxicant?

12.2 Risk Assessment for Chemical Hazards

12.3 Endogenous Toxicants

12.4 Naturally Occurring Toxicants

12.5 Synthetic Toxicants

Challenge!
Food Allergies and Food Intolerances

CHAPTER OBJECTIVES

After completing this chapter, you will be able to:
- List the three types of food toxicants, citing specific examples.
- Evaluate a dose-response curve.
- Explain the possibility of cyanide toxicity from eating certain vegetables.
- Describe the toxicity of domoic acid arising from shellfish consumption.
- Discuss the safety of herbal products.
- Explain the structure and mechanism of cholera toxin.
- Describe the problem of antibiotic resistance and how it relates to human health.
- Decide if growth promotants BST and DES are harmful and why.
- Explain how pesticides might be present in a fast food meal.
- Discuss the distinction between a food allergy and a food intolerance.

The coverage of food microbiology and food safety in previous chapters provided preparation for this chapter on food toxicology. The story regarding the safety of the foods we eat does not begin and end with microbiology. In recent years, the possible toxicity of foods has gained consumers' attention, in part because people have suffered ill effects from the uncontrolled use of medicinal herbs and plant extracts. Some plant products have been found to cause ailments ranging from hallucinations to seizures and even death. Certain consumers have also been concerned about the possible toxicity of certain food additives, such as MSG and aspartame. In addition, various microorganisms are capable of producing powerful toxins in food products, the consumption of which can lead to severe gastroenteritis or worse. In this chapter, we review the various types of toxicants, including the naturally occurring food toxicants, and look at specific toxin action in several food products. We also explore whether the health risk posed by these hazards can be quantified. Food allergies is the topic in the Challenge section.

12.1 WHAT IS A FOOD TOXICANT?

Simply stated, a **toxicant** is any chemical substance that can elicit a detrimental effect in a biological system. Food toxicants can be subdivided into three categories:

- endogenous
- naturally occurring
- synthetic

Endogenous toxicants are those produced by tissue cells in plants and other biological raw materials. These chemical compounds often serve the purpose of protecting plant tissue from pests, as well as from pathogenic organisms. Most endogenous toxicants are produced by plants, however, avoiding such plants may not be enough to avoid consuming the toxins. Transmission to people can easily occur through the consumption of toxic plants by animals that are then used for human food.

Naturally occurring toxicants are those that are produced by organisms that contaminate the food product. Microorganisms such as dinoflagellates, fungi, and bacteria can produce toxicants that, upon consumption, can produce disease. Not all microorganisms produce toxins, but some produce more than one type of toxin. Some toxin-producing organisms produce toxins in the food matrix (which can produce intoxication if consumed), while others produce toxins inside the victim (intoxification). These toxicants differ in their stability, with some able to withstand heating temperatures used in cooking, while others are able to tolerate extremes of pH without losing activity.

Synthetic toxicants are those that are synthetically produced, which find their way into our food supply through contamination of the food processing environment. Due to our ability to genetically engineer cells that can produce chemical substances

TABLE 12.1 Toxicants that might be found in a traditional holiday meal.

Food Item	Endogenous Toxicant	Natural Toxicant
Cream of mushroom soup	hydrazines	*Clostridium botulinum* toxin
Carrots, radishes, tomatoes	carotatoxin, myristicin, isoflavones, quercetin	*Clostridium botulinum* toxin
Roast turkey stuffing (onions, black pepper, mushrooms) and cranberry sauce, bread	heterocyclic amines, dihydrazines, psoralens, safrole	*Clostridium botulinum* toxin, mycotoxins
Red wine	tannins	

found in nature, production of natural toxicants can also be considered "synthetic" in certain situations.

Toxicants can be found in numerous types of foods. In the case of endogenous toxicants, fruits and vegetables represent the main category, followed by herbs. In addition, some foods inherently contain cyanide, a poison present in the seeds of peaches, plums, cherries, and other fruits. In the case of naturally occurring toxicants, any food that can become contaminated with toxin-producing microorganisms has the potential of being toxic. Organisms such as the bacteria *Staphylococcus aureus* and the enterotoxigenic strains of *Escherichia coli* can be found in fresh meats and vegetables. Mycotoxin-producing molds such as *Fusarium* species are typically found in grains and in some cheese products.

Synthetic toxicants can include drug residues in foods of animal origin. Anyone thinking that consumption of such toxicants is rare is challenged to look at the toxicants in Table 12.1 that might be found in a traditional holiday meal. So, the study of food toxicology is quite relevant and provides the basis for understanding the risk posed by the variety of toxicants to human health, and how to measure this risk.

12.2 RISK ASSESSMENT FOR CHEMICAL HAZARDS

The Challenge in Chapter 11 dealt with the topic of risk assessment for biological hazards. Here we reexamine this subject in food toxicology, since decisions regarding public health are made based upon gathering information regarding toxicants and disease. In 1970, the Environmental Protection Agency (EPA) was established as the agency responsible for regulation and assessment of environmental chemical hazards.

For the EPA to do its job, it became obvious that evaluation of the severity of the threat caused by each chemical was needed. This evaluation, known as **risk assessment**, is used to determine the toxicity of chemical hazards and the probability that humans will be exposed to those hazards. Risk assessment consists of two components:

- **Dose-response assessment**—the determination of the concentration of a particular toxicant needed to cause an unfavorable effect on a biological system (i.e., animals, people)
- **Exposure assessment**—the determination of the risk of exposure of a biological system to a toxicant.

In carrying out a dose-response assessment, we must consider that toxicants can produce acute and chronic effects on the consumer. **Acute effects** take place quickly, from minutes to days after exposure to the toxicant. **Chronic effects** appear after weeks, months, or even years of exposure. Acute toxicity is easy to measure with short-term studies providing sufficient data to determine the levels of a particular toxicant that will lead to an adverse effect. Chronic toxicity is not so easy to measure since it takes place after exposure to very low levels of a toxicant over a long period of time. Thus, in assessing the dose necessary to cause adverse chronic effects, a large number of subjects and a long time of exposure are needed.

TABLE 12.2 General toxicity categories.

Category	System Affected	Symptoms
Respiratory	Nose, trachea, lungs	Irritation, coughing, choking, tight chest
Gastrointestinal	Stomach, intestines	Nausea, vomiting, diarrhea
Renal	Kidney	Back pain, urination
Neurological	Brain, spinal cord	Headache, dizziness, depression, coma, convulsions, ocular effects, mouth effects
Hematological	Blood	Anemia, weakness
Dermatological	Skin, eyes	Rash, itching, swelling, redness
Reproductive	Ovaries, testes	Infertility, miscarriage

One of the principal aspects of risk assessment is gathering information about the toxicant. **Epidemiological** studies are those that provide us with information on the frequency of occurrence of a toxic effect in a human population. These studies also provide information on the geographic distribution of cases and other factors. Epidemiology ultimately seeks to link these two sets of information (frequency and distribution) with the toxicant in question to assess the risk to the population. One of the problems with epidemiological studies is that factors such as smoking, exposure to other toxicants, or diet can alter the effect of a toxicant. Thus, data obtained may not be open to unequivocal interpretation.

Besides epidemiological studies, animal studies, called **bioassays,** can also be used to gather information in risk assessment. They are used to predict the effects that toxicants would have on humans based on results obtained with animals. These studies use animals that exhibit a response to the toxicant in question that is very similar to that which occurs in humans. Similarly, **cell culture studies** can be used to observe the effects of toxicants on specific cell types.

Modes of Action of Toxicant

Toxicants work by changing the speed of certain body functions, such as increasing the heart rate or respiration rate, or by decreasing them. Typically, there is a progression from a biochemical effect to cellular and physiological effects. For example, an enzyme in the body may be inactivated by a toxicant. This may lead to an increase in the activity of nerve cells. This increase in activity may then result in labored breathing, twitching of muscles, or pain.

Depending on its type and its mechanism of action, a toxicant can have widespread effects throughout the body or can be confined to a specific target organ. Table 12.2 gives general toxicity categories and shows the variation in physiological symptoms. These symptoms can also be caused by other hazards, such as bacterial infection and viruses. Thus, to prove that a disease condition is caused by a toxicant, the toxicant must be detected in the body, and at levels known to cause illness.

Assessing Dose-Response

A well-known saying of food toxicologists is that the dose determines the poison. This principle was first championed by the chemist Paracelsus in the sixteenth century, who said: "solely the dose determines that a thing is not poison." In the book *The Third Defense*, Paracelsus attempted to defend his therapeutic use of mercury compounds, a powerful group of toxicants, to treat syphilis. His point was that a patient could be cured or killed, depending on whether a threshold value (above which poisoning would occur) was exceeded or not. This is the concept of **dose-response,** based on the principle that there is a relationship between a toxic reaction (the response) and the amount of toxicant received (the dose).

The dose-response relationship is critical in understanding the cause-and-effect relationship between exposure to a toxicant and an adverse effect. If the dose to which

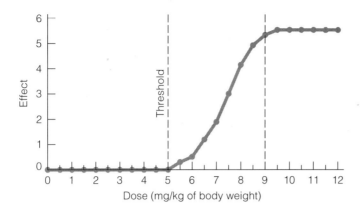

FIGURE 12.1 Dose-response curve.

the subject is exposed is below a threshold level, then no toxic effect is observed. This **threshold level** is the dose above which adverse effects are produced. Beyond the threshold level, the effect (or response) increases with dose, until a maximum effect is reached. Figure 12.1 illustrates this concept in a **dose-response curve.** No adverse effects are seen until the dose reaches the threshold dose of 5. Then the effect increases as the dose increases until the maximum effect is reached at a dose of 9.5.

In assessing the dose-response of a toxicant, concentrations are expressed as mg/kg of body weight of the human, or animal, in which the toxic response is measured. Potent toxicants may require fairly low doses to generate a response in a subject, expressed as μg/kg of body weight.

When animal studies are conducted to determine the dose-response of a toxicant, it is often expressed as the **ED$_{50}$**, or **effective dose** that will cause an adverse effect in 50 percent of the animals tested. Another commonly used measure of toxicity is the **LD$_{50}$**, or the **lethal dose** required to kill 50 percent of the animals tested. A toxicant that has a small LD$_{50}$, such as 100 μg/kg, is considered highly toxic, whereas a toxicant with a large LD$_{50}$, such as 1,000 mg/kg, is practically nontoxic.

Assessing the toxicity of a toxicant in animal studies is sometimes difficult. Factors such as temperature, food, light, stress, and the species, age, sex, health, and hormonal status of the animal can skew the results. Scientists consider all these variables in an attempt to control variation from experiment to experiment. In addition, since animal studies are the basis for recommendations regarding doses of particular toxicants that are allowed in food, several animal species may be tested when carrying out dose-response studies.

Carcinogen Testing

Carcinogens are chemical toxicants that are known to cause cancer. It is important to differentiate effects known to occur in animals versus those in humans. Testing, of course, is ethical only when done on animals or in cell culture. Animal studies do help to identify those substances carcinogenic to animals that are suspected of being human carcinogens as well. For this reason, these studies are crucial to the safety of people.

To assess the risk posed by a toxicant in terms of its ability to cause cancer, we must remember that just as with other adverse effects, the dose is critical. The generally accepted increase in the risk of cancer by exposure to environmental toxicants is one additional cancer in one million people. However, in the case of food toxicants, no amount of carcinogen is allowed. Of course, this refers only to toxicants intentionally added to foods and not to those that are endogenous. In addition, carcinogens that are produced by food contaminants, such as mycotoxins and marine toxins, are considered somewhat unavoidable and thus difficult to regulate.

To assess the dose that will lead to an increase in the risk of developing cancer, two types of studies are usually involved: epidemiological studies and bioassays. Animal experiments are easier to interpret than epidemiological studies because toxicants can be evaluated one at a time. In addition, very high doses can be administered, extraneous factors such as the effect of diet on the animal can be controlled, and the animals can be sacrificed during the course of the study. The latter enables researchers to conduct detailed tissue analysis during autopsies.

However, animal studies have two disadvantages: (1) it is difficult to decide how to apply these high-dose results to much lower dose exposures that happen in the real world, and (2) extrapolation of results from animals to humans may result in overestimation of the risk posed by a toxicant.

Knowing the basics of risk assessment of chemical toxicants and how dose-response measurements are made allows us to move forward to detail the specific toxicants found in foods. Our study of toxicants begins with those endogenous to foods, followed by those naturally produced by microorganisms, and finally the synthetic and intentionally added ones.

12.3 ENDOGENOUS TOXICANTS

We will look at six categories of endogenous toxicants: flavonoids, goitrogens, coumarins, cyanide compounds, herbal extracts, and mushroom toxins. The first three categories are compounds found in fruits and vegetables, but the quantities of these toxins that we consume in a normal varied diet have not been shown to be harmful. So, you can't use flavonoids as an excuse not to eat broccoli.

Flavonoids in Fruits and Vegetables

Fruits and vegetables inherently contain compounds known as **flavonoids,** which are organic molecules that impart color and sometimes flavor to these commodities (Table 12.3). Seven groups of flavonoids are found in foods:

flavones	leucoanthocyanins
flavonols	anthocyanins
flavanonols	catechins
flavanones	

Flavonoids can be found in plant tissues and cells, as well as in plant secretions. They can impart pigmentation to foods and can promote or inhibit plant growth. In addition, some have **antioxidant** activity—or the ability to inactivate toxic oxygen radicals. In the intestinal tract, flavonoids are metabolized by microorganisms into phenolic acids or lactones. They can then be absorbed and either excreted in bile or

TABLE 12.3 Common flavonoids found in fruits and vegetables.

Fruits	Flavonoid	Vegetables	Flavonoid
apple	quercetin, catechin	bell pepper	quercetin
apricot	quercetin, catechin	broccoli	quercetin
bilberry	quercetin	cabbage	quercetin
blackberry	quercetin	chive	quercetin
black currant	quercetin	endive	quercetin
blueberry	quercetin	garlic	quercetin
cherry	quercetin, catechin	horseradish	quercetin
cranberry	quercetin	kale	quercetin
gooseberry	quercetin	leek	quercetin
grape juice	quercetin	lettuce	quercetin
peach	quercetin, catechin	onion	quercetin
pear	quercetin	radish	quercetin
plum	quercetin, catechin	soybean	daidzein, genistein
prune	quercetin		
red currant	quercetin		

metabolized in the liver. The flavonoid content of plants varies according to the particular type of plant cultivar and maturity of the plant tissue.

Flavonoids are not considered harmful by most people but instead are recognized for their benefits to health as anticancer agents, and more recently, as nutraceutical substances. The benefits are due to their antioxidant effect, which inactivates cancer-promoting oxygen radicals commonly found in the human body. This antioxidant ability of flavonoids inhibits the development of rancidity in foods, caused by lipid oxidation.

However, some flavonoids, such as quercetin (one of the flavonols), can cause mutations in mouse tissue culture cells. Quercetin also causes intestinal and bladder tumors in experimental mice, rats, and hamsters and has been shown to reduce the lifespan of short-lived mice. Certain flavonoids are suspected of causing spontaneous abortions in cattle and may produce infertility in sheep. Scientific study is needed to fully understand the effects and toxic potential of flavonoid compounds in humans.

Most of the intake of flavonoids in the United States is attributed to the consumption of cocoa, cola, coffee, beer, wine, fruit juices, and spices, with a total daily consumption from all sources estimated at 1,000 mg/person.

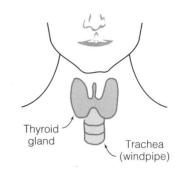

FIGURE 12.2 The thyroid gland.

Goitrogens in Cruciferous Vegetables

Cruciferous vegetables (family Cruciferae, genus *Brassica*) such as broccoli, cauliflower, and brussels sprouts contain several compounds, some of which are harmful to health. **Goitrogens** are toxicants found in these products, responsible for the development of goiter in humans and death in some farm animals. **Goiter** is a condition caused by enlargement and atrophy of the thyroid gland (Figure 12.2). This gland, located in the front of the neck, produces thyroid hormones that are important for human metabolism, development, and growth.

Thiocyanate is one of the most commonly found goitrogens in cabbage. Progoitrin, a goitrogen found in rapeseed, inhibits production of thyroid hormones in humans. Some goitrogens have been shown to have anticancer effects in animal studies. However, when the cancer-inducing compound (carcinogen) is administered before the consumption of cruciferous vegetables, higher cancer rates are seen in the animals, compared to animals given the carcinogen only. Thus, it appears that a tumor-promoting effect can be induced by consumption of cruciferous vegetables under some conditions. Hamsters fed cabbage at levels comparable to those of humans resulted in an increase in the number of gallbladder tumors. The principal cruciferic compound involved in these effects is *indole-3-carbinol*, also known as I3C.

Coumarins in Citrus

Coumarins are compounds primarily found in the peel of citrus fruits and are known for the pleasant citrus aroma they impart to oranges, grapefruit, lemons, and limes. Dermatitis, an inflammatory irritation of the skin, has been shown to occur in some people from simple contact with citrus fruits and their juices. The compound responsible is *bergapten*, a major coumarin found in limes. Others include psoralen, xanthotoxin, and isopimpinellin, which are *photosensitizing agents*, causing oversensitivity to light in humans.

Cyanide Compounds

Several plants produce compounds that contain **cyanogenic glycosides** (or glucosides). When ingested, these are converted to hydrogen cyanide (HCN), a powerful toxicant. This production of HCN is based on the breakdown of cyanogenic glycosides, either by enzymatic action of the enzyme *β-glucosidase*, or by a low pH—see Figure 12.3. This enzyme is produced by various organs, as well as by microorganisms that reside in the intestinal tract of humans and animals.

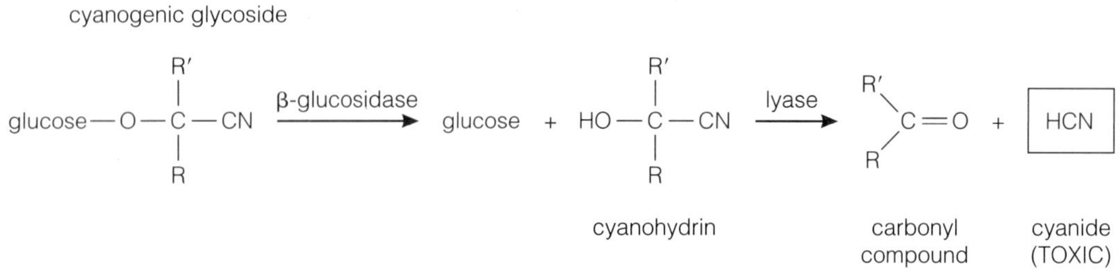

cyanogenic glycoside

$$\text{glucose}-O-\underset{R}{\overset{R'}{C}}-CN \xrightarrow{\beta\text{-glucosidase}} \text{glucose} + HO-\underset{R}{\overset{R'}{C}}-CN \xrightarrow{\text{lyase}} \underset{R}{\overset{R'}{C}}=O + \boxed{HCN}$$

cyanohydrin   carbonyl compound   cyanide (TOXIC)

FIGURE 12.3 Metabolism of cyanogenic glycosides to produce cyanide. *Breakdown of the glycoside produces a cyanohydrin, which liberates cyanide (HCN, also known as "prussic acid") plus the carbonyl product when acted upon by the appropriate enzymes. The enzymes are compartmentalized in the cyanogenic plants (certain legumes, stone fruits, and grasses). Chewing or mechanical force brings enzymes and substrate together, liberating the toxic agent, HCN.*

Cyanide-containing compounds inhibit respiration of body cells. Death of the person consuming the toxicant can occur if exposed to 200–500 mg/kg of body weight. Acute toxicity at small doses can cause headache, tightness in the throat and chest, and muscle weakness. Long-term exposure effects are not well known. Exposure to cyanide-containing foods has been linked to several conditions, such as anemia, degeneration of the optic nerve in people with a vitamin deficiency, and atrophy of the optic nerve.

The seeds of peaches, cherries, plums, almonds, apples, pears, apricots, and cassava contain cyanide. In addition, bamboo shoots, sorghum, peas, and beans can contain this compound. In Turkey, children were poisoned after eating apricot seeds, and in another case, consumption of a milkshake made with dried apricot kernels purchased at a health food store resulted in one death. Tragically, a woman suffering from cancer was poisoned after an attempt at self-treatment, which involved consuming several dozen apricot pits containing 4,100 μg of cyanide/g of apricot pit.

Cassava is a vegetable crop used extensively in the tropics for its high energy content. It is extensively consumed in developing countries, with adults in West Africa consuming 750 g/day. Cassava contains *linamarin*, a cyanide-based chemical, with bitter cassava crops containing higher levels than sweet cassava. For people in West Africa, the average daily exposure to HCN from consumption of cassava is about half of the dose required to kill a person. Such a high-cassava diet has been associated with bilateral deafness, Parkinson's disease, cerebellar degeneration, psychosis, and dementia. Table 12.4 contains information on content of various cyanogenic glucosides in plants.

Herbal Extracts

In many cultures, people have traditionally used herbs and herbal preparations to cure a variety of maladies. **Herbal extracts** are considered by many as superior to synthetic medicines, because they believe that herbs are "natural," and thus cannot be harmful (Figure 12.4). Many commercial drugs *do* contain one or more herbal components, however, some active ingredients of herbs can also cause detrimental health changes.

TABLE 12.4 Hydrogen cyanide content of various food plants.

Plant	Cyanogenic Glycoside	HCN content (mg/kg of plant weight)
Bitter almonds	amygdalin	2,500
Cassava root	linamarin	530
Lima bean	linamarin	100 to 3,120
Whole sorghum	dhurrin	2,500

FIGURE 12.4 Herbal products.
These products are unregulated, with no guarantee of safety, effectiveness, or purity.

Pyrrolizidine alkaloids are toxic compounds found in comfrey, Indian herbal teas, tomatoes, potatoes, and eggplant. They inhibit the enzyme *cholinesterase*, which serves to break down *acetylcholine*. Since acetylcholine is the chemical used by nerve cells to transmit signals to the rest of the body, stopping the action of cholinesterase prevents regulation of acetylcholine, resulting in overstimulation of nerve cells.

In addition to producing enhanced nerve activity, alkaloids can cause stomach pain, nausea, vomiting, rapid and difficult respiration, and death. Laboratory studies have found that alkaloids cause cirrhosis of the liver in rats and mutations in tissue culture cells. The amounts of toxic alkaloids consumed in one cup of comfrey root tea can be as much as 36 mg. A dose of 120 mg/kg of body weight of alkaloids given to rats causes a variety of pathological changes in the liver. Thus, drinking comfrey tea may not be as healthy a practice as advertised. Many health officials have recommended that comfrey be removed from the consumer market.

Chamomile is an herb used by many people in an herbal tea for adults or to treat childhood illnesses. Some allergic reactions to the tea have been observed, and continued use over time can cause diarrhea and seizures. Sassafras root can also be made into a medicinal tea. It contains 75 percent *safrole*, a known carcinogen, which used to be in root beer until it was banned in the 1960s.

Rosemary and sage are commonly used herbs. Plant extracts of these products have been found to be toxic to yeast cells, with *eucalyptol* being the main fungicidal component. Small amounts of rosemary and sage constituents can cause a variety of symptoms, including nausea, severe headache, and vomiting, due to these bioactive components. Table 12.5 lists these compounds, the amounts per gram of rosemary and sage, and the LD_{50} for each compound. Be assured that using normal small amounts of rosemary and sage in cooking is not going to kill you or your guests.

Along with cotton root bark, sage has been used as an abortifacient to induce abortion for unwanted pregnancies. An abortion usually ensues in 30 percent of the women 2 to 6 days after drinking one quart of tea made from these herbs. This practice is not recommended at all for many reasons, one of which is that birth defects can develop in babies that come to full term.

Some herbs are considered *psychoactive* for their ability to cause cerebral excitation and hallucinations in people. Carrots and nutmeg contain *myristicin*, a psychoactive substance. However, a person would have to consume 1,400 pounds of carrots or 20 g of nutmeg to produce such effects!

TABLE 12.5 Compounds commonly found in rosemary and sage.

Compound	Rosemary (μg/g of herb)	Sage (μg/g of herb)	LD_{50}*
Borneol	120–470	140–2,636	2,000 mg/kg
Camphor	539–2,910	28–1,410	810 mg/kg
Essential oil	4,000–19,000	7,000–10,000	unknown
Linalool	40–120	1,191–3,500	2,790 mg/kg
Safrole	32–95	0	1,950 mg/kg
Terpinen-4-ol	10–520	29–1,018	4,300 mg/kg
Ursolic acid	39,000	21,000	50 mg/kg

*Dose of compound required to kill 50% of experimental animals (rabbits). Recall from Section 12.2 that a toxicant with an LD_{50} over 1000 mg/kg is practically nontoxic.

Toxic Mushrooms

Mushrooms are very popular food items around the world, often consumed raw. All mushrooms contain at least one toxicant (hydrazine), including those grown for food and including prized wild species such as *Agaricus bisporus*, *Boletus edulis*, *Cantharellus cibarius*, and *Gyromitra esculenta*. Most of the toxicants cannot be eliminated by cooking, canning, freezing, or any means of processing. Thus, the only way to avoid the toxicants is to avoid eating the mushrooms. The good news is that the toxin levels are low enough that there are no health effects from consuming normal amounts of edible mushrooms.

Mushroom poisoning occurs when people eat known poisonous species. The effects of the toxicants in mushrooms are acute. Toxic mushrooms can be classified by the type of symptoms they produce. We will discuss three main types:

1. Protoplasmic poisoning
2. Neurotoxic poisoning
3. Gastrointestinal poisoning

Protoplasmic Poisoning. **Protoplasmic poison** mushrooms are those that affect the liver and kidneys. These produce one of three types of toxin: amanitins, hydrazines, and orellanines. Amanitins are produced by the mushrooms *Amanita phalloides*, *A. virosa*, *A. verna*, *Galerina autumnalis*, and others. Poisoning occurs after a period of 6 to 15 hours, during which no symptoms are detected. At the end of this period, sudden, severe seizures of abdominal pain, persistent vomiting and watery diarrhea, thirst, and lack of urination are manifested. The patient appears to recover for a short time, but then pain and loss of strength follow. Progressive, irreversible liver, kidney, heart, and muscle damage occurs, with development of jaundice. Convulsions and a period of coma follow, with death taking place within a few days due to necrosis (rotting) of the liver and kidney.

Hydrazines are produced by all mushrooms (Figure 12.5), but *Gyromitra esculenta* and *G. gigas* produce enough that the principal toxicant is called gyromitrin. Approximately six hours after ingestion, a feeling of fullness develops in the victim, along with vomiting and headache. The central nervous system is affected, as is the liver, although mortality is low.

The poison orellanine is produced by mushrooms in the genus *Cortinarius* and other species. This protoplasmic poisoning is characterized by a long period of 3 days to 2 weeks in which the patient suffers no symptoms. Nausea, headache, chills, muscle aches, and even loss of consciousness can then develop, with kidney failure in severe cases. Recovery may require several months.

Neurotoxic Poisoning. The second type of poisonous mushrooms are the **neurotoxic poison** mushrooms, which affect the central nervous system. Muscarine poisoning,

Shiitakes have a distinct, slightly smoky flavor.

Oyster mushrooms have a subtle flavor.

Enoki mushrooms have a mild, pleasant flavor that isn't particularly distinctive.

Button mushrooms are the most common variety sold; somewhat nutty and creamy when raw, they become earthy and rich when cooked.

Portabella mushrooms are simply cremini grown to gargantuan proportions; their flesh is denser and usually more fibrous.

Cremini are a darker variety of the standard button mushroom; they're firmer with a more intense flavor.

FIGURE 12.5 *All mushrooms, even harmless varieties like these grown for food, contain levels of hydrazines, known carcinogens. However, the levels are so small, no health effects are seen under normal levels of consumption.*

for instance, is caused by ingestion of species of mushroom belonging to the *Inocybe* and *Clitocybe* genera. The main symptom of the illness is profuse sweating, with an increase in salivation and lacrimation within 30 minutes of ingestion of the toxin. Severe nausea, diarrhea, abdominal pain, blurred vision, and even labored breathing can also occur if large doses are ingested. The symptoms typically disappear after a couple of hours.

Other neurotoxin-producing varieties of mushrooms are *Amanita muscaria* and *A. pantherina*, which produce the toxins ibotenic acid and muscimol. Abdominal discomfort usually takes place within 2 hours of ingestion, with drowsiness and sleepiness being the main symptoms. Interestingly, hyperactivity often follows, with delirium

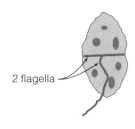

FIGURE 12.6 Structure of dinoflagellate.

being experienced by some patients, all of which fade within a few hours. Children can have severe symptoms if they ingest too much of these mushrooms, developing convulsions and even coma.

Neurotoxic mushrooms of the genera *Psilocybe, Panaeolus, Copelandia,* and *Pluteus* produce psilocin and psilocybin, toxins that cause symptoms that mimic drunkenness several hours after ingestion. Hallucinations can also take place; these mushrooms are purposely consumed by some Native American tribes as part of their religious ceremonies.

Gastrointestinal Poisoning. The third type of mushroom poisoning causes gastrointestinal irritation. The Green Gill, Gray Pinkgill, Tigertop, Jack O'Lantern, and Horse mushroom are some examples of common names given to those that cause gastrointestinal illness. Typical symptoms are nausea, vomiting, diarrhea, and abdominal cramps. The onset of illness is very rapid, which contrasts with the protoplasmic poisoning mushrooms, which take at least 6 hours for their poison to take effect. The symptoms caused by gastrointestinal irritants can last for several days, with replacement of bodily fluids being of uppermost importance for recovery to begin.

12.4 NATURALLY OCCURRING TOXICANTS

Marine Toxins

Several syndromes are caused by ingestion of shellfish and finfish containing toxins. The toxins are not produced by the fish but by living creatures that the fish themselves eat. These are **dinoflagellates,** members of the algae family that carry out metabolism through photosynthetic reactions (Figure 12.6). Dinoflagellates cluster on the surface of tropical seaweed, which fish consume as part of their diet.

There are at least five types of marine toxic poisonings:

1. Ciguatera
2. Paralytic shellfish poisoning (PSP)
3. Neurotoxic shellfish poisoning (NSP)
4. Diarrhetic shellfish poisoning (DSP)
5. Amnesic shellfish poisoning (ASP)

Ciguatera is caused by consumption of tropical herbivorous reef fish, such as grouper and red snapper, and the carnivorous fish that feed upon them. The source is the dinoflagellate *Gambierdiscus toxicus.* The symptoms develop very quickly, approximately 3 to 5 hours after ingestion. Diarrhea, nausea, vomiting, and abdominal pain are experienced first, with neurological symptoms developing 12 hours later. These include very strange phenomena, such as **hot/cold inversion,** in which patients drinking hot beverages believe them to be cold, and those drinking cold beverages believe them to be hot. Muscle aches, *paresthesia* (tingling), and *disesthesia* (numbness) of the lips and tongue are also experienced. *Pruritus* (itching), metallic taste, and dryness of the mouth can develop, along with anxiety, dizziness, chills, sweating, blurred vision, and temporary blindness. Reoccurrences can take place up to 25 years later, with alcohol and caffeine consumption bringing them on. The prognosis is usually good if the patient survives the first 24 hours. Ciguatera toxin is heat-stable, thus not easily destroyed by cooking.

Paralytic shellfish poisoning is caused by consumption of mussels, clams, and oysters, especially the dark meat, that come from cold waters between 5 and 8°C (41–46°F), which is the temperature range of the dinoflagellates reponsible for toxin production. Within 30 minutes of ingestion, tingling and numbness of the lips and fingertips develops. *Aphasia* (incoherent speech) and inability to understand speech are also experienced, with death due to respiratory paralysis within a few hours. The source of the illness is toxin produced by the dinoflagellate *Gonyaulax catenella,* which produces 12 **saxitoxins.** These saxitoxins bind to nerve cells, preventing the adequate transmission of signals from the brain to the muscles in the body. The toxins are also heat-stable, making them difficult to inactivate.

Neurotoxic shellfish poisoning results from ingestion of oysters and clams that have ingested the dinoflagellate *Ptychodiscus brevis*. These are usually found in the west coast of Florida, in the Caribbean, and in the Gulf of Mexico. The presence of these organisms is marked by red tide, in which sea water appears red wherever the organisms are found, visible for several miles. Tingling, muscle aches, vomiting, and diarrhea are typical symptoms, which pass after a couple of days. Respiratory irritation is experienced by people in the vicinity of the red bloom due to aerosols containing toxin. Several toxins called **brevitoxins** are produced by the dinoflagellate. They act on nerve cells, but instead of repressing activity they cause prolonged excitation.

Diarrhetic shellfish poisoning is seen mainly in Japan, Europe, and South America, in particular, Chile. It is caused by consumption of mussels and clams contaminated with the dinoflagellate *Dinophysis fortii*, as well as species of the genus *Prorocentrum*. The toxins formed resemble the brevitoxins, with the most widely distributed one being **okadaic acid**. The toxin inhibits protein phosphorylation by body cells and is believed to be a tumor promoter. Common symptoms include acute diarrhea with an onset of only a few minutes.

Amnesic shellfish poisoning is caused by mussels harvested from the eastern shore of Canada that are infected with the dinoflagellate *Nitschia*. The toxin produced is **domoic acid**, an analog of the amino acid glutamic acid (Figure 12.7). Thus, the toxin binds to glutamate receptors in the brain of the victim, exciting the brain cells until they die from overstimulation. The symptoms are quite interesting, including short-term memory loss and disorientation, both of which are irreversible. Some gastroenteritis, mainly vomiting and diarrhea, are also experienced.

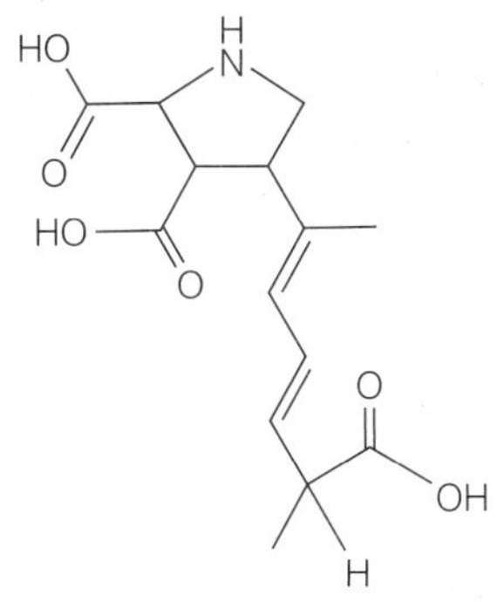

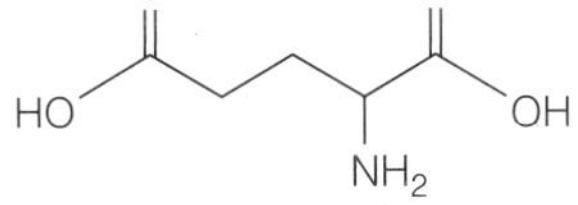

Besides these five syndromes, **scombroid poisoning** is another significant illness caused by consumption of marine species. However, it is not caused by ingestion of a toxin, but rather by ingestion of histamine produced by bacteria living in the fish, such as *Hafnia alvei*, *Morganella morganii*, and *Klebsiella pneumoniae*. Fish of the family Scombridae are implicated, such as tuna, mackerel, mahi-mahi, sardines, and anchovies. The onset of symptoms is from a few minutes to a few hours. They include rash, hives, localized inflammation, nausea, vomiting, diarrhea, abdominal cramps, and low blood pressure. Some neurological symptoms can also develop, such as headache, palpitations, tingling, flushing, burning sensation in the mouth, and itching. Facial flushing is a frequent occurrence as well as a metallic or peppery taste in the mouth. Scombroid poisoning is often confused with an allergic reaction because antihistamines, used to treat allergies, can be used to treat scombroid.

FIGURE 12.7 Comparison of domoic acid structure to the amino acid glutamic acid.

Microbial Toxins

Microbial toxins are those produced by microorganisms, either molds, in the case of mycotoxins, or by bacteria, in the case of bacterial toxins. The mycotoxins were discussed in Chapter 11, Food Safety. In this section, we will limit our discussion to the following bacterial toxins:

Staphylococcal enterotoxins
Cholera and cholera-like enterotoxins
Verocytotoxins
Clostridial toxins

Staphylococcal Enterotoxins. **Staphylococcal enterotoxins** are produced by the organism *Staphylococcus aureus*, a bacterium commonly found on the skin and mucous membranes of humans and animals. The cells produce seven enterotoxins—they affect the enteric, or intestinal, system. All of the toxins are pyrogenic, able to cause fever, and can suppress the immune system of the person affected. The toxin types are: A, B, C1, C2, C3, D, and E. Type A is the most commonly found in outbreaks of food poisoning caused by *S. aureus* in the United States, with the fewest cases being attributed to toxin E. Type B is rarely involved in outbreaks but is the most studied because, when found, it is produced in large quantities by the organism.

Toxins C1, C2, and C3 are very similar, chemically speaking, but are not identical. All three C types are heat-stable, being more stable than either toxins B, A, D, or E. To inactivate the toxins, food needs to be cooked at 100°C (212°F) for at least 30 minutes. The organism produces toxin in varying amounts, depending on growth conditions. For example, less toxin is produced if the conditions are anaerobic, or if the water activity (a_w) is low. The toxins are not absorbed by the body in the intestines. Rather, they act on the lining of the intestinal tract to induce vomiting by triggering a signal to the brain.

A typical set of circumstances that result in staphylococcal food poisoning is as follows: the organism contaminates a given food product, and the storage temperature is such that it allows the growth of the cells. During growth, they produce toxin, which upon ingestion, results in illness. Even if food is reheated, toxin remains active, unless it is heated sufficiently, as stated above.

Cholera and Cholera-like Enterotoxins. Cholera toxin was one of the first enterotoxins for which the mechanism was elucidated. This toxin was first recognized as being produced by the organism *Vibrio cholerae*. However, similar toxins have been found to be produced by certain serotypes of *Escherichia coli*, by *Campylobacter jejuni*, and others. The toxin is composed of 5 B subunits, each measuring 10.5 kilodaltons, and all arranged in a five-member ring. An A subunit in the middle is composed of two peptide chains termed A1 and A2 (see Figure 12.8).

The entire holotoxin, containing all 5 B subunits and the A1 and A2 subunits, binds to receptors on the surface of tissue cells lining the intestinal tract. A disulfide bond holds A1 and A2 together. This bond becomes reduced, allowing subunit A1 to enter the intestinal tissue cell. Upon entering, the A1 subunit transfers nicotinamide (NAD) to a protein residing in the cytoplasm of the tissue cell, known as the "G" protein. This modified G protein travels the length of the tissue cell and attaches to the enzyme **adenylate cyclase** at the base of the cell membrane. The enzyme cleaves two phosphates off an ATP molecule, producing AMP in a cyclical configuration, known as cAMP. This mechanism was presented earlier in Figure 11.2 (the intoxification mechanism of *Campylobacter jejuni*).

Increased levels of cAMP activate an enzyme known as protein kinase A, which adds a phosphate molecule to proteins. This protein phosphorylation leads to a decrease in the ability of the tissue cell to absorb NaCl, causing an imbalance in the amount of NaCl inside versus that outside the cell. This imbalance forces water molecules to go outside of the cell and into the intestinal lumen (Figure 12.9). The excess water mixes with fecal material, resulting in diarrhea.

Enterotoxigenic strains of *Escherichia coli* produce a cholera-like toxin, except that instead of forming cAMP, the toxin stimulates the enzyme **guanylate cyclase**, resulting in production of cyclic guanosine monophosphate (cGMP). A typical scenario of food poisoning caused by a cholera or cholera-like toxin involves ingestion of a toxin-producing bacterium; production of the toxin by the organism occurs inside the intestinal tract.

Verocytotoxins. **Verocytotoxins** are produced by enterohemorrhagic serotypes of the organism *E. coli*, with *E. coli* serotype O157:H7 being the most studied. The cells produce at least two toxins, VT1 and VT2, both of which are cytotoxic to African Green Monkey kidney cells in culture, also known as Vero cells. VT1 is usually found inside the bacterial cell, while VT2 is usually excreted by the organism. Both toxins consist of one A and one B subunit, with the A subunit being much larger (33 kilodaltons vs. 7 kilodaltons). The receptor for both toxins is found on kidney cells, thus the target organ is the kidney.

The toxins act by blocking protein synthesis, which manifests itself by a shutdown of the kidneys, or renal failure. The typical scenario for verotoxin poisoning involves the ingestion of cells of the causative organism. Once on the intestinal lining,

⚡ *A dalton is a mass unit to define the size of a biomolecule. One dalton is equivalent to one amu (atomic mass unit).*

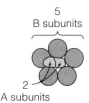

FIGURE 12.8 General structure of cholera toxin.

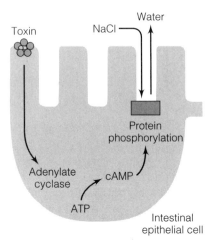

FIGURE 12.9 Mechanism of action of cholera toxin.

the cells produce either or both toxins, which are absorbed by the capillaries surrounding the intestinal wall and are quickly transported to the kidneys, where the toxins exert their effect.

Clostridial Toxins. **Clostridial toxins** are produced by different species of the genus *Clostridium. C. perfringens* produces an enterotoxin that is actually a structural component of the spore coat of this organism. The typical scenario involves the ingestion of cells of this organism in food. Once inside the body, the cells sporulate as they become attached to the intestinal tissue, producing toxin. The toxin causes increased permeability of water into the intestinal lumen, resulting in diarrhea.

The toxin produced by *C. botulinum* has been called the most potent natural toxin known. There is little doubt of this, with ingestion of the **botulinum toxin** resulting in death of most victims in as few as 10 days, unless antitoxin is provided. The toxin is produced intracellularly and released during growth, or as a result of cell lysis at the end of the stationary phase. The toxin has a molecular weight of 150 kilodaltons in the form of a hook. For the toxin to cause disease, it must be cleaved into two pieces, a 50 kilodalton and a 100 kilodalton piece, with both pieces remaining attached together by a disulfide bond (see Figure 12.10). Some strains of *C. botulinum* (notably type A) have a proteolytic enzyme that is released with the toxin, causing cleavage of the subunits to activate the toxin. Other strains, notably type E, do not activate the toxin, but rather the toxin is activated by proteolytic enzymes found in the environment.

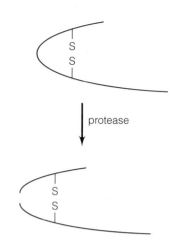

FIGURE 12.10 Activation of *Clostridium botulinum* toxin.

The toxin is so powerful that a typical culture of *C. botulinum* cells can have an activity of 10^4 to 10^5 MLD_{50} (or **mouse lethal dose**) per milliliter of culture. This means that in such a culture, 1.0 mL could kill 50 percent of 10,000–100,000 mice, or 5,000–50,000 rodents. Given that 3,500 MLD_{50} are needed to kill a person, 1.0 mL of such a bacterial culture could kill 5 to 50 people. The typical scenario of botulinum food poisoning involves the presence of *C. botulinum* spores in a food. The food is then improperly heat-canned, causing the spores to germinate as a result of the heat treatment. The germinated cells produce toxin in the canned product, due to the anaerobic environment of the sealed container. Once the toxin is ingested, illness develops.

12.5 SYNTHETIC TOXICANTS

Antimicrobial Agents

The animal husbandry industry depends on the careful use of certain drugs for the treatment of diseases that may strike the herd. Large numbers of animals raised in a high-density population under confined quarters are prone to acquiring infections, mainly due to increased stress and to the close proximity between animals. Antimicrobial agents are drugs used to prevent and to treat infections by microorganisms that can affect the health of the animal. When antimicrobials are used in animal husbandry, the potential exists for drug residues to be present in products derived from those animals, such as meat and milk.

Antimicrobials used in animal husbandry fall into one of three categories:

1. Therapeutic
2. Prophylactic
3. Modification of function

Therapeutic treatments are those in which drugs are used to cure or treat a certain disease condition. **Prophylactic** treatment involves the application of a drug to an animal in order to prevent disease, or to minimize its probability of occurrence. Drugs used for modification of function are those that will aid in promoting the growth of the animal, enhancing the efficiency in which feed is converted to muscle mass as rapidly as possible.

Antimicrobial drug residues can exceed the tolerance level due to misuse, illegal use, or negligence. The U.S. Food and Drug Administration, the Environmental Protection Agency, and the U.S. Department of Agriculture are the entities responsible for ensuring that only safe levels of drugs are present in raw meat and poultry, as well as in seafood grown in aquaculture. The typical antimicrobial agents used in animal husbandry are antiprotozoal drugs, dewormers, and antibiotics.

Antiprotozoal Drugs. Antiprotozoal drugs are used to eliminate protozoan infections, such as those caused by *Entamoeba tenella, E. necatrix,* or *E. acervulina* in chickens. Examples of these drugs are the nitroimidazoles such as dimetridazole, ipronidazole, metronidazole, and ronidazole. These compounds are used against bovine urogenital infections and for swine enteritis. At low doses, nitroimidazoles have growth-promoting activity. Concerns have been raised regarding their use because of data that indicate their **mutagenic** and **tumorigenic** properties.

Nitrofurans are another class of antiprotozoal agents used extensively in animal husbandry. In addition to eliminating protozoa, they are effective against bacterial infections such as caused by *Salmonella cholerasuis*. Nitrofurans are absorbed rapidly by the animal, with a propensity of being bound to tissue proteins. It is believed that they have mutagenic properties and that they are possibly carcinogenic.

Dewormers. Dewormers have been used since the 1960s to eliminate parasitic intestinal worms, with the general class of benzimidazoles being the most common. These compounds interfere with the metabolism of worms, preventing the synthesis of ATP, which results in death of the parasite. Upon exposure to benzimidazoles, humans have complained of loss of appetite, vomiting, nausea, diarrhea, drowsiness, headache, and dizziness.

Antibiotics. Antibiotics are typically used to treat and/or eliminate bacterial infectious agents of significance to animal health. A common class of these compounds are the aminoglycosides, of which streptomycin and gentamycin are examples. These compounds act by interfering with protein synthesis by bacteria. Streptomycin is effective against *Brucella, Salmonella, Klebsiella, Shigella,* and *Mycobacterium*. It is used in cattle to combat leptospirosis and in poultry to prevent chronic respiratory disease and infectious sinusitis. Aminoglycosides can cause kidney damage in humans, and can cause hearing loss.

A second class of antibiotics is chloramphenicol, which acts by inhibiting cell wall synthesis in bacteria. It is effective against bacterial pneumonia, brucellosis, and salmonellosis of various animals. Some people are sensitive to this antibiotic, developing **aplastic anemia,** a blood disorder that renders the patient unable to manufacture red blood cells.

Macrolides are another type of antibiotic, used as an alternative to other antibiotics when bacteria develop resistance. Erythromycin is a common example of a macrolide, used primarily as a bacteriostatic agent to inhibit growth, but not necessarily kill the organism. Human health concern about erythromycin use lies not in its possible toxic effects but rather that its use may enhance the development of resistance by bacteria. Similar concerns exist about the use of penicillins, a fourth class of antibiotic.

Sulfa drugs are perhaps one of the oldest antibiotics ever used. They are used to control urinary tract infections in cattle and swine. These antibiotics have been implicated as possible carcinogenic compounds. Tetracyclines are a group of antibiotics classified as **broad spectrum,** meaning that they are effective against a wide variety of bacteria. They are used in the treatment of chronic respiratory diseases in poultry, and in swine and cattle they are used to prevent bacterial enteritis. These antibiotics can pass to the milk of dairy cows. Some individuals in the population have been found to be sensitive to exposure to these agents.

Resistance to Antibiotics. One of the negative consequences of the use of antibiotics in animal husbandry is the possibility of development of resistance by microorganisms exposed to the antibiotics on the farm. Although the degree to which the use of antibiotics on the farm affects human health has not been proven, we should recognize that this can lead to resistant strains of pathogenic bacteria. Without an effective means to control their growth, such resistant pathogens could proliferate in the human body, causing serious illness.

Bacterial resistance to antibiotics is not something new—it was first discovered in the 1940s. Two of the scientists involved in the discovery of penicillin discovered strains of *Escherichia coli* that could survive exposure to this agent. This led many to believe that exposure to antibiotics imposes a selective pressure on bacterial populations such that those that can survive are selected and become the dominant strains.

However, in the 1960s bacteria resistant to the antibiotic tetracycline were found in the Solomon Islands, even though that antibiotic had never been used there. Thus, speculation began regarding the ability of some organisms to spontaneously develop resistance to antibiotics, even if not previously exposed to them. In the 1980s, microorganisms with resistance to multiple antibiotics were detected, and in the 1990s this trend continued, having been observed not only in bacteria commonly found in clinical settings, but also in pathogens transmitted via food.

The use of antibiotics has been increasing in the United States for several decades, including those that are used in animal husbandry. Thus, speculation is that overuse of these drugs in the rearing of food animals such as cattle and swine may play a significant role in the development of resistant bacteria of human health significance. Scientists believe there is a link in the transmission of antibiotic-resistant microorganisms from animals to humans, and vice versa.

Resistance in organisms associated with animals is believed to occur from the use of antibiotics on the farm, given to animals through feed and water. The organisms residing in the intestinal tract of animals become resistant due to selective pressure and are shed through the feces into lakes, streams, and soil. These resistant organisms can also end up as residues on meat and milk derived from the animals. Through consumption of these products, as well as through direct contact between animals and humans, and between human and animal feces, people acquire some resistant strains.

However, resistance can also be acquired through excessive administration of antibiotics by the medical community. Such microorganisms can therefore be associated with people in clinical health settings. These resistant bacteria can cause disease in the human host. They can also be shed in human sewage, which can contaminate the environment in some cases, coming into contact with animals and starting the cycle all over again.

Growth Promotants

Hormones are compounds that can enhance or suppress the growth of specific body cells. Certain hormones are used in animal husbandry to promote growth and increase feed efficiency. *Diethylstilbestrol*, or DES, is a synthetic hormone that causes salt retention, enhancing the ability of the animal to also retain nitrogen. This stimulates protein synthesis, and thus muscle build-up. However, its use was banned in 1979 by the FDA because it was classified as a carcinogen by the Environmental Protection Agency.

Bovine somatotropin (BST) is a hormone given to cows to increase milk production. Bacteria serve as "factories" to mass produce this product via recombinant technology. The hormone is technically a protein (peptide). It is injected into cows rather than given orally because peptide hormones do not withstand digestion in the gut. It differs enough structurally from human growth hormone, HST (human somatotropin), that it would not exert any growth effects in humans. The pasteurization processing of BST milk destroys about 90 percent of the BST, the remainder broken apart by the body's digestive enzymes.

Trenbolone acetate is another steroid used in cattle, sheep, goats, and swine for improving feed efficiency and increasing the rate of weight gain. It also acts by increasing nitrogen retention. The main concern about steroidal compounds is their ability to enhance the effects of *estrogen*, a sex hormone commonly found in females. Children

FIGURE 12.11 Dairy cow. *The hormone bovine somatotropin is given to cows to increase milk production.*

FIGURE 12.12 The flow of chemical toxicants through the food chain.
Pesticide build-up in fish results in the human consumer being exposed to a potentially high level of toxicants. Small fish consume tons of plants during the span of a year, and in turn, tons of these small fish are eaten by larger fish consumed as food by humans. If the sea plants that are fed on by small plant-eating fish are contaminated with toxic chemicals, such as pesticides, cadmium, lead, mercury, and PCB (polychlorinated biphenyl), then concentration up the food chain takes place.

Level 4
A 150-lb. person

Level 3
100-lb. of larger fish

Level 2
A few tons of plant-eating fish

Level 1
Several tons of plants

exposed to steroids in milk and meat can enter into puberty too early. In addition, cancer has been linked to the consumption of steroids. Examples of various steroid compounds used in animal husbandry are estradiol, hexaestrol, and progesterone.

Pesticides

Chemicals that can be used to prevent, control, or eliminate insects and other pests are known as **pesticides.** Another class of chemicals, **herbicides,** are used in agriculture to prevent, control, or eliminate unwanted plants such as weeds. Pesticides and herbicides have been used since the 1940s. Dichlorodiphenyltrichloroethane, better known as DDT, was first used to reduce malaria and typhus through killing of the carrier pest, the mosquito. The use of DDT also resulted in an increase in the yield of potato, tomato, and apple crops in the United States.

In 1954, the Pesticide Chemical Act was passed by Congress, mainly due to an increasing concern over the safety of these compounds, including concentration of pesticides and heavy metals up the food chain (Figure 12.12). This concern culminated in the 1960s with several reports regarding the safety of commonly used pesticides. Aminotriazole, used to disinfest cranberries, was reported to induce thyroid cancer, and Agent Orange (an herbicide used in the Vietnam War) were suspected of causing a myriad of serious maladies. Legislative actions resulted in passage of several environmental protection laws.

A new method of pest management was introduced in the 1980s, which is based on the minimal use of chemical pesticides to control insects. As much as 50 percent reduction in the amount of conventional pesticides are used, with the addition of alternative pest control measures such as trap-setting. This new system is called **Integrated Pest Management,** or IPM, because it seeks to integrate several techniques to control insects and other pests, instead of relying solely on chemicals.

Regulating Pesticides. The Food and Drug Administration is the government agency in charge of setting limits on the use of pesticides and herbicides. The goal is to minimize the concentration of residues that can be found in food produced from treated crops. **Maximum residue levels,** or MRLs, are set and kept as low as possible. The MRLs are so strict that the margin of safety is large, practically eliminating the risk to health from pesticides used in the production of food crops, milk, and meat. The main area of concern is not the domestic use of pesticides, but actually their use by other countries that export food to the United States. As much as 25 percent of incoming fruit has been reported to have higher residue levels than those allowed by the FDA (Figure 12.13).

Categories of Pesticides. Pesticides and herbicides of concern can be divided into the following categories:

1. Organochlorines
2. Organophosphates
3. Carbamates
4. Fumigants
5. Plant hormones
6. Herbicides

Organochlorines are lipid-soluble, highly stable chemicals. Of these, DDT is the most well known, however, it has been found to be carcinogenic, developing liver tumors in rats. Human cancer has not been directly linked to the use of DDT, although accidental ingestion of large quantities can lead to tremors and convulsions a few hours after exposure. Organochlorines can remain in the soil for months and can accumulate in root crops. One positive aspect is that these compounds are not easily absorbed through the skin, accounting for their good safety record.

Organophosphates do not persist in the environment but are easily absorbed through the skin. The well-known pesticides *malathion* and *paraquat* belong to this class, used primarily to prevent contamination of crops with soft-bodied insects like the boll weevil. Toxicity in humans is caused by inhibition of acetylcholinesterase at high doses, an enzyme necessary for nerve signaling. However, mild poisoning can lead to dermatitis, and impaired ability to concentrate. Some organophosphates have been shown to be **teratogenic** (induce birth defects) and mutagenic (induce DNA mutations), but studies have not been able to show their carcinogenic potential.

Carbamates are similar in mode of action to organophosphates in that they inhibit acetylcholinesterase. They have not been proven to be carcinogenic to humans, although they are highly toxic. *Aldicarb*, for instance, can be used only in a limited number of crops, and only at levels below 1 ppm (see Table 12.6). Exposure to high levels results in spasms, low blood pressure, and convulsions, and may lead to respiratory failure and cardiac arrest. Exposure to low levels can lead to headache, vomiting, diarrhea, confusion, blurred vision, and learning impairment.

TABLE 12.6 What is the meaning of "part per million"? *Pesticide residues in foods are permitted by the FDA only at extremely low levels. When you hear about pesticide residues in food, the amounts measured are either parts per million (ppm), parts per billion (ppb), or parts per trillion (ppt). The following comparisons show just how tiny these amounts are.*

1 ppm:	1 ppb:	1 ppt:
1 gram of residue in 1 million grams of food	1 gram of residue in 1 billion grams of food	1 gram of residue in 1 trillion grams of food
1 inch in 16 miles	1 inch in 16,000 miles	1 inch in 16 million miles
1 minute in 2 years	1 second in 32 years	1 second in 32,000 years
1 cent in $10,000	1 cent in $10 million	1 square foot on floor tile the size of Indiana

SOURCE: Adapted from International Food Information Council, *Pesticides and Food Safety* (Washington, DC: International Food Information Council, 1995), p. 2.

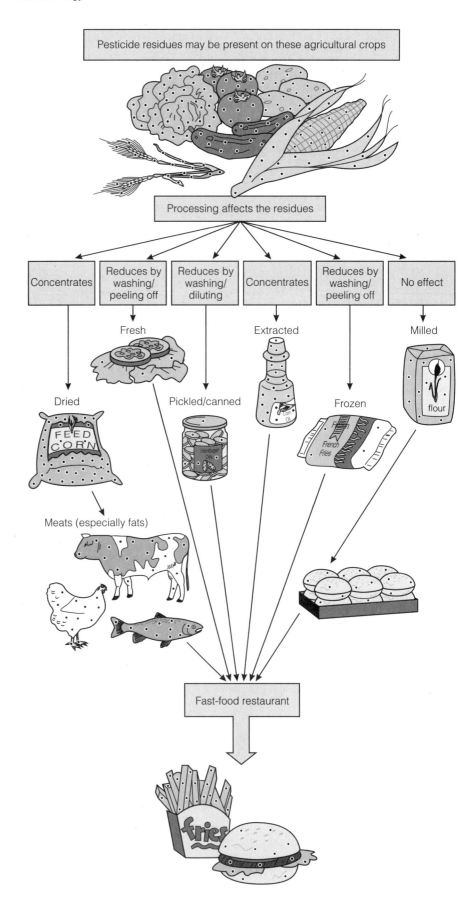

FIGURE 12.13 How pesticide residues could potentially end up in a fast food meal. *Small dots represent hypothetical pesticide residues on crops from spraying and postharvest application. While washing and peeling operations reduce pesticides in foods, other methods either have no effect or concentrate pesticide levels.*

Fumigants like *ethylene dibromide* (EDB) are used to eliminate insect pests from grain, soil, and spices. EDB was banned in 1984 because studies showed its carcinogenicity in experimental animals, as well as its mutagenicity to tissue culture cells. Phosphine is hydrogen phosphide, an insecticide used as an alternative to EDB.

Plant hormones like daminozide (better known as *Alar*) are used to reduce shoot formation and increase plant resistance to disease. Until 1989, it had been used on apples, cherries, and other fruit crops. To achieve exposure to doses that cause cancer in animals, it is estimated that a person would have to consume 50,000 pounds of apples every day for approximately 80 years. Ethylene oxide plus CO_2 in a 1:9 mixture acts as a nonresidual hormone and is considered safe.

Herbicides are used to eliminate weeds, disrupting metabolic processes that occur in these types of plants. They are not as toxic as pesticides. *Dinitrophenol*, also known as *dinoseb*, can cause headache, increased sweating, fever, and weight loss. It has been shown to cause birth defects in experimental animals.

As we have seen, chemicals are used in food production to prevent or minimize damage done to crops by parasites and insects and to improve yield through enhanced plant maturation. Without the use of these agents, agricultural production would not be as bountiful, nor the crops we eat be as high in quality or as inexpensive as consumers are accustomed to. Approval for the use of insecticides, plant hormones, and other chemicals is granted by the Food and Drug Administration, and depends upon experimental data, which may show whether the risk of damage to laboratory animals by the use of such an agent is below tolerable levels. Some of these chemicals have been shown to cause illness, and thus have been banned from use by the U.S. government. There are several manifestations of illness as a result of consumption of foods contaminated with toxic pesticides, ranging from vomiting and diarrhea, to more long-term effects such as birth defects and cancer. The Environmental Protection Agency and other government agencies are currently utilizing risk assessment models to better predict the effect of specific chemicals to public health, and thus be able to make science-based decisions regarding how the risk posed by some of these agents can be mitigated.

Key Points

- Toxicants can be endogenous, naturally occurring, or synthetic.
- Endogenous toxicants are produced by plant tissues and other biological systems.
- Naturally occurring toxicants are produced by microorganisms that reside on the food.
- Synthetic toxicants are those that are purposely added in the raising of food animals and that inadvertently end up in the food supply.
- Risk assessment is the process of determining the toxicity of a toxicant and the probability of humans being exposed to this hazard.
- As part of risk assessment, the dose-response of a toxicant is determined, either from epidemiological data, from animal studies, or from studies carried out with tissue culture cells.
- Endogenous toxicants include the flavonoids, goitrogens, coumarins, cyanide-containing compounds, herbal extracts, and toxins produced by some mushrooms.
- Naturally occurring toxicants include marine toxins such as ciguatera, produced by dinoflagellates, and bacterial toxins such as botulinum toxin.
- Synthetic toxicants include drugs used as antimicrobial agents, such as antibiotics; drugs used to promote animal growth, such as hormones; and pesticides and herbicides.

Study Questions

Multiple Choice

1. Toxicants that are found inherent in foods are
    a. endogenous.
    b. exogenous.
    c. insiders.
    d. inherents.
    e. endorphins.

2. Mycotoxins are produced by
    a. yeasts.
    b. molds.
    c. bacteria.
    d. viruses.
    e. parasites.

3. The concentration effect of a toxicant is measured by
    a. dose-response.
    b. dose-mapping.
    c. dose-delivery.
    d. dosing.
    e. overdose.

4. A method to determine toxicity that utilizes a living organism is
    a. bioassay.
    b. biorhythm.
    c. biofeedback.
    d. living toxic effect.
    e. biotoxicology.

5. The level above which a toxic effect is observed is
    a. baseline level.
    b. threshold level.
    c. evenness level.
    d. level 1.
    e. level a.

6. The dose of toxicant required to kill 50% of a population is the
    a. LT_{50}
    b. LD_{50}
    c. TD_{50}
    d. PD_{50}
    e. TT_{50}

7. A substance that causes cancer is a(n)
    a. tumorigen.
    b. mutagen.
    c. carcinogen.
    d. antigen.
    e. cancerinogen.

8. Substances that quench oxidative reactions are
    a. antioxigenins.
    b. antibiotics.
    c. antioxidants.
    d. anticarcinogens.
    e. antihistamines.

9. An example of a cruciferous vegetable is
    a. broccoli.
    b. celery.
    c. potato.
    d. peas.
    e. corn.

10. The substance that carries signals from the brain through nerve cells is
    a. neurotransferase.
    b. glucosidase.
    c. acetylcholine.
    d. adenylate cyclase.
    e. acetic acid.

11. NSP is caused by consumption of toxic
    a. shellfish.
    b. mushrooms.
    c. beef.
    d. cheese.
    e. liver.

12. Consumption of a toxicant that affects the digestive tract results in
    a. neurotoxic poisoning.
    b. hepatic poisoning.
    c. ataxia.
    d. gastrointestinal poisoning.
    e. pruritis.

13. Diarrhetic shellfish poisoning is caused by
    a. *Dinophysis*.
    b. *Daphnia*.
    c. *Deuteromyces*.
    d. *Deinonichus*.
    e. *Escherichia coli*.

14. Scombroid poisoning is caused by consumption of
    a. histamine.
    b. histidine.
    c. herpex.
    d. histocomplex.
    e. heparin.

15. The enzyme affected by cholera toxin is
    a. cholerase.
    b. guanylate esterase.
    c. adenine cyclase.
    d. adenylate cyclase.
    e. cholerine.

16. Toxins produced by *Escherichia coli* O157:H7 are
    a. virtual toxins.
    b. hela toxins.
    c. reverberant toxins.
    d. vero toxins.
    e. mycotoxins.

17. MLD_{50} is the term used for
    a. dose required to kill 50% of mycotoxins.
    b. dose required to kill 50 people.
    c. dose required to kill 50% of mice.
    d. minimum dose required.
    e. minimum lethal dose.

18. Mutagens are
    a. substances that cause mutations.
    b. substances that mutate.
    c. substances that cause muting.
    d. substances that inhibit speech.
    e. substances that inhibit mutations.

19. Which of the following is a mycotoxin?
    a. flavonoids
    b. histamine
    c. aflatoxins
    d. verotoxins
    e. cholera toxin

20. Which is a mushroom toxicant?
    a. hydrazine
    b. histamine
    c. aflatoxin
    d. verotoxin
    e. cholera toxin

21. Which toxicity would a person complaining of anemia and weakness likely be suffering from?
    a. neurological
    b. hematological
    c. renal
    d. dermatological
    e. cardiovascular

22. The cassava root toxin is produced by the breakdown of _____ by _____ enzymes.
    a. cyanohydrin compound, beta glucosidase
    b. cyanogenic glycoside, cyanohydrin
    c. glycoside compound, two
    d. HCN, beta glucosidase
    e. lyase, HCN

23. Which is a true statement?
    a. In the pesticide food chain, humans are at the middle.
    b. In the pesticide food chain, tons of plants are consumed annually by a person.
    c. In the pesticide food chain, tons of pesticides are consumed annually by fish.
    d. In the pesticide food chain, tons of pesticides are consumed annually by a person.
    e. The larger fish harbor fewer pesticides per individual fish than do the small plant-eating fish.

24. Since domoic acid resembles the amino acids glutamate and glutamic acid,
    a. it is an essential amino acid.
    b. it will produce diarrhea.
    c. it causes cAMP to produce ATP and overstimulates the thyroid.
    d. it inhibits phosphorylation of body cells and acts as a tumor promoter.
    e. it fools neurons into recognizing it as a neurotransmitter.

25. BST
    a. is banned by the FDA because of its carcinogenicity.
    b. cannot be digested.
    c. is a hormone but not a peptide.
    d. is destroyed by pasteurization.
    e. has a structure similar to HST, therefore it causes growth effects in humans.

References

Institute of Food Technologists. 1975. Naturally occurring toxicants in foods: a scientific status summary. *Journal of Food Science* 40:215.

Pierson, M. D., and Stern, N. J. 1986. *Foodborne Microorganisms and Their Toxins*. Marcel Dekker, New York.

Ragsdale, N. N., and Menzer, R. E. 1989. *Carcinogenicity and Pesticides: Principles, Issues, and Relationships*. American Chemical Society, Washington, DC.

Taylor, S. L. 1992. Chemistry and detection of food allergies. *Journal of Food Science* 46(5):148.

Ward, D., and Hackney, C. R. 1991. *Microbiology of Marine Food Products*. Van Nostrand Reinhold, New York.

Additional Resources

CAST. 1986. Agriculture and ground water. Report 103. Council for Agricultural Science and Technology, Ames, IA.

Glaziou, P., and Legrand, A. M. 1994. The epidemiology of ciguatera fish poisoning. *Toxicon* 32:863.

Hefle, S. L. 1996. The chemistry and biology of food allergens. *Journal of Food Science* 50(3):86.

Hui, Y. H., Gorham, J. R., Murrell, K. D., and Cliver, D. O. 1994. *Foodborne Disease Handbook: Diseases Caused by Hazardous Substances*, Vol. 3. Marcel Dekker, New York.

Web Sites

http://ace.orst.edu/info/extoxnet/faqs/
Extension Toxicology web site, ExtoxNet, maintained by the University of California at Davis, Cornell University, Oregon State University, University of Idaho, and Michigan State University. This address is replete with information on toxicants in the enviroment, including those of significance to food safety.

http://www.abvt.org/
American Board of Veterinary Toxicology. This site contains information on toxic fungi, endogenous toxicants, natural toxins, and much more.

http://ntp-server.niehs.nih.gov/
National Toxicology Program of the U.S. Department of Health and Human Services. Contains information on chemical health and safety, including topics in food safety of toxicants.

http://risk.lsd.ornl.gov/rap_hp.shtm
Risk Assessment Information System, a part of the U.S. Department of Energy Office of Environmental Management. Good information on risk assessment of toxicants.

http://www.nwfsc.noaa.gov/hab/biotoxins.htm
Northwest Fisheries Science Center's (NWFSC) Marine Biotoxins. This web site provides information and services to the public about marine biotoxins, including domoic acid.

http://www.foodallergy.org/index.html
The Food Allergy Network. This site includes important information on product alerts and recalls, as well as educational information for patients and families.

Challenge!

FOOD ALLERGIES AND FOOD INTOLERANCES

People often assume that an unpleasant reaction to a food is due to an allergy. However, true clinically proven food allergies are uncommon. The incidence of food allergy is about 1 percent of the total adult population.

The majority of cases of suspected food allergies turn out to be due to reactions called "food intolerances" rather than food allergies. A true **food allergy**, or hypersensitivity, is an abnormal response to a food that is triggered by the immune system. The immune system is not responsible for the symptoms of a food intolerance, even though these symptoms can resemble those of a food allergy.

The Mechanism of Food Allergic Response

The basic cause of food allergies is the reaction of the immune system of sensitive individuals to some component of the food, usually a protein. These components are termed antigens or food **allergens.** Allergens are usually not broken down by cooking or by stomach acid. Thus they easily enter the bloodstream, enabling them to cause allergic reactions in several parts of the body. Both whole foods, especially nuts, dairy products, eggs, wheat, and soy foods, as well as some food additives are thought to cause allergic reactions in sensitive people.

Common symptoms include several types of skin irritations, as well as gastrointestinal maladies like diarrhea and vomiting. Symptoms can vary in severity, time of onset, and duration, with all of these depending on the amount of food consumed.

The most severe symptom is *anaphylaxis*. This reaction is experienced by some allergic individuals in which more than one part of the body is engaged. Not only does the patient develop skin irritations, but also difficulty breathing and low blood pressure, with the potential for death due to suffocation. Immediate medical attention is required in these cases—injection of adrenaline to open up the airways is the most common treatment.

Allergic reactions involve the production of a certain type of *antibody*, called immunoglobulin E (IgE). These proteins are produced by specific blood cells that are part of the arsenal of weapons that the immune system utilizes to inactivate foreign substances. Some individuals in a population have a genetic predisposition to form IgE in response to exposure to certain foods. Thus, someone with allergic parents is likely to develop food allergies.

Another component of the immune system involved in food allergies is the *mast cell*. These cells are white blood cells typically found in the nose, throat, lungs, skin, and gastrointestinal tract, all areas that are involved in allergic reactions. They contain granules in their cytoplasm that harbor chemicals that can prompt diarrhea, skin rash, and other symptoms.

The specific series of events that take place during an allergic reaction begin by entry of the allergen into the circulatory system through absorption from the intestinal tract. Cells of the immune system known as *T helper cells* stimulate a group of blood cells, the *B cells*, to produce IgE antibody (see Figure 12.14). Mast cells have special receptors on their surface that bind to IgE. This triggers an increase in intracellular levels of calcium through alteration of the cell membrane that allows calcium to flow inside.

The excess of calcium elicits a *degranulation effect*, in which mast cells release the granules that contain chemicals that produce the symptoms associated with allergic reactions. For instance, histamine and heparin are released, both of which dilate blood vessels, resulting in swelling and skin rash. They can also increase mucus secretion in

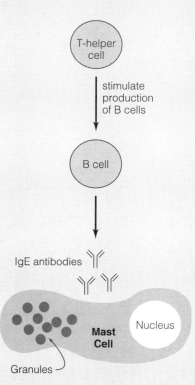

FIGURE 12.14 Antibody and Mast Cell Interaction.

the lungs, leading to bronchial congestion, and breathing difficulties that can asphyxiate the victim.

There is no cure for food allergies, other than avoiding the offending food. Most deaths occur because the person is unaware that a particular food contained the allergen. The principal problem in adults is consumption of restaurant food where listing the ingredients of dishes is impractical and difficult to enforce. For children, it is wise for parents to watch for signs of food dislikes, which may be clues to food insensitivities or true allergies. When a particular food needs to be excluded from the diet, especially dairy or egg products that are important sources of nutrients, care must be taken to find a nutritionally adequate replacement.

Food Intolerances

The following substances and reactions are not allergens, but some people are sensitive to them.

Aspartame. Aspartame, a commonly used sugar substitute, was approved for use in 1981 by the FDA. Controlled studies have shown that it is not an allergen. However, certain people who suffer from the disease phenylketonuria (PKU), as well as pregnant women with high levels of phenylalanine in the blood, cannot metabolize the amino acid phenylalanine, a major component of aspartame. For these people, consumption of aspartame can lead to brain damage.

Monosodium Glutamate. Monosodium glutamate, better known as MSG, is used in many restaurants, most notably those serving Chinese cuisine. Some people are sensitive to this additive, experiencing mild symptoms referred to as "Chinese restaurant syndrome."

Sulfites. Sulfites occur naturally in some foods and can be added to prevent discoloration of some fruits and vegetables, including potatoes, cut bananas, and dried apples. They are also used to inhibit microbial growth in wine. Sulfites in high concentrations can affect people who suffer from severe asthma. Sulfite-containing foods may release sulfur dioxide, which the asthmatic inhales while eating, which irritates the lungs and can precipitate a severe bronchospasm, a constriction of the lungs. As a result, the FDA banned sulfites as spray-on preservatives in fresh fruits and vegetables. However, they are still added to certain foods and are natural products of wine fermentation.

Gluten. Gluten is a class of proteins found especially in wheat, and by association, baked products such as bread. Gluten intolerance is associated with the disease called *gluten-sensitive enteropathy* or celiac disease. It is caused by an abnormal immune response to gluten, which is a component of wheat and some other grains.

Migraine Headaches. It is controversial to claim that migraine headaches are caused by food allergies. Some studies have shown that people who suffer from migraines can have their headaches brought on by histamines and other substances in foods. However, there is virtually no evidence that food allergies actually cause the migraines.

Child Hyperactivity. Some people believe hyperactivity in children is caused by food allergies. However, researchers have found that this behavior is only occasionally associated with food additives, and then only when consumed in large amounts. There is no evidence that a true food allergy can affect a child's activity except that if a child itches, sneezes, and wheezes in response to an additive, the child will be miserable and therefore more difficult to direct.

Challenge Questions

1. What is the key feature of a true food allergy that distinguishes it from other adverse reactions to foods?
   a. dilation of the pupils
   b. antigen production by the body
   c. migraine and hyperactivity
   d. an immune response
   e. granulation of the tongue

2. Which statement is correct?
   a. Aspartame is not an allergen.
   b. Sulfites are additives and are not naturally present in foods.
   c. Migraine headaches are due to food allergies.
   d. Individuals who suffer from PKU generate an immune response to aspartame.
   e. Mast cells produce IgE antibodies.

3. Which is the correct sequence?
   a. absorption, degranulation, antibody production, allergen ingestion
   b. IgE production, mucus secretion, stimulation of B cells
   c. calcium level increase, B cell stimulation, blood vessel dilation
   d. anaphylaxis, antigen released, antibody produced
   e. IgE binding, increased intracellular calcium, histamine release

4. Why must care be taken to find a nutritionally adequate replacement food when a particular item like milk has to be excluded from a child's diet due to suspected intolerance?
   a. Anaphylaxis can result.
   b. Child may lapse into a diabetic coma.
   c. Child is at risk of developing a nutrient deficiency.
   d. An allergy will develop.
   e. A hypersensitivity will develop.

5. Which is *not* a reason why food allergies are thought to be common?
   a. Foods can cause allergy-like symptoms.
   b. Foods or additives trigger antigen-antibody responses in about one-fourth of adults.
   c. Many people believe that headaches can be caused by foods or additives.
   d. Many people have experienced an adverse reaction to a food.
   e. Certain additives can cause allergy-like symptoms.

CHAPTER 13

Food Engineering

13.1 Food Engineering—Basic Terms and Principles

13.2 Deep-Fat Frying—An Illustration of Heat Transfer, Mass Transfer, and Boundary Layers

13.3 Food Materials Science—A Physicochemical Approach

13.4 Food Microstructure—Influencing Physical and Sensory Qualities

13.5 Psychrometrics—Looking at Air and Food Processing

13.6 Rheology—Studying Flow and Deformation

13.7 Extrusion Technology in the Food Industry

Challenge!
Food Packaging

CHAPTER OBJECTIVES

After completing this chapter, you will be able to:
- Define the broad scope of food engineering.
- List the thermal properties of foods.
- Explain the processes of heat transfer and mass transfer.
- Describe how materials science principles can be applied to foods.
- Explain the significance of the glass transition.
- Discuss the link between food microstructure and food quality.
- List the psychrometric properties of air.
- State the importance of the key rheological parameters.
- Explain the purpose of extrusion technology.
- List examples of food package types and the plastics used in their fabrication.

Food engineering overlaps food processing and preservation, as well as food chemistry and product development. While a food chemist studies food components from a molecular point of view, and a product development scientist makes use of knowledge regarding the chemical and physical nature of foods and ingredients, a food engineer focuses on the physical makeup of foods and how to measure the effect of processing on food's physical properties. The approaches complement each other because microstructure is based upon physical materials composed of molecules, and processing and ingredient formulation both act to change food properties and quality. Food engineering also encompasses the design and operation of food processing, storage, and analytical equipment. In this chapter, the rheological properties of foods and food materials science will be emphasized. However, time will be spent examining the importance of heat transfer, psychrometrics, and extrusion technology. A detailed introduction to food packaging technology in the Challenge section rounds out the chapter.

13.1 FOOD ENGINEERING—BASIC TERMS AND PRINCIPLES

Food engineering is the application of engineering principles to the manufacture of foods. A relatively new branch of engineering, it impacts the design of processes and systems extending from food materials handling through processing, packaging, and distribution. Topic areas important to the food engineer include heat transfer, mass and energy conservation, matter flow, and mechanical food properties. Food engineering is energized by the concepts of physics—including the study of mechanics and thermodynamics—and the principles of materials science.

Characteristics of Temperature and Heat

Temperature Scales. Three temperature scales measure and record temperature: Celsius, Fahrenheit, and Kelvin (or absolute)—see Figure 13.1. Food scientists use the Celsius (or centigrade) scale in research. However, the others are not ignored. For example, when food scientists develop guidelines for consumers regarding the safe cooking temperatures for meats, they use both Celsius and Fahrenheit. The following equations relate temperature in Celsius (°C) and Fahrenheit (°F):

$$T_c = \frac{5}{9}(T_f - 32°) \qquad T_f = \frac{9}{5}(T_c + 32°)$$

$$T_c = (T_f - 32°) \div 1.8 \qquad T_f = (T_c \times 1.8) + 32°$$

where T_f means Fahrenheit temperature and T_c means Celsius temperature.

Two basic concepts related to temperature are thermal contact and thermal equilibrium. If a carton of milk is removed from the refrigerator and placed on a counter at room temperature, it comes in *thermal contact* with the counter (and surrounding air). After a set amount of time, the counter and the milk achieve the same temperature as the environment and are at **thermal equilibrium.**

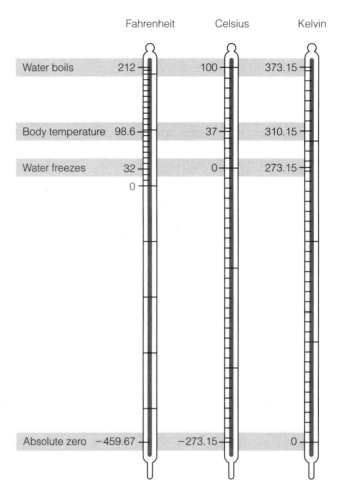

FIGURE 13.1 Temperature scales used to measure heat/cold intensity. *Every temperature on the Celsius scale corresponds to a particular temperature on the Fahrenheit scale. On the Celsius scale, the freezing point of water is 0°C and the boiling point of water is 100°C. On the Fahrenheit scale, the freezing point of water is 32°C and the boiling point of water is 212°C. Food scientists commonly use the Celsius (centigrade) scale. The simple formulas in the text make it easy to convert from Celsius into Fahrenheit and vice versa. To find a temperature on the Kelvin scale, add 273 to the Celsius temperature. For example, water freezes at 273 K (273+ 0°C), and boils at 373 K (273+ 100°C). Zero on the Kelvin scale is considered to be the point at which all molecular motion stops.*

Molecular Motion and Temperature. All food molecules are in motion due to the energy they possess. Molecular motion (the movement of food molecules) depends upon the temperature and the phase of the food material. Is there a temperature so cold that all motion stops?

Absolute zero temperature (measured as −273°C or 0°K) is the temperature at which all molecular motion stops. At this temperature, matter would possess no heat energy. But above this temperature, under typical, everyday conditions, the atoms and molecules in foods are in constant motion. But molecular motion varies in the individual states of matter, apart from temperature. In other words, at any particular temperature (e.g., room temperature, about 22°C), the molecules of solid, liquid, and gaseous substances have a different range of motion.

In a gas, molecules travel at high speeds and are relatively unrestricted to move about. There is much space between them, and gas molecules exert little self-attraction. In a liquid, molecular motion is more restricted because the molecules are closer together. These may touch and tumble over one another and exhibit self-attraction, which contributes to a definite shape. In a solid, molecular motion is the most restricted. Molecules

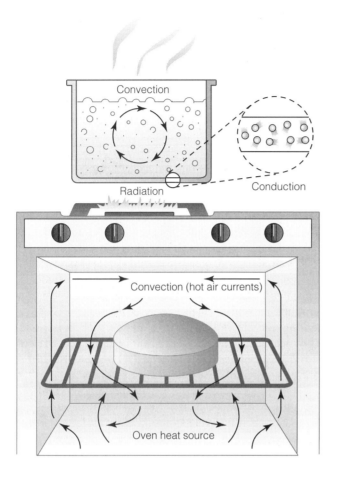

FIGURE 13.2 Heat transfer by conduction, convection, and radiation.

are locked into a specific location and shape, and the only kind of motion they have is **rotational** and **vibrational**. In spite of these differences, each state of matter will respond identically with the addition of heat energy: Their molecular motion increases, a result that is measurable as an increase in temperature.

Measuring Heat. The common unit to measure heat is the calorie. A **calorie** is defined as the amount of heat necessary to raise the temperature of 1 gram (g) of water by 1 Celsius degree. A kilocalorie (kcal) is the heat needed to raise the temperature of 1 kilogram (kg) of water by 1 Celsius degree.

The *joule* (J) is another unit used to measure heat. One joule equals 0.239 calories; conversely, 1 calorie = 4.18 joules. The *Btu* is still another heat term. One Btu is the amount of heat that will raise or lower the temperature of one pound of liquid water 1°F. What is the relationship between Btu, calorie, and the joule? One Btu = 252 calories = 1055 joules.

Conduction, Convection, and Radiation. Now let's discuss the transfer of heat energy to foods, which can involve several methods. These include conduction, convection, and radiation. Did you know that the common thermos bottle was created with the physics of heat transfer in mind? It is designed to minimize heat transfer from radiation, and from conduction and convection. The thermos is a cylindrical flask constructed of double-walled glass with silvered inner walls, which are separated by an evacuated space. This helps maintain the desired temperature of a liquid food or beverage (cold liquids remain cold, warm liquids remain warm, despite the ambient temperature).

Figure 13.2 illustrates the three kinds of heat transfer. *Conductive heat transfer* or **conduction** occurs at the molecular level. The molecules of an egg in a frying pan gain thermal energy from the stove, which causes them to vibrate more rapidly. These vibrations are passed along from molecule to molecule in the food, although there is no physical movement of the food material during conductive heat transfer. A *temperature gradient* (the simultaneous existence of hot and cold areas) between the stove, the pan, and the food causes heat transfer to be directed from the high temperature region to the lower temperature region.

When heat is exchanged between either liquid or gas and solid surfaces, it is called **convection,** or convective heating. The gas or liquid is considered a fluid substance, and the solid is considered nonfluid. To aid in understanding, consider a fire. The warmed mass of air that rises from above the flames is due to *natural convection*. The movement of the fluid material in this case—the air—is due to differences in density.

A form of convection called *forced convection* can be used in food processing. Forced convection involves the use of a fan or pump to create movement of the fluid material. In a *retort canner*, a large-scale pressurized cooker used in the canning industry to heat-process canned foods, heat energy is transferred by both conduction and convection. Canning inspectors periodically check the *cold point* of a can to make sure that the critical canning temperature has been achieved throughout the food contained by the can.

Radiation offers another form of heat transfer. We have all experienced the warming rays of the sun and appreciate solar (or radiant) energy as a heat source. This form of electromagnetic radiation is part of the electromagnetic energy spectrum, which spans from low energy (alternating current) to high energy sources (gamma and cosmic radiation). All materials, not only the sun, emit electromagnetic radiation based upon their surface temperature. In physical terms, the transfer of heat between two surfaces by radiation depends on two factors: the *emissivity* of the radiating surface and the *absorptivity* of the other surface. The shape and design of the object are also important.

Microwave Energy. Microwave ovens have achieved widespread use with consumers in the United States since their introduction in the 1970s. They are effective in heating foods because foods contain water. The dipolar molecule of water is the electrically active substance allowing microwaves to heat foods. *Microwaves* are part of the electromagnetic spectrum, with a characteristic wavelength and frequency. Microwaves are long in wavelength, high in frequency (2.45 gigahertz, or 2.45 billion cycles per second), and low in energy.

As photon particles, microwaves create a fluctuating electrical field inside a microwave oven that changes direction 2.45 billion times a second. This means that an electrically charged particle inside the oven, such as water, will be twisted back and forth that many times a second. This intense twisting causes adjacent individual water molecules to rub back and forth against each other. The rubbing causes thermal friction, and the molecules heat up. See Figure 13.3.

Heated water molecules in turn pass heat energy along to other molecules in the food, such as proteins, carbohydrates, and fats. Energy, then, arrives at the food item as microwave energy, not heat energy. Heat is generated internally within the food as a result of friction between molecules—this is different from conductive or convective heating. Microwaves penetrate (or are absorbed) to a depth of 5–7 cm, with the result that

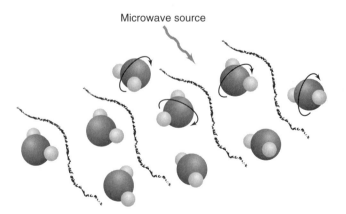

FIGURE 13.3 Microwave energy.
Friction of adjacent spinning water molecules in food within a microwave oven creates heat.

foods cook faster than in a conventional oven, since the slow process of first heating the outer surface, followed by inward conduction, does not have to occur.

Thermal conductivity (k) refers to the rate of heat that will be conducted through a unit thickness of material, such as any food material. In microwave energy heating, thermal conductivity depends on the amount of water in a food sample, as well as the other chemical constituents. The following empirical relationship can be used for certain foods:

$$k = 0.6W + 0.2P + 0.18F + 0.2C + 0.14A$$

where W, P, F, C, and A are volume fractions of water, protein, fat, carbohydrates, and ash in a food, and ($W + P + F + C + A = 1$).

As an example, suppose there are two different foods, I and II. The W, P, F, C, and A values for food I are given as 0.700, 0.050, 0.150, 0.099, and 0.001, while food II has these values: 0.600, 0.050, 0.100, 0.245, and 0.005. Which food, I or II, conducts heat at a higher rate (has the larger thermal conductivity)?

To answer, plug in the W, P, F, C, and A values separately for foods I and II, and solve for k. You should find that food I has a greater thermal conductivity value (k for food I is 0.4769, while k for food II is 0.4377). This means that food I conducts heat more quickly than food II. Why? It has to do with compositional differences between foods I and II. By looking at the input data, it should be clear that the difference in k values is due mainly to the greater volume fraction for water in food I.

Dielectric Properties. Although microwaves are in and of themselves not heat, materials like foods in a microwave oven convert the energy to heat. The food scientist investigating the microwave processing of foods must be familiar with the key electrical properties such as the **dielectric constant** (represented as K' or ϵ') and the **dielectric loss factor** (represented as K'' or ϵ''). Foods do not heat up equally fast in a microwave oven. Pork chops heat more slowly than diced carrots. Foods with "high loss" heat quickly and conveniently in a microwave, while "low loss" materials do not.

Dielectric properties ϵ' and ϵ'' reflect the ability of a material to store (ϵ') and to dissipate (ϵ'') electrical energy. What this relates to is a material's ability to act as an insulator. In this sense, foods tend to be *poor electrical insulators*, instead absorbing a signifcant fraction of energy from a microwave field.

The amount of microwave energy a food material can absorb is termed *absorptivity*. In general, the more polar a molecule, the greater its absorptivity, and the higher the ϵ'. Water, as the significant polar molecule in foods, is the substance interacting the most with microwaves to produce heat. This results in a more instantaneous heating than with conventional heating.

The dielectric properties of foods differ because of their chemical composition (such as percent moisture), physical structure, and temperature—see Table 13.1. The

TABLE 13.1 Dielectric properties of semisolid food products at 2,450 MHz. *This table compares dielectric data obtained (at two different temperatures, 25°C and 50°C) from three foods of varying moisture content heated in a microwave. Recall the dielectric constant (ϵ') and the dielectric loss factor (ϵ''), and that foods with "high loss" heat better in a microwave, than "low loss" materials. From the data, carrot and beef, while having the same moisture content, show differences in both dielectric properties.*

Product	Moisture	Temperature	ϵ'	ϵ''
Potato	76.4%	25°C	64	14
		50°C	58	13
Carrot	90.9%	25°C	72	15
		50°C	65	14
Beef	90.9%	25°C	61	17
		50°C	55	18

Source: Mudgett, 1982.

mathematics used to determine how food materials heat up is complex, and it is undetermined for some situations. Food mechanical factors such as product density and viscosity, as well as dielectric properties, and the thermal heat capacity each affect how rapidly a food heats up once microwaves are deposited in it.

Conservation of Mass

Mass is a measure of the amount of matter in an object and is the property that makes an object "reluctant" to change its state of motion (or nonmotion, if at rest). It is invariant, which means for example that a carrot has the same mass on Earth as it would on the moon. The force (gravity) that Earth exerts on an object of specific mass is called the object's *weight* on earth. Weight is not an invariant quantity. The weight of a carrot on Earth is different from the weight of the carrot on the moon, because of the gravitational differences.

The law of *conservation of mass* says when a reaction takes place, the total mass of reactants is equal to the total mass of products formed plus the mass of reactant remaining. All changes in matter are accompanied by a *flow of energy*. In addition to heat, energy exists in many forms, such as chemical, electrical, mechanical, and radiant energy. In the case of a food material being acted upon by heat energy in a closed container, there may be several consequences. Some of the matter may be consumed within the system and be converted into radiant energy, and a portion of the matter may generate new matter (product) or accumulate within the container boundary. For example, consider a container (system) used to reduce the moisture content of a food product, such as raw carrots. If there is no accumulation of mass within the system, the law of mass conservation simply asserts that the material input through the system equals the material output through the system. This defines a special situation called the *steady state*.

Steady state has to do with the nature of heat transfer and refers to system properties that do not change with time. For example, when a heat transfer process is operating at steady state, the temperature in the system remains constant in time, although it may vary from location to location within the system. Figure 13.4 illustrates this point, in which the energy being transferred within a hot chamber represents steady state heat transfer. Another example of steady state heat transfer occurs in a milk pasteurizer.

This steady state is in contrast to that of *transient heat transfer*. In transient heat transfer, system properties do change with time. This "unsteady state" is characterized by temperature changes with time. This form of heat transfer is common to food processing, in which the rate of heat transfer changes with time. Transient heat transfer also can mean that the temperature of a heating or cooling material is different at a different time. The cooling that occurs in a retorting operation following thermal processing or of products such as baked cookies at ambient temperature in a cooling environment are examples.

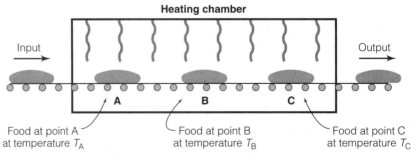

FIGURE 13.4 A steady state process. *When a food product at point A reaches point B within the heating chamber, its temperature is equal to T_B. Since the food enters the chamber "cold," it is clear that $T_A < T_B < T_C$. Upon leaving the chamber, all foods will achieve a temperature of T_C.*

Thermodynamics

A *system* is any object or set of objects (such as a food item) that we are studying; everything else is called the *environment*. Food items such as plants and animals are considered "open systems" while they are alive and exchanging nutrients, wastes, and gases such as CO_2, O_2, and H_2O with the environment. They are considered "closed systems" when they achieve constant mass (hence they no longer grow, respire, etc.).

Thermodynamics is the study of how natural processes are affected by changes in temperature. It embodies laws that seek to describe energy transformations in systems, involving heat, work, and energy. **Heat** is defined as a transfer in energy due to a difference in temperature, while *work* results from a transfer of energy that is not due to a temperature difference. Work consists of motions across the boundary of a system that are organized on a macroscopic scale, while heat involves motions at the molecular level.

The *first law of thermodynamics* states that a change in the internal energy of a system is equal to the heat added to the system minus the work done by the system. This is actually a restatement of *the conservation of energy*: Total energy is neither increased nor decreased in any process; energy can be transformed from one kind to another, but the total amount remains constant.

The *second law of thermodynamics* states that heat will spontaneously flow from a hot object to a cold one, but not the reverse, and that natural processes tend to move toward a greater state of disorder rather than order. To describe thermodynamic disorder, the term *entropy* is used. **Entropy** is a quantitative measure of disorder in a system.

Two important types of thermodynamic processes encountered by food engineers are called isothermal and adiabatic processes. An *isothermal* process is one carried out at constant temperature. An **adiabatic** process is one in which no heat is exchanged.

Heat Transfer. Heat Transfer refers to the manner in which heat energy is transported from a food's surroundings to the surface and interior of the food. The driving force for heat transfer is temperature difference. The prime mechanisms of heat transfer in foods are conduction, convection, and radiation. As mentioned earlier, conduction occurs when heat moves through a material due to molecular motion; convection is due to the movement of a heated fluid from hot regions to cold, and radiation occurs when heat is transferred directly between objects without an intervening medium.

Food *surface area* has an important effect on the speed at which foods will increase in temperature as heat energy is applied. A thin, 1-ounce slice of ham will heat up quicker than a 1-ounce *cube* of ham because the former has a greater surface area exposed to heat. In explaining heat transfer, foods generally heat up from the outer surface first. Molecules serving as the heat source transfer heat to the surface of foods by colliding with food molecules on the food surface. This causes the food surface to heat up and transfer energy inward throughout the food via molecular collisions with those molecules located interior to the surface.

The rate of heat transfer is influenced by food thermal properties such as specific heat and thermal conductivity and rheological properties such as viscosity, as well as by product density and thickness. The factors that affect conduction are presented in the following equation:

$$q = \frac{Q}{A} = \frac{\Delta t}{R}$$

where A = area, Q = heat transfer measured in watts, q = heat transfer per unit area, R = thermal resistance of material, and t = temperature. Knowledge of heat transfer is important for the food scientist—in the canning of foods in a retort, in the thermal drying of foods, and in the unit operation of deep-fat frying. We will focus on deep-fat frying in Section 13.2 to illustrate some of these basic heat transfer principles.

Heat Capacity and Specific Heat. The **heat capacity** of a material is the amount of heat energy necessary to speed up its molecules enough to raise the temperature of a unit mass of a material by 1 degree Celsius. **Specific heat** is a term that is often misused in place of heat capacity. Specific heat is actually the ratio of the heat capacity of a material to that of water.

Water has an average heat capacity of 1 kcal/kg · C°. The average specific heat of air is 0.25, of protein is 0.40, and of glass is 0.20. Substances with specific heat below 1.0 heat up more quickly than water. The lower the value, the more quickly the substance will heat up. The specific heat of a food fat or oil is approximately 0.5 units, compared to 1.0 for water. The specific heat value for a given food depends on its percentage composition of carbohydrate, protein, fat, and water.

A handy equation can be used to determine the specific heat for food products of known composition:

$$c_p = 1.424\, m_{carb} + 1.549\, m_{pro} + 1.675\, m_{fat} + 4.187\, m_{water}$$

In this equation, m_{carb} refers to the mass fraction of carbohydrate in the sample, while the other terms refer to the mass fractions of protein, fat, and water (as moisture). The units for specific heat determinations are in kJ/kg · C (Celsius) or K (Kelvin).

Thermal Conductivity. *Thermal conductivity* was defined in the section on microwave energy—it is the rate that heat will be conducted through a unit thickness of material. This is an important property to determine in food processing operations. The thermal conductivity (k) of high-moisture foods, such as fruits and vegetables, as well as meats, tends to be similar to that of water (k_{water} = 0.597 W/m · °C).

If the water content (as a %, symbolized by i) is known, then the following equations can be used to solve for thermal conductivity:

$$k = 0.148 + 0.00493i \text{ (for fruits and vegetables)}$$
$$k = 0.08 + 0.0052i \text{ (for fresh meats)}$$

Heating and Chemical Changes. Heat energy can cause chemical, physical, sensory, and nutritional changes in foods. For instance, heat energy in a convection oven creates chemical changes in starch during the baking of cakes. Cooking a steak on a grill produces not only physical changes, but chemical changes such as protein denaturation and gelation, and hydrolysis of food polymers.

As a general rule, the rate of a chemical reaction *doubles for every 10°C increase* in temperature. Are chemical reactions in foods desirable? Some are, because they account for good flavors, aromas, textures, and colors, but other kinds of chemical reactions are deleterious. This explains why food needs to be stored in a refrigerator or freezer—so that the enzyme-mediated, microbial, oxidative, and other potential chemical changes in foods are slowed.

Heat Exchangers

A **heat exchanger** is a specialized piece of equipment that is used in food processing and storage to either add or remove heat from food. The function of a heat exchanger is to transfer heat from one fluid to another, and the design often keeps the fluids physically separated by material that offers good conduction of heat to the product without fluid mixing (noncontact design). A simple type of arrangement is a double-pipe heat exchanger, consisting of one pipe inside of another pipe—see Figure 13.5.

Another type of heat exchanger used in the food industry is called a **scraped surface heat exchanger (SSHE)**. The main design implementation is the addition of a mechanical scraping device (blade) that "cleans" off any accumulating product (e.g., chocolate) on the pipe wall surfaces. Such a scraped surface maintains rapid heat transfer. A **plate heat exchanger** consists of closely spaced stainless steel plates or "fins" arranged in parallel, with fluid flow occurring between plates. The creation of turbulence within the product stream is accomplished by designing appropriate patterns that are pressed onto the plates, the result of which is improved heat transfer. Such equipment has been routinely used for decades in the dairy and beverage industries for pasteurization.

A wide spectrum of food products, ranging from canned vegetables, canned meats, fluid dairy foods, and dried fruits and grains receive heat exchange processing treatment.

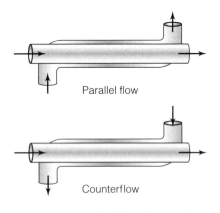

FIGURE 13.5 A double-pipe heat exchanger. *In a double-pipe heat exchanger, fluid flow can be described as parallel (in the same direction for both pipes) or counterflow (in opposite directions). Occasional fluid accumulation ("fouling") on pipe walls contributes to "drag," resistance to flow.*

Each offers unique challenges related to heat transfer. Although beyond the scope of this chapter, attempts to solve these problems involve quantitative approaches utilizing the mathematical concepts that deal with specific thermal properties of foods, such as specific heat and thermal conductivity calculations.

13.2 DEEP-FAT FRYING—AN ILLUSTRATION OF HEAT TRANSFER, MASS TRANSFER, AND BOUNDARY LAYERS

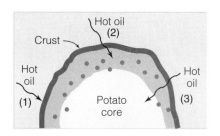

FIGURE 13.6 Boundary layers in a frying potato. *The layers are (1) the interface between the potato and the hot oil, (2) the crust layer; and (3) the inner core. The crust layer is porous and contains most of the transferred oil; oil penetration depends upon factors such as oil freshness, oil type, and frying time.*

In deep-fat frying, conduction occurs within a solid food material while convection occurs between the oil and the food material. It is common to describe the details of this process with respect to **boundary layers,** which represent dividing lines or fronts as two dissimilar materials, the heating oil and the food material, come in contact with each other (Figure 13.6). We should think of the existence of a stationary boundary layer at the food surface. Moving boundary layers within foods that separate the crumb (core) from outer crust regions in a fried food have also been postulated. At boundary layers interesting things happen, such as heat transfer and mass transfer.

Changes to Frying Food and Frying Oil

Changes take place within the food material as it fries, and changes also occur in the frying oil medium as a result of the food being there and the high frying temperature. After all, the oil is no longer in isolation; it is in contact with the moisture inherent in the food and with a variety of hydrophobic and hydrophilic food molecules. Heat, and the oxygen that is present, serve as initiators of physical and chemical changes to the oil. For instance, viscosity changes and lipid oxidation can occur, as well as thermal breakdown of the oil.

With respect to the food material, a frying potato, for example, will experience thermal changes to its constituent macromolecules of starch and protein, as gelatinization and denaturation, respectively. In addition, color and flavor development occur, as well as the loss of moisture due to vaporization and mass transfer, and oil uptake through mass transfer.

Mass Transfer

Mass transfer refers to the movement or migration of a liquid (such as frying oil) or a food component, either within one phase or between different phases, which is caused by physical conditions (i.e., concentration gradients) present in the liquid oil/food system. Mass transfer is characterized at the molecular level by mass diffusion, and macroscopically by bulk mass transport as material flow is directed by convection.

At this point, the similarities and differences between heat and mass transfer might become apparent. There are some common threads, including the existence of boundary layers, flow rates and resistance to flow, physical changes, and the ever-present adherence to the second law of thermodynamics, the tendency of systems to move toward equilibrium.

13.3 FOOD MATERIALS SCIENCE—A PHYSICOCHEMICAL APPROACH

Although "food" is a familiar material, it is nonetheless complex—much science is required to describe the nature of food. One method of description is the materials science approach. It seeks to categorize food materials the same way as nonfood materials, such as ceramics, metals, and plastics. These materials are characterized as possessing microscopic crystalline and amorphous (noncrystalline; without structure) character, which impacts physical parameters such as strength and density. In foods, this approach includes the notion of a structure-property-processing relationship, which impacts food quality.

For example, a manufactured bar of chocolate can be described at various stages of processing according to its microscopic crystalline and amorphous makeup. The finished product should have a uniform glossy brown appearance and a texture that produces a "snap" when the bar is bent. These properties result only when the microstructure is optimal with respect to amorphous and crystalline regions and when the fat crystals settle into the correct stable form due to tempering. A defect in chocolate (*bloom*—see Section 9.7) can be traced to a particular change in the microstructure—the fat crystals change from one type to another.

This materials science approach to foods enables the food scientist to construct three-dimensional models of food system structures, including gels, foams, and emulsions, and to follow changes to these systems created during processing and storage under various conditions (e.g., temperature).

Physicochemical States, Glass Transition, and Water Mobility

A materials science approach adapted to food molecules established in the early 1990s by American scientists Harry Levine and Louise Slade has influenced the view of food systems containing water. The mobility of water in foods and the relationships between water mobility, food ingredient and polymer mobility, and physicochemical phase transitions, are important factors that influence food quality.

The basic idea is that the behavior of the food macromolecules, which are thought of as polymers (especially polysaccharides and polypeptides), should be similar to that of the synthetic polymers studied in materials polymer science, such as ceramics, metals, and plastics. Of particular importance are the mobility and physicochemical state of food polymers and of certain food ingredients present in low-water-content foods and frozen foods. An understanding of what is called the glassy state, and the glass transition temperature, is central to the useful application of such an approach to food systems.

Four **physicochemical states** characterize the movement of water and ingredient molecules in food: (1) crystalline, (2) liquid, (3) amorphous rubbery, and (4) amorphous glassy—see Figure 13.7. Most foods are multiphasic, meaning they consist of both amorphous rubbery and glassy areas, plus water. In foods of varying water content, ingredients shift between these physicochemical states (or phases) as a function of temperature, concentration, and time. Water and molecules in foods may undergo physicochemical phase changes during processing and storage, these changes are significant because they influence food stability and quality. Freezing and evaporation are examples of conditions that promote such changes.

In polymer science, a **glass** is defined as any material that has solidified and become a rigid material without forming a regular crystal structure. A material in the glassy state

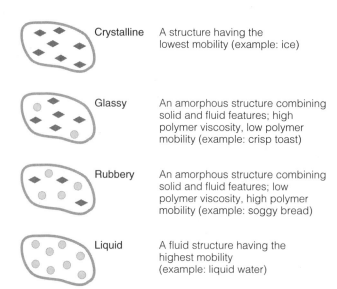

FIGURE 13.7 A simplified drawing of the four physicochemical states of food. *Viscosity describes the resistance of liquids to shear, agitation, or flow. It is commonly referred to as a fluid product's thickness. Viscosity is defined as the tangential force that a fluid flowing in one plane exerts on an adjacent plane.*

is described as brittle or crisp. A hard, dry sugar candy is a glass. When foods transition to the rubbery phase due to moisture migration, they lose their crispness. You can see this would be devastating to the eating quality of products such as snack chips or crackers.

The glassy and rubbery states are considered unstable because they do not persist indefinitely. Polymer mobility changes as a result of transitions from glassy to rubbery and vice versa. **Glass transition** (T_g) refers to the change in the physicochemical state of amorphous food materials between the solid glass and the liquid or rubbery states. The temperature range over which this transformation occurs can help predict the storage life and stability of foods. **Glass transition temperatures** (T_g') of food materials range from very low (water) to relatively high (high molecular weight food polymers such as starch). The actual glass transition temperature of such hydrophilic food components depends on the amount of water present. Experiments have shown that T_g' varies greatly among different ingredients as a function of concentration and time.

Water acts as a **plasticizer** in low-moisture and frozen foods, lowering the glass transition temperature of food polymer materials. Water-soluble polymers such as polysaccharides and polypeptides become plasticized by water and can then undergo the transitions between various states. If you think about water being available to microorganisms during storage, or water that alters texture during storage, you can see the link between physicochemical state, molecular mobility, and food storage stability, quality, and safety.

Mobility of Food Molecules and Ingredients

Food molecules and water are either free to move, are somewhat restricted in motion, or are immobilized in foods depending upon the conditions of storage and the interactions they exhibit toward each other. Ingredient molecules can become immobilized within crystalline structures when dehydrated or frozen. In liquid, ingredient molecules are free to disperse and to move about. In the amorphous rubbery and glassy states, ingredient molecules retain mobility but become increasingly constrained as their concentration increases.

For example, uncooked pasta is made up of starch, and the starch contains amylose and amylopectin polymers in an organized, hard, "glassy" condition. If this glassy pasta is added to water and heated, a transformation occurs in the organization of the starch because the water plasticizes the starch polymers. In simple terms, the pasta absorbs water as it cooks, and the glassy condition changes to the rubbery condition at the glass transition temperature of the polymers. The pasta is no longer dry and hard, but has become wet and soft.

Caution: Do not think that *all* food changes from a rigid to a more rubbery texture are due to water-mediated phase change. For example, consider a brick of cheese in the refrigerator at 4°F, compared to the same sample sitting out for an hour on the counter at 75°F. True, the warmer sample will seem more rubbery—but the reason is mainly the effect of temperature on the proportion of solid crystals to liquid oil in the lipid molecules in the cheese, not on polymer mobility due to water migration. Fats harden in the refrigerator and soften (solid crystals melt) at higher temperatures, but the change has nothing to do with water.

Food ingredients, as small molecules (e.g., glucose, maltose) or long polymers (e.g., amylose and pectin molecules), possess mobility. Polysaccharides can twist, rotate, and exhibit motion. As their concentration in a food increases, their mobility decreases but physical interactions (entanglement between polymers that touch one another) occur. This process of molecular entanglement and aggregation is what leads to an amorphous rubbery or glassy state. The highly dense, disordered nature of the amorphous state is quite distinct from the low-density, highly ordered structure of the crystalline state.

The rubbery and glassy conditions are types of amorphous states, each having specific characteristics. The least stable amorphous state for a food ingredient is the rubbery state. In this state, there is mobility and low viscosity—*viscosity* is the friction within a fluid that prevents it from flowing freely. Shelf stability and other quality problems of food ingredients may occur. On the other hand, the more stable amorphous state is the glassy

TABLE 13.2 Comparing the amorphous glassy and the amorphous rubbery physicochemical states. *The main differences in characteristics between them are due to solute: solvent interactions, which affect solute (polymer) mobility, water (solvent) mobility, and product shelf life.*

Amorphous rubbery state	Amorphous glassy state
Decreased product density	Increased product density
Decreased product viscosity	Increased product viscosity
Increased solute mobility	Decreased solute mobility
Increased polymer mobility	Decreased polymer mobility
Increased potential for microbial growth	Decreased potential for microbial growth
Decreased ingredient stability	Increased ingredient stability
Decreased shelf stability	Increased shelf stability

state. This comparatively dense state is characterized by minimum polymer mobility and maximum viscosity. Table 13.2 summarizes the differences between the two states.

Processing or storage of product well above the glass transition temperature may result in deteriorative food changes as the food shifts from the solid glassy state to the liquid-like rubbery state. On the other hand, the closer a food is to the glassy state transition temperature, the less likely it is to undergo dynamic phase shifts between liquid, crystalline, and rubbery.

A food's net T_g' value is an average, depending upon the individual T_g' values of its component molecules. For a food to remain stable, its transport, handling, and storage temperatures should not exceed its net T_g' value. Said another way, if a food is likely to experience temperatures above its net T_g', then a good strategy would be to reformulate if practical using an ingredient with as high a T_g' value as possible, such as soluble starch or maltodextrins. See Table 13.3 for a comparison of glass transition temperatures (T_g') for some food ingredients. This addition would effectively elevate the food product's net glass transition temperature and make it more stable at elevated storage temperatures. Improved stability results because the food would reside in the glassy amorphous state, where the mobility of water is most restricted, rather than the rubbery amorphous state.

To test your understanding, suppose you are involved in reformulating a food product that has exhibited a poor shelf life. You have the choice of selecting maltose (T_g' = −30°C) or maltodextrins (T_g' = −6°C) as an added ingredient to move the food system closer to more stable amorphous state (glassy). Which one would you select, and why? Maltodextrins would be the correct choice, because its addition would increase the net T_g' of the food product, since it has a *higher* T_g' than maltose (−6°C vs. −30°C). This ingredient should inhibit shifting to the amorphous rubbery state and result in greater product viscosity and lower polymer and solute mobility because of increased water binding and/or polymer entanglement. The water activity will also be decreased. The sum of these effects should be increased shelf life.

13.4 FOOD MICROSTRUCTURE—INFLUENCING PHYSICAL AND SENSORY QUALITIES

The concept of food microstructure is an important aspect of food engineering and complements the food materials science approach. As we have seen, certain food materials like proteins and starches are considered as polymers. Furthermore, food materials can be characterized at the microscopic level with the aid of microscopic techniques. They can be described, for instance, as having crystalline (highly ordered) or amorphous (less ordered) structure. **Food microstructure** refers to the organization of food structures at the microscopic level and their interactions to produce a food product's physical and sensory characteristics—see Figure 13.8. The study of food microstructure provides a window to the understanding of food textural properties and quality, since these depend upon food microstructural elements.

TABLE 13.3 A comparison of glass transition temperatures (T_g') for selected food ingredients measured using standard conditions.

Ingredient	Glass transition temperature (T_g')
Soluble starch	−3.5°C
Maltodextrins	−6°C
Wheat gluten	−6°C
Sodium caseinate	−10°C
Gelatin	−15°C
Corn syrup (20 DE)*	−15°C
Food gums and pectins	−25°C
Maltose	−30°C
Glucose (100 DE)*	−43°C

*DE means dextrose equivalent.
Source: Based upon Best, 1992.

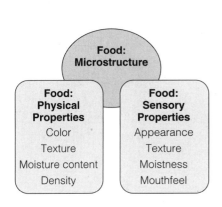

FIGURE 13.8 Food microstructure produces physical and sensory characteristics. *The formation of a gel in yogurt creates a semisolid texture that imparts characteristic moistness, flavor, and mouthfeel.*

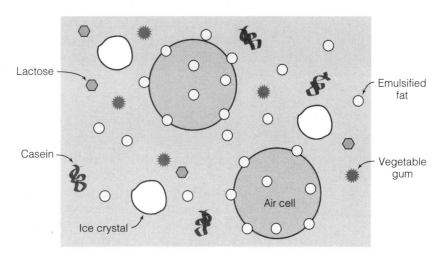

FIGURE 13.9 Important food microstructural elements in ice cream. *Ice cream is a complex colloidal dispersion having characteristics of a foam and emulsion. It consists of butterfat globules (emulsified fat), air, and a dispersion of ice crystals in a continuous "serum" of dissolved sugars, milk proteins, and stabilizers. Gums are important in preventing crystallization of lactose sugar and of water as ice crystals.*

For example, the microstructural appearance of a food fat can be viewed by polarized light microscopy, which shows the fat crystal microstructure as a three-dimensional network of solid and liquid components. Within the network are solid fat components of varying sizes. Smallest are the crystallites, while largest are the crystallite clusters of melted fat that have been subsequently cooled. The solid fat crystal components of the network are distributed throughout the liquid oil phase.

Texture/structure relationships are important considerations during food processing because the natural microstructure of foods can be altered, giving rise to new arrangements of molecules (as a "processed microstructure" in gel, emulsion, or low-moisture food systems). What can be termed food "destructuring" and food "restructuring" are often goals of food processing in the creation of specific textures. Such approaches contribute to the ultimate structure of a processed food, which in turn is sensed by the consumer as the typical and desired structure for that food product.

Food product engineers are food scientists specially trained to understand the connection between food structure and food properties. This structure-property way of thinking about foods, derived from a materials science approach, leads to an interesting and logical manner in which to perceive foods and their processing. This includes identifying macroscopic-microscopic structural levels of foods, and the process engineering (unit) operation that can be applied.

For example, consider the microstructure of ice cream—see Figure 13.9. Ice cream represents a highly complex food system. In terms of phases and systems, it is not only a foam (G/L), but a colloidal dispersion (S/L) and an emulsion (L/L). Lactose molecules and casein molecules are colloidally dispersed within an aqueous matrix containing frozen ice crystals, while air cells incorporated into the fat phase during mixing are surrounded by emulsified fat droplets. Although individual molecules of casein and lactose are too small for detection, the microstructural elements of ice cream that can be seen with a microscope include the air cells, fat, and ice crystals.

13.5 PSYCHROMETRICS—LOOKING AT AIR AND FOOD PROCESSING

The field of **psychrometrics** is concerned with evaluating the thermodynamic properties of the mixed gases present in air. At first glance, this may seem unrelated to the study of food science, but we shall see it has definite relevance. Air is a mixture of nitrogen, oxygen, carbon dioxide, and water vapor. At atmospheric pressure, what are known as "ideal gas relationships" are in effect. To describe these relationships, scientists have developed the "ideal gas law," which relates the pressure, volume, temperature, and amount of any gas under set conditions. This relationship is given by the equation:

$$PV = nRT$$

where P is pressure, V is the volume occupied by the gas, n is the amount of gas (in number of moles), R is something called the gas constant (0.0821 liter · atm/mole · °K) and T is absolute temperature in degrees K (Kelvin). The ideal gas law equation can be rewritten for a mixture of several gases—for instance, air. In the expression below, the concentrations of the various gases in air (carbon dioxide, nitrogen, oxygen) are summed as $n_1 + n_2$, etc.

$$P = (n_1 + n_2 + \cdots)\frac{RT}{V}$$

At this point, the focus on air will shift from the gas equation relationships to the use of psychrometric charts. But before discussing the actual properties expressed on the psychrometric chart, we define the concepts of sensible and latent heat.

Latent and Sensible Heating and Cooling

Recall that a change of phase occurs when foods change physical state (e.g., liquid to a solid in freezing). Consider a sample of ice at −30°C, exposed to heat. When plotting temperature versus heat input on a graph, five distinct effects would be observed—see Figure 13.10. In three regions of the plot, the addition of heat energy increases the temperature of the sample. This increase could be felt to the touch and is called *sensible heating*. However, in two other regions of the plot, even though heat was being added, a change in state rather than an increase in temperature resulted. The heat input that results in a change of state rather than a change in temperature is called *latent heat*.

Sensible heat is the heat energy in an airstream due to the temperature of the air. An airstream of 80°F/50% RH (relative humidity) has more sensible heat than an airstream of 70°F/50% RH. **Latent heat** is the heat energy in an air stream due to the moisture of the air. An airstream of 80°F/60% RH has more latent heat than an airstream of 80°F/50%.

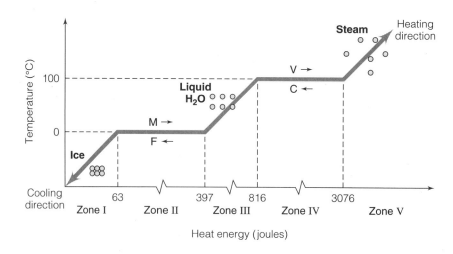

FIGURE 13.10 Latent and sensible heating and cooling curves for water. *Zones I, III, and V represent sensible heating portions of the curve, while zones II and IV represent latent heating portions of the curve. M = melting, F = freezing, V = vaporization, and C = condensation.*

Psychrometric Charts

Early in the twentieth century a German engineer Richard Mollier invented a graphic method of displaying the properties of the various mixtures of air and water vapor. This device is known as the Mollier diagram or psychrometric chart. In the United States the invention of the chart is assigned to Willis Carrier. It is a useful way of simplifying sensible (dry heat) and latent (wet heat) load calculations in systems. The chart describes all the

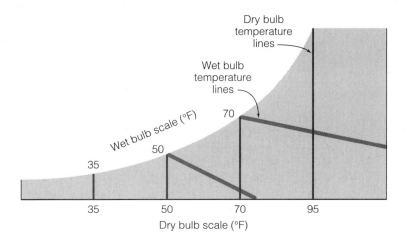

FIGURE 13.11 Simplified dry and wet bulb temperature psychrometric chart showing respective lines and scales.

possible combinations of temperature, moisture content, density, and heat content properties of air.

A **psychrometric chart** is a graph that represents the eight physical properties of air. In spite of its complicated appearance, the psychrometric chart is quite simple to use. By using the physical relationships between the eight properties of air, the psychrometric chart provides a graphical approach to problem solving. The usefulness of the psychometric chart is that given any two properties of air, the remaining properties can be determined graphically. A *psychrometer* is a device that measures dry bulb and wet bulb temperatures. These temperatures represent two of the eight key properties of air:

1. **Dry bulb temperature** (db) is the temperature (degrees F) that we measure with a standard thermometer that has no water on its surface. When people refer to air temperature, they are commonly referring to the dry bulb temperature. In Figure 13.11, dry bulb temperature is represented as vertical lines on the chart at the bottom, increasing from left to right.

2. **Wet bulb temperature** is the temperature of air in degrees F measured with a standard thermometer whose bulb is kept wet by covering it with a moistened cloth called a sock, which surrounds and cools the bulb. This cooling effect occurs because the air moving past the sock absorbs part of its moisture, causing a drop in temperature. The air before it passes across the sock has a certain moisture content, which influences cooling. The drier the air, the greater the rate of evaporation of water from the sock, and the greater the cooling effect upon the thermometer bulb. The only time that dry bulb and wet bulb temperatures will be the same is at saturation—100% relative humidity. Wet bulb temperature lines slant from the lower right toward the upper left corner of the chart in Figure 13.11.

3. **Relative humidity** (RH) is the ratio of the amount of water present in air to the maximum amount of water the air could contain if fully saturated with water. Therefore 0% RH represents fully dry air, while 100% RH represents fully saturated air. The amount of moisture the air can hold increases as the dry bulb temperature of the air increases. So if the weather forecaster refers to relative humidity, he or she should define the dry bulb temperature of the air referred to. In Figure 13.12, relative humidity lines curve from the lower lefthand portion of the chart up and to the right.

4. **Moisture content** (also called *specific humidity*) is a measure of the actual amount (mass) of water vapor present in the air. Represented as horizontal lines on a complete chart, it is normally given in "grains" of moisture per pound of dry air, with 7,000 grains being equal to one pound. Its scale is the *y*-axis.

5. **Dew point** is the temperature at which, as air is cooled, its moisture starts to form a visible condensate. Dew point (Dpt) corresponds to the dry bulb (or wet bulb) temperature at which air becomes completely saturated. If moist air is

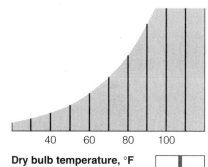

Dry bulb temperature, °F
Vertical line scale located at the base of the chart

Vertical

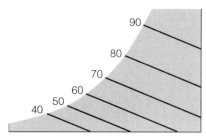

Wet bulb temperature, °F
Sloping line scales located along the curved upper left portion of the chart

Sloping

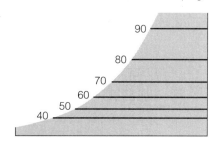

Dew point temperature, °F
Horizontal lines located along the curved upper left portion of the chart

Horizontal

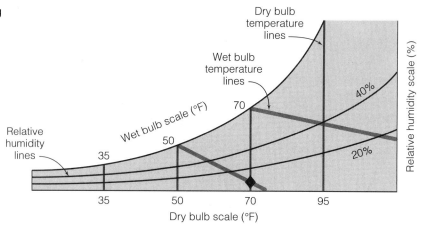

FIGURE 13.12 Psychrometric chart showing respective lines and scales for dry and wet bulb temperature plus relative humidity (state point given as ◆).

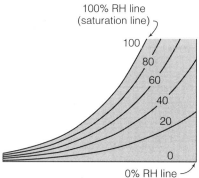

RH lines
These lines sweep from lower left to upper right of the chart

Upward sweep

cooled, it cannot hold the same amount of moisture. At some point, the moisture will condense out of the air onto any nearby surface. Dew point is reached as the dry bulb temperature is reduced and the specific humidity is held constant. Dew point is read along the saturation line (100% RH) on the psychrometric chart.

6. **Total heat content** (enthalpy) is the total heat energy in the air, expressed as Btu per pound of dry air. Three components contribute to this heat content: sensible heat from air temperature, latent heat required to convert liquid water to water vapor, and the heat energy of the water vapor. Enthalpy (h) is a useful but somewhat confusing concept. When air is hot, the enthalpy is high. Enthalpy is also high when air is moist. This is because it takes heat to evaporate moisture into the air. The more moisture in the air, the more heat is necessary to heat it and evaporate it. Conversely, more energy is required to reduce the amount of moisture in the air and cool it. Every pound of water requires approximately 1061 Btu for evaporation. Enthalpy can be thought of as the amount of useable heat (sensible and latent) in the air, expressed in Btu/lb of dry air. Enthalpy lines coincide with wet bulb temperature lines but have a separate linear scale to the left of the wet bulb scale on a full psychrometric chart (see the websites mentioned in the text and at the end of the chapter).

7. **Specific volume** refers to the space occupied by a pound of dry air at a standard atmospheric pressure (14.7 pounds per square inch, or psi), and expressed in cubic feet per pound of dry air (ft³/lb air). As air is heated, with the moisture content held constant, it occupies more space (volume). Noted in cubic feet per pound, specific volume lines slant from the lower righthand corner of the chart toward the upper left corner like wet bulb temperature lines but are more vertical.

8. **Vapor pressure**, in kilopascals (kPa), millibars (mb), or pounds per square inch (psi), can also be determined using the psychometric chart. Each water molecule in the air exerts pressure on the surrounding environment. The amount of vapor pressure at a certain moisture content is the sum of the pressures of all the water molecules. The vapor pressure scale appears to the left of the y-axis of the psychrometric chart.

Several different types of psychrometric charts are used, depending on the requirement. Charts are calibrated for what are called low, normal, or high temperature, and charts that are set up to read dry bulb temperature, wet bulb temperature, relative humidity, moisture, and so on. The key is that a comprehensive psychrometric chart would contain lines and scales for each property of interest.

Access *www.ianr.unl.edu/pubs/generalag/g626.htm* from the University of Nebraska at Lincoln, entitled "Air Properties—Temperature and Relative Humidity." Click on the provided charts and look at the example questions and answers. These will provide you with practice on using psychrometric charts.

To recap, psychrometric charts contain scales and lines for the air parameters of interest. If two independent parameters are known, other psychrometric parameters can be calculated using the chart. To use the chart properly, we must be aware of the difference between a line and a scale. The sample charts in Figures 13.11 and 13.12 contain scales numbering 35, 50, and so on, plus vertical dry bulb lines and somewhat diagonal wet bulb lines.

Locate the dry bulb temperature scale on the bottom of Figure 13.12. Move vertically up the dry bulb temperature line for the 70° dry bulb temperature. Find the wet bulb temperature scale in the curved portion and follow the 50° wet bulb temperature across and down to the right. It intersects the 70° dry bulb temperature at a specific point called the state point. The **state point** is the junction between any two air property lines.

The state point is useful in a more completely drawn psychrometric chart that shows, in addition to dry and wet bulb temperature lines and scales, the relative humidity lines and scales. In such a chart, the position of the state point would be compared to the relative humidity lines and scales, and an estimate of the relative humidity of air having the given dry and wet bulb temperatures could be obtained using the chart. Psychrometric processes can be represented by the movement of the state point on the psychrometric chart, and psychrometry is useful in determining air properties as applied to these common processes:

- Sensible cooling / sensible heating
- Cooling and dehumidification / heating and humidification
- Humidification / dehumidification
- Evaporative cooling / chemical dehydration

An even more complex chart which adds moisture, specific volume, dew point, and total heat lines and scales would enable the user to determine the values of the physical properties of air that is heated to a higher dry bulb temperature. For a closer look at psychrometric charts, the reader is referred to: virtual.clemson.edu/groups/psapublishing/Pages/FYD/HL237.pdf

Food Dehydration and Psychrometry

What does psychrometry have to do with foods? The answer has to do with the application of psychrometric charts to food processing. Knowledge of the eight properties of air can aid the design of equipment that regulates the storage temperature and humidity required by fresh agricultural products, such as cereal grains and produce.

Another use for the psychrometric chart is in dehydration processing. Psychrometrics studies the thermodynamic properties of gas-vapor mixes, most commonly, air-water vapor mixes. Understanding the psychrometric characteristics of such mixtures is important in the design and analysis of food processing systems such as a cereal grain dryer (Figure 13.13). When heated air is directed past moist grain or granular food products in a

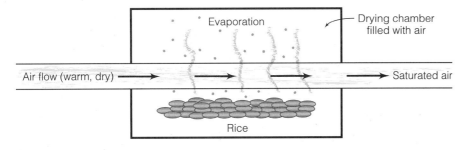

FIGURE 13.13 Adiabatic saturation in a cereal grain dryer. *Heated air forced to flow past moist rice grain is an example of a drying process as well as an adiabatic saturation process. Air-water vapor mixtures are part of the processing of fresh grains into dried product. In a well-insulated drying chamber (hence no loss nor gain of heat to the surroundings) the entering air causes evaporation of water vapor from the food product. In the process, part of the (sensible) heat of the entering air is converted to latent heat, and adiabatic saturation of the air results.*

drying apparatus, the drying process can be represented on a psychrometric chart. This situation is called an **adiabatic saturation process:** Heated air flows past a moist food product in a drying operation, with the air supplying the heat of evaporation to dry the product, with no change in enthalpy. Enthalpy refers to the internal energy content of the air, which varies with temperature and moisture (for example, at 0°C, the enthalpy of steam is zero).

As dry, warm air passes through the moist product (which contains water in the liquid state), the moisture moves from the product and is captured by the air. Sensible air heat is then converted to latent heat, with the uptake of water in the vapor state. The air exits the drying tunnel or chamber in a saturated condition. Although this is an adiabatic process and no heat is exchanged, the dry bulb temperature decreases during this process, while the enthalpy remains constant.

The amount of water (moisture) removed from the food product per kg of dry air can be calculated by using a psychrometric chart and knowing the temperature and relative humidity of the heated air. Such a chart plots dry bulb temperature (°C) versus humidity (kg water per kg dry air). Psychrometric charts can be useful in solving problems related to increasing the efficiency or energy conservation of food processing equipment such as food dryers.

13.6 RHEOLOGY—STUDYING FLOW AND DEFORMATION

Before we get to rheology, consider this statement: Foods, in the material sense, tend to display properties of fluids, or properties of solids, or properties of both (which could be called semisolid foods). Another term used to describe semisolids is "structured fluid." Suspensions, emulsions, gels, and foams are all types of structured fluids.

A **structured fluid** is a low modulus material having solid-like and liquid-like properties simultaneously due to its three-dimensional inner structure. A low modulus material has low stress to strain ratio. To understand what this means, consider a food gel like a wheat starch gel. A special test apparatus designed to twist a freestanding preformed cylinder shape of the gel will eventually cause the gel to rupture. The force needed to cause the rupture is called the *stress*, while the distance the gel will bend (move) before it actually breaks (fails in terms of its gel structure) is the *strain*. This mechanical form of stress is correctly termed a *shear stress*, and the corresponding strain, a *shear strain*.

A low stress to strain gel behaves in this way: A small force causes rupture after much movement by the gel. This is the situation in a *rubbery* gel. By comparison, a high stress to strain gel would move only a small distance before rupturing due to experiencing a large force. As an example of this, *brittle* gels exhibit high stress and low strain. What other stress and strain combinations and gel types are there? There are two more to consider. A *mushy* gel material would exhibit low stress and low strain, while a very strong, *tough* gel would demonstrate both high stress and strain at the point of structural failure.

Although this simplifies reality because a particular food can exhibit behaviors related to each category, often one type predominates. This approach reduces nicely: If a food is mainly fluid in nature, the parameter of interest is its *viscosity*. Viscosity can be thought of as friction within a fluid that prevents it from flowing freely. On the other hand, for solid foods their texture and elastic or plastic behavior are of prime importance. **Texture** represents both a physical property and a mechanical behavior of food that takes into account a food's size and shape dimensions, its density, and whether it has a porous or nonporous surface. Food texture can be measured by sensory or physical (rheological) methods. We will look at elastic and plastic solids shortly.

Rheology is related to **fluid mechanics,** a branch of mechanics that, for the food engineer, studies the flow characteristics of liquid and semisolid foods. What is the difference between rheology and viscosity? **Viscosity** is the principal parameter that characterizes the flow properties of foods, whereas **rheology** is the science that is concerned with the flow and deformation characteristics of food materials. Thus viscosity, or viscous flow, is a type of rheological measurement, whereas rheology itself is the area of scientific study involved in understanding such a measurement.

The study of rheology is relevant because (1) it can provide information about a food's structure, (2) it can aid in the design of food processing equipment, (3) it can provide vital information related to a food's shelf life, (4) rheological data can be correlated to sensory data, and (5) it can help in product development to fine tune foods to consumer's liking.

Mechanical and Rheological Measurements

Mechanics is the study of the *motion* of objects. Food engineers study the effects of external forces on foods, which cause them to move or to deform (become compressed, expand, bend, or break). The mechanical properties of solid foods form the basis of food *texture*, and influence the ways in which these foods are handled and processed. In developing a crispy snack cracker, a food scientist measures the amount of force that is required to snap the cracker in half. If too great a force is needed to break the cracker, or if it merely bends, then it probably will not be a hit with the consumer. The formulation may need to be reworked, or the processing of the cracker modified.

Rheological measurements include measures of viscosity, fluidity, elasticity, and plasticity of nonsolids. To measure these, we must talk about stresses, because the way a food responds to a stress puts it into either the fluid, elastic, or plastic category. How and why food materials deform and flow when they experience conditions of stress and strain is an area of active research in food engineering.

Stresses, Elastic Solids, and Plastic Solids

A **stress** (consider pressure as a stress, or mechanical action as a stress) can be considered as force per unit area of food material. Stress is distinct from **strain,** which is a result of stress, and may take the form of deformation in the case of a solid material. Stresses cause a variety of interesting and different effects, depending upon the direction of the force, whether a food is primarily a liquid or a solid material, and the food's viscosity component.

Fluid foods show continuous deformation at a flow rate that is proportional to the amount of stress applied to them. Fluids do not recover after a stress is removed. Instead, fluids respond to stresses by changing shape, which means movement or flow away from the stress. Solids respond to stress by exhibiting strain, deformation, or rupture. Figure 13.14 shows how plastic, elastic, and fluid foods respond to stress forces.

It is conventional to examine shear stress-strain relationships when dealing with solids. **Shear** refers to the relative motion of one surface with respect to another surface in parallel to it, which creates a zone of shearing action on any substance (fluid or solid) located between the moving surfaces. When a shear stress is applied to a food material and its response is a certain amount of deformation (the strain) away from the stress, but

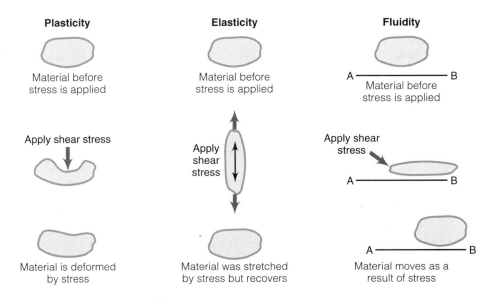

FIGURE 13.14 The response of plastic, fluid, and elastic food materials to shear stress forces.

without true flow movement, then we are dealing with an **elastic solid.** Such food materials will return to their original shape when the deforming stress is removed (Figure 13.14).

Elasticity is equated to "springiness," a term used to calculate the distance that a food recovers after being initially compressed. Bread dough exhibits such elastic behavior, although it really has a more complex response and is considered a viscoelastic food material. Let's examine this interesting property a bit further. **Viscoelastic** food behavior means that some foods are composed of both viscous and elastic materials, and in addition, they exhibit both flow and deformation. Bread dough is a good example of a viscoelastic food material. Recall that the key wheat gluten proteins, gliadin and glutenin, are responsible for dough viscosity and elasticity, respectively. If stretched, bread dough will spring back to its original shape. It exhibits a resistance to extension, which is termed **extensibility.** Extensibility can be measured as the distance extended in millimeters versus the force applied in newtons. If the force applied is sufficiently great, the extended dough will break or rupture. At the point of rupture, there is no longer any resistance to extension.

A **plastic solid** behaves differently when under the influence of a shear stress. Plastic solids show continuous deformation at a rate that is proportional to the amount of stress applied. Although there is some recovery after the stress is removed, there is not the level of return to original shape that elastic foods display (Figure 13.14). A material that exhibits plastic flow resists change in position until a minimum force is applied, at which point it can be made to flow. Cake frosting has this desirable characteristic, which as a performance criterion is termed spreadability. Certain soft cheeses are also examples of plastic foods.

Due to their wide diversity, food materials behave differently under various conditions. To understand food texture and the relationships of stress-strain and deformation to food texture and food quality, it helps to have standards or models for comparative purposes. Scientists have identified several representative models for study. We will not examine them in detail but list them with hints about their utility in parentheses:

- Bingham model (representative of an ideal plastic)
- Hooke solid (representative of an ideal solid)
- Kelvin-Voigt model (representative of a viscoelastic solid)
- Maxwell model (representative of a viscoelastic liquid)
- Newtonian fluid (representative of an ideal liquid).

Newtonian and Non-Newtonian Foods

Sir Isaac Newton first defined viscosity as "the tendency of a fluid to resist any internal motion." Materials evaluated for viscosity are classified today as either Newtonian or non-Newtonian, depending on their flow behavior. **Newtonian foods** are homogeneous mixtures that exhibit no change in viscosity as the rate of shear (applied mechanical force) is increased. **Non-Newtonian foods** are heterogeneous mixtures that exhibit a change in viscosity as the rate of shear (applied mechanical force) is increased.

Table 13.4 gives examples and the characteristics of Newtonian and non-Newtonian foods, including **shear thickening** and **shear thinning** examples. These can include fluids such as thickened solutions, concentrates, gels, purees, and other fluid food systems. The size of the dispersed molecules influences viscosity behavior in response to shear. A Newtonian fluid is one in which the viscosity is independent of the shear rate, alternately expressed that the ratio of shear rate (flow) to shear stress (force) is constant. A Newtonian fluid represents an "ideal" fluid whose viscosity and flow characteristics are readily predictable. Behavior of non-Newtonian fluids is not as simple to predict.

For solutions of small molecules such as sugars, the viscosity of the solution is independent of the rate at which the solution flows or is stirred (shear rate). The same viscosity value would be obtained whether the sugar solution was subjected to high or low shear rates. This is "ideal" or Newtonian viscosity behavior. For larger molecules such as poly-

TABLE 13.4 **Examples and characteristics of Newtonian and non-Newtonian foods.**

Viscosity Type	Characteristics and Food Examples
Newtonian	Viscosity is not affected by changes in shear rate and remains constant. Examples: water, fruit juice, carbonated beverages, milk, honey, vegetable oils.
Non-Newtonian Bingham plastic	A certain shear stress (called the yield point) is required to initiate flow; once flow starts, shear rate has no effect. Example: tomato paste.
Non-Newtonian plastic	A yield point must be achieved to cause product flow, at which point pseudoplastic, behavior is observed. Examples: mayonnaise, margarine.
Non-Newtonian pseudoplastic	"Shear-thinning" fluids decrease in viscosity as shear rate increases. Examples: applesauce, banana puree, orange juice concentrate, guar and xanthan gum thickened products.
Non-Newtonian dilatant	"Shear-thickening" fluids increase in viscosity as shear rate increases. Examples: corn starch suspensions, chocolate syrups.

mers (polysaccharides and polypeptides), viscosity can depend on shear rate. This is important in deciding whether a product should be processed in a way that subjects it to a high (example: valve homogenization) or low shear rate. On a more personal level, shear viscosity behavior influences our sensory response to a food as it is subjected to chewing (a form of shear force).

Shear thinning fluids are said to exhibit *pseudoplastic* behavior. What this means is that the viscosity of the fluid decreases (becomes thinner) as shear rate increases. Think of thick yogurt, for example. When you are not stirring it is hard to pour, but when you stir it, it seems relatively thin. The extent of shear thinning depends on the amount of polymer material. At moderate concentrations, the material is thought to stretch out in the direction of flow, and thereby not impede it. At higher levels, polymers are thought to become entangled but also to become disrupted by the shear effect. The latter process overtakes the former, and the net effect is a decrease in viscosity.

Shear thickening fluids are said to exhibit *dilatant* behavior. What this means is that the viscosity of the fluid increases (becomes thicker) as shear rate increases. The curve of the plot of shear stress versus shear rate is nonlinear, with shear stress increasing faster than the shear rate (Figure 13.15). The thickening effect is relatively rare in food systems, and is due to entanglement of polymers. One example is highly concentrated starch suspensions. If you put some dry corn flour or custard powder in a bowl and add just enough water to pour it out into another bowl you'll notice that it becomes more viscous as you mix it. This is an example of a shear thickening, or dilatant, fluid behavior.

Additives can alter a food's rheology. For example, adding a food gum (hydrocolloid) to a Newtonian food changes its rheological properties and can convert it to a non-Newtonian food. This is why food gums are used as thickening agents.

Viscosity in Food and in Processing

Consider ketchup (catsup); everyone knows the consequences of a low-quality batch of ketchup—it either runs out of the bottle all over your burger and fries, or it refuses to budge. What influences ketchup behavior? viscosity. What determines its viscosity?

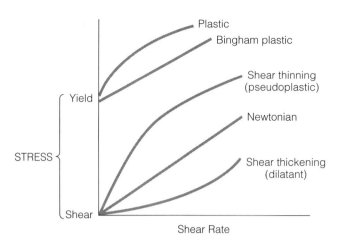

FIGURE 13.15 The relationship between shear stress and shear rate for typical time-independent fluids. *Newtonian foods are considered the simplest because shear stress is directly proportional (straight linear relationship) to shear rate. Among the non-Newtonian fluids, you can see that in the shear thickening type, shear stress increases faster than shear rate. Both Newtonian and non-Newtonian fluids are time-independent materials, meaning they respond by flowing immediately upon action by a stress. Yield stress is the minimum amount of stress that produces flow.*

- Formulation (tomato with its pectin content)
- Processing conditions (pumping/ shear rate effects)
- Package or container characteristics (wide mouth bottle vs. narrow mouth influence the ease in which the product will begin flowing)

In general, the factors that affect any food's viscosity include physical or chemical changes, pressure, temperature, sample homogeneity, and sample agitation (turbulence). Devices used to measure food viscosity are called viscometers.

Food engineers examine shear-stress–shear-rate relationships when dealing with fluids. The **absolute viscosity** (η) can be calculated in terms of shear stress, the force acting in the plane of a fluid, and shear rate, the velocity experienced by the fluid between moving plates:

$$\eta = \frac{\text{shear stress }(\tau)}{\text{shear rate }(D)}$$

By definition, absolute viscosity measurements are not under the influence of gravity during the measurement determination. The units of viscosity are typically mPa · sec (milli-Pascal seconds), or poise (one P equals 100 cP, or centipoise).

Absolute viscosity is different from apparent viscosity. **Apparent viscosity** refers to the viscosity of non-Newtonian food materials of both shear-thinning and shear-thickening types. The former decrease in apparent viscosity as shear rate increases, while the latter increase. Apparent viscosity is measured at a single shear rate.

Figure 13.15 shows the relationships between shear stress and shear rate for typical time-independent fluids.

The viscosity of food products is critical to many food processing operations, including mixing, pumping, filling, and forming. These unit operations are mechanical in nature, utilizing special food processing equipment. A result of these operations is to produce shear, which influences food product viscosity. Up until this moment the discussion has described time-independent flow behavior. In some, but not all cases, a fluid food's viscosity as a result of processing does not reach a constant value until the material is allowed to rest for a period of time following shear action. This indicates a time dependency.

Viscosity can decrease over time when a constant shear rate is applied—as in *pseudoplastic* behavior described in Table 13.4. **Thixotropic** behavior is of this type, except that it is time-dependent. Only after the passage of time from the cessation of shear action will the viscosity reach an unchanging value. Starch pastes are an example of thixotropic fluid foods. **Rheopectic** behavior refers to time-dependent shear thickening, which however is rarely seen in food materials.

One last point—viscosity is not related to surface tension. Viscosity is essentially a frictional force between different layers of fluid as they travel past one another. As such it represents a characteristic of fluids or liquids beneath their surfaces. The surface of a liquid can be thought to act like a stretched membrane under tension. This tension, acting parallel to the surface, exists due to the attractive forces between like (liquid) molecules. The effect is called **surface tension,** defined as the force per unit length that acts across any surface. Surface tension is a characteristic of emulsion systems.

In homogenized milk, the homogenization process alters the milk viscosity and the surface tension of the milk emulsion. There is a slight increase in viscosity compared to unhomogenized milk due to the increased volume of suspended matter and increased adsorption of casein to fat globule surfaces (emulsified fat). Homogenized milk has a slightly greater surface tension too since there are fewer surface active proteins present. (Why? homogenization affects milk proteins by pumping milk through small orifices at temperatures above 38°C, 100°F.) As a result some of the proteins experience structural changes that prevent them from functioning as emulsifiers.

13.7 EXTRUSION TECHNOLOGY IN THE FOOD INDUSTRY

Extrusion and Extruded Products

Extrusion devices, or extruders, once solely in use in the plastics industry, have become important in food production. In the early 1950s, extruders were developed to mix and melt plastic polymers. It took about twenty years for the technology to be applied to foods. Extruded food products range from ready-to-eat (RTE) breakfast cereals and snack foods to pet foods.

Extrusion processing is even useful to food ingredient suppliers. For instance, functional starch ingredients can be created by pregelatinizing the starch within the extruder through heating and steam mixing. The combination of heat, pressure, and shear conditions within the extruder can break the starch (amylose) molecules down into small fragments called short-chain starches. The significance of these physically rather than chemically modified starches is that they offer particular functional properties, such as increased solubility and dispersibility, which are applied to food product development.

How an Extruder Works

An **extruder** is a device that uses rotating screw technology within a tunnel (barrel) that has the ability to cook and shape certain foods much more quickly and efficiently than by conventional means. It accomplishes this through mechanical friction and shearing action as the screw churns and plasticizes food material, as well as by heat transfer and pressure increase generated by steam and temperature effects within the barrel. As such it offers HTST (high-temperature short-time) processing.

This technology boasts comparatively low cost, plus the ability to produce shaped foods of safe microbiological quality, with high nutrient retention. We need to remember that food microstructure and extrusion are related. The goal of thermal extrusion is to produce a desirable product microstructure, which results in a fabricated food with a texture that is highly acceptable to the consumer.

Extruders can be of several types: (1) single screw extruders, which are widely used to produce cereals, pasta, and pet foods; (2) twin (co-rotating) screw extruders, which are popular with snack food manufacturers and works better with viscous or sticky foods than does the single screw; (3) single reciprocating screw extruders, which unlike the others create only minor elevations of temperature and pressure due to screw and mixing action.

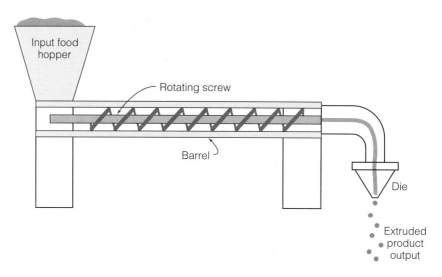

FIGURE 13.16 Simplified outline of an extruder device.

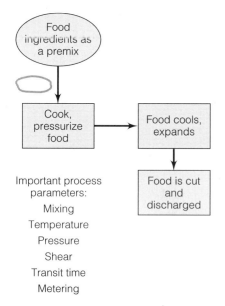

FIGURE 13.17 Steps to produce an extruded product.

The general idea of extrusion cooking is to force food material along a barrel under specific conditions of temperature, moisture, and pressure, so that the material expands and sets up a particular microstructure as it cooks (Figure 13.16).

A typical single screw extruder can be viewed as having three sections: (1) the input end where unprocessed food material is fed into the unit; (2) the transition section, in which product cooking and cooling are accomplished; and (3) the output end, which contains a die to create the final size and shape of the finished product and expel it from the extruder—see Figure 13.17. Common to each section is the screw rotating within the barrel length, which propels the food mixture along from the input to the output stage.

Suppose a manufacturer wants to produce a cookie or cracker, using flour as the basic ingredient. Figure 13.17 shows the basic steps. First the product flour premix is fed into the extruder. A dough results when water is added and mixed in. The water level in the product plasticizes the dough so that it can be mixed, shaped, and propelled along. Inside the extruder, the dough is rapidly cooked to achieve a certain moisture level and texture. The product moves to the output stage of the extruder, is expelled, and may be transferred to an external toaster oven to reduce the moisture content and provide additional color and flavor. Upon cooling, extruded products are packaged and shipped for distribution.

Key Points

- Heat energy applied to foods causes them to increase in molecular motion and temperature.
- Heat transfer is accomplished in foods via conductive, convective, or radiation modes of heat transfer.
- Mass transfer, characterized by diffusion and bulk motion, occurs in frying due to migration of liquid frying oil.
- Materials science principles are applied to the study of foods through a polymer science approach.
- Categorizing food systems according to four physicochemical states of matter helps explain water mobility and specific phase transition phenomena that alter quality in foods.
- The organization of food structures at the microscopic level and their interactions helps in the understanding of food textural properties and sensory quality.
- Psychrometric charts depict the relationships between eight physical properties of air and are applied to cooling, dehumidification, evaporative, and heating processes.
- The psychrometric properties of air and water vapor mixtures are important in the design and analysis of food processing equipment.
- Rheological parameters of fluid foods such as viscosity that contribute to consistency are influenced not only by formulation but also mechanical forces that create stress and strain relationships.
- Extrusion technology has been adopted by the food industry to produce foods such as ready-to-eat breakfast cereals and snack foods.

Study Questions

Multiple Choice

1. Which statement is *true*?
   a. Glass transition (T_g) refers to the time required to change between solid and liquid states.
   b. T_g' temperature range can help predict food storage life and stability of foods.
   c. T_g' values of food materials range from very high (water) to low (starch).
   d. Glass transition temperature of hydrophilic components is independent of water content.
   e. T_g' values for food ingredients do not vary as a function of concentration and time.

2. Which mechanisms are responsible for heat transfer in foods?
   a. irradiation, high pressure, canning
   b. conduction, convection, irradiation
   c. grinding, melting, evaporation
   d. conching, tempering, pressing
   e. radiation, conduction, convection

3. What is *true* regarding boundary layers in frying foods?
   a. Movement of heat boundary layers partition (separate) the core from crust regions.
   b. Heat transfer, but not mass transfer, occurs.
   c. Frying destroys boundary layers.
   d. They only form upon cooling of fried foods.
   e. Mass transfer, but not heat transfer, occurs.

4. A hot soup placed in storage at 5° C in the steady state
   a. will cool down to about 41°F.
   b. will cool down to about 15°F.
   c. will not achieve thermal equilibrium.
   d. transitions from liquid to crystalline.
   e. transitions from plastic to elastic.

5. Which thermal events occur during the frying of a potato slice in hot oil?
   a. gelatinization of protein
   b. color loss of oil
   c. denaturation of protein
   d. gelatinization of oil
   e. denaturation of oil

6. The movement or migration of frying oil caused by concentration gradients is called
   a. heat transfer.
   b. heat exchange.
   c. food microstructure.
   d. the second law of thermodynamics.
   e. mass transfer.

7. Palm oil is a plant oil containing saturated, monounsaturated, and polyunsaturated fatty acids. What is true about its microstructure?
   a. consists of a three-dimensional network
   b. consists of liquid fat crystals dissolved in solid oil
   c. it is related to the oil's macrostructure
   d. a and b
   e. a and c

8. In the manufacture of tomato ketchup, which unit operation does *not* correspond to the microstructural level?
   a. mixing
   b. homogenizing
   c. cleaning
   d. heating
   e. none of these

9. The key microstructural elements in ice cream are
   a. air, water, and fat.
   b. protein, sugar, and air.
   c. ice, sugar, egg, pectin, and lactose.
   d. casein, starch, sucrose, albumen, and cocoa butter.
   e. lactose, casein, ice, air, and fat.

10. A food scientist studying the effects of formulation and processing on the problem of ketchup viscosity is working in the area of
    a. thermodynamics.
    b. rheology.
    c. electromagnetics.
    d. specific heat.
    e. heat of fusion.

11. An _____ is a device that uses rotating screw technology within a barrel and has the ability to cook and shape foods.
    a. intruder
    b. extruder
    c. grinder
    d. concher
    e. extrusion

12. _____ show continuous deformation at a rate proportional to the amount of stress applied.
    a. Viscoelastic foods
    b. Structured fluids
    c. Adiabatic substances
    d. Plastic solids
    e. Liquids

13. Which statement is *false*?
    a. The glassy state is an amorphous state.
    b. The crystalline state is an amorphous state.
    c. The rubbery state is an amorphous state.
    d. Transitions from rubbery to glassy affect food quality.
    e. Generally, it is undesirable to maintain a food in the rubbery state during extended storage.

14. Peanut butter and process cheese are examples of structured fluids; which statement below is *false*?
    a. These products can be considered "soft solids."
    b. These products possess both hydrophilic and hydrophobic components dispersed uniformly in separate liquid phases.
    c. These products are unstable food systems.
    d. These products retain properties of both liquids and solids.
    e. These products are examples of food emulsions.

15. Which of the following statements is *incorrect*?
    a. Relative humidity is the ratio of the amount of water present in air to the maximum amount in fully saturated air.
    b. When air is hot, the enthalpy is high.
    c. Specific volume is the space occupied by a pound of dry air at standard atmospheric pressure.
    d. When air is moist, the enthalpy is high.
    e. It is not possible to determine vapor pressure using the psychometric chart.

16. The formula to calculate food viscosity is $\mu = \dfrac{\pi \, \Delta P \, R^4}{8 \, L \, V}$ where μ = viscosity, ΔP = measured pressure difference, π = 3.414, R = radius of a measuring device (capillary tube viscometer, L = tube length, and V = flow rate in the tube. Consider two fluid foods, A and B. The ΔP value of A is twice that of B. In measuring their viscosities (μ), what makes the most sense?
    a. Viscosity of A is double that of B.
    b. There is no effect of ΔP on μ.
    c. Viscosity of A and B are identical.
    d. The larger the ΔP, the smaller the μ value.
    e. Viscosity of B is double that of A.

17. What is the junction between any two air property lines?
    a. state point
    b. cold point
    c. target point
    d. dew point
    e. air plot value

18. Which is *not* a dielectric food property?
    a. ϵ''
    b. dielectric loss factor
    c. q
    d. K'
    e. ϵ'

19. When a potato slice is placed in hot frying oil,
    a. there will be no change in the color of the potato.
    b. oil uptake into the potato is through mass transfer.
    c. a crust layer develops in the potato that has greater moisture content than the core.
    d. oil uptake into the potato is due to heat transfer.
    e. the potato is the frying medium.

20. A metal shelf in a refrigerated storage area feels colder than the frozen vegetables resting upon it, although both are at thermal equilibrium. Why?
    a. The metal shelf has a lower k value than the vegetables.
    b. The vegetables would be about 5 degrees colder than the metal shelf.
    c. The metal shelf would be about 5 degrees colder than the vegetables.
    d. Metals have high thermal conductivities.
    e. Body heat is conducted away more quickly when touching the food compared to the shelf.

21. Consider two brands of ketchup, A and B. You work for the manufacturer of brand B, and are performing rheological testing of A, which is the current top-selling brand and comparing it to your company brand B. Upon analysis of brands A and B using identical test conditions and given the viscosity equation ($\mu = \pi \Delta P R^4 / 8 L V$), the V parameter for brand A is found to be a factor of 10 higher than the V value of brand B. What recommendation should you make to the company management? (μ = viscosity; ΔP = measured pressure difference; π = 3.414; R = radius; L = tube length; V = flow rate).
    a. Product B, the market leader, is more viscous (thicker) than A.
    b. Recommend decreasing the viscosity of B through formulation and/or processing modifications.
    c. Product A is much thicker out of the bottle than your company's ketchup.
    d. Recommend increasing the viscosity of B through formulation and/or processing modifications.
    e. Your company's product is too runny (not thick enough).

22. Which of the following would not be useful in determining the % RH if given a dry bulb temperature of 70°F and a wet bulb temperature of 50°F?
    a. determine if a low, normal, or high temperature psychrometric chart is required
    b. locate the wet and dry bulb, and % relative humidity scales and lines
    c. locate the state point
    d. locate the vertical line that represents the 70°F dry bulb temperature
    e. locate the vertical line that represents the 50°F wet bulb temperature

23. In designing a transport system for a semifluid food manufactured in a processing facility, such as a pipeline to pump the product from a bulk container into specific processing equipment, which is a product consideration rather than a process consideration that enters into the design?
    a. ambient temperature in the processing facility
    b. the target viscosity of the end product
    c. the force of the pumping process
    d. pumping velocity and pressure within the pipeline
    e. pipeline length and interior diameter dimensions

True/False

24. A "processed microstructure" is always identical to a food's natural microstructure.
25. A food processing example that describes the destruction of the natural microstrucure of foods is the reduction in particle size of wheat endosperm in the manufacture of flour.

References

Aguilera, J. M. 2000. Microstructure and food product engineering. *Food Technology* 54(11):56.

Aguilera, J. M., Stanley, D. W., and Baker, K. W. 2000. New dimensions in microstructure and food products. *Trends in Food Science and Technology* 11(1):3.

Best, D. 1992. New perspectives on water's role in formulation. *Prepared Foods* 161(4):59.

Brody, A. L. 2000. New food packaging polymer and processing techniques. *Food Technology* 54(12):72–74.

Buffler, C. R. 1993. *Microwave Cooking and Processing: Engineering Fundamentals for the Food Scientist*. Van Nostrand Reinhold, New York.

Hegenbart, S. 1995. Beating the big squeeze when creating extruded products. *Food Product Design* 5(11):81.

Kramer, F. 2000. Extruders not just for snack foods. *Food Processing* 61(5):180.

Kuntz, L. A. 1992. In the thick of it: a look at measuring viscosity. *Food Product Design* 2(9):70.

Kuntz, L. A. 1995. Dealing with shear. *Food Product Design* 5(11):69.

Mudgett, R. E. 1982. Electrical properties of foods in microwave processing. *Food Technology* 36(2):109.

Singh, R. P. 2000. Moving boundaries in food engineering. *Food Technology* 54(2):44.

Singh, R. P., and Heldman, D. R. 1993. *Introduction to Food Engineering*. Academic Press, San Diego, CA.

Additional Resources

Aguilera, J. M., and Stanley, D. W. 1990. *Microstructural Principles of Food Engineering and Food Processing*. Elsevier Science Publishers, New York.

Barbosa-Canovas, G. V., Ma, L., and Barletta, M. S. 1997. *Food Engineering Laboratory Manual*. Technomic Publishing, Lancaster, PA.

Brody, A. L. 1996. Integrating aseptic and modified atmosphere packaging to fulfill a vision of tomorrow. *Food Technology* 50(3):56.

Cooksey, K. 2000. Portable packaging. *Food Product Design* 10(6):137.

Forcinio, H. 1997. Special delivery. *Prepared Foods* 166(3):75.

Han, J. H. 2000. Antimicrobial food packaging. *Food Technology* 54(3):56.

Jenkins, W. A., and Harrington, J. P. 1991. *Packaging Foods with Plastics.* Technomic Publishing, Lancaster, PA.

Katz, F. 1997. Aseptic packaging accommodates particulates. *Food Technology* 51(8):132.

Lozano, J. E., Anon, C., Parada-Arias, E., and Barbosa-Canovas, G. 2000. *Trends in Food Engineering.* Technomic Publishing, Lancaster, PA.

Peleg, M. 1983. The semantics of rheology and texture. *Food Technology* 37(11):54.

Rao, M. A., and Hartel, R. W. 1998. *Phase/State Transitions in Foods: Chemical, Structural, and Rheological Changes.* Marcel Dekker, New York.

Roth, L. O., and Field, H. L. 1999. *Introduction to Agricultural Engineering.* Aspen Publishers, Gaithersburg, MD.

Thompson, L. J., Deniston, D. J., and Hoyer, C. W. 1994. Method for evaluating package-related flavors. *Food Technology*, 48(1).90.

Windhab, E. J. 1995. Rheology in food processing. In: *Physico-Chemical Aspects of Food Processing,* S. T. Beckett, ed. Chapman & Hall, London.

Web Sites

www.ift.org/divisions/food_eng/
IFT Food Engineering Division

http://howthingswork.virginia.edu/table_of_contents.html
Louis A. Bloomfield, Professor of Physics, The University of Virginia, has created a fine web site that can help anyone better understand physics.

http://howthingswork.virginia.edu/microwave_ovens.html
On the topic of microwaves, select Microwave Ovens as a topic from the list presented by Dr. Bloomfield at the site above, or go directly to this site.

www.foodengineeringmag.com
Food Engineering. This trade magazine provides new product information, special reports, news, and industry resources.

virtual.clemson.edu/groups/psapublishing/Pages/FYD/HL237.pdf
Clemson University's psychrometric charts

www.aimglobal.org/technologies/barcode/
Bar Code Technology

Challenge!

FOOD PACKAGING

Because foods are perishable commodities, various forms of containment and protection have been in use throughout history. Early examples included materials such as wood, clay, metal, glass, cloth, and even carved out gourds and dried calf stomachs. More recent technology has produced packaging materials specifically designed to meet standards of preservation, safety, and shelf life. To the consumer, the main function of packaging may be its convenience. Consumers may not be aware of the meaning of the expression "barrier properties," terms such as "thermoplastic" and "thermoset," or abbreviations such as LLDPE or PET.

Categories and Function

The primary functions of food packaging include (1) preservation, (2) protection against physical damage, (3) protection against chemicals, dirt, and biological contamination, (4) easing the distribution of foods from manufacturer to vendor and on to the consumer, and (5) to create or enhance visual (sales) appeal and customer recognition of product. Functional package components are referred to as primary, secondary, tertiary, and quaternary.

A **primary package** is the packaging material that comes in direct contact with the food product it is surrounding. For example, if you lift the unopened inner package of a cereal box out from the box, you see that you have two packages. The slightly opaque plastic package containing your cereal is the primary package, while the cardboard box with all of the brand graphics and nutrition information is the **secondary package.** The other two types of package components, tertiary and quaternary, have to do with food shipping and distribution activities. See Figure 13.18.

A corrugated (folded panel) box, or shipping container, serves as the most common form of *tertiary package*. The job of the tertiary container is to hold a group (perhaps a dozen) products together for ease in shipping from the manufacturing facility. Boxing is the last station of the production line, and boxed product is taken to the shipping and receiving area for distribution. However, boxes are not transported one at a time. Instead, they are compiled as a *quaternary package*. Typically, this means that a stack of tertiary boxes containing product is arranged onto a wooden pallet and is bundled with stretch plastic wrap to hold all the boxes together as a single unit during distribution.

Packaging foods has become a very complex enterprise, to the point where the large food companies utilize in-house packaging divisions. Since packaging is of such importance today, some universities offer degrees in package engineering. Although food scientists do not have to be experts in packaging, they may occasionally be called upon to assist with packaging decisions and problems, based upon knowledge of the food system being packaged.

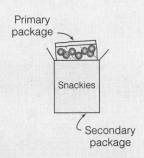

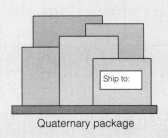

FIGURE 13.18 Primary, secondary, tertiary, and quaternary package components.

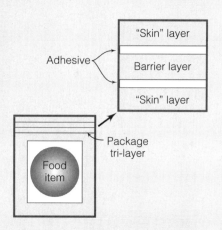

FIGURE 13.19 Diagram of a flexible pouch made of co-extruded tri-layer composite material.
Co-extrusion is the simultaneous combination of several layers of plastic using adhesives to promote bonding between the polymers, which remain discrete despite the extrusion process. The inner and outer layers can be composed of polyethylene, polypropylene, or polystyrene, while the middle layer may be composed of SARAN, the trade name for polyvinylidene chloride, PVDC.

To be effective, food packaging must meet several criteria. We might consider these the 10 Golden Rules for Food Packaging. Food packaging must (1) be nontoxic, (2) act as a barrier to moisture loss or gain, (3) provide a barrier to oxygen, (4) block microbial entry, (5) not allow product leakage or loss, (6) be easy to open and close, (7) be easy to dispose of, (8) not constitute an environmental burden, (9) meet specified size, shape, and weight requirements, and (10) be compatible with the food it will contain. With regard to Rule 10, since foods and their packaging come in direct contact with each other, interactions can occur. Volatile compounds in the packaging should not migrate into the food during heating and storage; migration could create undesirable flavors or odors. Continued research, safety testing, and federal regulations applied to this area keep the situation safe for the consumer.

Container Types

Containers offer not only containment, but also the convenience of portability, sized from single servings up to entire meals. The common primary container types include metal cans, flexible pouches, glass containers, and plastic containers.

Cans. Cans are considered the best overall container for heat-sterilized foods, due to the protection they offer, plus the speed of forming, filling, sealing, and processing. Cans are made of metals such as steel, and to a much lesser extent aluminum; steel cans may be coated with tin on the inside to reduce corrosion. All food cans are **hermetically sealed** by fusion, which means that they are completely sealed so that no gas (such as water vapor or oxygen) can enter or escape.

Flexible Pouches. Often made of composite material, meaning an outer layer of polyester, middle layer of aluminum for eliminating light and setting up a barrier to gas entry, and an inner layer of nylon or polypropylene, flexible pouches offer advantages over solid cans (see Figure 13.19). These include faster and greater heat penetration, no corrosion, easier opening, and light weight. However, there are disadvantages; unlike solid cans, flexible pouches are awkward to stack, provide no structural support for contents, and are somewhat susceptible to cutting and tearing.

Glass Containers. The use of glass bottles and jars permits product visibility, does not affect product taste, and, unlike cans, offers noncorrosive containment. Equally important is the type of closure used to hermetically seal the glass container. These include a cork, a crimped metal crown with cork liner, and an aluminum screw-cap or screw lid.

Microwavable Containers. These include containers made of high gas barrier materials such as polypropylene (abbreviated PP) bonded to an ethylene vinyl alcohol copolymer (abbreviated EVOH). This arrangement withstands microwave energy, is lightweight, unbreakable, easy to open, and able to be closed.

Edible Films. Edible films have been in use in the sausage industry as edible sausage casings. Another food example is the use of wax to coat vegetables and fruits. Food materials such as amylose from starch, and casein from milk protein, when solubilized, can be cast to make sheets of edible films upon drying. These films can also be fabricated into small packets to hold food ingredients. Edible coatings are a subfamily of edible films. The hard candy "shell" that makes a certain plain chocolate, chocolate-covered peanut, or chocolate-covered almond "melt in your mouth and not in your hands" is a famous example of an edible coating. The challenge there was more than to provide a moisture or gas barrier; it was to prevent chocolate melting and provide extended handling and shelf life at elevated temperatures. Edible coatings also serve as glazes to improve product appearance, or base material onto which snack chip or cracker flavors and seasonings adhere. All edible films must be approved by the FDA for human consumption.

Packaging Approaches

Controlled and Modified Atmosphere Packaging.
Controlled atmosphere storage involves controlling the amounts of gases in the food's environment. The shelf life of postharvest fruits and vegetables can be extended in this manner. Deterioration or senescence of fresh fruits and vegetables can be delayed during shipping or storage by depleting the storage atmosphere of O_2 and enriching with CO_2. This control can be accomplished by sealing the fruit or vegetable in an airtight package and allowing it, through normal respiration, to use up the oxygen in the package and to thus create its own modified atmosphere high in CO_2 content. The process wherein a respiring commodity creates its own atmosphere during storage is called *passive modification*. Also, unlike a true modified atmosphere (fixed gas) package, gas mixtures within controlled atmosphere packages (CAP) change during storage due to biochemical processes of living plant tissue.

Modified atmosphere packaging or MAP refers to the exchange of air in a processed food product package with a gas or gas mixture to exclude oxygen from the area within the package not occupied by the food product, known as the *headspace*. A synonym for this procedure is called gas flush packaging, or "active modification." Removal of oxygen and addition of N_2, CO_2, or a ratio of the two gases to the headspace of packaged instant coffee, dry milk, or baked products inhibits oxidation reactions in these products. MAP-packaged fresh-cut produce has become an important consumer item. In addition to its antimicrobial effect, here MAP is effective in inhibiting the normal respiratory activities of packaged fruits and vegetables. Figure 13.20 shows the product and headspace in food packages.

Occasionally, especially in the case of MAP-packaged baked products that are porous, a small packet containing reactive iron particles is added for the purpose of **oxygen scavenging**. Because porous foods are like sponges, with many crevices that can retain oxygen, the scavenger packet eliminates oxygen that resides within the MAP-packaged product. This is possible because over time, the movement of oxygen from within the food to the headspace of the package brings it into contact with the iron, causing a chemical reaction that removes oxygen from the headspace and product. Such a combination of MAP packaging plus oxygen absorber technology reduces oxygen levels to the vanishing point in the package, and greatly extends product shelf life.

Vacuum packaging refers to the removal of air from a packaged food. It has been applied to retail cuts of fresh meat, and is practiced both en route to the retail outlet and for use at the retail level. This type of modified atmosphere package prevents changes in myoglobin, thus enabling a fresh red color to the meat upon opening and exposure to air, and preventing the growth of aerobic spoilage organisms. Such packages may also be flushed with carbon dioxide or nitrogen as a means of maintaining quality.

Aseptic Packaging.
An aseptic process is designed to make a product free from pathogenic or spoilage microorganisms. The rapid heating of processed foods to high temperatures and their subsequent filling into presterilized containers that are hermetically sealed in a sterile environment is what constitutes traditional **aseptic packaging**. In addition to eliminating microbes, aseptically packaged products are notable for their freshness and superior color, texture, and flavor compared to traditional thermally processed foods. Some examples of aseptically packaged food products available to the consumer are juices, puddings, deli products, and soups.

The most widely used consumer aseptic packages are brick or block shaped, and consist of laminates (layers of nonextruded bonded materials) of paperboard, polyolefins, and aluminum foil as a gas and moisture barrier. In addition, bulk packages used to store fresh ingredients have been treated with chlorine dioxide to produce an aseptic package storage system. Holding foods in this manner to maintain freshness until needed is of particular interest to the foodservice industry.

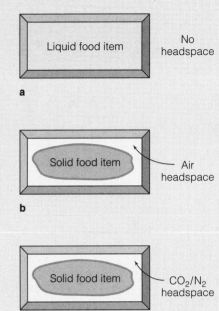

FIGURE 13.20 Inside view of a food package, showing product and headspace. *Food packaging systems may be (a) product and package without any headspace (e.g., packaged liquid foods), or (b) product and package with headspace present (flexible pouches, bottles, and cans). In a modified atmosphere package (c), a processed food is kept in a barrier package surrounded by inert gases.*

Plastics Used as Packaging Materials

Plastics can be thought of as polymers, the same way that proteins and polysaccharides are. In the case of plastics, the repeating monomeric units that are linked together are not amino acids or sugars, but either aromatic units containing the benzene ring, or are nonaromatic (aliphatic chain) units. More than one monomer type can coexist, and side branches may be present, which can crosslink and add structural rigidity to plastic packaging materials.

Terminology. A few important terms to define before continuing with this topic are resin, copolymer, thermosetting, and thermoplastic. A **resin** is a polymer in the form of small, individual pellets that are manufactured via melting and extrusion into plastic packaging sheets. The term **copolymer** refers to a plastics polymer formed by the reaction of at least two different comonomers. Ethylene-vinyl alcohol (EVOH) is a common example of a copolymer. **Thermoplastic** means that a plastic material can be melted and resolidified without degradation. The vast majority of plastics used for food and beverage packaging applications are thermoplastics. **Thermosetting** refers to certain plastics that harden when they are heated and set into a solid structure that will not remelt without decomposing. Epoxies and polyurethanes are examples of thermosetting polymers.

Thermoplastic resin molecules, although they are not food molecules, also exhibit a shift from a brittle material state to a softer more rubbery material. This is a shift from amorphous glassy to amorphous rubbery state, occurring at the glass transition temperature. However, what would be considered undesirable in a food polymer (rubbery toughness) can be an asset when considering plastic performance, which depends a lot on flexibility as well as strength. Resin polymers exhibit a range of T_g' values. Those with comparatively high values are brittle at refrigeration temperatures but exhibit good high temperature stability, whereas polymers with comparatively low values are tough and flexible at refrigeration temperatures. The addition of plasticizers, or the process of copolymerization, can extend the useful temperature range of polymer packaging materials.

Thermoplastic Polymers. Before listing some of the thermoplastic polymers most frequently used to make food packaging materials, cellophane deserves mention as a major food packaging material in use before 1970. **Cellophane** is a nonplastic film material derived from chemically treated wood pulp. Because of its great stiffness it cannot be shaped, for instance, into bottles, but can serve as packaging film. Cellophane popularity in the packaging industry waned when polyolefin polymers were introduced. A **polyolefin** is any polymer that is composed solely of straight-chain hydrocarbon units ("olefins") of covalently bonded carbon and hydrogen atoms.

Polyethylene (PE) is the most common plastic material for packaging and is used to make zippered plastic bags and a variety of plastic storage containers. It forms copolymers that act as excellent moisture barriers at low cost. Polyethylene and ethylene copolymers include **LDPE** (low-density polyethylene), **HDPE** (high-density polyethylene), and **LLDPE** (linear low-density polyethylene). LDPE is derived from ethylene ($H_2C = CH_2$) that has been subjected to high pressure and temperature in a reactor. During the reaction, many side chain branches form, which are a mixed blessing as they not only decrease density and improve clarity, but also reduce stiffness and impact toughness. **Vinyl acetate** (VA) is an unsaturated monomer containing the acetate ester group. It can copolymerize with ethylene to form **ethylene vinyl acetate** (EVA), which offers improved toughness compared to LDPE, and is commonly referred to as freezer wrap. EVA maintains its moisture-loss protection without becoming excessively brittle at low temperatures.

Table 13.5 gives examples of polymers used to manufacture food packaging plastics, including polypropylene (PP), polyvinyl chloride (PVC), polystyrene (PS), and polyethylene terephthalate film (PET). **Polypropylene (PP)** is used in retort pouches and as an inside layer of food packages subjected to high heat due to its high melting point and tensile strength. **Polyvinyl chloride (PVC)** refers to polymerized vinyl chloride, or more simply, vinyl. This material makes a good barrier to odors and moisture, and prevents

TABLE 13.5 Examples of polymers used to manufacture food packaging plastics.

Polymer	Chemical Formula
LDPE (low-density polyethylene)	$\cdots -CH_2-[CH_2-CH_2]_n-CH_2-\cdots$
VA (vinyl acetate)	$CH_2=CH-O-CO-CH_3$
PP (polypropylene)	$\cdots -CH_3CH-CH_2-\cdots$
PVC (polyvinyl chloride)	$\cdots -CH_2-ClCH-\cdots$
PS (polystyrene)	$\cdots -[CH_2-HC-\bigcirc]_n-\cdots$
PET (polyethylene terephthalate)	$\cdots -CH_2-CH_2-OCO-\bigcirc-OCO-\cdots$

freezer burn in packaged frozen foods. **Polystyrene (PS)** refers to polymerized styrene, which, when foamed, is known as expanded polystyrene, or styrofoam. PS provides thermal insulation and is used to produce egg cartons, hot/cold drinking cups, and "packaging peanuts" as shipping box filler material. **Polyethylene terephthalate (PET)** is a packaging film from the polyester family of materials. It is the most important polyester in plastic packaging.

An interesting distinction exists regarding the chemical nature of PET polymerization versus the polymerization of the polyethylene family (LDPE, HDPE, LLDPE), PP, PS, PVC, and EVOH. The latter group of polymers are made from unsaturated aliphatic hydrocarbons that bond together when a double bond is broken by a peroxide initiator catalyst to form a free radical. However, polyesters (e.g., PET) are condensation polymers. Condensation polymers are formed by the reaction of two molecules that eliminate a small molecule (water) when combining. The loss of water exposes binding sites that enable the reactants to polymerize into long chains of repeating units.

Properties of Plastics. When designing plastic packaging materials, certain properties stand out as most critical. These include heat-set properties, including the strength of the package seal immediately after it is formed, its strength after aging for a period of time, optical properties (gloss, haze, and percent light transmitted), bonding strength in the case of multilayer packaging materials, and gas and moisture transmission rates. The **water vapor transmission rate** (WVTR) is of prime consideration in the packaging of moisture-sensitive foods. This property is tested under specific conditions, such as 25°C and 90% RH, and is expressed in units as grams of water vapor transmitted per unit time per unit thickness per unit area. Other important packaging material properties are tensile and tear strength, impact strength, **flexural modulus** (stiffness), and flex life, which is the number of flexing cycles that a packaging film can endure before failure. This happens to be a key parameter in selecting packaging materials for products like processed meats.

Plastic Manufacture. As mentioned previously, extrusion technology gave rise to the plastics industry. Plastic films and sheets used for food packaging are manufactured by the extrusion process. A general overview of the extrusion manufacture of packaging materials is shown in Figure 13.21. A food package begins life as a rounded pellet. For instance, small pellets of LDPE are fed into an extruder. In the extrusion process, they are mixed and melted, and forced through an opening (the die). Here, they form a shape (sometimes called a "bubble") that can be immediately oriented by a special stretching process. Such **orientation** is critical—it refers to the alignment of individual polymer molecules within a fabricated film. After the stretching and casting of the melted, expanded polymer material, the stretched film is wound onto a roll and saved in preparation for further manufacture. For example, the roll of stretched film can be made into various pouches.

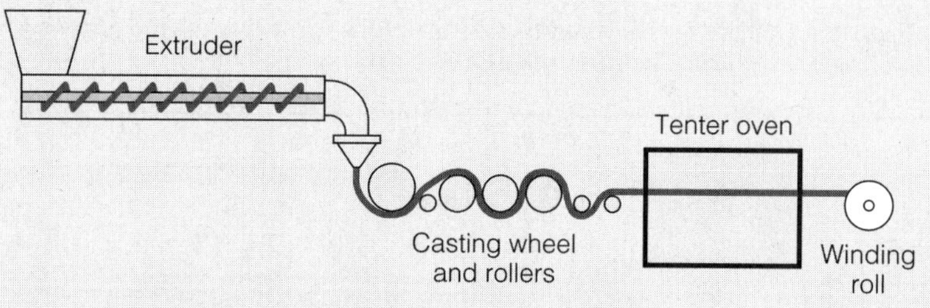

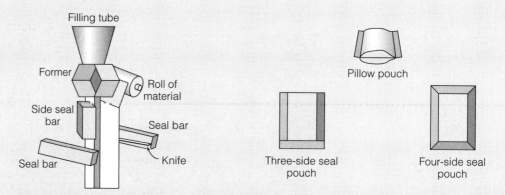

FIGURE 13.21 A general overview of the extrusion manufacture of packaging materials. *Pelleted plastic material is fed through the extruder, and is subsequently cast and sheeted by a roller and tenting oven. The newly formed stretched material is collected by winding upon a roll.*

FIGURE 13.22 A vertical form-fill-seal machine, and examples of pouches used in the food industry.

Packaging machinery exists that will take the roll of stretched extruded film and simultaneously form it into a pouch, fill it with product, and seal the open-fill side. Such equipment is called form-fill-seal machinery, and may be vertically or horizontally oriented. Typical pouch types made by this process are called the flat bottom bag, the four side-seal pouch, and the pillow pouch—see Figure 13.22. Foods packaged into these pouches include snack foods such as potato chips, individual ketchup and condiment portions, frozen food, milk, and various granular products.

Bar Code Technology

The last several decades has witnessed the development of bar code technology. **Bar code technology** was developed as a method of automating data collection. It encompasses the symbologies that encode data to be optically scanned, the printing technologies that produce machine-readable symbols, the scanners and decoders that capture visual images of the symbologies and convert them to computer-compatible digital data, and the verifiers that validate symbol quality. Bar code technology has the advantage of little if no human error.

Widespread use of bar code technology began in the supermarket industry and succeeded to the degree that virtually every food supplier now uses the UPC (universal product code) symbol on product packaging to enable point-of-sale (POS) scanning. Bar codes have accelerated the flow of products and information throughout the global business community. As a footnote to the food packaging topic, bar code technology has combined with modern packaging practices to offer the food industry a most efficient and effective means to contain, control, and track inventory.

Challenge Questions

1. Consider the statement: The most functional packaging materials are those made of metals, such as steel or aluminum. Which choice does not apply?
   a. Metals fulfill most of the "10 Golden Rule" requirements for effective packaging.
   b. Metals are disposable and exhibit low corrosion.
   c. Metal containers offer food preservation, protection against physical damage, and protection against contamination.
   d. Metal containers contribute to easy food product distribution and provide customer recognition of product.
   e. Metals act as functional microwave containers.

2. Which chemical reaction in baked products (e.g., cakes, bread) is inhibited by modified atmosphere packaging?
   a. moisture loss
   b. lipid oxidation (especially in cakes)
   c. enzymatic browning
   d. staling (especially in bread)
   e. polymerization

3. Oxygen scavenging technology
   a. is an example of a controlled atmosphere method.
   b. is an example of a modified atmosphere method.
   c. established increased package oxygen content.
   d. alters the carbon dioxide levels in a package.
   e. alters the nitrogen levels in a package.

4. Explain the chemical basis for the following statement: Plastics are organic molecules.
   a. Plastics are composed of aliphatic monomers.
   b. Plastics are composed of aromatic monomers.
   c. Plastics are composed of carbon-containing units linked together.
   d. Plastics are composed of high molecular weight polymer chains.
   e. all of these

5. Which statement is *false?*
   a. LDPE, LLDPE, PP, and HDPE are examples of polyolefins because they are straight-chain hydrocarbons composed only of carbon and hydrogen atoms.
   b. PS is a straight-chain hydrocarbon.
   c. PET is not an olefin because it contains oxygen.
   d. PS is not an olefin because it contains benzene rings.
   e. PVC and PVDC are not olefins because they contain chlorine.

CHAPTER 14

Food Biotechnology

14.1 What Is Food Biotechnology?

14.2 Genetic Engineering

14.3 Regulations Controlling the Application of Food Biotechnology

14.4 Improving Plant Products through Biotechnology

14.5 Improving Animal Products through Biotechnology

14.6 Improving Food Processing Aids through Biotechnology

14.7 Applying Biotechnology-Derived Foods in Food Safety

14.8 Major Concerns About Biotechnology-Derived Foods

Challenge!
Bioengineering of β-lactoglobulin

CHAPTER OBJECTIVES

After completing this chapter, you will be able to:
- Define biotechnology, and food biotechnology.
- List the benefits provided by biotechnology in food production.
- Explain the basics of genetic engineering techniques.
- List the issues related to the regulatory aspects of biotechnology-derived foods.
- List and explain the three categories of equivalence used in determining the safety of biotechnology-derived foods.
- Give examples of biotechnology-derived plant and animal products.
- Give examples of biotechnology-derived food processing aids.
- Describe the use of biotechnology in food safety applications.
- Discuss the concerns associated with biotechnology-derived foods.

The subject of biotechnology has gained much attention in recent years. The application of genetic engineering techniques has been shown to increase crop yields, a necessary step toward reducing world hunger and poverty. Biotechnology-derived foods can also be created with enhanced nutritional value, helping to improve the health of consumers. However, the application of biotechnology in food production has also raised concerns among some consumers, and the safety of genetically engineered products has been questioned. In addition, techniques such as cloning have many people wondering whether scientific advances in this field have taken into account any moral implications of their application.

In this chapter, the potential benefits of biotechnology-derived foods and genetic engineering techniques will be examined. The safety of biotechnology-derived foods will be addressed within the context of government regulations. Examples of foods and products made possible through biotechnology are reviewed. Consumer concerns are also presented.

14.1 WHAT IS FOOD BIOTECHNOLOGY?

In its simplest terms, **food biotechnology** refers to the technology of manipulating or modifying deoxyribonucleic acid—DNA (Figure 14.1) for the purpose of improving the quality and/or safety of foods. It employs the tools of genetic engineering to improve plants, animals, and microorganisms for food production. The term *biotechnology* was actually used first by Karl Ereky in the 1920s, who used the word to refer to the application of technology-intensive agricultural methods.

Surmounting the Species Barrier

For many years, people have selected seeds and cross-bred different species of plants to produce new varieties with increasingly desirable characteristics. Early bakers, brewers, and dairy farmers, such as the ancient Sumerians, utilized and manipulated different types of microorganisms to select for certain ones that would allow them to produce the highest quality, best-tasting breads, beers, wines, and cheeses. Thus, whether they knew it or not, our ancestors relied on genetics, the basis of biotechnology, to enhance or change the quality of food.

In present times, crops have been manipulated to develop desirable traits such as improved taste and resistance to diseases. However, there are limits to this traditional manipulation. **Cross-breeding** of plants, for instance, can take place only with species that are very closely related, which limits the results that can be obtained. Cross-breeding also results in unwanted characteristics being passed on to the new crop, along with the desired ones. Thus, plant breeders spend a great deal of time and effort **backcrossing** the newly developed crop with the parent crop, repeating this process many times to breed out the unwanted characteristics. This process is very slow, taking as long as 12 years for successful completion. In short, results from traditional cross-breeding technology are random, difficult to control, often imprecise, and time-consuming (Figure 14.2). Further,

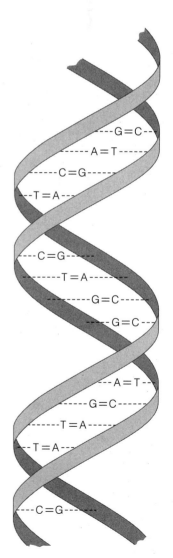

FIGURE 14.1 The DNA molecule.

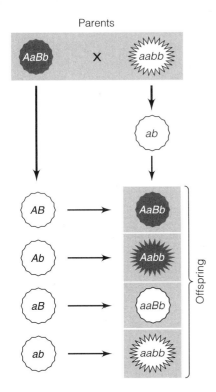

FIGURE 14.2 Backcrossing to produce a disease-resistant cultivar.
A backcross (or testcross) is the process by which an individual that displays a dominant characteristic (i.e., disease resistance) is crossed with an individual that displays a recessive characteristic. The dominant individual is phenotypically resistant to disease, but genetically it may be composed of a mix of dominant (A, B) and recessive (a, b) genes. Thus it is genotypically heterozygous (AaBb). The recessive individual is phenotypically susceptible to disease, and so it is composed of only recessive genes (a, b) and genotypically homozygous (aabb).

the possibility of expanding the choices of desirable characteristics is impossible, mainly because living cells of different species cannot be mated due to biological boundaries between species.

By surmounting the species barrier, biotechnology can be employed to improve the effectiveness and efficiency of transferring desirable genetic traits to food in a controlled manner. Generating a product with a certain characteristic no longer has to rely on chance for that characteristic to emerge from uncontrolled genetic crossings. Instead, breeders can select a specific characteristic, place the genetic code for it into a plant, and thus generate a product with the desired trait. A plant or animal modified by genetic engineering is called **transgenic** (Figure 14.3). Through transgenic biotechnology, the pool of available genetic traits can be greatly expanded, and the time for generating improved foods can be shortened significantly.

The Benefits of Food Biotechnology

Food biotechnology has the potential to address broad global food issues:

- Alleviating food shortages in developing countries
- Minimizing the impact of food production on the environment
- Improving the nutritional health of populations

In addressing these concerns, several options are available through biotechnology. First, much of the food that is produced in some countries is never consumed because of the losses that occur due to spoilage. In Russia alone, historically 15 to 50 percent of horticultural crops have been lost after harvest due to poor food processing and food distribution systems. Fruits and vegetables often ripen and spoil before they reach consumers because of long-distance shipping or because of the lack of efficient distribution and storage systems. Through biotechnology, foods that ripen slowly could be engineered, resulting in a product shipped to far-away markets without the concern of overripening before reaching the consumer. Even after ripening, these biotechnology-improved crops would retain their high quality for a longer period of time than conventional products, reducing the losses associated with food spoilage.

An added bonus from such biotechnology-derived crops could be superior taste to that of fruits and vegetables produced under conventional methods. To prevent spoilage, such commodities are traditionally harvested before they have properly ripened on the vine. Thus, ripening takes place enroute, resulting in poor-tasting crops. Production of fruits and vegetables that ripen slowly through biotechnology would permit growers to leave the commodities on the vine longer, allowing them to ripen properly. Given the slowing down of the maturation process, such products would not spoil during transport, ensuring their arrival to retail markets at their apex of quality.

Food shortages can also be alleviated by maximizing food production efficiency. Crops that could withstand extremes of cold or heat, or even droughts could be developed through biotechnology, providing greater productivity and self-reliance to developing nations.

Biotechnology can address the second issue of minimizing the impact of food production on the environment by reducing herbicide and pesticide use. Crops that are resistant to plant diseases could be developed, lessening the need for pesticide sprays to control them. In addition, new varieties of plants that can manufacture their own pesticide could also be produced.

Herbicides must decrease the encroachment of weeds and other undesirable plants within a planting field without harming the food crop. To accomplish this, farmers must use highly toxic herbicides that do not readily degrade, remaining in the environment for prolonged periods of time. Crops that are tolerant to herbicides could be generated through biotechnology, allowing farmers to use more "environmentally friendly" herbicides that still accomplish the job of weed eradication without damaging the food crop.

The third issue of concern is that of supplying highly nutritious foods. Problems of poverty among developing nations results in millions of undernourished people. Fruits and vegetables containing higher levels of certain nutrients, compared with conventional

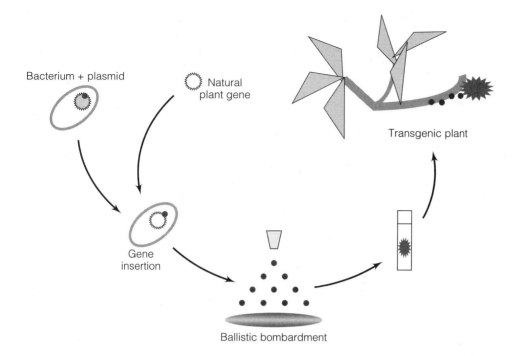

FIGURE 14.3 Production of transgenic plants by ballistic bombardment. *The details of genetic engineering will be discussed in Section 14.2.*

crops, can be developed. The incidence of chronic diseases such as cancer and heart disease could be lowered by consumption of foods with extra measures of phytochemical antioxidants including beta-carotene.

Many other benefits can be derived from biotechnology-produced foods. Specific improvements in plant products, animal products, and food processing aids will be discussed in Sections 14.4, 14.5, and 14.6. In addition, organisms can be made to produce natural chemicals that would inhibit foodborne pathogens, making such products safer (Section 14.7).

14.2 GENETIC ENGINEERING

Genetics is the field of study concerned with the elements in nature responsible for passing down traits from one generation to another. These elements are called genes, and they determine all inherited characteristics of plants, animals, and any other living form. Genes form part of the DNA (deoxyribonucleic acid) of a cell, most of which is packaged into the nucleus. The DNA molecule can be very long, containing millions of genes, all of which compose the **chromosome** of the cell.

Bacteria are notable exceptions, since they do not have nuclei, but rather house their DNA in a tightly wound-up coil called the "bacterial chromosome." They also have small circular rings of DNA called **plasmids,** containing just a few genes. Viruses can have either DNA or RNA (ribonucleic acid) as their genetic material, containing much fewer genes than those found in other organisms.

So, what is **genetic engineering?** It involves the manipulation of genes and the organisms containing them. **Biotechnology** then is the practice of genetic engineering to make useful biological products. The raw material necessary for genetic engineering is DNA. This molecule is typically found as a double helix of two intertwined strands (Figure 14.1). Each strand is simply a chain of sugar and phosphate molecules linked together alternatively. A **base** is attached to each sugar molecule, with weak bonds between bases joining the two strands of DNA into the double helix like rungs in a ladder. Four bases are found in DNA: adenine (A), thymine (T), cytosine (C), and guanine (G). A always pairs up with T, and C always pairs up with G (Figure 14.4).

Genes are made up of sequences of bases of a certain length. A sequence of three bases (codon) within a gene specifies the amino acid that will be made. A gene, which is

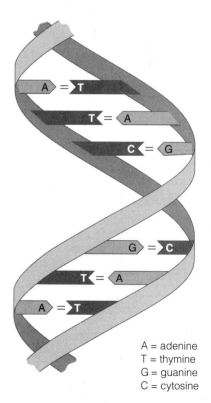

A = adenine
T = thymine
G = guanine
C = cytosine

FIGURE 14.4 Complementary base pairing in the DNA molecule.

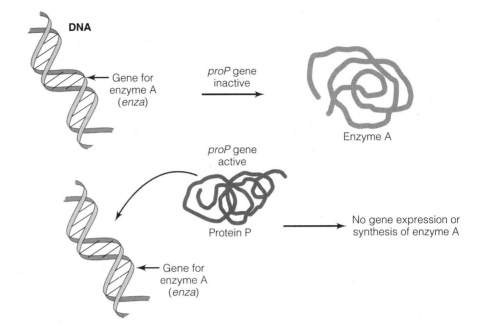

FIGURE 14.5 Inhibition of gene expression by protein P. *In this model, the protein being synthesized is enzyme A. The section of the organism's DNA that contains the specific sequence of bases necessary for production of enzyme A is the enzyme A gene. When this enzA gene is not inhibited, protein sythesis occurs, and enzyme A is produced. However, protein P acts to suppress enzyme A production when its proP gene is active.*

composed of many such three-base sequences, specifies all the amino acids that a protein will contain. For example, the sequence of bases TTT codes for the amino acid phenylalanine, while the sequence of bases TCA codes for the amino acid serine. A gene containing these sequences of bases possesses the genetic code needed to synthesize a protein containing these two amino acids.

Protein Synthesis

In addition to the sequence of bases necessary for the synthesis of amino acids (and ultimately of proteins), DNA contains the codes necessary for regulation of protein production. Depending on the needs of the cell, only certain proteins will be synthesized at any one time. The instructions for controlling such synthesis are found within the DNA molecule itself, coded by a specific gene.

For example, imagine a protein, enzyme A, which is necessary for the metabolism of a certain sugar by the cell. The DNA molecule of the organism contains a gene, *enzA*, which has the sequence of bases necessary for the synthesis of enzyme A. Suppose that the enzyme will not be synthesized in the presence of a protein, protein P, because it inhibits expression of the *enzA* gene. Protein P itself is coded by a gene, *proP*, within the DNA molecule. See Figure 14.5. Thus, expression of the *enzA* gene will be indirectly affected by expression of the *proP* gene.

These general concepts regarding DNA and gene expression are true for all living things. The genetic code is universal, with DNA molecules from all organisms being built with the same blocks—the A, T, C, and G bases. Thus, transferring genes from one organism, such as a bacterium, to another, such as a plant, is possible. Such transfers are what genetic engineering is all about. Now let's look at the techniques of genetic engineering.

Manipulating DNA in Food Production

To manipulate DNA to be useful in food production, the right tools must be available. A tool that allows cutting the molecule easily and at the right place in order to remove the desired gene is essential. Bacteria naturally contain such a tool in the form of **restriction enzymes.** These enzymes recognize particular base sequences within the DNA molecule, with specific enzymes being restricted in their ability to cut DNA only at specific sequences.

Restriction enzymes normally help bacteria cut the foreign DNA of invading viruses in order to fend off their attack. Hundreds of these enzymes are produced by bac-

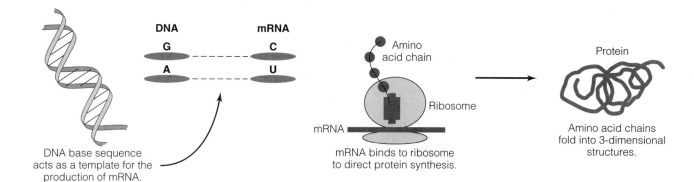

FIGURE 14.6 mRNA function in protein synthesis.

teria, most of which are available commercially. One important aspect of restriction enzymes is that when they cut, they do so wherever they find the particular sequence of bases along the DNA molecule. This leaves several unpaired bases at each point of the molecule where the enzyme has cut, which is significant.

Given the complementarity of bases (A with T, and C with G), the end of a DNA strand cut with a restriction enzyme will link to the end of another DNA strand that has been cut with the same enzyme. Besides restriction enzymes, another tool of the genetic engineer is the enzyme **DNA ligase**. This enzyme helps form chemical bonds that will join together two DNA strands, previously cut by restriction enzymes.

Using Messenger RNA in Genetic Engineering.

Often, the task of finding a specific gene, which a genetic engineer wishes to cut from the DNA molecule, can be quite daunting because of the enormous number of genes within the molecule. So, instead of manipulating DNA, engineers have discovered they can manipulate messenger RNA (mRNA).

There are many mRNA molecules, each containing the code for a specific protein. Thus, utilizing mRNA in genetic engineering is a much more precise way to manipulate the genes for specific proteins than working with the entire DNA genome. In addition, because mRNA molecules contain the sequence for only one protein, they are much smaller, therefore easier to manipulate, than DNA (See Figure 14.6).

Once the scientists find the desired genes in mRNA, they usually need to make DNA from the mRNA in order to use the restriction enzymes to cut out a gene of interest. These enzymes will not work on mRNA since this molecule contains the base uracil (U) in place of thymine (T), making it impossible for restriction enzymes to recognize their target sequence. To make DNA from mRNA, genetic engineers use the enzyme **reverse transcriptase**, which assembles a single strand of DNA from a strand of mRNA. To construct the complementary DNA strand so as to have a double helix, scientists use the enzyme **DNA polymerase**, which catalyzes the synthesis of DNA. This resulting DNA molecule, synthesized from mRNA, is usually referred to as copy, or cDNA:

mRNA $\xrightarrow{\text{reverse transcriptase}}$ one strand of DNA $\xrightarrow{\text{DNA polymerase}}$ double-stranded cDNA molecule

Inserting DNA into Bacteria.

Once a particular gene has been identified from mRNA, and the complementary cDNA has been synthesized, the cDNA is inserted into the target cell (microorganism, plant, or animal). If the objective is to insert DNA into a cell so as to manufacture large quantities of the gene product, the organism-of-choice is a bacterium, since it can be cultured relatively easily, and large quantities of material can be produced in a relatively short period of time. For insertion of DNA into microorganisms, plasmids are used as the vehicles, or vectors, to carry the gene into the cell.

Plasmids are short circular segments of DNA usually measuring 15,000 to 20,000 bases (15 to 20 kb) in length, and thus can carry only a few genes. They are naturally

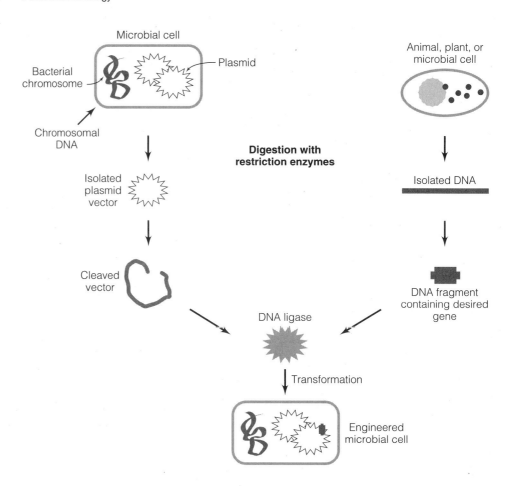

FIGURE 14.7 Inserting a gene into a simple cell system.

found in bacteria, typically being responsible for carrying genes that confer resistance to the organism to antibiotics. Plasmids are easily shuttled from one cell to another, this being the basis for transmission of antibiotic resistance among microorganisms in nature. They can be cut with restriction enzymes, allowing a piece of DNA containing a particular gene to be easily pasted into the plasmid. The result is a specialized plasmid that is readily inserted into a bacterial culture in order to manufacture the desired gene product. See Figure 14.7

Inserting DNA into Cells. If the purpose of the genetic engineer is to insert DNA into plant cells, **ballistic bombardment** can be used. DNA that is to be inserted is first pasted onto microscopic gold or tungsten particles. These particles, now coated with the desired gene, are shot into plant tissue at high velocity using an instrument dubbed the "metal gun." Another way to insert DNA into a plant cell, or a bacterium for that matter, is through a technique called **electroporation.** In this case, an electric field is applied to the target cell in a series of pulses. Pores momentarily appear on the membrane of the cell, allowing DNA (either in plasmid form or as a linear strand) to be inserted.

If the objective is to introduce DNA into animal cells, a different manipulation, called **microinjection,** is employed. DNA is injected into newly fertilized egg cells through a fine glass pipette. Egg cells are used so as to introduce the new genes at a very early stage in the animal's development. The eggs are then inserted into the uterus of a surrogate mother, allowing it to develop normally. The fetus will grow, and the mature animal, now possessing the inserted gene in its chromosome, can pass it along to its offspring in the future.

Screening Cells for Gene Insertion. The discussion has focused thus far on how particular genes are identified, how they are copied by producing cDNA, and how they can be inserted into a vector or directly into a target organism. In addition to these steps, genetic engineers realize that only a small number of cells that are treated in this way actually take up the desired gene. Thus, they need to verify whether it has in fact been successfully introduced. Two techniques for screening insertion of genes are used: inserting marker genes and using a combination of polymerase chain reaction and genetic probing.

The use of marker genes. **Marker genes** are genes that code for an easily recognizable product, and they are inserted into cells at the same time as the desired gene. If this gene is taken up, so is the marker gene. Thus, if the cell synthesizes the marker gene's product, one can assume that the cell also can synthesize the desired gene's product. As an example of the use of marker genes, consider Bt corn. The transgenic events that allow development of Bt corn include the insertion of a Bt gene, a promoter gene, and a marker gene. This lets corn breeders know which plants have the new genetic material. The promoter gene allows the Bt gene to be turned on, and different promoter genes may allow the Bt toxin to be expressed at different times of the year or different parts of the plant.

In screening bacterial cells for successful gene insertion, antibiotic resistance genes are typical markers. Introduction of these into a cell can be easily ascertained by inoculating the organism into a medium containing the antibiotic. Ability to grow in such a medium is only possible if the cell has acquired the necessary resistance through insertion of the marker gene.

In the case of plant or animal cells, marker genes other than those conferring antibiotic resistance have been used, such as genes that produce a compound with a particular color, or one that fluoresces when exposed to ultraviolet light.

The use of PCR and genetic probing. A more common way to screen insertion of genes into plant and animal cells is based on the coupling of two technologies: **polymerase chain reaction (PCR)** and **genetic probing**. PCR is used to generate many copies of DNA in a relatively short time. Once the DNA is amplified in this way, a genetic probe can be used to detect the presence of the DNA.

Simply stated, such a probe is made of a piece of DNA that contains a sequence of bases that is complementary to that of a target DNA. The probe usually also has a color marker molecule attached to one end. To verify whether a gene has been successfully inserted into a cell, the DNA is first extracted from the cell and then it is amplified, or copied, by PCR. A DNA probe containing the base sequence complementary to the desired gene is incubated with the amplified DNA. After incubation, the mixture is washed and the presence of color is measured. The presence of color indicates that the probe was able to bind to its target, which could only be possible if the target gene was present. When this happens, it means that the target gene must have been successfully inserted into the cell's DNA.

Gene Expression Through Genetic Switches. One last step before a genetic engineer can claim success is to make sure that the cell receiving the desired gene has the ability to express that gene and to synthesize the product coded by the gene. To confirm this, **genetic switches** are used—these are genes that are needed to trigger expression of other genes.

One example is the genetic switch for the *bgal* gene in some bacteria. This gene produces the enzyme beta-galactosidase, which enables the cell to metabolize the sugar lactose. The lactose content in dairy products can be reduced by treating it with this enzyme, which is helpful to lactose-intolerant consumers. The "genetic switch" gene responds to the presence of this sugar by turning on the *bgal* gene so it can produce beta-galactosidase, and thus be able to metabolize the sugar (Figure 14.8).

In genetic engineering, this genetic switch can be inserted into a target cell at the same time that a desired gene is inserted. The switch gene is usually placed several basepairs in front of the desired gene. When lactose is added to the medium in which the engineered cells are growing, the switch gene is triggered, which then turns on expression of the desired gene's product. In this case, the product is not beta-galactosidase, but whatever the desired gene codes for.

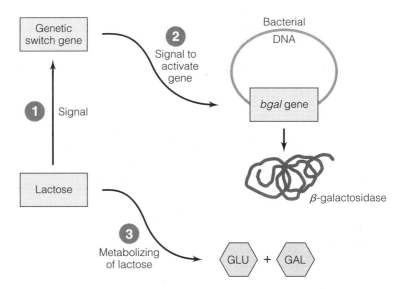

FIGURE 14.8 Genetic switch for bacterial β-galactosidase enzyme.

Propagating Growth of Genetically Engineered Cells. In addition to gene insertion, screening, and switching, biotechnologists require methods to propagate the growth of the genetically engineered cell. Individual cells such as bacteria can be grown in bioreactors or **fermentors,** typically large-volume containers in which cultures suspended in nutrient-rich broth are placed. Large quantities of the desired gene's product can be produced in a cost-effective and efficient manner. Temperature, pH, oxygen content, and other variables can be controlled and optimized to promote maximum synthesis of the desired product.

Genetically engineered plant tissue is usually grown in **cell culture** under controlled laboratory conditions. In this case, plantlets synthesize the desired gene product efficiently, since the plants are not subject to infection by viruses and pests. Some species of plants, usually those that do not have seeds, are partially grown in culture, and then directly planted in soil, allowing them to grow in a field under normal conditions.

Unfortunately, this method can result in variations in the quality of the gene product, as well as in the quantity being produced. An alternative is the use of "hairy root cultures," in which the plant tissue is first infected with the organism *Agrobacterium rhizogenes*. This bacterium promotes production of rootlike structures by the plant cells, allowing them to be grown almost indefinitely in media, eliminating the need for traditional field planting methods.

Animal cells can also be grown in culture, although only sheets of tissue can be produced, and not whole animals. For animal production, genetically modified embryos are inserted into surrogate mothers and allowed to mature. The result is a transgenic animal with modified genetic material. **Cloning** is a variation of this theme, in which the genetically modified embryo is split into eight cells. Each cell is implanted into a surrogate mother, resulting in eight identical animals that contain the desired genetic trait.

14.3 REGULATIONS CONTROLLING THE APPLICATION OF FOOD BIOTECHNOLOGY

Food regulations were presented in Chapter 7, so the focus here will be on regulations related to biotechnology. In the United States, biotechnology-derived foods are under the regulatory oversight of the Food and Drug Administration because the FDA's main charge is the enforcement of the Federal Food, Drug, and Cosmetic Act (FFDCA) of 1938, as amended. This agency is responsible for all foods, except for red meat, poultry, and egg products, which are under the purview of the United States Department of Agriculture.

FDA Policy

In 1992, the FDA published a statement describing its policy regarding these foods, to help companies involved with this technology comply with the FFDCA in the manufacture of their products. The following issues are discussed in their report:

- Genetic modification
- Toxicants
- Nutrients
- New substances
- Allergenicity
- Antibiotic resistance markers
- Animal feeds
- Labeling

Each of these topics will be discussed to gain a better understanding of the concerns of the FDA and the requirements the agency has of the food industry.

Genetic Modification. Regarding genetic modification itself, the FDA requires that the material or gene that is being introduced into a cell be highly characterized (discussed in Determining Safety), so that products resulting from its insertion are predictable. In addition, insertion of the gene into a cell's genome must be done in a way to ensure stability of the insertion, so that genetic rearrangements are minimized. These issues are important in ensuring that unexpected changes do not take place that will compromise the safety of these products.

Toxicants. Food biotechnology is not permitted to create toxic substances—the FDA stipulates that any gene that is inserted into a cell must not result in the production of harmful substances. This is especially true of toxins, which a plant may produce naturally, even if not genetically modified. If it is modified, then the plant must not produce these toxins at levels higher than those produced by the unmodified plant.

Nutrients. The FDA states that genetic engineering of plant tissues may change the levels of important nutrients produced by a modified plant, and that this change should be prevented so as not to detrimentally affect the nutritional quality of the food. Similarly, the agency believes it is important to ensure that the availability of nutrients, if not their levels, does not change with genetic modification.

New Substances. The FDA states that if new compounds not previously found in foods are generated by biotechnology, then these compounds must be considered new food additives. As such, they must undergo pre-market approval by the FDA just as all new food additives do.

Allergenicity. Along similar lines, by inserting genes from microorganisms, plants, or animals to produce food, a substance may be introduced into the food supply to which people may be allergic. If the source of the gene that is transferred is known to be allergenic, such as milk, nuts, and soybeans, then it is conceivable that the tissue in which the gene is introduced may result in production of an allergen, and thus cause an allergic reaction in sensitive individuals. Stability to heat, acid, and degradation by enzymes are often used as factors to determine whether a certain product is allergenic. Ultimately, immunological tests should be carried out to confirm if such products could cause allergies among the consuming population.

Antibiotic Resistance Markers. Recalling the concept of marker genes might help understand why the FDA strongly suggests that companies evaluate the use of antibiotic resistance genes as markers. The concern stems from the notion that having the product of such a gene in a food could result in inactivation of oral doses of antibiotics if prescribed to individuals consuming that food. In addition, there is the concern that an antibiotic

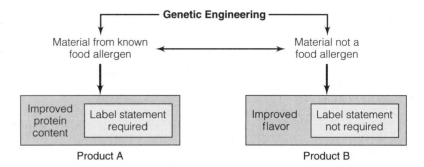

FIGURE 14.9 Example of labeling requirements of genetically engineered foods.

resistance gene present in plant or animal tissue may be transferred to microorganisms in the environment, thereby rendering them resistant to that agent. Exisiting evidence shows the possibility of this happening to be very small, but the FDA suggests that the use of marker genes be carefully evaluated.

Animal Feeds. Just as with foods destined for human consumption, the FDA regulates animal feeds, and these must meet the same safety standards under the FFDCA. Thus, it is important that companies determine whether biotechnology-derived plant products used in the manufacture of animal feed will result in any harm to the animals, whether by production of harmful substances or by reduction in the production or availability of essential nutrients in the plant.

Mandatory Labeling. Finally, the FDA requires that special labeling be used in genetically engineered products *if* such genetic manipulation affects the composition of the food to the extent of changing it significantly from its conventional counterpart (discussed further in Determining Safety). This is especially true if the biotechnology-derived food contains an allergenic substance not typically found in that food.

Food allergens can be transferred from one food to another by genetic engineering. For instance, a gene from Brazil nuts that codes for a protein nutrient has been introduced into soybeans. However, it also reproduced the same type of allergic response that Brazil nuts themselves do in people sensitive to them. As a result, it was not marketed. If it had been, it would be required by the FDA to carry information on the label to indicate it contained proteins identified as allergens to sensitive individuals (Figure 14.9).

If a genetically derived food differs significantly in composition from its natural counterpart, the name of the food must be changed or a qualifier added to make the difference apparent. Along with compositional differences, any differences in terms of usage of the food product, when compared with its natural counterpart, must also be listed on the label.

However, genetically engineered foods that do not differ from conventionally produced foods are not required to have any labeling since they are considered equivalent. This important topic of "equivalence" will be discussed in Determining Safety.

Voluntary Labeling. In addition to what is required of biotechnology-derived foods by the FDA, some food companies want to include other information on the label on a voluntary basis. One prime example of such action is the voluntary disclosure from some companies regarding the fact that they do *not* use biotechnology-derived ingredients and that their food products are not biotechnology-produced.

However, such claims must be substantiated. Because of the difficulty in proving the *absence* of impurities in foods, it is easier for food companies to claim that the ingredients used in their product were not bioengineered than to assert that the products themselves are free of bioengineered ingredients. For example, it is easier for milk companies to prove that they have not used genetically engineered, or recombinant, bovine somatotropin (rBST) than to claim the absence of rBST in their milk.

Based on this, the FDA has developed a guidance document, containing recommendations on the voluntary labeling of bioengineered foods, including the need for claims about the presence or absence of bioengineered ingredients.

Determining the Safety of Biotechnology-Derived Foods

In 1990, a joint committee formed by the Food and Agriculture Organization of the United Nations, and the World Health Organization published a report that set the standard for assessing the safety of foods in general. Basically, they stated that an essential element in considering the safety of any food is to compare it with a product having an acceptable standard of safety.

Thus, the international concept of **substantial equivalence** was born, which states that if a food or food component is found to be substantially equivalent to an existing food or food component, then it should be considered as safe as the conventional food or food component. This comparison between foods could be simple or take quite a bit of effort and time, depending on the availability of scientific data on the subject. The food companies have the burden of proof to conduct studies and present them to the FDA for review and approval.

Assessing substantial equivalence should be made comparing a food to one that is as close to the species level as possible. Thus, in the case of biotechnology-derived foods, characterization of the food at the DNA level, and comparisons with regards to expressed characteristics, key nutrients, toxicants, and allergens should be made. From such an assessment a genetically engineered organism, or the product from it, can be placed into one of three catagories.

First, it could be deemed substantially equivalent to a conventional counterpart available in the food supply. Second, certain differences may require that the product be declared substantially equivalent with these certain defined differences being stated. Third, substantial equivalence may not be attainable if no appropriate counterpart product exists in the market. See Figure 14.10. Biotechnology-derived food that is truly breaking new ground in terms of product development falls into this third catagory.

Substantially Equivalent Foods. Assessment of such products involves a two-component system. First, the genetically engineered organism from which the food is derived needs to be characterized. To do so, the biotechnology company must gather several facts about the host organism, the genetic modification being done, and the modified organism itself. Regarding the host, the company must determine its origin, classification, name, relationship to other organisms, history of use as a food source, production of toxins or allergenic compounds, and presence of factors that may negatively affect nutrients in foods.

The company must describe the method of insertion of DNA into the host, including the vector being used, the source of the DNA, insertion method used, and related

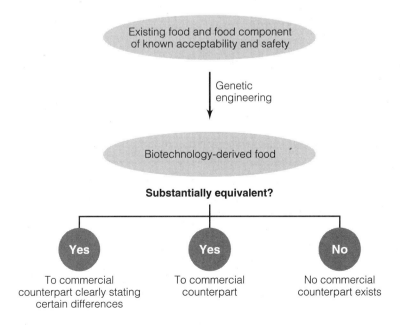

FIGURE 14.10 Substantial equivalence.

manipulations. Additionally, the company must provide to the FDA information on the modified organism, such as selection and screening methods, characteristics compared to those of the host, level of stability of expression of the new gene, number of copies of the gene within the organism's DNA, potential for rearrangement of the gene (gene interaction), and function of the gene.

Next, the food product resulting from the genetic manipulation must be characterized. This includes analysis of its composition, as well as its characteristics in comparison with an appropriate food already in the food supply, termed the **comparator.** In comparing composition, key nutrients and toxicants are examined. The more critical the nutrient or toxicant, the more care should be taken in making comparisons between the biotechnology-derived food and the comparator. In comparing characteristics, morphology, growth, yield, disease resistance, physiology, reproduction, and health are typically considered. Thus, substantial equivalence is proved to the FDA by demonstrating that the characteristics observed for the genetically engineered organism, or for the food derived from it, are equivalent to those of the comparator.

Substantially Equivalent Foods, Except for Defined Differences. Safety assessment of such products, beyond the initial evaluation, usually focuses on the defined differences only. These differences are the result of having introduced the new genetic material, and may be completely unintended. For instance, a genetically engineered organism may produce a certain desirable substance, the production of which was planned. This substance may lead to the production of an additional compound, completely unintentionally. Such compounds, whether planned or not, may be completely new, or may be the modified versions of existing molecules.

Products that are not exactly the same as existing foods must be assessed for safety by the company —and sometimes the FDA also commissions tests. If the bioengineered food is now able to produce a certain protein, the amino acid sequence of it is usually determined and compared with known proteins. If the protein is very similar in sequence to one known to be toxic, then toxicity is tested using animals or tissue culture cells. Another method to confirm the safety of proteins that are similar but not identical to known proteins is to analyze them for their susceptibility to degradation under low and neutral pH. Low pH degradation is performed to test the digestibility of the protein by stomach acid, and neutral pH degradation is performed to test the digestibility of the protein in the intestine. If the protein is not rapidly digested, other tests may be necessary.

For instance, a protein produced by introducing genetic material into an organism may perform adequately, but it has the additional ability to inhibit certain important functions such as absorption of nutrients by the body. In such a case, the effect of processing treatments to the food containing this protein must be assessed. If it turns out that this added function of the protein is inactivated after heating, then this protein may be allowed in the food, as long as the food is not consumed raw.

Foods Not Substantially Equivalent to Existing Foods or Products. Most probably, there are cases where foods manufactured through genetic engineering are substantially different from existing foods. Obviously, such foods cannot enter the marketplace, making it impossible to know how many such foods there are." However, there is always the possibility that such foods may be produced in the near future. This does not automatically make the food unsafe, but it will require extensive testing to ascertain its safety.

A product with no conventional counterpart must first be characterized in terms of describing the donor, as well as the host organism that received the genetic material, the type of genetic modification carried out, and the properties of the modified organism and/or food. Once this initial characterization is made, if no potential safety problems are discovered, the next step is to conduct testing with animal feeding studies. Human studies may also need to be conducted if the intended use of the new product is to have it become a major component of the diet. Of course, these studies should be conducted only when the animal testing deems the genetically engineered food to be safe.

14.4 IMPROVING PLANT PRODUCTS THROUGH BIOTECHNOLOGY

Efficient production of high-quality, nutritional crops is one of the keys to feeding the world's population. Biotechnology has helped in this arena, with the introduction of many new varieties of plant products, as well as with the engineering of microorganisms used to improve crop production.

Insect-, Virus-, and Drought-Resistant Plants

The bacterium *Bacillus thuringiensis* has been used as a natural insecticide for many years, capable of producing a protein that is toxic to various pest species such as beetles and soil worms. Through genetic engineering, variants of this organism have been developed that can overproduce the toxin. Moreover, the genes responsible for insecticide production have been inserted into other organisms like *E. coli*, which are more easily and cheaply grown than *Bacillus* species. Perhaps an even better alternative is that of inserting these genes directly into a variety of food plants such as cabbage, maize, potato, and tomato, essentially enabling the plants to have built-in resistance to pests.

Herbicide application is another area where food biotechnology can make a significant impact. Farmers typically apply herbicides as a preventive measure to their crops before weeds have an opportunity to grow. The problem is that some of these chemicals, those deemed to be safest in terms of being less damaging to the environment, actually affect the growth of desired food crops. Thus, farmers are left with having to use herbicides that are more environmentally toxic because these inhibit weeds while not affecting food crops. Plants are being developed that have the capability of tolerating the safest herbicides, enabling farmers to replace harmful chemicals with those more favorable to the environment.

Besides insecticide production and herbicide tolerance, biotechnology can impart other qualities to plants that enable farmers to grow better crops more efficiently. Resistance to plant viruses is one such quality that is highly desirable. These microorganisms are responsible for major reductions in yield and can affect the quality of some crops. There is no effective antiviral chemical, thus farmers have had to rely on conventional plant breeding methods to select for varieties that are naturally resistant.

One approach based on biotechnology is to insert genes into plants that code for viral proteins. The theory is that the presence of the genes would inhibit subsequent infection by that virus. Another application is currently being used to help plants fight fungal diseases such as soft rot. Genes that code for the production of the enzyme chitinase help the plant fight the fungal infection by digestion of the cell wall of the organism, which is primarily made of chitin.

Drought resistance would be of benefit to farmers, who often lose large amounts of their crops due to lack of sufficient rainfall. Dried yeast is an organism that can withstand dehydration without being harmed, requiring water to be revitalized. A plant called the "resurrection plant" can be found in a desert environment, which is also able to withstand prolonged periods of drought.

What imparts this ability to both the yeast and the plant? Answer: the sugar **trehalose**. This compound sequesters water for the cell, allowing the organism to survive when water is not available. It could be possible to insert the genes involved in the production of trehalose into other plants, such as potato and tomato, making them resistant to drought conditions.

Crop Improvement

The vast majority of the research being conducted in plant biotechnology is in the area of plant quality improvement. The research ranges from preventing spoilage to increasing sweetness and nutritional content.

Preventing Spoilage. Ripening of fruits and vegetables has been studied for many years due to the impact that it has on our ability to transport products to far-away markets, and to have out-of-season temperate crops available year-round. In the 1970s, scientists isolated the gene responsible for softening in tomatoes, which codes for the enzyme **polygalacturonase** (PG) (polygalacturonic acid). Using a technique called **antisensing**, researchers have been able to prevent expression of the gene coding for polygalacturonase, essentially inhibiting softening while not affecting the other aspects of plant ripening such as color and flavor development. The result is a tomato that looks and tastes as if it was ripened naturally, while still maintaining its firmness.

The technique of antisensing involves making a special copy of the DNA from the plant that codes for polygalacturonase, for instance. This "antisense" DNA gene has a sequence of bases that is complementary to that of the DNA normally produced by the plant. Once inserted, both DNA molecules will synthesize two mRNA molecules in the plant cell, one normal and the other an "antisense" mRNA. The idea is that the two complementary mRNA molecules will anneal or bind to each other, preventing translation of the genetic code that produces the enzyme polygalacturonase, thus preventing the softening process.

The FLAVR SAVR™ tomato developed by Calgene in 1994 can remain on the vine longer to ripen to full flavor before harvest and to remain firm after harvest because the PG enzyme has been suppressed (see Figure 14.11).

Antisense technology has also been used to inhibit production of ethylene by fruits and vegetables such as broccoli, raspberries, and melons. **Ethylene**, naturally found in many plants, is produced by the cells as part of the natural ripening process. Its inhibition impedes ripening, allowing fruits to be picked in the unripe state. At the desired time, storage of these products to an atmosphere containing ethylene would cause ripening to be resumed, but under controlled conditions.

Enhancing Quality Attributes. Biotechnology can also be used to produce fruits and vegetables with enhanced quality attributes. Sweeter crops like lettuce and tomatoes have been produced by inserting the genes for natural sweeteners, for example, **thaumatin**. This compound originates from tropical plants and has been calculated to be 3,000 times sweeter than sucrose.

Another example of quality improvement through biotechnology is the introduction of bacterial genes into potato plants, which increases the amount of starch in these tubers while reducing the water content. This increase in starch translates to less absorption of fat by the potato during frying, resulting in a more desirable product.

Increasing Nutritional Content. In improving quality, biotechnology can also be used to enhance the nutritional quality of foods. Potatoes have been developed in Israel that have improved amino acid content through genetic engineering. Given the overreliance by peoples in developing countries on potatoes for much of their diet, such enhanced products can help to minimize malnutrition in much of the world.

Oilseeds have been genetically modified by insertion of specific genes to produce a greater proportion of unsaturated to saturated fats. Such a change yields not only a more high-temperature oil, but also one that is considered by nutritionists to be a healthier alternative. Research involving rapeseed, soybeans, sunflowers, wheat, rice, and corn is directed at making low saturated varieties of oilseeds or varieties that process more efficiently or offer special phytonutrients. For instance, wheat bred to contain reduced levels of amylose and increased levels of amylopectin (termed "waxy starch") allows food processors to develop a wider range of products. Certain of these have exhibited improved freeze-thaw stability, which is important in frozen microwavable foods.

Preventing Production of Harmful Compounds. Genetic engineering can be used to prevent the production of certain harmful compounds, thereby increasing the wholesomeness of foods. One example is the manipulation of cassava, a plant product consumed in many African countries. The levels of cyanogenic compounds is fairly high

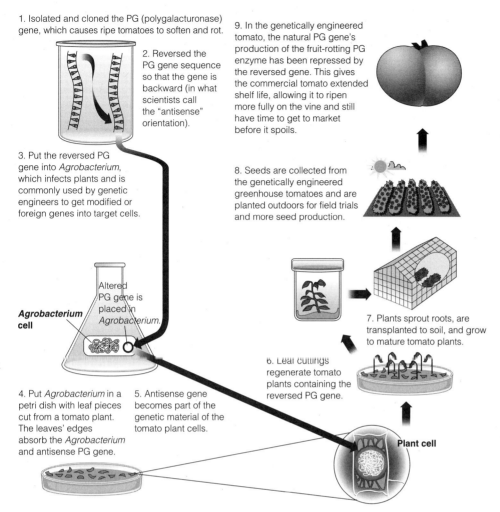

FIGURE 14.11 How the FLAVR SAVR tomato is created.

in these foods, leading to chronic illness if the product is not cooked thoroughly enough to inactivate these poisons (see Chapter 12, Food Toxicology). Genes can be manipulated by antisense technology and other methods to produce varieties with low levels of these potentially deadly compounds.

14.5 IMPROVING ANIMAL PRODUCTS THROUGH BIOTECHNOLOGY

Perhaps the best known application of genetic engineering in animal production is the manufacture of bovine somatotropin hormone, or **BST**. Back in the 1930s, scientists discovered that this hormone, produced by the pituitary gland of cows, increased milk production by 10–14 percent when injected into these animals. The problem with its use was that not enough BST could be extracted from one animal, with the pituitaries of many cows being needed to yield enough BST to make even one injection. The gene responsible for BST production was isolated in the 1970s, and it was successfully inserted into *Escherichia coli* (Figure 14.12). Now, BST can be mass-produced in large fermentors, enough to supplement many animals.

Similar advances have been made with porcine somatotropin, or PST, which is used in swine production to convert energy from feed into muscle instead of fat, resulting in

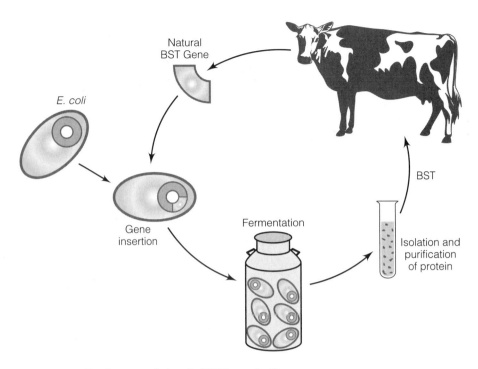

FIGURE 14.12 Bovine somatotropin (BST) production.

leaner pork. Similar techniques as those used for BST and PST production can be applied in the production of growth promotants and vaccines. In addition, biotechnology can be used to construct strains of bacteria, termed **probiotics,** that are known to help in the digestion of feed, improving animal health and efficiency of nutrient uptake.

The challenge is that probiotics have to compete with other microorganisms colonizing the intestinal tract for a place in that environment. Probiotic bacteria could be genetically engineered so that they could easily colonize the gastrointestinal tract of animals by producing inhibitory substances that would not allow the growth of other intestinal bacteria. Thus, through the use of genetically engineered probiotics one could essentially control the types of organisms living in the gut of food animals.

In addition to BST and probiotics, genetic engineering can be applied to actual animals to improve their health and the quality of products derived from them. In experiments in the late 1980s, pig embryos were injected with a gene that encoded for PST. The result was animals with very lean muscle tissue. Similarly, scientists have inserted an extra copy of the gene for a specific milk protein called **lactoferrin** into cow embryos. This protein helps fight bacterial infection in animals; it is active against organisms that cause mastitis. The implications of this research are that healthier animals can be produced, minimizing the need for antibiotic therapy on the farm. Similar experiments have been carried out with chickens and fish.

14.6 IMPROVING FOOD PROCESSING AIDS THROUGH BIOTECHNOLOGY

Many food processing aids, meaning specific food additives such as gums, sweeteners, natural coloring agents, and enzymes are necessary for the production of many of the foods we enjoy. Biotechnology can be used to manufacture these additives more economically and efficiently than conventional methods. The basis for the manufacture of food processing aids by this technology is through the use of genetically modified microorganisms. We will explore some examples of how bacteria and yeasts can be used to produce high-yielding, good-quality food additives.

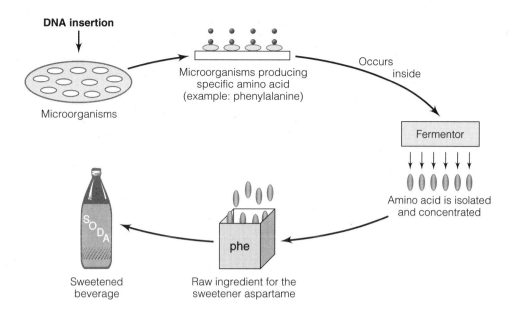

FIGURE 14.13 Use of microorganisms to mass-produce specific amino acids.

Amino Acids Used in Food Production

Amino acids are used in food production as flavor enhancers, nutritional additives, and seasoning agents. Microorganisms are first modified by inserting a gene that codes for the desired amino acid. Large fermentors are then used to grow the microorganism rapidly and efficiently, resulting in the release of large quantities of the desired product into the fermentation medium for easy harvesting. Examples of amino acids that can be manufactured in this way are glutamic acid, which is used as a flavor enhancer in the form of monosodium glutamate, and phenylalanine, which is used in the making of the sweetener aspartame (see Figure 14.13).

Gums Used in Food Production

Gums are processing aids that serve to thicken and/or emulsify foods, and some are used as fillers. Some bacteria are known to produce hydrocolloid substances that can be used for these purposes, such as curdlan, gellan, and xanthan gum. Through biotechnology, such organisms can be genetically altered to overproduce these polysaccharides, thereby manufacturing large quantities of a wide range of new gums. In a different application of biotechnology, bacteria can be modified to overproduce the enzyme galactosidase, which can convert cheap gums, such as guar gum, into more expensive locust bean gum.

Enzymes Used in Food Production

Enzymes of importance to the food industry can also be produced in large quantities by modified microorganisms in a fermentor. Examples are chymosin, used to coagulate milk proteins in the manufacture of cheese; pectinases, used to clarify wine and fruit juices; and proteases, used to tenderize meat and to modify proteins in dairy products. Similarly, biotechnology-derived citric, acetic, lactic, and ascorbic acids can be manufactured in large volumes by genetically engineered bacteria through fermentation.

Microorganisms Used in Food Production

In addition to bacteria, yeasts such as *Saccharomyces* are used in food production, typically in the manufacture of bread and beer products. The yeast *S. cerevisiae* is used traditionally, but since it cannot break down starches to simple sugars by itself, enzymes such as amylase must be added before sugar fermentation by the yeast can take place. This makes the process somewhat cumbersome and expensive.

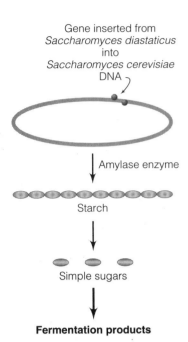

FIGURE 14.14 Insertion of gene from one yeast cell to another.

Another yeast, *S. diastaticus*, has been shown to produce amylase, while also having the ability to ferment simple sugars. The problem is that the organism also produces compounds that impart bitter flavors to foods. With biotechnology, it is now possible to take out the gene that codes for the amylase enzyme from *S. diastaticus* and insert it into *S. cerevisiae*, enabling the latter to accomplish both the breakdown of starch and the fermentation of the resulting sugars (Figure 14.14).

14.7 APPLYING BIOTECHNOLOGY-DERIVED FOODS IN FOOD SAFETY

Starter bacterial **cultures** used in the manufacture of yogurt and other dairy products have been studied for their ability to synthesize specific compounds that inhibit disease-causing organisms.

Nisin: Antimicrobial Agent

Starter organisms include the lactobacilli, some of which have been genetically engineered to overproduce the compound **nisin.** This product is considered an antimicrobial agent and is known to inhibit the pathogen *Listeria monocytogenes*. This organism is present in the environment and has been implicated in several outbreaks where postprocessing contamination occurred. Most notably, a large outbreak of foodborne illness from consumption of soft cheese contaminated with *L. monocytogenes* resulted in several deaths in the early 1980s. Addition of genetically modified bacteria capable of producing large quantities of nisin could thus improve the safety of dairy products that may be contaminated with this organism.

Diagnostic Biotechnology

Another application of biotechnology to food safety is in the area of diagnostics. Correct identification of pathogens isolated from foods by quality assurance personnel is imperative in the production of safe food products. Misidentification can result in risk to consumers if it is incorrectly deemed as harmless, or can result in unnecessary product recalls by the company if an organism is falsely identified as a pathogen.

An example of such a diagnostic test based on biotechnology is DNA probe analysis. With this technique, a probe containing a known sequence of bases belonging to a particular organism, for example, *Salmonella*, is allowed to react with a piece of DNA from the suspected organism. A positive reaction signifies that the two DNA molecules were able to anneal, or bind to each other, an event that can take place only if the bases contained in the piece of DNA from the organism in question match those of the probe.

14.8 MAJOR CONCERNS ABOUT BIOTECHNOLOGY-DERIVED FOODS

Biotechnology-derived foods are currently under public scrutiny. What specific issues are being raised, and what responses are being given?. When new products are introduced into the marketplace and consumer advocacy groups express concerns, it is up to science to provide accurate answers.

Concern #1: Are Biotech Foods Harmful?

Since biotechnology-derived foods are new and long-term effects have not been studied, some people fear that crops or new foods developed with biotechnology may result in harm to the consumer. Since there are several historical examples of approved substances later being shown to be harmful, this is a rational concern.

Some people are also worried about the unknown consequences of genetic engineering. What happens when genestically modified organisms mutate? What happens when genes from different species within a genetically modified organism interact over many generations?

Concern #2: Do Biotech Foods Harm the Ecosystem Balance?

Concern #2 is the fear that distributing biotechnology-derived entities into the environment may provide an unfair advantage to a particular sector of the ecosystem over another. For example, can using genetically engineered bacteria capable of producing herbicides to eliminate weeds ultimately result in the overgrowth of plants that are naturally resistant to these chemicals? Presence of these plants could cause a natural imbalance, resulting in elimination of certain animal species.

Concern #3: Will Big Companies Dictate Food Biotech Practices?

Concern #3 is that biotechnology may result in the concentration of economic power in the hands of a few corporations, who would dominate the marketplace, making it difficult for others to participate. This fear is expressed in many developing nations, who believe that their technological inadequacy may leave them prey to multinational corporations to fulfill their food supply needs.

Concern #4: Does Genetic Engineering Create Unnatural Consequences?

Concern #4 is that scientists will use genetic engineering not only to cross species barriers but to carry out unnatural acts such as the cloning of animals and humans and the cross-breeding of animal genes with those of plants. Some people view these actions as immoral, violating the sanctity of life on which we base our society.

Some Answers

Concerns about the safety of biotechnology-derived foods is often due to a lack of understanding on the part of consumers regarding the procedures necessary for these foods to be approved by the FDA. Concerns may also be due in part to a distrust of the FDA or other branches of government. Lack of labeling may contribute to the distrust. However, the extensive testing that is conducted in petitioning the government for approval of any new food or ingredient makes biotechnology-derived foods among the most thoroughly researched.

It is true that these crops and products have not been tested for 20 or 30 years. But foods produced by conventional methods are also not without health risks, yet in many cases their wholesomeness has not been as thoroughly studied as that of genetically engineered products. Whether they realize it or not, people have actually been consuming bioengineered plants and animals for years. Production of flavorful varieties of fruits, beef cattle able to withstand heat, and cows that produce large quantities of milk are some examples of successful bioengineering.

Concerns regarding the release or mutation of genetically engineered microorganisms, plants, or animals is a question of risk. How likely is it that such a practice will result in damage to the ecosystem, and ultimately to the consumer? Do the benefits of biotechnology outweigh the risks? Science can provide only *some* answers—it is up to the public and to policymakers to make these value judgments. However, consumers and policymakers should do so with knowledge that is based on science. Concerns about the potential for monopolies to develop among food production companies and about the moral implications of biotechnology should be discussed and debated nationwide.

The International Food Biotechnology Council

To address these issues, the International Food Biotechnology Council, a body composed of 30 major food processing and biotechnology companies, commissioned a report in 1990 to investigate whether the above-mentioned concerns have basis in fact. After more than two years, a panel composed of scientific experts from universities and member companies produced a document of findings and recommendations.

This document was subsequently reviewed by more than 150 additional scientists, regulators, and other experts from 13 countries. One of the major conclusions of the panel is that the necessary scientific knowledge already exists to ensure that these concerns are properly addressed. Further, they recommended that biotechnology-derived foods should be considered safe, provided they fulfill the following criteria:

1. The nutrients do not differ from those of traditional foods.
2. Any toxicant is within typical and acceptable levels found in traditional foods.
3. New products or constituents do not present an unacceptable risk.

For any concern regarding the use of biotechnology in food production to be properly addressed and discussed, public education is essential. The education of government regulators is also needed to ensure as much as possible that this technology will be used prudently and with minimal risk to consumers.

The StarLink Controversy

Beginning in the fall of 1999, the FDA received several consumer complaints of allergic reactions due to consumption of foods containing StarLink™ corn. This variety of corn was produced through biotechnology by Aventis Corporation and had been approved by the FDA only for use as animal feed. The consumers experienced vomiting, diarrhea, hives, runny nose, asthma, and anaphylaxis, apparently after consuming corn tortilla shells, chips, and corn cereal sold in grocery stores.

Approximately six dozen persons were affected, as reported by the Centers for Disease Control and Prevention. Specifically, the symptoms may have been caused by an allergic reaction to the protein Cry9C, contained within the genetically modified corn. As a result, millions of taco shells were pulled from grocery stores nationwide.

The controversy regarding this issue involves several items. First, since the tests used by FDA could detect only the presence of the gene that codes for this protein, but not the protein itself, the presence of the allergen was not proven. Second, Aventis data show that StarLink corn protein contains only 0.0129 percent Cry9C, thus the probability for allergic reactions is very small. In addition, Aventis claims that this protein is eliminated during the process of wet-milling and contends there is no risk from consumption of processed products. Third, given that some people are allergic to normal corn, the banning of StarLink corn may be an unfair overreaction on the part of the government.

Still, consumers have expressed a concern over the possible presence of this genetically modified corn in their food supply. Given that StarLink was not approved for human consumption by the FDA, their concern is quite valid. The reader can be the judge regarding what should be learned from this episode!

Key Points

- Genetic engineering is a tool that can be used to improve the quantity and quality of our food supply, forming the basis for the field of biotechnology.
- Electroporation can be used to introduce genetic material into a cell, with plasmids being used as the vectors to carry new genes.
- DNA can be inserted into plant cells through ballistic bombardment, and into animal cells through microinjection.
- Marker genes are used to verify whether DNA has been successfully introduced into the desired organism; the ability to express the gene is verified through gene switches.
- Once genetic material is successfully inserted into a cell, its product can be mass-produced by growing the microorganisms in a fermentor. In the case of plant cells, tissue culturing is used. With genetically modified animal cells, cloning is employed to insert the embryo into a surrogate mother.
- The Food and Drug Administration is the governmental agency that controls and regulates the application of biotechnology for food production.
- The concept of substantial equivalence is used globally to establish the standards by which biotechnology-derived products can be deemed safe.
- Examples of applications of biotechnology to increase or improve plant production are the development of drought-resistant, herbicide-resistant, or pesticide-tolerant plants.
- Application of biotechnology in animals include the increase of milk production by artificial use of BST and promotion of lean muscle tissue development in swine by PST.
- Food processing aids such as enzymes, gums, flavor enhancers and even antimicrobial agents can be produced by genetically altered microorganisms.

Study Questions

Multiple Choice

1. The mating of an offspring with its parent is known in plant breeding as
   a. cross-mating.
   b. cross-pollination.
   c. back-to-back mating.
   d. backcrossing.
   e. crisscrossing.

2. The term *transgenic* refers to plants or animals that are
   a. modified by transformation.
   b. modified by gene expression.
   c. modified by transmutation.
   d. modified by genetic mutations.
   e. modified by genetic engineering.

3. This gene is one that is inserted in reverse:
   a. nonsense
   b. antisense
   c. reversible
   d. backward
   e. negative

4. Genetic engineering is defined as
   a. the manipulation of chromosomal DNA.
   b. the manipulation of mRNA.
   c. the engineering of transfer DNA.
   d. the manipulation of genes.
   e. the manipulation of life.

5. _____ enzymes used to cut DNA in genetic engineering.
   a. Manipulation
   b. Convertible
   c. Restricted
   d. Restriction
   e. Conjugated

6. The use of an electric field to insert DNA into cells is
   a. electrostriction.
   b. electric shock.
   c. electrolysis.
   d. electron capture.
   e. electroporation.

7. Genes that code for a recognizable product, such as antibiotic resistance, are _____ genes.
   a. marker
   b. flag
   c. complete
   d. recognizable
   e. durable

8. The letters PCR stand for
   a. Polymerase Crucial Reaction.
   b. Polymerase Circular RNA.
   c. Polymerase Chain Reaction.
   d. Polymerase Correct Recycling.
   e. Polymerase Chain Routing.

9. Bacterial reactors used for mass production of specific products are called _____.
   a. reactors
   b. fermentors

c. vats
d. modulators
e. production vessels

10. The infection of plant tissue with *A. rhyzogenes* is
    a. root culturing.
    b. cell culture.
    c. hairy root culture.
    d. infective culture.
    e. culturing.

11. Cloning is defined as the splitting of an embryo into _____ cells.
    a. 2
    b. 4
    c. 6
    d. 8
    e. 10

12. Comparison of genetically modified foods with standard foods is called
    a. standard equivalence.
    b. standard equity.
    c. substantial equity.
    d. substantial effort.
    e. substantial equivalence.

13. The appropriate food that bioengineered foods are compared to are
    a. comparators.
    b. comparables.
    c. appropriates.
    d. compatriots.
    e. comparallels.

14. _____ is an organism known to produce a natural herbicide.
    a. *B. thuringiensis*
    b. *B. cereus*
    c. *B. thermophilus*
    d. *E. coli*
    e. *E. thermodurans*

15. The enzyme _____ breaks down fungus cell walls.
    a. polymerase
    b. chitinase
    c. cholerase
    d. pectinase
    e. endonuclease

16. Trehalose sequesters _____ from cells.
    a. sugars
    b. lipids
    c. proteins
    d. carbohydrates
    e. water

17. The enzyme responsible for softening in tomatoes is
    a. polymerase.
    b. polygalacturase.
    c. polygalacturonidase.
    d. polygalacturonase.
    e. polyase.

18. Fruits produce the following compound, which aids in the ripening process:
    a. galacturonic acid
    b. ethylene
    c. oxygen
    d. salt
    e. urea

19. Thaumatin is a natural _____ .
    a. antibiotic
    b. fat substitute
    c. sweetener
    d. acidulant
    e. antioxidant

20. Live cultures of bacteria that aid in food digestion are known as
    a. procultures.
    b. probiotics.
    c. biotins.
    d. prodigestins.
    e. bioproteins.

Critical Thinking

21. When bananas are cut, enzymatic browning causes the fruit to become brown and darken. This event can be prevented through the use of sulfates, but their use has been limited because of allergic reactions to them. Thus, pre-cut bananas are not usually made available to the public. Discuss how we could use biotechnology to solve this problem.

22. You work for a company that produces a new food processing aid called "berrylicious." This compound imparts a berry flavor when added to any food, while it also increases the vitamin C content of the food. What specific steps would you take to test the safety of this new product so you would satisfy the requirements set by the FAO/WHO?

23. Identify four concerns that some consumers have about biotechnology-derived foods, and address them using sound science.

24. Team Exercise: Work in teams to complete the following. Some people think that the safest, most nutritious, and overall best-tasting crops grown are those that are organically grown. What "organic" typically means is that the crops are grown without the addition of any chemical pesticides or chemical fertilizers. But what if a crop was produced through biotechnology? The *Bacillus thuringiensis* bacterium produces a protein, called Bt, that kills insect pests. By inserting the bacterium's genes that encode for this protein into corn, "Bt corn" results. The Bt corn crop would therefore have built-in protection (pest resistance) without the need of pesticide chemicals. Another way would be to grow cultures of the bacterium that produce Bt protein, and collect the protein, and make it into a spray for applying to crops instead of spraying with pesticides. The following scenario addresses whether genetically engineered crops can be called organic.

    Imagine that you are a farmer raising potatoes and that you have stopped using chemicals. Instead, you have switched to a Bt system of pest control, as supplied by a genetic engineering company. You have seen great results with the product. You want to market your produce as "organically grown" since you no longer use chemicals and can prove it with documentation.

    The two people who serve as your major distributors are Ms. Pro and Mr. Con. They differ in their opinion of whether you should try to market your potatoes through their retail outlets as "organic" produce: Ms. Pro says yes while Mr. Con says no.

    • Read the following statements, which could have been made by either Ms. Pro or Mr. Con, or even both, and identify which person you think made each statement.
    a. "A bioengineered potato is an organic food."
    b. "A bioengineered potato is not an organic food."

c. "Bt is a natural killer of Colorado potato beetles. If a beetle grub happens, while chomping on a leaf, to ingest Bt potato cells, the Bt poison will kill the grub."
d. "Organic farmers spray Bt to control beetles. It's not a poison; it's a natural enemy. It doesn't infect bees or worms or birds or people or anything but the particular beetles that are its host."
e. "The problem here is not that companies make mistakes—of course they do. The problem is that genetic engineering, like nuclear power, is not an arena where we can afford any mistakes."
f. "The Colorado potato beetle is second only to the green peach aphid in its acquired resistance to hard-core pesticides. But it is not yet resistant to Bt."
g. "Nature doesn't normally mix the genes of bacteria and potatoes, or frogs and lettuces, or pigs and people, and for very good reasons."
h. "Bt corn has proven to be a valuable pest management option for the corn producer. It provides nearly 100 percent control of the European corn borer, which protects the crop from a yield loss and helps reduce insecticide use."
i. "We don't know what ecosystems will do with genetically altered species."
j. "Throughout time, species have survived according to their ability to fit into changing ecosystems. In the hands of biotechnicians, the rate of evolution speeds up, and species are selected by their ability to fit into economic markets."

- For each of the ten statements (a–j) above, decide (1) which one of these six areas—food safety, pest resistance, altering the natural pace and selection mechanism, unknown natural consequences, human error, and breaking the species barrier—is the focal point of the statement, and (2) whether or not the statement supports genetic modification of food crops.
- What is your team's opinion—is a bioengineered potato an organic food?

References

Bills, D. D., and Kung, S. 1990. *Biotechnology and Food Safety*. Butterworth and Heinemann, Boston.

Bruhn, C. M. Consumer concerns and educational strategies: focus on biotechnology. *Food Technology* 46(3):80.

Cooperative Extension Service, University of Wyoming, and the Cooperative Extension at Washington State University. 1994. *Biotechnology, Food, and Agriculture*.

Gendel, S. M., Kline, A. D., Warren, D. M., and Yates, F. 1990. *Agricultural Bioethics: Implications of Agricultural Biotechnology*. Iowa State University Press, Ames.

Harlander, S. 1989. Introduction to biotechnology. *Food Technology* 43(7):44.

Harlander, S. 1991. Biotechnology—a means for improving our food supply. *Food Technology* 45(4) 89.

Knorr, D. 1987. *Food Biotechnology*. Marcel Dekker, New York.

O'Donnell, C. 1996. Bio-engineered proteins. *Prepared Foods* 165(1):43.

Otto, P. E. 1996. Biotech facts squelch seeds of controversy. *Food Product Design* 6(12):17.

USFDA. 1994. Biotechnology of food. USFDA, Center for Food Safety and Applied Nutrition. FDA Backgrounder, May 18, 1994.

Additional Resources

Babcock, B. C., and Francis, C. A. 2000. Solving global nutrition challenges requires more than new biotechnologies. *Journal of the American Dietetic Association* 100:1308.

Boyle, M. A. 2001. *Personal Nutrition*, 4th ed. Wadsworth/Thomson Learning, Belmont, CA.

Delves-Broughton, J. 1990. Nisin and its uses as a food preservative. *Food Technology* 44(11):100.

Henkel, J. 1995. Genetic engineering: fast forwarding to future foods. *FDA Consumer*, April 1995, p. 6.

Pendick, D. 1992. Better than the real thing: industry serves up the fruits of tomato biotechnology. *Science News* 142:376.

Raso, J. 1998. The biopharm revolution. *Priorities* 10:33.

Sizer, F. and Whitney, E. N. 2000. *Nutrition: Concepts and Controversies*, 8th ed. Wadsworth/Thomson Learning, Belmont, CA.

Whitney, E. N., and Rolfes, S. R. 1999. *Understanding Nutrition*, 8th ed. Wadsworth/Thomson Learning, Belmont, CA.

Zind, T. 1999. A new breed of ingredients. *Food Processing* 60(1):39.

Web Sites

The following web sites are useful resources for information on food biotechnology.

www.agnic.org
USDA Biotechnology Information Center (Agriculture Network Information Center)

www.whybiotech.com
Council for Biotechnology Information

www.ift.org/govtrelations/biotech/biotechnology.shtml
IFT biotech information

www.ific.org
Search for "biotechnology" at the International Food Information Council site.

www.greenpeaceusa.org
Greenpeace. For an antibiotechnology viewpoint go to the genetic engineering section.

Other web sites of biotechnology interest:

www.who.int/fsf/gmfood/index.htm

www.oecd.org

www.cast-science.org/biotechnology/index/html

BIOENGINEERING OF β-LACTOGLOBULIN

This challenge is based upon information in the article "Bio-engineered proteins" (O'Donnell, 1996). It focuses on a specific natural protein, β (beta)-lactoglobulin, a functional whey protein component of milk. β-lactoglobulin is an important protein biologically in living systems as well as functionally in foods. One of its roles in the body is to transport the vitamin A precursor molecule retinol, a hydrophobic compound. It comes as no surprise, then, that it can also function in foods to bind hydrophobic molecules, such as lipid-based flavorings.

This whey protein, like other food proteins, can form a gel under the proper conditions of concentration, pH, added emulsifier, and ion concentration. The fact that whey is rich in β-lactoglobulin accounts for its ability to act as a flavor carrier. As the whey substance gels, any flavor ingredient added to the system becomes entrapped in it. Typical lipid-based flavorings added to food gels include lemon oil, orange oil, citral, and benzaldehyde (Figure 14.15).

One of the challenges facing the food product developer in the use of whey proteins as gelling agents and flavor carriers is their stability. Being a protein, β-lactoglobulin is affected by heat because heat denatures proteins. Thus, gels made from whey protein, and the entrapment of any flavor in that gel, become unstable if temperatures reach a critical level. See Figure 14.16.

Thus, a protein's molecular binding properties, gelling ability, stability to heat, and flavor retention properties are all interrelated. What this means to the food biotechnologist is a new opportunity because it might be possible to change such properties by altering the amino acid sequence of a polypeptide, which would alter its shape and functioning.

As an example, although β-lactoglobulin is soluble under acidic conditions (pH about 3), when heated in excess of 85°C (185°F) it denatures and forms aggregates. Under certain conditions this can precede gel formation.

The application of genetic engineering would be to identify the critical amino acid sequence in the β-lactoglobulin structure that relates to specific functioning. Next, the protein could be modified through cloning experiments to express the complementary DNA (remember cDNA) using an organism like *Escherichia coli*. Later, if it were feasible, the more economical way to express the new bioengineered protein would be to insert the gene into cow DNA.

From the molecular point of view, the replacement of an amino acid or two (in this case, alanine or leucine) in the protein's natural sequence by a cysteine amino acid would

FIGURE 14.15 Chemical structures of limnolene, naringin, citral, and benzaldehyde, as examples of lipid-based flavorings.

Limnolene (lemon oil)   Naringin (orange oil)   Citral   Benzaldehyde

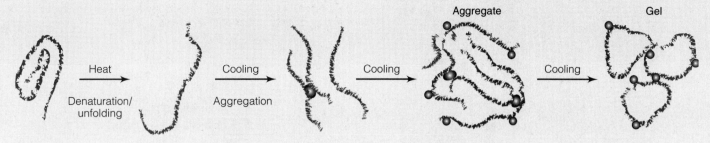

Figure 14.16 Thermal treatment of polypeptides. *When β-lactoglobulin is dispersed in heated water, hydration and denaturation occur. Allowed to cool, the dispersed polypeptides form aggregates through increased protein: protein associations. The warm sol system with time and continued cooling sets into a gel. Within the gel, water becomes entrapped into the three-dimensional gel network of protein.*

allow for the creation of a strong crosslink, via a cysteine-cysteine disulfide bond, in the modified protein.

Formation of a disulfide bond:

R--------SH HS--------R   R--------S-S----------R
cysteine residues ("thiol" groups)   disulfide bond between cysteines

According to researchers, this would create a molecule with more thermal stability and reduced gelling tendency. The application would be for use of this modified β-lactoglobulin as a flavor carrier in non-gel food systems, such as beverages. Furthermore, increasing β-lactoglobulin's stability to heat would permit the development of yogurt products that, although heat pasteurized, would retain the desired gel structure.

Challenge Questions

True/False

1. T/F   At a temperature of 90°C (194° F), unmodified β-lactoglobulin in solution would exhibit increased solute:solute rather than solute:solvent interactions.
2. T/F   A useful way to bioengineer β-lactoglobulin would be to modify its gene so that the amino acid content of the engineered protein would be higher in leucine and alanine, but lower in cysteine.
3. T/F   For modified β-lactoglobulin, increased thermal stability corresponds to decreased gelation.
4. T/F   Modified β-lactoglobulin would display decreased thermal stability if a greater number of disulfide linkages could be achieved.
5. T/F   Limnolene, citral, and benzaldehyde are examples of genetically engineered lipid-based flavorings commonly added to protein gels.

Critical Thinking

6. How would an emulsifier contribute to the structure of a flavored whey protein gel?
7. The interior of the β-lactoglobulin molecule is hydrophobic. Assume this protein is globular in shape, and sketch how the lipid-based flavoring citral would associate with it.
8. Figure 14.17 shows flavor release of benzaldehyde in whey protein isolate (WPI) gels exposed to various microwave heat treatments. What effect did the emulsifier have on the heat stability of the flavor in the whey protein isolate gel?

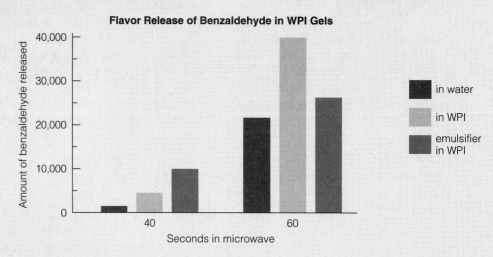

FIGURE 14.17 Flavor release of benzaldehyde from WPC gels *Adapted from O'Donnell, 1996.*

Multiple Choice

9. Which characteristic of beta lactoglobulin helps it carry lipid flavors?
   a. hydrophobicity.
   b. gelation functionality.
   c. ability to denature.
   d. water solubility.
   e. crosslinking potential.

10. What do limnolene, naringin, and citral have in common?
    a. dual benzine ring structure.
    b. presence of alcohol groups.
    c. presence of disulfide bonds.
    d. presence of aldehyde groups.
    e. lipid solubility.

11. Which is *not* a desired functional property of beta lactoglobulin in a beverage?
    a. act as a flavor carrier.
    b. increased thermal stability.
    c. stability to pasteurization.
    d. increased gelling ability.
    e. ability to be evenly dispersed.

CHAPTER 15

Sensory Evaluation and Food Product Development

15.1 What Is Sensory Evaluation?

15.2 Sensory Odor, Flavor, and Mouthfeel Perception

15.3 Sensory Texture and Color Perception

15.4 Responses Contributing to Sensory Perception

15.5 Sensory Tests

15.6 The Role of the Sensory Evaluation Specialist in Product Development

15.7 Product Development

15.8 The Role of Marketing in Food Product Development

15.9 Product Probability, Life Cycle, and ANN

Challenge!
Experimental Design in Product Development

CHAPTER OBJECTIVES

After completing this chapter, you will be able to:
- Discuss the meaning and value of sensory evaluation.
- Summarize the key sensory parameters of importance in sensory work.
- Classify sensory methods as discrimination, descriptive, and affective testing.
- Evaluate the need to obtain objective and subjective measurements in determining food quality.
- Explain the stages of product development.
- Assess the role of marketing in product development.
- Calculate the probability of success for a new food product.
- Define what is meant by a product's life cycle.

New food products are an essential part of the food industry. "New" may mean a previously unavailable product, a different package for a traditional product, or a formulation change to create an improved product. For example, a manufacturer has test marketed an alcoholic lemonade; another has studied incorporating long-chain PUFAs (arachidonic acid and DHA, docosahexaenoic acid) into infant formulas; into products traditionally made with hydrogenated fats; a third has introduced a juice called "Cold Buster 100," an orange juice blended with zinc, vitamins A and C, and the herb *Echinacea*. The research and development of these and other new food and beverage products are within the realm of product development, which involves the conceptualization, formulation, processing, testing, and marketing of food products. In product development, the use of sensory evaluation techniques, which obtain human responses to foods (called organoleptic responses), assumes great importance because sensory data provide information on both food product acceptability and quality.

In this chapter, the scientific basis of sensory evaluation will be presented, noting its link to product development. Since successful product development requires the food scientist to be competent in all of the areas covered in this text, including additive functionality, food safety, and regulations, it is fitting that the text conclude with a chapter devoted to the understanding of sensory evaluation and product development.

15.1 WHAT IS SENSORY EVALUATION?

Sensory evaluation is the assessment of all the qualities of a food item as perceived by the human senses. It is not merely food "tasting"; it can involve describing food color as well as texture, flavor, aftertaste, aroma, tactile response, and even auditory response. Sometimes *sensory analysis* is used interchangeably with *sensory evaluation*, although analysis is perhaps more correctly used regarding the statistical method applied to understand sensory data. The Institute of Food Technologists (IFT) defines sensory evaluation as:

> the scientific discipline used to evoke, measure, analyze, and interpret human reactions to those characteristics of foods and beverages as they are perceived by the senses of sight, smell, taste, touch, and hearing.

Figure 15.1 illustrates some of the sensing impressions provided by the five senses.

A Scientific Method

It is important to appreciate that sensory evaluation is a *scientific method* used to obtain, analyze, and interpret observations made through the senses of sight, smell, touch, taste, and hearing. As a scientific method, it incorporates the identification of a problem, the statement of a hypothesis, and an experimental strategy to investigate the problem. This strategy is accomplished through data collection and analysis, and a conclusion is reached that answers the original question. The conclusion either accepts or rejects the hypothesis according to the results of the study.

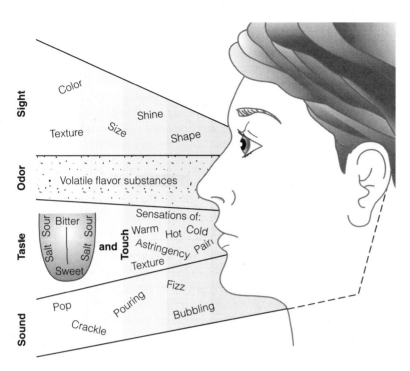

FIGURE 15.1 The role of the five senses in sensory evaluation.

A Quantitative Science

Sensory evaluation or sensory science is a quantitative science in which numerical data are collected to establish specific relationships between product characteristics and human perception. In sensory studies, human responses to stimuli are quantified. When sensory science is used in conjunction with product development, for instance, the proportion of a population of individuals that expresses a preference for one product over another can be determined.

In stating or defining the problem, the sensory specialist must define precisely what is to be studied and measured. The sensory specialist must choose the correct null hypothesis (that of no difference) and the correct alternative hypothesis. The test design must leave no room for subjectivity, must take into account known sources of bias, and minimize the amount of testing required to produce the desired accuracy of results. The nature of the test subjects ("instruments") must be determined; for example, are untrained consumers called for, or trained panelists? If trained, panelists must be selected on the basis of the ability to provide a reproducible verdict during training. The sensory specialist may only draw conclusions that are unambiguously supported by the results.

Sensory Science in the Food Industry

The main applications of sensory evaluation in the food industry are in quality assurance and product development. For maximum benefit, the sensory evaluation department or team within a food company interacts with other departments. The primary interaction is in support of product research and development, in the same way that marketing research supports a company's marketing efforts. In addition, sensory specialists interact with those who work in packaging and design, quality assurance, marketing research, and legal services. The latter deal with regulatory issues such as claim allowance and substantiation, and also advertising challenges.

15.2 SENSORY ODOR, FLAVOR, AND MOUTHFEEL PERCEPTION

Two types of senses are involved in sensory perception: chemical (taste and odor) and physical (sight, sound, and touch). However, the complex nature of flavor must be recognized. The important biological and physiological mechanisms of these senses will be briefly examined, and the sensory terminology relevant to describing foods as "character notes" will be set forth. **Character notes** are the sensory attributes of a food that define its appearance, flavor, texture, and aroma. Learning how food characteristics are perceived and influence acceptability is relevant because the food scientist applies it in food product development.

Taste

The organ of human taste (gustation) is the tongue. **Taste** may be defined as the sensation derived from food as interpreted through the tongue-to-brain sensory system. The four primary taste sensations of sweet, salty, sour, and bitter, as well as a fifth one, umami (meaning "delicious" along with meaty and savory suggestions) all trigger brain response. The receptor areas on the tongue shown in Figure 15.2 are those of maximum sensitivity for each of the four primary tastes.

Taste is thought of as a chemical sense. Taste involves the detection by the taste buds located primarily on the surface of the tongue of various food stimuli molecules that are perceived to have taste, called **tastants.** These molecules can possess the characteristics of sweet, bitter, and so forth. Contact between stimulus and taste bud is critical for the taste sensation to occur. The taste sensation is the result of a swarm of chemical signals inside the taste buds that culminates in the release of neurotransmitter molecules.

Taste buds are actually epithelial receptor cells organized into clusters of 50–150, which are embedded into structures called *papillae*. Humans possess several types of papillae: foliate, circumvallate, fungiform, and filiform. The taste receptor cells function to detect taste stimuli as well as transfer taste information to the brain. Newborns have between 8000 and 10,000 taste receptors. Adults possess 4000 to 6000 taste receptors, and the number declines with age to 2000 to 3000 in later life.

A chemical substance used in experiments to identify the detection threshold is PROP, 6-n-propylthiouracil. People differ in their taste sensitivity to this chemical. Three categories of tasters can be identified based upon response to PROP: "supertasters," "medium tasters," and "nontasters"—see Table 15.1. It is thought that individuals with greater numbers of fungiform papillae on the tongue are supertasters, while others are either medium tasters or nontasters.

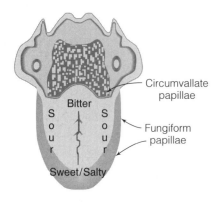

FIGURE 15.2 **The tongue, sensory organ of taste, showing areas of *maximum* sensitivity for each primary taste.** *All qualities of taste can be elicited from all the regions of the tongue that contain taste buds (papillae). The current evidence contradicts a long-held view of spatial segregation of sensitivities into specific zones or areas of the tongue.*

*Character notes differ in meaning from top notes. When researchers speak of "top notes" they mean the most immediate, prominent sensory flavors in a particular food or beverage.

TABLE 15.1 Characteristics of tasters: PROP categories.

Category	Characteristic
Supertasters	Have strong food dislikes
	Talk about hating a food
	Can detect specific ingredients in food samples
Medium tasters	Tend to like all foods
	Are fussy about preparation
	Think about food in an eager, positive way
Nontasters of PROP	Can taste other flavors
	Difficulty distinguishing different ingredients in a food
	Think of foods in terms of healthfulness rather than taste
	Attentive to the presentation of the food
	Attentive to the atmosphere of the place in which they are eating
	Are more dependent on smell than people of the other two groups
	Often prefer higher sweetness levels than other taster groups

SOURCE: "Taste, Flavor, and Food Choice" by V. Utermohlen.

Transduction and Sensitivity

The tongue is interlinked to the brain through the central nervous system (CNS). In explaining the brain's response to taste stimuli, scientists refer to a process called **taste transduction.** In the example of sweet tastant molecules, these interact with cell receptors through AH, B, X binding sites. One of the key events triggered by tastants is depolarization. **Depolarization** occurs when a positive (+) charge accumulates in the cell, making the membrane potential less negative than normal, and more positive. Accompanying depolarization is neuronal signaling—the release of neurotransmitters to the brain. In our example, the brain would interpret the neurotransmitter signal as "sweet." Salt and sour taste transductions are believed to involve similar mechanisms of depolarization and release of neurotransmitters. The transduction of bitter taste appears to be more complex, with several mechanisms proposed.

Odor

In addition to taste, smell plays a major role in terms of the enjoyment of food. The terms *odor, fragrance,* and *aroma* are often interchanged. However, in characterizing unpleasant or undesirable sensations, food scientists tend to describe off-odors rather than off-aromas or off-fragrances. **Odor** or aroma can be defined as the sensation derived from food as interpreted through the olfaction (sense of smell or odor perception) mechanism. Technically speaking, components of aroma can include either olfactory sensations (such as fruity or rancid) as perceived by the olfactory nerve, or nasal feelings (such as cool or pungent sensations) perceived by the nose's tactile nerves. The oral cavity, throat, and nasal chambers are interconnected, which means that aromas can be directly sniffed through the nostrils as well as passively sensed through the passage at the back of the mouth.

Olfaction strictly refers to the perception of odors by nerve cells in the nasal area. Airborne **odorants** (molecules that possess odor) are sensed by the **olfactory epithelium,** which is the outer layer of receptor cells located on the roof of the nasal cavity. Specifically, in what is still an incompletely understood phenomenon, odorant molecules are sensed by millions of tiny, hairlike **cilia** that cover this epithelium. The interaction between odorant and receptor cell initiates a cascade of events that, as in taste transduction, are related to the production of ion channel electrical signals. Optimal contact time between odorant and the nasal receptors occurs within 1 to 2 seconds. Scientists have demonstrated that exposure beyond this requires that the receptors readjust for about 20 seconds before a new, full-strength sensation can occur.

The sensitivity of olfactory receptors to stimulus molecules is much greater than in gustation. Two sensitivities are involved: being able to detect something at a low threshold of concentration and being able to differentiate one odor from another. Although humans do not have as acute a sense of smell, meaning as low a threshold of detection, as animals (e.g., a cat or dog), our ability to discriminate (differentiate odors) is equivalent. Humans can distinguish several thousand odors, but only a couple hundred taste variations. Individuals vary a great deal in both of these sensory abilities.

Flavor

Flavor results from the chemical stimulation of the tongue taste buds, the olfactory apparatus, and the organs of feeling present within the mouth, throat, and nose. Food **flavor** therefore is an overall impression combining taste, odor, mouthfeel factors, and trigeminal perception (Figure 15.3). The *trigeminal nerve* is an important neural region that runs through the entire facial area. Trigeminal nerve endings of importance are located especially in the nasal and oral cavities, and are responsible for what is alternately termed "trigeminal perception" or "somatosensory perception." Such perception refers to the sensation of astringency (from tannins in tea, which are polyphenols), burning (from isothiocyanates in mustard oil), cooling (from menthol in peppermint oil), and warmth (from capsaicin in paprika). (Refer back to Section 6.2.)

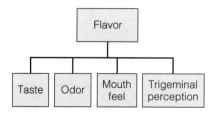

FIGURE 15.3 Flavor. *The overall impression of flavor is a combination of taste, odor, mouthfeel, and trigeminal perception.*

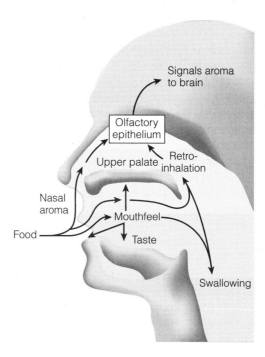

FIGURE 15.4 The sensory organs involved in the detection of aroma, mouthfeel, and taste.

Maltol is a flavor enhancer produced during high-temperature heating as well as commercially via fermentation to create enhanced sweetness.

Researchers believe food flavor is actually perceived at all areas of the tongue, although each of the four primary areas probably have maximum sensitivity to sweet, salty, sour, or bitter. Though the sensitivity to each taste varies, it is never zero at any area of this key taste organ. Flavor can also be influenced through the use of flavor enhancers or potentiators. These additives are substances such as monosodium glutamate, 5' nucleotides (e.g., inosine 5'-monophosphate), ibotenic acid, and maltol.

Aftertaste sensations are technically distinct from taste sensations. Aftertastes are perceived after the initial primary taste response and should be viewed as residual sensations that linger on the tongue after swallowing. Aroma/odor sensations contribute to flavor, as perceived by the olfactory nerve. What is called **retro-inhalation** plays a major role in flavor perception. It refers to the passage of flavor stimuli from the mouth through the pharynx up to the nose (Figure 15.4). Examples of olfactory character notes are fruity, vanilla, floral, and grassy.

Mouthfeel

Mouthfeel is the perceived sensation of food by the epithelial lining within the oral cavity, which includes tactile sensation as well as thermal response. The manner in which small movements of the tongue press food against the gums and palate perform an evaluation of viscosity and texture. Examples of thermal mouthfeel sensations include coolness and heat (e.g., menthol, and capsicum in cayenne pepper, respectively), metallic mouthfeel (food contaminated with iron and copper), and astringent mouthfeel (the presence of tannins in tea).

Gravy provides a good example on how taste, aroma, mouthfeel, and aftertaste all contribute to flavor perception. A typical brown gravy provides a certain savory flavor to complement a meat dish and also provides a characteristic aroma. The thickness of the gravy determines how it will coat and linger within the areas inside the mouth, and also its viscosity character. This ability to persist in the mouth can affect the aftertaste perception if the gravy particles persist in the oral cavity after swallowing. Also, gravy consistency can be smooth or lumpy in texture, a factor experienced as mouthfeel. The larger particles of lumpy gravy may release additional flavor during chewing compared to a smooth, more fluid gravy.

15.3 SENSORY TEXTURE AND COLOR PERCEPTION

Sensory Texture

Sensory texture has to do with the structure and composition of food, and it involves food molecules, their perception, and measurement. However, since memory, sight, hearing, and touch are automatically superimposed upon food texture attributes, the mechanisms of texture perception are incompletely understood, and food texture is challenging to classify.

Texture perception occurs whenever a food is chewed or beverages are swallowed. Texture measurement involves describing physical characteristics of food. Sensory texture encompasses several different types of sensations: mechanical, geometrical, and what is called "afterfeel." Food fat and moisture content also play a role in texture measurement and perception.

Sensory cells in the mouth, tongue, and throat perceive food particle shapes, sizes, thickness, and hardness, and transmit this information to the brain in a manner analogous to the taste and odor systems. In addition, people have a recollection of the texture of a particular food item they consumed in the past, and there is an expectation for the same texture if the item is consumed again.

The visual appearance of a plate of mashed potatoes prepares one for the texture that is anticipated. Breaking a Graham cracker into sections by the fingers produces a crisp snap. Both sight and sound create an expectation regarding how the texture will feel inside the mouth upon chewing. When a carbonated beverage is consumed, not only are gas bubbles heard popping, but their impact is felt as a tickle to the tactile nerve fibers of the mouth.

Important texture notes are also detected as food is manually handled in preparation for biting, chewing, and swallowing. Consider, for instance, a fruit like a melon or a pear. Although touching and pressing upon these foods with the hands is done outside of the mouth to evaluate their state of ripeness for eating, cues are being received by the tactile cells in the fingers and transmitted to the brain via the central nervous system.

Mechanical Sensory Characteristics. **Mechanical sensory characteristics** are generally due to attractive forces between the molecules in a food and the opposing force of disintegration. The sciences of physics and engineering come into play in the understanding of these character notes, as their definitions will demonstrate. The mechanical characteristics presented include the hardness, cohesiveness, viscosity, chewiness, and adhesiveness of food.

When food is chewed in the mouth, **hardness** is described by the amount of force required to compress the food between the teeth. Hardness intensities range from soft through firm and hard. Examples of particularly hard foods exhibiting this sensory characteristic include carrots, peanuts, crisp cookies, and hard candy.

Cohesiveness is described by the degree to which a food will deform or compress between the teeth before it breaks. Sensory cohesiveness intensities range from easy to disintegrate through difficult to disintegrate. Foods exhibit this sensory characteristic to varying degrees; how would you predict celery, grapes, toast, and potato chips to be judged?

Sensory viscosity is related to the force required to draw a fluid food from a spoon across the tongue at a steady rate. Sensory viscosity intensities range from thin to thick through viscous. Examples of fluid foods exhibiting differing viscosities include bottled water, honey, ketchup, and grapefruit juice.

Food **chewiness** is the length of time in seconds required to chew a food sample at a steady rate (one chew per second) with a constant force applied that results in a food

TABLE 15.2 Factors which contribute to a food's sensory texture.

Adhesiveness
Afterfeel
Chewiness
Cohesiveness
Fat
Hardness
Moisture
Particle shape
Particle size
Viscosity

consistency that is just ready to be swallowed. Sensory chewiness intensities range from tender to chewy through tough. Examples of foods of varying chewiness include meats, licorice, cheese, and chewing gum.

Food **adhesiveness** is a unique mechanical parameter because it is related to surface properties rather than forces of attraction and breakage. Adhesiveness is the force required to remove food material that attaches to the mouth (primarily the upper palate) during normal chewing. Sensory adhesiveness intensities range from sticky to tacky through gooey. Examples of foods of varying adhesiveness include broccoli, peanut butter, cream cheese, and sandwich bread.

Geometrical Food Characteristics. **Geometrical characteristics** of food also impact sensory texture. These characteristics are related to the size of discrete food particles present in the food, as well as to their shape and orientation. Sensory geometrical characteristics related to size and shape range from chalky to gritty to grainy and coarse. *Grainy* would indicate food-containing grains of a specified size, as for example cream of wheat cereal, wheat germ, and farina.

The designation *coarse* indicates relatively large particle size, as seen in cooked oatmeal. Particle shape and orientation can vary from fibrous to cellular and crystalline. *Fibrous* indicates foods composed of fibers, as in the meat fibers of cooked chicken or steak. *Crystalline* describes foods composed of crystals, such as granulated sugar.

Afterfeel. The term "afterfeel" is analogous to aftertaste in terms of food texture. **Afterfeel** indicates that after certain foods are chewed and eaten or beverages sipped, a "texture sensation residue" persists. Why? Perhaps on the molecular level, even though the food or beverage has been cleared from the mouth and throat, microscopic amounts of matter continue interacting with the sensory receptors and the oral lining, providing an ongoing stimulation and sensation of texture.

Fat and Moisture Content. The perception of the fat and moisture content in a food affects texture. Foods differ in fat content, and also in the manner in which fat is absorbed into them during preparation and processing and released into the mouth upon consumption. In addition, due to chemical differences, fats vary with respect to melting behaviors. As a result, foods such as egg rolls, chicken tenders, doughnuts, and onion rings will be variable relative to oiliness and greasiness. Moisture is also variable according to content, as well as to the rate and manner of its absorption and release. Depending on the specific food, moisture content may vary from dry (e.g., cereal grains) to moist (e.g., cake) to wet (tomato slice) and watery (canned fruit).

Sensory Interactions

It is clear to scientists who study sensations and perceptions that foods, being complex systems, exhibit interactions in terms of taste, odor, texture, and flavor. This interaction can generate a wide variety of effects. For instance, salt (sodium chloride) diminishes the perceived sourness of food acids, and it also enhances the sweetness of sucrose. Sucrose, on the other hand, counteracts the saltiness of salty foods.

This raises the point regarding competing levels of sweet, sour, bitter, and salty substances in foods or added to foods. Clearly, coffee is perceived as less bitter when sugar is added. An off-odor in a food will accompany and reinforce perception of off-flavor. Texture differences among samples of the same food item can influence taste and flavor. For example, viscosity differences can affect the distribution and concentration of flavor molecules within a food, which creates a variation in flavor intensity or perception.

When excessive sensory stimulation occurs, particularly with respect to taste (but also involving response to odor, aftertaste, and flavor), saturation of the sensory system results in **sensory overload.** This "sensory fatigue" can occur when an excessive number of samples are presented to panelists for evaluation. For instance, consider an overzealous researcher who is trying to determine the effect of different levels of alum on fermented

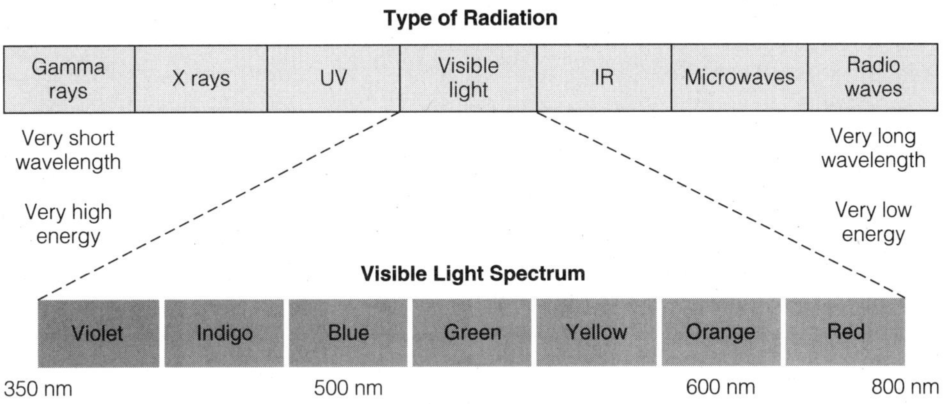

Type of Radiation

| Gamma rays | X rays | UV | Visible light | IR | Microwaves | Radio waves |

Very short wavelength · Very long wavelength

Very high energy · Very low energy

Visible Light Spectrum

| Violet | Indigo | Blue | Green | Yellow | Orange | Red |

350 nm   500 nm   600 nm   800 nm

FIGURE 15.5 **The portion of the electromagnetic spectrum associated with visible light and color.**

pickle texture. (Note: Alum is an additive that creates a "crunchy" pickle; it contains aluminum, a metal that can be toxic if ingested at a high enough level). If the investigator decides to serve fifteen samples to be evaluated at a single session, chances are very good that the panelists will become fatigued (and will have puckered expressions from all the acidity!). A better approach would be to plan a design that spreads out the samples over several sessions. Since sensory overload prevents sensory panelists from responding normally to food stimuli, the sensory specialist must exercise care to design experiments that do not result in sensory overload or fatigue.

Color

In addition to taste, flavor, and texture, color plays an important role in food quality and sensory acceptability. Readers are already aware of pigment molecules as the chemical basis of color. Foods possess color based upon wavelengths of reflected and absorbed visible light. From a physics standpoint, visible light is part of the electromagnetic spectrum, and visible light can be dissected into the spectrum of colors, from violet to red, each with a characteristic wavelength and energy level (Figure 15.5). As such, colors represent forms of radiation energy. However, color is not simply a physical phenomenon, because color energy not only exists, but is sensed, perceived, and responded to.

The reflected light energy of particular wavelengths in the range of 350 to 800 nanometers (nm) acts as the stimulus to the sensory cells in the eye, which communicate signals to the brain, enabling us to see the colors of foods and beverages. So again, we see commonality in the design of a sensory system, based upon incoming stimuli, receptor cells to receive them, and a manner to transmit the information to the brain for interpretation and response. The specific mechanism of color visualization requires study of the structure of the human eye and its specialized sensory cells.

The importance of sensory color in foods has been investigated. In consumer studies, the top three criteria used to define food quality included freshness, good appearance/color, and good taste. This conclusion conflicts with the notion that color is merely a cosmetic attribute of food. However, it is simple to realize the appeal that color holds for the consumer, and how it influences choice. Imagine being asked to select the highest quality food, let alone the most acceptable, from a group of three samples while blindfolded! Consider too how food flavor, texture, sweetness, and ripeness are all influenced by the color of a food. The altered appearance and color of rancid and spoiled foods have taught a valuable lesson regarding food color and its quality and even safety. Color can and has been used as an indicator of the economic value of products.

In the food industry, color is recognized as a quality parameter that standardizes a product. For instance, the USDA recognizes the importance of color in grading orange juice. The measurement of color can be accomplished through the analytical technique of tristimulus colorimetry. The three components of the tristimulus factor include the shade of darkness of a sample plus two hue components.

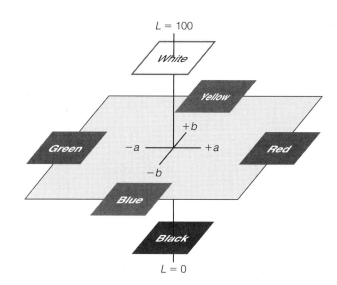

FIGURE 15.6 The Hunter L, a, b color solid. *True or false: A ripe, deep red tomato would be expected to have higher* L *and* +a *values than a pale, unripe tomato?*

As an example of this, consider the Hunter L, a, b system, or **Hunter tristimulus data,** widely used for food analysis. The Hunter color difference meter is a device used often in food research and testing. Based on the use of the Hunter Color Solid, the device has become the most widely used system for measuring color. Spaces within the solid are located from the values L, a, and b. The Hunter L value, located vertically on the scale, measures lightness or darkness. L values range from 100 (white) to 0 (black). Hue is measured by the Hunter a and b scales. Red is given by $+a$, green by $-a$, yellow by $+b$, and blue by $-b$. See Figure 15.6.

In this system, the three color dimensions match what is proposed to be occurring between the retina and brain regarding response to color and lightness. The L, a, and b values can be mathematically converged to produce a single color function to describe the color of any food or beverage sample. The following relationship is called the **Hunter color difference** (ΔE):

$$\Delta E = (\Delta L)^2 + (\Delta a)^2 + (\Delta b)^2$$

Objective evaluations are of no real value unless they correlate to what the eyes see as well as the value judgments and responses that perception provides. The Hunter system of measurement is useful and correlates well to visual color because it is patterned after the light-sensing principle of the eye retinal receptors. Color is but one of several possible subcategories of food **appearance.** Appearance also includes interior product characteristics, such as the crumb cell size of cake or bread, and external characteristics, such as a food's surface smoothness or roughness.

15.4 RESPONSES CONTRIBUTING TO SENSORY PERCEPTION

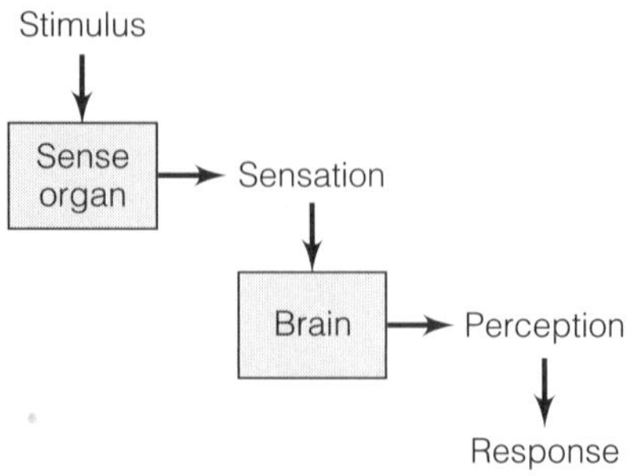

Sensory study involves quantifying a response to a stimulus. When sensory specialists study the relationship between a given physical stimulus and the human response, the single outcome actually is the end result of a multi-step process, with the actual response being one of taste, odor, and color. Here is a common sequence of events in sensory perception (Figure 15.7):

1. The stimulus interacts with the sense organ (tongue, nose, eye) and is converted into nerve signals that travel to the brain.
2. The brain interprets the incoming signals and organizes them into perceptions.
3. A response is elicited from the subject based upon these perceptions.

FIGURE 15.7 The sequence of sensory perception.

Objectivity and Subjectivity

The difference between **sensation** and **perception** can be viewed as the difference between regarding the brain as a machine ("objective brain function") versus regarding the brain as a thinking mental process ("subjective brain function") subject to conditioning, preconceptions, natural ability, and so forth. Using this analogy, sensations correspond to objective functioning, while perceptions are subjective.

Many features of food quality related to flavor, texture, color, and odor are measurable by **objective procedures,** using test equipment. For example, an instrumental device can measure the amount of force needed to shear a food sample into pieces, which might then be used as an index to tenderness, or a color meter can be used to determine the product's color. Objectivity can be a good thing, because it eliminates bias. However, instrumental methods alone are insufficient indicators of product quality, since they do not measure human **subjective response.** Since minor components in food formulations can create major differences in subjective results, it is important to assess the quality of foods by the intended consumer, using humans as sensory "test equipment."

Is the human subjective mental process at odds with the goal of "objective" sensory evaluation? The answer is no, because people can respond with increased objectivity with appropriate training. With focused training and the use of reference products, the mental process can be shaped to the point that all individuals on a sensory panel move toward showing the same approach to a given stimulus. This is not saying that all panelists respond identically, but panelists understand what is being evaluated and how to evaluate, and rely on their personal, innate ability to respond to food product character note intensities.

As an example, imagine a manufacturer needing to demonstrate the sensory acceptability of a bagel formulated to have extended shelf life. Suppose a reference bagel, which was subject to Maillard browning and the development of off-flavors and odors, was used in panel training. Then, the panelists could be trained to detect the presence or absence of those specific changes in reformulated bagels. Also, the success or failure of several test formulations aimed to minimize the effects of Maillard reaction could be assessed. Using the reference bagel as a "worst case scenario" would provide the panelists a valuable frame of reference from which to make specific acceptance evaluations.

Intensity

Intensity has to do with the product stimulus, while sensitivity has to do with panelist ability to sense. **Intensity** is the degree to which a character note is present, as determined by its product concentration and perception by a person. So, the intensity of a stimulus refers to its perceived strength. Intensities for product character notes can be quantified. The measurement of sensory attributes evaluated over a period of time following an initial exposure is called **time-intensity measurement.** This type of data is obtained when information related to time-dependent product attribute rates of change or duration in intensity is desired. A time-intensity curve shows the total duration of a taste, the maximum intensity, the time required to achieve maximum, and the rate of intensity decline to a minimum value (Figure 15.8).

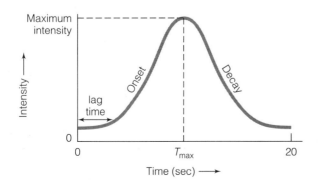

FIGURE 15.8 A time-intensity curve.

TABLE 15.3 Relative potency of tastant molecules. *Each tastant representing the four taste groupings receives a potency index based upon a reference compound, which has the maximum potency index of 1.0.*

Tastant Substance	Potency Index
Sour Compounds	
Hydrochloric acid reference	1.0
Lactic acid	0.85
Malic acid	0.60
Acetic acid	0.55
Citric acid	0.46
Bitter Compounds	
Quinine reference	1.0
Caffeine	0.40
Sweet Compounds	
Sucrose reference	1.0
Saccharin	675
Aspartame	150
Fructose	1.7
Glucose	0.8
Salty Compounds	
Sodium chloride reference	1.0
Sodium iodide	0.35
Potassium chloride	0.6

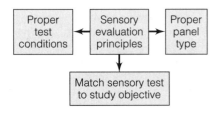

FIGURE 15.9 Central principles of sensory analysis.

Threshold

A **threshold** for sensitivity refers to distinct points of transition for sensory judgments corresponding to concentrations of stimuli. Each person has a lower limit at which he or she first begins to respond to a stimulus, and an extended range for the ability to differentiate low, medium, and high levels of a stimulus, before a saturation point is reached and further distinctions are not possible. **Detection threshold** is the point at which a person initially responds to a stimulus, the dividing line between lack of sensation to sensation.

A stimulus molecule's potency is connected to how it is perceived. Sour acids, such as lactic acid in dairy products, and citric acid in fruits, present differing potencies, which in turn affect taste detection thresholds. The same is true of bitter, sweet, and salty substances, and these differences have been quantified—see Table 15.3.

In addition to detection threshold, other thresholds are of significance. The **recognition threshold** is the point at which the identity of a stimulus is made. **Difference threshold** refers to the minimum amount of stimulus change that results in a change of sensation. **Terminal threshold** is the amount of stimulus above which any increase in intensity cannot be detected without the use of reference standards.

15.5 SENSORY TESTS

Sensory testing involves a large number of decisions, guided by a group of important principles. Although a specific sensory test cannot provide definitive answers to all key questions, it is a key part of a larger sequence of information gathering during the product development process. The sensory specialist must have expert knowledge of the variety of sensory tests available and be able to apply the correct test to address any given problem. To carry out effective and meaningful sensory research, the main guiding principle is that the objective of the specific project determines the sensory technique and test chosen by the specialist (Figure 15.9). Other important principles are listed below:

- Trained panelists should not be used to obtain acceptability judgments.
- Consumer panels should not be used to obtain accurate descriptive information.
- Individuals who are involved with the R&D aspects of products should not serve on panels.
- Samples should be labeled with random 3-digit codes to minimize bias.
- Sample order should be randomized to avoid artifacts due to the order of presentation.
- Panelist interaction during sensory evaluation is forbidden because it can bias the data.

Classification of Test Methods

Sensory tests are typically classified according to their primary purpose and most valid use. As stated before, matching the test method to the objectives of the project is critically important. Three classes of sensory tests are most commonly used, each having a different goal and selection criteria for panelists (Figure 15.10).

    Affective   Discrimination   Descriptive

Affective Test Methods

Affective tests attempt to quantify the degree of liking or disliking of one product over another, and include preference, hedonic, and consumer acceptance tests. With food products, the two main approaches to using consumers in sensory evaluation are in measuring preference and measuring acceptance. **Preference** allows the consumer a choice and asks for it to be made between products on a hedonic (like, dislike) basis. **Ranking** is a form of preference testing that uses more than two samples. It is not the same as rating.

In ranking, panelists order a group of products with respect to either their degree of liking or their perceived intensity of a sensory attribute. **Rating** is a term meaning the assigning of numbers by panelists to characterize products in various response categories based on specific sensory attributes, and is a type of acceptance test.

In **consumer acceptance testing,** consumers rate their liking or disliking on a scale. This generates an average value from all the participant data, which can be linked to a food's acceptability. The scale used may be implied rather than obvious. However, the sensory analyst would know to convert a "like extremely" response to a score of 9, and "dislike extremely" to a score of 1. A comparative sample is not needed. Consumer acceptance tests require large numbers of panelists (50–100 or more) who are selected as representative of the typical end user of a product.

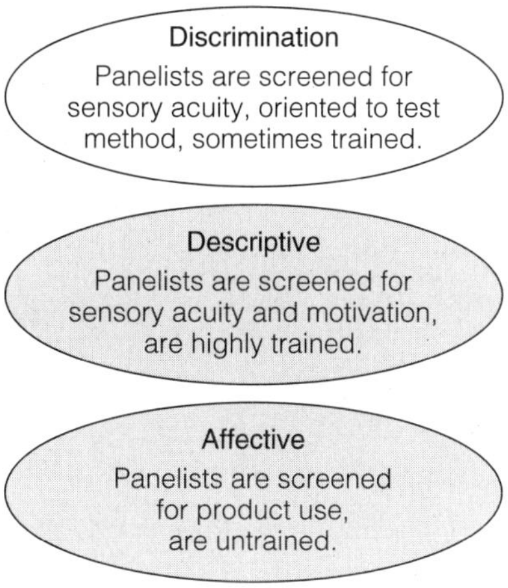

Discrimination Test Methods

Discrimination (or difference) **tests** simply attempt to answer whether any difference at all exists between two types of products. Typically, the test is between a control and a test sample. However, the nature of the difference is not determined through discrimination testing. The analysis of the data is based upon statistics of frequencies and proportions (the number of right versus the number of wrong answers). The ability of a panelist to discriminate is determined when a critical number of consistent correct judgments are made above the level expected by chance guessing.

Examples of difference tests include the difference-from-control, duo-trio, paired comparison, and triangle test. These are used to determine whether panelists can discriminate between experimental treatments. Table 15.4 gives an overview of the various types of sensory tests.

The sensory specialist decides which discrimination test to use based upon various criteria. For instance, if the difference between the samples is known to be a chemical one, a paired comparison test is usually selected. If on the other hand the difference is a qualitative one, then a **triangle test** is commonly used. In a triangle test of two products, three samples, two of one product and one of the other, are presented to the panelists in a specific order. As an example, if A and B are used to represent each product, there are six possible orders of presentation: AAB, ABA, BAA, BBA, BAB, ABB. If, after sufficient trials, panelists can consistently pick out the different (sometimes called "odd") sample in each set, then a statistically significant or "true" sensory difference is said to exist. Notice that AAA and BBB presentation are never used in triangle testing because of the definition of the "triangle."

Descriptive Test Methods

Descriptive tests seek to describe specific product attributes related to flavor, texture, mouthfeel, and so forth, by quantifying the perceived intensities of specified sensory characteristics. Descriptive analysis, which requires highly trained panelists familiar with attribute scales, has proved to be the most comprehensive and information-sensitive evaluation tool available to the sensory specialist. The wording used in descriptive testing is very precise and careful, so as not to be ambiguous. Scorecards for use in descriptive sensory studies include information on each specific characteristic to be evaluated, and may be developed by the researcher in certain cases, or by the panelists in others.

Quantitative descriptive analysis (QDA) is a test developed at the Stanford Research Institute that uses line scales, replicated experimental designs, descriptive terminology, and analysis of variance. QDA generated data are often represented by spiderweb or radar plots, in which the intensity of each sensory attribute increases outward radially from a central point—see Figure 15.11.

Flavor and texture profiling are frequently used descriptive analysis methods. They employ very detailed descriptions of products as developed by trained panelists. In the **flavor profile method,** odor, taste, flavor, and mouthfeel character notes are determined by the panelists. Sensations are quantified in terms of intensity, order of perception, and duration, from the initial sensation through those lingering after samples are swallowed. In

FIGURE 15.10 Classification of sensory tests and the panelists used for each.

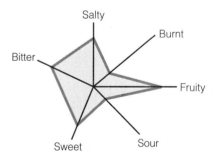

FIGURE 15.11 QDA spiderweb plot. *The panelist responses create a polygon shape, in this case, having six sides. The more a product elicits a response from panelists, the further from the origin each attribute point will be. It is relatively simple to visually inspect the differences in polygon shapes obtained for control versus test samples to see what differences exist.*

TABLE 15.4 Overview of sensory tests. *Affective testing obtains information regarding panelist preferences. Discrimination testing and descriptive testing are analytical test approaches that provide information about product differences.*

Analytical Tests Detectable Differences	Affective Tests Individual Preferences
Descriptive Tests—Quantify the Differences **QDA** Quantitative descriptive analysis is a form of analytical testing using trained sensory panelists to quantify and describe specific product parameters. **Flavor or Texture Profile** Used to detail the specific flavors (garlic, vanilla, caramel, boiled milk) or textures (smoothness, springiness, moistness) of a food or beverage. ***Discriminative Tests—Discernible Differences*** These tests fall into two basic categories: difference and sensitivity tests. ***Difference—Differentiate Between Samples*** **Triangle.** Three samples are presented simultaneously—two are the same and one is different. Panelists are asked to identify the odd sample. **Duo-trio.** Three samples are presented at the same time, but a standard is designated, and the participant is asked to select the one most similar to the standard. **Paired Comparison.** Two samples are presented, and the taster is asked to select the one that has the most of a particular characteristic (sweet, sour, thick, thin, etc.). **Ranking.** More than two samples are presented and compared by ranking them from lowest to highest for the intensity of a specific characteristic (flavor, odor, color). **Ordinal.** A scale that usually uses words like "weak, moderate, strong" to describe samples that differ in magnitude of an attribute.	***Hedonic Tests—Personal Preference*** Consumers are tested for their preferences (hedonic means "relating to pleasure") rather than their evaluation of differences. They are often provided a form to record their likes or dislikes of a particular sample. **Example #1:** Product Score Sheet Product Date Instructions: Rank the food from 1 (very poor) to 5 (very good) in each of the following categories. Total each column and then the final Total row for a complete score Very Poor 1, Poor 2, Fair 3, Good 4, Very Good 5 Taste Odor Mouthfeel (texture) Appearance Total = Total Score = **Example #2:** "Smiley" or "frowny" faces can be used for children.

texture profiling, the mechanical, geometrical, fat and moisture, and afterfeel character notes of food samples are described and quantified in a like manner.

Selection of Test Method

The appropriate selection of a sensory test method involves a large number of decisions. These decisions apply to the nature of the food product and the objective of the project, panelist selection and training, experimental design, and statistical analysis. Figure 15.12 gives a decision flowchart for test selection.

15.6 THE ROLE OF THE SENSORY EVALUATION SPECIALIST IN PRODUCT DEVELOPMENT

A well-functioning sensory program is useful in helping the food industry not only meet consumer expectations but enhance the potential for a product's marketplace success. The correct application of the sensory tests just presented, coupled to proper training and use of panelists, and the awareness of factors such as interactions, help define the role and importance of the sensory specialist. It is not difficult to see the value of sensory information in reducing the risk in decisions about product development and strategies for meeting consumer needs.

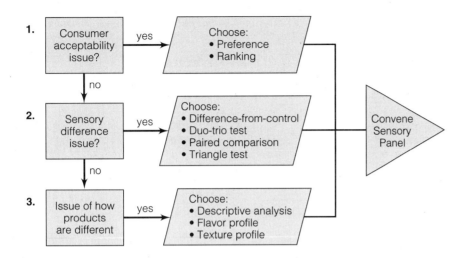

FIGURE 15.12 Sensory decision flowchart.

The Sensory Specialist

The sensory specialist is more than simply a provider of a specialized testing service. By playing an active, critical role in developing the project objective, and by collaborating in the development of the experimental designs that will best answer the questions posed in a product development project, the sensory specialist functions as an enabler of product development. Experimental design, as well as information about the sensory environment, is presented in this chapter's Challenge section. The sensory specialist performs seven key roles in product development.

1. Determine the project objective through understanding the needs of the project leader in order to apply the appropriate sensory test and properly analyze the data. Define treatments and variables; determine if it is a new product or a product modification.
2. Determine the test objective; this means deciding whether it is to determine difference, preference, or acceptability. Before starting the study, record in writing the project objective, the test objective, and a brief statement of how the test results will be obtained, analyzed, and applied.
3. Screen the samples during the first discussion phase of the project. The sensory specialist should examine all of the sensory properties of the samples to be tested to determine whether any sensory biases (such as color or thickness differences) are introduced by the samples.
4. Design the test, which involves: selection of the test technique, score sheet design, selection and training of panelists, the criteria for sample preparation and presentation, and the manner of data analysis. Panelist criteria are summarized in Table 15.5.
5. Conduct the test, with the sensory specialist being responsible for seeing that all requirements of the design are met by the sensory staff.
6. Analyze the data, using the procedure for data analysis that was determined at the test design stage. The appropriate statistical programs and expertise need to be in place to begin data analysis as soon as the test is completed. Data is often analyzed for the main treatment effect (test objective) and other test variables such as order of presentation, day effects, and subject variables (such as panelist age, gender, and geographical area).
7. Interpret and report results; it is always desirable to review the results and express them in terms of the stated objectives. An initially clear statement of the project and test objectives enables the sensory specialist to accomplish this. The written sensory report should summarize the data, describe the samples, provide information regarding the panelists, identify the key test findings, and make informed recommendations.

TABLE 15.5 **Panelist criteria.** *Panelist selection and training depend upon various factors, according to the sensory test application.*

Criteria	Test Applied To
Healthy, no illness, no food allergies	All
Socioeconomic, geographical, ethnic, and cultural factors	Affective
Basic taste threshold response ability	Discrimination
	Descriptive
Ability to recognize duplicate samples (using triangle test)	All
Consistent panelist response during trial runs	All
Ability to correctly rank intensity of basic taste samples (as levels in water solutions)	Descriptive

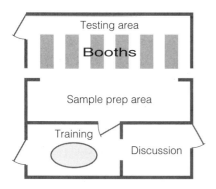

FIGURE 15.13 **Sensory testing environment.** *The sensory evaluation environment consists of sample preparation, panel discussion and training, and testing areas.*

The Sensory Environment

The layout and environment of the sensory testing area must be carefully planned to minimize distractions so that the panelists can focus their attention on sample evaluations. See Figure 15.13. This is usually accomplished through the use of individual sensory booth seating, with partitions between panelists. The color of the booth is a neutral shade such as gray or beige, and the temperature is controlled to a comfortable setting. Two-way hatches or sliding doors between scientist and panelist can be used to pass samples and retrieve score sheets. It is important to separate the panelists from the sample preparation area, and provide adequate ventilation in the testing room to keep out competing odors. Since talking can not only distract but also bias panelists, it is necessary to instruct panelists to avoid any communication with each other. The inclusion of computer terminals in the booths allow panelists to input data as they perform their evaluations, which simplifies the work of the sensory scientist.

15.7 PRODUCT DEVELOPMENT

Product development is a process in which new food product ideas are generated, and the products themselves are created and marketed. It involves the conceptualization, formulation, processing, testing, and marketing of food products. The products may be totally novel or improvement modifications of existing products. Through product development, a food manufacturer expands the variety of products offered to consumers. New product ideas present challenges to the manufacturer in terms of development cost, formulation, testing, and marketing strategies.

Thousands of new product ideas are generated annually, but the vast majority of them fail for one reason or another. As an example, a major manufacturer believed it had a winning idea regarding a soup/sandwich combination several years back. Only after a costly full-scale production launch was it realized that the excessive amount of packaging that the product required made it unpopular with consumers, who viewed it as not only inconvenient to access, but an environmental nightmare as well. Besides, the consumer already knew how to prepare a quick sandwich and heat up some soup the old-fashioned way.

New product development can be expensive. As both a costly and risky undertaking, new product development needs to be as organized and efficient a process as possible. One way to accomplish this is to employ the scientific method in product research and development, and to understand and follow a defined sequence of steps that facilitate the introduction of the new product to the consumer.

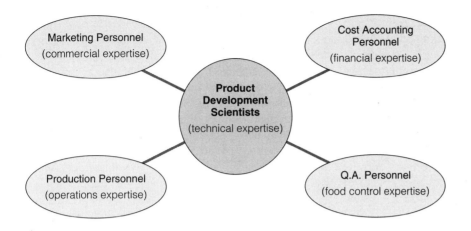

FIGURE 15.14 The product development team.

The **product development team** refers to all of the individuals employed by a food manufacturer who play a role in the product development process. Included are all technical, marketing, financial, production, and quality assurance personnel (Figure 15.14). Panels of consumers take part in product evaluation during market testing, and are representative of the market segment for which the product was created.

The Scientific Method in Product Development

As members of the product development team, product development scientists use elements of the scientific method in their approach to developing ideas and solving problems. The scientific method, as you might recall, is a procedure in which a problem or question (the "challenge") is identified and a sequence of steps is employed to find, solve, or answer the challenge. While in actual practice each phase of product development does not rely on the scientific method, a product development scientist is trained to:

1. Define a problem or ask the question: "What new food product are we trying to create?"
2. Solve the problem or answer the question by arriving at a hypothesis: "This is our proposal regarding the formulation and processing required to make the new food product."
3. Test the hypothesis through carefully designed and controlled experiments in which observations are made and data recorded.
4. Accept or reject the hypothesis based upon analysis and interpretation of the results of the experiments.

If a fat-free cookie needs to be developed, or a shelf-stable meat snack is required, food scientists apply their knowledge of food formulation, processing, and the basic sciences of chemistry, physics, and microbiology to these challenges. Science and the scientific method become part of the overall approach to developing these new products. An important skill required of product development scientists is the selection and application of food ingredients, such as those presented in our chapter on food additives. In addition, new and state-of-the-art processing and packaging can be employed to create **value-added food products.** To be considered value-added, foods must offer the consumer some improvement in terms of quality or convenience over the traditional or previously available foods. An instant pudding is an example of a value-added food because it offers the convenience of less preparation time.

The Stages of Product Development

A number of phases or stages, each accompanied by specific activities, are required to successfully carry out new food product development. These stages involve not only creativity and expertise on the part of research and development scientists but also cooperation with other team members to progress from the initial product idea or concept through marketing. Before we discuss specific stages and activities, it is necessary to realize the underlying basis for product development. The true starting point for product work is called the **corporate mission.** This mission embodies the foundational aspects of any corporation: who it is, what are its products, and who are its customers. From this mission, one or more corporate objectives are generated. As an example, a company might have as an objective to become the number one manufacturer of frozen french-fried potatoes in the nation.

A **product strategy** is a plan that is established to accomplish a company's mission and objectives. In the example of a frozen french-fried potato manufacturer, the strategy might be to market steak fries, crinkle-cut fries, and shoestring fries under a single brand name. Although distinct from each other, these products represent a single **product line** because they are basically similar products serving the same market. If, however, the manufacturer wants to produce frozen french-fried potatoes in both conventional and microwave formats, two product lines would result. In many instances, the realistic practicality of a new product variety or line is discussed in terms of its **feasibility.** In the above example this would pertain to whether or not a microwave frozen french-fry product would be feasible in terms of actual development (e.g., are formulation, equipment, and packaging costs prohibitive?).

A **line extension** is a new form of an established product or family of products. For instance, a new flavor snack chip, a new ready-to-serve canned soup variety, and a low-fat peanut butter are examples of product line extensions. In general, the more creativity, innovation, and technical complexity required in a new product, the greater the research and development time, hence cost. However, one factor to appreciate regarding line extensions is that they are relatively inexpensive to create because they take comparatively little effort and time for development compared to products made "from scratch." If a breakfast cereal manufacturer already markets a puffed rice cereal, it is relatively simple to formulate and process a honey-flavored variant of it as a line extension. However, a goal to introduce not only a new flavor but also a new product shape and package type to appeal to a particular age group would represent a more complex (and costly) line extension.

From product strategy and feasibility, a product development project objective emerges, which is related to various stages and activities. Since it is a source of possible confusion, let's distinguish the stages in product development from the activities in product development.

For convenience, the whole of product development can be divided into three general stages, under which specific activities fall:

1. the idea stage
2. the development stage
3. the commercial stage

During the development stage, a variety of testing and production-related activities must take place that are distinct from those occurring in other stages. See Figure 15.15.

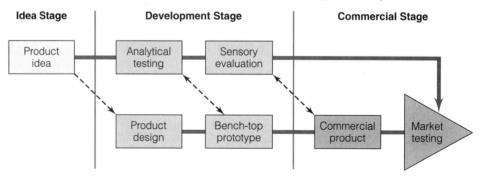

FIGURE 15.15 Several activities involved in product development. *Adapted from Fuller, 1998.*

The Idea Stage. The initial stage of product development is usually referred to as the concept or **idea stage.** This stage identifies the idea, concept, problem, or specific challenge that the product development team works with. The concept should be framed to meet corporate objectives and should be economically and otherwise feasible. This stage requires the input of not only food scientists and engineers, but also the marketing and management individuals of a food company. It is critical to address certain questions during the idea stage, related to the product concept goal, such as:

- What is the anticipated demand for this product, and who is the competition?
- Does the product meet a specific consumer need?
- Which ingredients, processing, and packaging are required to formulate and manufacture the product, and are they available at reasonable cost?
- What are the desired characteristics of the new product?
- How much time and money will it take to bring the product to the commercial stage?
- What are the profit objective and profit potential of the new product—how much money will it make for the company?

The Development Stage. The **development stage** refers to the work involved in creating the design and nature of the product, from a compositional standpoint as well as a processing, packaging, and marketing one. The majority of the activities undertaken in product development occur at this stage. From formulation to process fine-tuning, to the use of sensory and analytical testing, all of these activities must flow from the idea stage as a single, connected process.

One of the first accomplishments of the developmental stage is to formulate a **bench-top prototype** product. This prototype is usually made in the laboratory as a small scale (a few pounds in weight) batch that can be easily and reproducibly formulated for repeated testing without too much expense. Prototype products of batch sizes ranging up to one hundred pounds are made in facilities called pilot plants. These are identical in nature to actual commercial production facilities except they are designed for smaller output of product and require less floor space, equipment, personnel, and lower operating cost.

Questions often addressed during the development stage include:

- Is the formulation reproducible, or are changes needed?
- How will quality and cost be affected by altering ingredients and processing conditions?
- What are the storage requirements of the product and its shelf life?
- Will spoilage due to yeasts, molds, or bacteria be a problem?
- Does the product require any special packaging?
- Does sensory analysis indicate the product is meeting its concept goal?
- What is required to scale up for commercial production of this product?

The Commercial Stage. For our purposes, the **commercial stage** refers to production scale-up from pilot plant to commercial plant, plus market testing and the subsequent introduction of the new product to a nationwide market. In this stage product development scientists interact with manufacturing and operations personnel in the actual commercial production plant. A further aspect of this stage has to do with obtaining information from the end user of a product—the consumer—through market testing and market research.

Market testing and market research both rely on consumer data, but are distinct activities. Market testing occurs after the successful production of the product in a commercial production plant. It involves a **test market,** which is the location where the new product is sold to consumers. In market testing, feedback regarding products is obtained from consumers by phone or survey form after the products are used in their homes.

Market research is an organized inquiry that seeks to identify and measure the factors that influence the consumer marketplace, so as to guide marketing professionals in decision making regarding new products. It is important to recognize that the consumer, much more than the product idea or the application of technology, will determine the success of a new food product. Equally important is the understanding of who consumers are, what brings them satisfaction, what is their history regarding purchasing certain products, and how do they and their needs change with time.

Questions that must be addressed during the commercial stage:

- How much operating time must be devoted to new product manufacture at the manufacturing plant?
- How much cost is involved in large commercial batch production runs?
- Are there any concerns regarding the bulk storage of raw materials in the commercial plant?
- Does the product have to be returned to the pilot plant for retesting for any reason, or is it ready to be test marketed?
- What, if any, image should the new product project?
- Where will the product be test marketed?
- How will consumer attention be directed toward the new product?
- Is the institutional market (schools, hospitals, restaurants, etc.) a potential user of this product?

15.8 THE ROLE OF MARKETING IN FOOD PRODUCT DEVELOPMENT

It is not a simple two-step process to have scientists create tasty new food products in the lab any particular day and offer the foods for sale in the supermarket the next. The difficulties related to new product development can be envisioned as a three-dimensional challenge. The first relates to the technical expertise required of product development scientists. Added to this are the complexity of the marketplace, as influenced by competitors and the economy, and difficult-to-predict consumer factors, especially their changing needs versus wants behavior ("volatility"), which influences the market. Figure 15.16 shows these factors on three axes.

Early on, the marketing staff of a food manufacturer must address the basic question: "What is the specific need for this product?" This is equivalent to the question: "What is the current and future market need, and how can it be met?" The marketing department must be attuned to the desires of the consumer and be ready to communicate these to upper management. Through the process of **brainstorming,** the marketing group examines consumer trends and uses the information to speculate on what new products would meet those trends. The importance of marketing brainstorming to actual manufacture and production is to identify something that is needed and can be produced, rather than simply making a new product to sell. In this way, it is correct to state that new product development is market-driven.

Marketing Steps

Most marketing ideas or concepts come from marketing professionals. The next step after brainstorming for an idea is to decide on its market potential. One of the most valuable steps in creating and improving products is to discover precisely which factors—specific ingredients and processes—have the greatest impact on consumer response. Marketing puts together discussion groups called focus groups for this discovery process. The use of **focus groups** is a form of consumer testing for product development and usually involves assessing the attitudes of 8 to 12 carefully selected consumers and their acceptance of a concept or product. Each focus group is made up of consumers selected as representative of the target consumer for the new product. To be of benefit, the individuals who participate in focus groups must be articulate, honest, and uninhibited with regard to expressing their opinions.

Focus groups meet and interact for several hours with a trained moderator. The group is provided information regarding the product concept, usually with the aid of a

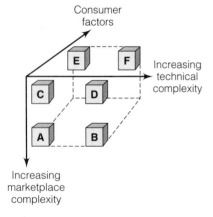

FIGURE 15.16 Three-dimensionality of new product development.
Product A is part of a more complex marketplace. Product D offers greater technical complexity. Product E is part of a familiar marketplace, is technically simple, but is to be positioned in an untried consumer market segment. Adapted from Fuller, 1998.

FIGURE 15.17 The consumer is the final judge in test marketing.

product mock-up or prop, in order to obtain their detailed attitudes regarding the concept. They are often asked to judge the desired features of the new product against existing products, its ability to perform in meeting specific needs (i.e., what do they want in a product), and the likelihood of purchase. This approach speeds up product development, reduces biases inherent to quantitative testing protocols, and generates information regarding the concept, the product, and the goodness-of-fit between them. It also accesses the beliefs and attitudes of the consumers and uncovers potential product functionalities (e.g., health benefits, convenience of preparation, sensory pleasure). Focus group information is not used as the basis for business decisions but has definite value regarding the direction of new product ideas.

The next step after the focus group is to actually evaluate or test the concept. The product is made into a prototype through bench-top and pilot plant formulation and production. In-house sensory testing of the prototype guides its development. A given product may need several cycles of such testing and modification to bring it to the stage of matching not only the original product concept but also satisfying consumers. Consumer sensory testing must also be done, with the product presented to a group of approximately 200 consumers for evaluation. Through such **consumer response testing**, consumers are asked questions regarding the likelihood they would purchase the product, how much they would pay for it, and how frequently they anticipate using the product. Affective testing is often a component of consumer response testing.

This information is brought before corporate management, who decide whether to proceed with the product based upon the consumer response. The criteria for recommending project continuance by management varies from company to company, although it is safe to say that most would frown upon a likely-to-purchase score below 50 percent. As with in-house testing, consumer concept evaluations influence the development of the product prototype. If comments received from a majority of consumers indicated a formulation or packaging flaw, for instance, this would be addressed, and the prototype modified accordingly.

If the product survives prototype testing, it is ready to be test marketed, which involves the operations end of the company, such as engineering. A basic question that should have already been answered before test marketing begins is "can we produce enough product in our existing facility?" Market testing is expensive, typically costing companies millions of dollars. However, this amount of money is small compared to the debt incurred were a product to be introduced nationally without the benefit of adequate market testing, only to fail. Figure 15.18 shows the marketing steps in product development through test marketing.

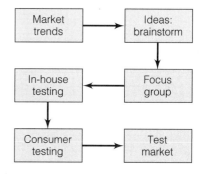

FIGURE 15.18 The marketing steps in product development through test marketing.

One value of test marketing is that success at that level, although not a total guarantee, suggests success in the national market. But even when a product is introduced nationally, having proven itself in all previous evaluations, there is still only a 1 or 2 percent chance of success. This is a sobering thought, considering that the expense to a company in bringing a product before the national market is on the order of $25 million or more. Success would be measured according to a product life cycle duration that earns the company a predetermined profit. Product life cycle, which we will define in detail shortly, is affected by various factors, but is especially subject to consumer wants and needs.

Meeting Market Need and Product Marketability

As we have seen, the market need refers to the need by the consumer for the product. This fact establishes the import of identifying the targeted user of the product, not only in terms of **demographics** such as the age, gender, income, and geographical location, but as related to factors such as shopping habits, level of education, and motivation to buy. Consumer habits change with time, and trends with respect to what is and is not a desirable product also change.

In a specific hypothetical case, a manufacturer might market a traditional as well as squeezable or spray butter product and ask the marketing department to keep watch for signs of change in consumer attitudes to these products. In fact, marketing's primary role is probably to monitor the marketplace for changes that might influence the course of new product development. Value-added products can be developed to meet particular market needs. When marketing considers what constitutes a value-added product in the eyes of the consumer, the answer can be something as simple as the functionality of a product container.

For example, in the case of the butter product available in a squeezable or spray bottle, each container offers more convenience than the traditional package, thus offering added value. From the product development standpoint, we should appreciate the formulation challenge to create the proper performance characteristics of such a squeeze or spray product under specified conditions of use. The technical expertise would require an understanding of emulsion chemistry, fat and oil processing, melting point, and perhaps fat fractionation to achieve the desired end product.

For a spray butter product for which a market need has been identified, a number of tasks would be completed, usually in an integrated fashion: market justification, development of a prototype, laboratory testing of the prototype, modification of the prototype, preference testing in the marketplace, test marketing, market research, sales analysis and forecasting, stop or go manufacturing decisions, and possible market expansion. Advertising strategies would be created, and with the assistance of the technologists, product label design, label statements or claims, label art work, and media promotions would be devised by marketing and sales staff.

15.9 PRODUCT PROBABILITY, LIFE CYCLE, AND ANN

Product Probability

Before we can adequately discuss a mathematical concept like probability, we need a working definition for it. **Probability** is conventionally defined as the ratio of the number of favorable outcomes to the number of total possible outcomes (both favorable and unfavorable). The range of values for probabilities is from zero to 1.00 inclusive. Zero means that there is no possibility of a favorable outcome, whereas 1.00 means that a favorable outcome is certain. As you can see, probability is related to the ability to forecast or predict information, such as the likelihood of success for a hypothetical food product.

The process for determining the overall probability of success for a hypothetical food product can be determined if the individual probability values of each event in the entire sequence are known. This approach can be simplified to a sequence of A→B→C→D→P, in which A is the starting ingredient and P is the final product. Letters

B, C, and D can represent product intermediates, with the arrows indicating processing steps or product development stages. Suppose the probabilities are given as:

$$A \rightarrow B = 0.7 \quad B \rightarrow C = 0.5 \quad C \rightarrow D = 0.4 \quad D \rightarrow P = 0.8$$

The overall probability of product success is found by multiplying the individual probabilities times one hundred:

$$(0.7) \times (0.5) \times (0.4) \times (0.8) \times 100 = 11\%$$

Thus, the probability of producing P is only 11 percent, which is quite poor. A favorable probability would be a higher number, preferably in excess of the 0.50 (50%) level.

In addition to being able to calculate the overall probability of success in going from A to P, it is also possible to calculate the total time and cost required in producing P if the data for the appropriate data pertaining to the previous steps are known. In each case, the answer is merely estimated to be the sum of the individual times and costs. These figures are useful in providing a basis of judgment regarding project continuation or termination.

With this basic understanding regarding predicting the probability of success in developing a new food product, let's progress to an understanding of what is experienced by all successfully developed food products that make it to the consumer market: that of a life cycle.

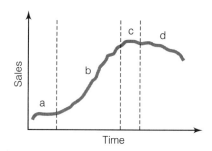

FIGURE 15.19 The association between product life cycle and sales. *Several distinct phases of a product life cycle are shown: a = lag phase when product is first introduced, b = strong growth, c = peak sales, non-growth, and d = decline in sales.*

Product Life Cycle

A **product life cycle** refers to the birth of a new product, its duration as a good selling item, and the time required to see sales decline and the product's removal from the marketplace. The cycle can be represented as having distinct phases. The initial period or lag phase is associated with low sales but high advertising and promotion costs during the introduction of the product. A strong follow-up growth phase is indicative of successful promotion and consumer acceptance, with both new and repeat purchases driving sales. As new markets open, growth continues until the static phase is reached, when sales peak but no longer increase over time. This is typically followed by the start of a sales decline, which can accelerate rapidly, especially if competitors anticipate and capitalize on new market trends. Figure 15.19 shows these phases in the product life cycle.

The fact that food products generally are introduced, experience sales spurts, reach a point of maximum growth, and then decline in sales, points to the need for continual replenishment of product via ongoing product development programs. The successful food company always considers the uncertainty and unpredictability of the future. To maintain profits and to keep ahead of competitors, it will have new replacement products ready before a product life cycle decline is seen for each of its product lines. This ongoing new product development will prevent losing ground to the market competition, positions that, once lost, may prove difficult or even impossible to regain.

Artificial Neural Network

Product development is a creative task, and its multistep process is time-consuming and costly. A technology that would simplify the process and result in cost savings by shortening the development cycle is an attractive prospect. The development of an **artificial neural network (ANN)**, a form of computer intelligence that uses electronic sensors for instant pattern recognition, is touted as an important part of the future of sensory evaluation and product information processing.

In adopting ANN, the task of sensory evaluation will remain the same, except that multifunctional sensors and optical sensors could increasingly be used in conjunction with human sensory testing. High-performance computer microchips are able to store the combined sensor- and human-generated data, which increases the total product data storage capability. Neural networks accomplish this through the design and implementation of many thousands of input connections, through an internal layering scheme that mimics that of biological neurons (Figure 15.20).

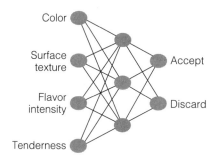

FIGURE 15.20 Artificial neural network. *A neural network can rapidly calculate stimulus data and has the potential to indicate how a product is accepted or rejected. The figure represents the activity direction flow, from left to right. Electronic sensors that are designed to respond to color, surface texture, flavor, and tenderness transfer the data to a "brain," where the information is evaluated. From this point a decision to accept or reject (discard) the product is generated. (Adapted from Bomio, 1998).*

An instrument called the "electronic nose" has been developed using this technology. Similar to the human nose, the electronic nose contains three basic components: sensors, conversions of sensor outputs, and data analysis. It uses a sensor array that responds to each chemical stimulus in a slightly different way. It transmits signals to the "brain"—a computer with sophisticated software—for analysis. In the electronic nose, the ANN is self-learning, meaning the more data presented, the more discriminating the instrument becomes.

Does this mean that machines for sensory evaluation will replace people? No, not by any means. Humans will always play the critical role in determining consumer preference and acceptance of food products. However, the product developer will have a new ally, a sensitive instrument not subject to fatigue or emotion, which can provide reliable data.

Key Points

- Sensory evaluation is a quantitative science involving the measurement and analysis of consumer reactions to foods as they are perceived by the senses.
- Chemical and physical sensations perceive food color, flavor, odor, mouthfeel, and texture.
- Chemical or physical stimuli interact with the senses, which transmit sensations as electrical signals to the brain for interpretation, perception, and response.
- Test equipment can provide objective measurements of sensory parameters, while panelists provide subjective responses.
- Carrying out sensory evaluation involves affective, discrimination, and difference test methods.
- In product development, new products are conceptualized, formulated, processed, tested, marketed, and evaluated.
- Three key aspects to product development work are the idea, development, and commercial stages.
- Test marketing and market research rely on consumer data and play an essential role in food product development and ultimate product success.
- A product's life cycle comprises the time interval from its conception and initial sales, market duration, and sales decline, to its removal from market.
- Artificial neural networks involves the use of electronic and optical sensors linked to a computer to provide data supplemental to human sensory testing.

Study Questions

Multiple Choice

1. Which is a characteristic of supertasters?
   a. insensitive to PROP (6-n-propylthiouracil)
   b. often prefer intense sweetness
   c. tend to like all foods
   d. difficulty distinguishing ingredients present in foods
   e. often have strong food dislikes

2. The main applications of sensory evaluation in the food industry are
   a. establishing SOPs and GOPs.
   b. in quality assurance and product development.
   c. in market research and sales.
   d. in formulation and purchasing.
   e. in quality assurance and integrated resource management.

3. The sensory attributes that define a food's appearance, flavor, texture, and aroma are
   a. affectations.
   b. top notes.
   c. character notes.
   d. primary taste response.
   e. gustation factor.

4. These cells function to detect taste stimuli as well as transfer taste information to the brain.
   a. basal cells.
   b. taste receptor cells.
   c. umami cells.
   d. circumference papillae cells.
   e. depolarizing cells.

5. Flavor combines all of the following *except*
   a. mouthfeel.
   b. odor.
   c. taste.
   d. residual sensations.
   e. appearance.

6. Which mechanical texture characteristic refers to the degree to which a food deforms or compresses between the teeth before breaking?
   a. cohesiveness
   b. compression
   c. chewiness
   d. hardness
   e. viscosity

7. Which one is *not* a geometrical food characteristic?
   a. chalky
   b. adhesive
   c. coarse
   d. gritty
   e. particle size

8. What negative result occurs if a sensory panelist is presented with too many samples?
   a. sensory fatigue
   b. sensory interactions
   c. sensitivity enhancement
   d. threshold detection
   e. subjective perception enhancement

9. Which statement is *not* true regarding line extensions?
   a. can refer to a new form of an already established product
   b. require the same amount of time and money to develop as totally new products
   c. typically involve concept, idea, development, and commercialization
   d. can vary in terms of complexity and cost
   e. are comparatively inexpensive to create

10. What is the meaning of *tristimulus*?
    a. the a, b, c color space
    b. relationship between color, appearance, and gloss
    c. link between color, flavor, and odor
    d. simultaneous exposure to three sensory stimuli
    e. three color dimensions

11. Which pairing is *incorrect*?
    a. color, shine
    b. astringency, pain
    c. crackle, volatile flavors
    d. salt, bitter
    e. bubbling, fizz

12. Which panelist selection and training criterion should *not* be applied to all sensory test situations?
    a. consistent correct responses
    b. healthy personal status
    c. lack of food allergies
    d. socioeconomic level and cultural background
    e. ability to recognize duplicate samples

13. In this form of sensory test, the mechanical, geometrical, fat and moisture, and afterfeel characteristics of food are described and quantified.
    a. consumer acceptance
    b. QDA
    c. triangle test
    d. texture profiling
    e. flavor profile method

14. Which corresponds most to affective testing?
    a. triangle test
    b. defining product attributes
    c. hedonic/acceptance/preference
    d. the use of highly trained panelists
    e. detecting a difference between two samples

15. What is ANN, and what key advantage does it offer?
    a. artificial neural network; time savings
    b. accelerated neural network; time savings
    c. accelerated neural network; accuracy
    d. artificial neural network; increased sales
    e. advanced neural network; precision

16. To maintain profits, a successful food company
    a. buys out all competitors.
    b. will almost never introduce a new product.
    c. anticipates the possibility of bankruptcy.
    d. puts all of its resources into products that have sold well.
    e. considers the uncertainty of the future.

17. Which generally is *not* associated with a food product's life cycle?
    a. strong growth followed by peak sales
    b. lag phase at product introduction
    c. decline in sales
    d. high advertising expense during decline phase
    e. sales growth as new markets open

18. Suppose you are working as a product development scientist. Your company is committed to developing a high soy isoflavone beverage. The question "Where should the product be test marketed?" addresses which activity stage in product development?
    a. brainstorming stage
    b. commercial stage
    c. idea stage
    d. development stage
    e. consumer stage

19. Given: the individual probabilities for success of a food product are A→B (0.9), B→C (0.8), C→D (0.9), and D→P (0.8). What is the correct conclusion?
    a. The product will definitely fail.
    b. The product should not be developed beyond the level of C.
    c. The probability of product success is about 52%.
    d. The product should not be developed beyond the level of B.
    e. The probability of product success is about 85%.

20. Which pairing is *incorrect*?
    a. aspartame, fructose
    b. caffeine, saccharin
    c. lactic acid, malic acid
    d. sodium iodide, potassium chloride
    e. quinine, caffeine

21. The Hunter color data for a light green vegetable would have values that are _____ compared to a dark green vegetable.
    a. lower in L
    b. lower in L and lower in $+a$
    c. lower in L and higher in $-a$
    d. higher in L
    e. higher in $+a$

22. Which does *not* apply to the triangle test?
    a. One of the sample presentation orders is AAA.
    b. It is in the same category of test as paired comparison.
    c. One of the sample presentation orders is ABA.
    d. It can be used in panelists selection and training.
    e. One of the sample presentation orders is BAA.
23. T/F   Both odorant and tastant molecules interact through similar transduction mechanisms.
24. T/F   Regarding sensory test methods, discrimination tests are used to see if panelists can distinguish an overall difference between two products.
25. T/F   In adopting ANN technology, the need for human sensory tests is eliminated.

References

Bomio, M. 1998. Neural networks and the future of sensory evaluation. *Food Technology* 52(8): 62–63.

Clydesdale, F. 1991. Color perception and food quality. *Journal of Food Quality* 14: 61–74.

Fuller, G. W. 1998. *New Food Product Development: From Concept to Marketplace*. CRC Press, Boca Raton, FL.

Goldman, A. 1994. Predicting product performance in the marketplace by immediate- and extended-use sensory testing. *Food Technology* 48(10):103–106.

Hegenbart, S. 1992. Fostering creativity and innovation in food product design. *Food Product Design* 1(9)26–32.

Hegenbart, S. 1992. Sensory analysis: maximizing effectiveness in product development. *Food Product Design* 48(5):51–60; 65.

Hollingsworth, P. 1994. The perils of product development. *Food Technology* 48 (6): 81–88.

Institute of Food Technologists. 1990. Taking the gamble out of product development. *Food Technology* 44(6): 110–117.

Lawless, H. T., and Heymann, H. 1998. *Sensory Evaluation of Food, Principles and Practices*. Chapman & Hall, New York.

McLaughlin, S., and Margoleskee, R. F. 1994. The sense of taste. *American Scientist* 82(6):538–545.

Murano, P. S., and Johnson, J. M. 1998. Volume and sensory properties of yellow cakes as affected by high fructose corn syrup and corn oil. *Journal of Food Science* 63(6): 1088–1092.

O'Mahony, M. 1995. Sensory measurement in food science: fitting methods to goals. *Food Technology* 49(10):72–82.

Schutz, H. G. 1998. Evolution of the sensory science discipline. *Food Technology* 52(8):42–46.

Sidel, J. L., and Stone, H. 1976. Experimental design and analysis of sensory tests. *Food Technology* 30 (11): 33–38.

Utermohlen, V. Taste, flavor, and food choice. http://www.cce.cornell.edu/food/expfiles/topics/utermohlen/utermohlenqanda.html

Additional Resources

Baker, R. C., Wong Hahn, P., and Robbins, K. R. 1988. *Fundamentals of New Product Development*. Elsevier Publishing Company, New York.

Cohen, J. C. 1990. Applications of qualitative research for sensory analysis and product development. *Food Technology* 44(11):164–174.

Demetrakakes, P. 1999. Playing to win. *Food Processing* 60(1):15–20.

Erickson, P. 1992. Comparing product design strategies in the United States and abroad. *Food Product Design*, 1(9) 59–65.

Jeremiah, L. E., Gibson, L. L., and Burwash, K. L. 1997. Descriptive sensory analysis: the profiling approach. Research Branch, Agriculture and Agri-Food Canada, Technical Bulletin 1997-2E.

Larmond, E. 1977. *Laboratory Methods for Sensory Evaluation of Food*. Canadian Govt. Publ. Center, Ottawa, Ontario.

Meilgaard, M., Civille, G. V., and Carr, B. T. 1991. *Sensory Evaluation Techniques*, 2nd ed. CRC Press, Boca Raton, FL.

Munoz, A. M., Civille, G. V., and Carr, B. T. 1992. *Sensory Evaluation in Quality Control*. Van Nostrand Reinhold, New York.

Stone, H. W., and Sidel, J. L. 1998. Quantitative descriptive analysis: developments, applications, and the future. *Food Technology* 30 (11): 48–52.

Szczesniak, A. S. 1998. Sensory texture profiling—historical and scientific perspectives. *Food Technology* 30 (11): 54–57.

Whitfield, P., and Stoddart, M. 1985. *Hearing, Taste, and Smell: Pathways of Perception*. Torstar Books, New York.

Web Sites

www.cce.cornell.edu/food/expfiles/topics/utermohlen/utermohlenqanda.html
"Taste, Flavor, and Food Choice" by V. Utermohlen. Includes characteristics of tasters: PROP categories.

www.ifrn.bbsrc.ac.uk/public/set99/map.html
Institute of Food Research: Dispelling the Tongue Map Myth! This site challenges the tongue map proposed by Hanig, which described the distribution of taste sensitivity around the perimeter of the tongue.

www.umds.ac.uk/physiology/jim/tasteolf.htm
The United Medical and Dental Schools, London. Taste and smell are differentiated and physiological basis explained.

http://www.ift.org/divisions/marketing/
IFT Marketing and Management Division supports the development of the scientist/business person in the area of business tactics and skills, consumer trends, marketing strategies, organizational development, management techniques, and leadership.

http://www.ift.org/divisions/prod_dev/
IFT Product Development Division

http://www.pdlab.com/
PD Lab. This is a central site for product developers in food, beverage, cosmetics, and pharmaceutical industries.

Challenge!

EXPERIMENTAL DESIGN IN PRODUCT DEVELOPMENT

Years ago during the 1950s and 1960s, the words "made in Japan" on a manufactured product meant an inferior product. Today, the single most important factor associated with Japan's economic success is its near obsession with quality. Quality! Other countries, including the United States, have been taught a lesson. If a serious share of the market is desired, a company must produce top quality products. And that goes for food products, too. To be competitive, a commitment to quality is imperative.

But what is **quality?** Quality may not be such an easy word to define. Quality can be thought of as the sum total of a product's characteristics that satisfy the consumer under specified conditions of use. However, quality embodies concepts; it is more than a simple definition. The first of these concepts is that quality must be consistent throughout the product. The second is that consumer perceptions are needed to determine and judge key quality parameters. A third concept is that quality is thought of as multidimensional. The multidimensional nature of quality refers to product performance, features, aesthetics, cost, and reputation.

The Relationship Between Quality and Experimental Design

Quality should be designed into every food product that is developed because quality influences the consumer's buying decision. Additionally, defects in quality most frequently can be traced back to the early stages of product design. The genius of the Japanese was in their attention to the design of experiments before initiating production, which resulted in improving product quality.

So where did the concept of experimental design originate? An Englishman named Sir Roland Fisher introduced the notion as an application of statistical design to agricultural research during the 1920s. However, not much was done with it until the 1950s.

In the 1950s, W. Edwards Deming proposed that business processes should be analyzed and measured to identify sources of variations that cause products to deviate from customer requirements. Deming's design of experiments (DOE) is an essential method for developing competitive world-class products. He recommended that business processes be continually evaluated so that managers could provide needed improvements. He believed that every organization could gain significantly from increasing its understanding of product variation and the use of experimental design and statistical methods.

Deming is famous for creating the PDCA cycle for Plan, Do, Check, and Act. "Plan" means pre-experimental planning to improve results; "do" means implementing the plan and measuring its performance; "check" means assessing the measurements and reporting the results to decision makers; and "act" requires a decision regarding changes needed for improvement. Deming stressed the importance of checking one's experimental results by conducting confirmation experiments.

What is **experimental design?** It is a tool used by food scientists to determine the effects of experimental treatments upon a particular population. A population can imply people (e.g., consumers) or products (e.g., different batches of low-fat blueberry muffins). Experimental design can refer to a deliberate plan to vary factor levels (such as an ingredient level used in a formulation) to study the effect on specific product attributes.

An understanding of experimental design and becoming accustomed to using experimental design as a part of the product development process is important for the prospective food scientist. Experimental design is effective and efficient, saving time and

General Case:

	Levels		
Test variable A	1	2	3
Test variable B	1	2	3

Specific Case:

	Levels (mg)		
Citric acid	0.1	0.4	0.8
Maltodextrins	1.0	2.0	3.0

FIGURE 15.21 A 2 × 3 factorial experimental design.

money. Results obtained from a well-designed experiment will be precise enough to detect important differences in product characteristics.

Names of Experimental Designs

Completely randomized, factorial, incomplete block balanced, Latin square, randomized block, response surface, and split-plot—these are all names of types of experimental designs. We will briefly define a few of these within the appropriate context of sensory evaluation in product development. In a **complete block experimental design,** all product treatment (i.e., ingredient) levels are assigned to one block (where block means the panel). This corresponds to each panelist evaluating every version of a product or having all products evaluated within a single session. When only certain levels of a variable being studied are assigned for evaluation during a single session, an **incomplete block experimental design** results.

Factorial experimental designs treat all levels of one test variable with all levels of another test variable. For example, a product might be formulated with two specific additives, an emulsifier and a modified starch, at 0%, 0.5%, and 1.0% addition levels of each. Two variables with three levels of each results in a 2×3 factorial design (see Figure 15.21). With proper application, any of the above designs meets the criteria of a good experimental design. Good designs are kept as simple as possible, should answer the key questions and meet the established objectives of the study, and minimize experimental error.

In product development work, experimental design often refers to all of the main components of a sensory experiment. Since human-derived data is often highly variable, it is desirable for the sensory specialist to control or eliminate as much variability in the data obtained as possible. Certain biases or influences are controlled by having well-defined objectives, and through the use of individualized sensory booths, randomized numbers, randomized presentation order, and so forth:

Objectives refer to the project objective (e.g., "to reduce the sodium content of pretzel snacks") and the test objective (e.g., "to determine whether the control and treatment (reduced sodium) pretzels are acceptable to consumers").

Testing environment is the physical setting where the sensory testing is carried out. The sensory room must be isolated from the sample preparation area, and it must be controlled with respect to distractions, noise, light, temperature, comfort, use of color, and layout because the physical conditions of the test facility can affect the results of a sensory test. It is best to have panelists perform noninteractive evaluations in separate testing booths.

Panelists (alternatively referred to as participants, judges, or tasters) are screened for applicability to a given sensory test. They may be untrained or trained, as required by the test objective. The number of panelists is determined in part by practicality and in part by the type of sensory test, the number chosen yielding sufficient data able to be interpreted by statistical methods.

Samples or treatments, including an appropriate control, are developed to specifically meet the project and test objectives. They must be easy to prepare and to evaluate. All samples must be standardized in appearance, size, and temperature of serving. The amount of sample tested by each panelist must be within a limit that offers both statistical validity and comfort to the panelists. Samples must be coded with random 3-digit numbers, such as "715," without the use of letters such as A, B, C, etc. to prevent psychological bias effects.

Response forms (evaluation forms, scorecards, or score sheets) are frequently written to provide information regarding a specific product attribute, such as flavor intensity or juiciness. Or the panelist may be instructed to select a preference, rate product acceptability, or identify a difference, according to the goal of the research. The form of the response should be matched to the use of a specific statistical test that will be applied in data analysis in order to test the hypothesis of the experiment.

Serving procedure of the samples takes into account physical condition (sample dimensions, weight, shape, and temperature) as well as the presentation sequence (some-

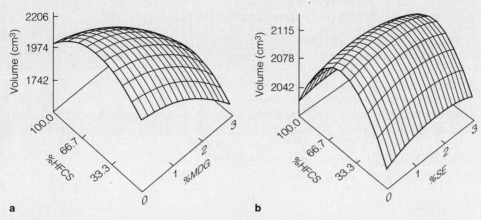

FIGURE 15.22 Response surface plots. *These three-dimensional plots show how two emulsifier ingredients, mono- and diglycerides (MDG) and sucrose esters (SE), in combination with high fructose corn syrup (HFCS), affect cake volume. (From Murano and Johnson, 1998.)*

times called the test plan). The serving procedure can affect the choice of data analysis method.

A **balanced design** is a common type of sensory design, in which interaction effects between panelist and sample (due to sample position on serving tray, sample order, frequency of appearance of a sample number) are balanced. A balanced design minimizes the possibility of generating meaningless data and experimental error.

Data analysis is accomplished by the selection of the appropriate statistical test method for the experimental design. Data means, standard deviations, correlation coefficient determinations, analysis of variance (ANOVA), mean separation procedures, and response surface methodology are frequently employed to meet the objectives of an experiment and answer the basic research questions posed by the study. Variables may also be related to processing conditions. For example, the processing time and temperature of a dried meat product might be varied to determine the effect on product tenderness.

The use of statistics in data analysis and good experimental design is essential to draw meaningful conclusions in product development work. A statistically designed experiment is always divided into two parts. One is the **screening design,** which is used to determine which variables are of primary importance. The other is **optimization,** where experimentation finds the best operating conditions for each of the important variables.

Response surface design methodology is used to identify optimum levels of ingredients or optimum product processing conditions in product development. It allows the product developer the opportunity to vary levels of several ingredients (the independent variables) simultaneously and check their influence on a particular response (the dependent variable). It becomes a relatively simple matter to optimize ingredient levels and combinations to create the best products. As an example, various emulsifier and corn syrup levels can be added to a cake formulation to determine their effects on cake volume—see Figure 15.22.

The **Edisonian approach** to experimentation is to attempt to study all variables and approaches until a solution to a problem is found. But this approach is too time consuming and incorporates none of the advantages inherent in experimental design strategies. Experimental design has made the Edisonian approach obsolete and impractical. One of the main reasons that experimental design is more successful than Edisonian experiments is that the design takes into account the effect of interactions. **Interaction** in the experimental sense is what product developers commonly know as synergy. These terms signify that the relationship between the response and one variable depends on the level of a second and/or third variable. Understanding the interaction allows the product developer to accurately predict what will happen in a system when one or more of the variables are altered.

Table 15.6 lists the steps in experimental design.

TABLE 15.6 Experimental design involves several steps.

1. Make a clear statement of the objectives.
2. List the objectives of others on the product development team, including Marketing, Management, Engineering.
3. Put the objectives of the experiment in writing, and allow for comment.
4. List the variables that might affect the measured responses.
5. Determine which responses are going to be measured. For example, product greasiness, fat content, and sensory acceptance could be measured.
6. Identify the types of variables, whether they are discrete or continuous.
7. Determine which variables can be controlled and which cannot.
8. State the hypothesis based upon expectations.

Variables and Error Considerations

We want next to present the topic of error but need first to discuss the nature of the variables in an experiment. There are two types of experimental variables, discrete and continuous. **Discrete variables** are those that can take on only certain values. For example, if the effects of three brands of starch in a baked product are under investigation, the discrete levels would be brand A, brand B, and brand C. **Continuous variables** are those that may take on an infinite number of levels. If the percentage of starch in a formula is being studied, it could range anywhere from 0 percent to over 50 percent.

Precision refers to the repeatability of a test result. On repeat testing, slight variations in data observation are typically due to error variance. This error variance is kept low by isolating the sensory response to the factor of interest, minimizing extraneous influences, controlling sample preparation and presentation, or, when needed, screening and training panelists. **Accuracy** is the ability of a test instrument (the sensory panel) to produce a value that is close to the "true" value (if knowable), such as that defined by an independent instrument that has been appropriately calibrated.

In sensory testing, the **validity** of a test relates to its ability to measure what it was intended to measure. The test results should reflect the perceptions of consumers who might use the product and should generally be applicable to the larger population of users. A good sensory test minimizes errors in measurement, in conclusions, and in decisions. Of primary concern in sensory testing is the sensitivity of a test to differences among products. A test should not miss important differences that are present, nor should it find any that aren't real. "Missing a difference" that is present or "finding a false difference" that is not implies an error due to an incorrect or insensitive test procedure.

In statistical language, detecting true differences means avoiding Type I and Type II errors. **Type I error** is detection of a false difference, which means concluding that a difference exists when there is none. **Type II error** is missing a difference and concluding that a difference does not exist when there is one. Each kind of error can be devastating in product development, which places a statistical burden upon sensory specialists.

In summary, experimental design makes a lot of sense in product development work. It ensures proper experimental set-up, data collection, and data analysis. It minimizes error and the potential for incorrect decision making. It is cost effective. Fewer experiments need to be performed, which adds up to fewer costs, with reduced labor, time, and raw material needs. The product developer who follows the steps of a good experimental design can trust the data obtained through experiments and be confident in the interpretations and conclusions drawn. Perhaps most important of all, experimental design properly carried out provides the gateway to the establishment of consistent product quality.

Challenge Questions

1. An experiment wherein the effects of adding aspartame, methylcellulose, and citric acid to a food product, each at three levels, are studied is
    a. a randomized design.

b. a 3×3 factorial design.
   c. a complete block design.
   d. a 3×3 factorial design.
   e. an incomplete block design.

2. Which refers to detecting a false difference?
   a. Type I error
   b. validity
   c. accuracy
   d. Type II error
   e. lack of precision

3. Consider an orange juice beverage. Which terms, in proper order, refer to the multidimensional quality aspects of aesthetics, performance, and nutrition?
   a. potassium, sweetness, color
   b. brand name, tartness, potassium
   c. package, viscosity, pulp content
   d. brand name, container, vitamin C
   e. container, orange flavor intensity, vitamin C

4. Evaluate the following pairings and identify which one is incorrect
   a. screening design/identity of key variable
   b. samples/meet test objectives
   c. ANOVA/example of complete block design
   d. serving procedure/important in test plan
   e. response surface design/identify optimum ingredient levels

5. In sensory work, it is possible to control variability in the data and minimize error. All of the following help accomplish this *except*
   a. randomization of sample presentation order.
   b. proper experimental design selection to allow assessment of interaction effects.
   c. layout of sensory evaluation area open to food preparation area.
   d. use of individualized sensory booths of neutral color and proper lighting.
   e. randomization of sample presentation order.

Critical Thinking

Read the following scenario of a hypotheical cake mix devolpment project, and assess the skill of the sensory specialist in charge of the project.

Scenario: A food product development team has re-formulated a pre-existing cake mix to make it low fat. In the sensory evaluation of the product, eight untrained consumers were asked to perform descriptive analysis of the standard mix and the low fat mix in order to look for differences between the products. When the results indicated that the products were equally described, and only four panelists detected differences, the sensory specialist concluded that the new low fat treatment was a success, and recommended progress to the commerical stage.

Evaluate to following statements, and agree or disagree with each.

- The proper sensory test was not selected.
- Sensory testing did not match the objective.
- Consumers should not participate on descriptive panels (descriptive panels must use highly trained judges).
- An acceptable experimental design approach was not employed.
- Proper statistical analysis of the data was not performed.
- The conclusion was not based upon a sound experimental approach, suggesting that the researcher committed a Type II error.

GLOSSARY

Absolute viscosity Viscosity measurements obtained via methods not under the influence of gravity during the determination.

Absorption Digested nutrients entering the bloodstream through the capillaries of the small intestine.

Accuracy Ability to produce a value that is close to the "true" value, such as that defined by an independent instrument that has been appropriately calibrated.

Acetylcholine Chemical produced by nerve cells that helps transmit electrical signals from nerve cells to muscle.

Acid food A food with a natural pH of 4.6 or below.

Acid value A measure of the number of free fatty acids present in a fat.

Acidified food Low-acid food to which acids are added, having water activity values exceeding 0.85, and a final equilibrium pH of 4.6 or below.

Acidity control Controlling the pH of a food through the use of a combination of acidification, heat treatment, and/or refrigeration storage.

Acidulant Food additive that functions to lower pH; examples of acidulants include citric acid, phosphoric acid, and benzoic acid.

Actinomyosin Crosslinked contractile muscle tissue proteins actin and myosin.

Action level A level for contamination of a food, below which no court enforcement action is necessary.

Activation energy The amount of energy needed to convert substrate molecules from the ground or baseline energy state to the ES complex. Under proper reaction conditions, it represents the greatest barrier to enzyme product formation.

Active site Region on the surface of an enzyme where catalytic activity occurs.

Acute effect The rapid response of a biological system to a toxicant, lasting a short time and due to one exposure event.

Addition reaction Reaction of two or more substances to give a third substance.

Additive Substance added to foods for a specific reason, such as a coloring agent, a flavor ingredient, or a thickener.

Adenosine triphosphate (ATP) A molecule necessary for cell-building reactions through the transfer of high-energy phosphate.

Adenylate cyclase Enzyme that catalyzes the production of cyclic adenosine monophosphate (cAMP) from adenosine triphosphate (ATP).

Adhesiveness Mechanical mouthfeel parameter equal to the force required to remove food material that attaches to the mouth (primarily the upper palate) during normal chewing.

Adiabatic A drying situation occurring when warm air provides the heat of evaporation to dry a food product with no heat gained or lost to the surroundings.

Adiabatic saturation process Heated air flowing past a moist food product in a drying operation, with the air supplying the heat of evaporation to dry the product, with no change in enthalpy.

Adsorbed water (structural water) Water that associates in layers via intermolecular hydrogen bonds around hydrophilic food molecules.

Adulteration Purposeful introduction of foreign material into food, especially those aesthetically objectionable, indicative of unsanitary or unscrupulous manufacturing practices.

Aerobe Organism that requires oxygen for survival.

Aerosols Suspended droplets of liquid containing microorganisms.

Affective tests Sensory test methods that attempt to quantify the degree of liking or disliking of one product over another; include preference, hedonic, and consumer acceptance tests.

Aflatoxin A powerful mycotoxin produced by the *Aspergillus* mold, which is associated with crops such as peanuts, corn, rice, cottonseed meal, oats, hay, barley, sorghum, cassava, and millet that are stressed by drought.

Afterfeel A texture sensation residue that persists in the oral cavity following consumption.

Aftertaste Taste sensation perceived after the initial primary taste response, viewed as residual sensations that linger on the tongue after swallowing.

Agricultural Marketing Service (USDA/AMS) Branch of the USDA that administers the Egg Products Inspection Act, offers grading of meat, poultry, fruits, vegetables, and dairy products, and administers the National Organic Program.

Agricultural Research Service (USDA/ARS) Branch of the USDA that leads the national agricultural research agenda to ensure high-quality, safe food and other agricultural products, and to sustain a competitive agricultural economy.

Alcohol Organic compounds containing hydroxyl (OH) functional groups.

Aldehyde Organic molecule in which oxygen is double-bonded to carbon.

Ale A type of beer with a pH of 3.8.

Alimentary pastes Cooked and dried pasta products.

Alimentary toxic aleukia Disease condition characterized by three stages of progression, and caused by the mold *Fusarium sporotrichioides* that produces the causative trichothecene toxin.

Allergens Antigen susbstances (usually proteins) foreign to the body that elicit an immune system response and trigger inflammation and other symptoms.

Alternative sweetener Substance other than sucrose (table sugar) used to sweeten foods and beverages; examples are high fructose corn syrup, saccharin, and aspartame.

Ames test A simple, rapid test to identify the mutagenic potential of chemical substances including chemical food additives.

Amine Compound that contains an amino (NH_2) functional group.

Amino acids Component subunits of proteins.

Amnesic shellfish poisoning Poisoning caused by consuming mussels infected with the dinoflagellate *Nitschia*, containing the toxin domoic acid.

Amphiphilic molecules or amphiphiles   Molecules that contain polar (hydrophilic) and nonpolar (hydrophobic) regions in their structure.

AN/TN ratio   Measure of the relative amount of amino nitrogen (AN) present in a hydrolysate relative to the total amount of nitrogen (TN) in the intact protein substrate.

Anaerobe   Organism that cannot survive or grow in the presence of oxygen.

Annatto   Coloring compound that imparts the characteristic yellow-orange color to cheddar cheese.

Anthocyanins   Water-soluble flavonoid compounds that range in color from deep purple to orange-red.

Antibiotic   Drug used to treat bacterial infections in animals and humans.

Anticaking and free-flowing agents   Substances that keep ingredients in a powder form for ease of incorporation into formulations during product manufacture.

Antimicrobial agents   Additives that act to inhibit the growth of bacteria, yeasts, and molds and thus function as preservatives.

Antioxidant   Additive that acts to inhibit the oxidation of fats and pigments by molecular oxygen to prevent product rancidity and altered color.

Antiprotozoal   Drug utilized to eliminate or treat protozoan infections in animals and humans.

Antisensing   The technique of making a copy of DNA with the complementary sequence, resulting in production of an antisense mRNA molecule.

Aplastic anemia   Blood disorder that renders the patient unable to manufacture red blood cells, caused by oversensitivity to chloramphenicol antibiotics.

Apparent viscosity   Viscosity of non-Newtonian food materials of both shear-thinning and shear-thickening types measured at a single shear rate.

Appearance   Exterior as well as interior product characteristics; for example, the crumb cell size of cake or bread, and external or surface characteristics, such as a food's smoothness or roughness.

Artificial neural network (ANN)   Form of computer intelligence that uses electronic sensors for instant pattern recognition and product information processing.

Aseptic canning   The separate sterilization of food and canning container, and then bringing them together in a sterile environment where the container is filled and sealed.

Aseptic packaging   Presterilized containers that are hot-filled with product and hermetically sealed.

Autolyzed yeast   Organisms that have been centrifuged to remove cell wall material to concentrate such flavors and flavor precursor substances as free amino acids, peptides, monosaccharides, and Maillard reaction products generated during thermal processing.

Avidin   Egg white protein that binds to biotin, making it unavailable nutritionally.

Backcrossing   A breeding technique of crossing the first generation with the parent crop and repeating this process many times to eliminate unwanted characteristics.

Bacteria   Unicellular organisms, measuring about 1 micron in length, which stain either gram-positive or gram-negative.

Bactericidal   Able to kill bacterial cells.

Bacteriostatic   Able to inhibit the growth of but not kill bacterial cells.

Balanced design   Common type of sensory design, in which interaction effects between panelist and sample (due to sample position on serving tray, sample order, frequency of appearance of a sample number) are balanced.

Ballistic bombardment   A means to insert ("shoot") DNA into cells at high velocity using microscopic gold or tungsten particles coated with the desired gene.

Bar code technology   A method of automating data collection using symbologies to encode data that is optically scanned and computer processed.

Barley malt   A compound that breaks down starch from barley into individual sugars during beer production.

Base   A nitrogen-containing compound found in the DNA molecule (adenine, thymine, cytosine, guanine).

Bench-top prototype   Product made in the laboratory as a small scale (a few pounds in weight) batch that can be easily and reproducibly formulated for repeated testing without too much expense.

Beta-glucans   Polysaccharides of glucose similar to cellulose, but less linear, occurring in oats, barley, and yeast, useful as a fat replacer.

β (beta)-lactoglobulin   A whey protein component of milk.

Betalains   Group of two types of water-soluble plant pigments: betacyanins and betaxanthins.

Beverage   Drinkable liquid that is consumed for its ability to quench thirst, for its stimulant effect, for its alcohol content, for its health value, or for enjoyment.

Bioassay   Controlled study utilizing laboratory animals.

Bioavailability   Degree to which nutrients are able to be digested by human enzymes in the digestive tract and absorbed by the body.

Biological changes   Changes in foods due to microbial fermentation of carbohydrates into small organic molecules such as acids and alcohols, and gases such as hydrogen sulfide and carbon dioxide.

Biological hazards   Bacteria, molds, viruses, and parasites such as protozoa, flatworms, and roundworms that cause foodborne disease.

Biological value (BV)   Measure of protein quality equal to the amount of nitrogen derived from food protein used in the body to promote growth; expressed as a ratio of the nitrogen retained to the amount of nitrogen absorbed from a food.

Biologically active water   Water available to participate in chemical reactions and accessible to microorganisms in foods.

Biotechnology   The practice of genetic engineering to make useful products.

Biotechnology policy of the FDA   The food products of biotechnology were to be viewed no differently than ordinary food by the FDA, except in cases in which the genetic engineering process added a new substance to the food that is not GRAS or that is significantly different in structure, function, or amount than substances found ordinarily in food.

Blanching   The rapid application of mild heat processing to deactivate browning (polyphenol oxidase) and tissue softening (pectinase) enzymes present in fruits and vegetables that are to be further processed.

Bleaching   Processing to remove colored substances from an oil, through heating and use of adsorbants.

Botanical   Herbal medicine or herbal pharmaceutical used in traditional medicine.

Bottled water   As distinguished from ordinary drinking (tap) water, includes natural mineral waters, carbonated waters, and sweetened, flavored waters that are typically carbonated.

Botulinum toxin Powerful neurotoxin produced by the bacterium *Clostridium botulinum*.

Botulism Disease caused by consumption of toxin produced by the bacterium *Clostridium botulinum*.

Boundary layer Interface between two dissimilar materials, such as frying oil and a food material, that come in contact with each other and experience the phenomena of heat and mass transfer.

Bound water (water of hydration) Water in a tight chemically bound arrangement in food that does not exhibit the typical properties of water.

Bovine spongiform encephalopathy (BSE) A deadly condition thought to be transmitted by a protein termed a prion; "mad cow disease."

Brainstorming In marketing, examining consumer trends to speculate on what new products would meet those trends.

Braising Physical meat tenderizing treatment that involves a slow, moist heating process, causing the pulling apart of the collagen strands in tough meat.

Bran Outermost layer in a grain kernel, high in fiber content (cellulose and hemicellulose), and also containing protein, B vitamins (niacin, thiamin, and riboflavin), and iron.

Bread The product of baking a 3:1 mixture of flour and water, with added salt, yeast, and other ingredients.

Breaking The process of shell breaking to separate the egg white from the egg yolk.

Brevitoxins Toxins produced by the dinoflagellate *Ptychodiscus*, responsible for neurotoxic shellfish poisoning.

Broad spectrum antibiotic Antibiotic effective against a wide variety of bacterial infections.

BST (bovine somatotropin hormone) A naturally produced substance that increases milk production in dairy cows.

Buffer A solution of a weak acid and its salt at a pH where the solution has the ability to maintain that pH when quantities of base are added.

Buffering effect The prevention of changes in pH.

Butter Dairy spread made from either sweet or sour cream, as a water-in-oil emulsion.

Cadaverine Compound produced from the decarboxylation of lysine by some microorganisms.

Calorie Amount of heat needed to raise the temperature of 1 gram of water by 1 degree Celsius.

Candling Rotating an egg in front of a light source to examine the size and position of the air cell, clearness of the white, yolk position and mobility, and shell condition and to check for staleness, blood clots, embryo development, and quality grade.

Cane sugar The sucrose product obtained from sugarcane, generally produced in two stages after harvesting.

Canning Food preservation method achieved by filling food into sealed containers and heating to destroy spoilage microorganisms (bacteria, yeast, and molds).

Capsaicinoids Pungent alkaloid compounds that occur in chiles.

Caramel color An example of a natural colorant exempt from certification, and produced commercially by heating sugar and other carbohydrates under strictly controlled conditions.

Caramelization Formation of brown caramel pigments as a result of applying heat energy to sugars.

Carbamate Pesticide similar in mode of action to organophosphates; not considered carcinogenic.

Carbonation The saturation of water with carbon dioxide under pressure in which the CO_2 gas dissolved in the water becomes carbonic acid.

Carboxylic acid Acid containing the COOH functional group.

Carcinogens Chemical toxicants that are known to cause cancer in laboratory animals or humans.

Carnauba wax Exudate from palm tree leaves that functions as a coating in chewing gum, sauces, fruits and vegetables, and confections.

Carotenoids Class of fat-soluble plant pigments that consists of carotenes and xanthophylls that are responsible for the yellow, orange, and red colors of fruits and vegetables.

Casein micelles Large colloid particles in milk composed of calcium phosphate complexed to casein.

Caseins Proteins found in milk.

Cell culture A population of tissue cells grown under controlled in vitro conditions.

Cell culture study Study that uses tissue culture cells in place of animals or people.

Cellophane Film material derived from chemically treated wood pulp having a high flexural modulus.

Cellulase Enzyme produced by microorganisms that break down cellulose.

Cellulose Polysaccharide found in the cell wall of plants, composed of glucose molecules, a type of insoluble fiber that quickens the movement of food through the intestine.

Cereal grain Any grain used for food, such as corn, rice, wheat, barley, millet, rye, sorghum, and oats.

Character notes Sensory attributes of a food that define its appearance, flavor, texture, and aroma.

Cheddaring Repeated cutting of the cheese curd, and draining of the whey.

Cheese A concentrated dairy food defined as the fresh or matured product obtained by draining the whey (the moist serum from the original milk) after coagulation of casein.

Chemical acidulants Substances added to lower food pH as an alternative to fermentation by microorganisms.

Chemical bonds Forces that hold atoms and molecules together.

Chemical changes The action of microbial enzymes called proteases and lipases on food protein and lipid molecules, respectively, that result in altered food texture, flavor, and odor.

Chemical hazards Chemical substances that are either naturally occurring in foods, such as plant toxins, as well as those that are added to food, such as antibiotics, that cause foodborne disease.

Chemogastric Revolution The decades (1940–1970) when the processed food industry saw a dramatic surge in growth.

Chemotrophic organism Organism that requires chemicals for metabolism.

Chewiness Mechanical texture characteristic equal to length of time in seconds required to chew a food sample at a steady rate (one chew per second) with a constant force applied that results in a food consistency that is just ready to be swallowed.

Chilling injury Damage to fruits and vegetables characterized by pitting and browning of the surface and the flesh, due to severe cooling treatment.

Chocolate A natural food derived from cacao beans from the tree *Theobroma cacao*.

Cholera Disease caused by consumption of cells of the bacterium *Vibrio cholerae*, which produces a toxin inside the intestinal tract responsible for most of the symptoms.

Chroma The clarity and purity of a color.

Chromophoretic compounds in foods A structurally diverse group of pigment molecules.

Chromosome An internal cell component that packages the cell's functional DNA units.

Chronic effect The slow response of a biological system to a toxicant, due to repeated long-term exposure.

Chymosin Enzyme that breaks down casein proteins in milk.

Ciguatera Disease caused by consumption of tropical fish that feed upon dinoflagellates of the genus *Gambierdiscus*.

Cilia Hairlike projections on olfactory epithelial (receptor) cells that sense odorant molecules.

Cleaners Amphiphilic compounds having hydrophilic and hydrophobic structure that interact with water and dirt/debris by suspending these particles in solution.

Cleaning Removal of surface dirt, debris, and associated bacteria by washing with water and detergent.

Climacteric fruits Fruits in which ripening is accompanied by increased respiration (e.g., apples, banana).

Cloning The splitting of an embryo cell into eight individual cells, with each cell containing the genetic information necessary to produce an identical animal.

Clostridial toxins Toxins produced by members of the bacterial genus *Clostridium*.

Co-extrusion Simultaneous combination of several layers of plastic using adhesives to promote bonding between the polymers, which remain discrete despite the extrusion process.

Cochineal extract A natural color additive derived from the dried bodies of cochineal insects.

Cocoa butter Fat obtained from cacao beans that supplies most of the fat calories in chocolates.

Code of Federal Regulations (CFR) A publication of the Office of the Federal Register, National Archives and Records Administration, in Washington, DC, that includes food standardization information and nutrition labeling regulations.

Codex Alimentarius A publication that establishes international food quality standards to facilitate world trade.

Codex Alimentarius Commission A UN organization established in 1962 by the FAO and WHO to develop international standards and safety practices for food and agricultural products.

Coenzymes Specific B vitamins that function with enzymes to facilitate the metabolism of the energy nutrients.

Cohesiveness Mechanical texture characteristic—the degree to which a food will deform or compress between the teeth before it breaks.

Cold-break process Processing of tomatoes without heat, which permits the enzyme pectinase to lower the viscosity of the product.

Cold point In retort processing, the last area within a can to reach processing temperature; it is measured to determine if heat has penetrated a container of food and completely heated all the food particles to the desired temperature.

Collagen A tough fibrous protein that comprises the majority of the connective tissue surrounding muscle fiber bundles.

Colloids Surface active ingredients such as fatty acids, glycerides, phospholipids, polysaccharides, and proteins too large to dissolve and become the dispersed phase of a true solution.

Color A perception of the physical appearance of food that includes chroma, hue, and intensity.

Color additive Any dye, pigment, lake, or substance that can impart color when added or applied to a food.

Color additive certification The FDA approval process for synthetic food color additives that assures the safety, quality, consistency, and strength of a color.

Color Additives Amendments A 1960 amendment to the FFDCA that requires dyes used in foods, drugs, cosmetic, and certain medical devices to be approved by the FDA prior to their marketing.

Colorants, or food colors Natural or synthetic additives that produce or retain attractive colors in foods and beverages.

Commercial stage The third stage in product development—production scale-up from pilot plant to commercial plant, plus market testing and the subsequent introduction of the new product to a national market.

Commercially sterile The degree of sterilization at which all pathogenic and toxin-producing organisms are destroyed, as well as any spoilage organisms.

Comminution Meat particle size reduction; the unit operation by which muscle tissue is chopped, diced, emulsified, ground, and transformed into minute particles for incorporation into a sausage.

Comminuted meat emulsions Finely chopped meat mixed with water, fat, and sometimes additives such as preservatives and water-binding agents.

Comparator A food already present in the food supply against which genetically engineered foods are compared with respect to analysis of composition and appropriate characteristics.

Complete block experimental design All product treatment (i.e., ingredient) levels are assigned to one block (where block means the panel). This corresponds to having each panelist evaluate every version of a product or to having all products evaluated within a single session.

Composition reaction Chemical reaction in which two or more substances are combined into a single product.

Composting A process applied to organic material, decomposing it in order to release nutrients and eliminate pathogens.

Compound Two or more elements chemically bonded together in definite proportions by weight.

Condensation Reaction in which separate reactant molecules are linked together by special chemical bonds, sometimes through the action of enzymes.

Conduction The movement of molecules or heat energy through a medium from the more energetic particles to the less energetic ones.

Conjugated double bonds Alternating single and double carbon-to-carbon bonds.

Connective tissue Tissue composed of a watery dispersion of stromal protein matrix and includes elastin and collagen.

Consumer acceptance testing Affective test method in which consumers rate their liking or disliking of a product on a scale.

Consumer response testing Asking consumers questions regarding the likelihood they would purchase a product, how much they would pay for it, and how frequently they anticipate using the product. Affective testing is also a component of consumer response testing.

Continuous variables   Variables that may take on an infinite number of levels.

Controlled atmosphere storage   Storage that controls the amounts of gases in the food environment in order to extend shelf life.

Convection   The movement of heat energy due to density differences in a product caused by temperature gradients within the system, and characterized by random molecular motion (diffusion) and bulk motion.

Convulsive ergotism   Disease condition characterized by hallucinations and convulsive seizures due to a mycotoxin produced by the genus *Claviceps*.

Copolymer   A plastics polymer formed by the reaction of at least two different comonomers.

Coprophagy   Direct ingestion of feces, which are often infected with parasite eggs.

Corporate mission   Statement that embodies the foundational aspects of any corporation: who it is, what are its products, and who are its customers.

Corrective actions   In a HACCP plan, specific outlined steps that dictate what is to be done to correct the cause of the violation, and what is to be done with the product that was produced while the critical limit was violated.

Coumarins   Chemical compounds found in the peel of citrus fruits, some of which are toxic.

Covalent bond   A bond in which the filling of valence shells occurs through the sharing of electrons.

Cream   The high-fat, liquid product that is separated from whole milk; according to federal Standards of Identity, cream must be at least 18% milkfat.

Creaming   In milk systems, large fat globules clump together and rise to layer on the top of the water phase.

Critical control points (CCPs)   Steps during the food production process for which control is essential in order to produce the safest food possible.

Critical limits   Numerical values that, if exceeded or not met, will result in the loss of control of the particular CCP.

Cross-breeding   Breeding of two closely related plant species for the purpose of producing crops with desirable characteristics.

Cross-contamination   Critical control point that occurs when microorganisms are transferred from one food to another.

Cruciferous vegetable   Vegetable in the family Cruciferae, such as broccoli, cauliflower, brussels sprouts.

Crude oil   Lipid material obtained via pressing and rendering processing procedures.

Crystal   Solid made up of units in a repeating pattern.

Crystallization   Formation of a crystalline structure, an organized three-dimensional array of unit cells into a solid form.

Curd   The coagulate that forms when milk proteins precipitate out of solution during fermentation.

Cured fish   Fish processed by salting, smoking, or pickling.

Curing   Addition of salt, sugar, and sodium nitrite to meats for the purposes of color development, flavor enhancement, preservation, and safety.

Curing agents   Additives used in meats, including sodium nitrite, which helps retain the pink color of cured meats, as well as acts as a preservative.

Cyanogenic glucosides   Compounds prodced in some plants that when consumed can be converted to hydrogen cyanide, a powerful toxicant.

Cyclamate   An alternative sweetener that, in 1968, was banned as a food ingredient when testing showed that a saccharin/cyclamate mixture caused cancer in experimental laboratory mice.

Cyst   Calcified wall or capsular enclosure surrounding the larval stage of a roundworm organism which is found in certain meat muscle tissue.

Danger zone   Temperature range between 40 and 140°F in which pathogenic microorganisms can proliferate.

Data analysis   Obtaining of data means, standard deviations, correlation coefficient determinations, and using analysis of variance (ANOVA), mean separation procedures, and response surface methodology to meet the objectives of an experiment.

Decimal reduction time or D value   The time required for a bacterial population to pass through one log cycle, in which 90% of the organisms have been killed.

Decomposition reaction   Chemical reaction in which food macromolecules come apart during processing conditions or by enzyme action.

Definitive host   The host in which a parasite actually reproduces to multiply in number.

Degrees Brix (° Brix)   A measure of a beverage or liquid's sugar concentration, equal to the weight percent of sucrose in solution.

Degumming   Separating the hydrated phospholipids from oils by washing in water and separating by centrifugation.

Dehydration   A form of food preservation in which moisture is driven off by the application of heat, resulting in a stable food that has a moisture content below that at which microorganisms can grow, thus preventing microbial and enzymatic deterioration.

Delaney Clause   A legislative provision in the Food Additive Amendment that prohibited the approval of an additive if it was found to cause cancer in humans or animals.

Demographics   Information such as the age, gender, income, and geographical location of consumers and factors such as shopping habits, level of education, and motivation to buy.

Denaturation   An unfolding of protein structure (due to H bonds breaking) without disrupting protein covalent bonds.

Deodorization   Application of evaporation and vacuum to separate certain odorous low molecular weight compounds from an oil.

Depolarization   Accumulation of a positive (+) charge in the cell, making the membrane potential less negative than normal, and more positive. Accompanying depolarization is neuronal signaling—the release of neurotransmitters to the brain.

Descriptive tests   Sensory test methods that attempt to describe specific product attributes related to flavor, texture, mouthfeel, and so forth, by quantifying the perceived intensities of specified sensory characteristics.

Detection threshold   Point at which a person initially responds to a stimulus—the dividing line between lack of sensation to sensation.

Development stage   The second stage in product development involved in creating the design and nature of the product, from a compositional standpoint as well as a processing, packaging, and marketing one.

Dew point Temperature at which, as air is cooled, its moisture starts to form a visible condensate.

Dewormers Drugs used to eliminate or treat parasitic infections in animals.

Dextrans Polysaccharides produced metabolically by certain bacteria and yeast composed of glucose units connected by α-1,6 glycosidic bonds.

Dextrins Linear arrays of glucose units bound by α-1,4 glycosidic linkages, produced by industrial starch hydrolysis.

Dextrose equivalent Measure of the percentage of glycosidic bonds hydrolyzed in disaccharides and polysaccharides, which indicates the level of reducing sugar present.

Diarrhetic shellfish poisoning Poisoning from consumption of clams containing toxin produced by the dinoflagellates *Dinophysis* and *Prorocentrum*.

Dielectric constant (K' or ϵ') Electric property and indicates the ability of a food to store electrical energy.

Dielectric loss factor (K'' or ϵ'') Electrical property that describes a material's ability to lose heat.

Dietary fiber Residue of plants left undigested after consumption of edible fiber.

Dietary supplement A product intended to supplement the diet that contains one or more of the following dietary ingredients: a vitamin, a mineral, an herb or other botanical, an amino acid, or other dietary substance used as a concentrate, metabolite, constituent, extract, or combination of these ingredients.

Dietary Supplement Health and Education Act (DSHEA) of 1994 Legislation that redefined and regulated dietary supplements.

Difference threshold Minimum amount of stimulus change that results in a change of sensation.

Dinoflagellates Members of the algae family, photosynthetic, consumed by certain types of fish; produce toxins that affect humans that consume the infected fish.

Direct expansion refrigeration Cold storage method that pumps a gaseous refrigerant through a coil to achieve temperature reduction.

Disaccharide Two monosaccharides joined together by a glycosidic linkage.

Discrete variables Quantifiable factors that can take on only certain values.

Discrimination tests or difference tests Sensory test methods that attempt to answer whether any difference at all exists between two types of products. Typically, the test is between a control and a test sample.

Disintegrating Unit operation that refers to particle size reduction of foods.

Disulfide bonds Bonds between sulfur atoms ($-S-S-$) between cysteine amino acids in flour dough proteins that tighten dough structure.

DNA ligase Enzyme that catalyzes the formation of the final (phosphodiester) bond in DNA replication.

DNA polymerase Enzyme that catalyzes the synthesis of DNA.

Domoic acid Toxin produced by the dinoflagellate *Nitschia*, responsible for amnesic shellfish poisoning.

Dose rate The dose accumulated per unit of time, such as 1 kGy per hour.

Dose-response The concept that there is a relationship between a toxic reaction (the response) and the amount of toxicant received (the dose); the direct response that can be measured as a result of administration of a particular dose of a particular chemical or agent to a live subject.

Dose-response assessment Determination of the concentration of a particular toxicant needed to cause an adverse effect on a biological system.

Double-helix structure Arrangement of two polynucleotide chains to form DNA.

Dough strengtheners Substances such as SSL (sodium stearoyl lactylate) used to improve the machinability of bread dough during processing.

Drinking water Water intended for human consumption; may be manufactured into sealed bottles or other containers.

Drum drying A form of drying using rotating drums to dry potatoes into flakes and create other products.

Dry bulb temperature (db) Temperature of air in degrees Fahrenheit as measured with a standard thermometer.

Drying An extensive approach to moisture removal in which product moisture is reduced to a few percent.

Dye Water-soluble chemical used to color food products, for example lollipops, throughout.

Edisonian approach Attempting to study all variables and approaches until a solution to a problem is found.

Effective dose (ED_{50}) Dose that will cause an adverse effect in 50 percent of the animals tested.

Egg substitute A very low-fat egg product based on egg white (albumen) rather than egg yolk.

Elastic solid Food material that deforms without true flow movement, and that returns to its original shape when the deforming stress is removed.

Electrolytes Specific mineral elements inside and outside of body cells that conduct electricity, such as sodium and potassium.

Electromagnetic spectrum An organized scale of electromagnetic radiation, which includes gamma radiation, radio waves, visible light, and microwaves based upon frequency, wavelength, and energy value.

Electronegative Designation for an element or molecule with a strong attraction for electrons.

Electron volt (eV) Energy possessed by an electron; 1 eV is the amount of kinetic energy gained by an electron as it accelerates through an electric potential difference of 1 volt.

Electroporation Technique for inserting DNA into a cell that applies an electric field to the target cell as a series of pulses, which creates pores on the cell membrane, allowing DNA to be inserted.

Electropositive Indicates an element or molecule with a weak attraction for electrons.

Element Simplest type of pure substance that cannot be broken down without splitting atoms.

Emulsification The process in digestion by which bile surrounds fat molecules and converts them into smaller particles for digestion.

Emulsifiers Substances that keep fat globules dispersed in water or water droplets dispersed in fat.

Emulsifying salts Substances that function to enhance natural emulsifier activity in food systems such as process cheese.

Emulsion Colloidal dispersion of two liquids, usually oil and water, that are immiscible (not mixable).

Encapsulation A technique applied to flavors to accomplish convenience, stability, and timed release.

Endogenous toxicant Toxicant produced by tissue cells in plants and other biological entities.

Endosperm Major layer in a grain kernel, made up of starch-storing parenchyma cells, and having a content of B vitamins and protein.

Energy metabolism How the body uses the energy nutrients (carbohydrates, fats, and proteins) as fuel sources.

Enology Study of wines and winemaking.

Enrichment The addition of nutrients to a food to meet a specific standard.

Enthalpy In psychrometrics, the internal energy content of air, which varies with temperature and moisture.

Entropy Measurement of randomness or disorder.

Enzymatic hydrolysis Fragmentation of large food molecules by the action of hydrolytic enzymes such as carbohydrases (sucrase, lactase, maltase, amylase), lipases, and proteases.

Enzymatic oxidation Reactions that occur when oxygen is added to, or hydrogen or electrons are removed from, food molecules in the presence of an active enzyme.

Enzymatic reduction The gaining of one or more electrons, gaining hydrogen, or losing oxygen from the structure of a food molecule in the presence of an active enzyme.

Enzyme Specialized protein molecule composed of amino acids that speeds up chemical reactions, such as oxidation/reduction or hydrolysis/condensation.

Epidemiology Study of the effects that consuming a particular food has on the development of a toxic effect or disease on a specific human population.

Equation The written description of a chemical reaction, using chemical symbols and formulas.

Ergogenic aid Substance ingested in an attempt to improve athletic performance

Ergoty Descriptive term to denote grains that contain greater than 0.3% of *Claviceps* mold by weight.

Ester Organic compound having a carbonyl-oxygen-carbon system, present in fruits and responsible for characteristic flavors and aromas.

Ethylene Naturally occurring substance in plants, important in the ripening process.

Ethylene vinyl acetate (EVA) A packaging material formed from the copolymerization of vinyl acetate (VA) and ethylene.

Eucaryotes (from the Greek meaning "true nucleus") Organisms having a nucleus, and include the fungi, protozoa, plants, and animals.

Evaporation Removal of moisture from a food to concentrate its solids content.

Experimental design Research strategy employing experimental setup and statistics used to understand the effects of treatments upon a particular population of products or product users.

Exposure assessment Determination of the risk of exposure of a biological system to a toxicant.

Extensibility Resistance to extension, which is measured as the distance extended in millimeters versus the force applied in newtons.

Extruder Device that uses rotating screw technology within a barrel to cook and shape foods quickly and efficiently.

Factorial experimental design Design that treats all levels of one test variable with all levels of another test variable.

Facultative anaerobe Organism that prefers oxygen for growth but can also grow in the absence of oxygen.

Fair Packaging and Labeling Act of 1966 Legislation passed to counter problems with underweighing of products, particularly cereals; it required food labels to indicate product identification, the name and place of business of manufacturer, the net quantity of contents, and the net quantity of a serving when the number of servings is represented.

Fat mimetic Substance that does not possess all of the true fat physical properties, such as flavor and flavor release, but can imitate some of them, such as creaminess.

Fat substitute Food substance or combination of ingredients formulated into food products to replace all or part of the natural fat content.

FDA Modernization and Accountability Act of 1997 (FDAMAA) Legislation specifically aimed at reducing the bureaucracy inherent in the health claim approval process.

Feasibility Determination of whether a new product can be developed in terms of formulation, equipment, and costs.

Federal Food, Drug, and Cosmetic Act (FFDCA) of 1938 Legislation that gave the FDA authority over food and food ingredients, and defined requirements for truthful labeling of ingredients.

Fermentation A metabolic process carried out under anaerobic conditions.

Fermented food Low-acid food subjected to the action of certain microorganisms.

Fermentor Large container in which bacteria can be grown anaerobically in a fermentation reaction, allowing production of large quantities of certain chemicals by the organisms under controlled conditions.

Fiber Nonstarch polysaccharide carbohydrate portion of plants that helps maintain structural rigidity.

Fibrous proteins Parallel strands of amino acids linked by their side chains; examples are actin, myosin, collagen, and elastin.

Finfish Fish having a backbone and fins (e.g., freshwater trout and catfish; saltwater salmon, tuna, and cod).

Flash pasteurization A high-temperature, short-time (HTST) treatment in which pourable products, such as juices, are heated for 3–15 seconds to a temperature that destroys pathogenic microorganisms.

Flavonoids Organic molecules, some of which are toxic, that impart color and flavor to fruits and vegetables.

Flavor Overall impression of food combining taste, odor, mouthfeel factors, and trigeminal perception.

Flavor enhancer (and flavor potentiator) Substance that does not itself impart flavor but rather intensifies in some manner the flavors that are naturally present or are added to food.

Flavor profile method Descriptive test method in which flavor sensations are quantified in terms of intensity, order of perception, and duration, from the initial sensation through those lingering after samples are swallowed.

Flavoring Natural or synthetic substance added to foods for flavor production or modification.

Flexural modulus The stiffness of a plastic packaging film or sheet.

Fluid mechanics Branch of mechanics which, for the food engineer, studies the flow characteristics of liquid and semisolid foods.

Foam Colloidal dispersion in which the dispersed phase is a gas, within a liquid continuous phase.

Focus group Interactive panel composed of about 10 consumers plus a trained moderator to obtain detailed attitudes regarding the concept of a proposed new product.

Food Additives Amendment to the FFDCA A 1958 amendment that required FDA approval for the use of an additive prior to its inclusion in food, and also required the manufacturer to prove an additive's safety for the ways it would be used.

Food allergy or hypersensitivity An abnormal response to a food that is triggered by the immune system.

Food and Drugs Act of 1906 A law that prohibited interstate commerce in misbranded and adulterated foods, beverages, and drugs.

Food biosecurity or food insecurity The potential of a food supply to fall victim to planned contamination with disease-producing or toxic hazards as a result of malicious and criminal intent.

Food biotechnology The practice or application of genetic engineering for improving the quality and/or safety of foods.

Foodborne illness Any illness resulting from the ingestion of food.

Food chemistry Application of the principles of chemistry to the study of foods and food molecules.

Food engineering Application of the science of engineering principles to the manufacture of foods; impacts the design of processes and systems extending from food materials handling through processing, packaging, and distribution.

Food Guide Pyramid (USDA) System created for the consumer to graphically illustrate the Dietary Guidelines; recommended serving sizes are presented for six food categories within the framework of a pyramid design.

Food irradiation Application of radiation, as ionizing energy, to foods, as a nonthermal processing method that offers a means of preservation to destroy microorganisms without heating the food material.

Food microbiology Study of the microorganisms that cause food spoilage, those that make foods unsafe by producing disease, and those that are used to make fermented foods.

Food microstructure The organization of food structures at the microscopic level and their interactions to produce a food product's physical and sensory characteristics.

Food preservation Inhibiting spoilage microorganisms in food to extend their freshness by manipulating environmental conditions such as moisture level, pH, temperature, and solute concentration through the use of additives, canning, dehydration, freezing, or other suitable processing techniques.

Food processing The steps involved in the manufacture of food products; this encompasses specific unit operations within the food industry such as materials handling, cleaning and separating as well as the application of technologies (canning, freezing, packaging) prior to consumer purchase.

Food Safety and Inspection Service (FSIS) Branch of the USDA responsible for the safety, wholesomeness, and labeling of meat and poultry products.

Food science Scientific study of raw food materials and their behavior during formulation, processing, packaging, storage, and evaluation as consumer food products.

Food scientist or food technologist A person who applies scientific knowledge and technological principles to the study of foods and their components.

Food system Dispersion containing two phases: a continuous phase and a dispersed phase.

Food technology Application of food science to the selection, preservation, processing, packaging, distribution, and use of safe, nutritious, and wholesome food.

Forming Creating specific food shapes during processing.

Formula The symbols of all of the elements present in a compound.

Fortification The addition of nutrients, either absent or present in insignificant amounts, to restore nutrients lost during food processing or to prevent or correct a particular nutrient deficiency in a population.

Fractionation A chemical reaction that splits an oil into its higher melting point components, such as stearic acid, and lower melting point components, such as oleic acid.

Free radical Atom or molecule with an unpaired electron.

Free water Lightly entrapped food water that acts as a dispersing agent and solvent, and can be removed by drying.

Freeze-drying Food preservation method that requires low-pressure chilling under a vacuum, allowing moisture to form ice crystals slowly, which are evaporated into the gas phase via sublimation, without passing through the liquid phase.

Freezer storage (-18°C, 0°F) low-temperature form of meat preservation causes cessation of microbial growth and reproduction.

Fructooligosaccharides (FOS) Naturally occurring sugars consisting of multiple units of sucrose joined to one, two, or three fructose molecules via glycosidic linkage to the fructose portion of the sucrose molecule.

Fructose (levulose or fruit sugar) Monosaccharide commonly found in fruits and other plant foods.

Fruit The sweet, edible, fleshy seed-bearing or reproductive part of flowering plants.

Fumigant Chemical used to disinfest grain, soil, and spices.

Functional food A food with demonstrable beneficial effects on health due to the presence of one or more biologically active components.

Functional groups Arrangements of just a few atoms that create particular properties of molecules.

Functional properties Physical and chemical properties of food molecules that affect their behavior and produce desired effects in foods during formulation, processing, preparation, and storage.

Galactose Monosaccharide that exists bound to another monosaccharide (glucose) in forming the structure of the disaccharide lactose.

Gangrenous ergotism Disease condition characterized by a burning sensation in the extremities and caused by a mycotoxin produced by the genus *Claviceps*.

Gel Two-phase system in which a liquid is dispersed in a solid.

Gelatin Protein that originates from heated collagen.

Gelatinization Irreversible disruption of the molecular organization of starch granules due to heat and water.

Gelation (starch) The formation of a gel from a cooled paste.

Genetic engineering The manipulation of genes, and the organisms containing them.

Genetic probing The use of a piece of DNA with a particular sequence of bases for the purpose of allowing it to anneal, or bind, to another piece of DNA.

Genetic switches Genes that are used to trigger expression of other genes.

Geometrical characteristics The size, shape, and orientation of discrete food particles present in food.

Germ Layer in a grain kernel that is rich in unsaturated fat, specific B vitamins (niacin, thiamin, and riboflavin), and iron.

Glass Any material that has solidified and become a rigid material without forming a regular crystal structure.

Glass transition (T_g) The change in the physicochemical state of amorphous food materials between the solid glass and the liquid or rubbery states.

Glass transition temperature (T_g') The temperature at which a food polymer or ingredient molecule achieves transition to or from the amorphous glassy physiochemical state.

Glassy state The most stable amorphous physicochemical state for a food molecule during storage; a comparatively dense state characterized by minimum polymer mobility and maximum viscosity.

Globular proteins Irregular polypeptide structures twisted into somewhat spherical shapes; examples include egg white albumen, myoglobin, and enzymes.

Glucose Monosaccharide used as the preferred body fuel.

Glutamine peptide (GP) Example of a protein hydrolysate purported to provide specific physiological benefits.

Glycemic effect Degree to which a food causes an increase in blood glucose.

Glycerophospholipids Polar lipids in foods, such as the emulsifier lecithin.

Glycolysis The splitting apart of glucose during anaerobic metabolism to obtain energy.

Glycophore The sweet portion of a tastant, consisting of AH and B units together with a third unit called γ (gamma), to form a tripartite structure.

GMP Good manufacturing practices.

Goiter Enlargement and atrophy of the thyroid gland.

Goitrogens Toxicants found in cruciferous vegetables, responsible for the development of goiter in humans.

Good Manufacturing Practices (GMPs) Regulations that limit the amount of food and color additives used in foods.

Grain A small hard seed produced by plants that are grasses.

Grain (in meat) The direction and organization of muscle fibers in muscle held together by connective tissue into bundles; fine-grain is more tender and has smaller bundles, whereas coarse-grain meat is tougher and has larger bundles.

Gram staining Method used to differentiate bacteria, with those having a thin cell wall and an outer membrane being classified as gram-negative, and those having a thick cell wall and no outer membrane being classified as gram-positive.

GRAS substances Additives that have been adequately shown to be safe in food prior to January 1, 1958, through either scientific procedures or experience based on common use in food.

Guanylate cyclase Enzyme that catalyzes the production of cyclic guanosine monophosphate (CGMP) from guanosine triphosphate (GTP).

Hardness Mechanical texture characteristic—the amount of force required to compress the food between the teeth.

Harvesting The collecting of fruits and vegetables to sell them at the specific time of peak quality of color, texture, and flavor.

Haugh unit An expression relating egg weight to the height of the thick white.

Hazard Analysis and Critical Control Point (HACCP) Food safety protocol for food processors and manufacturers based upon the principles of hazard analysis and risk assessment and the determination and monitoring of what are termed critical control points.

Hazard Any biological, chemical, or physical entity that can harm the consumer.

HDPE High-density polyethylene.

Heat Energy transferred between a system and its surroundings due to temperature difference.

Heat capacity Amount of energy needed to speed up molecular motion to cause the temperature of one unit of mass to increase by 1 degree Celsius.

Heat exchange The application or removal of heat from a food.

Heat exchanger Specialized piece of equipment used in food processing and storage to either add or remove heat from food.

Heat resistance The ability of an organism to survive thermal processing of a particular time and temperature combination that destroys non–heat-resistant organisms.

Heat transfer The movement of heat energy from an outside heat source to food particles in a container, such as a can, bottle, or pouch, typically a result of conduction, convection, and radiant energy.

Heat treatment Food processing to achieve preservation; includes pasteurization, blanching, baking, canning to achieve commercial sterility, extrusion cooking, and microwave cooking.

Hemolytic uremic syndrome Condition that results in kidney failure, due to damage to the convoluted tubules of the kidneys; caused by toxins produced after infection by hemorrhagic *Escherichia coli* serotypes, such as *E. coli* O157:H7.

Hemorrhagic colitis Condition that results in profuse bleeding due to inflammation of the large intestine; caused by infection with pathogenic bacteria, such as *Escherichia coli* serotype O157:H7.

Hepatitis type A Viral disease transmitted by food, associated with shellfish and characterized by malaise, headache, anorexia, nausea, abdominal discomfort, and jaundice.

Herb (medicinal) Plant or plant extract containing pharmacologically active substances.

Herbal extract A water-soluble or lipid-soluble compound that is extracted from herbs through distillation (water-soluble) or through solvent extraction (lipid-soluble).

Herbicides Chemicals used to prevent, control, or eliminate unwanted plants such as weeds.

Hermetic seal Manner by which cans are sealed by fusion, which means that they are completely sealed so that no gas can enter or escape the can.

Hermetically sealed container Any package, regardless of its composition (i.e., metal, glass, plastic, polyethylene-lined cardboard), that is capable of maintaining the commercial sterility of its contents without refrigeration after processing, due to its airtight seal.

Heterofermentor Microorganism that produces more than one compound, such as lactic acid and diacetyl, during fermentation.

High-acid foods Foods naturally low in pH (<4.6) due to the presence of organic acids such as citric, malic, or tartaric acid.

HM pectin (high methoxyl) Defined as galacturonic acid polymer having more than 50% of the carboxylic acids esterified with methanol.

Homeostasis Body's tendency to maintain a state of chemical and metabolic equilibrium.

Homofermentor Microorganism that produces one compound, such as lactic acid, during fermentation.

Homogenization Process used to decrease the size of fat globules dispersed in milk.

Homogenous The uniform distribution of matter (i.e., solutes, microorganisms, etc.) within a sample.

Hops Plant flowers that impart bitterness and flavor in the fermentation of grains into beer.

Hormone Substance secreted by endocrine gland that influences functioning of another organ—can enhance or suppress growth of specific body cells.

Hot-break process Use of heat in production of tomato paste to destroy pectinase enzyme, producing a thicker (higher viscosity) product.

Hot/cold inversion Neurological condition in which hot beverages appear cold and cold beverages appear hot.

Hue The actual color name (e.g., red, blue, and green).

Humectant A substance that attracts water within a food product, which may lower the product's water activity.

Hunter color difference (ΔE) A measure of the distance in color space between two colors, as determined by the formula $\Delta E = (\Delta L)^2 + (\Delta a)^2 + (\Delta b)^2$.

Hunter tristimulus data In food color analysis, the three values of L (lightness), a (red to green color dimension), and b (yellow to blue color dimension), as represented by a three-dimensional color solid.

Hurdle A stress placed on a microorganism that it must overcome to survive, grow, and reproduce in food; hurdles are equivalent to the factors affecting microbial growth: water activity, pH, temperature, pressure, and chemical antimicrobials.

Hurdle technology Approach that creates a combination of suboptimal growth conditions using a combination of hurdles to control microorganisms in foods.

Hydration Process by which water molecules surround and interact with solutes by acting as a solvent.

Hydrogenation The forced addition of hydrogen atoms to the unsaturated bonds in an unsaturated fat.

Hydrogen bond Intermolecular covalent bond characterized by unequal sharing of electrons between the bonding atoms of separate molecules.

Hydrogen sulfide Substance produced from the breakdown of cysteine by certain bacteria.

Hydrogenization Processing to saturate double bonds and harden an oil, and make it resistant to oxidative rancidity.

Hydrolysis A reaction in which a water molecule enters the region of the functional group of a larger molecule and splits it off.

Hydrolytic rancidity Off-flavor resulting from chemical reactions that liberate free fatty acids by water hydrolysis and enzyme action.

Hydrolyzed vegetable proteins (HVP) Proteins derived from soybeans used for flavor and as alternate sources of protein to meat and dairy products.

Hydroperoxide (ROOH) The result of the oxidation of an unsaturated fatty acid.

Hydrophobic ("water-fearing") Inability of a substance to dissolve in water; fats are hydrophobic compounds.

Hygroscopicity Ability of a substance to attract moisture.

Hypobaric storage Low-pressure storage (40–80 mm of mercury) and temperatures of 5°C (40°F) to reduce ethylene production and respiration in order to maintain quality in fruits for up to several months.

Ice cream Complex colloidal system of dispersed and clustered fat droplets in a concentrated mixture of sugars and proteins, with the additional presence of dispersed ice crystals, a polysaccharide gel, and air bubbles.

Idea stage Initial stage of product development.

Incomplete block experimental design Design in which only certain levels of a variable being studied are assigned during a single evaluation session.

Indirect additives Food contaminants, or substances that accidentally occur in a food product during its production, processing, or packaging.

Infection Foodborne illness due to ingestion of food contaminated with large numbers of microorganisms that colonize the intestine and damage the intestinal lining.

Infusion Food processing step that utilizes heat and pressure to integrate substances into foods, to add flavor, and color, and to achieve a certain texture and moisture content.

Integrated Pest Management (IPM) System of pest control that integrates the use of pesticides and alternative nonchemical measures, such as trapping.

Integrated Resource Management (IRM) Management of resources within a processing manufacturing environment.

Intensity Degree to which a character note is present, as determined by its product concentration and via human perception.

Intentional additives Substances purposely added to foods to perform specific functions; they include such substances as sugar, salt, corn syrup, baking soda, citric acid, and vegetable coloring.

Interaction or synergy The relationship between the response and one variable depends on the level of a second and/or third variable.

Interesterification A process that removes fatty acids from glycerol and causes their subsequent rearrangement or recombination into numerous configurations, most of which differ from the original fat molecule.

Intoxication A foodborne illness due to ingestion of food contaminated with microorganisms that have produced toxin in the food.

Intoxification A foodborne illness due to ingestion of bacteria that produce toxins once inside the small intestine.

Intramolecular bonding The bonding between a hydrogen atom and an oxygen atom within the same molecule.

Inulin A nondigestible fructooligosaccharide that functions as a soluble dietary fiber and a prebiotic.

Invert sugar A mixture of glucose plus fructose produced by hydrolysis of sucrose.

Iodine value A measure of the degree of unsaturation of a fat or oil.

Iodophores Iodine-based sanitizers.

Ion An atom that carries either negative or positive charge, depending on whether it has gained or lost electrons.

Ionic bond A bond in which the filling of valence shells occurs through the transfer of electrons between reactants.

Ionization constant (Ka) A measure of the tendency of an acid to lose protons via chemical dissociation.

Ionizing energy A form of energy in the electromagnetic spectrum.

IQF foods Foods created by the Individual Quick Frozen (IQF) process using carbon dioxide cryogenic equipment to quickly cool cooked foods from 71°C (160°F) to −23°C (−10°F) in 10 minutes.

Irradiation The process of applying radiation to matter.

Isoelectric point (pI) The pH value at which a protein molecule (e.g., casein) loses its net electrical charge and is easily denatured and precipitated.

Isoflavones Compound classified as flavonoids. Research has suggested that soy isoflavones may be effective in the prevention and treatment of cancer and certain chronic diseases.

Ketone Organic compound in which an interior carbon atom is double-bonded to an oxygen atom.

Lactacel A commercial starter culture used in fermentation of meat.

Lactic acid bacteria Organisms able to ferment lactose to form lactic acid under anaerobic conditions.

Lactoferrin A milk protein with antibacterial properties, as a natural defense to mastitis.

Lactose (milk sugar) Disaccharide composed of galactose bound to glucose.

Lager A type of beer with a pH of 4.2.

Lake An insoluble dispersion derived from dyes, used to color the surface of foods or fat-based products, including chocolates.

Latent heat The heat energy in an airstream due to the moisture of the air; in latent heating, energy input results in a change of state of water rather than an increase in temperature.

Leavening Production of gas by yeast fermentation or the production of gas caused by the reaction of an acid with baking soda, in batter and dough products that contributes to the volume achieved during baking and to the final aerated texture.

Leavening acid Substance that generates hydrogen ions and facilitates the release of carbon dioxide from baking soda (sodium bicarbonate), causing the expansion of a baking dough or batter product.

Leavening agents Additives such as baking powder that improve the volume of dough in baked products.

Legumes Edible seeds and pods of certain flowering plants, such as beans, lentils, soybeans, and peas.

Lethal dose (LD_{50}) Dose required to kill 50 percent of the animals tested.

Levans Polysaccharides composed of fructose units produced by certain microorganisms.

Line extension New form of an established product or family of products.

Linoleic acid Fatty acid having two double bonds, part of the omega family of fatty acids.

Linolenic acid Fatty acid having three double bonds, part of the omega family of fatty acids.

Lipids Food molecules that do not mix or dissolve in water, and include all types of fats, oils, phospholipids, sterols, and waxes.

Lipoprotein Substance composed of lipid plus protein material.

Lipoxidation The chemical mechanism in which heat, light, or metals trigger a chain reaction within stored fats and oils, resulting in fat rancidity.

Lipoxygenase Enzyme present in soybeans that needs to be deactivated by heat, otherwise it creates objectionable "beany" off-flavors in soy products.

Listeriosis Foodborne illness caused by consumption of *Listeria monocytogenes* cells.

Lithotrophic organism Chemotroph that requires inorganic compounds such as minerals.

LLDPE Linear low-density polyethylene.

LM pectin (low methoxyl) Defined as galacturonic acid polymer having less than 50% of the carboxylic acids esterified with methanol.

Low-acid food Food having an equilibrium pH greater than 4.6 and a water activity (a_w) greater than 0.85.

Low-density polyethylene (LDPE) A plastic polymer derived from ethylene.

Low-temperature treatment Cold storage, including refrigeration and freezing, to achieve the preservation of seafood, meats, poultry, and dairy products.

Macronutrients Nutrients humans must consume in the largest amounts, including carbohydrate, lipid, and protein.

Maillard browning The browning of foods as a result of the Maillard reaction.

Maillard reaction Nonenzymatic sequence of chemical reactions involving sugars and amino acids.

Malnutrition Nutritional imbalance, either too much or too little intake of essential nutrients or calories, resulting in poor health.

Maltodextrins Polysaccharide fragments derived from starch hydrolysis and defined as having a dextrose equivalent of less than 20.

Maltose Disaccharide composed of two glucose units bonded together.

Margarine Water-in-oil emulsion made from various fat ingredients (e.g., animal fat, soy, or cottonseed oils) that are churned with cultured, pasteurized skim milk or whey.

Marker genes Genes that code for an easily recognizable product, which are inserted into cells at the same time as the desired gene.

Market research Organized inquiry that seeks to identify and measure the factors that influence the consumer marketplace, so as to guide marketing professionals in decision making regarding new products.

Mass transfer Movement or migration of a liquid (such as frying oil) or a food component, either within one phase or between different phases, which is caused by physical conditions (i.e., concentration gradients) present in the liquid oil/food system.

Materials handling The manner in which raw commodities such as crops and animals or animal by-products are harvested and transported to a food processing facility.

Maturity The condition of a fruit when it is picked at or just before the ripened stage.

Maximum growth temperature The temperature at which enzyme and structure inactivation outbalances the enhanced ability to synthesize new cell material.

Maximum residue levels (MRL) Maximum concentration of pesticides and herbicides allowed in foods.

Meat Inspection Act The first law to regulate meat quality and safety, passed on June 30, 1906.

Meat   Edible animal flesh, to include processed or manufactured products derived from such tissue (beef, chicken and other poultry, pork, fish, lobster, lamb).
Mechanical sensory characteristics   Texture characteristics due to attractive forces between the molecules in a food and the opposing force of disintegration.
Mechanically deboned meat   Poultry meat, or any other meat, from which the bones are removed by a mechanical device.
Mesophile   Organism that cannot tolerate extremes of temperature.
Metabolism   The chemical reactions that take place within the human body.
Methyl group   The CH_3 group.
Metmyoglobin   Myoglobin derivative associated with aged meat exposed to air, producing a grayish or brown meat color.
Micelles   Small, spherical complexes that are products of lipid digestion, composed of monoglycerides, long-chain fatty acids, cholesterol, and phospholipid; in a micelle the hydrophobic groups are directed away from the water while the polar (charged) groups are exposed on the external surface.
Microaerophilic organism   Organism that cannot survive in the presence of atmospheric levels of oxygen but requires more reduced oxygen conditions, typically 6% O_2.
Microinjection   Manipulation to introduce DNA into newly fertilized animal egg cells.
Micronutrients   Nutrients the body requires in lesser amounts, including the vitamins and minerals.
Microorganisms   Living entities that are too small to be seen with the naked eye—the bacteria, viruses, protozoa, and fungi such as yeasts and molds.
Microparticulation   A processing technique used to reduce the particle size of fat replacers like proteins in order to create a creamy sensation in the mouth similar to fat.
Microwave pasteurization   Method to heat foods using microwave energy to destroy microorganisms in beverages such as milk and wine.
Minimally processed vegetables   Vegetables processed without preservatives and heating or chemical treatment, but instead relying on various types of packaging and gas environments to maintain product freshness.
Minimum growth temperature   The temperature at which the generation time is highest.
Mixing   Blending of food ingredients to create a food product.
Mixture   physical rather than chemical combination of two or more substances.
Modified atmosphere packaging (MAP)   Packaging in a modified gas atmosphere to prevent oxidation reactions in foods.
Moisture   Amount of water present in a food, as a component, relative to all the solid constituents, such as proteins, carbohydrates, and any nonwater liquid (i.e., oils).
Moisture content   Measure of the actual amount of water vapor present in the air.
Moisture removal   Removing water in a product through drying, dehydration, evaporative concentration, and intermediate moisture processing.
Moisture sorption isotherms (MSI)   Graphs of data that interrelate the water (moisture) content of a food with its water activity at a constant temperature.

Molds   Multi- or unicellular organisms that grow on organic matter.
Monitoring   In HACCP consists of conducting procedures that enable the determination of whether a critical limit is being maintained or not.
Monosaccharide   Single unit carbohydrate, also called a simple sugar.
Mouse lethal dose (MLD_{50})   Dose required to kill 50 percent of mice.
Mouthfeel   Perceived sensation of food by the epithelial lining within the oral cavity, which includes tactile sensation as well as thermal response.
mRNA (messenger RNA)   Molecule utilized by DNA to carry the genetic code from the nucleus of the cell to the cytoplasm for protein synthesis.
Mutagen   Substance that causes a change (mutation) in the base sequence of a cell's DNA.
Mycelium   The tangled mass of macroscopic mold growth that spreads readily and may cover several inches within 3 days.
Mycotoxins   Toxic substances produced by molds.
Myofibrils   Small structures (about 2,000 per cell) that compose each muscle cell or muscle fiber.
Myoglobin   Protein pigment molecule that binds oxygen and provides color to red meat.
Naturally occurring toxicant   Toxicant produced by organisms that contaminate foods.
Needle stitch pumping   Common pickle curing method in which a single needle with multiple openings or multiple needles inject brine directly into meat tissue.
Neurotoxic poisoning   Poisoning caused by microbes, plants, or animals that affects the victim's nervous system.
Neurotoxic shellfish poisoning   Poisoning from consumption of oysters and clams containing toxin produced by the dinoflagellate *Ptychodiscus*.
Neutralization   Processing a fat with strong alkali solution to remove free fatty acids.
Newtonian foods   Homogeneous mixtures that exhibit no change in viscosity as the rate of shear (applied mechanical force) is increased.
Nisin   Antimicrobial agent known to inhibit the pathogen *Listeria monocytogenes*.
Nitric oxide (NO)   A derivative of nitrite that acts as a reductant, and accelerates color development in cured meats.
Nitric oxide myoglobin or nitrosylmyoglobin   Bright pink-red myoglobin reacted with nitric oxide.
Noncarbonated soft drinks   Soft drinks similar in ingredients and processing to carbonated soft drinks, but rather than use carbonation, they are usually pasteurized to destroy microorganisms.
Nonclimacteric fruits   Fruits in which postharvest ripening is not accompanied by increased respiration (e.g., grape, strawberry).
Noncovalent interactions   Hydrogen bonding (intermolecular H bonds), ionic interactions (ions in water), and hydrophobic interactions (micelle structure).
Non-Newtonian foods   Heterogeneous mixtures that exhibit a change in viscosity as the rate of shear (applied mechanical force) is increased.
Nonnutritive sweeteners   Compounds that provide sweetness without calories due to greater sweetness intensity per amount

when compared to sucrose, examples are aspartame, acesulfame potassium (acesulfame-K), and saccharin.

Nonthermal processing Methods that do not rely on heat to create preservation, including irradiation, high pressure, and pulses of light and electric fields.

Nutraceutical Proposed new regulatory category of food components that may be considered a food (in whole or in part) that provides health benefits beyond nutrition.

Nutraceutical beverages Drinks formulated with special functional ingredients that promote some aspect of health or reduce the risk of certain diseases.

Nutrition Study of the nutrient substances in food, their contribution to health and disease, and the processes by which the body digests, absorbs, and utilizes them.

Nutrition Labeling and Education Act (NLEA) Legislation that mandated a change in the label of almost every food product in the United States. NLEA authorized certain health claims for foods, the food ingredient panel on labels was standardized, serving sizes were standardized, and terms such as *low-fat* and *light* were standardized.

Nutritional additives Nutrients included in foods to boost nutrient intake and provide for a more balanced diet.

Nutritive sweeteners Products such as sucrose, fructose, maltose, lactose, xylitol, sorbitol, molasses, and honey, which provide significant calories from carbohydrates in addition to a level of sweetness intensity.

Objective In product development identifies the overall goals of a project; the test objective refers to the specific goal of a sensory experiment.

Objective procedures Testing procedures using analytical equipment.

Odor or aroma The sensation derived from food as interpreted through the olfaction (sense of smell or odor perception) mechanism.

Odorants Molecules that possess odor.

Off-flavors Unwanted flavor development in foods.

Ohmic heating or resistance heating a method in which a food product is subjected to an alternating current passed through a salt brine.

Oilseeds Legumes that are higher in fat, such as soybeans, cottonseed, sesame seed, sunflower seed, and peanut seed.

Okadaic acid Toxin produced by the dinoflagellate *Dinophysis fortit*, causing diarrhetic shellfish poisoning.

Olfaction Perception of odors by nerve cells in the nasal passage or cavity.

Olfactory epithelium Outer layer of receptor cells located on the roof of the nasal cavity.

Oligosaccharides Complex carbohydrates of 10 or fewer (typically nonreducing) sugar units.

Omega (φ) fatty acid Type of unsaturated fatty acid; both omega-3 and omega-6 designations refer to the first double-bond position from the methyl end of the molecule; comprise 30% of fatty acids in fish oil.

Optimization The use of statistics in data analysis and experimentation to find the best operating conditions for each of the important variables.

Optimum growth temperature The temperature at which the time it takes for the cells to divide is the shortest.

Organic alcohol A molecule that contains carbon atoms attached to —OH (alcohol) groups.

Organic salts Compounds formed from organic acids in which the hydrogen atom of the acid group, COOH, is replaced by a metal ion such as sodium, calcium, or potassium.

Organochlorines Lipid-soluble, highly stable, chlorine-containing pesticides.

Organophosphate Phosphate-containing pesticides, easily absorbed through the skin.

Organotrophic organism Chemotroph that requires organic compounds such as carbohydrates.

Orientation The alignment of individual polymer molecules within a fabricated film.

Overnutrition Excessive intake of one or more nutrients, which may lead to toxic response and overdose disease.

Oxidation The addition of oxygen in a chemical reaction.

Oxidative rancidity Off-flavor resulting from chemical reactions between unsaturated fatty acids and oxygen, producing hydroperoxides and their odorous breakdown products.

Oxidized environment Environments in which molecules have a relatively high affinity for electrons.

Oxidizing agent Substance that causes another substance to become oxidized, while the oxidizing agent itself becomes reduced in a reaction.

Oxygen scavenging Technology using iron oxide material to eliminate headspace oxygen that resides within packaged foods.

Oxymyoglobin Myoglobin derivative responsible for a bright red color.

Ozone An unstable, colorless gas used in the bottled water industry as a powerful oxidizer and a potent germicide.

Packaging A unit operation that offers a contained product protection from the environment plus convenience for retailers as well as consumers.

Pancreatin Digestive juice secreted by the pancreas containing a mixture of two proteases, chymotrypsin and trypsin, which is useful in the production of protein hydrolysates.

Panelists Sensory participants, judges, or tasters.

Paralytic shellfish poisoning Poisoning caused by consumption of mussels, clams, and oysters containing toxin produced by the dinoflagellate *Gonyaulax*.

Pasteurization A food preservation process that heats liquids to 160°F (71°C) for 15 seconds, or 143°F (62°C) for 30 minutes, to kill bacteria, yeasts, and molds.

Patent flour The purest flour, selected from the purest flour streams released in the mill.

PCR (polymerase chain reaction) Technique used to generate many copies of DNA (DNA amplification) in a relatively short time.

Pectic acid A short-chain demethylated derivative of pectinic acid associated with overripe fruit.

Pectic substances High molecular weight polysaccharides found in plant cell wall middle lamellae, composed of galacturonic acid units joined by α-1,4 glycosidic linkages.

Pectin Polysaccharide found in many fruits, composed of galacturonic acid molecules.

Pectin esterase An enzyme produced by some microorganisms that can break down pectin.

Pectin gel A system containing pectin polymers and a large volume of water arranged within a three-dimensional solid network.

Pectinase (pectin methyl esterase) A plant enzyme that degrades pectin substrate and softens texture.

Pectinic acid A methylated galacturonic acid polymer produced during fruit ripening.

Pectins or pectic substances Soluble fibers found in plants.

Peptide bonds Chemical bonds joining amino acids in protein (polypeptide) structures.

Perception Interpretation or subjective response to a stimulus; perceptions are subject to conditioning, preconceptions, and natural ability.

Peroxide value (PV) A measure of lipid oxidation, using potassium iodide, to assess an oil's quality.

Pesticide Residue Amendment FDDCA amendment that split jurisdiction for pesticides between the USDA, which could register pesticides for specific uses, and the FDA, which was charged with setting tolerances for residues of pesticides in raw agricultural products.

Pesticides Chemicals used to prevent, control, or eliminate insects and other pests.

pH The hydrogen ion concentration, expressed on a logarithmic scale, of the free or dissociated hydrogen ions in a product.

pH control agents Acidulants, which lower food pH, and alkaline compounds, which increase food pH.

Pheophytin Gray-green and olive-green degradation product of chlorophylls *a* and *b*, respectively.

Phosphates Salts of phosphoric acid, H_3PO_4.

Phospholipid Type of fat that contains phosphorus, as well as fatty acid residues, in its structure.

Phototrophic organism Organism that requires energy in the form of light to live.

Physical changes Loss of moisture in foods due to evaporation and separation of phases, such as oil and water.

Physical hazards Bone, metal, plastic and any other foreign matter that can cause damage to the consumer upon ingestion.

Physicochemical states Four states that characterize the movement of water in food upon food polymers and ingredients: (1) crystalline, (2) liquid, (3) amorphous rubbery, and (4) amorphous glassy.

Phytochemical Plant ("phyto")-derived chemical that is biologically active and thought to function in the body to prevent certain disease processes; considered nonnutritive.

Plant hormones Chemicals used to reduce shoot formation and increase plant resistance to disease.

Plasmids Short circular segments of DNA usually measuring 15 to 20 kilobases in length, used by scientists as vectors to carry genes into cells.

Plasticity The physical property of a fat that describes its softness at a given temperature.

Plasticizer Substance that lowers the glass transition temperature when added to a polymer food system.

Plasticizing Softening a hard fat, which changes the fat's consistency.

Plastic solid Food material showing continuous deformation at a rate that is proportional to the amount of stress applied, exhibiting some recovery upon removal of the stress.

Plate drying Method in which concentrated egg white is sprayed onto metal plates after fermentation and dried at 45 to 50°C (113–122°F) for a minimum of 48 hours to produce egg white crystals.

Plate heat exchanger Exchanger consisting of closely spaced stainless steel plates (fins) arranged in parallel, with fluid flow occurring between plates.

Poising effect The prevention of changes in redox potential in a medium.

Polyethylene terephthalate (PET) A packaging film from the polyester family of materials; represents a condensation polymer.

Polygalacturonase Enzyme that causes softening in tomatoes.

Polymer High molecular weight molecule created by the repetitive reaction of hundreds or thousands of low molecular weight units.

Polymerization Reaction in which many molecular units (monomers) are joined together end to end to form very large molecules called polymers.

Polymorphism The property of a fat such as cocoa butter that can exist in multiple crystal forms ("polymorphs") depending upon conditions of tempering and crystallization.

Polyolefin Any polymer that is composed solely of straight-chain hydrocarbon units ("olefins") of covalently bonded carbon and hydrogen atoms.

Polyols (sugar alcohols) Polyhydric alcohol counterparts of the sugars maltose, mannose, sucrose, and xylose called maltitol, mannitol, sorbitol, and xylitol.

Polypropylene (PP) Packaging polymer used in retort pouches and as an inside layer of food packages subjected to high heat due to its high melting point and tensile strength.

Polysaccharides Complex carbohydrates; long chain-like linkages of sugar units.

Polystyrene (PS) Polymerized styrene that, when foamed, is known as expanded polystyrene, or styrofoam.

Polyvinyl chloride (PVC) Polymerized vinyl chloride, or more simply, vinyl.

Polyvinylidene chloride (PVDC) SARAN.

Powdered soft drinks Dry powder bases made by blending flavoring material with such ingredients as dry acids, gums, and artificial color, to which water must be added.

Prebiotics Substances that promote the growth of probiotic bacteria.

Precision Repeatability of a test result.

Preference Consumer choice on an affective test, based on a hedonic (like, dislike) basis.

Pregelatinized starch Example of a chemically modified starch that increases product thickness with minimal heating.

Preservative Substance that functions to maintain or preserve a food product's freshness.

Pressing Mechanical squeezing of oil from oilseeds.

Primary package Type of containment barrier that comes in direct contact with a food item and may be a can, a jar, or a pouch.

Primary structure of protein A covalently bonded backbone chain of —C—C—N—C—C—N— atoms in sequence derived from amino acids joined by peptide bonds.

Prior-sanctioned substances Substances that the FDA or the USDA had determined were safe for use in specific foods prior to the 1958 amendment, such as sodium nitrite and potassium nitrite.

Probability Ratio of the number of favorable outcomes to the number of total possible outcomes (both favorable and unfavorable).

Probiotics Bacterial organisms believed to be beneficial to health.

Procaryotes (from the Greek meaning "before nucleus") Microorganisms with no nucleus, and include the bacteria.

Process cheese Form of cheese produced from a mixture of natural cheeses plus emulsifier, which are blended together during controlled heating.

Process flavor substances Substances obtained by heating a mixture of ingredients, not necessarily themselves having flavor properties, of which at least one contains nitrogen and another is a reducing sugar.

Processed meat A whole muscle product that has been transformed into a manufactured product by chemical, enzymatic, or mechanical treatment.

Processing aids Acidulants and alkalis, buffers, and phosphates, which are added to help maintain a constant pH in a food.

Product development Process in which new food product ideas are generated and the products themselves are created and marketed. It involves the conceptualization, formulation, processing, testing, and marketing of food products.

Product development team All of the individuals employed by a food manufacturer who play a role in the product development process.

Product life cycle Birth of a new product, its duration as a good selling item, and the time required to see sales decline and the product's removal from the marketplace.

Product line Similar products that differ according to some characteristic but that serve the same market.

Product strategy Plan that is established to accomplish a company's mission and objectives.

Prophylactic Used to prevent disease or minimize the probability of its occurrence.

Protein complementation The practice of combining specific nonmeat and nondairy foods to create a high-quality protein meal.

Protein hydrolysate A protein breakdown product obtained via enzyme or chemical (acid, alkali) action.

Protein synthesis Reassembling digested food amino acids into specific proteins that the body uses for specific functions.

Protopectin Nonmethylated galacturonic acid polymers found in immature fruit.

Protoplasmic poisoning Toxic mushroom poisoning that affects the liver and kidneys.

Protozoa Single-celled eucaryotes.

Psychrometer Instrument used to measure dry bulb and wet bulb temperatures.

Psychrometric chart Graph that represents the eight physical properties of air.

Psychrometrics Field of study concerned with evaluating the thermodynamic properties of the mixed gases present in air.

Psychrophile Literally, "cold loving"—organism that thrives in cold environments (down to $-5°C$).

Psychrotroph Literally, "cold feeder"—organism that lives and feeds at low temperature.

Psychrotrophic Microorganisms adapted to living and multiplying in cold environments and are able to cause spoilage of refrigerated foods.

Pulsed light technology Nonthermal processing that utilizes brief bursts of high-intensity light energy to cause selective damage to the cell membranes of bacterial cells without the disruption of food tissue.

Pumping Mechanical method of moving food material from one point to another during processing with the use of a centrifugal pump.

Pungency A warming or hot sensation in the mouth and lips caused by pungent molecules.

Pure Food Congress An event held in Washington in 1898 that focused national attention on the growing movement to enact federal legislation against the misbranding and adulteration of foods.

Purge Juices that flow from the meat, as water, soluble material, and extracellular fluid.

Putrescine Compound produced by the decarboxylation of ornithine by certain microorganisms.

Pyrrolizidine alkaloids Toxic compounds found in comfrey, herbal teas, tomatoes, potatoes, and eggplant.

Quality assurance System for assuring that commercial products such as foods meet certain standards of identity, specific sanitary manufacturing procedures, fill of container, and safety considerations.

Quality (of food) The sum of a product's qualitative and quantitative characteristics that provide consumer satisfaction.

Quantitative descriptive analysis (QDA) Descriptive test developed at the Stanford Research Institute that uses line scales, replicated experimental designs, descriptive terminology, and analysis of variance of trained panel data in order to understand and quantify product differences.

Radappertization Dose above 10 kGy, such as 20 to 30, which can sterilize a food product, and considered a high processing dose.

Radiation The emission and propagation of energy through matter or space by electromagnetic disturbances called photons.

Radicidation Dose of irradiation less than 1.0 kGy, considered a low processing dose.

Radiolytic products Unstable atoms or molecules derived from substances naturally present in foods treated by ionizing energy.

Radura The international symbol for irradiation.

Radurization Dose of 1 to 10 kGy, considered a medium processing dose.

Ranking A form of preference testing that uses more than two samples in which panelists order a group of products with respect to either their degree of liking or their perceived intensity of a sensory attribute.

Rating Assigning of numbers by panelists in order to characterize products in various response categories based on specific sensory attributes, as a type of acceptance test.

Reaction flavors Flavors produced by chemical reactions taking place under controlled conditions. Typically, reactants such as sugars and proteins are combined at specified pH, and temperature until the desired product results.

Reaction technology Method used to produce what are referred to as reaction flavors, such as a cheesy or meaty process flavor.

Recognition threshold Point at which the identity of a stimulus is made by a panelist.

Red Book An FDA publication officially known as *Toxicological Principles for the Safety Assessment of Direct Food Additives and Color Additives Used in Food*.

Redox potential The ratio of total oxidizing molecules to the total reducing molecules in a medium.

Reduced environment Environment in which molecules have a relatively low affinity for electrons.

Reducing agent Substance that causes another substance to become reduced, while the reducing agent itself becomes oxidized in a reaction.

Reducing sugars Sugars containing the aldehyde or ketone carbonyl group; they function as reducing agents.

Reduction The gain of hydrogen in a chemical reaction.

Refining Removal of impurities from an extracted fat or oil.

Refrigerated storage The most common method of meat preservation, utilizing temperatures of about 3–4°C (38°F).

Refrigeration Cooling process to slow the rates of reactions in foods, to prolong their storage.

Relative humidity (RH) Ratio of the amount of water present in air to the maximum amount of water the air could contain if fully saturated with water.

Rendering Heating of fatty meat scraps in water, which allows the fat to melt and rise to the surface to be separated from the tissue and water.

Rennet coagulation Method of producing lactic acid that uses chymosin to break down casein in milk.

Rennin (or rennet) Enzyme used to coagulate milk, derived from the gastric juice of the fourth stomach of calves.

Representative sampling Collection of enough samples from a lot or batch of food from which conclusions can be drawn about the entire lot.

Resin Polymer in the form of small, individual pellets that are manufactured via melting and extrusion into plastic packaging sheets.

Respiration The biological oxidation of organic molecules to produce energy plus carbon dioxide and water, which plays an essential role in postharvest fruit and vegetable ripening and quality.

Response forms Written evaluation forms, scorecards, or score sheets used by panelists to provide information regarding a specific product attribute, such as flavor intensity or juiciness.

Response surface design Methodology used to identify optimum levels of ingredients or optimum product processing conditions in product development.

Restriction enzymes Hydrolytic enzymes that break bonds at specific DNA base sequences.

Retort processing Utilizing a chamber with steam valve jets that can be set to allow steam to enter the chamber for precise temperature control, used to heat sealed cans to destroy bacteria and spores.

Retrogradation Result of heating and cooling starch in water, with reassociation of especially amylose polymers into an ordered structure.

Retro-inhalation Inhalation of air carrying volatile food odorants already present in the mouth after a food item is tasted.

Reverse osmosis A membrane separation system that allows certain components within a fluid food to pass more easily through the membrane.

Reverse transcriptase Enzyme that catalyzes the synthesis of DNA on an rnRNA template.

Reversion flavor Mild off-flavor developed by refined oils that have become exposed to oxygen.

Rheology Science that is concerned with the flow and deformation characteristics of food materials.

Rigor mortis Transient postmortem physical and biochemical event that takes place in animal muscle tissue causing a loss of extensibility of the tissue.

Ripeness The optimum or peak condition of flavor, color, and texture for a particular fruit.

Ripening (in cheese) Changes in physical and chemical composition of cheese that take place between the time of curd precipitation and the time when cheese develops the desired characteristics of its own particular type.

Ripening (of fruit) The chemical and physical transformation of a fruit from an immature stage to one having optimal color, flavor, texture, firmness, and aroma, due to natural internal chemical changes under enzyme and hormone control.

Risk assessment Evaluation of the severity of the threat posed by a health risk to a population.

Rotational motion Spinning motion of molecules in response to relatively weak energy (e.g., microwaves).

Rubbery state Least stable amorphous physicochemical state for a food molecule, in which shelf instability and other quality problems may occur.

Saccharin An alternative sweetener that was removed from the GRAS list in 1972.

Saccharin Study and Labeling Act of 1977 Legislation that mandated a warning label on all products containing saccharin, with Congress passing a moratorium on the ban to keep saccharin-containing foods in the marketplace.

Salmonellosis Disease caused by consuming bacterial cells of the genus *Salmonella*.

Samples Experimental treatments, including controls, which are developed to specifically meet the project and test objectives.

Sanitizer Chemical agent that can kill biological hazards such as bacteria and molds.

Saponification value Test that gives the average molecular weight of the fatty acids in a fat.

Saturated fat Fatty acid chain that does not contain any carbon-to-carbon double bonds.

Saxitoxins Class of 12 toxins produced by the dinoflagellate *Gonyaulax*, responsible for paralytic shellfish poisoning.

Scientific agriculture The input of specially selected chemicals, technological advances, and various forms of energy into the agricultural system to improve crop yields.

Scientific method A sequence of steps that scientists apply when solving problems and during experimentation.

Scientific principle General understanding of cause and effect involving careful observation, testing, and the development of conclusions.

Scombroid poisoning Poisoning caused by ingestion of histamine produced by bacteria living in scombroid fish.

Scoville heat unit (Shu) Unit of measure of the hotness of chiles.

Scraped surface heat exchanger (SSHE) Exchanger having the addition of a mechanical scraping device (blade) that "cleans" off accumulating product on the pipe wall surfaces to maintain rapid heat transfer.

Screening design Part of a statistically designed experiment used to determine which variables are of primary importance.

Secondary package or container The outer box or wrap that does not contact the food directly but allows for protection and the organized shipping and distribution of the food.

Secondary structure of protein Hydrogen bonding between —NH— and C=O groups above and below polypeptide helical structure, resulting in areas of regular, repeating patterns such as an alpha-helix or a beta-sheet.

Semi-dry sausage A fermented meat product, containing 50% moisture.

Semolina A functional grain ingredient obtained by milling hard wheat grains, and especially useful in the processing of pasta.

Senescence Quality decline in stored, respiring fruits and vegetables that occurs after harvesting; the phase associated with deteriorative processes, leading to aging and death of tissue, and quality decline in fruits.

Sensation The objective response to a stimulus; sensations are objectively sensed.

Sensible heat The heat energy in an airstream due to the temperature of the air; in sensible heating, energy input results in a change of temperature of water rather than a change of state.

Sensory evaluation Assessment of all the qualities of a food item as perceived by the human senses.

Sensory overload Excessive sensory stimulation, saturating the sensory system.

Sensory viscosity Force required to draw a fluid food from a spoon across the tongue at a steady rate.

Separating Unit operation that isolates a desirable part of a food raw material from another part.

Sequestrants Additives that act to combine with metal elements, such as copper and iron, to inhibit the development of off-flavors and odors due to oxidation.

Serving procedure of sensory samples Encompasses physical condition (sample dimensions, weight, shape, temperature) as well as the presentation sequence (sometimes called the test plan).

Serving size According to NLEA (the Nutrition Labeling and Education Act) serving size is the amount of food customarily eaten at one time.

Shear The relative motion of one surface with respect to another surface in parallel to it, which creates a zone of shearing action on any substance (fluid or solid) located between the moving surfaces.

Shear thickening An increase in viscosity as shear rate increases.

Shear thinning A decrease in viscosity as shear rate increases.

Shelf life The length of time that a food remains safe and useful to the consumer under specified conditions of storage.

Shellfish Crustaceans with a hard upper shell and a soft under shell (e.g., crab, shrimp, lobster) or mollusks with two enclosing shells (e.g., clams, oysters, scallops).

Significant hazard A hazard that is reasonably likely to occur, or whose severity in terms of its impact on public health is high enough that even a low probability of occurrence does not diminish its significance.

Smart packages Packages that meet the special needs of certain foods, such as packages incorporating oxygen-absorbing materials such as iron to remove oxygen or packaging containing antimicrobial agents.

Smoke point Temperature at which smoke emanates continuously from the surface of a lipid heated under standard conditions.

Soft drinks Nonalcoholic carbonated or noncarbonated beverages, usually containing a sweetening agent, edible acids, and natural or artificial flavors.

Sol Two-phase system in which a solid is dispersed in a liquid.

Solid fat index Proportion of solid fat crystals to liquid oil in a food lipid sample at a particular temperature.

Solubility The maximum amount of solute that dissolves in a specified volume of solvent at a specified temperature.

Soluble fiber Fiber that can be dispersed in water and in general delay intestinal transit time; examples are oat bran, guar gum.

Solution Homogeneous mixture in which one substance (the *solute*) is dissolved in another (the *solvent*).

Solvent extraction Separation of oil from cracked seeds using a nontoxic fat solvent such as hexane.

Sometribove A recombinant bovine somatotropin (BST) product for increasing milk production in dairy cows.

Soy concentrates Defatted soy products containing about 70% protein.

Soy isolates Defatted soy product concentrates containing over 90% protein.

Specific heat The ratio of the heat capacity of a material to that of water.

Specific volume Space occupied by a pound of dry air at a standard atmospheric pressure (14.7 pounds per square inch, or psi), and expressed in cubic feet per pound of dry air (ft^3/lb air).

Spoilage The loss of food quality as a result of specific biological, chemical, and physical changes.

Spore Protective coating composed of several layers which certain types of bacteria develop when conditions are not conducive to growth, such as limited nutrient availability and low moisture.

Spray drying Method in which a fine stream of the liquid product is dried very quickly in midair by heat.

SSOP Sanitation Standard Operating Procedures; the written procedures that describe step-by-step what an operator must do to carry out a sanitation procedure, such as in cleaning and sanitizing food contact surfaces, equipment.

Stabilizer Substance that functions in a food system to keep it from changing its form or chemical nature.

Stabilizers and thickeners Additives that combine with water in foods to increase product viscosity, to form gels, and to prevent product crystallization.

Standard of Identity Legal standard defined by the FDA for a limited number of staple foods, regarding a food's minimum quality specifications, including permitted ingredients and processing requirements, if any.

Standard Operating Procedures (SOPs) Step-by-step procedures to comply with GMP requirements.

Staphylococcal enterotoxins Toxins produced by the bacterium *Staphylococcus aureus*; heat-resistant.

Starch gel Rigid, thickened starch and water mixture that has the properties of a solid.

Starch granule Naturally occurring spherical aggregate of amylose and amylopectin in plants.

Starch paste Viscoelastic starch and water system that possesses both thick liquid-like (viscous) and solid-like (elastic) properties.

Starter culture A culture of one or more microorganisms that can ferment organic molecules in foods, yielding one or more products such as lactic acid.

State point The junction between any two air property lines on a psychrometric chart.

Sterile Devoid of microbial contaminants.

Sterilization The complete destruction of microorganisms, usually requiring 121°C (250°F) of wet heat for 15 minutes, or equivalent.

Sterilization procedures Specific time and temperature treatments needed to destroy microorganisms in foods; a more severe treatment than pasteurization.

Strain Response of a material to a stress, which may take the form of a relative deformation.

Strecker degradation A chemical reaction between carbonyl compounds and proteins to form flavor and aroma compounds.

Stress Pressure or mechanical action defined as force per unit area of food material.

Strong gluten flours Flours with a relatively high protein content, with an elastic gluten suited to breadmaking.

Structured fluid Low modulus material (low ratio of stress to strain) simultaneously having solid-like and liquid-like properties.

Subjective response Human subjective response, as opposed to objective instrumental response; the use of humans for sensory testing of food products.

Substantial equivalence Term used to determine whether a biotechnology-derived food should be considered the same as an existing food or component, thereby deeming it as safe as that food or component.

Sucrose Disaccharide composed of one glucose molecule attached to one fructose molecule.

Sugar refining The production of high-quality sugars from remelted raw cane sugars.

Sulfhydryl group The SH group.

Sun drying Form of natural drying used to produce dried fruits and nuts.

Superchlorination A process wherein water is exposed to a high concentration of chlorine to inactivate microorganisms.

Superoxide radical Highly reactive derivative of oxygen.

Surface active agents, or surfactants Additives that act as wetting agents, lubricants, dispersing agents, and emulsifiers, by affecting the surface tension of materials present in food systems.

Surimi Popular form of processed fish flesh, or mincemeat, to which sugar or sorbitol is added as a cryoprotectant to extend shelf life.

Sweet taste transduction Biochemical pathway in which a chemical stimulus, the tastant molecule, is converted into electrical impulse interpreted as a "sweet" sensation by the brain.

Symbiotic A relationship between two organisms in which their growth depends on each other.

Symbols Abbreviations representing elements (e.g., H for hydrogen) or elements as part of a compound or molecule (e.g., the C, H, and O atoms in CH_3COOH, acetic acid).

Synapse Site or juncture where two neurons interconnect.

Syneresis Increased tendency to release water from a gel.

Synthetic toxicant Toxicant synthetically or naturally produced but found as residue in foods.

Tastants Molecules that possess the primary sensory characteristics of sweet, bitter, salt, and sour as sensed by the tongue.

Taste Sensation derived from food as interpreted through the tongue-to-brain sensory system.

Taste buds Epithelial receptor cells organized into clusters embedded into tongue structures called papillae.

Taste transduction Mechanism of cell-to-cell communication through pores in cell membranes (ion channels) that function as "gates" between the tongue and the central nervous system (CNS).

Teamsmanship Ability to work well with others as part of a group or team toward a common goal.

Temperate virus Virus that inserts its nucleic acid into the host's DNA, leaving the host cell intact; most food viruses are temperate.

Tempering (of cocoa butter) The manipulation of temperature to heat and cool the sample into the desired stable beta crystal form.

Teratogen Substance that causes abnormal fetal development and birth defects.

Teratogenic Chemical able to induce birth defects in animals and/or humans.

Terminal threshold Amount of stimulus above which any increase in intensity cannot be detected without the use of reference standards.

Terpenoids A group of compounds that include a variety of lipid flavor molecules from plants, such as limnolene and citral.

Tertiary structure of protein The three-dimensional shape created by the spatial arrangement all the amino acids linked together in a protein molecule.

Test market Location where the new product is sold to consumers.

Testing environment Physical setting where the sensory testing is carried out.

Texture Measure of the physical characteristics of food, encompassing four different types of sensations: mechanical, geometrical, fat and moisture content, and afterfeel; takes into account a food's size and shape dimensions, its density, and whether it has a porous or nonporous surface; food texture can be measured by sensory or physical (rheological) methods.

Texture profiling Descriptive test method that describes and quantifies the mechanical, geometrical, fat and moisture, and afterfeel character notes of food samples.

Texturized soy protein Extrusion-produced fiberlike soy protein products that offer excellent protein quality and low fat from soy flour and are functional as meat extenders/replacers.

Texturizing agents Starches, nonstarch polysaccharides, and proteins that promote viscosity, increase firmness, and cause gelation in food systems.

Thaumatin protein Compound originating from tropical plants that has been calculated to be 3,000 times sweeter than sucrose.

Therapeutic Used for the treatment of disease.

Thermal conductivity Rate of heat that will be conducted through a unit thickness of material.

Thermal death time (TDT) A combination of heating time and temperature that kills various spore concentrations.

Thermal equilibrium A state of equalized temperature achieved between a food and its surrounding environment.

Thermodynamics Study of energy interchanges in chemical and physical processes.

Thermophiles Organism that prefers high temperatures, typically above 40°C, to grow.

Thermoplastic Plastic material that can be melted and resolidified without degradation.

Thermosetting Certain plastics that harden when they are heated and set into a solid structure that will not remelt without decomposing.

Thermotroph Organism that can grow at high temperatures.

Thiols Compounds containing the SH group.

Thixotropic A time-dependent shear-thinning behavior of a fluid food.

Three-class attribute sampling A sampling plan in which a product is deemed acceptable, marginal, or unacceptable.

Threshold for sensitivity Distinct points of transition for sensory judgments corresponding to concentrations of stimuli.

Threshold level Dose above which adverse effects are produced in biological systems.

Time-intensity measurement Measurement of sensory attributes evaluated over a period of time following an initial exposure.

Titratable acidity Measure of the total acidity in a sample, both as free hydrogen ions and as hydrogen ions still bound to undissociated acids.

Top note The predominant initial aroma or flavor characteristic for a substance.

Total heat content of air The total heat energy in the air, expressed as btu per pound of dry air, derived from sensible heat from air temperature, latent heat required to convert liquid water to water vapor, and the heat energy of the water vapor.

Toxicant Any chemical substance that can elicit a detrimental effect on a biological system.

Toxicity A potentially fatal physiological response of an organism to a biological, chemical, or physical agent.

Transgenic Modified by genetic engineering.

Transpiration A process of moisture loss through pores in fruit tissue.

Transport The circulatory system's delivery of nutrients and oxygen to the body cells.

Trehalose Sugar that sequesters water for plant cells, allowing them to survive when water is not available.

Triacylglycerols or triglycerides Esters of glycerol and three fatty acids.

Triangle test Simple discrimination or difference test in which three samples, two of one product and one of the other, are presented to the panelists in a specific order.

Trichinosis Condition in which consumption of muscle tissue, or meat, contaminated with cysts of the roundworm *Trichinella spiralis* results in germination of the cysts inside the body resulting in severe muscle pain, diarrhea, vomiting, and fever.

Trimethylamine (TMA) Phospholipid molecule derivative produced by fish or bacterial enzymes, which results in a strong fishy odor.

Trimming Detaching superfluous plant parts from fruits and vegetables.

Trophozoites Motile form of some parasites, able to penetrate the small intestine of the infected animal; postcyst forms of the protozoan genus *Giardia*, which penetrate the intestinal wall causing cramps, nausea, weight loss, severe diarrhea, vomiting, and flatulence.

Tumorigenic Chemical able to induce tumor formation in animals and/or humans.

Turgor The structural rigidity of plant cells due to water content, an important factor in determining the texture of fruits and vegetables.

12D concept Processing for a time-temperature combination that results in twelve log cycle reductions to establish a safety margin for thermally processed foods.

Two-class attribute sampling A sampling plan in which a product is deemed either acceptable or unacceptable according to whether it exceeds a limit or not.

Type I error Detection of a false difference, which means concluding that a difference exists when there is none.

Type II error Missing a difference, and concluding that a difference does not exist when there is one.

Ultrafiltration Special membrane system to separate sugars and proteins at the molecular level from food salts, acids, bases, and water; an example of a food industry unit operation called separating.

Undernutrition Dietary deficit of one or more nutrients, resulting in deficiency diseases.

Unit operations Categories of common food processing operations in practice in the food industry; the number of operations vary by product and application.

Unsaturated fat Fatty acid chain that contains carbon-to-carbon double bonds.

Vacuum packaging Removal of air from a packaged food by using a vacuum draw.

Valence shell The outermost orbital (or high-energy level) in which electrons occur.

Validity of a test A test's ability to measure what it was intended to measure.

Value-added food products Products that offer the consumer some improvement in terms of quality or convenience over the traditional or previously available foods.

Vapor pressure Pressure exerted by a vapor in equilibrium with the solid or liquid phase of the same substance; the partial pressure of the substance in the atmosphere above the solid or the liquid.

Vegetable Herbaceous plant containing an edible portion such as a leaf, shoot, root, tuber, flower, or stem.

Vegetable gums Plant hydrocolloid substances that distribute in water as colloidal dispersions.

Vegetative cells Bacterial cells that are actively growing and carrying on metabolic activities.

Verification Actions designed to confirm whether procedures designed to control hazards are correctly being carried out. For example, visual inspection of monitoring records is a way of confirming that monitoring activities are being carried out. Another example is the calibration of a thermometer, which is a verification action designed to confirm that a thermometer used to measure cooking temperature is accurate.

Verocytotoxins Toxins produced by enterohemorrhagic serotypes of *Escherichia coli*.

Vibrational motion Rapid back and forth motion of molecules in response to relatively strong energy (e.g., infrared energy).

Vinyl acetate (VA) Unsaturated monomer containing the acetate ester group; it can copolymerize with ethylene to form ethylene vinyl acetate (EVA).

Viscoelastic Type of food behavior exhibiting both flow and deformation due to thickness (viscosity) and elasticity (ability to deform).

Viscosity Principal parameter that characterizes the flow properties of foods; the friction within a fluid that prevents it from flowing freely.

Warmed-over flavor (WOF) An unpleasant, stale taste in reheated meats.

Washing Using water as a soaking medium or as a direct spray (pressurized water) on fruits and vegetables.

Water activity (a_w or A_w) The availability of water molecules to enter into microbial, enzymatic, or chemical reactions.

Water holding capacity (WHC) Ability of certain food tissues to retain water molecules within their cells as intracellular water.

Water Vapor Transmission Rate (WVTR) Amount of water vapor transmitted per unit time per unit thickness per unit area in packaging material.

Waxes Esters of fatty acids and even-numbered long carbon chain alcohols, occurring in nature as low melting point solids that coat plant leaves and fruits.

Weak gluten flours Flours low in protein, which produce a softer, more fragile dough for cakes and biscuits.

Wet bulb temperature (wb) Temperature of air in degrees F measured with a standard thermometer whose bulb is kept wet by covering it with a moistened cloth called a "sock."

Whey A component of milk, containing unique proteins.

Whey protein The specific milk proteins lactalbumin, lactoglobulin, lactoferrin, lysozyme, and lactoperoxidase.

White House Conference on Food, Nutrition and Health (1969) Conference that led to major changes in federal food policies, including the food stamp program and various FDA regulations, including the first nutrition labeling rules, which were subsequently established by the FDA in 1973.

Winterizing Low-temperature treatment of oils to eliminate short-chain or highly unsaturated triglycerides.

Wort The liquid resulting after mashing of malted barley and cereal grains in the manufacture of beer.

Yeasts A type of fungus, unicellular in structure, that does not grow in the form of mycelia, but rather as single cells that are spherical or oval.

Yersiniosis The disease caused by consumption of cells of the bacterium *Yersinia enterocolitica*.

Zero-order reaction Enzymatic reaction in which the progress of the reaction is determined solely by enzyme concentration.

INDEX

absolute zero, 358
absorption, 61–62, 361
accuracy, 448
acesulfame-K (Sunette), 78, 80, 258
acetic acid, 108, 217
acetylcholine, 337
acidity. *See* food acids; pH
acidosis, 88
acids. *See* food acids
acid strength, 106–107
acid value, 253
acrolein, 138
act, definition of, 183
actin, 36, 38
actinomyosin, 35
action levels, FDA, 187
activation energy, 97
active sites, 97
acute effects, 331
addition reactions, 98
additives. *See* food additives
adenylate cyclase, 342
adhesiveness, 426
adiabatic processes, 363, 374
adulteration of foods, 7–8
aeration, fats and, 140
aerobes, 286
aerosols, 288
affective tests, 430–431
aflatoxins, 187–188, 195, 250, 312
afterfeel, 426
aftertaste, 424
Agent Orange, 346
aglycone, 159
Agricultural Marketing Service (AMS), 208
Agrobacterium rhizogenes, 400
AH,B Sweetness Theory of Shallenberger, 175–176
air moisture relationships, 369–373
Alar, 349
alcohol functional groups, 99
alcoholic beverages, 25
aldehyde functional groups, 100
ale, 297
aleukia, 311
algae, 340
alimentary pastes, 262
alimentary toxic aleukia, 311
alitame (Novasweet), 80
allergenicity of bioengineered foods, 401–402, 412
allergens, 353
allicin, 52, 163
Allura Red, 160–161

almonds, 34
alpha-linolenic acid, 71–72
amanitins, 338
Ames test, 194–195
amines, 100
amino acids
  bitterness, 281
  in cereal grains, 28
  classification, 142
  essential, 73
  genetically modified, 409
  in infant formulas, 17
  limiting, 74
  Maillard reactions with sugars, 123–125
  nutritive claims, 84
  structure, 73, 142–143
  sweetness, 176
amino functional groups, 100
aminotriazole, 346
ammonium bicarbonate, 28
amnesic shellfish poisoning (ASP), 341
amphiphilic molecules (amphiphiles), 103, 117–118
amphoteric substances, 145
amygdalin, 34
amylopectin, 130–131
amylose, 130–131
anabolic reactions, 82
anabolic steroids, 84
anaphylaxis, 353
animal feeds, genetically engineered, 402
animals, antimicrobial agents for, 343–345
anions, 95, 102, 108
annatto, 46, 161, 294
anthocyanins, 158–159
anthoxanthins, 158–159
antibiotic resistance markers, 401
antibiotics, 344–345
antibodies, 353
anticaking agents, 181
antimicrobial agents, 181, 343–345
antioxidants, 139, 181, 218, 233, 334–335
antiprotozoal drugs, 344
antisensing, 406
AN/TN ratio, 280
APHIS (Animal and Plant Health Inspection Service), 193
aplastic anemia, 344
apparent viscosity, 378
appearance, 428
Appert, Nicolas, 6
applesauce, 268
artificial neural network (ANN), 441–442
artificial sweeteners, 78–80

aseptic packaging, 387
aseptic processing, 216
aspartame, 78, 80, 258, 354, 409
Aspergillus mold, 187, 311
astringency, 162–163
Atkins diet, 88–90
ATP (adenosine triphosphate), 292
autolyzed yeast, 165–166
avidin, 40

baby foods, nutrient labeling for, 198
Bacillus cereus, 221, 307
Bacillus megaterium, 291–292
Bacillus mesentericus, 291
Bacillus stearothermophilus, 221
Bacillus thuringiensis, 405
backcrossing, 393–394
bacteria
  antibiotic resistance, 344–345
  in cheesemaking, 45
  definition, 283
  foodborne infection, 305–306
  foodborne intoxication, 305–307
  foodborne intoxification, 306, 308–309
  gram-negative/gram-positive, 213, 283
  heat resistance, 221
  hurdle technology against, 223–224, 288
  illness from *E. coli*, 310
  illness from *Listeria monocytogenes*, 310–311
  illness from *Salmonella*, 308–310
  inserting DNA into, 397–398
  moisture content of foods and, 215, 285–286
  in muscle meat, 289–290
  probiotics, 128, 259, 408
  spore life cycle, 284
  toxins from, 341–343
  in water, 289
bactericidal treatments, 320
bacteriostatic methods, 320
baked products, 28, 29
baking powder, 28, 109
baking soda, 28, 109
ballistic bombardment, 398
bar code technology, 390
bases, nucleic, 395–396
B cells, 353
bee pollen, 84
beer, 297
bench-top prototypes, 437
Benecol, 205–206
benzimidazoles, 344
benzoic acid, 217

I–1

beta-carotene, 159
beta fat crystals, 140
beta-glucans, 127
betalains, 158–159
beverages
  alcoholic, 25
  composition of, 25
  definition, 24
  degrees Brix, 26
  degrees Brix/acid ratio, 26–27
  nutraceuticals, 258–259
  nutrient density, 24, 26
  ozone sanitization, 256
  processing, 255–258
  soft drinks, 25–26, 256–258
  stimulant, 25
  thirst-quenching, 25
  water beverages, 255–256
  water content, 26
bgal gene, 399
BHT/BHA, 218
bile, role in digestion, 70
Bingham plastics, 377
bioassays, 332
bioavailability, 19, 28, 61
"Bioavailability for rats of vitamin E from fortified breakfast cereal," 19
bioengineered foods. *See* biotechnology
biological changes, 220
biological hazards, 304–308, 327–328
biologically active water, 212
biological value (BV), 28, 74
biology, food science and, 4
biotechnology
  animal products, 407–408
  benefits of, 394–395
  β-lactoglobin, 416–418
  Bt corn, 399
  crop improvement through, 405–407
  definition, 393
  FDA policy, 189–190, 206–208, 400–404
  food processing aids, 408–410
  food safety applications, 410
  genetic engineering, 395–400
  insect-, virus-, and drought-resistant plants, 405
  International Food Biotechnology Council, 412
  labeling requirements, 207, 402
  major concerns about, 410–412
  new substances, 401
  StarLink controversy, 412
  substantial equivalence concept, 403–404
  surmounting the species barrier, 393–394
biotin, 76
bitterness, of amino acids, 281
black cohosh, 53
blanching, 216, 265
bleaching, of oils, 251
bloom, in chocolate, 273–274
Borden, Gail, 6
botanicals, 51. *See also* phytochemicals

botulism, 311, 343
boundary layers, 365
bovine somatotropin (BST), 189–190, 345, 402, 407–408
bradyzoites, 313–314
brainstorming, 438
braising, 230
bread, 164, 260–261
breakfast cereals, 19, 27–28, 261
brevitoxins, 341
broccoli, 335
bromelin, 233
browning, nonenzymatic. *See* Maillard browning reactions
browning of fruits and vegetables, 98
Bt corn, 399
Btu (unit), 359
buffering effect, 288
buffers, 108, 145, 288
butter, 44, 116
butyric acid, 41

cadaverine, 292
caffeine, 84, 274
calcium, 78
calcium caseinate, 42
calories, 63, 82–83, 196, 359
Campylobacter jejuni, 284, 306, 308–309
cancer risks, 186–187, 194–195, 333–334
candling, 41, 228
candy. *See* confectionery
cane sugar, 253. *See also* sucrose
canning of foods
  aseptic, 232–233, 268
  color changes during, 160
  fruits and vegetables, 267–268
  heat transfer, 219–220
  heat treatment, 216
  origin of, 6, 215
  red meat, 232–233
  retort processing, 219, 360
  vacuum, 220
cans, 386
caprenin, 82, 141
capsaicinoids, 163
capsicum, 53
caramel color, 161–162
caramelen, 124
caramelization, 124, 164
carbamate pesticides, 347
carbohydrates, 121–133
  complex, 67–68
  dietary fiber, 67–68, 128–129
  digestion of, 62
  glycemic effect, 68–69
  irradiation of, 244
  metabolized or produced by microbial contamination, 291–292
  nutritional overview, 66
  oligosaccharides, 126
  polysaccharides, 126–131
  simple, 67

carbonated beverages, 256–258
carbonation, 257
carbonyl groups, 101
carboxylic acid functional groups, 100–101, 106–109
carcinogen testing, 333–334
carnitine, 84
carotenoids, 52, 159
carrageenan, 169
Carrier, Willis, 370
casein, 42, 151–153, 169, 226, 294
casein glycomacropeptide, 42
casein micelles, 151–153
cassava, 336, 406–407
catabolic reactions, 82
cations, 95, 102, 108
celiac disease, 354
cell culture, 332, 400
cellophane, 388
cells, organelles in, 4
cellulase, 291
cellulose, 127, 291
Center for Science in the Public Interest (CSPI), 205
cereal, definition of, 158
cereal grains
  amino acid content, 28
  breadmaking, 260–261
  breakfast cereals, 19, 27–28, 261
  corn chips, 262–263
  definition, 27, 258
  fermentation, 297
  food composition of, 27–28, 259
  oils, 259
  pasta, 260, 262
  processing, 258–262
CFSAN (Center for Food Safety and Applied Nutrition), 193
Challenge!
  bioengineering of β-lactoglobin, 416–418
  chemistry of sweeteners and sweetness, 175–177
  enzymes in food processing—what are protein hydrolysates?, 279–281
  experimental design in product development, 445–449
  food allergies and food intolerances, 353–355
  food irradiation, 241–248
  food packaging, 385–391
  food systems, 115–119
  high-protein, high fat diets, 88–90
  microbial sampling to verify food quality, 301–302
  milk protein chemistry, 151–153
  phytochemicals as food components, 51–56
  reading the research, 17–20
  regulation of functional, bioengineered, and organic foods, 205–209
  risk assessment for biological hazards, 327–328

chamomile, 337
character notes, 422
cheddaring, 46
cheese, 42, 45–46, 151–152, 226, 294
chemical acidulants, 296
chemical browning, 164
chemical changes, 220
chemical hazards, 304–305
chemistry. *See also* food chemistry
   activation energy, 97
   chemical bonds, 93–96
   classification of matter, 92
   decomposition/composition reactions, 96
   electron orbits and chemical bonds, 93–94
   enzymatic reactions, 96–98
   food science and, 4–5
   functional groups, 99–102
   homogeneous matter, 115
   nonenzymatic reactions, 98–99
   protein, 151–153
   solutions, 115–116
   of sweeteners and sweetness, 175–177
   symbols, formulas, and equations, 92–93
chemogastric revolution, 187
chemotrophs, 285
chewiness, 425–426
chicken, 236, 290–291
Chick, Harriet, 7
chiles, 163
chilling injury, 266
chitinase, 405
chloramphenicol, 344
chloride, 78
chlorine, as sanitizer, 318
chlorophylls, 159–160
chocolate, 46–47, 272–274
cholecystokinin (CCK), 62
cholera, 311, 342
cholesterol
   Benecol, 205
   Cholestin and, 201
   essential fatty acids and, 72
   lipoproteins and, 72
   plant sterol and stanol esters and, 199
   soluble oat fiber and, 199
   soy protein and, 199
Cholestin, 201
cholinesterase, 337
chroma, 156
chromium, 79
chromium picolonate, 84
chromophoretic compounds, 156
chromosomes, 395
chronic effects, 331
chymosin, 42, 294
ciguatera, 340
cilia, 423
cis forms of fatty acids, 135–136
citric acid, 111, 217
Claviceps purpurea, 311
cleaning, in food processing, 212–213

climacteric fruits, 263–264
cloning, 400
Clostridium botulinum
   botulism from, 311
   heat resistance, 221–222
   intoxication from, 216, 305, 307
   sodium nitrite against, 320
   sources, 331
   toxins, 343
Clostridium perfringens, 307, 343
coagulation, 145–146
cobalt, 79
Coca-Cola, 257
cochineal, 161
cocoa butter, 46, 168, 272–273
Code of Federal Regulations (CFR), 183–185
Codex Alimentarius Commission, 190–191
coenzymes, 75
coffee, 25
cohesiveness, 425
cold-break process, 269–270
cold point, 219–220, 360
cold shortening, 231
collagen, 36, 170–171
colloid, 116
colloidal dispersions, 44, 115–118
colon, digestion in, 62
color
   definition, 115–116
   food colorants, 160–162, 181
   of fruits and vegetables, 31, 98, 158–160
   Hunter tristimulus data, 428
   lipid pigments, 137
   natural pigments, 156, 158–160
   nitrates in meats, 233
   red meat, 156–158, 233
   sensory evaluation, 427–428
color additive certification, 181
Color Additives Amendments of 1960, 186
colorants, 160–162, 181
comfrey, 337
commercially sterile, 215–216
comminuted meat emulsions, 37–38
comminution, 234
commodities, definition of, 23
comparators, 403
complete block experimental design, 446
composition reactions, 96
compounds, definition of, 92
Compton effect, 243
conching, 273
condensation reactions, 98–99
conduction, 219–220, 359–360
confectionery, 47, 125
conjugated double bonds, 135, 158
conjugated linoleic acid (CLA), 72
connective tissue, 36, 230
conservation of energy, 363
conservation of mass, 362
consumer acceptance testing, 431
consumer response testing, 439
contamination of foods, 317–318

continuous variables, 448
controlled atmosphere storage, 266, 387
convection, 219–220, 360
convulsive ergotism, 311
CoolPure process, 224
copolymers, 388
copper, 79
coprophagy, 314
corn chips, 262–263
corn, genetically modified, 399, 412
corn syrup, high-fructose (HFCS), 145
corn syrup solids, 128
coronary heart disease, 199
corporate mission, 436
corrective actions, 322
cottage cheese, 45
cottonseed oil, 250
coumarins, 335
covalent bonds, 94–95
cream, 44
cream cheese, 45
creaming, 117
creatine, 84
Creutzfeldt–Jakob disease (CJD), 315–317
critical control points, 321
critical limits, 321
cross-breeding of plants, 393–394
cross-contamination, 318
cruciferous vegetables, 335
Cruess, W. V., 11
crustaceans, 39
Cryptosporidium parvum (Cryptosporidiosis), 313–314
crystallization, 140, 426
cultured milk, 42–43
curds, 42–43, 294
curing, 181, 234, 235
cyanide compounds, 335–336
cyanogenic glycosides, 335–336
cyclamates, 80, 188
Cyclospora, 313–314
cysteine (CYS), 17
Cysticercus bovis, 315
Cysticercus cellulosae, 315
cysts, 315

Dactylopus coccus costa, 161
Daily Reference Values (DRV), 63
Daily Values, on food labels, 63, 196
dairy products. *See* milk and dairy products
danger zone, 318–319
data analysis, 447
DDT, 346
decimal reduction time (D values), 221, 242
decomposition reactions, 96
deep-fat frying, 365
definitive hosts, 314
degranulation effect, 353
degree of esterification (DE), 129
degrees Brix, 26, 32
degrees Brix/acid ratio, 26–27

degumming, 251
dehydration, 32, 65–66, 269, 373–374
Delaney Clause, 186–187
Deming, W. Edwards, 445
demographics, 440
denaturation, of proteins, 144–146
deodorization, 251
depolarization, 423
descriptive test methods, 431–432
designer foods. *See* functional foods
design of experiments (DOE), 445–449
detection threshold, 430
dewormers, 344
dew point (Dpt), 371–372
dextrans, 128, 291
dextrins, 127–128
dextrose equivalents (DE), 123, 128
DHEA, 84
diarrhetic shellfish poisoning (DSP), 341
dielectric constant, 361–362
dielectric loss factor, 361
dietary fiber. *See* fiber, dietary
dietary goals, in calories, 83
The Dietary Guidelines for Americans, 59
Dietary Supplement Health and Education Act (DSHEA) of 1994, 200–201
dietary supplements, 200–201. *See also* vitamins
diethylstilbestrol (DES), 345
difference threshold, 430
digestion, 60–62
diglycerides, 98, 252
dilatant behavior, 377
dinoflagellates, 340–341
Dinophysis fortit, 341
dinoseb, 349
dipoles, 70, 175
direct expansion refrigeration, 216
directive, definition of, 183
disaccharides, 67, 98, 122
discrete variables, 448
discrimination test methods, 431–432
disintegrating, in food processing, 213
disulfide bonds, 107
diuresis, 88
DNA, 393, 396–400, 406
DNA ligase, 397
DNA polymerase, 397
docosahexaenoic acid (DHA), 39, 71–72
domoic acid, 341
Donkin, Brian, 6
dose rate, 242
dose-response, 327, 331–334
dough sheeting, 260–261
dough softening, 107
dough strengtheners, 181
drought-resistant plants, 405
drum drying, 214
dry bulb temperature, 371
drying, in food processing, 214–215, 229
D values, 221, 242
dyes, 161–162. *See also* colorants

echinacea, 53
E. coli
  antibiotic resistance, 345
  characteristics of, 306–307, 309
  cholera-like enterotoxins, 342
  mechanism of infection, 310
  outbreak of O157:H7, 267, 304
  sources, 331
edible films, 386–387
Edisonian approach, 447
Edwards Committee, 191
effective dose (ED_{50}), 333
efficacy, 180
eggs
  denaturation and coagulation of whites, 146
  egg substitute, 230
  Haugh units, 40–41
  nutrient composition, 39–41
  processing, 227–230
  protein quality, 74
  quality classes, 228
  structure of, 40
eicosapentaenoic acid (EPA), 39, 71–72
Eijkman, Christiaan, 7
elasticity, 375–376
elastic solids, 376
electrolytes, 65–66
electromagnetic spectrum, 427
electronegative/electropositive compounds, 175
electron orbits, 93–96
electron volts (eV), 242
electroporation, 398
elements, chemical, 92–93
emulsification. *See also* emulsions
  by bile, 70
  of fats, 140–141, 252
  in food processing, 214
  proteins in, 145, 151
emulsifiers, 117–118, 181, 252
emulsifying salts, 181
emulsions
  definition, 116
  emulsifiers, 117–118, 181, 252
  ice cream, 369
  meat, 37–38
  oil-in-water *vs.* water-in-oil, 37, 116–118
  process cheese, 46
  stability, 19, 117
  unprocessed milk, 117
  water in, 104
encapsulation, flavor, 166
endogenous toxicants, 330
endosperm, 28
energy levels, in atoms, 94
energy metabolism, 82–83
enology, 11
enrichment, 182
enterogastrones, 62
enterotoxins, 341–342
enthalpy, 372, 374

entropy, 363
Environmental Protection Agency (E.P.A.), 193
enzymatic browning, 98
enzymes
  activation energy, 97
  chlorophyll degradation, 159–160
  in digestion, 60–61
  enzymatic hydrolysis, 97–98, 279–281
  enzyme-produced flavors, 165–166
  factors affecting activity, 146
  as food additives, 181
  genetically modified, 409
  high-fructose corn syrup production, 145
  oxidation/reduction, 98
  polymerization, 98
  protein hydrolysates, 279–281
  reactions in foods, 96–98
  restriction, 396–398
  synthesis of, 396
  tabulation of, 147
  tenderizers, 171, 231
EPG (esterified propoxylated gycerol), 141
epidemiology, 332
equations, chemical, 93
Ereky, Karl, 393
ergogenic substances, 84
ergotism, 311
erythromycin, 344
erythropoesis, 75
Escherichia coli. *See E. coli*
essential amino acids, 73
essential fatty acids, 68
ester bonds, 98
esters, 101
estrogen, animal steroids and, 345–346
ethanol, 99, 291
ethylene, 406
ethylene dibromide (EDB), 349
ethylene vinyl acetate (EVA), 388
eucalyptol, 337
eucaryotes, 283
evaporated milk, 42, 43
evaporation, in food processing, 214
experimental design, 445–449
exposure assessments, 331
extensibility, 376
extrusion technology, 379–380, 386, 389–390

factorial experiment designs, 446
facultative anaerobes, 286
Fair Packaging and Labeling Act of 1966, 187
fat mimetics, 169
fat replacers, 78–82, 141, 169–170
fats. *See also* lipids
  chemical and physical testing of, 253
  classification, 69
  cocoa butter, 46
  content in foods, 72
  digestion of, 62, 70

emulsification, 140–141, 252
fatty acids and health, 70–72
as flavor compounds, 136, 141
in the Food Guide Pyramid, 60
fractionation, 137
genetical engineering of, 406
high-protein, high fat diets, 88–90
hydrogenation, 138
hydrolysis, 138
interesterification, 138–139
lipid composition, 69–70
melting points, 136
microstructure, 369
milkfat, 41
nutritional overview of lipids, 69
omega-3 fatty acids, 39, 71–72, 135
oxidation, 139
perception of, 426
polymorphism, 273
processing of, 250–253
rancid, 5, 39
saturated and unsaturated, 71, 134–135
solvent extraction, 251
structure of, 134
texture and, 168–169
triglycerides interesterification, 252
fatty acids
 acid value, 253
 cis- and trans-fatty acids, 72, 135, 138
 essential, 68, 70–71
 omega-3/omega-6, 39, 71–72, 135
 polymerization, 140
 rancidity, 167
 saponification value, 253
 saturated/unsaturated, 70, 134–135
 structure, 71, 134
FCCC (Food Chemicals Codex Committee), 193
FDA Modernization and Accountability Act of 1997, 191–192
FD&C colorants, 160–161, 181
feasibility, 436
Federal Food, Drug, and Cosmetic Act (FFDCA) of 1938, 185, 400
Federal Register, 183
Federal Trade Commission (FTC), 193, 206
fermentation reactions
 bread, 164, 261
 cereal grains, 297
 chocolate, 273
 definition, 292
 enzymatic hydrolysis, 97–98
 growth of genetically engineered cells, 400
 meat, 235, 295–296
 milk, 111, 293–295
 overview, 292–293
 pH, 110–111
 vegetables and fruits, 268, 296
feverfew, 53
fiber, dietary
 benefits and sources, 68

definition, 67
digestion of, 62
inulin, 128–129
soluble oat fiber and heart disease, 199
types of, 67, 128
ficin, 233
filtration, 267
fish. *See* seafood
fish oil, 39
flash freezing, 233–234
flash pasteurization, 216
flatworms, 308, 313–315
flavonoids, 51, 52, 158, 334–335
flavor
 chemical structure and, 162–164
 chemistry of sweeteners and sweetness, 175–177
 cooling sensation, 164
 definition of, 162
 deterioration, 166–167
 encapsulation, 166
 enzyme-produced flavors, 165–166
 fats, 136, 141
 flavor enhancers, 166, 181
 glossary of terms, 165
 process and reaction flavors, 164–166
 protein hydrolysates, 166
 pungency, 163–164
 taste mechanisms, 422–424
flavorings, 181
flavor profile method, 431–432
Flavr Savr tomato, 190, 406–407
flexible pouches, 386, 390
flexural modulus, 389
fluidity, 375
fluid mechanics, 374–379
fluoride, 79
foams, 118, 145–146, 151, 369
focus groups, 438–439
folate, 76
food acids. *See also* pH
 acid control during processing, 217
 acid foods, 111, 217
 acid strength, 106–108
 buffers, 108, 145, 288
 pH and pH scale, 110
 salts of organic acids, 107–108
 as sanitizers, 318
 structures, 106
food additives, 178–209
 Ames test, 194–195
 approval process for, 193–195
 in candy, 125
 definitions, 179
 food regulatory agencies, 190–193
 hyperactivity and, 181, 354
 intentional *vs.* indirect, 179
 irradiation, 179, 246
 laws and regulations on, 183–190
 major types of, 180–183
 processed meat, 233
 safety testing, 194

uses of, 180
zero-cancer-risk standard, 186–187
Food Additives Amendment (FAA) of 1958, 185–186
food allergies and intolerances, 353–355, 412
Food and Drug Act of 1906, 185
Food and Drug Administration (FDA)
 acceptable levels of filth, 192
 bioengineered foods policy, 189–190, 206–208, 401–402
 FDA Modernization and Accountability Act of 1997, 191–192
 formation of, 184–185
 functional foods, 205–206
 health claims approval, 192, 198–199
 organic foods, 208
 recalls, 190–191
 Red Book, 189
 responsibilities, 190–192
 supplements *vs.* drugs, 201
food biosecurity, 304
food biotechnology. *See* biotechnology
foodborne illness
 from bacteria, 305–307, 308–311
 definition, 304–305
 Hazard Analysis and Critical Control Points, 12, 267, 304, 320–322
 mad cow disease, 315–317
 mechanism of foodborne intoxification, 307
 from molds, 306, 311–312
 from parasites, 307–309, 313–315
 preventing contamination, 317–318
 preventing proliferation of microorganisms, 318–320
 reducing biological hazards, 320
 risk assessment for biological hazards, 327–328
 from viruses, 307, 312
food categories, 23
food chain, flow of chemical toxicants through, 346, 348
food chemistry. *See also* chemistry
 colorants, 160–162
 color of fruits and vegetables, 158–160
 color of red meat, 156–158
 enzymatic reactions, 96–98
 flavor, 162–167
 food acids, 106–109
 functional groups, 99–102
 HPLC analysis, 17
 leavening agents, 108–109
 nonenzymatic reactions, 98–99
 study of, 12
food components
 in baked goods, 29
 beverages, 25–26
 cereal grains, 27–28
 chocolate and confectionery, 46–47
 food composition tables, 22–24
 Food Guide Pyramid, 10–11

food components (continued)
  fruits and vegetables, 30, 32
  legumes, 33
  nuts, 35
  phytochemicals as, 51–56
food composition tables, 22–24
food engineering
  conservation of mass, 362
  deep-fat frying, 365
  extrusion technology, 379–380
  food dehydration, 373–374
  food microstructure, 368–369
  heat exchangers, 364–365
  microwave energy, 360–361
  optimizing low fat peanut butter, 18
  psychrometrics, 369–374
  rheology, 374–379
  study of, 12, 357
  tasks of engineers, 369
  temperature and heat characteristics, 359–362
  temperature scales, 357–358
  thermodynamics, 363–364
Food Guide Pyramid, 10–11, 23–24, 58–60
food gums, 19. *See also* vegetable gums
food irradiation. *See* irradiation
food labels, 62–65, 196–198
food laws. *See* laws and regulations
food materials science, 365–368
food microbiology. *See* microorganisms
food microstructure, 368–369
food packaging. *See* packaging
food poisoning. *See* food toxicology; toxicants/toxins
food preservation. *See also* irradiation
  decimal reduction time (D values), 221
  definition, 211
  hurdle technology, 223–224, 288
  innovative nonthermal methods, 223–225
  moisture removal, 214–215
  ohmic heating, 223
  thermal death time, 222
  thermal processing, 220–223
  12D concept, 222
food processing, 211–248. *See also* canning of foods
  acidity control, 217
  antimicrobial chemical preservatives, 217–218
  aseptic, 216
  definition, 5, 211
  heat treatment, 215–216
  history of, 5–8
  low-temperature treatment, 216
  moisture removal, 214–215
  nonthermal processing innovations, 219
  unit operations of, 212–214
food quality, 9, 40–41, 230–231, 244–245, 301–302, 448
food safety
  bacterial hazards, 305–306, 308–311
  foodborne illness, 304–305, 317–320

HACCP approach, 320–322
mad cow disease, 315–317
mold, 306, 311–312
parasites, 313–315
viruses, 285, 307, 312
Food Safety and Inspection Service (FSIS), 190
food science, 2–5
food science degree, 11–13
food science education, 11–14
food scientists/technologists, 8–10
food systems, 115–119
Food Technology, 9
food technology, definition of, 2–3
food toxicology. *See also* toxicants/toxins
  carcinogen testing, 333–334
  endogenous toxicants, 334–340
  food allergies and intolerances, 353–355, 412
  growth promotants, 345–346
  microbial toxins, 341–343
  mushroom poisoning, 338–340
  pesticides, 346–349
  risk assessments for chemical hazards, 331–334
  seafood toxins, 340–341
  synthetic antimicrobials, 343–345
  toxicants, 330–331
forming, in food processing, 214
formulas, chemical, 93
fortification, 182
fractionation, of fats, 252
frankfurters, 19–20, 37–38
free radicals, 139, 243–244
free water, 103
freeze-drying, 214–215, 234–235
freezer storage, 233, 266–267
freeze–thaw cycles, 105
French Paradox, 30–31
"fresh" designation, 198
fructooligosaccharides (FOS), 128
fructose, 67, 121–122
fruits
  acidity, 32, 111
  climacteric, 263–264
  color chemistry, 158–160
  coumarins in, 335
  definition, 28–29, 262
  dehydration of, 32, 269
  fermentation, 296
  flavonoids in, 334–335
  in the Food Guide Pyramid, 60
  freezing, 266–267
  hot-break/cold-break processing, 269–270
  infused, 32
  juices, 25–26, 267
  microbial contamination, 290
  nutrients compared to grains, 259
  nutrition labeling, 195
  packaging, 265
  processing, 264–265

quantitative and qualitative quality, 31–32
recommended intake of, 30–31
ripeness and maturity, 31, 263–264
storage, 266
fumaric acid, 106–107
fumigants, 349
functional foods, 51–52, 54–55, 205–206, 258–259
functional groups, 99–102
functional properties, 42, 124
fungi, 284
Fusarium moniliforme, 311
Fusarium sporotrichioides, 311

galactomannan, 259
galactose, 67, 122
gallbladder, 62, 70
Gambierdiscus toxicus, 340
gamma-linolenic acid (GLA), 72
gangrenous ergotism, 311
garlic, 53, 163
gastrin, 62
gastrointestinal poisoning, from mushrooms, 340
gelatin, 169–171
gelatinization, 126, 129–130
gelation, 131–133, 146, 151, 416–417
gels, 118, 132, 169
genes, 395–396, 399
genetically modified organisms (GMOs), 206–208
genetic engineering. *See also* biotechnology
  DNA in food production, 396–400
  FDA policy on, 189–190, 206–208, 401–402
  genetic switches, 399
  inserting DNA into bacteria, 397–398
  inserting DNA into cells, 398
  marker genes, 398–399
  messenger RNA, 397
  overview of, 395–396
  PCR and genetic probing, 399
  propagating growth of cells, 400
  protein synthesis, 396
  soybeans, 272
genetic probing, 399
genetic switches, 399
genistein, 51
germ, cereal, 28
Giardia lamblia, 284, 313
ginger, 53, 163
ginko, 53
ginseng, 84
glass and glassy state, 366–368
glass containers, 386
glass transition temperature, 105, 367–368
glucono delta lactone (GDL), 296
glucose, 67, 121–122, 176
glucoside, 159
glutamine peptide (GP), 280
gluten, 259–261, 354

glycemic effect, 68–69
glycerol, 99, 133
glycerophospholipids, 136–137
glycolysis, 292
glycophores, 175–176
glycosidic bonds, 98, 122–123, 127
goiter, 335
goitrogens, 335
Gonyaulax catenella, 340
Good Manufacturing Practices (GMPs), 191, 318, 322
gossypol, 250
G protein, 342
grains. *See* cereal grains
grainy texture, 426
gram staining, 283
GRAS (generally recognized as safe) substances, 186, 188
guanylate cyclase, 342
gums, genetically modified, 409
gyromitrin, 338

hairy root cultures, 400
hardness, 425
Harrison, James, 7
harvesting, definition of, 31, 263
Haugh units, 40–41
Hazard Analysis and Critical Control Points (HACCP), 12, 267, 304, 320–322
hazards, 304, 321
hazelnuts, 34
HDPE (high-density polyethylene), 388
health claims on labels, 192, 198–199, 200, 205–206
"healthy" designation, 198
heart disease, 199
heat
    content, 372
    definition of, 363
    exchange, 214, 364–365
    heat capacity, 363–364
    measurement of, 82–83, 359
    resistance, 221
heat transfer
    within a can, 219–220
    chemical reaction rates and, 364
    conduction, convection, and radiation, 359–360
    conservation of mass, 362
    definition, 219
    fats and, 141
    heat capacity and specific heat, 363–364
    heat exchangers, 214, 364–365
    latent and sensible heat, 370
    microwave energy, 360–361
    in a retort canner, 219, 360
    surface area and, 363
    thermal conductivity, 361, 364
    thermodynamics of, 363
    transient, 362
    vacuum canning, 220
    water and, 105

heat treatment, 215–216, 220–223, 268, 320
hedonic tests, 432
hemolytic uremic syndrome, 310
hemorrhagic colitis, 310
hepatitis A, 307, 312
herbicides, 346
herbs, 51, 53–54, 336–338. *See also* phytochemicals
hermetically sealed containers, 220, 268, 386
heterofermentors, 293–294
high acid foods, 217
high-density lipoproteins (HDL), 72
high-fructose corn syrup (HFCS), 145
high-pressure processing (HPP), 224
history of the food industry, 6–8
HMB, 84
HM pectin, 129–130
holotoxins, 342
homeostasis, 61
homofermentors, 292–294
homogeneous distribution, 302
homogenization, 117, 214, 226, 379
hops, 297
hormones, 61–62, 345–346, 349
horseradish, 163
hot-break process, 269–270
hot/cold inversion, 340
hot dogs, 19–20, 37–38
"HPLC method for cysteine and methionine in infant formulas," 17
hue, 156
humectants, 125–126, 164, 181–182
Hunter color difference, 428
Hunter tristimulus data, 428
hurdle technology, 223–224, 288
hydration, 103
hydrazines, 338–339
hydrocooling, 266
hydrogen bonds, 95–96, 176
hydrogenization, 251
hydrogen sulfide, 292
hydrolysis reactions, 60, 97–99, 138, 147, 166, 279–281
hydrolytic rancidity, 138, 167
hydrolyzed vegetable protein (HVP), 34, 166
hydroperoxide, 139
hydrophobic/hydrophilic compounds, 70
hygroscopicity, 106
hyperactivity, food additives and, 181, 354
hypobaric storage, 266

ice cream, 44, 226–227, 369
ideal gas law, 370
immune system, 353
immunoglobulins, 151, 353
incomplete block experimental design, 446
Individual Quick Frozen (IQF) process, 216, 266
indole-3-carbinol (I3C), 335

indoles, 52
infant formulas, 17
infusion, 32
ingredient lists, 64, 197
insect-resistant plants, 405
Institute of Food Science and Technology (IFST), 2
Institute of Food Technologists (IFT), 2, 14
insulin, 72
Integrated Pest Management (IPM), 346–347
intensity, of color, 156
interesterification, 138–139, 252
intermediate moisture foods (IMF), 214
International Food Biotechnology Council, 412
intestines, digestion in, 62
intoxication, foodborne, 305
intoxification, foodborne, 306, 308–309
intramolecular bonding, 96, 103
inulin, 128–129
invert sugars, 126
iodine, 79, 318
iodine value, 253
iodophores, 213
ionic bonding, 95
ionization constant, 106
ions, 95, 102
iron, 79
irradiation
    approved uses, 245, 247
    of crab meat products, 18–19
    description of, 235, 241
    dose and dose rate, 242
    effect on microorganisms, 18–19, 243
    facility safety, 245
    as food additive, 179, 246
    food quality and, 242–245
    future of, 246–247
    of mail at postal sorting centers, 245
    radiation sources, 241–242
    radiolytic products from, 245
    radura symbol, 246
isoelectric point, 45–46, 151
isoflavones, 33, 51, 52
isolated soy protein (ISP), 19, 81
isoprenoids, 52
isothiocyanates, 52
iterative processes, 8

Joint FAO/WHO Expert Committee on Food Additives (JECFA), 190
joules, 359
Journal of Food Science, 9
juices, 25–26, 267
The Jungle (Sinclair), 8, 185

kava, 53
Kelvin temperature scale, 357–358
ketones, 101
ketosis, 88
kwashiorkor, 75

labeling of foods, 62–65, 195–199, 207–208, 402
Lactacel, 295
lactic acid, 108, 111, 217, 293
lactic acid bacteria, 290, 292–294
Lactobacillus, 227, 291, 294, 295
Lactococcus lactis, 294
lactoferrin, 408
lactose, 67, 122–123
lactose intolerance, 41
lager, 297
lakes, 161–162. *See also* colorants
latent heat, 370
law, definition of, 183
laws and regulations
   on biotechnology, 189–190, 206–208, 400–404
   cyclamates, 188
   definitions, 183
   Delaney Clause, 186–187
   early events and legislation, 184–185
   FDA Red Book, 189
   FFDCA of 1938 and Amendments, 185–187
   international food agencies, 190
   laws *vs.* regulations, 183–184
   national food regulatory agencies, 190–193
   nutritional labeling, 189, 195–199
   pesticides and toxicants, 187–188
   processed foods innovations, 187
   proposed *vs.* finalized rules, 184
   saccharin, 188–189
   timeline of selected, 185
LDPE (low-density polyethylene), 388–389
leavening, 28–29, 108–109, 182
leavening acids, 108–109
lecithin, 40, 137, 273
legumes, 32–33
lethal dose (LD_{50}), 333
Leuconostoc mesenteroides, 291
levans, 292
Levine, Harry, 366
LFTT (lean finely textured tissue), 19
Liebig, Justus von, 7
light, color and, 115–116
lignans, 52
linamarin, 336
line extension, 436
linoleic acid (LA), 72, 134
linolenic acid, 39, 71, 134
lipids. *See also* fats
   chemical reactions, 137–140
   flavorings, 416
   functional properties, 140–143
   irradiation of, 244
   lipoxidation, 139
   overview, 68–72
   pigments, 137
   polar, 136–137
   processing, 250–253

   structures and types, 133–137
   waxes, 137
lipoproteins, 72
lipoxidation, 139
lipoxygenase, 33
Listeria monocytogenes, 306, 310–311, 410
listeriosis, 311
lithotrophs, 285
liver, 62, 72
LLDPE (linear low density polyethylene), 388
LM pectin, 129–130
locust bean gum, 409
low-acid foods, 111
low-density lipoprotein (LDL), 72
low-temperature treatment, 216
lycopene, 159
lyophilization, 214–215
lysine, 28

macadamia nuts, 34
macrolides, 344
macronutrients. *See also under names of individual macronutrients*
   carbohydrates, 66–68
   definition, 58
   lipids, 68–72
   protein, 73–75
   water, 64–66
mad cow disease, 315–317
magnesium, 78
Maillard browning reactions, 123–125, 164
malathion, 347
malic acid, 106, 111, 217
malnutrition, 58, 75–79
maltitol, 126
maltodextrins, 81, 128
maltol, 424
maltose, 67, 122–123, 127
manganese, 79
mannitol, 126
margarine, 44, 116–117, 138, 205
margin of safety concept, 188
marker genes, 399
marketing, in product development, 438–440
market research, 437–438
marshmallows, 118
mass transfer, 365
mast cells, 353
materials handling, in food processing, 212
matrix metalloproteases, 231
matter. *See* chemistry
maturity, of fruits, 31, 263
maximum growth temperature, 287
maximum residue levels (MRLs), 347
meat and meat products
   acceptable levels of contamination, 192
   canning, 232–233
   chemical additives, 233
   chemistry of red meat color, 156–158

   cold shortening, 231
   cold storage, 233–234
   comminution, 234
   curing, 234
   definition, 230
   drying, 234–235
   emulsions, 37–38, 116
   fermentation, 235, 295–296
   flavor development, 164–165
   in the Food Guide Pyramid, 60
   grain of, 230
   inspection of, 185, 190
   irradiation, 235
   mad cow disease, 315–317
   microbial contamination, 289–292
   nutrient composition, 37
   nutrition labeling requirements, 195
   packaging, 235
   processing methods, 35, 231–235
   quality, 230–231
   rigor mortis, 230–231
   smoking, 235
   structure of muscle tissue, 35–36
   tenderizers, 165, 171, 231
   warmed-over flavor, 167
meat emulsions, 37–38, 116
Meat Inspection Act of 1906, 185
mechanical deboning, 290
melanoidins, 123
melting points, 136
menthol, 164
mesophiles, 287
messenger RNA, 397
metabolism, 82–83
methionine (MET), 17, 28
methyl groups, 101
metmyoglobin, 157
mevinolin, 201
micelles, 103
Michaelis–Menton equation, 97
microaerophilic, 286
microinjection, 398
micronutrients, 58, 75–80
microorganisms. *See also* bacteria
   in dairy products, 290–291
   description of, 283–285
   effect of irradiation on, 18–19, 243
   food effects, 287–288, 291–292
   fruits and vegetable contamination, 290
   genetically modified, 409–410
   molds, 215, 226, 284–286, 292, 306, 311–312
   in muscle foods, 289–290
   nutrient requirements, 285–287
   representative sampling, 301–302
   sources in foods, 288–289
   study of, 12
   viruses, 285, 307–308
   water activity, 285
microparticulation, 80–81, 146
microwavable containers, 386

microwave energy, 360–361
middlings, 260
migraine headaches, 354
milk and dairy products
    β-lactoglobin bioengineering, 416–418
    bovine somatotropin, 189–190, 345
    butter and cream, 41, 44
    cheese, 42, 45, 226, 294
    fat processing, 252
    fermentation, 293–295
    as food group, 60
    homogenization, 226, 379
    ice cream, 44, 226–227, 369
    lactic acid, 108, 111
    lactose intolerance, 41
    microbial contamination, 290–291
    nutrient composition, 41
    nutrient density, 26
    pasteurization, 215–216, 225
    processing, 225–227
    proteins, 42–43, 151–153
    recombinant bovine somatotropin in, 189–190
    souring, 110
    standards of identity, 43, 186
    varieties of, 42
    yogurt, 227–228, 295
milk thistle, 53
minerals, essential, 75, 77–79, 259
mineral water, 256
minimum growth temperature, 287
mixing, in food processing, 213–214
mixtures, definition of, 92
mobility, of food molecules, 367–368
modified atmosphere packaging (MAP), 218, 265, 387
moisture content, 371, 426. See also water activity
moisture, definition, 103
moisture removal, 214–215, 296
moisture sorption isotherms, 104–105
molasses, 254
molds
    in cheese, 226
    definition, 284
    ergotism and aleukia from, 311–312
    microbial degradation by, 292
    moisture content of foods and, 215, 285–286
    mycotoxins from, 306
Mollier diagrams, 370–373
Mollier, Richard, 370
molybdenum, 79
monoglycerides, 252
monosaccharides, 67, 121–122
monosodium glutamate (MSG), 354, 409
monounsaturated fatty acids, 71
mouse lethal dose (MLD_{50}), 343
mouth, digestion in, 62
mouthfeel, 141, 424–425
muscle tissue, 35–36

mushrooms, 338–340
mustard, 163
mutations, Ames test for, 194–195
mutual supplementation, of protein, 74
mycotoxins, 306
myofibrils, 35–36, 38
myoglobin, 156–158
myosin, 36, 38
myristicin, 337

National Food Processors Association (NFPA), 205
National Marine Fisheries Service (NMFS), 193
National Nutrient Data Bank (NDB), 22
National Organic Program (NOP), 208
naturally occurring toxicants, 330
needle stitch pumping, 234
neurotoxic poisoning, 338–340
neurotoxic shellfish poisoning (NSP), 341
neutralization process, 251, 254
neutralizing value (NV), 109
Newtonian foods, 376–378
niacin, 76
nisin, 410
nitrate-cured meats, 157, 233
nitric oxide (NO), 233
nitric oxide myoglobin, 157
nitrofurans, 344
nitrogen balance, 74–75
nitroimidazoles, 344
Nitschia, 341
noncarbonated soft drinks, 258
noncovalent interactions, 103
nonfat dry milk (NFDM), 42–43
non-Newtonian foods, 376–378
nonnutritive sweeteners, 182, 258
nonthermal processing, 223
Norwalk-like viruses, 312
nutraceuticals, 9–10, 51, 258–259. See also functional foods
nutrient density, 24, 26
nutrients. See food components
nutrition
    definition, 3, 58
    digestion, absorption, and transport, 60–62
    energy metabolism, 82–83
    ergogenic substances, 84
    food composition tables, 22–24
    Food Guide Pyramid, 10–11, 23–24, 58–60
    food labels, 62–65, 195–199
    high-protein, high fat diets, 88–90
    legislation, 195–199
    macronutrients, 64–75
    micronutrients, 75–79
    sugar and fat substitutes, 77–82
    vitamin E fortification, 19
nutritional additives, 182
nutrition facts labels, 62–65, 195–199

Nutrition Labeling and Education Act (NLEA) of 1990, 189, 195–199
nutritive sweeteners, 182
nuts, 33–35

oat fiber, heart disease and, 199
Oatrim, 80–81, 127
odor, 423–424
odorants, 423
off-flavors, 166–167
ohmic heating, 223
oils. See also fats; lipids
    bleaching, 271
    from cereal grains, 259
    crude, 250
    as flavorings, 416
    fractionation, 137
    genetic engineering, 406
    processing of, 250–253
    soybean, 271–272
oilseeds, 33
okadaic acid, 341
oleic acid, 134, 207, 272
Olestra (OLEAN), 81–82, 141
olfaction, 423
olfactory epithelium, 423
oligosaccharides, 126
omega-3/omega-6 fatty acids, 39, 71–72, 135
oncospheres, 314
onions, 163
oocysts, 314
optimization, 447
"Optimizing low fat peanut butter spread containing sucrose polyester," 18
optimum growth temperature, 287
ordinance, definition of, 183
orellanine, 338
organic acids. See food acids
organic alcohols, 121. See also sugars
organic foods, 208
organic salts, 107–108
organochlorine pesticides, 347
organophosphate pesticides, 347
organotrophs, 285
oscillating magnetic fields (OMF), 224
overnutrition, 58
oxidation/reduction reactions, 98–99, 126
oxidative rancidity, 167
oxidizing agents, 99, 182
oxidizing environments, 286
oxygen, microbial growth and, 286–287
oxygen scavenging, in packaging, 387
oxymyoglobin, 157
ozone, 256

packaging
    bar code technology, 390
    categories and functions, 385–386
    container types, 386–387
    controlled and modified atmosphere, 387

packaging *(continued)*
  in food processing, 214, 218–219
  fruits and vegetables, 265
  plastics in, 388–390
  pulsed light technology for, 224
  vacuum, 218, 235, 265, 387
pancreas, 62
pancreatin, in protein hydrolysis, 279
panelists, 446
pantothenic acid, 76
papain, 171, 233
paralytic shellfish poisoning (PSP), 340–341
paraquat, 347
parasites, 307–309, 313–315
particle size, 115
part per million (ppm), 347
Pascalization, 224
passive modification, 387
pasta, 260, 262
pasteurization, 215–216, 225
patent flour, 260
PDCA cycle, 445
peanut butter, 5, 18
peanuts, 33
pectic acid, 129
pectic substances, 128–130
pectin esterase (pectinase), 269–270, 291
pectinic acid, 129
pectinolytic organisms, 285
Pediococcus cerevisiae, 295–296
Pekelharing, A. C., 7
Penicillium roqueforti, 45, 226
pepper, 163
peptide bonds, 73, 98, 143
"percent fat free" designation, 198
peroxide value (PV), 139, 253
Pesticide Residue Amendment of 1938, 187
pesticides, 346–349
PET (polyethylene terephthalate), 389
pH. *See also* food acids
  acidity control in food processing, 217
  acid/low-acid foods, 111, 217
  buffers, 108, 145, 288
  in cheesemaking, 45–46
  control agents, 182
  definition, 110
  fermented foods, 111
  of fruits, 32, 111
  microorganisms and, 286
  pigment colors, 158
  proteins and, 145, 152
  scale, 110
  of selected foods, 109
  titratable acidity, 32, 110–111
  titration curve for acetic acid, 108
pharmfoods. *See* functional foods
phenylketonuria (PKU), 354
pheophytin, 160
phosphate groups, 101–102
phosphate salts, 84
phosphoric acid, 217
phosphorus, 78

phototrophs, 285
physical changes, 220
physical hazards, 305
physicochemical states of food, 365–367
phytochemicals
  consumer safety, 54
  definition, 10, 51
  as food components, 51–56
  functions of, 52
  herbal preparations, 51, 53–54, 336–338
  role of food scientist in product development, 54–55
phytoestrogen, 51
phytonutrients, 51
phytotherapy, 51
pickle curing of meats, 234
pickles, 111, 268–269
pigments, 137. *See also* color
pilot plants, 437
pistachios, 34
plant hormones, 349
plant sterol and stanol esters, heart disease and, 199
plasmids, 395, 397–398
plastic containers, 256, 388–390
plastic fats, 136
plasticity, 141, 375
plasticizers, 105–106, 251, 367
plastic solids, 376
plate drying, 229
plate heat exchangers, 364
poising effect, 288
polygalacturonase (PG), 406
polymerase chain reaction (PCR), 399
polymerization reactions, 98, 140, 170–171
polymers, 98, 388–389
polymorphism, 140, 273
polyolefin, 388
polyols (sugar alcohols), 77, 126, 164
polypeptides, 143, 417
polyphenol oxidase (PPO), 98
polypropylene (PP), 388–389
polysaccharides
  beta-glucans, 127
  cellulose, 127, 291
  dextrins and maltodextrins, 127–128
  as fat replacers, 80
  fructooligosaccharides, 128
  functional properties, 126–127
  inulin, 128–129
  overview, 67
  pectic substances, 128–130
  starch, 129–133
  as texturizing agents, 169
  vegetable gums, 132–133
polystyrene (PS), 389
polyunsaturated fatty acids, 71, 135
polyvinyl chloride (PVC), 388–389
porphyrin rings, 157–158
potassium, 78
potatoes, 269, 406
potency index, 430

poultry, 236, 290–291
powdered soft drinks, 258
prebiotics, 128
precision, 448
preferences, 430
pregelatinized starch, 131
pre-market reviews, 206–207
Prescott, Samuel C., 11
preservatives, antimicrobial chemical, 217–218, 320
pressing, 251
prions, 316–317
prior-sanctioned substances, 186
probability of risk, 327
probability, product, 440–441
probiotics, 128, 259, 408
procaryotes, 283
processed commodities, 23
processed meat, 231–235
process flavor substances, 165
processing aids, 182
product development
  artificial neural network, 441–442
  commercial stage, 437–438
  definition, 434
  development stage, 437
  experimental design in, 445–449
  idea stage, 436–437
  overview, 8–9
  product development team, 435
  product life cycle, 441
  product probability, 440–442
  role of marketing, 438–440
  role of sensory evaluation specialist in, 432–434
  scientific method in, 435
product life cycle, 441
product line, 436
product recalls, 190–191
product strategy, 436
progoitrin, 335
PROP, 422
propionic acid, 94, 217
prostaglandins, 71
proteases, 147, 165, 220, 231, 279
protein energy malnutrition (P.E.M.), 75
protein hydrolysates, 166, 279–281
protein kinase A, 342
proteins
  AN/TN ratio, 280
  biological value, 28, 74
  β-lactoglobin bioengineering, 416–418
  casein and whey, 42–43, 151–153, 169, 226, 294
  in comminuted meat emulsions, 38
  complementation, 74
  conjugated *vs.* nonconjugated, 143
  digestion of, 62
  egg, 40
  enzyme hydrolysis of, 220
  functional properties, 145–147
  G, 342

high-protein, high fat diets, 88–90
hydrolysates, 279–281
hydrolyzed vegetable, 34, 166
irradiation of, 244
microbial metabolism of, 292
nitrogen balance, 74–75
nutritional overview, 73
prions, 316–317
soy, 33, 271–272
structure and shape, 142–144
synthesis, 74, 396–397
tertiary structure, 144
water-holding capacity, 147
protopectin, 129
protoplasmic poisoning, 338
protozoa, 284–285, 308–309, 313–314
prussic acid (cyanide), 335–336
PSE meat, 231
pseudoplastic behavior, 377
psilocybin, 340
psychrometrics, 369–374
psychrophiles, 287
psychrotrophs, 38, 287
psyllium, 53
Ptychodiscus brevis, 341
pulsed electric fields (PEF), 224
pulsed light technology (PLT), 224–225
pumping, in food processing, 213
pungency, 163–164
PureBright process, 224
Pure Food Congress of 1898, 185
purge, in meat processing, 234
putrescine, 292
pyrrolizidine alkaloids, 337
pyruvate, 84, 286

quality assurance (QA), 9, 40–41, 448
quality, definition of, 445
quantitative descriptive analysis (QDA), 431–432
quaternary ammonium compounds, as sanitizers, 318
QUATS, 213
quercetin, 334–335

radapperization, 242
radiation, 241, 360. *See also* irradiation
radicidation, 242
radiolysis, 244–245
radurization, 242
rancid fats, 5, 39
rancidity, hydrolytic *vs.* oxidative, 167
ranking, 430, 432
rating, 431
RBD oil, 250
recalls, 190–191
recognition threshold, 430
Recommended Dietary Allowance (RDA), 63, 197
Red Book, 189
redox potential, 286
red tide, 341

reducing agents, 99, 123
reducing environments, 286
reducing sugars, 123
reduction reactions, 98–99
red wine, 30–31
Reference Daily Intake (RDI), 63
reference dose, 188
refining, definition of, 250
refrigerant gases, 267
refrigerated storage, 233, 319
refrigeration technology, 7
regulation, definition of, 183. *See also* laws and regulations
relative humidity, 104, 371
rendering, 251
rennet, 147, 294
rennin, 42, 45, 152
research, basic, 8
resonance electrons, 158
respiration, 264
response forms, 446
response surface design, 447
restriction enzymes, 396–398
restructuring, 236
resveratrol, 30–31
Retinol Activity Equivalents (RAE), 197
retort processing, 219, 360
retrogradation, 132
retro-inhalation, 424
reverse transcriptase, 397
reversion flavor, 136
rheology, 374–379
rheopectic behavior, 379
riboflavin, 76
rickets, vitamin D and, 7
rigor mortis, 230–231
ripeness, 31, 263
ripening, 263–264, 406
risk assessments, 327–328, 331–334
risk–benefit relationships, 189
RNA, messenger, 397
rosemary, 337–338
rotaviruses, 312
roundworms, 313
rule, definition of, 183
Russell, H. L., 11

saccharin, 78, 80, 176, 188–189, 258
Saccharin Study and Labeling Act of 1977, 188–189
Saccharomyces cerevisiae, 285, 291, 297, 409–410
Saccharomyces diastaticus, 410
safrole, 195, 337
sage, 337–338
St. John's wort, 54
Salatrim (Benefat), 81–82, 141
Salmonella spp., 228, 267, 306, 308–310, 410
Salmonella enteritidis, 310
Salmonella typhimurium, 194, 242, 309
salmonellosis, 310
salts, 104, 107–108

samples, 446
sampling, to verify food quality, 301–302
sanitizing, in food processing, 212–213, 318
santitation standard operating procedure (SSOP), 322
saponification value, 253
saponins, 52
sarcomere units, 36
sassafras root, 337
saturated fats, 71, 134–135
sausage processing, 234, 295–296. *See also* meat and meat products
saw palmetto, 54
saxitoxins, 340
scientific agriculture, 6
scientific method, 420, 435
scientific principles, definition of, 9
scombroid poisoning, 341
Scoville heat units, 163
scraped surface heat exchanger (SSHE), 364–365
scrapie, 316–317
screening design, 447
seafood
  canning, 232–233, 235
  curing, 235–236
  nutrient composition, 37–39
  perishability of, 38–39
  poisoning from, 340–341
  surimi, 236
  trimethylamine and fish odor, 39
secretin, 62
selenium, 79
semolina, 260, 262
senescence, 31, 264
sensible heat, 370, 374
"Sensory and microbial quality of irradiated crab meat products," 18–19
sensory evaluation
  color, 427–428
  data analysis, 447
  decision flowchart, 433
  definition, 12–13, 420
  experimental design, 446–447
  flavor, 423–424
  of frankfurters, 19–20
  intensity, 429
  mechanical sensory characteristics, 425–426
  mouthfeel, 424–425
  objectivity and subjectivity, 429
  odor, 423
  quantitative, 421, 431–432
  role in food industry, 421
  as scientific method, 420
  sensory environment, 434, 446
  sensory overload, 426–427
  sensory texture, 425
  sequence of perception, 428
  specialists, 432–434
  taste, 422

sensory evaluation *(continued)*
   test methods, 430–432
   threshold, 430
   transduction and sensitivity, 423
   variables and error considerations, 448
sensory overload, 426–427
sensory viscosity, 425
separating, in food processing, 212
sequestrants, 182
serving procedures, 446–447
serving sizes, 23–24
shear stresses, 375–378
shear thickening/thinning, 376–377
Shigella, 306
significant hazards, definition of, 321
Simplesse, 81, 146
Sinclair, Upton, 8
Slade, Louis, 366
smart packages, 219
smoked meats, 235, 320
smoke point, 140, 253
smouch, 7
soapmaking, 251
sodas, 25–26, 256–258
sodium, 65–66, 78
sodium bicarbonate, 84
sodium chloride, 104
sodium tripolyphosphate (STPP), 19–20
soft drinks, 25–26, 256–258
solid fat index, 140
sols, 119
solubility, 104, 116, 147, 151
solutions, 115–116
solvent extraction, 251
sometribove, 189–190
sorbestrin, 141
sorbitol, 126, 164
sour cream, standard of identity for, 186
soy products
   flour, 271
   heart disease and, 199
   isoflavone content, 33, 259
   meal, 270–271
   nutrient composition, 33
   oil, 271–272
   processing, 270–272
   soy milk, 271
   soy protein concentrates, 272
   soy protein isolates, 271–272
   texturized soy protein, 271–272
spaghetti, 262
sparkling water, 256
specific heat, 363–364
specific volume, 372
spores, 284
sporozoites, 314
spray drying, 214, 229
stabilizers, emulsion, 117, 182
standard of identity, 42–43, 185–186
standard operating procedures (SOPs), 322
stanol esters, heart disease and, 199, 205
staphylococcal enterotoxins, 341–342

Staphylococcus aureus
   effect of pulsed light on, 225
   heat resistance of, 221
   intoxication from, 305, 307
   meat fermentation and, 295
   sources, 331
   staphylococcal enterotoxins from, 341–342
   water activity and, 215, 285
starches. *See also* carbohydrates
   digestion of, 62
   gelatinization, 126, 129–130, 133
   pasting, 131–133
   retrogradation, 132–133
   starch gels, 132–133
StarLink controversy, 412
starter cultures, 293, 410
state point, 373
statute, definition of, 183
steady state, 362
Stepp, Wilhelm, 7
stereoisomers, 176
sterilization, 215
steroid alcohols, 137
stomach, digestion in, 62
storage, 233–234, 266, 319, 387, 406
strain, 375–376
strawberries, 246
Strecker degradation, 164
Streptococcus lactis, 45
Streptococcus pyogenes, 306
Streptococcus salivarius spp. *thermophilus*, 227, 295
streptomycin, 344
stress forces, 375–376
structured fluids, 374
substantial equivalence, 403
sucralose, 78–80, 258
sucrose, 26–27, 67, 122–123, 253–255. *See also* sugars
sucrose inversion, 26
sucrose polyester (SPE), 18
sugar alcohols, 78, 126, 164
sugars. *See also* polysaccharides; sweeteners
   browning reactions, 123–125
   caramelization, 124, 164
   chemistry of, 121–123, 127, 175–177
   crystallization, 125
   digestion of, 62
   humectancy, 125–126
   invert, 126
   oxidation/reduction, 126
   processing of, 253–255
   raw, 255
   reducing, 123
   refining, 255
   relative sweetness, 126
   as texturizers, 170
sulfa drugs, 344
sulfhydryl groups, 102
sulfites, 354

sulfur, 78
sun drying, 214
superchlorination, 257
supplements, 200–201
surface active agents (surfactants), 183
surface area, heat transfer and, 363
surface tension, 379
surimi, 236
suspensoids, 118
sweetened condensed milk, 42–43
sweeteners. *See also* sugars
   alternative, 78–80
   chemistry of, 175–177
   cyclamates, 80, 188
   degrees Brix measurement, 26
   digestion of, 62
   nutritive *vs.* nonnutritive, 182, 258
   saccharin, 78, 80, 176–177, 188–189
   simple sugars, 126
   in soft drinks, 257
   sucrose inversion, 26–27
sweetness, theory of, 175–176, 423
sweet taste transduction, 175
symbols, chemical, 92
syneresis, 132
synthesis reactions, 98
synthetic toxicants, 330–331

tachyzoites, 313–314
Taenia saginata, 313–314
Taenia solium, 313–314
tannins, 163
tapeworms, 308, 313–315
tartaric acid, 111, 217
tastants, 422, 430
taste, 422–424. *See also* flavor
taste buds, 422
taste transduction, 423
TBH, 218
tea, 25
teamsmanship, 14
temperate viruses, 285
temperature
   absolute zero, 358
   of blanching, 265
   Celsius, Kelvin and Fahrenheit scales, 357–358
   cold storage, 266
   collagen and gelatin, 171
   danger zone, 318–319
   dew point, 371–372
   glass transition, 105, 367
   heat treatment of foods, 215–216, 220–223
   low-temperature treatment of foods, 216
   microbial growth and, 287
   molecular motion and, 358–359
   rate of chemical reactions and, 364
   solubility and, 116
   water activity and, 104
   wet bulb/dry bulb, 371

tempering of fats, 141, 273–274
tenderizers, 143, 165, 171, 231
teratogenic compounds, 347
terminal threshold, 430
terpenes, 52, 141
terrorism, biological, 304
tertiary structure of proteins, 144
test market, 437
tetracyclines, 344–345
texture
   classification, 168
   definition, 167–168
   influence of fats and water, 168–169
   rheology and, 374–375
   sensory, 425
   stability, 115
   sugars, 126
   texturizing agents, 169–171
texture profile analysis (TPA), 18, 432
texturized soy protein, 271–272
texturizing agents, 169–171
thaumatin, 406
thaw rigor, 231
T helper cells, 353
theobromine, 273
thermal conductivity, 361, 364
thermal contact, 357
thermal death time (TDT), 216, 222
thermal equilibrium, 357
thermal processing, 215–216, 220–223, 268, 320
thermodynamics, 363
thermophiles, 287
thermoplastic, 388
thermos bottles, 359
thermosetting, 388
thermotrophs, 287
thiamin, 76
thickening agents, 117, 182, 376–377
thiocyanate, 335
thiols, 102
thixotropic behavior, 378–379
three-class representative sampling, 302
threshold level, 333
thyroid, 335
time–intensity measurement, 429
titratable acidity, 32, 110–111
tomatoes and tomato products, 190, 269–270, 406–407
top notes, 165
tortilla dough, 107
total heat content, 372
toxicants/toxins
   categories of, 332
   cholera, 332
   clostridial, 343
   coumarins, 335
   cyanide compounds, 335–336
   definition, 330–331
   endogenous, 334–340
   FDA policy on biotechnology of, 401
   flavonoids, 334–335

   flow through the food chain, 346
   goitrogens, 335
   herbal extracts, 336–338
   marine, 340–341
   microbial toxins, 341–343
   mineral toxicity, 77–78
   mode of action of, 332
   naturally occurring, 340–343
   toxic mushrooms, 338
   verocytotoxins, 342–343
   vitamin toxicity, 76–77
Toxoplasma gondi (toxoplasmosis), 313–314
trans-fatty acids, 72, 135–136, 138
transgenic organisms, 394–395, 400
transmissible spongiform encephalopathies (TSEs), 315–317
transpiration, 264
transport, of nutrients, 61
trehalose, 405
trenbolone acetate, 345–346
triacylglycerols, 133
triangle tests, 431–432
Trichinella spiralis (trichinosis), 243, 313, 315
trigeminal perception, 423
triglycerides, 69–70, 72, 133–134, 136, 140, 169. *See also* fatty acids
trimethylamine, fish odor and, 39, 235
trimming, 264
trophozoites, 284, 313–314
tropocollagen, 170
turgor, 269
turkey. *See* poultry
turkey x disease, 187
12D concept, 222
two-class attribute sampling, 301
Type I and Type II error, 448

umami, 162–163, 422
undernutrition, 58
United Nations food agencies, 190
United States Department of Agriculture (USDA), 190–191
unit operations, 212. *See also* food processing
unsaturated fats, 134–135

vacuum canning, 220
vacuum cooling, 266
vacuum packaging, 218, 235, 265, 387
valence shells, 94
valerian, 54
validity of tests, 448
value-added food products, 435
vapor pressure, 372
vegetable gums, 132–133
vegetables
   color chemistry, 158–160
   definition, 29, 262
   dehydration of, 269
   fermentation, 296
   flavonoids in, 334–335
   freezing, 266–267

   goitrogens in, 335
   hot-break/cold-break processing, 269–270
   microbial contamination, 290
   nutrients compared to grains, 259
   nutrition labeling, 195
   packaging, 265
   pickles, 111, 268–269
   processing, 264–265
   recommended intake of, 30–31
   ripening, 263–264
   storage, 266
vegetables food group, 59
vegetarian diets, 74
vegetative cells, 284
verification, in HACCP program, 322
verocytotoxins, 342–343
very low-density lipoprotein (VLDL), 72
Vibrio cholerae, 309, 311, 342
Vibrio parahaemolyticus, 306
Vibrio vulnificus, 306
vinyl acetate (VA), 388, 389
viral gastroenteritis, 312
viruses, 285, 307–308, 312
virus-resistant plants, 405
viscoelastic foods, 376
viscosity, 31, 367, 374, 376–379, 425
visible spectrum, 155, 427
vitamins
   discovery of, 7
   guide to, 75–77
   irradiation of, 244–245
   on nutrition labels, 197
   vitamin A, 76, 197, 244
   vitamin B_1 (thiamin), 244
   vitamin B_6 (pyridoxine), 76
   vitamin B_{12} (cobalamin), 76
   vitamin C (ascorbic acid), 76, 244
   vitamin D (cholecalciferol), 7, 77
   vitamin E, 19, 77, 244
   vitamin K, 77
   vitamin O, 206

walnuts, 34
warmed-over flavor (WOF), 167
washing, 264
water
   activity, 103–104, 168, 215, 285
   biologically active, 212
   bottled, 255–256
   bound, 103, 168
   content in foods, 104–105
   covalent bonding in, 95
   in digestion, 62
   dipoles in, 70
   electrolytes and dehydration, 65–66
   in emulsions, 104
   functional properties in foods, 102
   heat capacity, 364
   heat transfer and, 105
   hydrogen bonding in, 95–96
   irradiation of, 244

water *(continued)*
   metabolic, 65
   microbial contamination, 288–289
   moisture sorption isotherms, 104–105
   molecular structure, 102–103
   nutritional role, 64–66
   as plasticizer, 105–106, 367
   recommended intake, 89
   role in texture, 168
   solvation and dispersing action, 103
   treatment for use in soft drinks, 257
water activity, 103–104, 168, 215, 285
water-holding capacity (WHC), 147
water vapor transmission rate (WVTR), 389
waxes, 137
Weigand, Ernest, 11
weight loss diets, 88–90
wet bulb temperature, 371
wheat milling, 259–260
whey, 43, 151–153, 226, 294, 416–418
White House Conference on Food, Nutrition and Health (1969), 188
winterizing, 251
work, definition of, 363

xylitol, 164

yeasts, 28, 165–166, 215, 284, 286, 405
Yersinia enterocolitica, 306, 311
yield point, 377
yield stress, 378
yogurt, 227–228, 295

zero-cancer-risk standard, 186–187
zero-order reaction, 97
zinc, 79
Z-trim, 81